Probability and Statistics for Data Science

This self-contained guide introduces two pillars of data science, probability theory, and statistics, side by side, in order to illuminate the connections between statistical techniques and the probabilistic concepts they are based on. The topics covered in the book include random variables, nonparametric and parametric models, correlation, estimation of population parameters, hypothesis testing, principal component analysis, and both linear and nonlinear methods for regression and classification. Examples throughout the book draw from real-world datasets to demonstrate concepts in practice and confront readers with fundamental challenges in data science, such as overfitting, the curse of dimensionality, and causal inference. Code in Python reproducing these examples is available on the book's website, along with videos, slides, and solutions to exercises.

This accessible book is ideal for undergraduate and graduate students, data science practitioners, and others interested in the theoretical concepts underlying data science methods.

CARLOS FERNANDEZ-GRANDA is Associate Professor of mathematics and data science at New York University, where he has taught probability and statistics to data science students since 2015. The goal of his research is to design and analyze data science methodology, with a focus on machine learning, artificial intelligence, and their application to medicine, climate science, biology, and other scientific domains.

"Fernandez-Granda's *Probability and Statistics for Data Science* is a comprehensive yet approachable treatment of the fundamentals required of all aspiring data scientists – whether they be in academia, industry, or elsewhere. The language is clear and precise, and it is one of the best-organized treatments of this material I have ever seen. With lucid examples and helpful exercises, it deserves to be the leading text for these topics among undergraduate and graduate students in this technical, fast-moving discipline. Instructors take note!"

—Arthur Spirling, Princeton University

"If you're mathematically inclined and want to master the foundations of data science in one go, this book is for you. It covers a broad range of essential modern topics – including nonparametric methods, causal inference, latent variable models, Bayesian approaches, and a thorough introduction to machine learning – all illustrated with an abundance of figures and real-world data examples. Highly recommended."

—David Rosenberg, Office of the CTO, Bloomberg

Probability and Statistics for Data Science

CARLOS FERNANDEZ-GRANDA

New York University

CAMBRIDGE
UNIVERSITY PRESS

CAMBRIDGE
UNIVERSITY PRESS

Shaftesbury Road, Cambridge CB2 8EA, United Kingdom

One Liberty Plaza, 20th Floor, New York, NY 10006, USA

477 Williamstown Road, Port Melbourne, VIC 3207, Australia

314–321, 3rd Floor, Plot 3, Splendor Forum, Jasola District Centre,
New Delhi – 110025, India

103 Penang Road, #05–06/07, Visioncrest Commercial, Singapore 238467

Cambridge University Press is part of Cambridge University Press & Assessment,
a department of the University of Cambridge.

We share the University's mission to contribute to society through the pursuit of
education, learning and research at the highest international levels of excellence.

www.cambridge.org
Information on this title: www.cambridge.org/9781009180085

DOI: 10.1017/9781009180108

First published 2025

Cover image: philipp igumnov/Moment via Getty Images

A catalogue record for this publication is available from the British Library

A Cataloging-in-Publication data record for this book is available from the Library of Congress

ISBN 978-1-009-18008-5 Hardback
ISBN 978-1-009-18009-2 Paperback

Para Irida, Iliana y Chrysa

Contents

Preface   *page* xi
Book Website   xiii

Introduction and Overview   1

1   Probability   6
1.1   Intuitive Properties of Probability   6
1.2   Mathematical Definition of Probability   9
1.3   Conditional Probability   13
1.4   Estimating Probabilities from Data   19
1.5   Independence   22
1.6   Conditional Independence   26
1.7   The Monte Carlo Method   29
       Exercises   33

2   Discrete Variables   37
2.1   Discrete Random Variables   37
2.2   The Empirical Probability Mass Function   43
2.3   Discrete Parametric Distributions   46
2.4   Maximum-Likelihood Estimation   52
2.5   Comparing Parametric and Nonparametric Models   57
       Exercises   63

3   Continuous Variables   66
3.1   Continuous Random Variables   66
3.2   The Cumulative Distribution Function   69
3.3   The Probability Density Function   76
3.4   Functions of Random Variables   81
3.5   Nonparametric Probability-Density Estimation   84
3.6   Continuous Parametric Distributions   90
3.7   Maximum-Likelihood Estimation   96
3.8   Inverse-Transform Sampling   101
       Exercises   104

4   Multiple Discrete Variables   109
4.1   Multivariate Discrete Random Variables   109
4.2   Marginal Distributions   115

4.3	Conditional Distributions	118
4.4	Independence	122
4.5	Conditional Independence	125
4.6	Causal Inference	127
4.7	The Curse of Dimensionality	137
4.8	Classification via Naive Bayes	138
4.9	Markov Chains	141
	Exercises	152
5	**Multiple Continuous Variables**	**161**
5.1	Joint Distribution of Continuous Random Variables	161
5.2	Joint Cumulative Distribution Function	163
5.3	Joint Probability Density Function	166
5.4	Multidimensional Density Estimation	170
5.5	Marginal Distributions	171
5.6	Conditional Distributions	174
5.7	Independence	177
5.8	Conditional Independence	180
5.9	Jointly Simulating Multiple Random Variables	184
5.10	Gaussian Random Vectors	186
	Exercises	196
6	**Discrete and Continuous Variables**	**202**
6.1	Joint Distribution of Discrete and Continuous Variables	202
6.2	Conditional Distribution of Continuous Variables Given Discrete Variables	204
6.3	Conditional Distribution of Discrete Variables Given Continuous Variables	207
6.4	Independence	211
6.5	Classification via Gaussian Discriminant Analysis	212
6.6	Clustering via Gaussian Mixture Models	217
6.7	Bayesian Parametric Modeling	225
	Exercises	235
7	**Averaging**	**241**
7.1	The Mean	241
7.2	Linearity of Expectation	248
7.3	Independent Random Variables	249
7.4	Mean of Parametric Distributions	251
7.5	Sensitivity of the Mean to Extreme Values	254
7.6	The Mean Square	255
7.7	The Variance	257
7.8	The Conditional Mean	262
7.9	The Average Treatment Effect	274
	Exercises	280

8	**Correlation**	284
8.1	Correlation between Standardized Quantities	284
8.2	Correlation and Covariance	287
8.3	Estimating Correlation from Data	290
8.4	Simple Linear Regression	293
8.5	Properties of the Correlation Coefficient	299
8.6	Uncorrelation and Independence	306
8.7	A Geometric Analysis of Correlation	310
8.8	Correlation (Usually) Does Not Imply Causation	315
	Exercises	321
9	**Estimation of Population Parameters**	325
9.1	Random Sampling	325
9.2	The Bias	329
9.3	The Standard Error	332
9.4	Deviation Bounds: Markov's and Chebyshev's Inequalities	337
9.5	The Law of Large Numbers	339
9.6	Some Averages Are Not to Be Trusted	344
9.7	The Central Limit Theorem	349
9.8	Confidence Intervals	363
9.9	The Bootstrap	370
	Exercises	384
10	**Hypothesis Testing**	390
10.1	Selecting a Hypothesis	390
10.2	The Null Hypothesis and the Test Statistic	391
10.3	Parametric Testing and the P-Value	392
10.4	Two-Sample Tests	395
10.5	Statistical Significance	399
10.6	The Power	402
10.7	Nonparametric Testing: The Permutation Test	408
10.8	Multiple Testing	417
10.9	Hypothesis Testing and Causal Inference	423
10.10	P-Value Abuse	425
	Exercises	429
11	**Principal Component Analysis and Low-Rank Models**	433
11.1	The Mean in Multiple Dimensions	433
11.2	The Covariance Matrix	436
11.3	The Sample Covariance Matrix	441
11.4	Principal Component Analysis	445
11.5	Dimensionality Reduction	457
11.6	Low-Rank Models	464
11.7	Matrix Completion for Collaborative Filtering	483
	Exercises	491

12   Regression and Classification   495

12.1   Linear Regression   495

12.2   Generalization and Overfitting in Linear Regression   514

12.3   Ridge Regression   531

12.4   Sparse Regression   543

12.5   Logistic Regression   546

12.6   Softmax Regression   552

12.7   Tree-Based Models   556

12.8   Neural Networks and Deep Learning   577

12.9   Evaluation of Classification Models   584

     Exercises   590

Appendix: Datasets   599

References   602

Index   604

Preface

I started teaching probability and statistics to data science students in 2015. At first, I followed a traditional approach. I would begin with probability theory and then eventually move on to statistics and data analysis. It did not work. Students found it hard to understand the connection between the two halves of the course. By the time we reached statistics, they would struggle to recall the relevant concepts from probability theory, and I would often have to review them all over again. Motivated by this, I decided to completely restructure the material and write this book.

The book is a self-contained, rigorous guide to probability and statistics for data science. It intertwines the material on probability and statistics, explaining how to estimate each probabilistic object from data *right after its mathematical definition*. For example, we present nonparametric and parametric models just after defining random variables. This makes it possible to provide examples with real data from the very beginning. Throughout the book, these examples illustrate the connection between the material and practical data analysis, confronting students with fundamental challenges in data science, such as overfitting, the curse of dimensionality, and causal inference.

I began to think about how to best teach probability and statistics as a teaching assistant for Ayfer Özgür and Abbas El Gamal in graduate school. Abbas gave me the opportunity to teach my first course and encouraged me to use real-data examples, one of which survives in this book (the call-center example in Chapter 2). The book is also influenced by several other great teachers I had the luck to learn from in my studies, including Julian Manley, Eduardo Lorenzo Pigueiras, Francis Bach, Dan Boneh, Stephen Boyd, and Andrea Montanari.

While writing the book, I received invaluable feedback from my students at the Center for Data Science in New York University[1]. I had the good fortune to co-teach this material with Brett Bernstein and Ilias Zadik and to have the help of my wonderful teaching assistants Annika Brundyn, Nadine Hussami, Vlad Kobzar, Nhung Le, Mu Li, Xintong Li, Sheng Liu, Taro Makino, Sreyas Mohan, Tinatin Nikvashvili, Jonas Peeters, Lisa Ren, Levent Sagun, Michael Stanley, Zhiyuan Wang, and Weicheng Zhu. I am extremely grateful to Diego Herrero Quevedo and Louis Mittel, who proofread the manuscript and provided many insightful comments, and also to Juan Argote, Afonso Bandeira, Chris Chen, Shi-ang Chi, Daniel Climent, Rajesh Ranganath, Alvaro Reig, and David Rosenberg for their helpful suggestions.

In addition, I would like to thank Diana Gillooly, who originally suggested that I publish the book, my editor Lauren Cowles, for her help and extreme patience, my mentors Emmanuel Candès, Julia Kempe, Bob Kohn, and Eero Simoncelli, for their guidance, and

---

[1] My favorite feedback from a student evaluation: *The instructor is like Spanish wine. I don't like Spanish wine.*

xi

the Division of Mathematical Sciences of the National Science Foundation, for their support through grants 1616340 and 2009752.

Finally, I cannot end without thanking my in-laws Dimitra and Daniil, my parents Yolanda and Carlos, my daughter Irida (for constantly sharing the books she was writing in kindergarten to motivate me), my daughter Iliana (for showing up as I was writing the book, and making things more fun), and my wife Chrysa (for everything).

Book Website

Supporting material for the book, including code, videos, slides, and solutions to all exercises, can be found on the book website:

www.ps4ds.net

Introduction and Overview

This book provides a self-contained guide to probability and statistics, and their application to data science. We present probability theory intertwined with statistics, as opposed to first covering probability and then statistics, as in most existing texts. The goal is to highlight the connections between probabilistic concepts and the statistical techniques used to estimate them from data. Throughout the text, and from the very beginning, computational examples illustrate the application of the material to real-world data, extracted from the datasets listed in the Appendix. Code in Python reproducing these examples is available on the book website (www.ps4ds.net). The website also contains additional supporting material, including videos, slides, and solutions to all exercises. In the remainder of this section, we provide an overview of the contents of each chapter.

Chapter 1 introduces **probability**. We begin with an informal definition, which enables us to build intuition about the properties of probability. Then, we present a more rigorous definition, based on the mathematical framework of probability spaces. Next, we describe conditional probability, which makes it possible to update probabilities when additional information is revealed. In our first encounter with statistics, we explain how to estimate probabilities and conditional probabilities from data, as illustrated by an analysis of votes in the United States Congress. Building upon the concept of conditional probability, we define independence and conditional independence, which are critical concepts in probabilistic modeling. The chapter ends with a surprising twist: In practice, probabilities are often impossible to compute analytically! Fortunately, the Monte Carlo method provides a pragmatic solution to this challenge, allowing us to approximate probabilities very accurately using computer simulations. We apply the method to model a 3x3 basketball tournament from the 2020 Tokyo Olympics.

Chapter 2 introduces **random variables**, and explains how to use them to model uncertain numerical quantities that are **discrete**. We first provide a mathematical definition of random variables, building upon the framework of probability spaces. Then, we explain how to manipulate discrete random variables in practice, using their probability mass function (pmf), and describe the main properties of the pmf. Motivated by an example where we analyze Kevin Durant's free-throw shooting, we define the empirical pmf, a nonparametric estimator of the pmf that does not make strong assumptions about the data. Next, we define several popular discrete parametric distributions (Bernoulli, binomial, geometric, and Poisson), which yield parametric estimators of the pmf, and explain how to fit them to data via maximum-likelihood estimation. We conclude the chapter by comparing the advantages and disadvantages of nonparametric and parametric models, illustrated by a real-data example, where we model the number of calls arriving at a call center.

Chapter 3 introduces **continuous random variables**, which enable us to model uncertain continuous quantities. We again begin with a formal definition, but quickly move on to describe how to manipulate continuous random variables in practice. We define the cumulative distribution function and quantiles (including the median), and explain how to estimate them from data. We then introduce the concept of probability density and describe its main properties. We present two approaches to obtain nonparametric models of probability densities from data: The histogram and kernel density estimation. Next, we define two celebrated continuous parametric distributions – the exponential and the Gaussian – and show how to fit them to data using maximum-likelihood estimation. We use these distributions to model the interarrival time of calls at a call center, and height in a population, respectively. Finally, we discuss how to simulate continuous random variables via inverse transform sampling.

Chapter 4 describes how to jointly model the interactions between several uncertain discrete quantities. Mathematically, this is achieved by representing the quantities as **multiple discrete random variables** within the same probability space. In practice, such variables are characterized using their joint pmf. We explain how to estimate the joint pmf from data and use it to model precipitation at three locations in Oregon. Then, we introduce marginal and conditional distributions, utilizing the real-world Oregon precipitation data as a running example. Marginal distributions describe the individual behavior of each variable in a model. Conditional distributions describe the behavior of a variable, when the values of other variables are fixed. Next, we generalize the concepts of independence and conditional independence to random variables. At this point, we discuss the problem of causal inference, which seeks to identify causal relationships between variables. Causal inference enables us to understand why a relatively unknown NBA player can have a better three-point shooting percentage than the best shooter in history. We then turn our attention to a fundamental challenge in statistics and data science: It is impossible to completely characterize the dependence between the variables of a probabilistic model, unless they are very few. This phenomenon, known as the curse of dimensionality, is the reason why independence and conditional independence assumptions are needed to make probabilistic models tractable. We conclude the chapter by describing two popular models based on such assumptions: Naive Bayes and Markov chains.

In Chapter 5 we describe how to jointly model continuous quantities, by representing them as **multiple continuous random variables** within the same probability space. We define the joint cumulative distribution function and the joint probability density function, and explain how to estimate the latter from data using a multivariate generalization of kernel density estimation. Next, we introduce marginal and conditional distributions of continuous variables and also discuss independence and conditional independence. Throughout, we model real-world temperature data as a running example. Then, we explain how to jointly simulate multiple random variables, in order to correctly account for the dependence between them. Finally, we define Gaussian random vectors, which are the most popular multidimensional parametric models for continuous data, and apply them to model anthropometric data.

Chapter 6 discusses how to build probabilistic models that include both **discrete and continuous variables**. Mathematically, this is achieved by defining them as random variables within the same probability space. In practice, the variables are manipulated using their marginal and conditional distributions. We define the conditional pmf of a discrete random variable given a continuous variable, and the conditional probability density of a continuous random variable given a discrete variable. We use these objects to build mixture models and apply them to model height in a population. Next, we describe Gaussian discriminant analysis, a classification method based on mixture models with Gaussian conditional distributions, and apply it to diagnose Alzheimer's disease. Then, we explain how to perform clustering using Gaussian mixture models and leverage the approach to cluster NBA players. Finally, we introduce the framework of Bayesian statistics, which enables us to explicitly encode our uncertainty about model parameters, and use it to analyze poll data from the 2020 United States presidential election.

Chapter 7 focuses on **averaging**, which is a fundamental operation in probability and statistics. We begin by defining an averaging procedure for random variables, known as the mean. We show that the mean is linear, and also that the mean of the product of independent variables equals the product of their means. Then, we derive the mean of popular parametric distributions. Next, we caution that the mean can be severely distorted by extreme values, as illustrated by an analysis of NBA salaries. In addition, we define the mean square, which is the average squared value of a random variable, and the variance, which is the mean square deviation from the mean. We explain how to estimate the variance from data and use it to describe temperature variability at different geographic locations. Then, we define the conditional mean, which represents the average of a variable when other variables are fixed. We prove that the conditional mean is an optimal solution to the problem of regression, where the goal is to estimate a quantity of interest as a function of other variables. We end the chapter by studying how to estimate average causal effects, motivated by two real-world causal-inference questions: *Do all-caps titles attract more views on YouTube?* and *Do private lessons improve students' grades?*

Chapter 8 focuses on **correlation**, a key metric in data science, which quantifies to what extent two quantities are linearly related. We begin by defining the correlation between normalized and centered random variables. Then, we generalize the definition to all random variables and introduce the concept of covariance, which measures the average joint variation of two random variables. Next, we explain how to estimate correlation from data and analyze the correlation between the height of NBA players and different basketball stats. In addition, we study the connection between correlation and simple linear regression. We then discuss the differences between uncorrelation and independence. In order to gain better intuition about the properties of correlation, we provide a geometric interpretation of correlation, where the covariance is an inner product between random variables. Finally, we show that correlation does not imply causation, as illustrated by the spurious correlation between temperature and unemployment in Spain.

Chapter 9 explains how to **estimate population parameters** from data. As running examples, we consider the problems of estimating the mean height in a population and the prevalence of COVID-19 in New York City. We begin by introducing random sampling, a simple yet powerful approach that enables us to obtain accurate estimates from limited data. We then define the bias and the standard error, which quantify the average error of an estimator and how much it varies, respectively. In order to gain a deeper understanding of the properties of random sampling, we derive deviation bounds, which characterize the probabilistic behavior of a random variable just based on its mean and variance. We use these bounds to prove the celebrated law of large numbers, which states that averaging many independent samples from a distribution yields an accurate estimate of its mean. An important consequence of this law is that random sampling provides an arbitrarily precise estimate of means and proportions, as the number of data grows. However, we also caution that this is not necessarily the case, if the underlying data contain extreme values, as demonstrated by a real-world economic dataset. Next, we discuss another fundamental mathematical phenomenon, the central limit theorem (CLT), according to which averages of independent quantities tend to have Gaussian distributions. We again provide a cautionary tale, inspired by the 2008 Financial Crisis, warning that the CLT does not hold in the absence of independence. Then, we explain how to use the CLT to build confidence intervals, which quantify the uncertainty of estimates obtained from finite data. Finally, we introduce the bootstrap, a popular computational technique to estimate standard errors and build confidence intervals.

Chapter 10 presents the framework of **hypothesis testing**, which can be used to evaluate whether the available data provide sufficient evidence to support a certain hypothesis. We consider two questions as running examples: *Is a toy die unfair?* and *Is Giannis Antetokounmpo's free-throw shooting worse in away games than in home games?* The main idea behind hypothesis testing is to play devil's advocate and assume a null hypothesis, which contradicts our hypothesis of interest. We explain how to use parametric modeling to implement this idea and define the p-value. A small p-value indicates that the data cannot be explained by the null hypothesis, which is evidence in favor of the original hypothesis. We prove that thresholding the p-value is guaranteed to control the probability of endorsing a false finding. In addition, we define the power of a test, which quantifies the test's ability to identify positive findings. Next, we show how to perform hypothesis testing without a parametric model, focusing on the permutation test. Then, we discuss multiple testing, a setting of great practical interest where many tests are performed simultaneously. Using real data from NBA players, we demonstrate that avoiding false findings in such situations is very challenging, but can be achieved by adjusting the p-value threshold. To end the chapter, we provide three reasons why hypothesis testing should not be used as the only stamp of approval for scientific discoveries. First, hypothesis testing does not necessarily identify causal effects; it is complementary to causal inference. Second, small p-values do not imply practical significance. Third, relying on p-values to validate findings produces a strong incentive to cherry-pick results, a practice known as p-hacking.

Chapter 11 covers **principal component analysis and low-rank models**, which are popular techniques to process high-dimensional datasets with many features. We begin by defining the mean of random vectors and random matrices. Then, we introduce the covariance matrix, which encodes the variance of any linear combination of the entries in a random vector, and explain how to estimate it from data. We model the geographic location of Canadian cities as a running example. Next, we present principal component analysis (PCA), a method to extract the directions of maximum variance in a dataset. We explain how to use PCA to find optimal low-dimensional representations of high-dimensional data, and apply it to a dataset of human faces. Then, we introduce low-rank models for matrix-valued data and describe how to fit them using the singular-value decomposition. We show that this approach is able to automatically identify meaningful patterns in real-world weather data. Finally, we explain how to estimate missing entries in a matrix under a low-rank assumption and apply this methodology to predict movie ratings via collaborative filtering.

Chapter 12 delves deeper into the problems of **regression and classification**, where the goal is to estimate a certain quantity of interest (the response) from observed features. In regression, the response is modeled as a numerical variable. In classification, the response belongs to a finite set of predetermined classes. We begin with a comprehensive description of linear regression models, which are ubiquitous in data science and statistics, because of their simplicity and interpretability. As a running example, we build a linear model of premature mortality in United States counties. Then, we discuss how to leverage linear regression to perform causal inference. In addition, we explain under what conditions linear models tend to overfit, and under what conditions they generalize robustly to held-out data. Motivated by the threat of overfitting, we introduce the concept of regularization. First, we provide a theoretical analysis of ℓ_2-norm regularization (a.k.a. ridge regression) and show that it can mitigate overfitting in practice. Second, we explain how to leverage ℓ_1-norm regularization (a.k.a. the lasso) to perform sparse regression, where the goal is to fit a linear model that only depends on a small subset of the available features. Next, we introduce two popular linear models for binary and multiclass classification: Logistic and softmax regression. We apply these methods to several classification tasks involving real data: Diagnosis of Alzheimer's disease, digit recognition, and identification of wheat varieties. At this point, we turn our attention to nonlinear models, which are cornerstones of modern machine learning. First, we present regression and classification trees and explain how to combine them to build complex nonlinear models via bagging, random forests, and boosting. Second, we describe the framework of deep learning and explain how to train neural networks to perform regression and classification. Finally, we end the chapter (and the book) by discussing how to evaluate classification models.

1

Probability

Overview

Who will win the next presidential election? What will be the price of a certain stock tomorrow? Will the New York Knicks win the NBA championship next season? There is no definite answer to these questions, because they pertain to *uncertain* phenomena with different possible outcomes. To describe an uncertain phenomenon, we interpret it as a repeatable experiment, which enables us to define the *probability* of different events associated with the phenomenon. This simple idea is a fundamental underpinning of statistics and data science. In Section 1.1, we provide an intuitive definition of probability and describe its main properties. Building upon this intuition, Section 1.2 introduces the mathematical framework of probability spaces. Section 1.3 defines conditional probability, which allows us to update probabilities when additional information is revealed. In Section 1.4, we explain how to estimate probabilities from data. Sections 1.5 and 1.6 introduce the key concepts of independence and conditional independence, respectively. Finally, Section 1.7 describes the Monte Carlo method, which enables us to approximate probabilities using computer simulations.

1.1 Intuitive Properties of Probability

In order to define probabilities associated with an uncertain phenomenon, we interpret the phenomenon as an *experiment* with multiple possible outcomes. The set of all possible outcomes is called the *sample space*, usually denoted by Ω. As the following examples show, the sample space can be discrete or continuous.

Example 1.1 (Die roll: sample space). If we roll a six-sided die, there are six possible results that are mutually exclusive (the die cannot land on two numbers at the same time). These six outcomes form the sample space $\Omega := \{1, 2, 3, 4, 5, 6\}$ associated with the die roll. In this case, the sample space is a finite set.
··

Example 1.2 (Rolling a die until it lands on a six: sample space). Imagine that we roll a six-sided die repeatedly until it lands on a six. Modeling the outcomes for this situation is not as straightforward as in Example 1.1. If we are just interested in the number of rolls that occur, we can set the outcome to equal that number. In that case, the sample space is the set of natural numbers $\Omega_1 := \mathbb{N}$. If we are interested in the actual values of the rolls, then we can set the outcome to equal the sequence of rolls (e.g. if we roll a four, a one, and a six, the outcome is $4 \to 1 \to 6$). The sample space Ω_2 is then the (infinite) set of all such sequences. Either way, the sample space is discrete, but countably infinite.
··

Example 1.3 (Weather in New York: sample space). If we want to model the weather in New York, then there are a lot of choices to make! To simplify matters, let us assume that we are only interested in the temperature in Washington Square Park at noon. We define the outcome to be that temperature, represented as a real number. The sample space containing all possible outcomes is the real line $\Omega := \mathbb{R}$.[1] In this case, the sample space is continuous, and the number of outcomes is uncountable.

....

Once we have defined the sample space, we quantify the uncertainty about our phenomenon of interest by determining how likely it is for the outcome to belong to different subsets of the sample space. We call these subsets *events*. Events can consist of several outcomes, a single outcome, the whole sample space, or no outcomes at all. An event occurs when the outcome of the experiment belongs to the event, as illustrated by the following examples.

Example 1.4 (Die roll: events). Possible events associated with the sample space in Example 1.1 include:

- Rolling a five, $A := \{5\}$.
- Rolling an even number, $B := \{2, 4, 6\}$.
- Rolling any number, $C := \{1, 2, 3, 4, 5, 6\}$.

If the roll is a four, then events B and C occur, but A does not.

....

Example 1.5 (Rolling a die until it lands on a six: events). In Example 1.2, the structure of the events depends on the choice of sample space. For example, the event *Rolling twice to obtain a six* contains a single outcome $\{2\}$, if the sample space is Ω_1, and five outcomes $(1 \to 6, 2 \to 6, 3 \to 6, 4 \to 6, 5 \to 6)$, if the sample space is Ω_2.

....

Example 1.6 (Weather in New York). If we model the temperature in Washington Square Park at noon and fix the sample space to be the real numbers ($\Omega := \mathbb{R}$), then possible events include:

- The temperature is above 30°, $A := [30, \infty)$.
- The temperature is equal to 35°, $B := 35$.
- The temperature is any number, $C := \mathbb{R}$.

If the temperature turns out to be 40°, then A and C occur, but B does not.

....

In order to quantify how likely an event is, we assign it a number, which we call a *probability*. The key idea behind the concept of probability is to interpret the uncertain phenomenon of interest as an experiment, which *can be repeated over and over*. Of course, this is just an abstraction. Many uncertain phenomena, such as the next presidential election, will occur only once. However, thinking of them as repeatable experiments enables us to quantify our uncertainty about them. The probability $P(A)$ of an event A represents the fraction of times that the event occurs (i.e., the outcome of the experiment belongs to the event), when we repeat the experiment an arbitrarily large number of times:

[1] Strictly speaking, temperatures cannot be lower than absolute zero, but we use the whole real line for convenience.

$$P(\text{event}) := \frac{\text{times event occurs}}{\text{total repetitions}}. \tag{1.1}$$

Notice that the probability is between zero and one because the number of times the event occurs must be between zero and the total number of repetitions. This is an informal definition of probability, which enables us to build intuition about its properties. We provide a formal definition in Section 1.2.

When determining the probabilities associated with a sample space, we do not need to assign a probability to every subset of the sample space. In fact, when the sample space is continuous, it is usually not possible to do this in a consistent manner. We refer the interested reader to any textbook on measure theory for more details. In any case, we definitely want to assign probabilities to *some* events. In the remainder of this section, we discuss what these events should be and derive their associated probabilities using our informal definition of probability.

1.1.1 Probability of the Sample Space

We should definitely assign a probability to the event that *anything at all* happens. This event contains all possible outcomes, so it is equal to the sample space Ω. Every time we repeat the experiment, we obtain an outcome that must be in Ω, so by our informal definition of probability,

$$P(\Omega) = \frac{\text{times } \Omega \text{ occurs}}{\text{total repetitions}} \tag{1.2}$$

$$= \frac{\text{total repetitions}}{\text{total repetitions}} \tag{1.3}$$

$$= 1. \tag{1.4}$$

Therefore, the probability assigned to the sample space should always equal one.

1.1.2 Probability of Unions and Intersections of Events

If we assign a probability to two events, we should also assign probabilities to their union and intersection. The union of two events is the event that *either* of them occurs. The intersection of two events is the event that *both* of them occur simultaneously. We begin by considering *disjoint* events, which are events that do not have any outcomes in common, so their intersection is empty. In Example 1.4, the events A and B are disjoint because no outcome is in both events, but A and C are not disjoint because the outcome 5 belongs to both of them. If two events D_1 and D_2 are disjoint, our informal definition of probability implies

$$P(D_1 \cup D_2) = \frac{\text{times } D_1 \text{ or } D_2 \text{ occur}}{\text{total repetitions}} \tag{1.5}$$

$$= \frac{\text{times } D_1 \text{ occurs} + \text{times } D_2 \text{ occurs}}{\text{total repetitions}} \tag{1.6}$$

$$= \frac{\text{times } D_1 \text{ occurs}}{\text{total repetitions}} + \frac{\text{times } D_2 \text{ occurs}}{\text{total repetitions}} \tag{1.7}$$

$$= P(D_1) + P(D_2). \tag{1.8}$$

Therefore, the probability of the union of disjoint events should equal the sum of their individual probabilities.

If two events E_1 and E_2 are not disjoint, then their intersection is not empty. As a result, according to our informal definition, the probability of their union equals

$$P\left(E_1 \cup E_2\right) = \frac{\text{times } E_1 \text{ or } E_2 \text{ occur}}{\text{total repetitions}} \tag{1.9}$$

$$= \frac{\text{times } E_1 \text{ occurs } + \text{ times } E_2 \text{ occurs } - \text{ times } E_1 \text{ and } E_2 \text{ occur}}{\text{total repetitions}}$$

$$= \frac{\text{times } E_1 \text{ occurs}}{\text{total repetitions}} + \frac{\text{times } E_2 \text{ occurs}}{\text{total repetitions}} - \frac{\text{times } E_1 \text{ and } E_2 \text{ occur}}{\text{total repetitions}}$$

$$= P\left(E_1\right) + P\left(E_2\right) - P\left(E_1 \cap E_2\right). \tag{1.10}$$

We subtract the probability of the intersection to avoid counting its outcomes twice.

From (1.10) we obtain a formula for the probability of the intersection of two events:

$$P\left(E_1 \cap E_2\right) = P\left(E_1\right) + P\left(E_2\right) - P\left(E_1 \cup E_2\right). \tag{1.11}$$

1.1.3 Probability of the Complement of an Event

If we assign a probability to an event, we should also assign a probability to its complement, that is, to the event *not* occurring. Mathematically, the complement is the set of all the outcomes that are *not* in the event. In Example 1.4, the complement of A is $\{1, 2, 3, 4, 6\}$ and the complement of B is $\{1, 3, 5\}$. For any event E, the union of E and its complement E^c is equal to the whole sample space (every outcome is either in E or in its complement). In addition, E and E^c are disjoint by definition (no outcome can be in both events). By our informal definition of probability, this implies

$$P\left(E\right) + P\left(E^c\right) = P\left(E \cup E^c\right) \tag{1.12}$$

$$= P(\Omega) \tag{1.13}$$

$$= 1, \tag{1.14}$$

so to compute the probability of the complement of E, we just need to subtract its probability from one, $P\left(E^c\right) = 1 - P\left(E\right)$. An intuitive consequence is that if an event is very likely (probability close to one), its complement should be unlikely (probability close to zero), and vice versa.

1.2 Mathematical Definition of Probability

In this section, we present the mathematical framework of probability spaces, which allows us to characterize uncertain phenomena using probabilities. A probability space has three components. First, a sample space containing the mutually exclusive outcomes associated with the phenomenon. Second, a collection containing the events that are assigned probabilities. Third, a probability measure, which is a function that assigns a probability to each event in the collection.

The collection of events in a probability space must satisfy the conditions in the following definition.

Definition 1.7 (Collection of events). *When defining a probability space based on a sample space Ω, we assign probabilities to a collection of events (a set of subsets of Ω) denoted by C, such that:*

1 If an event belongs to the collection, $A \in C$, then its complement belongs to the collection, $A^c \in C$.

2 If two events A and B belong to the collection, $A, B \in C$, then their union belongs to the collection, $A \cup B \in C$. This also holds for infinite sequences; if $A_1, A_2, A_3, \ldots \in C$ then $\cup_{i=1}^{\infty} A_i \in C$.

3 The sample space is in the collection, $\Omega \in C$.

A collection satisfying Definition 1.7 is called a σ-algebra in mathematical jargon, which may sound somewhat intimidating. However, the definition just implements the intuitive properties discussed in Section 1.1. If we assign probabilities to certain events, then *we should also assign probabilities to their complements, unions, and intersections.* Although the definition does not mention intersections explicitly, it implies that intersections of events in C also belong to C. This follows from the fact that $A \cap B = (A^c \cup B^c)^c$ (a consequence of De Morgan's laws) combined with Conditions 1 and 2. The empty set $\emptyset$ always belongs to a valid collection because it is the complement of Ω. The simplest possible collection satisfying the conditions is $\{\Omega, \emptyset\}$, but this is not a very interesting collection; usually we want to consider more events.

Example 1.8 (Die roll: collection of events). A natural choice for the collection of events in our six-sided die example (Example 1.1) is the *power set* of the sample space $\Omega := \{1, 2, 3, 4, 5, 6\}$, which is the set of all $2^6 = 64$ subsets of Ω. However, other choices are possible. For example, we may want to consider the smallest possible collection containing the event $A := \{5\}$. In that case, the collection must also contain $A^c = \{1, 2, 3, 4, 6\}$ by Condition 1 in Definition 1.7, Ω by Condition 3, and the empty set $\emptyset$ by Conditions 1 and 3. This is enough. You can check that the collection $\{\emptyset, A, A^c, \Omega\}$ satisfies Definition 1.7.
. . . .

Once we have defined a sample space and a corresponding collection of events, the final ingredient to define a probability space is a probability measure that assigns probabilities to the events in the collection. The probability measure must satisfy the following axioms, which encode the intuitive properties of probability derived in Section 1.1.

Definition 1.9 (Probability measure). *Given a sample space Ω, let C be a collection of events satisfying the conditions in Definition 1.7. A probability measure P is a function, which maps events in C to a number between 0 and 1, satisfying the following axioms:*

1 *All probabilities are nonnegative, $P(A) \geq 0$ for any event $A \in C$.*

2 *The probability of the sample space is one, $P(\Omega) = 1$.*

3 *If the events $A_1, A_2, \ldots, A_n \in C$ are disjoint (i.e., $A_i \cap A_j = \emptyset$ for $i \neq j$), then the probability of their union equals the sum of their individual probabilities,*

$$P\left(\cup_{i=1}^{n} A_i\right) = \sum_{i=1}^{n} P(A_i). \tag{1.15}$$

Similarly, for a countably infinite sequence of disjoint events $A_1, A_2, \ldots \in C$,

$$P\left(\lim_{n \to \infty} \cup_{i=1}^{n} A_i\right) = \lim_{n \to \infty} \sum_{i=1}^{n} P(A_i). \tag{1.16}$$

Axiom 3 in Definition 1.9 implies the formula (1.11) for the probability of the intersection of two events, derived informally in Section 1.1.2.

Lemma 1.10 (Probability of the intersection). *For any probability measure* P *satisfying the axioms in Definition 1.9, and any events* A *and* B *in the corresponding collection of events,*

$$P(A \cap B) = P(A) + P(B) - P(A \cup B). \tag{1.17}$$

Proof First, we decompose A into the union of $A \cap B$ and $A \cap B^c$, which are disjoint events, so that by Axiom 3 in Definition 1.9,

$$P(A) = P(A \cap B) + P(A \cap B^c). \tag{1.18}$$

Similarly,

$$P(B) = P(A \cap B) + P(A^c \cap B). \tag{1.19}$$

Then, we decompose $A \cup B$ into the union of $A \cap B$, $A \cap B^c$, and $A^c \cap B$, which are all disjoint, so that

$$P(A \cup B) = P(A \cap B^c) + P(A^c \cap B) + P(A \cap B) \tag{1.20}$$
$$= P(A) + P(B) - P(A \cap B), \tag{1.21}$$

where the last equality follows from (1.18) and (1.19). ∎

The formula for the probability of the complement of an event derived in Section 1.1.3 is also a direct consequence of Definition 1.9. The proof follows from the same argument as in Section 1.1.3, so we omit it here.

Lemma 1.11 (Probability of the complement of an event). *For any probability measure* P *satisfying the conditions in Definition 1.9, and any event* A,

$$P(A^c) = 1 - P(A). \tag{1.22}$$

Another important consequence of Definition 1.9 is that, if an event B contains another event A, then the probability of B cannot be smaller than the probability of A.

Lemma 1.12 (Subset of an event). *For any probability measure* P *satisfying the conditions in Definition 1.9, assume that there exist two events* A *and* B *in the corresponding collection of events, such that* $A \subseteq B$. *Then* $P(A) \leq P(B)$.

Proof We can express B as the union of the two disjoint events $A \cap B$ and $A^c \cap B$. Since $A \subseteq B$, $A \cap B = A$, so that by Axiom 3 in Definition 1.9

$$P(B) = P(A) + P(A^c \cap B) \tag{1.23}$$
$$\geq P(A) \tag{1.24}$$

because $P(A^c \cap B) \geq 0$ by Axiom 1 in Definition 1.9. ∎

A caveat to Lemma 1.12 is that it is possible for a subset of an event in the collection to *not* belong to the collection, which means that its probability is not defined. Consider, for instance, the collection $\{\emptyset, A, A^c, \Omega\}$ in Example 1.8, where $A := \{5\}$. The event $\{2\}$ is a subset of A^c, but it does not belong to the collection, so there is no probability assigned to it.

Probability measures have similar properties to other measures such as mass, length, area, or volume. For example, the area of the union of two disjoint two-dimensional (2D) shapes is the sum of their individual areas. This motivates the use of Venn diagrams to visualize probability spaces. In a Venn diagram, the outcomes in the sample space are represented

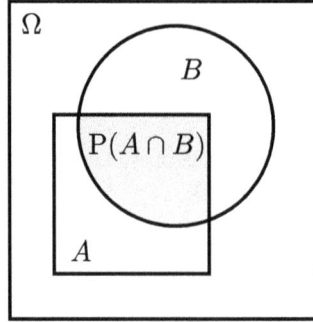

Figure 1.1 Venn diagram of a probability space. The Venn diagram represents a probability space. The sample space Ω is the big square that encompasses everything. The small square and the circle represent two events A and B. Their areas are equal to their respective probabilities. The probability of their intersection $A \cap B$ is equal to the area of the intersection between the two shapes, which is shaded.

as points in two dimensions. Events are sets of points, depicted as regions delimited by closed curves. The probability of each event is equal to the area of the corresponding region. The region representing the sample space must have area one and contain all outcomes. Figure 1.1 shows an example.

Example 1.13 (Simple probability measure). Consider the collection of events $\{\emptyset, A, A^c, \Omega\}$, where A is an arbitrary event. To define a valid probability measure, we just need to assign a probability to A, $\mathrm{P}(A) = \theta$. The probability θ can be any number between zero and one. Once that value is fixed, the probabilities of the remaining events are determined by the conditions in Definition 1.9. By Lemma 1.11, $\mathrm{P}(A^c) = 1 - \theta$. By Axiom 2, $\mathrm{P}(\Omega) = 1$, which implies $\mathrm{P}(\emptyset) = 0$, also by Lemma 1.11.

....

Example 1.14 (Die roll: probability measure). As explained in Example 1.8, a reasonable choice for the collection of events associated with the single six-sided die roll is the power set of the sample space $\Omega := \{1, 2, 3, 4, 5, 6\}$. At first, it may seem daunting to define the probability measure, given that there are 64 events in the collection (all possible subsets of Ω). However, we can apply a simple strategy: we divide the sample space into a *partition* of events and assign probabilities to the events in the partition.

A partition of the sample space Ω is any collection of disjoint sets $A_1, A_2, \ldots$ that covers Ω, meaning that $\Omega = \cup_i A_i$. In this case, we choose $A_i := \{i\}$, for i in $\{1, 2, 3, 4, 5, 6\}$. These six events are disjoint and their union equals Ω. We assign a probability to each of them,

$$P(A_i) = \theta_i, \tag{1.25}$$

where $\theta_1, \theta_2, \ldots, \theta_6$ are numbers between zero and one. The careful reader may have noticed that these numbers cannot be completely arbitrary. The sum of the probabilities must equal one, due to Axioms 2 and 3 in Definition 1.9:

$$\sum_{i=1}^{6} \theta_i = \sum_{i=1}^{6} \mathrm{P}(A_i) \tag{1.26}$$

$$= \mathrm{P}(\cup_{i=1}^{6} A_i) \tag{1.27}$$
$$= P(\Omega) \tag{1.28}$$
$$= 1. \tag{1.29}$$

Let us assume that this condition is satisfied. Then we are actually done! Any event of the collection in the probability space can be decomposed as a union of events in the partition. Since these events are disjoint, we can add their individual probabilities to compute the probability of the event, leveraging Axiom 3 in Definition 1.9. For instance, the probability of the event *the roll is even* ($\{2, 4, 6\}$) equals

$$\mathrm{P}\left(\{2, 4, 6\}\right) = \mathrm{P}\left(\cup_{i \in \{2,4,6\}} A_i\right) \tag{1.30}$$
$$= \sum_{i \in \{2,4,6\}} \mathrm{P}\left(A_i\right) \tag{1.31}$$
$$= \theta_2 + \theta_4 + \theta_6. \tag{1.32}$$

Note that for our strategy to work, the partition needs to be granular enough. The events $\{1\}$ and $\{2, 3, 4, 5, 6\}$ are also a partition of Ω, but we cannot express $\{2, 4, 6\}$ as a union of events in this partition.
. . . .

We have now rigorously defined all the elements of a probability space. This yields the following formal definition.

Definition 1.15 (Probability space). *A probability space is a triple $(\Omega, \mathcal{C}, \mathrm{P})$ consisting of:*

- *A sample space Ω, which contains all possible outcomes of the experiment.*
- *A collection of events $\mathcal{C}$, which satisfies the conditions in Definition 1.7.*
- *A probability measure P assigning probabilities to the events in $\mathcal{C}$, which satisfies the axioms in Definition 1.9.*

At this point, you may feel that this probability-space business sounds pretty complicated. We have explained how to choose a sample space, a collection of events, and a probability measure for a very simple example (the single die roll), and even that was not very straightforward. Imagine doing it for more complex phenomena! The good news is that, in practice, we never construct probability spaces like this. Instead, we use random variables to define probability spaces implicitly, without worrying about the gory mathematical details. We discuss random variables at length in Chapters 2–6.

1.3 Conditional Probability

1.3.1 Definition

Conditional probability is a crucial concept in probabilistic modeling. It allows us to update models, when additional information is revealed. Imagine that we are interested in the probability that an airplane is late, if it rains. We define a probability space where the collection of events contains the event R (*it rains*), the event L (*the airplane is late*), and all their complements, unions, and intersections. Let us assume that we have estimated the following probabilities from past data, as described in Section 1.4:

$$P\left(L \cap R^{c}\right)=\frac{2}{20}, \qquad P\left(L^{c} \cap R^{c}\right)=\frac{14}{20},$$

$$P\left(L \cap R\right)=\frac{3}{20}, \qquad P\left(L^{c} \cap R\right)=\frac{1}{20}. \tag{1.33}$$

By Axiom 3 in Definition 1.9, the probability that the plane is late equals

$$P(L) = P(L \cap R^{c}) + P(L \cap R) \tag{1.34}$$

$$= \frac{1}{4} \tag{1.35}$$

because $L = (L \cap R^{c}) \cup (L \cap R)$, and the events $L \cap R^{c}$ and $L \cap R$ are disjoint. However, this is not the probability we are interested in! According to our intuitive definition of probability in (1.1), we can interpret $P(L)$ as

$$P(L) = \frac{\text{times airplane is late}}{\text{total repetitions}}, \tag{1.36}$$

where we imagine that the flight is an experiment that can be repeated many times. This is not what we want. Our goal is to determine the probability that the plane is late *if it rains*, which can be captured by modifying (1.36) to equal the fraction of late arrivals *out of the times it rains*. This yields the *conditional probability*

$$P(L \mid R) = \frac{\text{times airplane is late and it rains}}{\text{times it rains}}. \tag{1.37}$$

Now, to express this quantity in terms of our known probabilities, we multiply and divide by the total repetitions,

$$P(L \mid R) = \frac{\text{times airplane is late and it rains}}{\text{total repetitions}} \cdot \frac{\text{total repetitions}}{\text{times it rains}} \tag{1.38}$$

$$= \frac{P(L \cap R)}{P(R)}. \tag{1.39}$$

Since $P(L \cap R) = 3/20$ and $P(R) = P(L \cap R) + P(L^{c} \cap R) = 1/5$ (by Axiom 3 in Definition 1.9),

$$P(L \mid R) = 3/4. \tag{1.40}$$

The conditional probability that the plane is late, given that it rains, is three times larger than $P(L)$.

Inspired by our intuitive reasoning, we define conditional probability more formally. Let $(\Omega, \mathcal{C}, P)$ be a probability space, and let $A \in \mathcal{C}$ be an event with nonzero probability. In order to condition on A, we build a new probability space $(\Omega_{A}, \mathcal{C}_{A}, P(\cdot \mid A))$ that preserves the properties of the original probability space as much as possible, but where *all outcomes are in A*. We denote the new probability measure $P(\cdot \mid A)$ to indicate that we are conditioning on A. In the new probability space, every outcome belongs to A, so it is natural to set the new sample space equal to A, $\Omega_{A} := A$. If an outcome in the new probability space belongs to an event B in $\mathcal{C}$, then it must lie in $A \cap B$. We therefore define the new collection of events $\mathcal{C}_{A}$ as the collection containing the intersections of A with each event in $\mathcal{C}$ (in Exercise 1.1, we check that this satisfies the conditions in Definition 1.7).

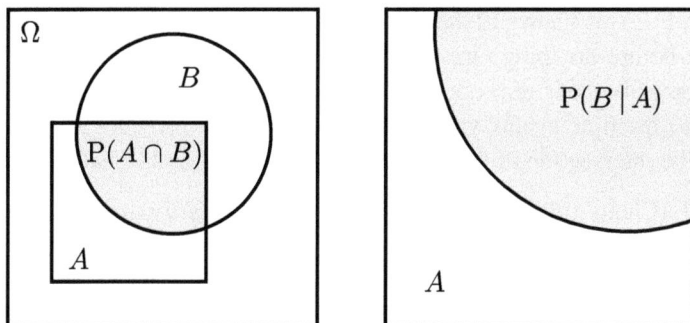

Figure 1.2 Conditional probability. The Venn diagram on the left depicts a probability space where the sample space Ω is a square with area one. The shaded area is the intersection $A \cap B$ of events A (represented by a square) and B (represented by a circle). In order to condition on A, we update the probability space as shown on the right. We set the sample space to equal A and discard the rest of Ω. In addition, we *blow up* the area assigned to A in the Venn diagram by a factor of $1/\mathrm{P}(A)$ to ensure that the new sample space has unit area. This increases the area assigned to $A \cap B$ from $\mathrm{P}(A \cap B)$ to $\frac{\mathrm{P}(A \cap B)}{\mathrm{P}(A)} := \mathrm{P}(B \mid A)$, which is the conditional probability of B given A by Definition 1.16.

All we have left to do is to define the probability measure $\mathrm{P}(\cdot \mid A)$. We could be tempted to use the same probability measure P as in the original probability space. Any event in $\mathcal{C}_A$ is of the form $A \cap B$ for some $B \in \mathcal{C}$, so it also belongs to $\mathcal{C}$ and is assigned the probability $\mathrm{P}(A \cap B)$ by P. However, this does not yield a valid probability measure for our new probability space. The probability of the whole sample space would equal $\mathrm{P}(A)$ instead of 1! The problem is that P assigns nonzero probability to A^c, which cannot occur in the new probability space. To correct for this, we divide all the probabilities by $\mathrm{P}(A)$. This normalizes the conditional probabilities and ensures that the conditional probability of A given A equals 1. The resulting definition of conditional probability coincides with our intuitive definition (1.39).

Definition 1.16 (Conditional probability). *Let A and B be events in a probability space $(\Omega, \mathcal{C}, \mathrm{P})$ and assume $\mathrm{P}(A) \neq 0$. The conditional probability of B given A is defined as*

$$\mathrm{P}(B \mid A) := \frac{\mathrm{P}(B \cap A)}{\mathrm{P}(A)}. \tag{1.41}$$

Defined in this way, $\mathrm{P}_A(B \cap A) := \mathrm{P}(\cdot \mid A)$ is a valid probability measure for the probability space $(A, \mathcal{C}_A, \mathrm{P}_A)$. In Exercise 1.1, we check that it satisfies the axioms in Definition 1.9. Figure 1.2 uses a Venn diagram to illustrate the definition of conditional probability.

1.3.2 The Chain Rule

By Definition 1.16, we can express the probability of the intersection of two events A and B as follows:

$$\mathrm{P}(A \cap B) = \mathrm{P}(A)\,\mathrm{P}(B \mid A) \tag{1.42}$$

$$= \mathrm{P}(B)\,\mathrm{P}(A \mid B). \tag{1.43}$$

In this formula, $P(A)$ is known as the *prior* probability of A because it describes our uncertainty about A before anything else is revealed. $P(B \mid A)$ is the *posterior* probability of B because it describes our uncertainty about B once we know that A occurred. Generalizing (1.42) to multiple events yields the *chain rule* that provides a factorization of the probability of the intersection of the events as a product of conditional probabilities.

Theorem 1.17 (Chain rule). *Let* $(\Omega, \mathcal{C}, P)$ *be a probability space. For any* n *events* $A_1, A_2, \ldots, A_n$ *belonging to the collection* $\mathcal{C}$,

$$P(\cap_i A_i) = P(A_1) P(A_2 \mid A_1) P(A_3 \mid A_1 \cap A_2) \cdots P(A_n \mid A_1 \cap \cdots \cap A_{n-1}) \quad (1.44)$$

$$= P(A_1) \prod_{i=2}^{n} P(A_i \mid \cap_{j=1}^{i-1} A_j). \quad (1.45)$$

Proof   We prove the result for three events A_1, A_2, and A_3. The argument can be easily extended by induction to any countable number of events. We apply (1.43) twice. Setting $A := A_3$ and $B := A_1 \cap A_2$ yields

$$P(A_1 \cap A_2 \cap A_3) = P(A_1 \cap A_2) P(A_3 \mid A_1 \cap A_2). \quad (1.46)$$

Setting $A := A_2$ and $B := A_1$, this implies

$$P(A_1 \cap A_2 \cap A_3) = P(A_1) P(A_2 \mid A_1) P(A_3 \mid A_1 \cap A_2). \quad (1.47)$$

∎

Note that the order in which we condition when applying the chain rule is *completely arbitrary*. For example, for three events A, B, and C, we have six possible factorizations,

$$P(A \cap B \cap C) = P(A) P(B \mid A) P(C \mid A \cap B) \quad (1.48)$$
$$= P(A) P(C \mid A) P(B \mid A \cap C) \quad (1.49)$$
$$= P(B) P(A \mid B) P(C \mid A \cap B) \quad (1.50)$$
$$= P(B) P(C \mid B) P(A \mid B \cap C) \quad (1.51)$$
$$= P(C) P(A \mid C) P(B \mid A \cap C) \quad (1.52)$$
$$= P(C) P(B \mid C) P(A \mid B \cap C). \quad (1.53)$$

In probabilistic modeling (and homework problems), it is often crucial to choose the order wisely in order to exploit the information that we have available.

To alleviate notation, in the rest of the book, we often use a comma instead of the symbol $\cap$ to describe intersections of events. For example, we write $P(A, B, C)$ instead of $P(A \cap B \cap C)$.

1.3.3 Law of Total Probability

Sometimes, estimating the probability of a certain event is more difficult than estimating its probability conditioned on simpler events. The law of total probability, illustrated in Figure 1.3, allows us to compute the probability of the event by combining such conditional probabilities.

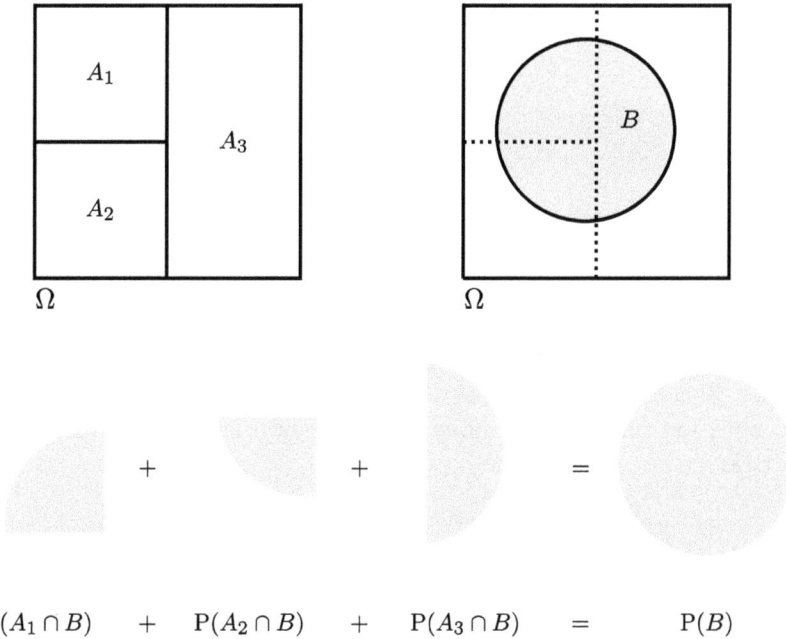

$$\text{P}(A_1 \cap B) \quad + \quad \text{P}(A_2 \cap B) \quad + \quad \text{P}(A_3 \cap B) \quad = \quad \text{P}(B)$$

Figure 1.3 The law of total probability. The Venn diagram in the upper left corner shows a partition of Ω consisting of three events A_1, A_2, and A_3. The Venn diagram in the upper right corner shows another event B. B can be decomposed into the union of three smaller events, each equal to its intersection with one of the events of the partition. These events are disjoint, so the sum of the areas of the 2D regions representing them in the Venn diagram is equal to the area of the 2D region representing B, as depicted by the graphic equation underneath the Venn diagrams. This is consistent with Theorem 1.18 that establishes that the probability of B is equal to the sum of the probabilities of $A_1 \cap B$, $A_2 \cap B$, and $A_3 \cap B$.

Theorem 1.18 (Law of total probability). *Let $(\Omega, \mathcal{C}, \text{P})$ be a probability space, and let A_1, $A_2, \ldots \in \mathcal{C}$ be a partition of the sample space Ω (meaning that the events are disjoint and $\cup_i A_i = \Omega$). For any event B belonging to the collection $\mathcal{C}$,*

$$\text{P}(B) = \sum_i \text{P}(B \cap A_i) \tag{1.54}$$

$$= \sum_i \text{P}(A_i) \text{P}(B \mid A_i). \tag{1.55}$$

Proof Consider the intersections between B and the events in the partition $B \cap A_1$, $B \cap A_2, \ldots$ (illustrated in Figure 1.3). Their union $\cup_i (B \cap A_i)$ is equal to B. To prove this, we show that the two events contain each other. $\cup_i (B \cap A_i) \subseteq B$ because every element of $\cup_i (B \cap A_i)$ is in $B \cap A_i$ for some i, and consequently in B. Conversely, $B \subseteq \cup_i (B \cap A_i)$ because every element of B is in A_i, and consequently in $B \cap A_i$, for one value of i (because $A_1, A_2, \ldots$ form a partition). Since $B \cap A_1, B \cap A_2, \ldots$ are disjoint (because $A_1, A_2, \ldots$ are disjoint), by Axiom 3 in Definition 1.9, and the chain rule (Theorem 1.17),

$$\text{P}(B) = \text{P}(\cup_i (B \cap A_i)) \tag{1.56}$$

$$= \sum_i \mathrm{P}\left(B \cap A_i\right) \qquad (1.57)$$

$$= \sum_i \mathrm{P}\left(A_i\right)\mathrm{P}\left(B \mid A_i\right). \qquad (1.58)$$

∎

Example 1.19 (Flight delay and rain). Imagine that in our flight delay example, we only have access to the conditional probability that the flight is late given rain and no rain, but we are interested in the probability that the flight is late. Specifically, we know that $\mathrm{P}\left(L \mid R\right) = 0.75$ and $\mathrm{P}\left(L \mid R^c\right) = 0.125$, and we want to compute $\mathrm{P}\left(L\right)$. The events R and R^c are disjoint and cover the whole sample space, so they form a partition of the sample space. Consequently, as long as we know their probabilities, we can apply the law of total probability to obtain the probability that the flight is late. Assuming $\mathrm{P}\left(R\right) = 0.2$, by Theorem 1.18,

$$\mathrm{P}\left(L\right) = \mathrm{P}\left(L \mid R\right)\mathrm{P}\left(R\right) + \mathrm{P}\left(L \mid R^c\right)\mathrm{P}\left(R^c\right) \qquad (1.59)$$

$$= 0.75 \cdot 0.2 + 0.125 \cdot 0.8 = 0.25. \qquad (1.60)$$

The probability that the flight is delayed is $1/4$.

1.3.4 Bayes' Rule

It is important to realize that in general $\mathrm{P}\left(A \mid B\right)$ is not necessarily equal to $\mathrm{P}\left(B \mid A\right)$. For example, most players in the NBA probably own a basketball: $\mathrm{P}\left(\text{owns ball} \mid \text{NBA}\right)$ is very high. However, most people who own basketballs (including myself) are not in the NBA: $\mathrm{P}\left(\text{NBA} \mid \text{owns ball}\right)$ is very low. The reason is that the prior probabilities are very different: $\mathrm{P}\left(\text{NBA}\right)$ is much smaller than $\mathrm{P}\left(\text{owns ball}\right)$. This is illustrated in Figure 1.4. Consequently, in order to compute $\mathrm{P}\left(A \mid B\right)$ from $\mathrm{P}\left(B \mid A\right)$, we need to take into account the prior probability of each event, as dictated by Bayes' rule.

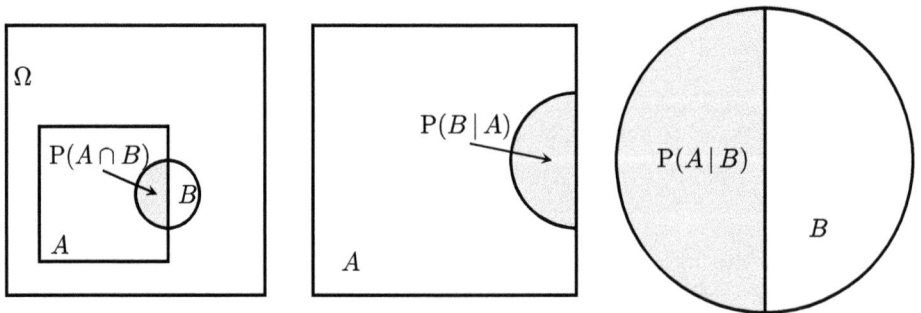

Figure 1.4 $\mathbf{P}(B \mid A) \neq \mathbf{P}(A \mid B)$. The Venn diagram on the left depicts a probability space where the sample space Ω is a square with area one. The shaded area is the intersection $A \cap B$ of the events A (represented by a square) and B (represented by a circle). In the middle diagram, we condition on A by setting the sample space to equal A, and expanding its area by a factor of $1/\mathrm{P}(A)$ (as in Figure 1.2). In the right diagram, we condition on B by setting the sample space to equal B, and expanding its area by a factor of $1/\mathrm{P}(B)$. Since $\mathrm{P}(A) \neq \mathrm{P}(B)$, the area corresponding to $A \cap B$ is enlarged to different extents in each case, and therefore $\mathrm{P}(A \mid B) \neq \mathrm{P}(B \mid A)$.

Theorem 1.20 (Bayes' rule). *For any events A and B in a probability space,*

$$P(A \mid B) = \frac{P(A) P(B \mid A)}{P(B)}, \tag{1.61}$$

as long as $P(B) > 0$.

Proof By the definition of conditional probability (Definition 1.16) and the chain rule (Theorem 1.17),

$$P(A \mid B) := \frac{P(A, B)}{P(B)} \tag{1.62}$$

$$= \frac{P(A) P(B \mid A)}{P(B)}. \tag{1.63}$$

■

Example 1.21 (Conditional probability of rain given flight delay). Imagine that the flight in Example 1.19 was late, but you don't know whether it rained or not because you spent the day indoors studying probability spaces. You decide to use your newly acquired knowledge to estimate the probability that it rained. The prior probability of rain is $P(R) = 0.2$, but since we know the flight was late, we should update the estimate. Applying Bayes' rule and the law of total probability:

$$P(R \mid L) = \frac{P(L \mid R) P(R)}{P(L)} \tag{1.64}$$

$$= \frac{P(L \mid R) P(R)}{P(L \mid R) P(R) + P(L \mid R^c) P(R^c)} \tag{1.65}$$

$$= \frac{0.75 \cdot 0.2}{0.75 \cdot 0.2 + 0.125 \cdot 0.8} = 0.6. \tag{1.66}$$

The posterior probability of rain conditioned on the flight delay is much higher than the prior probability of rain.

··

1.4 Estimating Probabilities from Data

Sections 1.1–1.3 describe the machinery of probability spaces, which provides a set of rules to define and manipulate probabilities. Now, we ask a question that takes us beyond probability theory into the realm of statistics and data science: *How do we estimate the probability of an event from data?*

In statistics, a rule for estimating a certain quantity of interest is called an *estimator*. In order to design an estimator for the probability of an event, we seek inspiration in our intuitive definition of probability (1.1). Assume that we have access to a dataset where each data point can be modeled as an outcome in a probability space. Since the probability of an event represents the fraction of times the event occurs, it seems natural to use the observed fraction of occurrences as an estimate of the probability.

Definition 1.22 (Empirical probability). *Let Ω denote a sample space, and A an event within that sample space, $A \subseteq \Omega$. Let $X := \{x_1, x_2, \ldots, x_n\}$ denote a dataset with values in Ω. The empirical probability of A is defined as the fraction of elements of X that belong to A,*

$$P_X(A) := \frac{1}{n} \sum_{i=1}^{n} 1(x_i \in A), \tag{1.67}$$

where $1(x_i \in A)$ is an indicator function that is equal to one, if $x_i \in A$, and to zero otherwise.

In words, the empirical probability of an event is the fraction of times we observe it in the data. In Exercise 1.2, we check that the empirical probability is a valid probability measure that satisfies the axioms in Definition 1.9.

Example 1.23 (Unfair die). In books about probability, six-sided dice are often assumed to be fair, meaning that there is an equal chance of rolling every number. However, this may not be the case for real dice. My daughter has a toy six-sided die, which I suspect is not fair. In order to resolve this question scientifically, I rolled it 60 times and recorded the results. Let $n_j, j \in \{1, 2, 3, 4, 5, 6\}$, denote the number of times that a roll with value j was observed. According to the data,

$$n_1 := 10, \quad n_2 := 8, \quad n_3 := 18, \quad n_4 := 7, \quad n_5 := 7, \quad n_6 := 10. \tag{1.68}$$

Following Example 1.14, we model the die roll using a probability space where the collection of events is the power set of the outcomes, and we define the probability measure by assigning a probability to each event $A_j := \{j\}$, for j in $\{1, 2, 3, 4, 5, 6\}$. We denote the data by

$$X := \{x_1, x_2, \ldots, x_{60}\}, \tag{1.69}$$

where x_i indicates the value of the ith roll. By Definition 1.22, the empirical probability of A_j is

$$P_X(A_j) := \frac{1}{60} \sum_{i=1}^{60} 1(x_i = j) \tag{1.70}$$

$$= \frac{n_j}{60}, \tag{1.71}$$

which yields

$$P_X(A_1) = \frac{10}{60}, \qquad P_X(A_2) = \frac{8}{60}, \qquad P_X(A_3) = \frac{18}{60},$$

$$P_X(A_4) = \frac{7}{60}, \qquad P_X(A_5) = \frac{7}{60}, \qquad P_X(A_6) = \frac{10}{60}. \tag{1.72}$$

From the results, it looks like the die may not be fair, in line with my suspicions. In Chapter 10, we evaluate this conjecture rigorously, using the framework of hypothesis testing.

When computing empirical probabilities, we interpret each data point as the result of repeating an experiment that represents the phenomenon of interest. Mathematically, we assume that the data are *independent and identically distributed* (i.i.d.), which means that the value of each data point only depends on the corresponding probability, and not on the values of the other data. We provide a more formal definition of the i.i.d. assumption in Example 2.18 and Definition 2.23.

Table 1.1 **Empirical probability of a coin toss.** *The table shows ten different estimates of the probability of heads, obtained by simulating a fair coin flip twenty times and then computing the empirical probability as described in Definition 1.22. Most of the empirical probabilities are different from the true underlying probability, equal to 0.5.*

Heads (out of 20)	15	13	10	9	9	8	9	9	12	8
Empirical probability	0.75	0.65	0.5	0.45	0.45	0.4	0.45	0.45	0.6	0.4

In most cases, the i.i.d. assumption is just an approximation, but even if it were to hold exactly, empirical probabilities cannot be expected to be completely accurate. This is illustrated in Table 1.1, where we compute empirical probabilities in an idealized situation where the true underlying probabilities are known. We perform the following procedure ten times: We simulate twenty flips from a fair coin, for which the probability of heads is 0.5, and compute the empirical probability of heads. The empirical probability is equal to the true probability only once. In fact, if we use twenty-one flips instead, the empirical probability cannot be correct (we would need to observe ten and a half heads). This is our first encounter with a fundamental challenge in statistical estimation: Estimates based on finite data are almost never exact. However, the empirical-probability estimator approximates the true probability with arbitrary precision if the number of data is large enough (under certain reasonable assumptions), as established in Theorem 9.24 (see also Example 2.18).

Empirical probabilities can also be used to estimate conditional probabilities from data. Inspired by (1.39) we define the empirical conditional probability of an event B given another event A as the fraction of times we observe B within the subset of data where A occurs.

Definition 1.24 (Empirical conditional probability). *Let Ω be a sample space, and $X :=$ $\{x_1, x_2, \ldots, x_n\}$ a dataset with values in Ω. For any two subsets A and B of Ω, $A, B \subseteq \Omega$, the empirical conditional probability of B given A is the fraction of the elements of X in A, which also belong to B,*

$$P_X(B \mid A) := \frac{\sum_{i=1}^{n} 1(x_i \in A \cap B)}{\sum_{i=1}^{n} 1(x_i \in A)}, \tag{1.73}$$

where $1(x_i \in S)$ is an indicator function that is equal to one if $x_i \in S$ and to zero otherwise, for any event $S \subseteq \Omega$.

Example 1.25 (House of Representatives: Empirical probabilities). In this example, we model the voting behavior of congressmen in the US House of Representatives using data extracted from Dataset 1. We consider votes on two issues: Adoption of the budget resolution and duty-free exports. Table 1.2 shows the voting records. For simplicity, we ignore absences and abstentions. We would like to understand the relationship between the two issues. If a representative votes Yes for the budget, are they more likely to vote Yes for duty-free exports? To answer such questions, we build a probabilistic model, in which the voting process is interpreted as a repeatable experiment.

The outcome of the experiment consists of the votes on both issues. The sample space contains the four possible outcomes: *Yes-Yes, Yes-No, No-Yes,* and *No-No*. We define the events B and E to represent positive votes on the budget and on the duty-free exports issue, respectively. Since we do not consider absences or abstentions, B^c and E^c represent negative

Table 1.2 **Voting data from the US House of Representatives.** *Number of representatives who voted Yes or No on the adoption of the budget resolution, and on duty-free exports, in Dataset 1.*

		Duty-free exports	
		Yes	No
Budget	Yes	151	88
	No	21	140

votes. To estimate the probability of B and E, we divide the positive votes for each issue by the total votes, following Definition 1.22:

$$P(B) = \frac{239}{400} = 0.598, \tag{1.74}$$

$$P(E) = \frac{172}{400} = 0.43. \tag{1.75}$$

To estimate the conditional probability of E given B, we only consider outcomes in B (i.e., representatives who voted Yes on the budget) and compute what fraction of them that are also in E, following Definition 1.24:

$$P(E \mid B) = \frac{151}{239} = 0.632. \tag{1.76}$$

Similarly,

$$P(E \mid B^c) = \frac{21}{161} = 0.130. \tag{1.77}$$

Our analysis shows that if we know nothing about a representative, they are slightly more likely to vote No on the duty-free issue because $P(E)$ is smaller than 1/2. However, if we know that they have voted Yes on the budget, then they are more likely to also vote Yes on the duty-free issue because $P(E \mid B)$ is larger than 1/2. If we know that they voted No on the budget, then they are very likely to also vote No on the duty-free issue because $P(E^c \mid B^c) = 0.870$.

1.5 Independence

Conditional probabilities quantify the extent to which the occurrence of an event affects the probability of another event. In some cases, it makes no difference: The events are *independent*. More formally, two events A and B are independent if and only if

$$P(A \mid B) = P(A). \tag{1.78}$$

This definition is not valid if $P(B) = 0$. We usually use the following definition, which is equivalent to (1.78) by the chain rule (Theorem 1.17), but can also be applied when the probability of one of the events is zero.

Definition 1.26 (Independence of two events). *Let* $(\Omega, \mathcal{C}, \mathrm{P})$ *be a probability space. Two events* $A, B \in \mathcal{C}$ *are independent if and only if*

$$\mathrm{P}(A \cap B) = \mathrm{P}(A)\,\mathrm{P}(B). \tag{1.79}$$

The following example shows that when we consider more than two events, pairwise independence does not necessarily imply a lack of dependence between the events.

Example 1.27 (Two coin flips). Let $(\Omega, \mathcal{C}, \mathrm{P})$ be a probability space representing two fair coin flips. The sample space Ω contains four outcomes: *heads-heads*, *heads-tails*, *tails-heads*, and *tails-tails*. The collection $\mathcal{C}$ is the power set (all possible subsets) of Ω. The probability measure assigns

$$\mathrm{P}(\{\text{heads-heads}\}) = \mathrm{P}(\{\text{heads-tails}\}) = \mathrm{P}(\{\text{tails-heads}\})$$
$$= \mathrm{P}(\{\text{tails-tails}\}) = \frac{1}{4}.$$

We are interested in the following events:

$$A := \{\text{heads-heads, heads-tails}\} \quad \text{(first flip is heads)}, \tag{1.80}$$
$$B := \{\text{heads-heads, tails-heads}\} \quad \text{(second flip is heads)}, \tag{1.81}$$
$$C := \{\text{heads-heads, tails-tails}\} \quad \text{(flips are the same)}. \tag{1.82}$$

By Axiom 3 in Definition 1.9,

$$\mathrm{P}(A) = \mathrm{P}(\{\text{heads-heads}\} \cup \{\text{heads-tails}\}) \tag{1.83}$$
$$= \mathrm{P}(\{\text{heads-heads}\}) + \mathrm{P}(\{\text{heads-tails}\}) \tag{1.84}$$
$$= \frac{1}{2} \tag{1.85}$$

since the individual events are disjoint. Similarly,

$$\mathrm{P}(B) = \mathrm{P}(\{\text{heads-heads}\} \cup \{\text{tails-heads}\}) = \frac{1}{2}, \tag{1.86}$$

$$\mathrm{P}(C) = \mathrm{P}(\{\text{heads-heads}\} \cup \{\text{tails-tails}\}) = \frac{1}{2}. \tag{1.87}$$

By Definition 1.26, A, B, and C are pairwise independent:

$$\mathrm{P}(A, B) = \mathrm{P}(\{\text{heads-heads}\}) = \frac{1}{4} = \mathrm{P}(A)\mathrm{P}(B), \tag{1.88}$$

$$\mathrm{P}(A, C) = \mathrm{P}(\{\text{heads-heads}\}) = \frac{1}{4} = \mathrm{P}(A)\mathrm{P}(C), \tag{1.89}$$

$$\mathrm{P}(B, C) = \mathrm{P}(\{\text{heads-heads}\}) = \frac{1}{4} = \mathrm{P}(B)\mathrm{P}(C). \tag{1.90}$$

This makes sense. Revealing the result of the first flip provides no information about the result of the second flip. However, does this imply there is *no dependence between the three events*? Not at all! Notice that $A \cap B \cap C = \{\text{heads-heads}\} = A \cap B$, so the conditional probability of C given $A \cap B$ is

$$P(C \mid A, B) = \frac{P(A, B, C)}{P(A, B)} \tag{1.91}$$

$$= \frac{P(\{\text{heads-heads}\})}{P(\{\text{heads-heads}\})} \tag{1.92}$$

$$= 1, \tag{1.93}$$

which is definitely not equal to $P(C)$. Indeed, if we know that the first flip is heads and also that the second flip is heads, we can be sure that the two flips are the same! The three events are therefore not independent, despite being pairwise independent.

··

Motivated by this example, we extend the definition of independence to more than two events.

Definition 1.28 (Mutual independence of multiple events). *Let* $(\Omega, \mathcal{C}, P)$ *be a probability space. The events* $A_1, A_2, \ldots, A_n \in \mathcal{C}$ *are mutually independent if and only if for any possible subset of* m *events* $A_{i_1}, A_{i_2}, \ldots, A_{i_m}, \{i_1, i_2, \ldots, i_m\} \subseteq \{1, 2, \ldots, n\}$,

$$P\left(\cap_{j=1}^m A_{i_j}\right) = \prod_{j=1}^m P\left(A_{i_j}\right). \tag{1.94}$$

Definition 1.28 guarantees complete independence because if (1.94) holds for all possible subsets of events, then all conditional probabilities of each event A_i conditioned on any subset of the remaining events equal $P(A_i)$. For example,

$$P(A_3 \mid A_1, A_2) = \frac{P(A_1, A_2, A_3)}{P(A_1, A_2)} \tag{1.95}$$

$$= \frac{P(A_1)P(A_2)P(A_3)}{P(A_1)P(A_2)} \tag{1.96}$$

$$= P(A_3). \tag{1.97}$$

The following example investigates independence between events using real data.

Example 1.29 (House of Representatives: Vote dependence). Based on the empirical probabilities computed in Example 1.25, the events B and E are clearly not independent, since $P(E)$ is very different from $P(E \mid B)$. Here, we repeat the same analysis for two other issues. Voting Yes on an anti-satellite test ban is represented by the event A, and voting Yes on an immigration issue is represented by the event I. Table 1.3 shows the data, which

Table 1.3 *Voting data from the US House of Representatives. Number of representatives who voted Yes or No on an immigration issue, and on an anti-satellite test ban, in Dataset 1.*

		Immigration	
		Yes	No
Anti-satellite test ban	Yes	124	113
	No	89	93

are also extracted from Dataset 1. We again ignore absences and abstentions. To determine whether the events are independent, we compute the empirical probabilities following Definition 1.22,

$$P(A, I) = \frac{124}{419} = 0.296, \tag{1.98}$$

$$P(A) = \frac{237}{419} = 0.566, \tag{1.99}$$

$$P(I) = \frac{213}{419} = 0.508, \tag{1.100}$$

and verify that

$$P(A)P(I) = 0.288 \approx 0.296 = P(A, I). \tag{1.101}$$

This seems to indicate that the events are almost independent, which is also reflected in the conditional probabilities. By Definition 1.24,

$$P(A \mid I) = \frac{124}{213} = 0.582 \approx 0.566 = P(A). \tag{1.102}$$

In our model, the probability that a representative voted Yes on the anti-satellite test ban barely changes, if we find out that they voted Yes on the immigration issue.
....

You may be a bit uneasy about our conclusion in Example 1.29. Strictly speaking, the events A and I are not independent because this requires equality to hold exactly in (1.101). However, in practice, we *cannot expect to observe exact equality* for empirical probabilities computed from data. The reason is that these probabilities are extremely unlikely to be completely accurate, as discussed in Section 1.4 (see Table 1.1). The following example illustrates this in a situation, where we are pretty sure that the events are independent.

Example 1.30 (Tom Brady and Category 5 hurricanes). Table 1.4 shows in what years Tom Brady won the Super Bowl (top row) and in what years there was at least one Category 5

Table 1.4 **Tom Brady and Category 5 hurricanes.** *The table shows in what years Tom Brady won the Super Bowl (top row) and in what years there was at least one Category 5 hurricane in the North Atlantic Ocean (bottom row) between 2002 and 2021.*

Year	02	03	04	05	06	07	08	09	10	11
Brady wins	✓	✗	✓	✓	✗	✗	✗	✗	✗	✗
Hurricane	✗	✓	✓	✓	✗	✓	✗	✗	✗	✗

Year	12	13	14	15	16	17	18	19	20	21
Brady wins	✗	✗	✗	✓	✗	✓	✗	✓	✗	✓
Hurricane	✗	✗	✗	✗	✓	✓	✓	✓	✗	✗

hurricane in the North Atlantic Ocean (bottom row) between 2002 and 2021. By Definition 1.22, the empirical probability of a hurricane, represented by the event H, is

$$P(H) = \frac{8}{20} = 0.4. \tag{1.103}$$

We denote the event that Tom Brady wins the Super Bowl by T. Conditioned on this event, by Definition 1.24, the empirical probability of a hurricane is

$$P(H \mid T) = \frac{4}{7} = 0.571, \tag{1.104}$$

which is very different from $P(H)$. Is this proof that the two events are not independent? No, we simply don't have enough data. In fact, if Brady had won the 2012 Super Bowl and lost in 2017,[2] then $P(H \mid T)$ would equal 0.429, which is almost equal to $P(H)$. In Example 10.25, we examine these data from the perspective of hypothesis testing. This example may seem a bit silly, but many sport news articles have been written with flimsier quantitative evidence.
· ·

1.6 Conditional Independence

The dependence between two events in a probability space can change completely, when we condition on a third event. In particular, conditioning may render the two events independent from each other. This is captured by the concept of conditional independence. Two events A and B are conditionally independent given a third event C, if and only if

$$P(B \mid A, C) = P(B \mid C), \tag{1.105}$$

where $P(B \mid A, C) := P(B \mid A \cap C)$. Intuitively, this means that the probability of B is not affected by whether A occurs or not, *as long as C is known to occur*. The chain rule (Theorem 1.17) holds for the probability measure $P(\cdot \mid C)$ (as it is a valid probability measure, see Exercise 1.1), so (1.105) is equivalent to

$$P(A, B \mid C) = P(A \mid C) P(B \mid A, C) \tag{1.106}$$
$$= P(A \mid C) P(B \mid C). \tag{1.107}$$

Definition 1.31 (Conditional independence). *Let $(\Omega, \mathcal{C}, P)$ be a probability space. Two events $A, B \in \mathcal{C}$ are conditionally independent given a third event $C \in \mathcal{C}$ if and only if*

$$P(A \cap B \mid C) = P(A \mid C) P(B \mid C). \tag{1.108}$$

The events $A_1, A_2, \ldots, A_n \in \mathcal{C}$ are mutually conditionally independent given another event C if and only if for any possible subset of m events $A_{i_1}, A_{i_2}, \ldots, A_{i_m}, \{i_1, i_2, \ldots, i_m\} \subseteq \{1, 2, \ldots, n\}$,

$$P\left(\cap_{j=1}^m A_{i_j} \mid C\right) = \prod_{j=1}^m P\left(A_{i_j} \mid C\right). \tag{1.109}$$

The following examples show that independence does not imply conditional independence or vice versa.

---

[2] If you follow American football, you might know that this was very close to happening.

Example 1.32 (Conditional independence does not imply independence). Let us consider the probability space in Example 1.19, extended to include the event T, representing that a taxi is available when the flight arrives. Assume that

$$P\left(T \mid R\right) = 0.1, \quad P\left(T \mid R^c\right) = 0.6, \tag{1.110}$$

where R denotes the event that it rains. We model the events L (*flight is late*) and T as conditionally independent given the events R and R^c,

$$P\left(T, L \mid R\right) = P\left(T \mid R\right) P\left(L \mid R\right), \tag{1.111}$$

$$P\left(T, L \mid R^c\right) = P\left(T \mid R^c\right) P\left(L \mid R^c\right). \tag{1.112}$$

We are assuming that the availability of taxis is unrelated to flight delay, as long as we know whether it rains or not. Does this imply that they are also unrelated *if we don't know whether it rains*? More formally, are T and R independent?

They are not. By the law of total probability (Theorem 1.18) and the chain rule (Theorem 1.17), since R and R^c form a partition of the sample space,

$$P\left(T\right) = P\left(T, R\right) + P\left(T, R^c\right) \tag{1.113}$$

$$= P\left(T \mid R\right) P\left(R\right) + P\left(T \mid R^c\right) P\left(R^c\right) \tag{1.114}$$

$$= 0.1 \cdot 0.2 + 0.6 \cdot 0.8 = 0.5, \tag{1.115}$$

$$P\left(T \mid L\right) = \frac{P\left(T, L\right)}{P\left(L\right)} \tag{1.116}$$

$$= \frac{P\left(T, L, R\right) + P\left(T, L, R^c\right)}{P\left(L\right)} \tag{1.117}$$

$$= \frac{P\left(T \mid R\right) P\left(L \mid R\right) P\left(R\right) + P\left(T \mid R^c\right) P\left(L \mid R^c\right) P\left(R^c\right)}{P\left(L\right)}$$

$$= \frac{0.1 \cdot 0.75 \cdot 0.2 + 0.6 \cdot 0.125 \cdot 0.8}{0.25} = 0.3. \tag{1.118}$$

$P\left(T\right)$ and $P\left(T \mid L\right)$ are very different, so the events are *not* independent. This makes intuitive sense. The events L and T are connected through R. L provides information about R (if a flight is delayed, then it is more likely that it rained) and R provides information about T (taxis are more difficult to find when it rains). Consequently, L provides information about T: If a flight is delayed, taxis are more difficult to find because it is more likely that it rained. We conclude that conditional independence does not imply independence.
· ·

Example 1.33 (Independence does not imply conditional independence). Flight delays are sometimes caused by mechanical problems in the airplane. We incorporate another event M into our model, which represents a mechanical problem. We model the events M and R as independent, which implies that M and R^c are also independent (see Exercise 1.3),

$$P\left(M\right) = P\left(M \mid R\right) = P\left(M \mid R^c\right). \tag{1.119}$$

In addition, we assume

$$P\left(M\right) = 0.1 \quad P\left(L \mid M\right) = 0.7, \quad P\left(L \mid M^c\right) = 0.2, \quad P\left(L \mid R^c, M\right) = 0.5.$$

Now, imagine that we are waiting for a flight on a sunny day, and we are wondering whether there could be a mechanical problem. The fact that it is not raining is of no use to us because M and R^c are independent. Without any further information, the probability of M is 0.1.

Suddenly, they announce that the flight is late. Now, what is the probability that there is a mechanical problem? We may be tempted to answer $\mathrm{P}(M \mid L) = 0.28$, which can be derived by the same reasoning as in Example 1.21. However, the actual conditional probability is $\mathrm{P}(M \mid L, R^c)$ because the information that it is sunny is now relevant. It implies that the rain is not responsible for the delay, so intuitively a mechanical problem should be more likely. Indeed, by the definition of conditional probability (Definition 1.16), the chain rule (Theorem 1.17) and our assumptions,

$$\mathrm{P}(M \mid L, R^c) = \frac{\mathrm{P}(L, R^c, M)}{\mathrm{P}(L, R^c)} \tag{1.120}$$

$$= \frac{\mathrm{P}(L \mid R^c, M)\,\mathrm{P}(M \mid R^c)\,\mathrm{P}(R^c)}{\mathrm{P}(L \mid R^c)\,\mathrm{P}(R^c)} \tag{1.121}$$

$$= \frac{\mathrm{P}(L \mid R^c, M)\,\mathrm{P}(M)}{\mathrm{P}(L \mid R^c)} \tag{1.122}$$

$$= \frac{0.5 \cdot 0.1}{0.125} = 0.4, \tag{1.123}$$

which confirms that if the flight is late, a mechanical problem is indeed more likely when it is not raining. Formally, $\mathrm{P}(M \mid L, R^c) \neq \mathrm{P}(M \mid L)$, so the events M and R^c are *not* conditionally independent given the event L. We conclude that independence does not imply conditional independence.

· ·

Example 1.34 (House of Representatives: Conditioning on political affiliation). A key factor that determines how politicians vote in congress is political affiliation. In Example 1.25, we observe that the events B and E are not independent. Is it possible that the dependence is mainly due to political affiliation? In that case, the two events would be conditionally independent given political affiliation. To investigate this, we incorporate affiliation into our model by defining an event R, which indicates that the politician is a Republican. Conversely, R^c means that they are a Democrat.

From the data on the left of Table 1.5, we compute the empirical conditional probabilities given R, following Definition 1.24,

$$\mathrm{P}(B, E \mid R) = \frac{7}{155} = 0.045, \tag{1.124}$$

$$\mathrm{P}(B \mid R) = \frac{22}{155} = 0.142, \tag{1.125}$$

$$\mathrm{P}(E \mid R) = \frac{14}{155} = 0.090, \tag{1.126}$$

and verify that

$$\mathrm{P}(B \mid R)\mathrm{P}(E \mid R) = 0.013 \tag{1.127}$$

is quite different from $\mathrm{P}(B, E \mid R)$. In addition, the conditional probability

$$\mathrm{P}(B \mid R, E) = \frac{7}{14} = 0.5 \tag{1.128}$$

Table 1.5 **Voting and political affiliation.** *Number of Republicans (left) and Democrats (right) who voted Yes or No on the adoption of the budget resolution, and on duty-free exports, in Dataset 1.*

Republicans		Duty-free exports	
		Yes	No
Budget	Yes	7	15
	No	7	126

Democrats		Duty-free exports	
		Yes	No
Budget	Yes	144	73
	No	14	14

is very different from $P(B \mid R)$. We conclude that the events B and E are not conditionally independent given R.

Now, let us condition on the representative being a Democrat. The empirical conditional probabilities (computed from the data on the right of Table 1.5) equal

$$P(B, E \mid R^c) = \frac{144}{245} = 0.588, \qquad (1.129)$$

$$P(B \mid R^c) = \frac{217}{245} = 0.886, \qquad (1.130)$$

$$P(E \mid R^c) = \frac{158}{245} = 0.645, \qquad (1.131)$$

so that

$$P(B \mid R^c)P(E \mid R^c) = 0.571 \approx P(B, E \mid R^c). \qquad (1.132)$$

Therefore, B and E are approximately conditionally independent given R^c. This is reflected in the conditional probability

$$P(B \mid R^c, E) = \frac{144}{158} = 0.911, \qquad (1.133)$$

which is close to $P(B \mid R^c)$. According to the data, if we are interested in whether a Democrat has voted Yes on the budget, then knowing that they voted Yes on the duty-free exports provides very little information. This is not the case if we do not know the affiliation of the representative, or if they are a Republican. As illustrated by this example, conditioning on different events can completely change the dependence structure of a probabilistic model.

1.7 The Monte Carlo Method

When performing probabilistic modeling in practice, one quickly comes to a shocking realization: It is often intractable to compute the probability of some events, even if we have all the necessary information! Example 1.36 illustrates this through a probabilistic analysis of a basketball tournament. Even if we know the probability of any team beating any other team, computing the probability that a team wins the tournament requires keeping track of an enormous number of possible results. Unfortunately, such combinatorial explosions are commonplace in probabilistic modeling. The Monte Carlo method provides a pragmatic solution to this problem, inspired in our intuitive definition of probability (1.1): We *simulate* a large number of outcomes and compute the empirical probability of the event of interest.

The Monte Carlo method was developed in the context of nuclear-weapon research in the 1940s, pioneered by Stanislaw Ulam and John von Neumann. The name *Monte Carlo* was a code name inspired by the Monte Carlo Casino in Monaco. Ulam came up with the idea motivated by a game of cards. In his own words, as reported in Eckhardt (1987):

The first thoughts and attempts I made to practice (the Monte Carlo Method) were suggested by a question which occurred to me in 1946 as I was convalescing from an illness and playing solitaires. The question was what are the chances that a Canfield solitaire laid out with 52 cards will come out successfully? After spending a lot of time trying to estimate them by pure combinatorial calculations, I wondered whether a more practical method than "abstract thinking" might not be to lay it out say one hundred times and simply observe and count the number of successful plays. This was already possible to envisage with the beginning of the new era of fast computers...

Definition 1.35 (Monte Carlo method for estimating the probability of an event). *Given a probability space* $(\Omega, \mathcal{C}, \mathrm{P})$, *let us assume that we can repeatedly generate outcomes from* Ω *according to the probability measure* P. *To approximate the probability of any event A in the collection $\mathcal{C}$, we:*

1 *Generate n simulated outcomes:* $s_1, s_2, \ldots, s_n \in \Omega$.
2 *Compute the fraction of the outcomes in A,*

$$\mathrm{P}_{\mathrm{MC}}(A) := \frac{\sum_{i=1}^{n} 1(s_i \in A)}{n}, \tag{1.134}$$

where $1(s_i \in A)$ is an indicator function that is equal to one, if $s_i \in A$, and to zero otherwise, for any event $A \in \mathcal{C}$.

Similarly, to approximate the conditional probability of any event $B \in \mathcal{C}$ conditioned on A, we:

1 *Generate n simulated outcomes:* $s_1, s_2, \ldots, s_n \in \Omega$.
2 *Compute the fraction of the outcomes in A that are also in B,*

$$\mathrm{P}_{\mathrm{MC}}(B \mid A) := \frac{\sum_{i=1}^{n} 1(s_i \in A \cap B)}{\sum_{i=1}^{n} 1(s_i \in A)}. \tag{1.135}$$

Example 1.36 (3 × 3 Olympic basketball tournament). The 2020 Tokyo Olympics were the first to include 3 × 3 basketball. Eight teams participated: Belgium, China, Japan, Latvia, the Netherlands, Poland, the Russian Olympic Committee (ROC), and Serbia. Here, we imagine that the tournament has not happened yet, and we want to estimate the probability of each participant winning a gold, silver, or bronze medal based on the ranking points of each individual player. These ranking points reflect the players' performance in the previous 12 months. The left column in Table 1.6 shows the total points of the four players in each team before the tournament, gathered from the official FIBA website.

We begin by using the ranking points to determine the probability of each team beating every other team. Consider two teams A and B. The higher the total sum of the ranking points

Table 1.6 **Predicting the 3 × 3 basketball Olympic tournament.** *The table shows the probability that each team wins a gold, silver, or bronze medal, or wins the group stage in the 3 × 3 basketball tournament of the 2020 Tokyo Olympics according to the model described in Example 1.36. The probabilities are estimated using 10^4 Monte Carlo simulations.*

Country	Ranking points	Probability of winning (%)			
		Gold	Silver	Bronze	Group
Serbia	2,997,304	43.2	27.1	19.6	43.3
Latvia	2,959,152	42.0	28.0	18.9	42.9
ROC	970,438	6.3	14.9	18.9	5.6
Netherlands	768,134	3.6	10.3	14.4	3.2
Belgium	664,381	2.2	8.5	11.4	2.4
Poland	654,908	2.2	7.7	11.3	2.1
China	356,522	0.3	1.7	3.1	0.4
Japan	334,018	0.2	1.7	2.5	0.2

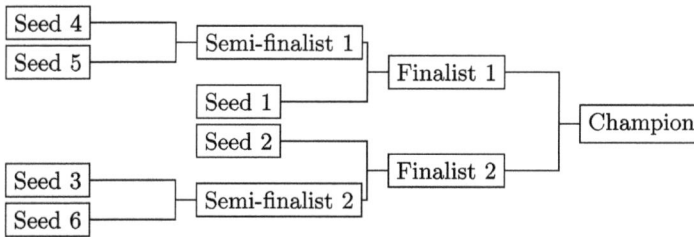

Figure 1.5 3 × 3 basketball bracket in the 2020 Tokyo Olympics. The eight participant teams were seeded according to the group stage. The first and second teams qualified directly for the semi-finals. The third, fourth, fifth, and sixth teams played the quarter-finals. The two last teams were eliminated.

of A with respect to B, the more likely A is to win a game between them. Consequently, a reasonable estimate for the probability that team A beats team B is

$$P(\text{team A beats team B}) = \frac{\text{ranking points of A}}{\text{ranking points of A } + \text{ ranking points of B}}. \quad (1.136)$$

This yields, for example,

$$P(\text{Belgium beats Poland}) = \frac{664381}{664381 + 654908} \quad (1.137)$$
$$= 0.504. \quad (1.138)$$

Our estimator is a simple heuristic that can probably be improved, but let us assume that we are happy with it. Now, how do we use these probabilities to derive the probability that a team wins the gold, silver or bronze medal?

We need to take into account the logistics of the tournament, which consisted of a group stage followed by playoffs. In the group stage, the eight participant teams played each other once, for a total of $\binom{8}{2} = 28$ games. The results determined the seeding for a playoff bracket with six more games: The five games in Figure 1.5, and the bronze-medal game. In order to compute the probability that a team wins a medal, we need to sum the probabilities of all the ways in which this can happen. This requires considering all 2^{34} possible results of

Table 1.7 **Conditioning on a rare event.** *The table shows predictions for the 3 × 3 basketball tournament in the 2020 Tokyo Olympics conditioned on the event that Serbia is eliminated in the group stage, according to the model described in Example 1.36. Each column shows the results for a different number of Monte Carlo simulations. Serbia was a heavy favorite, so they are unlikely to be eliminated, as is apparent from the last row, which shows the number of simulations where this happens. 10^4 total simulations are too few, but for 10^6 we observe enough relevant simulations for the conditional-probability estimates to be accurate.*

Country	Probability of gold conditioned on the event *Serbia does not reach bracket* (%)		
	10^4 sims	10^6 sims	10^7 sims
Latvia	68.6	63.5	63.4
ROC	10.0	13.3	13.2
Netherlands	7.1	8.5	8.6
Belgium	10.0	6.5	6.3
Poland	4.3	6.2	6.1
China	0	1.2	1.3
Japan	0	0.8	1.1
Serbia	0	0	0
Sims where Serbia does not reach bracket	70	5,539	55,719

the group stage and the bracket, which are more than ten billion! With modern computing, this is not intractable, but would take some time. However, in many practical situations, the number of possibilities makes exact computation completely impossible. For example, March Madness (the American college basketball championship) has sixty-seven games, which results in more than 10^{20} possible results, and the Wimbledon tennis tournament or the Premier League soccer championship have even more games.

Fortunately, the Monte Carlo method enables us to approximate our probabilities of interest. Following Definition 1.35, we repeatedly simulate the tournament using the probabilities in (1.136) and then compute the fraction of outcomes for which each event of interest occurs. Table 1.6 shows the results. Our model suggests that Latvia and Serbia are heavy favorites. Out of the 10^4 simulations of the tournament, they each won the gold medal about 40% of the time. Overall, the predictions of the model are quite reasonable. In the actual tournament, Latvia ended up winning gold, beating ROC in the final. Serbia won bronze, beating Belgium in the bronze-medal game.

The accuracy of the Monte Carlo method depends on the number of simulations that we perform. For example, an event with probability 0.01 only occurs in (approximately) 1 out of every 100 simulations, so we better consider at least a few hundred, or ideally a few thousand, simulations in order to estimate its probability. In practice, it is crucial to quantify the uncertainty associated with the probability estimates obtained via the Monte Carlo method, which can be achieved using confidence intervals, as explained in Example 9.46.

When approximating conditional probabilities, the number of *relevant* simulations can easily dwindle if the event we are considering is rare. To illustrate this, we consider the problem of predicting the tournament results if Serbia happens to be eliminated in the group stage. Following Definition 1.35, we simulate the tournament 10^4 times. Then, we select the outcomes in which Serbia ends up seventh or eighth during the group stage (resulting in elimination), and compute the fraction of these outcomes that are in each of the events of interest (e.g. Latvia wins the gold medal). If we don't pay attention, we could be fooled into thinking that this yields an accurate approximation. After all, we are using 10^4 simulations. However, the probabilities are estimated based exclusively on the subset of simulations in which Serbia drops out after the group stage, but this occurs only seventy times! Consequently, the estimated conditional probabilities, reported in Table 1.7, are not very precise. Increasing the number of simulations to 10^6, yields more than $5,000$ relevant instances in which Serbia is eliminated in the group stage, resulting in very different estimates. For instance, the conditional probability of Belgium winning gold drops from 10% to 6.5%. Further increasing the number simulations to 10^7 barely changes the estimates, suggesting that the approximation is accurate.

· ·

Exercises

1.1 (Conditional probability space) Let $(\Omega, \mathcal{C}, \mathrm{P})$ be a probability space, and let A be an event in the collection $\mathcal{C}$, such that $\mathrm{P}(A) \neq 0$. As explained in Section 1.3, in order to condition on A, we define $\mathcal{C}_A$ as the collection containing the intersection of A with all the events in $\mathcal{C}$:

$$\mathcal{C}_A = \{A \cap S \colon S \in \mathcal{C}\}. \tag{1.139}$$

If we consider a new sample space $\Omega_A := A$, prove that $\mathcal{C}_A$ is a valid collection, and also that the conditional probability measure

$$\mathrm{P}_A(S \cap A) := \frac{\mathrm{P}(S \cap A)}{\mathrm{P}(A)}, \tag{1.140}$$

where $S \in \mathcal{C}$, is a valid probability measure on $\mathcal{C}_A$.

1.2 (Empirical probability measure) We have available n data points $x_1, \ldots, x_n$ taking values in a discrete set Ω. We define a probability space where the sample space is Ω and the collection of events is the power set of Ω. The probability measure is defined following Definition 1.22. For each subset $S \subseteq \Omega$,

$$\mathrm{P}(S) := \frac{1}{n} \sum_{i=1}^{n} 1(x_i \in S), \tag{1.141}$$

where $1(x_i \in S)$ is an indicator function that is equal to one if $x_i \in S$ and to zero otherwise. As an example, suppose the data are coin flips where heads are represented by 1 and tails by 0, so $\Omega := \{0, 1\}$. If $n = 10$ and the data are 6 heads and 4 tails, then

$$\mathrm{P}(\emptyset) = 0, \quad \mathrm{P}(\{1\}) = 0.6, \quad \mathrm{P}(\{0\}) = 0.4, \quad \text{and} \quad \mathrm{P}(\{0, 1\}) = 1. \tag{1.142}$$

Prove that this is a valid probability measure.

1.3 (Independence and complements) Prove that if two events A and B in the same probability space are independent, then A^c and B are also independent.

1.4 (Conditional independence and complements) If two events A and B in the same probability space are conditionally independent given another event C in the same probability space, are A and B also conditionally independent given C^c? Prove that they are or provide a counterexample.

1.5 (Partition and independence) Show that events in a partition cannot be independent. Assume that every event in the partition has nonzero probability.

1.6 (Conditional probability and complements) Let A and B be two events in the same probability space. If $\mathrm{P}(A \mid B) = 1$, is it true that $\mathrm{P}(B^c \mid A^c) = 1$?

1.7 (The Linda problem) In (Tversky and Kahneman, 1983) Amos Tversky and Daniel Kahneman provided the following description to a group of survey respondents:

Linda is 31 years old, single, outspoken, and very bright. She majored in philosophy. As a student, she was deeply concerned with issues of discrimination and social justice, and also participated in anti-nuclear demonstrations.

They then asked the respondents which of the following options they considered more likely:

- Linda is a bank teller.
- Linda is a bank teller and is active in the feminist movement.

Most respondents chose the second option. Show that this contradicts the axioms of probability.

1.8 (Quiz) A school teacher gives a weekly quiz to her students consisting of two questions. The following table shows the questions that a student answered correctly (✓) or wrong (✗):

Week	1	2	3	4	5	6	7	8	9	10
Question 1	✓	✓	✓	✓	✗	✗	✓	✓	✓	✓
Question 2	✗	✗	✓	✓	✓	✗	✗	✓	✓	✗

Use the empirical-probability estimator to estimate the conditional probability that the student gets the second question right given that (1) they got the first question right, and (2) they got the first question wrong. Does this suggest that the answers to both questions could be independent?

1.9 (Baby name) Anna is having a baby, which will be a boy or a girl with probability $1/2$. When the baby is born, she will call her aunt Margaret to tell her whether the baby is a boy or a girl. Margaret is a bit deaf; she will misunderstand and think the baby is the wrong sex with probability 0.2. Then Margaret will tell her neighbor Bob, who is also a bit deaf. He will misunderstand what Margaret says (i.e., he will think that the baby is the opposite sex of what she says) with probability 0.1.

a What is the probability that both Margaret and Bob think that the baby is a girl?

b If Bob thinks the baby is a girl, what is the probability that he is right?

1.10 (Cake) Milena is preparing a birthday cake. To finish on time she requires some help, so she asks Scott and Antonis for help. If nobody helps, she won't finish on time. If both of them help, she finishes on time. If only one helps (no matter who), she finishes on time with probability 0.5. The probability that Scott helps is 0.4. The probability that Antonis helps is 0.8. They decide to help or not independently from each other.

a What is the conditional probability that Scott helps if we know that Milena finishes on time?

b What is the conditional probability that Scott helps given that Antonis helps and Milena finishes on time? Is the event *Scott helps* conditionally independent from the event *Antonis helps* given the event *Milena finishes on time*?

1.11 (Baby sleep) A babysitter is taking care of a baby. They give her some food and put her to sleep. Assume the following:

- The probability that the food is bad is 0.1.
- If a baby eats food that is bad, they will wake up in the middle of the night. If the food is not bad, they may still wake up (with a probability that depends on whether they are good or bad sleepers).
- All babies can be classified into *good sleepers* or *bad sleepers*. The probability that a baby that is a *good sleeper* wakes up in the middle of the night is 0.1. The probability for a baby that is a *bad sleeper* is 0.8.
- A baby is a *good sleeper* with probability 0.6.

Answer the following questions indicating any (reasonable) assumptions you make about independence between events.

a What is the probability that the baby wakes up in the middle of the night?

b If the baby wakes up in the middle of the night, what is the probability that the food is bad?

c Compute the conditional probability that the food is bad given that a good sleeper wakes up.

d Under our assumptions, are the events *baby is a good sleeper* and *food is bad* conditionally independent given the event *wakes up in the middle of the night*? Justify your answer mathematically and explain it intuitively.

1.12 (COVID-19 tests) A company with 10 employees decides to test them for COVID-19 before they go back to work in person. From available data, they determine that the probability of each employee being ill is 0.01. The employees have not been in contact with each other for a while, so the events *employee i is ill*, for $1 \leq i \leq 10$, are modeled as independent. If an employee is ill, the test is positive with probability 0.98. If they are not ill, the test is positive with probability 0.05.

a Is it reasonable to model the events *test i is positive*, for $1 \leq i \leq 10$, as mutually independent? From now on, model them as mutually independent whether you think it is reasonable or not, and also model the set of complements of these events as mutually independent.

b The company tests all employees. What is the probability that there is at least one positive test?

c If there is at least one positive test, what is the probability that nobody is ill? If you make any independence or conditional independence assumptions, please explain why you think they are reasonable.

1.13 (Boxing championship) In the boxing world championship, the challengers Manny and Saul will first fight one another, and the winner will fight the reigning champion Floyd. The probability that Manny beats Saul is only 0.4, but he has a higher probability of beating Floyd due to their different fighting styles. The probability that Manny and Saul beat Floyd are 0.25 and 0.1, respectively. We assume there cannot be any ties.

a If Floyd loses, what is the conditional probability that Manny becomes champion?

b Estimate the conditional probability that Manny becomes champion given that Floyd loses, from the following 20 simulations, generated according to our assumptions (F means that Floyd won, M that Manny won and S that Saul won):

Simulation	1	2	3	4	5	6	7	8	9	10
Challenger fight	S	M	S	M	M	S	M	M	S	S
Championship fight	F	F	F	M	M	F	F	F	F	F

Simulation	11	12	13	14	15	16	17	18	19	20
Challenger fight	M	M	M	S	M	S	S	S	S	S
Championship fight	F	F	F	F	M	F	F	F	F	S

Why is it not surprising that you obtain a different answer?

1.14 (Videogame) In a videogame, Silvio faces three fighters: first Honda, then Zangief, and then Blanka. He must defeat all three fighters to win the game. The probability that they defeat each fighter is 0.8 for Honda, 0.5 for Zangief and 0.4 for Blanka. If Silvio loses a fight, he faces the same fighter again, but if he loses the second fight, the game stops and he loses. We assume all fights are independent.

a What is the probability that the player wins the game?

b Use the following independent simulated fights to obtain independent simulations of the game:

Honda: L W L W W W W W L W W W L L W W W W W W L
Zangief: W L L L W L W L W W L W L L W L W L W W
Blanka: L W L L L L W L W W L L L W L L L L L L

W indicates that Silvio wins the fight, L that he loses. Report the results of the simulated games, and use them to obtain a Monte Carlo estimate of the probability that Silvio wins the game. Is the result accurate? How can you improve it?

1.15 (Rare event) If we apply the Monte Carlo method to estimate the probability of an event A, how many independent simulations should we perform to make sure that the estimated probability is nonzero with probability at least 0.99? Derive the number of simulations as a function of the probability of the event $P(A)$, and compute it for $P(A) := 0.01$.

1.16 (Streak of heads) In this problem, we consider the problem of testing whether a randomly generated sequence is truly random. A certain computer program is supposed to generate independent fair coin flips. When you try it out, you are surprised that it contains long streaks of heads. In particular, you generate a sequence of length 200, which turns out to contain a sequence of 8 heads in a row.

a Compute the probability that the longest streak of heads that you observe has length x for $x \in \{1, 2, 3, 4, 5\}$ when you flip a fair coin 5 times, and the flips are independent.

b Estimate these probabilities using the Monte Carlo method.

c What is the estimated probability that the longest streak of heads has length 8 or more for 200 flips? Is the sequence of 8 ones evidence that the program may not be generating truly random sequences?

2

Discrete Variables

Overview

In this chapter, we explain how to model uncertain numerical quantities that are *discrete*, which means that they take a finite or countably infinite number of values. Examples include the number of students attending a class, the goals scored in a soccer game, or the earthquakes occurring in San Francisco over the next 20 years. Section 2.1 introduces the machinery of discrete random variables, which allows us to represent and manipulate such quantities within the mathematical framework of probability spaces. Section 2.2 describes how to build discrete nonparametric models. Section 2.3 defines several popular discrete parametric distributions (Bernoulli, binomial, geometric, and Poisson). Section 2.4 introduces maximum-likelihood estimation, which enables us to fit parametric models to data. Finally, in Section 2.5, we discuss the advantages and disadvantages of nonparametric and parametric models, and how to evaluate them in practice.

2.1 Discrete Random Variables

Random variables are mathematical objects designed to represent uncertain quantities. They do not have a fixed value. Instead, they can take multiple values with different probabilities. We do not make statements such as *the random variable $\tilde{a}$ equals 3*, but rather *the probability that the random variable $\tilde{a}$ equals 3 is 0.5*. We use a tilde to indicate that variables are random (e.g. $\tilde{a}$, $\tilde{b}$, $\tilde{x}$, $\tilde{y}$), in order to distinguish them from nonrandom or *deterministic* variables (a, b, x, y) representing quantities that are not uncertain.

In Section 2.1.1, we define random variables as functions that map outcomes in a probability space to real numbers. However, in data science we don't often think of random variables as functions, unless we need to prove mathematical statements about them. Instead, we treat them as uncertain numerical quantities with associated probabilities. In fact, we usually define and manipulate the random variables exclusively through these probabilities, as explained in Sections 2.1.2 and 2.1.3.

2.1.1 Random Variables as Functions of Outcomes

In order to define random variables mathematically, we build upon the framework of probability spaces, as described in Chapter 1. A random variable is defined as a function of the outcomes in the sample space of the probability space.

Example 2.1 (Rolling a die twice). Consider a probability space representing two rolls of a six-sided die. Each possible outcome can be encoded as a vector with two entries,

$$\omega := \begin{bmatrix} \omega_1 \\ \omega_2 \end{bmatrix}, \quad \omega_1, \omega_2 \in \{1, 2, 3, 4, 5, 6\}. \tag{2.1}$$

The first entry ω_1 is the result of the first roll; the second entry ω_2 is the result of the second roll. The sample space Ω has 36 possible outcomes.

We are interested in modeling the result of the first roll. This is an uncertain numerical quantity that can take a discrete number of values (six). The quantity can be represented as a function of the outcome of the probability space:

$$\tilde{a}(\omega) := \omega_1. \tag{2.2}$$

We call such functions random variables. Since a random variable is a function, we refer to the set of values it can take as its *range*. The range of $\tilde{a}$ is $\{1, 2, 3, 4, 5, 6\}$. The random variable allows us to describe events related to the roll of the first die very succinctly. The event

$$\textit{First roll equals one} := \left\{ \begin{bmatrix} 1 \\ 1 \end{bmatrix}, \begin{bmatrix} 1 \\ 2 \end{bmatrix}, \begin{bmatrix} 1 \\ 3 \end{bmatrix}, \begin{bmatrix} 1 \\ 4 \end{bmatrix}, \begin{bmatrix} 1 \\ 5 \end{bmatrix}, \begin{bmatrix} 1 \\ 6 \end{bmatrix} \right\} \tag{2.3}$$

can be expressed as $\{\omega : \tilde{a}(\omega) = 1\}$ or just $\tilde{a} = 1$.

Similarly, we can define random variables to represent the second roll,

$$\tilde{b}(\omega) := \omega_2, \tag{2.4}$$

or the sum of the two rolls

$$\tilde{c}(\omega) := \omega_1 + \omega_2. \tag{2.5}$$

....

The definition of a random variable as a function concentrates all the uncertainty in the sample space. Selecting an outcome simultaneously determines the value of all random variables associated with the probability space. This is called *sampling* the random variables. The resulting values are known as *realizations* or *samples*. In Example 2.1, if $\omega = \begin{bmatrix} 3 \\ 1 \end{bmatrix}$, then the realizations corresponding to the different random variables defined in the probability space are automatically fixed to equal $\tilde{a}(\omega) = 3$, $\tilde{b}(\omega) = 1$, and $\tilde{c}(\omega) = 4$. Crucially, this allows us to model the dependence between multiple uncertain quantities, as we discuss in more detail in Chapter 4.

2.1.2 The Probability Mass Function

Describing a random variable by explicitly defining its associated probability space is very cumbersome. For this reason, we usually manipulate random variables through their *probability mass function* (pmf). The pmf encodes the information that we are really interested in: The probability that the random variable is equal to any element of its range. Probability mass is just another name for probability, motivated by the fact that it is a measure like mass (see Section 1.2). The *probability distribution* of a random variable $\tilde{a}$, meaning its probabilistic behavior, is completely determined by its pmf $p_{\tilde{a}}$ (we make this more precise in what follows). Because of this, we often say that a random variable $\tilde{a}$ is *distributed* according to a certain pmf $p_{\tilde{a}}$.

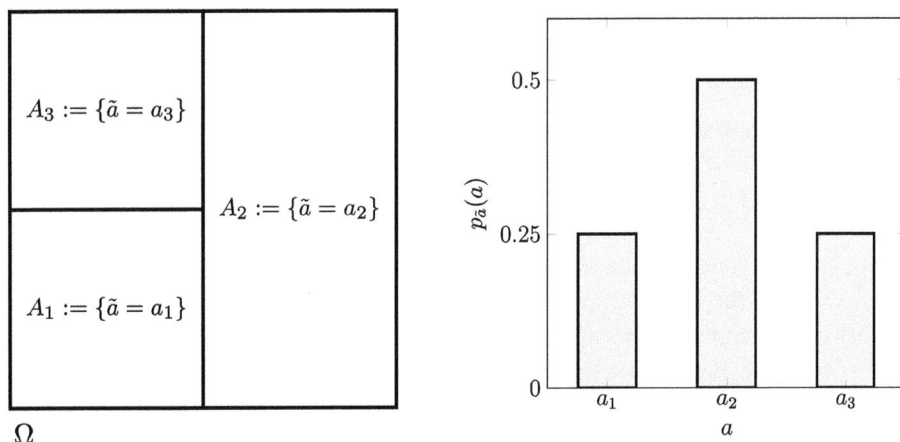

Figure 2.1 Probability mass function of a discrete random variable. The discrete random variable $\tilde{a}$ maps the outcomes in the sample space Ω to three possible values: a_1, a_2 and a_3. The three events A_1, A_2, and A_3 contain the outcomes that map to a_1, a_2 and a_3, respectively. A_1, A_2, and A_3 form a partition, as shown in the Venn diagram on the left. The graph on the right depicts the pmf of $\tilde{a}$. For $i \in \{1, 2, 3\}$, the pmf maps a_i to the probability of A_i, $p_{\tilde{a}}(a_i) = P(A_i)$, which is equal to the area of A_i in the Venn diagram.

Definition 2.2 (Probability mass function). *Let $\tilde{a}$ be a discrete random variable with discrete range $\{a_1, a_2, \ldots\}$, defined on a probability space $(\Omega, \mathcal{C}, \mathrm{P})$. The probability mass function (pmf) $p_{\tilde{a}} : \mathbb{R} \to [0, 1]$ of $\tilde{a}$ encodes the probability that $\tilde{a}$ equals each element of its range:*

$$p_{\tilde{a}}(a_i) := \mathrm{P}(A_i), \qquad i = 1, 2, \ldots, \tag{2.6}$$

where

$$A_i := \{\omega \colon \tilde{a}(\omega) = a_i\}. \tag{2.7}$$

Figure 2.1 illustrates Definition 2.2 with a simple example. The more mathematically oriented readers might be wondering whether the definition is sound. *How do we know that the probability $\mathrm{P}(A_i)$ of the events in (2.6) exists?* The answer is that we impose it explicitly in our formal definition of discrete random variables. For the probability to exist, the event A_i must belong to the collection of events of the probability space, so that the probability measure assigns it a probability. In that case, we say that the event A_i is *measurable*. We require that all such events be measurable for a discrete random variable to be well defined.

Definition 2.3 (Formal definition of discrete random variable). *Let $(\Omega, \mathcal{C}, \mathrm{P})$ be a probability space. Let $\tilde{a}$ be a function mapping elements in the sample space Ω to a finite or countably infinite set of values $\{a_1, a_2, \ldots\}$. The function $\tilde{a}$ is a discrete random variable if the events*

$$A_i := \{\omega \colon \tilde{a}(\omega) = a_i\}, \qquad i = 1, 2, \ldots, \tag{2.8}$$

containing the outcomes mapping to each element a_i of the range are all measurable. This implies that for all i, A_i belongs to the collection $\mathcal{C}$ and is assigned a probability by the probability measure P, so that

$$\mathrm{P}(\tilde{a} = a_i) := \mathrm{P}(A_i) \tag{2.9}$$

is well defined.

When I first came across the formal definition of random variables, I found it quite daunting. It seems to suggest that, in order to model an uncertain quantity, we need to: (1) build a probability space consisting of a sample space, a collection of events, and a probability measure, (2) define the random variable representing the quantity as a function from the sample space to the desired range, (3) ensure that the event mapping to every element of the range is in the collection. This sounds very difficult to do! Thankfully, in practice *we never actually do it*.

The probability space associated with a random variable is a mathematical abstraction, which ensures that the random variable is properly defined. However, we only describe this probability space explicitly in simple pedagogical examples, such as Example 2.1. Otherwise, we manipulate the random variable using only its pmf, which completely characterizes its behavior, as established in Theorem 2.5. You can think of the pmf as the *user interface* of a random variable. By contrast, the formal definition of the random variable as a function is its mathematical *implementation*, which we don't need to worry about most of the time. Indeed, when using discrete random variables to model data, we estimate their pmf directly, without ever defining the underlying probability space. This is mathematically legitimate because there always exists a valid underlying probability space, as we show in Theorem 2.10.

2.1.3 Properties of the Probability Mass Function

In Figure 2.1, the events A_1, A_2, and A_3 mapping to a_1, a_2 and a_3 form a partition of the sample space. This is no coincidence; it is a direct consequence of the definition of discrete random variables.

Lemma 2.4. *Let $\tilde{a} \colon \Omega \to \mathbb{R}$ be a discrete random variable with range $\{a_1, a_2, \ldots\}$ associated with a probability space $(\Omega, \mathcal{C}, \mathrm{P})$. The events*

$$A_i := \{\omega \colon \tilde{a}(\omega) = a_i\}, \qquad i = 1, 2, \ldots, \tag{2.10}$$

form a partition of the sample space Ω, i.e. they are disjoint and cover all of Ω (i.e. $\Omega = \cup_i A_i$).

Proof Recall that a function assigns a unique value to every element of its domain. Consequently, the events are disjoint because if there existed $\omega \in \Omega$ such that $\omega \in A_i$ and $\omega \in A_j$ for $a_i \neq a_j$, then $\tilde{a}(\omega)$ would be equal to both a_i and a_j. In addition, functions must assign a value to every element of their domain, so every $\omega \in \Omega$ is mapped to an element of $\{a_1, a_2, \ldots\}$ by $\tilde{a}$ and therefore belongs to A_i for some i, which implies that the events cover Ω. ∎

In Example 1.14, we build a partition of events associated with a die roll explicitly, in order to define the probability measure of the corresponding probability space. By Lemma 2.4, when we define a discrete random variable and its pmf, we are implicitly applying the same strategy! For instance, you can check that in Example 2.1 each random variable is implicitly associated with a different partition of the sample space.

Lemma 2.4 implies that the pmf can be used to determine the probability of a random variable belonging to any subset of its range, as established in Theorem 2.5. This frees us from having to worry about the underlying probability space! In fact, we usually refer to events of the form $\{\omega \colon \tilde{a}(\omega) \in S\}$ using the simplified notation $\tilde{a} \in S$, without even mentioning the sample space.

Theorem 2.5. *Let* $\tilde{a} \colon \Omega \to \mathbb{R}$ *be a discrete random variable with range* R *associated with a probability space* $(\Omega, \mathcal{C}, \mathrm{P})$. *The probability that* $\tilde{a}$ *belongs to any subset* $S \subseteq R$ *of its range equals,*

$$\mathrm{P}\left(\tilde{a} \in S\right) := \mathrm{P}\left(\{\omega \colon \tilde{a}\left(\omega\right) \in S\}\right) \tag{2.11}$$

$$= \sum_{a \in S} p_{\tilde{a}}\left(a\right). \tag{2.12}$$

Proof Let us denote the elements of the range by a_1, a_2, ... By Lemma 2.4, the events mapping to each element

$$A_i := \{\omega \colon \tilde{a}\left(\omega\right) = a_i\}, \qquad i = 1, 2, \ldots, \tag{2.13}$$

are disjoint. Consequently, by Axiom 3 in Definition 1.9 and Definition 2.2,

$$\mathrm{P}\left(\tilde{a} \in S\right) := \mathrm{P}\left(\{\omega \colon \tilde{a}\left(\omega\right) \in S\}\right) \tag{2.14}$$

$$= \mathrm{P}\left(\cup_{a_i \in S} A_i\right) \tag{2.15}$$

$$= \sum_{a_i \in S} \mathrm{P}\left(A_i\right) \tag{2.16}$$

$$= \sum_{a_i \in S} p_{\tilde{a}}\left(a_i\right). \tag{2.17}$$

■

A direct consequence of Theorem 2.5 is that the sum of all the entries of a pmf must equal one.

Lemma 2.6 (Pmfs sum to one). *For any discrete random variable* $\tilde{a} \colon \Omega \to \mathbb{R}$ *with range* R,

$$\sum_{a \in R} p_{\tilde{a}}(a) = 1. \tag{2.18}$$

Proof By Theorem 2.5 and Axiom 2 in Definition 1.9,

$$\sum_{a \in R} p_{\tilde{a}}(a_i) = \mathrm{P}\left(\tilde{a} \in R\right) \tag{2.19}$$

$$= \mathrm{P}(\{\omega \colon \tilde{a}\left(\omega\right) \in R\}) \tag{2.20}$$

$$= \mathrm{P}(\Omega) \tag{2.21}$$

$$= 1. \tag{2.22}$$

■

Example 2.7 (Goal difference in soccer game). Let us use a discrete random variable $\tilde{g}$ to model the goal difference in a soccer game between Barcelona and Atlético de Madrid. Figure 2.2 shows the probability mass function of $\tilde{g}$. You can check that its entries sum to one, as required by Lemma 2.6. We can compute the probability that $\tilde{g}$ belongs to different sets via Theorem 2.5. The probability of the difference being exactly two in favor of either team is

$$\mathrm{P}\left(\tilde{g} \in \{-2, 2\}\right) = p_{\tilde{g}}\left(-2\right) + p_{\tilde{g}}\left(2\right) = 0.2. \tag{2.23}$$

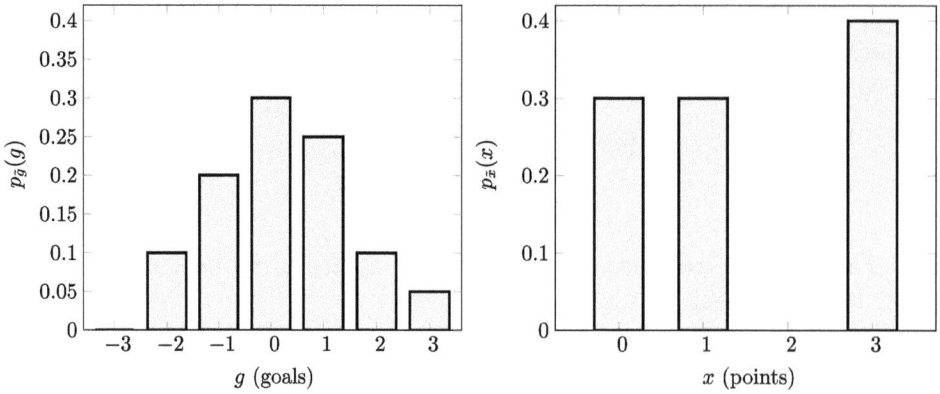

Figure 2.2 Soccer game. The figure shows the pmfs of the random variables $\tilde{g}$ and $\tilde{x}$ from Examples 2.7 and 2.9. The random variable $\tilde{g}$ represents goal difference in a soccer game. The random variable $\tilde{x}$ represents the corresponding number of points earned in the same soccer game.

The probability of Barcelona winning by more than one goal is

$$\mathrm{P}\left(\tilde{g} > 1\right) = p_{\tilde{g}}\left(2\right) + p_{\tilde{g}}\left(3\right) = 0.15. \tag{2.24}$$

. . . .

2.1.4 Functions of Random Variables

In probabilistic modeling, we often encounter deterministic functions of uncertain quantities. For instance, in Example 2.7, we might be interested in whether a team wins or not, which is a deterministic function of the goal difference (if a team scores more goals, they win). Functions of discrete random variables are discrete random variables themselves. Intuitively, if the input to a function is uncertain, then the output is also uncertain, so it makes sense for it to be represented as a random variable. The following lemma makes this mathematically rigorous. It also provides a simple formula for the pmf of a function of a random variable. To compute the probability of an output, we just add up the probabilities of all the inputs mapping to that output.

Theorem 2.8 (Function of a discrete random variable). *Let $\tilde{a}\colon \Omega \to R_{\tilde{a}}$ be a discrete random variable with range $R_{\tilde{a}} := \{a_1, a_2, \ldots\}$ associated with the probability space $(\Omega, \mathcal{C}, \mathrm{P})$, and let $h\colon R_{\tilde{a}} \to \mathbb{R}$ be a deterministic function. Then $\tilde{b} := h \circ \tilde{a}$, also denoted by $h(\tilde{a})$, is a discrete random variable, and its pmf is given by*

$$p_{\tilde{b}}\left(b\right) = \sum_{\{a:h(a)=b\}} p_{\tilde{a}}\left(a\right) \tag{2.25}$$

for any b such that $h(a) = b$ for some $a \in R_{\tilde{a}}$.

Proof　Let us denote the range of $\tilde{b}$ by $R_{\tilde{b}} := \{b_1, b_2, \ldots\}$. $R_{\tilde{b}}$ is discrete because it has at most as many elements as $R_{\tilde{a}}$, which is discrete. By Definition 2.3, to prove that $\tilde{b}$ is a valid discrete random variable, we need to show that the event

$$B := \{\omega\colon \tilde{b}\left(\omega\right) = b\} \tag{2.26}$$

is measurable for any $b \in R_{\tilde{b}}$. We express B as the union of the events mapped by $\tilde{a}$ to the elements of $R_{\tilde{a}}$, which are in turn mapped to b by h:

$$B = \cup_{\{a:h(a)=b\}} \{\omega : \tilde{a}(\omega) = a\}. \tag{2.27}$$

These events are all in $\mathcal{C}$, and hence are measurable because $\tilde{a}$ is a discrete random variable. Consequently, their union belongs to $\mathcal{C}$ by Condition 2 in Definition 1.7, which means that B is measurable and $\mathrm{P}(B)$ is well defined. Finally, by Theorem 2.5

$$p_{\tilde{b}}(b) = \mathrm{P}(\tilde{b} = b) \tag{2.28}$$

$$= \mathrm{P}(h(\tilde{a}) = b) \tag{2.29}$$

$$= \mathrm{P}\left(\tilde{a} \in \cup_{\{a:h(a)=b\}} \{a\}\right) \tag{2.30}$$

$$= \sum_{\{a:h(a)=b\}} p_{\tilde{a}}(a). \tag{2.31}$$

■

Example 2.9 (Converting goal difference to points). In most soccer leagues, a win is worth three points, a draw one point, and a loss zero points. Imagine that we want to model the number of points obtained by Barcelona in the soccer game of Example 2.7, but only have access to the pmf of the random variable $\tilde{g}$ modeling the goal difference. We need to derive the pmf of a new random variable $\tilde{x} := h(\tilde{g})$ representing the points, where h is the deterministic function

$$h(g) := \begin{cases} 0 & \text{if } g < 0, \\ 1 & \text{if } g = 0, \\ 3 & \text{if } g > 0. \end{cases} \tag{2.32}$$

The values of g such that $h(g) = 0$ are -2 and and -1, so by Theorem 2.8,

$$p_{\tilde{x}}(0) = p_{\tilde{g}}(-2) + p_{\tilde{g}}(-1) \tag{2.33}$$

$$= 0.3. \tag{2.34}$$

By the same argument, $p_{\tilde{x}}(1) = 0.3$, $p_{\tilde{x}}(3) = 0.4$, and $p_{\tilde{x}}(x) = 0$ for any other value of x. Figure 2.2 shows a plot of the pmf, which reassuringly adds up to one.
....

2.2 The Empirical Probability Mass Function

In Section 2.1, we define discrete random variables formally and show that their pmf completely characterizes their behavior. We now explain how to apply these concepts to model discrete data in practice. The idea is to interpret the data as realizations obtained by repeatedly sampling a random variable. To estimate the probability distribution of this random variable from the data, all we need to do is estimate its pmf. We do *not* need to define the underlying probability space. As established in the following theorem, as long as the pmf is nonnegative and adds up to one, we can rest assured that a valid probability space exists.

Theorem 2.10. *Any function* $p \colon A \to [0, 1]$ *that maps a discrete set* A *to the unit interval* $[0, 1]$ *can be interpreted as a valid pmf of a random variable, as long as*

$$\sum_{a \in A} p(a) = 1. \tag{2.35}$$

Proof We need to define a valid underlying probability space $(\Omega, \mathcal{C}, \mathrm{P})$. We set the sample space to be $\Omega := A$, the collection of events to be the power set (all possible subsets) of A, and the probability measure to equal the pmf p. If we define a random variable to be the identity function, then p is a valid pmf, as it is the probability that the random variable equals any element of A. We leave it as an exercise to check that this yields a probability space satisfying Definition 1.15 and a discrete random variable satisfying Definition 2.3. ■

In Section 1.4, we explain how to compute the empirical probability of an event from data. The pmf of a discrete random variable encodes the probability that the random variable equals each element of its range. It is therefore natural to estimate pmfs using empirical probabilities. The resulting statistical estimator is known as the *empirical pmf*.

Definition 2.11 (Empirical probability mass function)**.** *Let* $X := \{x_1, x_2, \ldots, x_n\}$ *denote a dataset with values in a discrete set* A*. The empirical probability mass function* $p_X \colon A \to [0, 1]$ *maps each value* $a \in A$ *to the fraction of data that equal* a,

$$p_X(a) := \frac{1}{n} \sum_{i=1}^{n} 1(x_i = a), \tag{2.36}$$

where $1(x_i = a)$ *is an indicator function that is equal to one, if* $x_i = a$, *and to zero otherwise.*

By Theorem 2.10 and the following simple lemma, the empirical pmf is guaranteed to be a valid pmf.

Lemma 2.12 (The empirical pmf adds up to one)**.** *Let* $p_X \colon A \to [0, 1]$ *be the empirical pmf of a dataset* $X := \{x_1, x_2, \ldots, x_n\}$ *taking values in a discrete set* A*. Then,* p_X *sums to one,*

$$\sum_{a \in A} p_X(a) = 1. \tag{2.37}$$

Proof Each $x_i \in X$ is equal to one (and only one) of the elements in A, so $\sum_{a \in A} 1(x_i = a) = 1$, which implies

$$\sum_{a \in A} p_X(a) = \frac{1}{n} \sum_{i=1}^{n} \sum_{a \in A} 1(x_i = a) \tag{2.38}$$

$$= \frac{n}{n} = 1. \tag{2.39}$$

■

The empirical pmf is closely related to the *histogram*, a widely used technique to visualize datasets.

Definition 2.13 (Histogram of discrete data)**.** *Let* $X := \{x_1, x_2, \ldots, x_n\}$ *be a dataset with values in a discrete set* A*. To build a histogram of the data, we assign a bin or bucket to every element in* A*. We then count how many data points are in each bin.*

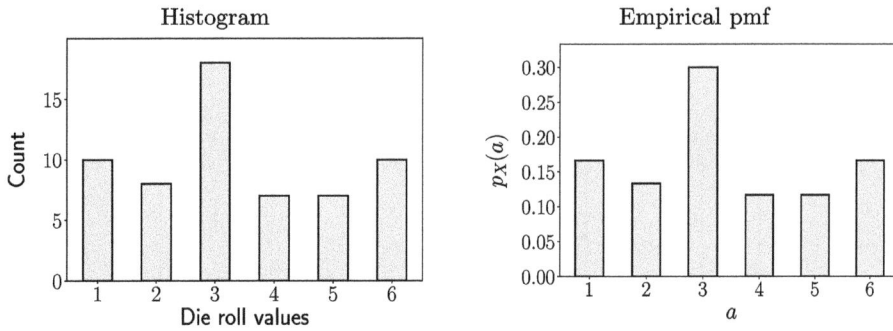

Figure 2.3 Histogram and empirical pmf for a die roll. The left plot shows the histogram of the data in Example 1.23. The empirical pmf of a random variable $\tilde{a}$ representing the die roll, shown on the right, is obtained by dividing the counts in each bin by the total number of data.

The empirical pmf is just a normalized histogram. The left plot of Figure 2.3 shows the histogram of the data in Example 1.23, which consists of rolls from a six-sided die. The count in each bin is the number of observed die rolls equal to the corresponding value. We obtain the empirical pmf of a discrete random variable representing the die roll by normalizing the histogram, dividing each count by the total number of data, as depicted in the right plot of Figure 2.3.

As we discuss in Section 1.4, empirical probabilities computed from a finite number of data are only approximations. As a result, empirical pmfs cannot be expected to be completely accurate. This is illustrated in Figure 2.4, where we show three different empirical pmf estimates obtained from 60 realizations of a fair die roll. Due to the limited number of data, there are substantial fluctuations in the estimated pmfs.

Example 2.14 (Free-throw streaks). Kevin Durant is an NBA player, known for his shooting accuracy. In this example, we model the length of his streaks of consecutive made free throws, which we represent by a random variable $\tilde{a}$. We compute the empirical pmf of $\tilde{a}$ using the first 3,015 free throws shot by Durant in his career, extracted from Dataset 2. To compute the empirical pmf, the 3,015 free throws are divided into 377 different streaks of varying length. For example, Durant made 2 free throws in a row (followed by a miss) 42 times. The value of the empirical pmf at 2 is therefore estimated to equal $42/377 = 0.11$, following Definition 2.11. Figure 2.5 shows the empirical pmf.

The general shape of the pmf estimated in Example 2.14 looks quite reasonable (the probability of a streak decreases as its length increases). However, a closer look reveals imprecisions due to the small number of data. The probability of a streak of length 7 is lower than the probability of a streak of length 12, even though the general trend is for longer streaks to be less likely. It seems plausible that, by mere chance, streaks of length 7 are underrepresented in the data, or streaks of length 12 are overrepresented.

Similarly, the probability of a streak of length 25 is zero because no such streaks were observed. However, there are five streaks of length 24 and two streaks of length 26, which suggests that this is again due to random chance. Indeed, imagine that the true probability of a streak of length 25 is around 0.01. Since the total number of streaks that we observe is less than 400, we would expect to see about four instances of such streaks in the data. However, if streaks with similar lengths (23, 24, 26, 27, etc.) occur with similar probabilities, we are bound to get unlucky for one of them, and not observe any instances at all.

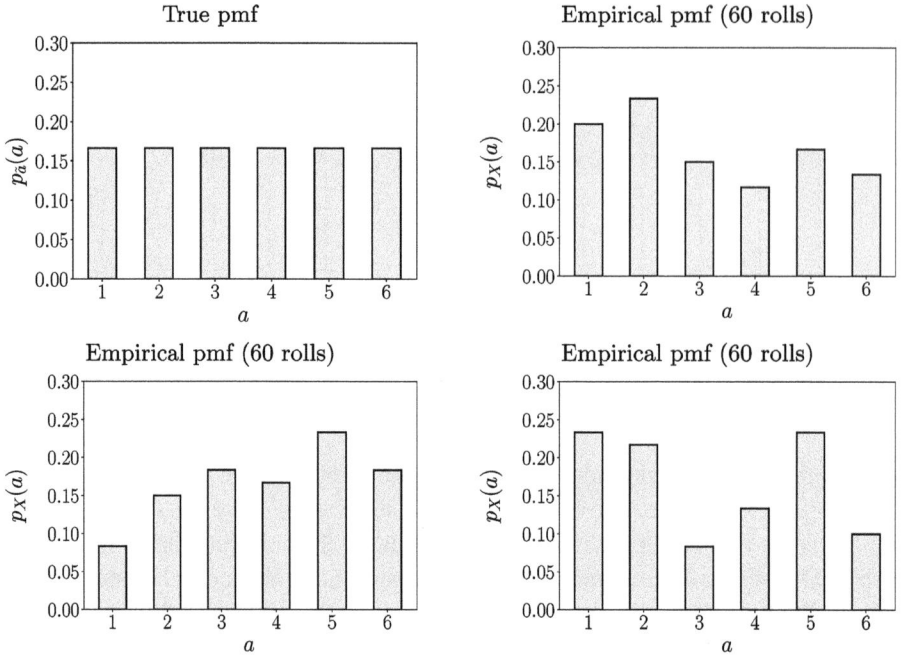

Figure 2.4 True and empirical pmf for a fair die roll. The upper left plot shows the true pmf of a random variable $\tilde{a}$ representing a fair die roll. The remaining three plots show the empirical pmfs estimated from 60 independent and identically distributed rolls (see Definition 2.23). Due to the limited number of data, there are substantial fluctuations in the empirical pmfs.

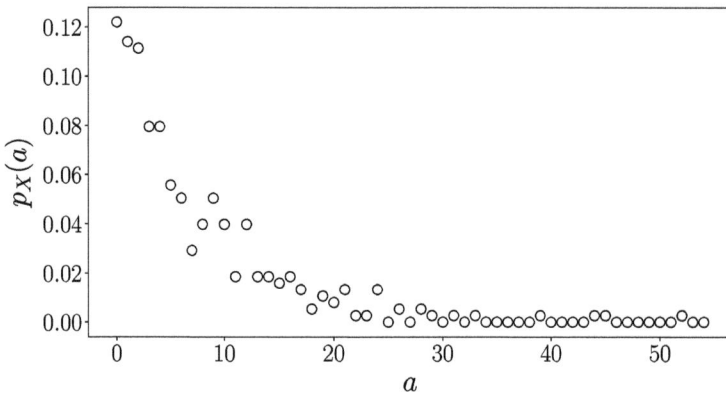

Figure 2.5 Empirical probability mass function of free-throw streaks. The graph shows the empirical pmf of the length of Kevin Durant's free-throw streaks, which encodes the probability that he makes a certain number of consecutive free throws before missing. The empirical pmf is obtained using the first 3,015 free throws in his career.

2.3 Discrete Parametric Distributions

As illustrated by Example 2.14, when the available data are limited, the empirical pmf can be an inaccurate estimator of the pmf of a random variable. Parametric modeling is a strategy to address this issue by incorporating additional assumptions. The idea is to design a pmf based on such assumptions, which only depends on a small number of parameters, and hence can be estimated robustly from limited data (following the procedure described in Section 2.4). In this section, we introduce several discrete distributions that are widely used for parametric modeling: Bernoulli (Section 2.3.1), geometric (Section 2.3.3), binomial (Section 2.3.2), and Poisson (Section 2.3.4).

2.3.1 The Bernoulli Distribution

The Bernoulli distribution is used to model uncertain phenomena that have two possible results, such as a basketball game (win or loss) or a driving test (pass or fail). By convention, we often represent one of the results by 0 and the other by 1. A canonical example is flipping a biased coin, where the probability of obtaining heads is θ. If we encode heads as 1 and tails as 0, then the result of the coin flip corresponds to a Bernoulli random variable with parameter θ.

Definition 2.15 (Bernoulli distribution). *The pmf of a Bernoulli random variable $\tilde{a}$ with parameter $\theta \in [0, 1]$ is*

$$p_{\tilde{a}}(0) = 1 - \theta, \tag{2.40}$$
$$p_{\tilde{a}}(1) = \theta. \tag{2.41}$$

2.3.2 The Binomial Distribution

We motivate the definition of the binomial distribution with a simple example involving coin flips.

Example 2.16 (Coin flips). We flip a coin n times. If the flips are independent and the probability of heads is θ, what is the probability of obtaining a heads?

Let us first consider the case where $n = 3$ and $a = 2$ to gain some intuition. We are interested in the probability of the event *obtaining two heads*, which occurs if the sequence of flips equals *heads-heads-tails*, *heads-tails-heads*, or *tails-heads-heads*. Since these individual events are disjoint (the sequence cannot be *heads-tails-heads* and *tails-heads-heads* at the same time), we can add their probabilities to compute the probability of their union by Axiom 3 in Definition 1.9. This suggests a strategy to answer the question: Compute the probabilities of obtaining a heads and $n - a$ tails *in every possible order* and then add them up.

By Definition 1.28, if the coin flips are independent, the probability of first obtaining a heads and then $n - a$ tails equals

$$\begin{aligned} &\mathrm{P}\left(\text{flip } 1 = h, \ldots, \text{ flip } a = h, \text{ flip } a + 1 = t, \ldots, \text{ flip } n = t\right) \\ &= \mathrm{P}\left(\text{flip } 1 = h\right) \cdots \mathrm{P}\left(\text{flip } a = h\right) \mathrm{P}\left(\text{flip } a + 1 = t\right) \cdots \mathrm{P}\left(\text{flip } n = t\right) \\ &= \theta^a \left(1 - \theta\right)^{n-a}, \end{aligned} \tag{2.42}$$

where h represents heads and t tails. By the same reasoning, this is also the probability of obtaining a heads and $n - a$ tails in any other order. By basic combinatorics, the number of possible orders is equal to the binomial coefficient

$$\binom{n}{a} := \frac{n!}{a! \, (n - a)!}. \tag{2.43}$$

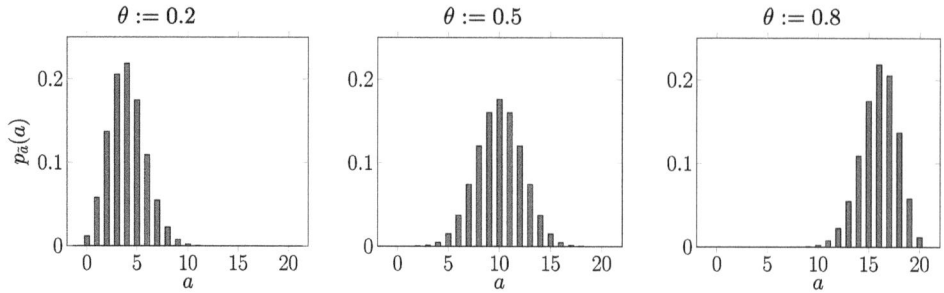

$\theta := 0.2$ $\theta := 0.5$ $\theta := 0.8$

Figure 2.6 Binomial pmf. The figure shows the pmf of a binomial random variable $\tilde{a}$ for $n := 20$ and $\theta := 0.2, 0.5, 0.8$.

We conclude that

$$P\left(a \text{ heads out of } n \text{ flips}\right) = \binom{n}{a} \theta^a \left(1 - \theta\right)^{n-a}. \tag{2.44}$$

The distribution derived in Example 2.16 is known as the binomial distribution. It is often used to model situations where there are n independent trials with a binary result that has constant probability. Figure 2.6 shows binomial pmfs with different values of θ.

Definition 2.17 (Binomial distribution). *The pmf of a binomial random variable $\tilde{a}$ with parameters n and θ is*

$$p_{\tilde{a}}(a) = \binom{n}{a} \theta^a \left(1 - \theta\right)^{n-a}, \quad a = 0, 1, \dots, n. \tag{2.45}$$

The binomial distribution enables us to analyze the behavior of the empirical-probability estimator. As we discuss in Section 1.4, whenever we compute an empirical probability using a finite number of data, the estimate is subject to random fluctuations. In the following example, we analyze these random fluctuations under the assumption that the data are sampled independently according to a fixed underlying probability. This illustrates how we can use probability theory to analyze the behavior of statistical estimators.

Example 2.18 (Analysis of the empirical-probability estimator). We are interested in estimating the probability of an event A from n data points $x_1, x_2, \dots, x_n$. To compute the empirical probability of A, we count how many data points are in A and divide by n, following Definition 1.22. In order to analyze the resulting estimator, we model the data probabilistically. Let B_i denote the event that the ith data point belongs to A. We assume

$$P\left(B_i\right) = \theta_{\text{true}}, \quad \text{for } 1 \le i \le n, \tag{2.46}$$

where θ_{true} represents the true probability of A. We also assume that the data are obtained independently, so that these events and their complements are all independent. To be more precise, for any sequence $S_1, S_2, \dots, S_n$, where S_i is either B_i or B_i^c, the events in the sequence are mutually independent.

Under our assumptions, each data point is analogous to a coin flip in Example 2.16, if we set $\theta := \theta_{\text{true}}$. In this analogy, heads corresponds to the data point being in A, and tails to

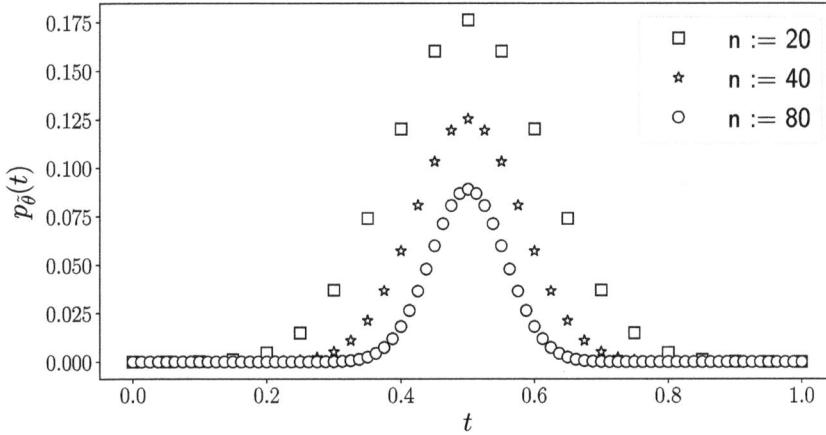

Figure 2.7 Analysis of the empirical-probability estimator. Pmfs of the empirical-probability estimator under the assumptions of Example 2.18, when the true probability $\theta_{\text{true}} := 0.5$ and the number of data n is 20, 40, and 80. As n increases, the estimator concentrates around the true probability.

the data point not being in A. Consequently, if we represent the number of data points in A as a random variable $\tilde{c}$, then $\tilde{c}$ is binomial with parameters n and θ_{true}, so its pmf is

$$p_{\tilde{c}}(c) = \binom{n}{c} \theta_{\text{true}}^c (1 - \theta_{\text{true}})^{n-c}, \quad c = 0, 1, 2, \ldots, n. \tag{2.47}$$

This allows us to derive the pmf of the empirical-probability estimator $\tilde{\theta} := \tilde{c}/n$, which equals the number of data in A divided by the total number of data:

$$p_{\tilde{\theta}}(t) := \mathrm{P}\left(\tilde{\theta} = t\right) \tag{2.48}$$

$$= \mathrm{P}(\tilde{c} = nt) \tag{2.49}$$

$$= \binom{n}{nt} \theta_{\text{true}}^{nt} (1 - \theta_{\text{true}})^{n-nt}, \quad t = 0, \frac{1}{n}, \frac{2}{n}, \ldots, 1. \tag{2.50}$$

Figure 2.7 shows the pmf of $\tilde{\theta}$ when $\theta_{\text{true}} := 0.5$ and n is 20, 40, and 80. As we gather more data, the pmf of the empirical-probability estimator concentrates around the true probability θ_{true}. Let us compute the probability of making an error of less than 0.1 when approximating θ_{true}. By Theorem 2.5,

$$\mathrm{P}(|\tilde{\theta} - \theta_{\text{true}}| \leq 0.1) = \sum_{t \in [\theta_{\text{true}}-0.1, \theta_{\text{true}}+0.1]} p_{\tilde{\theta}}(t) \tag{2.51}$$

$$= \sum_{k \in [n\theta_{\text{true}}-0.1n, n\theta_{\text{true}}+0.1n]} \binom{n}{k} \theta_{\text{true}}^k (1 - \theta_{\text{true}})^{n-k}. \tag{2.52}$$

For $\theta_{\text{true}} := 0.5$, this probability equals 0.737 ($n := 20$), 0.846 ($n := 40$), and 0.943 ($n := 80$). Reassuringly, the probability increases with the number of data, and is already very high for $n := 80$; the error is more than 0.1 only about 5% of the time. Theorem 9.24 establishes that this is not a coincidence: The empirical-probability estimator is guaranteed to converge in probability to the true probability of the event of interest, as long as it is computed using independent data.

. . . .

2.3.3 The Geometric Distribution

In the following example, we derive a parametric model for the free-throw data in Example 2.14.

Example 2.19 (Parametric model for free-throw streaks)**.** To obtain a parametric model, we make two simplifying assumptions:

1 The free throws are all independent.
2 The probability of making each individual free throw is equal to a fixed constant θ.

Let $\tilde{s}$ be a discrete random variable representing the length of a streak of made free throws. Under our assumptions, by Definition 1.28, the value of the pmf of $\tilde{s}$ equals

$$p_{\tilde{s}}(s) = \mathrm{P}(s \text{ free throws are made, followed by a miss}) \tag{2.53}$$
$$= \mathrm{P}(\text{1st made})\mathrm{P}(\text{2nd made})\cdots\mathrm{P}(s\text{th made})\mathrm{P}(s+1\text{th missed}) \tag{2.54}$$
$$= \theta^s(1-\theta) \tag{2.55}$$

for $s \in \{0, 1, 2, \dots\}$. This parametric pmf depends on a single parameter (θ).

. . . .

You may be questioning the assumptions used to derive the model in Example 2.19, and you would be right to do so. The probability of making a free throw probably changes depending on the type of game (e.g. regular season vs. playoffs), the time at which it is taken (e.g. beginning of the game vs. end of the game), and the state of mind of the player (e.g. if the game is close, contract negotiations are coming up, etc.). This illustrates how, in practice, we are usually forced to make assumptions that are not entirely correct in order to design parametric models. The key, paraphrasing George Box, is to ensure that the model is *wrong, but useful*. The simple model derived in Example 2.19 is definitely useful; in Section 2.5, we show that it achieves good performance.

The distribution derived in Example 2.19 is equivalent to a *geometric* distribution. This is a popular parametric distribution, which can be used, for example, to model how many times we need to flip a coin until we obtain heads (see Exercise 2.1), or the number of times we use a washing machine before it breaks down. The following standard definition of the

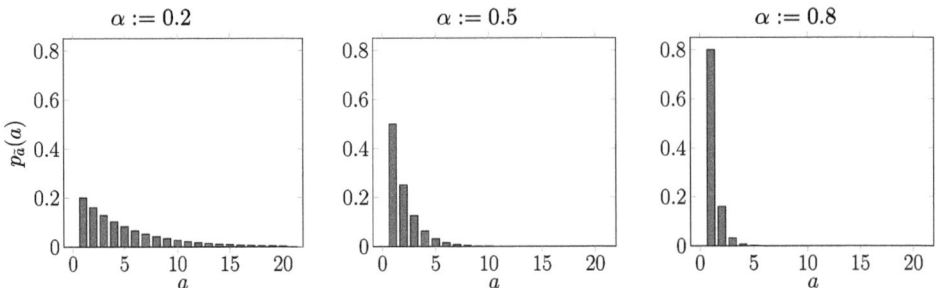

Figure 2.8 Geometric pmf. The figure shows the first twenty entries of the pmf of a geometric random variable $\tilde{a}$ for $\alpha := 0.2, 0.5$, and 0.8.

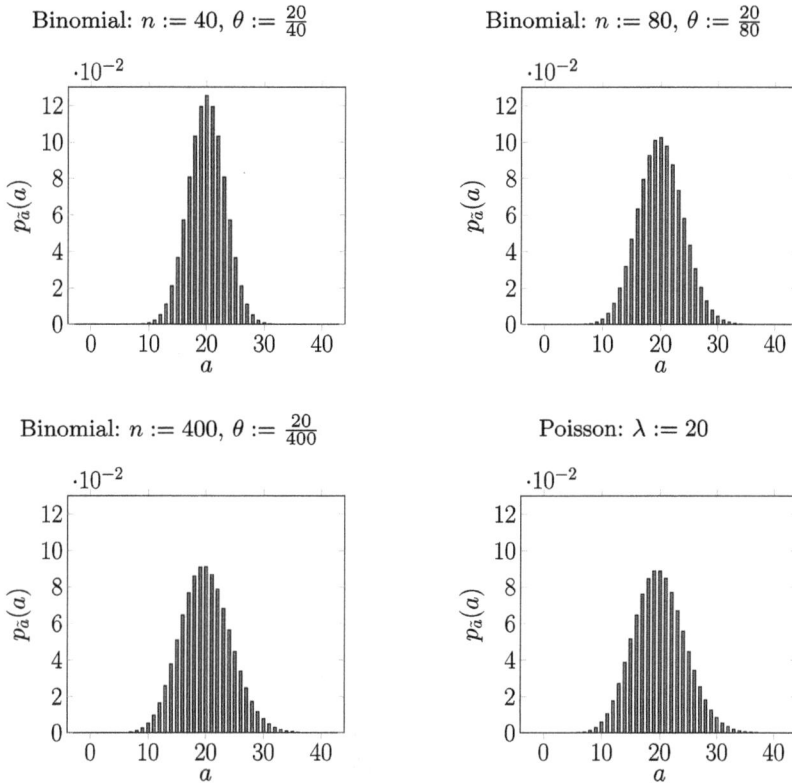

Figure 2.10 Convergence of the binomial distribution to a Poisson distribution. As n increases, the binomial pmf with parameters n and $\theta := \lambda/n$ converges to a Poisson pmf with parameter λ.

For any $a \in A$, $p_\theta(a)$ encodes the probability of observing a according to the model. Assume that we only have access to one data point x. Then a reasonable choice for θ is the one that maximizes the probability of observing that particular data point, $p_\theta(x)$. Note the change of perspective: We are suddenly thinking of $p_\theta(x)$ *as a function of* θ. This function is known as the *likelihood* of the parametric model. Maximizing the likelihood to fit the parameters of a parametric model is known as *maximum-likelihood estimation*.

Of course, we won't ever fit a model with a single data point. In order to extend the concept of likelihood to multiple data, we need to make some assumptions. A reasonable choice is to interpret the data as realizations of independent, identically distributed (i.i.d.) random variables. This means that the value of each data point is generated according to the same distribution and is independent of the rest of the data.

Definition 2.23 (Independent identically distributed random variables). *Let $\tilde{a}_1$, $\tilde{a}_2$, ..., $\tilde{a}_n$ be discrete random variables defined on the same probability space. The random variables are identically distributed, if they have the same pmf, so that for each element a in their range A:*

$$P(\tilde{a}_1 = a) = P(\tilde{a}_2 = a) = \cdots = P(\tilde{a}_n = a). \qquad (2.67)$$

The random variables are independent, if for any choice of a_1, a_2, ..., a_n belonging to A, the events $\tilde{a}_1 = a_1$, $\tilde{a}_2 = a_2$, ..., $\tilde{a}_n = a_n$ are mutually independent.

Let $x_1, x_2, \ldots, x_n$ be the available data. If we model them as realizations of i.i.d. discrete random variables with a parametric pmf p_θ, the probability of observing precisely those values equals

$$P(\tilde{a}_1 = x_1, \tilde{a}_2 = x_2, \ldots, \tilde{a}_n = x_n) = P(\tilde{a}_1 = x_1)P(\tilde{a}_2 = x_2) \cdots P(\tilde{a}_n = x_n) \quad (2.68)$$

$$= \prod_{i=1}^{n} p_\theta(x_i), \quad (2.69)$$

by Definition 1.28. Interpreting this probability as a function of θ yields the likelihood of the parametric model. Notice that the likelihood is a product of probabilities, which are numbers between zero and one. As a result, it tends to become extremely small when n is large. To avoid numerical instability, we often consider the logarithm of the likelihood, or *log-likelihood*, instead.

Definition 2.24 (Likelihood function under i.i.d. assumptions). *For some discrete set A, let $p_\theta \colon A \to [0, 1]$ be a parametric pmf model dependent on a parameter vector θ, and $X := \{x_1, x_2, \ldots, x_n\}$ a dataset with values in A. The likelihood of the model given the data under i.i.d. assumptions is*

$$\mathcal{L}_X(\theta) := \prod_{i=1}^{n} p_\theta(x_i). \quad (2.70)$$

The log-likelihood function is the logarithm of the likelihood function,

$$\log \mathcal{L}_X(\theta) = \sum_{i=1}^{n} \log p_\theta(x_i). \quad (2.71)$$

Maximum-likelihood estimation selects the parameters that maximize the likelihood or the log-likelihood, and therefore the probability of the observed data according to the parametric model. Maximizing the likelihood is equivalent to maximizing the log-likelihood because the logarithm is a strictly monotone function.

Definition 2.25 (Maximum-likelihood estimator). *For some discrete set A, let $p_\theta \colon A \to [0, 1]$ be a parametric pmf dependent on a parameter vector θ, $X := \{x_1, x_2, \ldots, x_n\}$ a dataset with values in A, and S the set of parameters for which p_θ is a valid pmf. The maximum-likelihood estimator of θ is*

$$\theta_{\mathrm{ML}} := \arg\max_{\theta \in S} \mathcal{L}_X(\theta) \quad (2.72)$$

$$= \arg\max_{\theta \in S} \log \mathcal{L}_X(\theta). \quad (2.73)$$

Example 2.26 (Maximum-likelihood estimator for the Bernoulli distribution). Let $X := \{x_1, \ldots, x_n\}$ be n data points equal to zero or one, representing the occurrence of some event of interest. Assuming that the data are i.i.d., we decide to fit a Bernoulli model with parameter θ (see Definition 2.15). According to the parametric model, for the ith data point x_i, $p_\theta(x_i) = \theta$, if $x_i = 1$, and $p_\theta(x_i) = 1 - \theta$, if $x_i = 0$. Consequently, by Definition 2.24, the likelihood function is

$$\mathcal{L}_X(\theta) = \prod_{i=1}^{n} p_\theta(x_i) \quad (2.74)$$

$$= \theta^{n_1} (1 - \theta)^{n_0}, \quad (2.75)$$

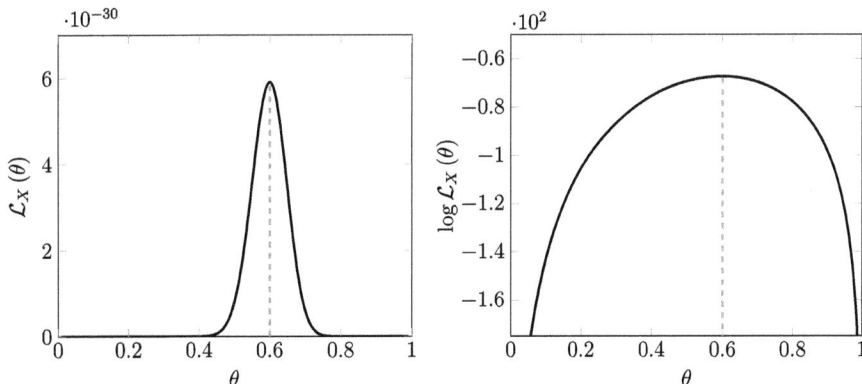

Figure 2.11 Likelihood and log-likelihood functions of a Bernoulli parametric model. Likelihood (left) and log-likelihood (right) functions of the Bernoulli parametric model derived in Example 2.26 for a dataset, where there are 60 data points equal to one and 40 data points equal to zero. The values of the likelihood are extremely small. The maximum of both functions occurs at 0.6, which is the maximum-likelihood estimate of the Bernoulli parameter by Definition 2.25.

where n_0 and n_1 are the number of observations equal to zero and one, respectively. The log-likelihood function equals

$$\log \mathcal{L}_X (\theta) = n_1 \log \theta + n_0 \log (1 - \theta). \tag{2.76}$$

Figure 2.11 shows the likelihood and log-likelihood functions for $n_0 = 40$ and $n_1 = 60$. Following Definition 2.25, we compute the maximum-likelihood estimator of the parameter θ by maximizing the log-likelihood function,

$$\theta_{\mathrm{ML}} := \arg \max_{\theta} \log \mathcal{L}_X (\theta) \tag{2.77}$$

$$= \arg \max_{\theta} n_1 \log \theta + n_0 \log (1 - \theta). \tag{2.78}$$

The derivative and second derivative of the log-likelihood function equal

$$\frac{\mathrm{d} \log \mathcal{L}_X (\theta)}{\mathrm{d}\theta} = \frac{n_1}{\theta} - \frac{n_0}{1 - \theta}, \tag{2.79}$$

$$\frac{\mathrm{d}^2 \log \mathcal{L}_X (\theta)}{\mathrm{d}\theta^2} = -\frac{n_1}{\theta^2} - \frac{n_0}{(1 - \theta)^2}. \tag{2.80}$$

The function is concave, as the second derivative is negative for all $\theta \in [0, 1]$. This is good news because it means that there is a single maximum within that interval. The maximum is at the point where the first derivative equals zero:

$$\theta_{\mathrm{ML}} = \frac{n_1}{n_0 + n_1}. \tag{2.81}$$

We conclude that the maximum-likelihood estimator is the fraction of data points that equal one, which is the empirical probability of observing a one. This is a very reasonable way to estimate the parameter θ, since θ represents the probability that the Bernoulli random variable is equal to one.

. . . .

We are now ready to fit the parametric model derived in Example 2.19 to the data in Example 2.14.

Example 2.27 (Fitting the parametric model for free-throw streaks). Let $X := \{x_1, x_2, \ldots, x_n\}$ contain the lengths of the $n := 377$ observed free-throw streaks. Since $p_\theta(a) = \theta^a(1 - \theta)$, by Definition 2.24, the log-likelihood equals

$$\log \mathcal{L}_X(\theta) = \sum_{i=1}^n \log p_\theta(x_i) \tag{2.82}$$

$$= \sum_{i=1}^n \log \left(\theta^{x_i}(1 - \theta) \right) \tag{2.83}$$

$$= \sum_{i=1}^n \left(x_i \log \theta + \log (1 - \theta) \right) \tag{2.84}$$

$$= \left(\sum_{i=1}^n x_i \right) \log \theta + n \log (1 - \theta) \tag{2.85}$$

$$= n_{\text{made}} \log \theta + n_{\text{missed}} \log (1 - \theta), \tag{2.86}$$

where $n_{\text{made}} := \sum_{i=1}^n x_i$ is the sum of the lengths of all streaks, which equals the number of made free throws, and n_{missed} is the number of missed free throws, which equals the number of total streaks n because every streak ends with a miss. The log-likelihood is exactly the same as the one derived for the Bernoulli model in Example 2.26. This makes sense: Under our assumptions, the individual free throws can be interpreted as realizations of i.i.d. Bernoulli random variables with parameter θ. Consequently, by the same argument as in Example 2.26, the maximum-likelihood estimator of θ equals the fraction of made free throws,

$$\theta_{\text{ML}} := \arg \max_\theta \log \mathcal{L}_X(\theta) \tag{2.87}$$

$$= \frac{n_{\text{made}}}{n_{\text{missed}} + n_{\text{made}}} \tag{2.88}$$

$$= 0.875. \tag{2.89}$$

The corresponding pmf is shown in Figure 2.12, which also shows the pmfs corresponding to two other values of θ. The pmf corresponding to the maximum-likelihood estimator achieves a much better approximation to the empirical pmf of the observed data. This is not a surprise, since maximum-likelihood estimation yields the parametric pmf according to which the data are most likely.

· ·

Figure 2.13 depicts the behavior of the maximum-likelihood estimator for the parametric model derived in Example 2.27, when the parametric assumption holds. The left column shows log-likelihood functions computed from three simulated datasets of 3,015 i.i.d. samples, drawn from the parametric distribution for θ equal to $\theta_{\text{true}} := 0.875$. The log-likelihood function depends on the observed data, so it varies for the different datasets, but its maximum θ_{ML} tends to be close to θ_{true} (the maximum-likelihood estimates are 0.878, 0.871, and 0.863). The right column shows the resulting pmf estimates $p_{\theta_{\text{ML}}}$, which achieve a much better approximation to the true pmf $p_{\theta_{\text{true}}}$ than the empirical-pmf estimator (computed following Definition 2.11). Of course, the comparison is not fair: The data are generated exactly according to the parametric model, which is never the case in reality.

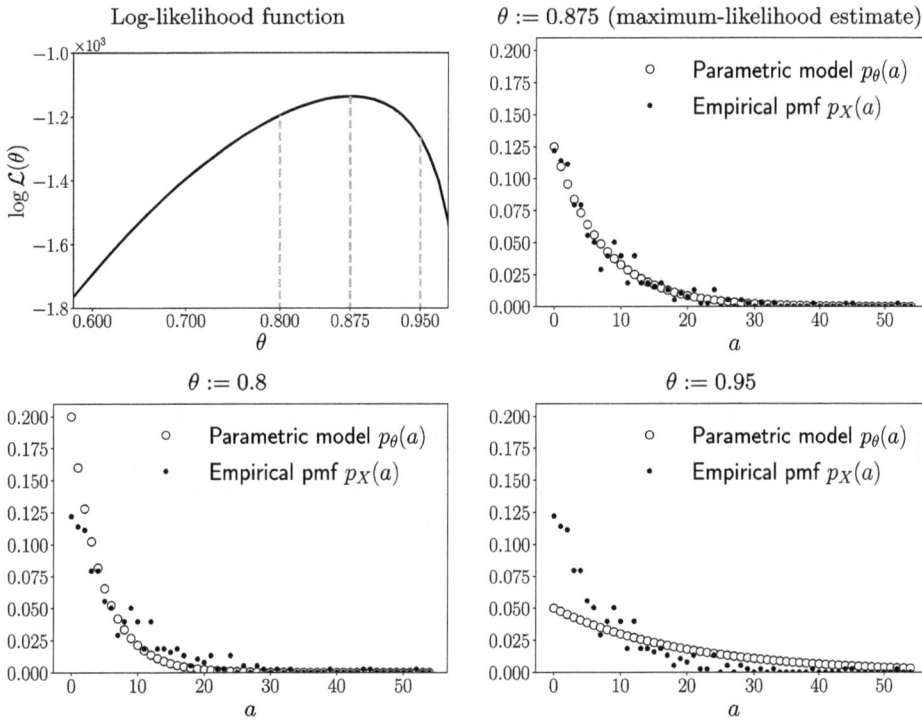

Figure 2.12 Parametric model for consecutive made free throws. The graph in the top left shows the log-likelihood function of the parametric model derived in Example 2.19. The corresponding pmf, obtained via maximum-likelihood estimation, is shown in the top right along with the empirical pmf of the data. The two bottom graphs show the pmfs corresponding to two other choices of the parameter θ, which produce a worse fit to the data.

2.5 Comparing Parametric and Nonparametric Models

In this chapter, we describe two approaches to estimate pmfs of discrete random variables from data. The first is to approximate the pmf directly using empirical probabilities, as explained in Section 2.2. This is known as *nonparametric* estimation. The second approach is *parametric* estimation, where we fit a pre-determined parametric distribution to the data, as explained in Section 2.4. In this section, we discuss the advantages and disadvantages of the two strategies.

Let us focus on the free-throw dataset from Example 2.14. We could be tempted to evaluate the pmf models based on how well they fit the data. To assess the fit of a given pmf estimate p_{est}, we compare the estimated probability $p_{est}(\ell)$ to the corresponding empirical probability of observing a streak of length ℓ. This probability is exactly equal to the empirical pmf at ℓ, by Definition 2.11. Consequently, according to this metric, the empirical-pmf estimator produces a perfect fit! Of course, this is too good to be true. As we explain in Example 2.14, the empirical pmf seems to capture spurious patterns due to the limited number of data.

The problem here is that we are using *the same data to build the model and evaluate it.* This is like evaluating a student on a test, after they have seen the solutions. The evaluation is meaningless because the student can mindlessly replicate the solutions without having

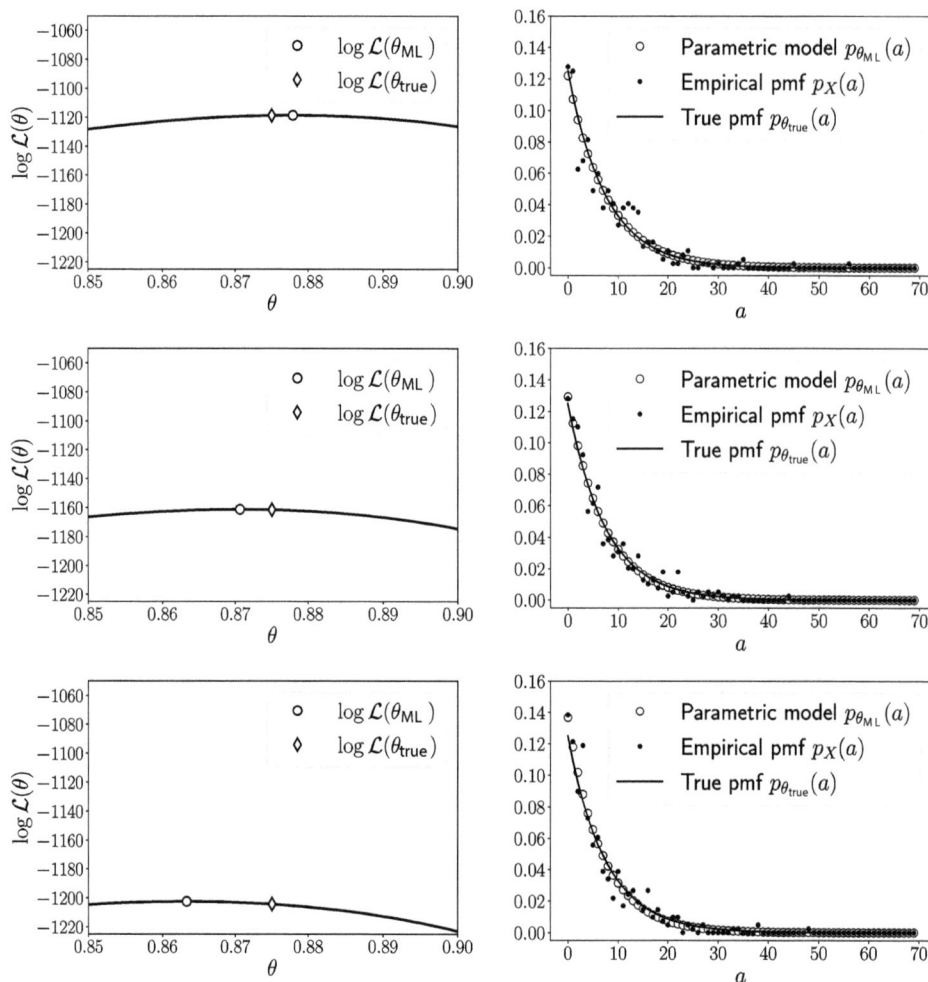

Figure 2.13 Maximum-likelihood estimation applied to simulated data. The left column shows the log-likelihood function of the parametric model derived in Example 2.19 computed from three synthetic datasets, each consisting of 3,015 i.i.d. samples from the assumed parametric distribution with a fixed parameter equal to $\theta_{\text{true}} := 0.875$. The maximum-likelihood estimate θ_{ML} of the model parameter is close to θ_{true} in all three cases. The corresponding log-likelihood values are indicated by a circle and diamond marker, respectively. The right column shows the parametric pmf corresponding to θ_{ML} (white circle markers), compared to the true pmf (black curve) and the empirical pmf of the data (black dots).

learned anything. Instead, we need to evaluate models by checking whether they *generalize* to *held-out data*, which have *not* been used to fit them.

In order to perform rigorous evaluation of a statistical estimator, we select a subset of the available data beforehand and reserve it exclusively for evaluation. We call this the *test set*. The data used to fit the model is known as the *training set* (*training* a model means fitting it to data). In Example 2.14, we use the first 3,015 free throws in Durant's NBA career as a training set to fit a nonparametric empirical-pmf model. In Example 2.19, we use the same training data to fit a parametric pmf based on the geometric distribution. To compare the generalization ability of both models, we leverage a test set consisting of the following

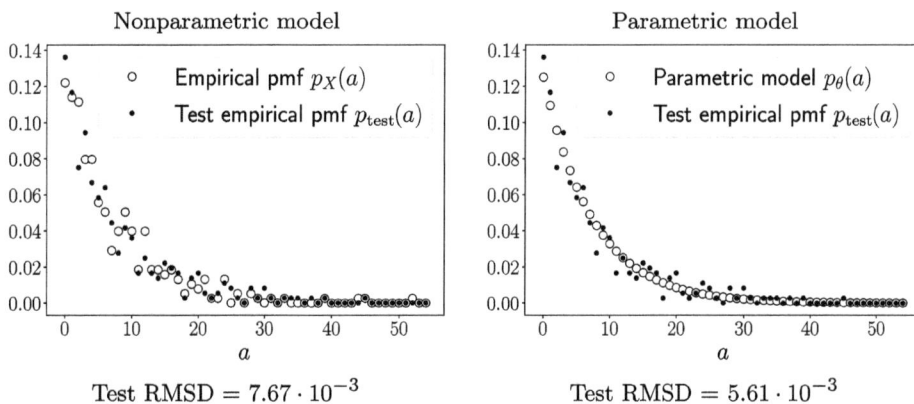

Nonparametric model   Parametric model

Test RMSD = $7.67 \cdot 10^{-3}$ Test RMSD = $5.61 \cdot 10^{-3}$

Figure 2.14 Parametric vs. nonparametric modeling of free-throw streaks. Probability of Kevin Durant making a certain number of consecutive free throws before missing, obtained by fitting a nonparametric empirical-pmf model (left graph, see Example 2.14) and a parametric geometric model (right graph, see Example 2.19). Both models are fit on a training set consisting of the first 3,015 free throws in Durant's NBA career. In order to evaluate the pmf estimates, we compare them to the empirical pmf of a test set corresponding to the following 3,015 free throws. The parametric model approximates the test pmf better, as quantified by the root mean square deviation, reported underneath the graphs.

3,015 free throws shot by Durant. Our evaluation metric is the root mean square deviation (RMSD) between each estimated pmf p_{est} and the empirical pmf of free-throw streak length in the test data p_{test}:

$$\text{RMSD}(p_{\text{est}}) := \sqrt{\frac{1}{L} \sum_{\ell=0}^{L} (p_{\text{est}}(\ell) - p_{\text{test}}(\ell))^2}, \tag{2.90}$$

where L is the maximum streak length, which we set equal to 55. The nonparametric model has a test RMSD of $7.67 \cdot 10^{-3}$. The parametric model has a smaller test RMSD error: $5.61 \cdot 10^{-3}$. Figure 2.14 shows both estimated pmfs superimposed on the empirical pmf of the test data.

As we had suspected, the fluctuations in the nonparametric empirical-pmf estimate (e.g. a streak of length 7 having lower probability than a streak of length 12, see Example 2.14) are not reproduced in the test data. This suggests that they are *noise*, a term we use to describe unpredictable structure in the data which does not carry any useful information. By contrast, the meaningful information that we are trying to extract from the data is called the *signal*. Estimators that fit the noise in the training data are said to *overfit* because they approximate the training set too closely. A training error that is much better than the test error is a symptom of overfitting. Here, the training error of the nonparametric empirical-pmf model is zero. By contrast, the training error of the parametric model is $5.46 \cdot 10^{-3}$, which is very similar to the test error, indicating that it does not overfit.

Complicated models with a large number of parameters are more prone to overfit than simpler models because they are able to fit noisy fluctuations in the training data. Even though the empirical pmf is called a *nonparametric* estimator, it also requires estimating some *parameters* from the data, namely each entry of the pmf. In the case of the free-throw data, the number of parameters to be estimated is 55.[1] By contrast, the parametric model only requires estimating one parameter! This constrains the shape of the pmf, preventing it from overfitting the noise in the training data.

Parametric models are not always superior to nonparametric models. The assumptions underlying parametric models never hold exactly. This is not very problematic when the training data are limited. In fact, as illustrated by the free-throw data, it can be a crucial advantage: The simplicity of the parametric model in Example 2.19 avoids overfitting. However, if training data are more plentiful, inaccurate parametric assumptions may result in an overly constrained model, which is not able to capture meaningful complex structure. This is known as *underfitting*. The following example evaluates a nonparametric and parametric model fit with training sets of different sizes. As the number of training data increases, the performance of the nonparametric model improves, eventually surpassing the performance of the parametric model, as shown in Figure 2.16.

Example 2.28 (Call center). In this example, we analyze Dataset 3, which consists of telephone data recorded at the call center of an anonymous bank in Israel in 1999. Our goal is to model the distribution of the number of calls that arrive between 6 am and 7 am on weekdays. We model the number of calls as a discrete random variable and estimate its pmf using two approaches: A nonparametric model based on the empirical pmf (see Definition 2.11), and a parametric model based on the Poisson distribution (see Section 2.3.4). We set the maximum number of calls to 24, so fitting the nonparametric model requires estimating 24 values. By contrast, the Poisson model has a single parameter, which we fit via maximum-likelihood estimation.

We begin by deriving the maximum-likelihood estimator of the parameter in the Poisson model. Let X denote the training data. Each data point x_i, $1 \leq i \leq n$, is the number of calls recorded in a specific day between 6 am and 7 am. By Definition 2.24, the likelihood is

$$\mathcal{L}_X(\lambda) = \prod_{i=1}^{n} p_\lambda(x_i) \tag{2.91}$$

$$= \prod_{i=1}^{n} \frac{\lambda^{x_i} e^{-\lambda}}{x_i!}, \tag{2.92}$$

where p_λ denotes a Poisson pmf with parameter λ, so the log-likelihood equals

$$\log \mathcal{L}_X(\lambda) = \sum_{i=1}^{n} (x_i \log \lambda - \lambda - \log(x_i!)). \tag{2.93}$$

The derivative and second derivative of the log-likelihood are

$$\frac{d \log \mathcal{L}_X(\lambda)}{d\lambda} = \sum_{i=1}^{n} \left(\frac{x_i}{\lambda} - 1 \right), \tag{2.94}$$

$$\frac{d^2 \log \mathcal{L}_X(\lambda)}{d\lambda^2} = -\sum_{i=1}^{n} \frac{x_i}{\lambda^2}. \tag{2.95}$$

---

[1] The maximum possible streak length is set to 55, so the possible lengths go from 0 to 55. From these 56 numbers, only 55 need to be estimated from the data because the pmf must sum to 1.

The function is concave, as the second derivative is negative (as long as at least one data point is nonzero). The maximum-likelihood estimator of the parameter (see Definition 2.25) is consequently at the point where the first derivative equals zero, namely

$$\lambda_{\mathrm{ML}} := \arg \max_{\lambda} \log \mathcal{L}_X(\lambda) \tag{2.96}$$

$$= \frac{1}{n} \sum_{i=1}^{n} x_i, \tag{2.97}$$

which is the average number of calls received between 6 am and 7 am, or equivalently the rate of calls per hour in that period. This is consistent with the interpretation of the parameter as a rate in our derivation of the Poisson pmf (see Example 2.21).

In order to compare the nonparametric and parametric models, we utilize training sets of different sizes to fit them. The training sets consist of the number of calls received during weekdays between 6 am and 7 am in a period from month 1 (January) to month m, where we vary m from one (just January) to six (January to June). We use the remaining months (July to December) as a test set. Figure 2.15 shows the resulting estimated pmfs, along with the empirical pmf of the test set. Figure 2.16 shows the test RMSD, as defined in (2.90), of the nonparametric and parametric models. When $m = 1$, the number of data is around 20, which is very small compared to the number of estimated values required by the empirical pmf. As a result, the nonparametric model overfits the training data, resulting in a test RMSD that is double that of the Poisson model. As we increase the number of months used for training, the nonparametric model improves substantially, eventually overtaking the parametric model. This is quite typical: Parametric models (with reasonable assumptions) tend to perform well when the training data are limited, whereas nonparametric models can be extremely effective, if we have enough training data to exploit their flexibility without overfitting.

Let us take a closer look at the pmf estimates in Figure 2.15. It turns out that January has an unusual number of days when there are no calls, possibly due to public holidays when the bank is closed. If the training set only contains January (top row in Figure 2.15), the empirical pmf assigns a high probability to zero. This results in a large test error because the fraction of data with zero calls in the test set is much lower (maybe because there are less holidays between July and December). When we include data from other months in the training set, the empirical pmf assigns less and less probability to zero because the additional training data contain less days with zero calls.

By contrast, the parametric model does not overfit the unusual number of days with zero calls, even if we only use January to train the model. The reason is that the shape of the Poisson pmf cannot assign a large probability to zero and at the same time fit the rest of the data. In this case, this results in effective generalization to the test data, but it could also have hurt it. In fact, in Figure 2.15, we see that the Poisson model systematically underestimates the probability of zero calls. This would have been even more apparent if there had been a month in the test data with a lot of holidays (for example, January of the following year). The problem is that the observed data violates the assumptions underlying the Poisson model, so the parametric model underfits. To address this, we cannot just use more training data; the Poisson model is too simple to account for the additional structure, no matter how much data we use to fit it. Instead, we need to use a more complex model.

Months   Nonparametric model   Parametric model

Test RMSD $= 5.92 \cdot 10^{-2}$ Test RMSD $= 3.00 \cdot 10^{-2}$

Test RMSD $= 3.22 \cdot 10^{-2}$ Test RMSD $= 2.23 \cdot 10^{-2}$

Test RMSD $= 2.20 \cdot 10^{-2}$ Test RMSD $= 2.22 \cdot 10^{-2}$

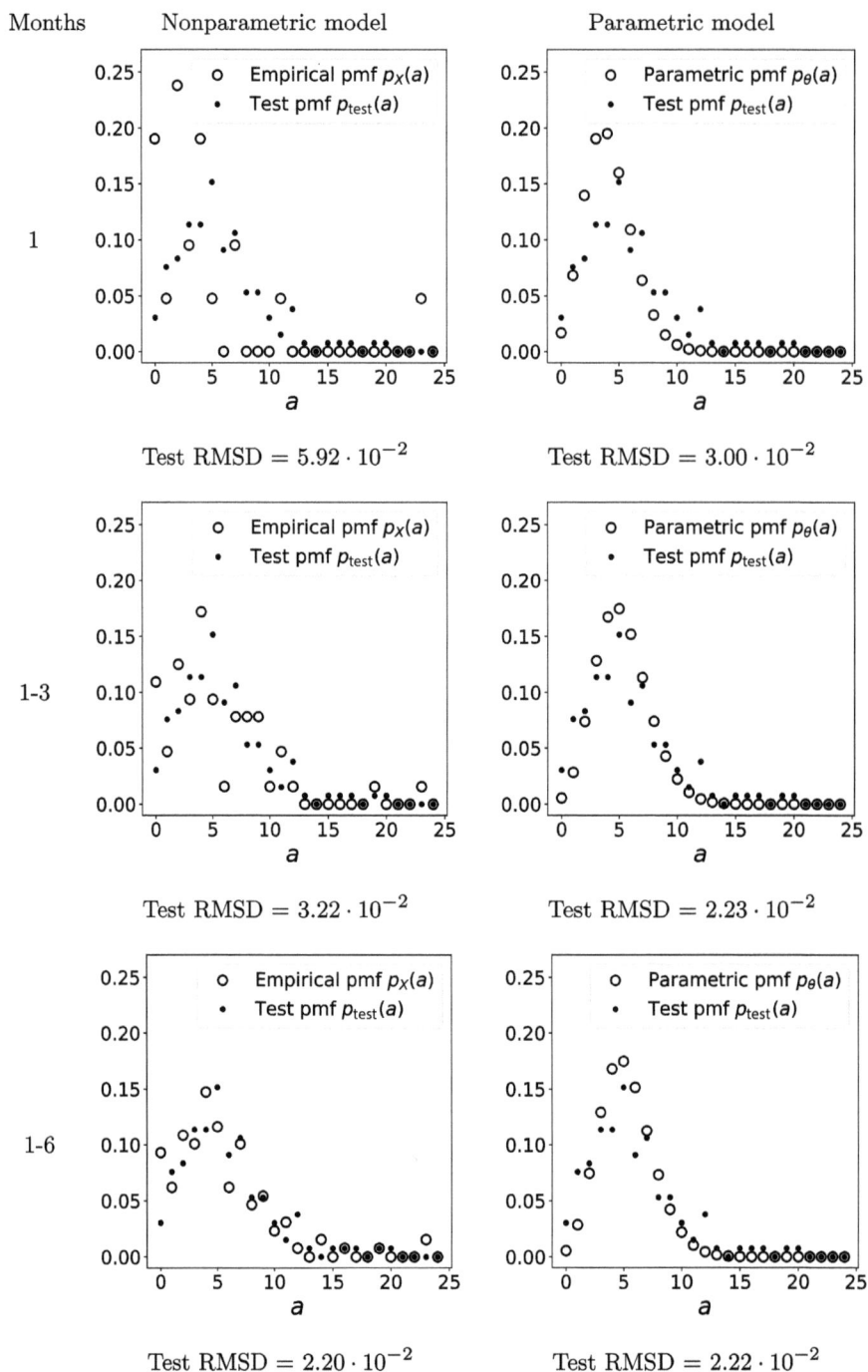

Figure 2.15 Parametric vs. nonparametric modeling of calls arriving at a callling center.
The graphs compare a nonparametric model based on the empirical pmf (left column) and
a parametric Poisson model (right column) fit to calling-center data, as explained in Example 2.28. Each row corresponds to a training set containing a different number of months: January
(first row), January–March (second row), and January–June (third row). The pmf estimates are
compared to the empirical pmf of the test data, which contains the months from July to December.

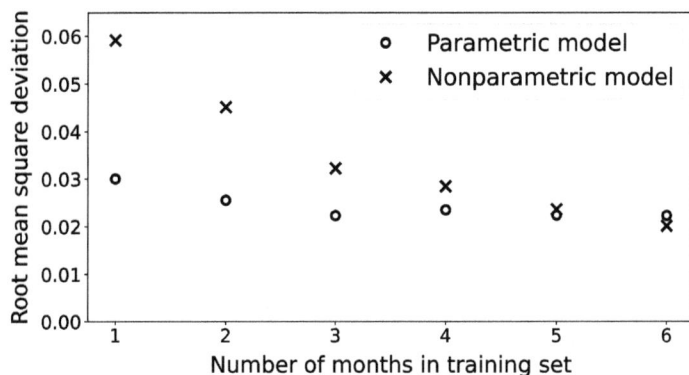

Figure 2.16 Comparison of parametric and nonparametric models for different amounts of data. The graph shows the test root mean square deviation of the parametric and the nonparametric model in Example 2.28. The training sets contain data from month 1 (January) to month m, where we vary m from 1 (just January) to 6 (January–June). The test set contains data from July to December. The parametric approach is clearly superior when the number of training data is small. However, the nonparametric model improves dramatically as we increase the training data, eventually outperforming the parametric model.

Exercises

2.1 (Flipping until heads) Imagine that we flip a coin until we obtain heads. If α is the probability of heads, and the flips are independent, show that the number of flips is a geometric random variable with parameter α, as defined in Definition 2.20.

2.2 (Geometric pmf) Prove that the pmf of a geometric random variable (Definition 2.20) is a valid pmf.

2.3 (Poisson pmf) Show that the pmf of a Poisson random variable (Definition 2.22) is a valid pmf.

2.4 (Memoryless property of the geometric distribution) Let $\tilde{a}$ be a geometric random variable with parameter α. What is the probability that $\tilde{a}$ equals a, for $a = 1, 2, 3, \ldots$, if we condition on the event $\tilde{a} > 5$? Justify your answer mathematically, and also explain why it makes sense intuitively, using coin flips as an example.

2.5 (Old car) Dani and Felix have a 2000 Ford Taurus. Every time they drive it, the car breaks down with probability $1/10$.

    a What distribution could they use to model the number of times that the car breaks down if they drive it 50 times? What assumption do they need to make?

    b What is the probability that the car breaks down at most one time in the first 50 drives? Compute the probability exactly under the assumption you suggested in the previous question and also approximate it using a Poisson distribution. Justify the approximation and compare the results.

    c What is the probability that the kth time the car breaks down is in the nth drive?

2.6 (Oil prospector) An oil prospector is analyzing an unexplored region. From past data, she determines that there are three possible types of regions according to their abundance of oil. In *oil-rich* regions every time you drill, you find oil with a probability of 1/2. In *standard* regions the probability is 1/10. In *oil-poor* regions the probability is 1/100. From a geological study, she determines that the new region is oil rich with probability 1/10, standard with probability 8/10, and oil poor with probability 1/10. Assuming that drilling attempts are all conditionally independent given the type of region, what is the pmf of the number of times the prospector has to drill to find oil in the new region?

2.7   (Darts) In a game of darts, a player needs to hit a certain number k times. Assume that all attempts are mutually independent, and that the probability of success in each attempt is θ. Derive the pmf of a random variable representing the number of required attempts.

2.8   (Binomial random variable) Let $\tilde{a}$ be a binomial random variable with parameters n and θ. What is the conditional probability that $\tilde{a} = 1$ given $\tilde{a} \leq 1$?

2.9   (Intersection) Cars can go left, right, or straight at an intersection. We model their decision as i.i.d. samples from a random variable $\tilde{c}$ that can equal -1 (left), 1 (right), or 0 (straight). We assume that the probability of going right is the same as the probability of going left. We design a parametric model for the pmf of the random variable, where the only parameter θ represents the probability of going right. Find the maximum-likelihood estimate of θ given the following data: $-1, 1, 0, 0, 1, 1$.

2.10 (Bad apples) A farmer is interested in determining the fraction θ of bad apples in her orchard. Her strategy is to pick apples uniformly at random with replacement (so every apple has the same probability of being picked each time) and test them. The test result indicates whether the apple is bad or not. The test is not perfect, it correctly identifies bad apples with probability 0.9. It also incorrectly identifies good apples as bad with probability 0.4.

    a Let $\tilde{t}$ be a Bernoulli random variable that equals 1, if a test is positive, and 0, if it is negative. Derive the pmf of $\tilde{t}$ as a function of θ.

    b What is the maximum-likelihood estimator of θ, if the farmer picks 10 apples and 5 are declared to be bad by the test? Interpret the data as i.i.d. samples from the random variable $\tilde{t}$.

2.11 (False positives) A company has developed a test for a new disease. They want to estimate the probability of a false positive θ. They evaluate it on individuals selected from a population where the disease prevalence is 10%, so the probability that each individual has the disease is 0.1. They assume that the probability of a false negative is zero.

    a Let $\tilde{t}$ be a Bernoulli random variable that equals 1, if a test is positive, and 0, if it is negative. Derive the pmf of $\tilde{t}$ as a function of the probability of false positives θ.

    b The company tests n people from the population and counts the number of positive tests n_{pos}. Derive the maximum-likelihood estimator of θ as a function of the fraction of observed positive tests $\phi := n_{\mathrm{pos}}/n$. Interpret the data as i.i.d. samples from the random variable $\tilde{t}$. (Hint: The estimator should be between zero and one because θ represents a probability.)

2.12 (Chess games) Garry and Anish decide to play 10 chess games. Garry wins 4, they draw 4, and Anish wins 2. We decide to model the games probabilistically, assuming that they are independent. In each game, Garry has a probability θ of winning and Anish has a probability α of winning.

    a Plot the log-likelihood function of the parametric model.

    b What are the maximum-likelihood estimates of θ and α?

    c Model the data as realizations of a discrete random variable and compute its empirical pmf. Compare this nonparametric model to the parametric model from the previous questions.

2.13 (Maximum-likelihood estimator for the geometric distribution) Prove that the maximum-likelihood estimator of the parameter α of a geometric distribution, given a dataset $x_1, \ldots, x_n$, where x_i is an integer greater than or equal to one for all i, is

$$\alpha_{\mathrm{ML}} = \left(\frac{1}{n} \sum_{i=1}^{n} x_i \right)^{-1}. \tag{2.98}$$

2.14 (Multinoulli distribution) The multinoulli or categorical distribution is used to model quantities that can take multiple different values. The pmf of a multinoulli random variable $\tilde{a}$ associated with the values $v_1, \ldots, v_m$ equals

$$p_{\tilde{a}}(v_k) = \theta_k, \qquad 1 \le k \le m, \tag{2.99}$$

and zero otherwise.

a What constraints must the parameters $\theta_1, \ldots, \theta_m$ satisfy?

b Given a dataset $x_1, \ldots, x_n$ taking values in $\{v_1, \ldots, v_m\}$, prove that the maximum-likelihood estimate of θ_k is the fraction of data points equal to v_k. Assume that the log-likelihood is concave, which can be proved by showing that its Hessian is negative semidefinite.

c Compare the parametric model to the nonparametric empirical-pmf estimator.

2.15 (Adaptive test) A test with two questions is designed so that the difficulty of the second question depends on whether the first question is answered correctly. If the first answer is correct, then the second question is twice as difficult as the first one, otherwise the difficulty is the same.

a Derive a parametric pmf for the number of correct answers. The parameter θ represents the probability of answering the first question correctly. Assume that the conditional probability of the second answer being correct is $\theta/2$, if the first answer is right, and θ, if the first answer is wrong.

b The number of correct answers for a group of three students is 2, 1, 1. What is the maximum-likelihood estimate of θ, if we interpret the data as i.i.d. samples from the parametric distribution? Assume that the log-likelihood is concave for $\theta \in [0, 1]$.

2.16 (Air traffic) Download the data in https://github.com/cfgranda/ps4ds/tree/main/data/airtraffic, which record the numbers of flights arriving at London Heathrow airport in 10-minute intervals between 18:00 and 19:30 (extracted from Dataset 4). The training dataset contains flights from June-December 2010, and the test set from January to June 2011. We model the number of arriving flights in a 10-minute interval as a discrete random variable.

a Estimate the pmf of the random variable by fitting a parametric model and a nonparametric model to the training set. Explain any assumptions you make to choose the parametric model. Plot the nonparametric and parametric pmfs.

b Evaluate the performance of your two models, based on the root mean square deviation between the estimated pmfs and the empirical pmf of the test set. Which model performs better?

3

Continuous Variables

Overview

Physical quantities such as temperature, time, or distance are typically considered to be continuous, meaning that they are not constrained to take discretized values. In this chapter, we explain how to model uncertain continuous quantities probabilistically, using random variables. We define continuous random variables in Section 3.1. Section 3.2 describes the cumulative distribution function and quantiles and explains how to estimate them from data. Section 3.3 introduces the probability density, a fundamental tool to describe and manipulate continuous random variables. Section 3.4 discusses functions of continuous random variables and explains how to derive their probability density. In Section 3.5, we present two nonparametric approaches to estimate probability densities: the histogram and kernel density estimation. Section 3.6 describes two popular continuous parametric distributions: the exponential and the Gaussian. Section 3.7 explains how to fit continuous parametric models to data using maximum-likelihood estimation. Finally, in Section 3.8, we discuss how to simulate continuous random variables via inverse-transform sampling.

3.1 Continuous Random Variables

In Chapter 2, we explain how to model discrete uncertain quantities using random variables. Mathematically, discrete random variables are functions in a probability space. We define continuous random variables analogously, as functions mapping outcomes in the sample space of a probability space to the real line. The only difference is that the range of the functions is continuous (and therefore uncountably infinite), as opposed to discrete.

In practice, we manipulate discrete random variables using their probability mass function, which encodes the probability that the random variable equals any specific value. Similarly, when using continuous random variables for probabilistic modeling, we do not define the underlying probability space explicitly. Instead, we describe the random variable in terms of the probability that it belongs to different intervals of the real line. In order for this probability to be well defined, every interval must belong to the collection of events in the probability space associated with the random variable. This is imposed in the mathematical definition of continuous random variables.

Definition 3.1 (Mathematical definition of a continuous random variable). *Let $(\Omega, \mathcal{C}, \mathrm{P})$ be a probability space and $\tilde{a} \colon \Omega \to \mathbb{R}$ a function mapping elements in the sample space Ω to the real line $\mathbb{R}$. The function $\tilde{a}$ is a valid random variable if for any interval $\mathcal{I} := [a, b] \subseteq \mathbb{R}$, $a \leq b$, the event*

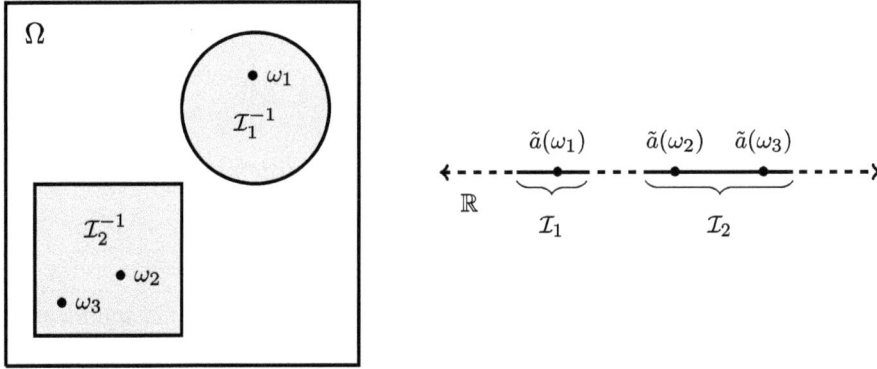

Figure 3.1 Mathematical definition of a continuous random variable. The continuous random variable $\tilde{a}$ maps outcomes in the sample space Ω, represented by the Venn diagram on the left, to the real line $\mathbb{R}$ depicted on the right. The events $\mathcal{I}_1^{-1}$ and $\mathcal{I}_2^{-1}$ contain all outcomes mapping to the two intervals $\mathcal{I}_1$ and $\mathcal{I}_2$, respectively. For $i = 1, 2$, the probability that $\tilde{a}$ belongs to $\mathcal{I}_i$ is equal to $P(\mathcal{I}_i^{-1})$, represented by the area of $\mathcal{I}_i^{-1}$ in the Venn diagram. If $\mathcal{I}_1$ and $\mathcal{I}_2$ are disjoint, then so are $\mathcal{I}_1^{-1}$ and $\mathcal{I}_2^{-1}$, as established in the proof of Theorem 3.2.

$$\mathcal{I}^{-1} := \{\omega : \tilde{a}(\omega) \in \mathcal{I}\}, \tag{3.1}$$

containing the outcomes mapping to $\mathcal{I}$, belongs to the collection $\mathcal{C}$. This means that the probability

$$P(\tilde{a} \in \mathcal{I}) = P(\mathcal{I}^{-1}) \tag{3.2}$$

is well defined. Such functions are called measurable. The random variable is said to be continuous, if the probability that it equals any single point $a \in \mathbb{R}$ is zero,

$$P(\tilde{a} = a) = 0. \tag{3.3}$$

We often denote the event that a random variable $\tilde{a}$ belongs to a set S by $\tilde{a} \in S$, although strictly speaking we mean

$$\tilde{a} \in S := \{\omega : \tilde{a}(\omega) \in S\}. \tag{3.4}$$

Figure 3.1 illustrates the mathematical definition of continuous random variables. It depicts two events of the underlying probability space, each mapping to an interval.

Notice that the probability that a continuous random variable is equal to a specific value is zero. Although this might seem a bit strange at first, it is a natural consequence of modeling a quantity as being continuous. A single point on a line has zero length. A single point inside an object has zero mass. In fact, it is impossible to assign nonzero probability to every point in an interval of the real line because the number of points in any interval is uncountably infinite, so the sum of the probabilities would explode to infinity (see Exercise 3.1).

As a consequence of Definition 3.1, we can decompose the probability that a continuous random variable belongs to any union of disjoint intervals into the sum of the probabilities that it belongs to each of the intervals.

Theorem 3.2. *Let $\tilde{a}$ be a continuous random variable satisfying Definition 3.1. For any n disjoint intervals $\mathcal{I}_1, \mathcal{I}_2, \ldots, \mathcal{I}_n$ of the real line $\mathbb{R}$,*

$$P(\tilde{a} \in \cup_{i=1}^n \mathcal{I}_i) = \sum_{i=1}^n P(\tilde{a} \in \mathcal{I}_i). \tag{3.5}$$

Similarly, if $\mathcal{I}_1, \mathcal{I}_2, \ldots$ is a countably infinite sequence of disjoint intervals of $\mathbb{R}$,

$$P(\tilde{a} \in \cup_{i=1}^\infty \mathcal{I}_i) = \sum_{i=1}^\infty P(\tilde{a} \in \mathcal{I}_i). \tag{3.6}$$

Proof   We prove the finite case, the infinite case follows from the same argument. For all i, let us denote by

$$\mathcal{I}_i^{-1} := \{\omega \colon \tilde{a}(\omega) \in \mathcal{I}_i\} \tag{3.7}$$

the event containing the outcomes mapped to $\mathcal{I}_i$. The key insight is that if $\mathcal{I}_1, \ldots, \mathcal{I}_n$ are disjoint, then so are $\mathcal{I}_1^{-1}, \ldots, \mathcal{I}_n^{-1}$. This is illustrated by Figure 3.1: Since $\mathcal{I}_1$ and $\mathcal{I}_2$ are disjoint, $\mathcal{I}_1^{-1}$ and $\mathcal{I}_2^{-1}$ must also be disjoint because otherwise there would be an outcome $\omega \in \mathcal{I}_1^{-1} \cap \mathcal{I}_2^{-1}$, which would map to a point $\tilde{a}(\omega) \in \mathcal{I}_1 \cap \mathcal{I}_2$, which is impossible if the intervals are disjoint. In addition, the union of all outcomes mapping to $\cup_{i=1}^n \mathcal{I}_i$, denoted by $(\cup_{i=1}^n \mathcal{I}_i)^{-1}$, is equal to $\cup_{i=1}^n \mathcal{I}_i^{-1}$ (to prove this formally, you can show that the two sets contain each other). Consequently, by Axiom 3 in Definition 1.9

$$P(\tilde{a} \in \cup_{i=1}^n \mathcal{I}_i) := P(\{\omega \colon \tilde{a}(\omega) \in \cup_{i=1}^n \mathcal{I}_i\}) \tag{3.8}$$

$$= P(\{\omega \colon \omega \in (\cup_{i=1}^n \mathcal{I}_i)^{-1}\}) \tag{3.9}$$

$$= P(\{\omega \colon \omega \in \cup_{i=1}^n \mathcal{I}_i^{-1}\}) \tag{3.10}$$

$$= \sum_{i=1}^n P(\{\omega \colon \omega \in \mathcal{I}_i^{-1}\}) \tag{3.11}$$

$$= \sum_{i=1}^n P(\tilde{a} \in \mathcal{I}_i). \tag{3.12}$$

$\blacksquare$

Theorem 3.2 is a crucial result because it liberates us from having to consider the probability space associated with a random variable, in order to describe its behavior. Instead, we just need to determine the probability that the random variable belongs to any interval of the real line. This information is encoded in the cumulative distribution function defined in Section 3.2 (see Lemma 3.5), and in the probability density function defined in Section 3.3 (see Theorem 3.14).

Since the probability that a continuous random variable is equal to any single point is zero, it does not matter whether we consider closed or open intervals when describing continuous random variables. For any continuous random variable $\tilde{a}$ and any interval $[a, b] \subseteq \mathbb{R}$, $a \leq b$,

$$P(\tilde{a} \in [a, b]) = P(\tilde{a} = a) + P(\tilde{a} \in (a, b)) + P(\tilde{a} = b) \tag{3.13}$$

$$= P(\tilde{a} \in (a, b)). \tag{3.14}$$

Given our definition of continuous random variables, we can only consider the probability that they belong to *Borel sets*, which are sets that can be described as countable unions of intervals. This is not restrictive in practice, although there do exist subsets of $\mathbb{R}$ that are not

Borel sets. However, it is unlikely that you will ever come across them, unless you happen to do very theoretical mathematical research.

3.2 The Cumulative Distribution Function

3.2.1 Definition

According to Theorem 3.2, in order to describe the behavior of a continuous random variable, we need to keep track of the probability that it belongs to any interval. To this end, we define the *cumulative distribution function* (cdf), which encodes the probability that the random variable is less than or equal to any real value.

Definition 3.3 (Cumulative distribution function). *Let $(\Omega, \mathcal{C}, \mathrm{P})$ be a probability space and $\tilde{a} \colon \Omega \to \mathbb{R}$ a random variable. The cumulative distribution function of $\tilde{a}$ is defined as*

$$F_{\tilde{a}}(a) := \mathrm{P}(\tilde{a} \leq a). \tag{3.15}$$

In words, $F_{\tilde{a}}(a)$ is the probability that $\tilde{a}$ is less than or equal to a.

As established in the following lemma, the cdf always has the same overall structure. It starts at zero because the probability that the random variable is smaller than an arbitrarily small number is zero. It ends at one because the probability that the random variable is smaller than an arbitrarily large number is one. It is also nondecreasing.

Lemma 3.4 (Properties of the cdf). *For any random variable $\tilde{a}$ with cdf $F_{\tilde{a}}$*

$$\lim_{a \to \infty} F_{\tilde{a}}(a) = 1, \tag{3.16}$$

$$\lim_{a \to -\infty} F_{\tilde{a}}(a) = 0. \tag{3.17}$$

In addition, $F_{\tilde{a}}$ is nondecreasing,

$$F_{\tilde{a}}(b) \geq F_{\tilde{a}}(a), \quad if\ b > a. \tag{3.18}$$

Proof To prove (3.16), we assume that a is positive (as we are going to take the limit when $a \to \infty$), we set $n := \lfloor a \rfloor$ (the largest integer smaller than a) and define the sequence of intervals

$$\mathcal{I}_0 := (\infty, 0), \tag{3.19}$$

$$\mathcal{I}_i := [i - 1, i), \quad 1 \leq i \leq n. \tag{3.20}$$

The intervals are disjoint and their union is included in the interval $(-\infty, a]$. As a result, by Definition 3.3, Lemma 1.12, and Theorem 3.2

$$\lim_{a \to \infty} F_{\tilde{a}}(a) = \lim_{a \to \infty} \mathrm{P}(\tilde{a} \leq a) \tag{3.21}$$

$$\geq \lim_{n \to \infty} \mathrm{P}(\tilde{a} \in \cup_{i=0}^{n} \mathcal{I}_i) \tag{3.22}$$

$$= \lim_{n \to \infty} \sum_{i=0}^{n} \mathrm{P}(\tilde{a} \in \mathcal{I}_i). \tag{3.23}$$

Since the intervals are disjoint, the events that map to them are also disjoint (see the proof of Theorem 3.2). As $n \to \infty$, the union of the intervals covers the whole real line, so the union of the events mapping to them must equal the whole sample space because every outcome is mapped to some real number by $\tilde{a}$. Consequently, by Axioms 3 and 2 in Definition 1.9,

$$\lim_{a \to \infty} F_{\tilde{a}}(a) \geq \lim_{n \to \infty} \sum_{i=0}^{n} P(\tilde{a} \in \mathcal{I}_i) \tag{3.24}$$

$$= \lim_{n \to \infty} \sum_{i=0}^{n} P(\{\omega : \omega \in \mathcal{I}_i^{-1}\}) \tag{3.25}$$

$$= P\left(\lim_{n \to \infty} \cup_{i=0}^{n} \{\omega : \omega \in \mathcal{I}_i^{-1}\}\right) \tag{3.26}$$

$$= P(\Omega) = 1. \tag{3.27}$$

This establishes (3.16) because $F_{\tilde{a}}(a) \leq 1$ for any a (due to Definition 1.9, since the cdf is a probability), so $\lim_{a \to \infty} F_{\tilde{a}}(a) \leq 1$. The proof of (3.17) follows from a similar argument.

The inequality (3.18) is a consequence of Theorem 3.2. The intervals $(-\infty, a]$ and $(a, b]$ are disjoint, so

$$F_{\tilde{a}}(b) = P(\tilde{a} \leq b) \tag{3.28}$$

$$= P(\tilde{a} \leq a) + P(a < \tilde{a} \leq b) \tag{3.29}$$

$$\geq P(\tilde{a} \leq a) = F_{\tilde{a}}(a) \tag{3.30}$$

because $P(a < \tilde{a} \leq b)$ is nonnegative by Axiom 1 in Definition 1.9. ∎

We can compute the probability that a random variable belongs to any interval from its cdf, as established in the following lemma. Consequently, the cdf completely determines the behavior of a continuous random variable, just like the pmf completely determines the behavior of a discrete random variable.

Lemma 3.5. *For any a, $b \in \mathbb{R}$, $a \leq b$, and any continuous random variable $\tilde{a}$,*

$$P(a < \tilde{a} \leq b) = F_{\tilde{a}}(b) - F_{\tilde{a}}(a). \tag{3.31}$$

Proof   By Definition 3.3 and Theorem 3.2,

$$F_{\tilde{a}}(b) := P(\tilde{a} \leq b) \tag{3.32}$$

$$= P(\tilde{a} \in (-\infty, a] \cup (a, b]) \tag{3.33}$$

$$= P(\tilde{a} \leq a) + P(a < \tilde{a} \leq b) \tag{3.34}$$

$$= F_{\tilde{a}}(a) + P(a < \tilde{a} \leq b). \tag{3.35}$$

∎

Example 3.6 (Using the cdf to compute probabilities). Consider a continuous random variable $\tilde{a}$ with the following cdf:

$$F_{\tilde{a}}(a) := \begin{cases} 0 & \text{for } a < 0, \\ 0.5\,a & \text{for } 0 \leq a \leq 1, \\ 0.5 & \text{for } 1 \leq a \leq 2, \\ 0.5\left(1 + (a-2)^2\right) & \text{for } 2 \leq a \leq 3, \\ 1 & \text{for } a > 3. \end{cases} \tag{3.36}$$

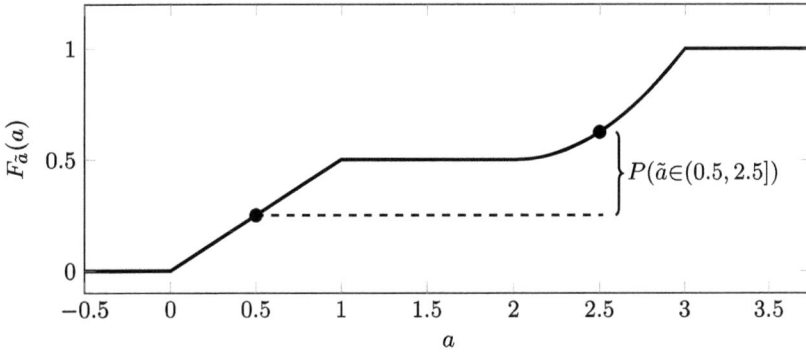

Figure 3.2 Example of cumulative distribution function. The graph shows the cdf of the random variable $\tilde{a}$ in Example 3.6 and illustrates Lemma 3.4. The probability that $\tilde{a}$ belongs to an interval is equal to the difference between the values of the cdf at the end and the beginning of the interval.

Figure 3.2 shows the cdf. You can check that it satisfies the properties in Lemma 3.4. We can apply Lemma 3.5 to compute the probability that $\tilde{a}$ belongs to any interval. For example,

$$\text{P}\left(0.5 < \tilde{a} \leq 2.5\right) = F_{\tilde{a}}\left(2.5\right) - F_{\tilde{a}}\left(0.5\right) = 0.375, \tag{3.37}$$

as illustrated in Figure 3.2.

. . . .

It is often useful to model quantities that are *uniformly* distributed in an interval, meaning that the probability that they are in any subinterval is proportional to the length of the subinterval. The following example shows that random variables with linear cdfs have this property. We define such uniform random variables more formally in Definition 3.17 using their probability density.

Example 3.7 (Linear cdf). Consider the continuous random variable $\tilde{u}$ with cdf

$$F_{\tilde{u}}\left(u\right) := \begin{cases} 0 & \text{for } u < 0, \\ u & \text{for } 0 \leq u \leq 1, \\ 1 & \text{for } u > 1. \end{cases} \tag{3.38}$$

The cdf is shown in Figure 3.3. By Lemma 3.5, the probability that $\tilde{u}$ belongs to any interval $[a, b] \subseteq [0, 1]$ is equal to the length of the interval,

$$\text{P}\left(a < \tilde{u} \leq b\right) = F_{\tilde{u}}\left(b\right) - F_{\tilde{u}}\left(a\right) \tag{3.39}$$
$$= b - a. \tag{3.40}$$

. . . .

We mainly use the cdf to manipulate continuous random variables, but discrete random variables also have cdfs, as we show in the following example.

Example 3.8 (Cumulative distribution function of a discrete random variable). We consider the discrete random variable $\tilde{g}$ from Example 2.7. By definition, the cdf of the random variable equals the probability that it is less than or equal to any real number. To compute these probabilities, we apply Theorem 2.5, which yields

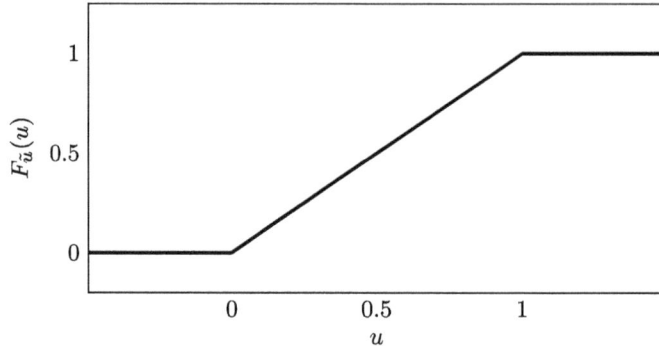

Figure 3.3 Uniform probability in the unit interval. Cumulative distribution function of a random variable $\tilde{u}$ that is uniformly distributed in $[0, 1]$. Due to the linear shape of the cdf, the probability that $\tilde{u}$ belongs to any subinterval of $[0, 1]$ is equal to the length of the subinterval, as established in Example 3.7.

$$F_{\tilde{g}}(g) := \mathrm{P}(\tilde{g} \leq g) = \begin{cases} 0 & \text{for } g < -2, \\ 0.1 & \text{for } -2 \leq g < -1, \\ 0.3 & \text{for } -1 \leq g < 0, \\ 0.6 & \text{for } 0 \leq g < 1, \\ 0.85 & \text{for } 1 \leq g < 2, \\ 0.95 & \text{for } 2 \leq g < 3, \\ 1 & \text{for } g \geq 3. \end{cases} \tag{3.41}$$

The cdf is piecewise constant because there are only six values where the probability is nonzero. The cdf has a jump discontinuity at each of these values. The amplitude of the discontinuity is equal to the probability that $\tilde{g}$ equals that value, as illustrated in Figure 3.4.

. . . .

As illustrated by Example 3.8, the cdf of any discrete random variable is discontinuous because the probability that the random variable equals individual values is nonzero. By contrast, for continuous random variables, the cdf is continuous because the probability of each single point is zero (see Exercise 3.2 for a formal proof).

3.2.2 Quantiles

The distribution of a continuous random variable can be described in terms of its *quantiles*. Quantiles divide the real line into regions, such that the random variable is in each region with equal probability. The cdf makes it very simple to define quantiles.

Definition 3.9 (Quantiles). *Let m be an integer and $\tilde{a}$ a random variable. The m-quantiles of $\tilde{a}$ are $m - 1$ points $q_1, q_2, \ldots, q_{m-1}$ such that*

$$\mathrm{P}(\tilde{a} \leq q_1) = \mathrm{P}(q_i < \tilde{a} \leq q_{i+1}), \qquad \text{for } 1 \leq i \leq m - 1, \tag{3.42}$$

$$= \mathrm{P}(\tilde{a} > q_{m-1}) \tag{3.43}$$

$$= \frac{1}{m}, \tag{3.44}$$

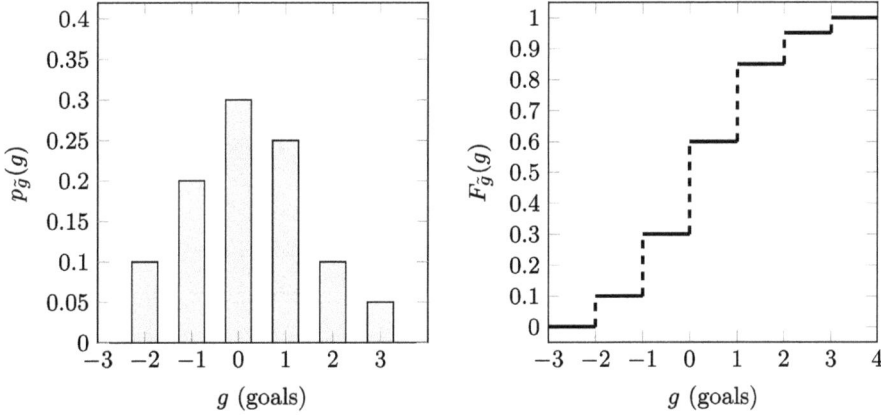

Figure 3.4 Cumulative distribution function of a discrete random variable. The graphs show the pmf (left) and the cdf (right) of the random variable $\tilde{g}$ from Example 2.7. The random variable is discrete: There are only six values where the probability is nonzero. Consequently, the cdf is piecewise constant and has jump discontinuities at these values. The amplitude of each discontinuity is equal to the probability that $\tilde{g}$ equals the corresponding value.

or equivalently,

$$F_{\tilde{a}}(q_i) = \mathrm{P}(\tilde{a} \leq q_i) \tag{3.45}$$

$$= \frac{i}{m}, \qquad \text{for } 1 \leq i \leq m - 1. \tag{3.46}$$

In the case that several points satisfy the equality, we usually choose their midpoint. In words, the quantiles partition the real line into m intervals, so that $\tilde{a}$ has the same probability (equal to $1/m$) of being in each interval. When $m := 4$, the quantiles are called quartiles. When $m := 10$, the quantiles are called deciles. When $m := 100$, they are called percentiles.

The three quartiles q_1, q_2, and q_3 are widely used to provide a succinct description of the distribution of random variables. The difference between the third and first quartile $q_3 - q_1$ is called the *interquartile range*. It quantifies the spread of the distribution. The second quartile q_2 is called the *median*. It separates the range of the random variable into two intervals with equal probability, and can therefore be interpreted as the midpoint of the distribution.

Definition 3.10 (Median). *The median q_2 of a random variable $\tilde{a}$ satisfies*

$$\mathrm{P}(\tilde{a} \leq q_2) = \mathrm{P}(\tilde{a} > q_2) = \frac{1}{2}, \tag{3.47}$$

or equivalently,

$$F_{\tilde{a}}(q_2) = \frac{1}{2}. \tag{3.48}$$

Figure 3.5 shows the quartiles of the random variables in Examples 3.6 and 3.7, illustrating their connection to the cdf.

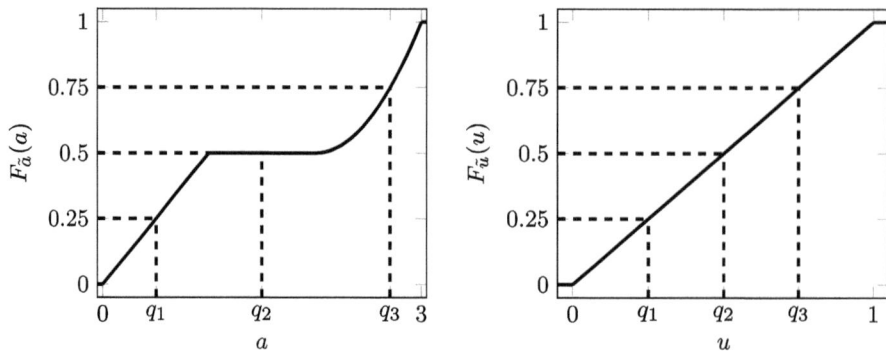

Figure 3.5 Quartiles. The graph shows the quartiles of the random variables in Examples 3.6 (left plot) and 3.7 (right plot). They can be found by *inverting* the cdf. To find the median, for example, we look for the point where the cdf equals 1/2. For the cdf on the left, any point between 1 and 2 could be considered the median; we arbitrarily choose 1.5.

3.2.3 Estimating the CDF and Quantiles from Data

In order to model an uncertain quantity as a continuous random variable, we can estimate its cdf from data. Since the cdf represents probabilities, it is natural to approximate it using the empirical-probability estimator, introduced in Definition 1.22. The resulting statistical estimator is known as the empirical cdf.

Definition 3.11 (Empirical cdf). *Let $X := \{x_1, x_2, \ldots, x_n\}$ denote a real-valued dataset. The empirical cumulative distribution function $F_X : \mathbb{R} \to [0, 1]$ maps each value $a \in \mathbb{R}$ to the fraction of data that are smaller than or equal to a,*

$$F_X(a) := \frac{1}{n} \sum_{i=1}^{n} 1(x_i \leq a), \qquad (3.49)$$

where $1(x_i \leq a)$ is an indicator function that is equal to one, if $x_i \leq a$, and to zero otherwise.

Figure 3.6 shows empirical cdfs computed from two real-world datasets. One contains heights of men in the US Army, extracted from Dataset 5, and the other national gross domestic products (GDPs) of countries, extracted from Dataset 6.

The following definition explains how to estimate the quantiles of a dataset.

Definition 3.12 (Quantile estimation). *Let $X := \{x_1, x_2, \ldots, x_n\}$ denote a real-valued dataset, interpreted as realizations from a random variable $\tilde{a}$. The m-quantiles of the data are $m - 1$ points $\hat{q}_1, \hat{q}_2, \ldots, \hat{q}_{m-1}$ such that*

$$F_X(\hat{q}_i) = \frac{i}{m}, \qquad i = 1, 2, \ldots, m - 1, \qquad (3.50)$$

where F_X is the empirical cdf (see Definition 3.11). In words, the fraction of data points smaller than or equal to the ith m-quantile $\hat{q}_i$ is i/m, i.e. one quarter of the data is smaller than or equal to the first quartile, one half is smaller than or equal to the median, and three thirds are smaller than or equal to the third quartile.

Note that there may be several points that satisfy (3.50). It is also possible that no point satisfies the equality because the empirical cdf is discontinuous, as illustrated in Figure 3.12. In that case, we can choose $\hat{q}_i$ so that $F_X(\hat{q}_i)$ is as close as possible to i/m.

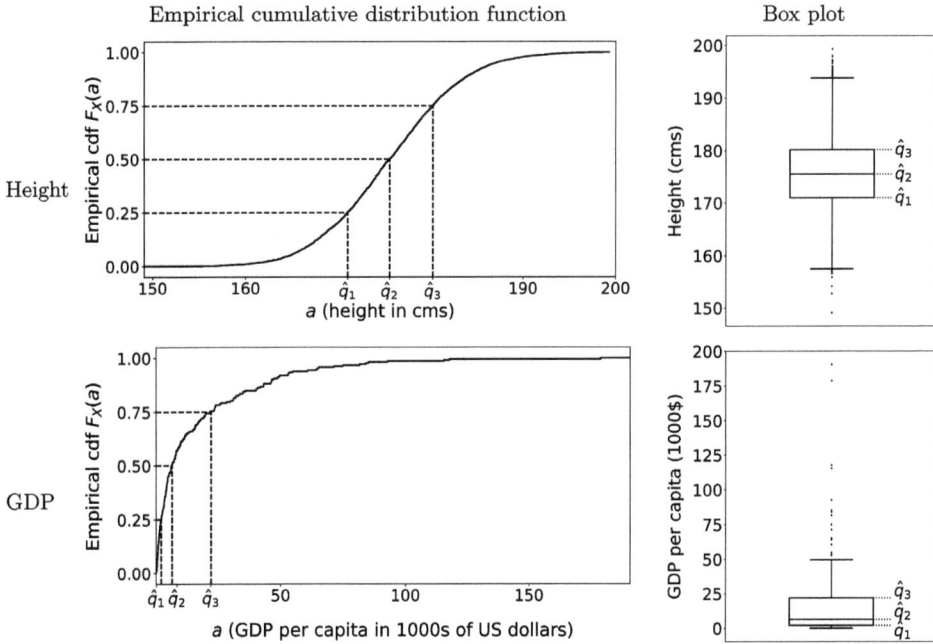

Figure 3.6 Empirical cdf and box plot. The top row shows the empirical cdf (left) of the heights of 4,082 men in the United States Army, extracted from Dataset 5, and the corresponding box plot (right). Both plots are annotated to indicate the three quartile estimates $\hat{q}_1$ (171.0 cm), $\hat{q}_2$ (175.5 cm), and $\hat{q}_3$ (180.2 cm). The bottom row shows the empirical cdf (left) of the gross domestic product per capita of 212 countries in 2019, extracted from Dataset 6, and the corresponding box plot (right). Both plots are annotated to indicate the three quartile estimates $\hat{q}_1$ ($2,181), $\hat{q}_2$ ($6,520), and $\hat{q}_3$ ($21,988).

The quantiles can be computed efficiently by sorting the data: the position of the ith m-quantile $\hat{q}_i$ is the closest integer to in/m. For example, the median is the middle element of the sorted dataset (or any point between the two middle elements, if the number of data is even).

As a result of the definition, if the number of data n is not too small,

$$\mathrm{P}_X(\tilde{a} \le \hat{q}_1) \approx \mathrm{P}_X(\hat{q}_i < \tilde{a} \le \hat{q}_{i+1}), \qquad \text{for } 1 \le i \le m-1, \tag{3.51}$$

$$\approx \mathrm{P}_X(\tilde{a} > \hat{q}_{m-1}) \tag{3.52}$$

$$\approx \frac{1}{m}, \tag{3.53}$$

where P_X denotes the empirical probability, computed from X following Definition 1.22. In words, the m-quantiles partition the data into m subsets with equal empirical probability.

The quartiles of a dataset provide a very concise description of the overall distribution of the data. They can be visualized using a *box plot*, which shows the median $\hat{q}_2$ of the data enclosed in a box. The bottom and top edges of the box are the first quartile $\hat{q}_1$ and the third quartile $\hat{q}_3$, respectively. This way of visualizing a dataset was proposed by the mathematician John Tukey. Tukey's box plot also includes *whiskers*, which depend on the interquartile range (IQR, defined as $\hat{q}_3 - \hat{q}_1$). The lower whisker is a line extending from the bottom of the box to the smallest value within 1.5 IQR of the first quartile. The upper

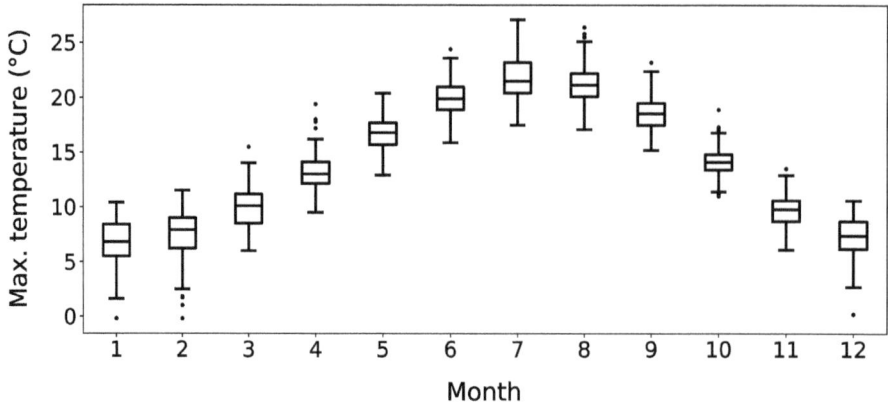

Figure 3.7 Temperature box plots. Box plots of the maximum monthly temperatures recorded at a weather station in Oxford over 150 years, extracted from Dataset 7.

whisker extends from the top of the box to the highest value within 1.5 IQR of the third quartile. Values beyond the whiskers are considered *outliers* and are plotted separately.

The box plots in Figure 3.6 reveal the differences between the distributions of the two datasets. The height data is evenly spread out, so the box plot is almost symmetric around the median. In stark contrast, the GDP dataset is very skewed. The number of countries decreases dramatically as the GDP per capita increases. The difference between the median and the first quartile is around \$2,400, whereas the difference between the median and the third quartile is around \$15,500, more than six times larger! Some outliers such as Monaco and Liechstenstein have GDPs per capita above \$175,000.

Figure 3.7 illustrates how box plots can be used to compare multiple continuous quantities. Each box plot in the figure represents the distribution of the maximum temperature in Oxford during a specific month. The data, extracted from Dataset 7, cover a period of 150 years. By comparing the box plots, we see that the summer months are warmer, the winter months are colder, and there is more intra-month variation in the summer and winter, than in the spring and fall.

3.3 The Probability Density Function

As its name indicates, the cdf provides a *cumulative* description of the distribution of a random variable. The cdf $F_{\tilde{a}}(a)$ of a continuous random variable $\tilde{a}$ at a certain point a does not encode how likely $\tilde{a}$ is to be close to a, but rather how likely $\tilde{a}$ is to be *smaller than or equal* to a. This makes it challenging to determine the *local* behavior of the random variable directly from the cdf. In this section, we show that the rate of change of the cdf, which we call the *probability density*, provides a more informative description of the local behavior of a random variable.

Consider a small interval $(a - \epsilon, a]$ that includes a point of interest a, for some small constant $\epsilon > 0$. By Lemma 3.5, the probability that the random variable $\tilde{a}$ belongs to the interval is equal to the change in the cdf $F_{\tilde{a}}$ over the interval,

$$\mathrm{P}\left(a - \epsilon < \tilde{a} \leq a\right) = F_{\tilde{a}}(a) - F_{\tilde{a}}(a - \epsilon). \tag{3.54}$$

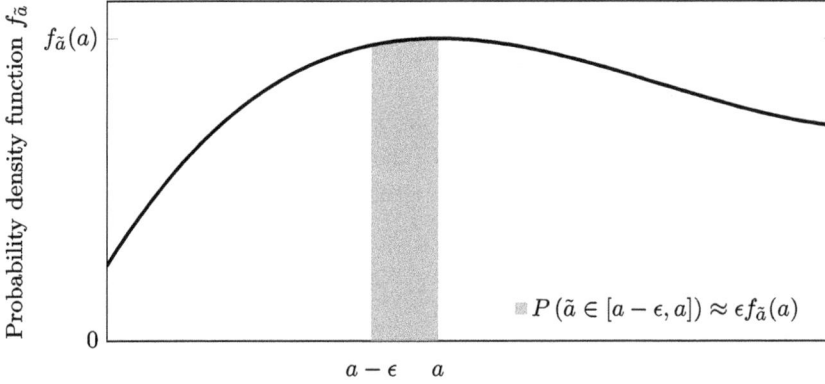

Figure 3.8 Probability density. The probability density $f_{\tilde{a}}(a)$ of a random variable $\tilde{a}$ at a is equal to the probability $\mathrm{P}(a - \epsilon \leq \tilde{a} \leq a)$ that $\tilde{a}$ belongs to the interval $[a - \epsilon, a]$, divided by the length of the interval ϵ, in the limit where $\epsilon \to 0$. Consequently, the probability that $\tilde{a}$ is in the interval is approximately equal to $\epsilon f_{\tilde{a}}(a)$. We represent the probability by the area of the shaded region, which can be approximated by a rectangle of dimensions $\epsilon \times f_{\tilde{a}}(a)$ for small ϵ.

If we normalize this probability by the length of the interval ϵ, and take the limit $\epsilon \to 0$, so that the interval shrinks to a, we obtain the probability density of $\tilde{a}$ at a,

$$f_{\tilde{a}}(a) := \lim_{\epsilon \to 0} \frac{\mathrm{P}(a - \epsilon \leq \tilde{a} \leq a)}{\epsilon}. \tag{3.55}$$

Notice that the probability of the interval converges to zero when $\epsilon \to 0$ because $\tilde{a}$ is continuous, so the probability that it equals a is zero, but the density can remain nonzero.

The density is a measure of probability per unit length. For a sufficiently small value of ϵ,

$$\mathrm{P}(a - \epsilon < \tilde{a} \leq a) \approx f_{\tilde{a}}(a)\epsilon, \tag{3.56}$$

as illustrated in Figure 3.8. Therefore, the higher the density is at a, the more likely it is for the random variable to be close to a.

By the definition of derivative and (3.54), the probability density is equal to the derivative of the cdf:

$$f_{\tilde{a}}(a) = \lim_{\epsilon \to 0} \frac{F_{\tilde{a}}(a) - F_{\tilde{a}}(a - \epsilon)}{\epsilon} \tag{3.57}$$

$$= \frac{\mathrm{d}F_{\tilde{a}}(a)}{\mathrm{d}a}. \tag{3.58}$$

Thus, the density is only defined at points where the cdf is differentiable. We often describe random variables with differentiable cdfs using their probability density function (pdf), which encodes their probability density at every point.

Definition 3.13 (Probability density function). *Let $\tilde{a} \colon \Omega \to \mathbb{R}$ be a random variable with cdf $F_{\tilde{a}}$. If $F_{\tilde{a}}$ is differentiable, then the probability density function of $\tilde{a}$ is the derivative of its cdf,*

$$f_{\tilde{a}}(a) := \lim_{\epsilon \to 0} \frac{\mathrm{P}(a - \epsilon \leq \tilde{a} \leq a)}{\epsilon} \tag{3.59}$$

$$= \frac{\mathrm{d}F_{\tilde{a}}(a)}{\mathrm{d}a}. \tag{3.60}$$

The pdf of a random variable $\tilde{a}$ can be integrated to obtain the probability that $\tilde{a}$ belongs to any union of intervals B (or more formally, to any Borel set). To explain why, let us decompose B into a partition of very small disjoint intervals of length ϵ delimited by a grid of points $a_1, \ldots, a_n$. By Theorem 3.2, the probability that $\tilde{a}$ belongs to B is the sum of the probabilities that it is in each interval of the partition. By (3.56), if ϵ is small, these probabilities can be approximated using the probability density at the points on the grid. This yields

$$P(\tilde{a} \in B) = \sum_{i=1}^{n} P(\tilde{a} \in [a_i - \epsilon, a_i]) \tag{3.61}$$

$$\approx \sum_{i=1}^{n} f(a_i)\epsilon. \tag{3.62}$$

If the cdf of $\tilde{a}$ is differentiable, the approximation becomes exact when we take the limit $\epsilon \to 0$. In that case, the pdf is integrable because it is the derivative of the cdf. Consequently, as $\epsilon \to 0$, the Riemann sum in (3.62) converges to an integral, so that

$$P(\tilde{a} \in B) = \int_{a \in B} f(a) \, \mathrm{d}a. \tag{3.63}$$

This means that the probability that $\tilde{a}$ belongs to B is the area under the corresponding segment of the pdf curve, as illustrated in Figure 3.9. This further justifies why we refer to the probability density as a *density*: We integrate the probability density to compute probabilities, just like we integrate the density of objects to compute their mass. We now provide a more formal proof based on the fundamental theorem of calculus.

Theorem 3.14. *Let $\tilde{a}$ be a continuous random variable with pdf $f_{\tilde{a}}$. For any Borel set or countable union of intervals $B \subseteq \mathbb{R}$,*

$$P(\tilde{a} \in B) = \int_B f_{\tilde{a}}(a) \, \mathrm{d}a. \tag{3.64}$$

Proof   By Lemma 3.5, the fundamental theorem of calculus and the definition of pdf, for any interval $\mathcal{I} := (a, b]$, $a \le b$,

$$P(a < \tilde{a} \le b) = F_{\tilde{a}}(b) - F_{\tilde{a}}(a) \tag{3.65}$$

$$= \int_{\mathcal{I}} f_{\tilde{a}}(a) \, \mathrm{d}a. \tag{3.66}$$

Since B is a Borel set, it can be decomposed into a countable union $B = \cup_i \mathcal{I}_i$ of disjoint intervals $\mathcal{I}_1, \mathcal{I}_2, \ldots$ (we can merge any intervals that are not disjoint). By Theorem 3.2 and (3.66),

$$P(\tilde{a} \in B) = P(\tilde{a} \in \cup_{i=1}^{n} \mathcal{I}_i) \tag{3.67}$$

$$= \sum_{i=1}^{n} P(\tilde{a} \in \mathcal{I}_i) \tag{3.68}$$

$$= \sum_{i=1}^{n} \int_{\mathcal{I}_i} f_{\tilde{a}}(a) \, \mathrm{d}a \tag{3.69}$$

$$= \int_B f_{\tilde{a}}(a) \, \mathrm{d}a. \tag{3.70}$$

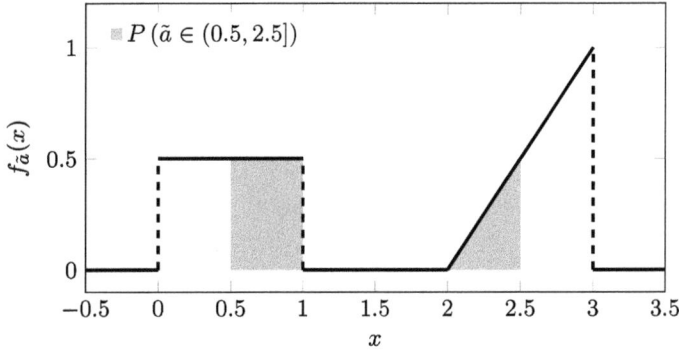

Figure 3.9 Example of probability density function. The plot shows the pdf of the random variable in Examples 3.6 and 3.15. The probability that the random variable belongs to the set $(0.5, 2.5]$ is equal to the shaded area under the pdf curve.

■

Example 3.15 (Using the pdf to compute probabilities). To compute the pdf of the random variable in Example 3.6, we differentiate its cdf:

$$f_{\tilde{a}}(a) = \begin{cases} 0 & \text{for } a < 0, \\ 0.5 & \text{for } 0 \leq a \leq 1, \\ 0 & \text{for } 1 \leq a \leq 2, \\ a - 2 & \text{for } 2 \leq a \leq 3, \\ 0 & \text{for } a > 3. \end{cases} \tag{3.71}$$

Figure 3.9 shows the pdf. To determine the probability that $\tilde{a}$ is between 0.5 and 2.5, we integrate over that interval to obtain the same answer as in Example 3.6,

$$P(0.5 < \tilde{a} \leq 2.5) = \int_{a=0.5}^{2.5} f_{\tilde{a}}(a) \, da \tag{3.72}$$

$$= \int_{a=0.5}^{1} 0.5 \, da + \int_{a=2}^{2.5} (a - 2) \, da \tag{3.73}$$

$$= 0.375. \tag{3.74}$$

The pdf provides a more intuitive description of the local behavior of the random variable than the cdf. From the pdf, we immediately see that $\tilde{a}$ can only be in the intervals $[0, 1]$, or $[2, 3]$. In the remaining regions the density is zero, so the probability that the random variable belongs to any subintervals within those regions is zero. The density is constant in $[0, 1]$, so the probability that $\tilde{a}$ is in any fixed-length subinterval within $[0, 1]$ is the same. By contrast, the density increases over the interval $[2, 3]$, so $\tilde{a}$ is more likely to belong to fixed-length subintervals that are closer to 3 than to 2. It is more difficult to glean this kind of insights by looking at the cdf, shown in Figure 3.2.

Just like the pmf completely characterizes the probabilistic behavior of a discrete random variable, the pdf completely characterizes the behavior of a continuous random variable. Indeed, by Theorem 3.14, it encodes the probability that the random variable belongs to any finite or countable union of intervals of the real line. Because of this, we often define random variables by stating that they are distributed according to a certain pdf, without ever mentioning the underlying probability space.

The following theorem shows that pdfs are always nonnegative and their integral over $\mathbb{R}$ equals one, and also that any function satisfying these properties is a valid pdf.

Theorem 3.16 (Properties of the pdf). *A function $f : \mathbb{R} \to \mathbb{R}$ is the pdf of a continuous random variable, if and only if it is nonnegative and*

$$\int_{\mathbb{R}} f(a) \, \mathrm{d}a = 1. \tag{3.75}$$

Proof If f is the pdf of a continuous random variable $\tilde{a}$, then it must be nonnegative because the cdf is nondecreasing by Lemma 3.4, and the pdf is the derivative of the cdf. In addition, since all points in the sample space Ω of the underlying probability space are mapped to a real number by $\tilde{a}$, Theorem 3.14 and Axiom 2 in Definition 1.9 imply

$$\int_{\mathbb{R}} f(a) \, \mathrm{d}a = \mathrm{P}\left(\tilde{a} \in \mathbb{R}\right) \tag{3.76}$$

$$= \mathrm{P}\left(\omega \in \Omega\right) \tag{3.77}$$

$$= 1. \tag{3.78}$$

To prove that every nonnegative function f, which integrates to one, can be interpreted as the pdf of a random variable, we need to build an underlying probability space. We choose the sample space to be the real line, and set the collection of events to be the collection of all Borel sets (which can be easily shown to satisfy Definition 1.7). Then, we define the probability measure based on f. For any Borel set B, we set $\mathrm{P}\left(B\right) := \int_B f(a) \, \mathrm{d}a$. You can check that the resulting probability measure satisfies Definition 1.9. Finally, we define the continuous random variable associated with the pdf as the identity function. This random variable is valid according to Definition 3.1 because the probability that it belongs to any interval is well defined. ∎

In Example 3.7, we consider a random variable with a linear cdf over the unit interval $[0, 1]$. The pdf of the random variable is constant because it is the derivative of a linear function, as shown in Figure 3.10. This means that any subinterval of equal length within $[0, 1]$ has the same probability of containing the random variable. We say that the variable is

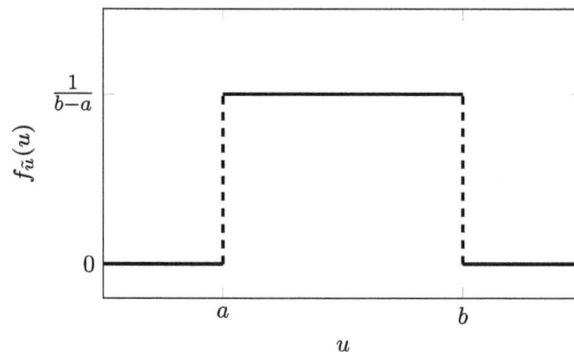

Figure 3.10 Uniform distribution. Probability density function of a uniform random variable $\tilde{u}$ in the interval $[a, b]$.

uniformly distributed over the unit interval. The following definition generalizes this concept to arbitrary intervals.

Definition 3.17 (Uniform distribution). *A random variable $\tilde{u}$ is uniformly distributed on the interval $[a, b]$, if its pdf is constant within the interval,*

$$f_{\tilde{u}}(u) = \frac{1}{b - a}, \qquad if \ a \leq u \leq b, \tag{3.79}$$

and zero otherwise.

The pdf of a continuous random variable can be larger than one. For example, the pdf of a uniform distribution in the interval $[0, 0.5]$ equals 2. This reminds us that the pdf does not encode a probability, like the pmf or the cdf, but rather a probability density.

3.4 Functions of Random Variables

As discussed in Section 2.1.4, in probabilistic modeling, it is often useful to characterize the behavior of functions of random variables. In contrast to the discrete case, not all deterministic functions applied to a random variable yield a valid random variable. Any function of a random variable is obviously a function itself, but to satisfy Definition 3.1, we need to guarantee that the set of outcomes mapping to any interval is assigned a probability, as stated formally in the following lemma.

Lemma 3.18 (Function of a continuous random variable). *Let $(\Omega, \mathcal{C}, \mathrm{P})$ be a probability space, and $\tilde{a} \colon \Omega \to \mathbb{R}$ a continuous random variable associated with the probability space. Given a deterministic function $g \colon \mathbb{R} \to \mathbb{R}$, $\tilde{b} := g \circ \tilde{a}$, also denoted by $g(\tilde{a})$, is a random variable as long as for any interval (a, b), $a \leq b$, the set $\{x \colon g(x) \in (a, b)\}$ of real values mapped by g to the interval is a Borel set (i.e. a countable union of intervals).*

Proof We need to show that for any interval (a, b), the event

$$A := \{\omega \colon g(\tilde{a}(\omega)) \in (a, b)\}, \tag{3.80}$$

which is mapped to (a, b) by $\tilde{b}$, is in the collection $\mathcal{C}$. In that case, the function $\tilde{b} := g \circ \tilde{a}$ satisfies Definition 3.1. The random variable $\tilde{a}$ maps A to $\{x \colon g(x) \in (a, b)\}$. Under the assumptions of the lemma, the latter set is a Borel set, so A must be in $\mathcal{C}$ by Definition 3.1 because $\tilde{a}$ is a valid random variable. ∎

Although there exist functions of continuous random variables that are not valid random variables, this is rarely the case in practice. Most functions, including all continuous functions, do result in valid random variables. We refer the interested reader to Section 1.3 in Durrett (2019) for more details.

The following theorem characterizes how the distribution of a random variable is transformed, when we scale it by a multiplicative constant. The pdf is stretched proportionally to the scaling factor, as illustrated in Example 3.20.

Theorem 3.19 (Multiplying a random variable by a constant). *Let $\tilde{a}$ be a continuous random variable with pdf $f_{\tilde{a}}$. For any positive constant $\alpha > 0$, the pdf of the scaled random variable $\tilde{b} = \alpha \tilde{a}$ is*

$$f_{\tilde{b}}(b) = \frac{1}{\alpha} f_{\tilde{a}} \left(\frac{b}{\alpha} \right). \tag{3.81}$$

Proof To derive the pdf of $\tilde{b}$, we compute its cdf and then differentiate it. The cdf can be computed by expressing it in terms of the cdf of $\tilde{a}$. By Definition 3.3,

$$F_{\tilde{b}}(b) := P \left(\tilde{b} \leq b \right) \tag{3.82}$$

$$= P \left(\alpha \tilde{a} \leq b \right) \tag{3.83}$$

$$= P \left(\tilde{a} \leq \frac{b}{\alpha} \right) \tag{3.84}$$

$$= F_{\tilde{a}} \left(\frac{b}{\alpha} \right). \tag{3.85}$$

The result then follows from differentiating $F_{\tilde{b}}(b)$ with respect to b, which yields $f_{\tilde{b}}(b)$ by Definition 3.13.   ∎

Example 3.20 (Current and voltage). Mariano, who is an electrical engineer, is modeling the voltage across a resistor in a certain device. He wants to know how the pdf of the voltage, represented by the random variable $\tilde{v}$, depends on the pdf $f_{\tilde{c}}$ of the current, represented by the random variable $\tilde{c}$. Since the voltage is equal to the current multiplied by the resistance r, which is a deterministic quantity, we have $\tilde{v} = r\tilde{c}$. By Theorem 3.19, the pdf of the voltage equals

$$f_{\tilde{v}}(v) = \frac{1}{r} f_{\tilde{c}} \left(\frac{v}{r} \right). \tag{3.86}$$

To test out the formula, Mariano uses a source that generates current uniformly at random between -1 and 1 amperes (the pdf is shown in the left plot of Figure 3.11). The corresponding pdf of the voltage, depicted in the middle plot of Figure 3.11, is uniform between $-r$ and r volts, as predicted by (3.86).

. . . .

In the proof of Theorem 3.19, we derive the pdf of a function of a random variable by first deriving its cdf, and then differentiating it. This is an effective strategy in general, as illustrated by the following example.

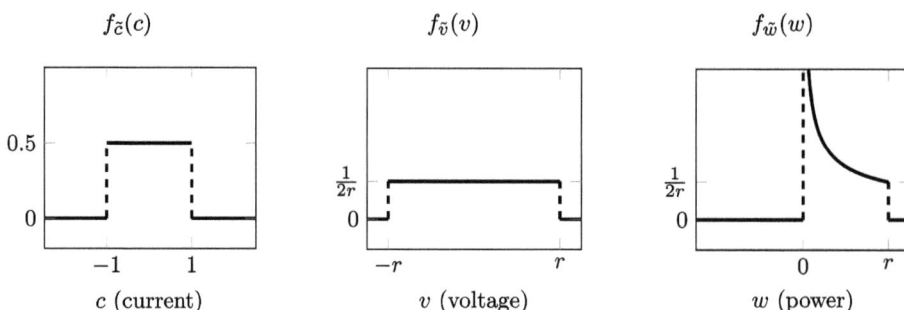

Figure 3.11 Current, voltage, and power. As derived in Example 3.20, if the pdf of the current across a resistor is uniform between -1 and 1 amperes (left), then the pdf of the voltage is uniform between $-r$ and r (center), where r denotes the resistance. The pdf of the power (right) is not uniform because the power is a quadratic function of the current. Instead, it exhibits a square-root decay between 0 and r, as derived in Example 3.21.

Example 3.21 (Current and power). Mariano now wants to derive the pdf of the power dissipated in the resistor, modeled as a random variable $\tilde{w}$. The power is equal to the square of the current multiplied by the resistance r, so $\tilde{w} = r\tilde{c}^2$. For $w < 0$, $F_{\tilde{w}}(w) = P(\tilde{w} \le w) = 0$ because the power is nonnegative. For $w \ge 0$,

$$F_{\tilde{w}}(w) := P(\tilde{w} \le w) \tag{3.87}$$
$$= P\left(r\tilde{c}^2 \le w\right) \tag{3.88}$$
$$= P\left(-\sqrt{\frac{w}{r}} \le \tilde{c} \le \sqrt{\frac{w}{r}}\right) \tag{3.89}$$
$$= F_{\tilde{c}}\left(\sqrt{\frac{w}{r}}\right) - F_{\tilde{c}}\left(-\sqrt{\frac{w}{r}}\right). \tag{3.90}$$

To compute the pdf, we differentiate the cdf, following Definition 3.13. We conclude that the pdf equals

$$f_{\tilde{w}}(w) = \frac{\mathrm{d}}{\mathrm{d}w}\left(F_{\tilde{c}}\left(\sqrt{\frac{w}{r}}\right) - F_{\tilde{c}}\left(-\sqrt{\frac{w}{r}}\right)\right) \tag{3.91}$$
$$= \frac{1}{2\sqrt{rw}}\left(f_{\tilde{c}}\left(\sqrt{\frac{w}{r}}\right) + f_{\tilde{c}}\left(-\sqrt{\frac{w}{r}}\right)\right) \tag{3.92}$$

if $w \ge 0$, and 0 otherwise.

The right plot in Figure 3.11 shows the pdf of the power, when the current is uniformly distributed between -1 and 1 amperes. In contrast to the voltage, the power is not uniformly distributed. By (3.92), it equals

$$f_{\tilde{w}}(w) = \frac{1}{2\sqrt{rw}}, \tag{3.93}$$

between 0 and r, because $f_{\tilde{c}}\left(\sqrt{\frac{w}{r}}\right)$ and $f_{\tilde{c}}\left(-\sqrt{\frac{w}{r}}\right)$ equal $1/2$ for $0 \le w \le r$.
· ·

To end the section, we study an interesting phenomenon. If we feed a continuous random variable $\tilde{a}$ into its cdf $F_{\tilde{a}}$, the resulting random variable $F_{\tilde{a}}(\tilde{a})$ always has the same distribution. Regardless of the distribution of $\tilde{a}$, $F_{\tilde{a}}(\tilde{a})$ is uniformly distributed in the unit interval. This is known as the *probability integral transform*, which turns out to be very useful for the analysis of p-values in hypothesis testing (see Theorem 10.13).

Theorem 3.22 (Probability integral transform). *Let $\tilde{a}$ be a continuous random variable with cdf $F_{\tilde{a}}$. The random variable*

$$\tilde{b} := F_{\tilde{a}}(\tilde{a}) \tag{3.94}$$

is uniformly distributed in the unit interval $[0, 1]$.

Proof By Definition 3.3, the cdf of $\tilde{b}$ equals

$$F_{\tilde{b}}(b) := P\left(\tilde{b} \le b\right) \tag{3.95}$$
$$= P\left(F_{\tilde{a}}(\tilde{a}) \le b\right). \tag{3.96}$$

Let us assume for a moment that $F_{\tilde{a}}$ is invertible. Since cdfs are nondecreasing by Lemma 3.4, $a \leq b$ is equivalent to $F_{\tilde{a}}(a) \leq F_{\tilde{a}}(b)$. Consequently, the event $F_{\tilde{a}}(\tilde{a}) \leq b$ is the same as the event $\tilde{a} \leq F_{\tilde{a}}^{-1}(b)$, which implies

$$F_{\tilde{b}}(b) = \mathrm{P}\left(\tilde{a} \leq F_{\tilde{a}}^{-1}(b)\right) \tag{3.97}$$
$$= F_{\tilde{a}}\left(F_{\tilde{a}}^{-1}(b)\right) \tag{3.98}$$
$$= b. \tag{3.99}$$

For any $0 \leq b \leq 1$. Differentiating with respect to b, following Definition 3.13, establishes that the pdf of $\tilde{b}$ is equal to one in $[0, 1]$, so the random variable is indeed uniformly distributed in $[0, 1]$.

If $F_{\tilde{a}}$ is not invertible, we can define the generalized inverse of the cdf as

$$F_{\tilde{a}}^{-1}(b) := \min_{x}\left\{x\colon F_{\tilde{a}}(x) = b\right\}. \tag{3.100}$$

The generalized inverse is well defined, as long as the random variable is continuous, because its cdf is continuous (otherwise the set $\{x\colon F_{\tilde{a}}(x) = b\}$ could be empty). Since (3.97) and (3.99) hold for the generalized inverse, $\tilde{b}$ is uniformly distributed in $[0, 1]$, even if the cdf of $\tilde{a}$ is not invertible. ∎

3.5 Nonparametric Probability-Density Estimation

In this section, we describe two popular *nonparametric* approaches for estimating pdfs from data, which do not assume that the data follow a predefined parametric model.

3.5.1 The Histogram

In Section 3.2.3, we show how to estimate the cdf from data using the empirical cdf. In order to estimate the corresponding pdf, we might be tempted to differentiate the empirical cdf, but unfortunately this is impossible. The empirical cdf is piecewise constant, as illustrated in Figure 3.12, and therefore not differentiable.

To derive a statistical estimator for the pdf, let us consider its intuitive definition (3.56). The pdf $f_{\tilde{a}}$ of a continuous random variable $\tilde{a}$ is approximately equal to

$$f_{\tilde{a}}(a) \approx \frac{\mathrm{P}(a - \epsilon < \tilde{a} \leq a)}{\epsilon}, \tag{3.101}$$

when ϵ is small enough. Inspired by this, we can apply the following strategy:

1 Divide the range of the random variable into short segments, which we call bins or buckets.
2 Approximate the probability that the random variable is in each bin via the empirical-probability estimator from Definition 1.22.
3 Compute the density in each bin dividing the estimated probability by the bin width, assuming that the density is approximately constant within each bin.

This strategy amounts to computing the histogram of the data and then normalizing it, in order to ensure that it integrates to one and is therefore a valid pdf.

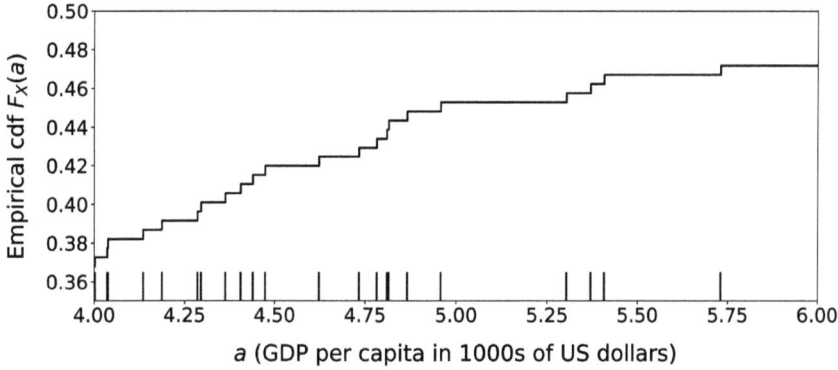

Figure 3.12 The empirical cdf is piecewise constant. The graph is a zoomed-in version depiction of the empirical cdf of gross domestic product per capita in the bottom row of Figure 3.6. The rug plot at the bottom shows the observed data that are within the range of the horizontal axis. The empirical cdf has jump discontinuities at each data point because its value at a point a is equal to the fraction of data smaller than or equal to a by Definition 3.11.

Definition 3.23 (Histogram of continuous data). *Let $X := \{x_1, x_2, \ldots, x_n\}$ be a dataset with values in an interval $[m, m + \ell] \subseteq \mathbb{R}$ of the real line, where m and ℓ are real-valued constants. To build a histogram of the data, we divide the interval into b bins or buckets of length ℓ/b:*

$$\mathcal{B}_i := \left[m + \frac{(i-1)\ell}{b}, m + \frac{i\ell}{b} \right), \quad 1 \leq i \leq b. \tag{3.102}$$

We then count how many elements of X are in each bin. The count for the ith bin equals

$$c_i := \sum_{j=1}^{n} 1 \left(x_j \in \mathcal{B}_i \right), \quad 1 \leq i \leq b, \tag{3.103}$$

where $1 \left(x_j \in \mathcal{B}_i \right)$ is an indicator function that is equal to one, if x_j is in $\mathcal{B}_i$, and to zero otherwise.

The histogram can be normalized to provide an estimate of the pdf of the data f_{hist}. For any $t \in \mathcal{B}_i$ and any $1 \leq i \leq b$,

$$f_{\text{hist}}(t) := \frac{b}{n\ell} \sum_{j=1}^{n} 1 \left(x_j \in \mathcal{B}_i \right). \tag{3.104}$$

The following lemma confirms that we have chosen the right normalization, so that the integral of the normalized histogram equals one, which implies that it is a valid pdf.

Lemma 3.24. *The normalized histogram in Definition 3.23 is a valid pdf.*

Proof From the definition, the estimated pdf f_{hist} is nonnegative, so by Theorem 3.16, we only need to check that it integrates to one. The width of each bin $\int_{t \in \mathcal{B}_i} \mathrm{d}t$ is ℓ/b, which implies

$$\int_{t\in\mathbb{R}} f_{\text{hist}}(t)\ \mathrm{d}t = \sum_{i=1}^{b}\int_{t\in\mathcal{B}_i}\frac{b}{n\ell}\sum_{j=1}^{n}\mathbb{1}\left(x_j\in\mathcal{B}_i\right)\mathrm{d}t \tag{3.105}$$

$$= \frac{b}{n\ell}\sum_{i=1}^{b}\sum_{j=1}^{n}\mathbb{1}\left(x_j\in\mathcal{B}_i\right)\int_{t\in\mathcal{B}_i}\mathrm{d}t \tag{3.106}$$

$$= \frac{1}{n}\sum_{j=1}^{n}\sum_{i=1}^{b}\mathbb{1}\left(x_j\in\mathcal{B}_i\right) \tag{3.107}$$

$$= \frac{1}{n}\sum_{i=1}^{n}1 = 1, \tag{3.108}$$

where we have used the fact that $\sum_{i=1}^{b}\mathbb{1}\left(x_j\in\mathcal{B}_i\right) = 1$, because each data point is in exactly one of the bins.   ∎

A key consideration when building a histogram is how to choose the number and width of the bins. If there are only a few wide bins, then the density estimate will be very coarse, and therefore not very informative. If there are many narrow bins, then each bin may contain very few data points, resulting in a poor estimate of the probability. The tradeoff is illustrated by the top two rows in Figures 3.13 and 3.14, which show histogram-based estimates of the pdf of height in the United States Army and of GDP per capita of countries, obtained from data extracted from Datasets 5 and 6.

3.5.2 Kernel Density Estimation

Given a dataset $X := \{x_1, x_2, \ldots, x_n\}$, we can decompose the normalized histogram in Definition 3.23 as follows:

$$f_{\text{hist}}(t) = \sum_{j=1}^{n}\Pi_j(t). \tag{3.109}$$

$\Pi_j(t)$ is a function representing the contribution of the jth data point. The function is only nonzero within the bin $\mathcal{B}_{\text{ind}(x_j)}$, where $\text{ind}(x_j)$ is the index of the bin that contains the data point x_j,

$$\Pi_j(t) = \begin{cases} \frac{b}{n\ell} & \text{for } t \in \mathcal{B}_{\text{ind}(x_j)}, \\ 0 & \text{otherwise.} \end{cases} \tag{3.110}$$

The amplitude of the function within $\mathcal{B}_{\text{ind}(x_j)}$ is chosen so that its integral equals

$$\int_{t\in\mathbb{R}} \Pi_j(t)\ \mathrm{d}t = \int_{t\in\mathcal{B}_{\text{ind}(x_j)}}\frac{b}{n\ell}\ \mathrm{d}t \tag{3.111}$$

$$= \frac{1}{n} \tag{3.112}$$

since the width of the bin is ℓ/b. This ensures that the integral of the whole histogram, which consists of the sum of these n nonnegative rectangular functions (one for each data point), equals to one. You can check that (3.109) is exactly equivalent to Definition 3.23. In words, the normalized histogram is a superposition of small rectangles centered at the bin centers.

Histogram

50 bins

20 bins

10 bins

Normalized histogram

50 bins

20 bins

10 bins

Kernel density estimate

Bandwidth := 0.25

Bandwidth := 1.5

Bandwidth := 5

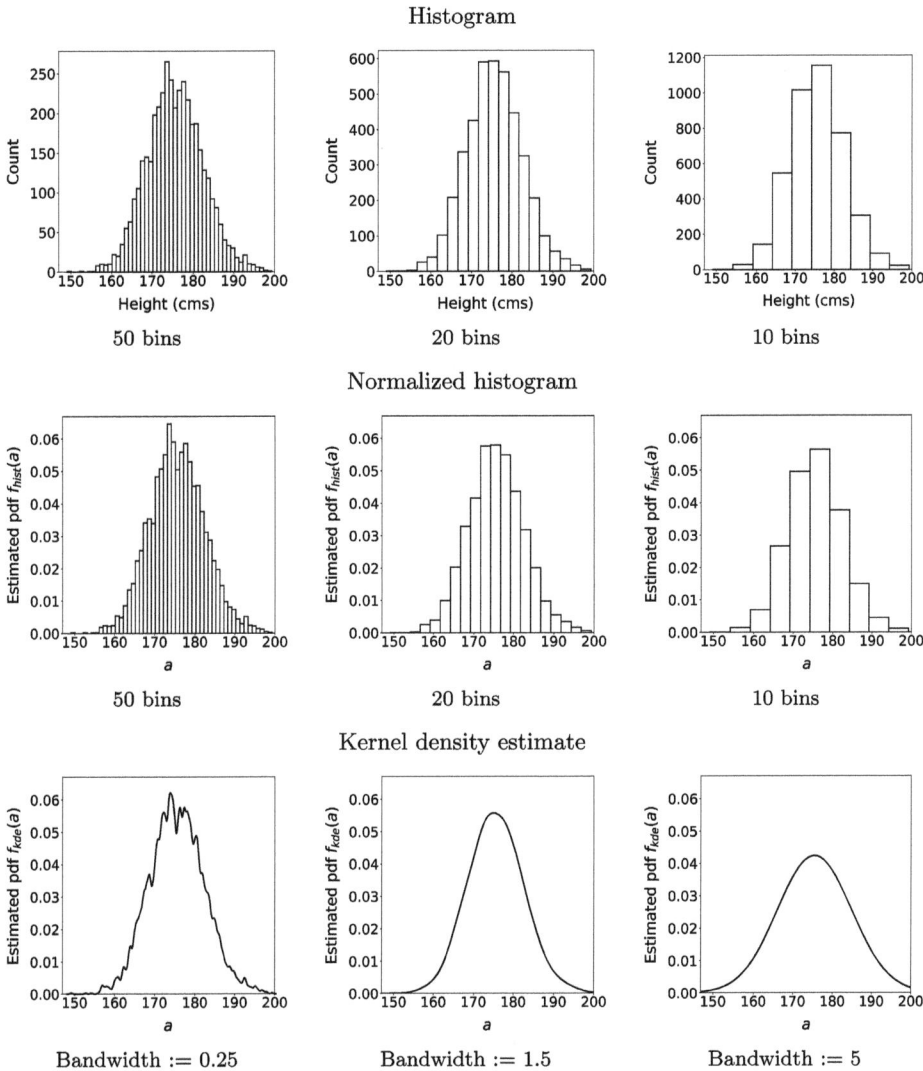

Figure 3.13 Probability density function of height in the US army. The top row shows histograms with different numbers of bins, computed using the heights of 4,082 men in the United States Army, extracted from Dataset 5. The pdf estimates on the second row are obtained by normalizing the histograms, following Definition 3.23, so their integral is equal to one. The third row shows pdf estimates obtained via kernel density estimation (see Definition 3.25) using Gaussian kernels with different bandwidths. Choosing narrow bins in the histogram, or a small bandwidth for kernel density estimation, results in noisy estimates (left column). Increasing the bin width or the bandwidth smooths the pdf estimates (middle column), but may also eliminate informative structure (right column).

This analysis of the histogram reveals a shortcoming. The rectangular function $\Pi_j(t)$ is centered at the center of $\mathcal{B}_{\mathrm{ind}(x_j)}$. This is problematic when x_j is near the edge of the bin because then the data point contributes to the density estimate at points that are far from it (but within the bin) on one side, and it does not contribute to points that are close (but outside the bin) on the other side. A simple solution to remedy this is to center $\Pi_j(t)$ at the

Histogram

40 bins   20 bins   10 bins

Normalized histogram

40 bins   20 bins   10 bins

Kernel density estimate

Bandwidth := 0.15   Bandwidth := 0.4   Bandwidth := 1

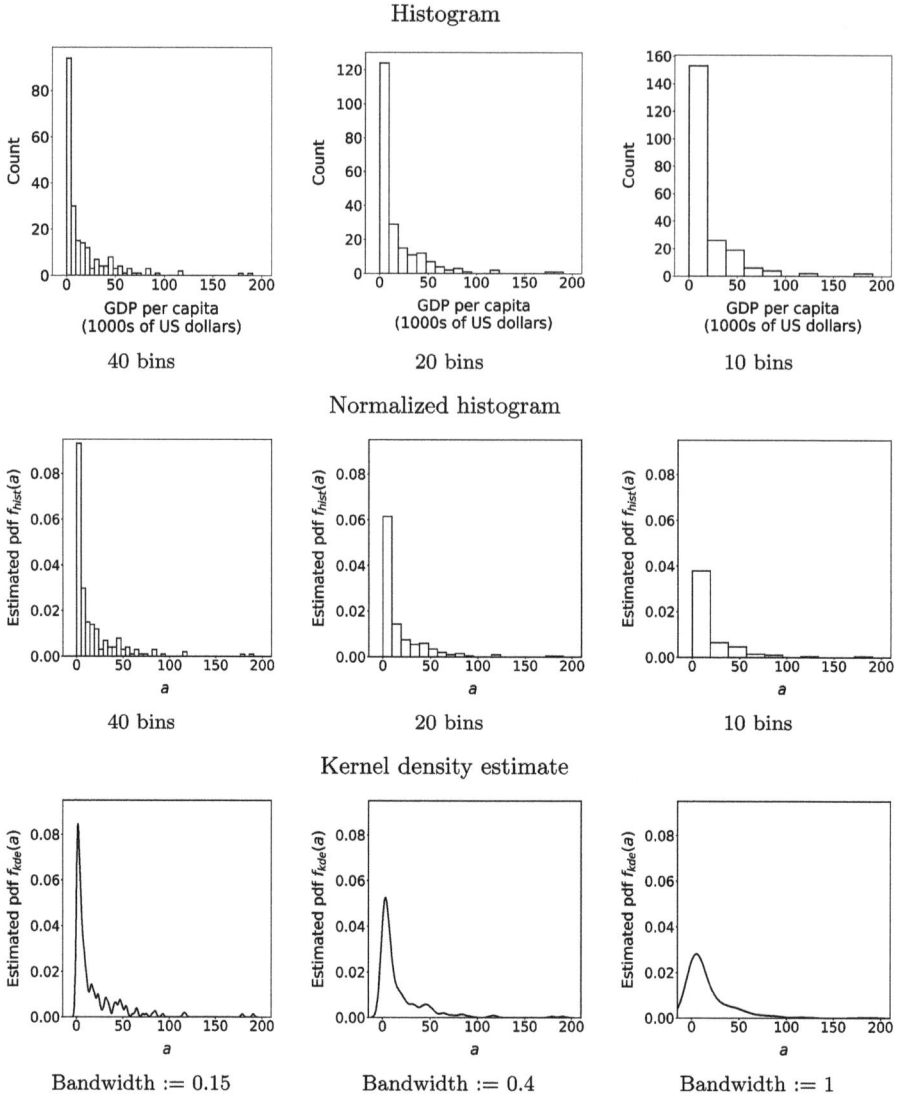

Figure 3.14 Probability density function of GDP per capita. The top row shows histograms with different numbers of bins, computed using the gross domestic product per capita of 212 countries in 2019, extracted from Dataset 6. The pdf estimates on the second row are obtained by normalizing the histograms, following Definition 3.23, so that their integral is equal to one. The third row shows pdf estimates obtained via kernel density estimation (see Definition 3.25) using Gaussian kernels with different bandwidths. Choosing narrow bins in the histogram, or a small bandwidth for kernel density estimation, results in noisy estimates (left column). Increasing the bin width or the bandwidth smooths the pdf estimates (middle column), but may also eliminate informative structure (right column).

data point x_j. This approach is known as *kernel density estimation*. Here *kernel* refers to the function Π_j representing the contribution of each data point. There is no need for the kernel to be rectangular. In fact, it makes sense for it to decay away from its center, so that the influence of each data point on the density estimate is higher, the closer we are to the point.

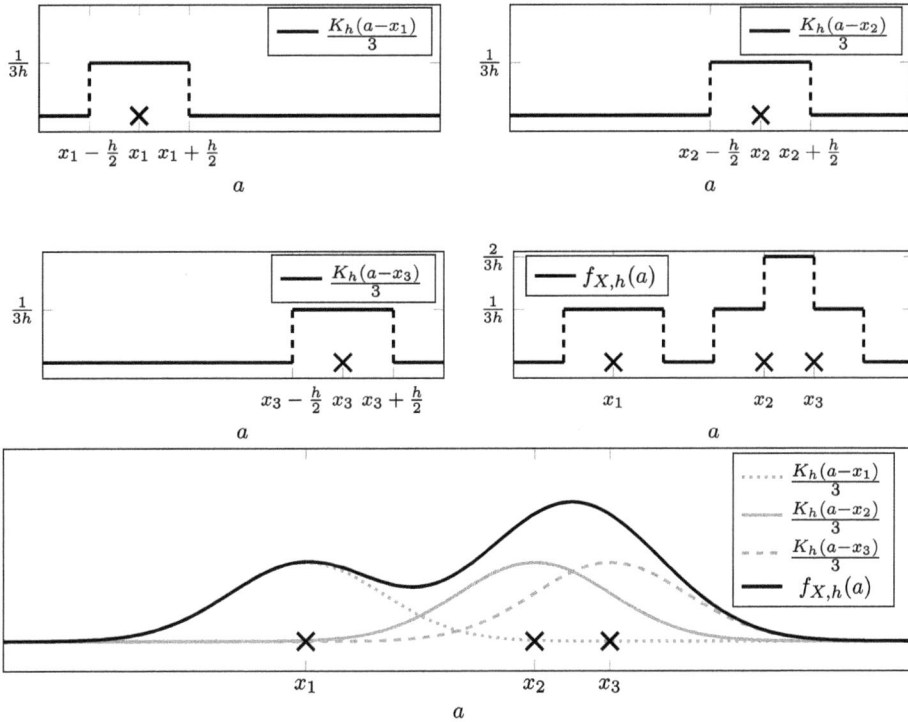

Figure 3.15 Kernel density estimation. The figure shows pdfs estimated via kernel density estimation with different kernels (see Definition 3.25) from a dataset $X := \{x_1, x_2, x_3\}$. The three first plots show individual rectangular kernels, centered at each data point. The fourth plot shows their sum, which is the resulting pdf estimate. The kernel density estimate depicted in the bottom graph is based on a Gaussian kernel instead, which results in a smooth pdf estimate.

Definition 3.25 (Kernel density estimator). *Let* $X := \{x_1, x_2, \ldots, x_n\}$ *denote a real-valued dataset. The corresponding kernel density estimate at a point* $a \in \mathbb{R}$ *is*

$$f_{X,h}(a) := \frac{1}{n\,h} \sum_{i=1}^{n} K\left(\frac{a - x_i}{h}\right), \tag{3.113}$$

where $K \colon \mathbb{R} \to \mathbb{R}$ *is a kernel function, centered at the origin, which satisfies*

$$K(a) \geq 0, \quad \text{for all } a \in \mathbb{R}, \tag{3.114}$$

$$\int_{\mathbb{R}} K(a)\,\mathrm{d}a = 1. \tag{3.115}$$

The bandwidth h *is a positive constant that determines the width or spread of the kernel function.*

A popular choice for the kernel is the Gaussian function

$$K(a) := \frac{1}{\sqrt{2\pi}} \exp\left(-\frac{a^2}{2}\right), \tag{3.116}$$

which is smooth and decays rapidly away from its center. Figure 3.15 shows examples of kernel density estimates with rectangular and Gaussian kernels for a simple dataset.

In summary, kernel density estimation approximates the pdf using a weighted local average of the data. An important consideration is the spread of the kernel, governed by its bandwidth h. If the bandwidth is very small, the density estimate at each point is dominated by the closest data. This captures fine-scale structure, but may also overfit spurious fluctuations, if the number of data is limited. Increasing h amplifies the influence of distant data points, smoothing out such fluctuations. However, a large h may over-smooth the estimate, eliminating meaningful structure. The effect of varying the bandwidth h on the pdf estimate is illustrated in the bottom row of Figures 3.13 and 3.14.

3.6 Continuous Parametric Distributions

When the number of available data are limited, it may be challenging to estimate the cdf or the pdf via the nonparametric approaches in Sections 3.2.3 and 3.5. In such cases, it is often advisable to instead leverage an appropriate parametric model with a small number of parameters. In this section, we describe two of the most popular continuous parametric distributions: the exponential distribution (Section 3.6.1) and the Gaussian distribution (Section 3.6.2).

3.6.1 The Exponential Distribution

Exponential parametric models are often used to represent the time between intermittent phenomena such as earthquakes, telephone calls, radioactive decay of particles, or neuronal action potentials. We describe the assumptions underlying the exponential model and derive its pdf in the following example.

Example 3.26 (Interarrival time between earthquakes). Let us consider the problem of modeling the distribution of the interarrival time between earthquakes in Example 2.21, under the same modeling assumptions.

The first assumption is that, for small enough ϵ, the probability of an earthquake occurring in an interval of length ϵ is equal to $\lambda\epsilon$, where λ is a fixed parameter quantifying the rate at which earthquakes occur. We can interpret the assumption in the following way: Given that no earthquake occurs by time t, the conditional probability that the earthquake occurs between t and $t + \epsilon$ is $\lambda\epsilon$. More formally, if $\tilde{t}$ is a continuous random variable representing the interarrival time,

$$\mathrm{P}\left(t \leq \tilde{t} \leq t + \epsilon \,|\, \tilde{t} > t\right) \approx \lambda\epsilon, \tag{3.117}$$

where the approximation becomes exact as $\epsilon \to 0$. Let us express the conditional probability in terms of the *survival function* $S(t) := 1 - F_{\tilde{t}}(t)$. By Definition 1.16 and Lemma 3.5,

$$\mathrm{P}\left(t < \tilde{t} \leq t + \epsilon \,|\, \tilde{t} > t\right) = \frac{\mathrm{P}\left(t < \tilde{t} \leq t + \epsilon\right)}{\mathrm{P}\left(\tilde{t} > t\right)} \tag{3.118}$$

$$= \frac{F_{\tilde{t}}(t + \epsilon) - F_{\tilde{t}}(t)}{1 - F_{\tilde{t}}(t)} \tag{3.119}$$

$$= \frac{S(t) - S(t + \epsilon)}{S(t)} \approx \lambda\epsilon. \tag{3.120}$$

Reordering the terms in the equation and taking the limit when $\epsilon \to 0$ yields the differential equation,

$$-\lambda = \frac{1}{S(t)} \lim_{\epsilon \to 0} \frac{S(t+\epsilon) - S(t)}{\epsilon} \tag{3.121}$$

$$= \frac{1}{S(t)} \frac{\mathrm{d}}{\mathrm{d}t} S(t) \tag{3.122}$$

$$= \frac{\mathrm{d}}{\mathrm{d}t} \log S(t). \tag{3.123}$$

Integrating, and then applying the exponential function on both sides, we obtain

$$c \exp(-\lambda t) = S(t) \tag{3.124}$$

$$= 1 - F_{\tilde{t}}(t) \tag{3.125}$$

for some constant c. Since the interarrival time cannot be negative, we impose the condition $F_{\tilde{t}}(0) = \mathrm{P}(\tilde{t} \leq 0) = 0$, which implies $c = 1$. We conclude that the cdf of the interarrival time is

$$F_{\tilde{t}}(t) = 1 - \exp(-\lambda t), \tag{3.126}$$

so, by Definition 3.13, its pdf equals

$$f_{\tilde{t}}(t) = \frac{\mathrm{d}F_{\tilde{t}}(t)}{\mathrm{d}t} \tag{3.127}$$

$$= \lambda \exp(-\lambda t). \tag{3.128}$$

. . . .

Random variables with the pdf derived in Example 3.26 are called exponential random variables. Figure 3.16 shows several examples.

Definition 3.27 (Exponential distribution). *An exponential random variable $\tilde{t}$ with parameter $\lambda > 0$ has a pdf of the form*

$$f_{\tilde{t}}(t) = \begin{cases} \lambda \exp\left(-\lambda t\right), & if \ t \geq 0, \\ 0, & otherwise. \end{cases} \tag{3.129}$$

Example 3.28 (Interarrival times at a call center). In this example, we model the interarrival times between telephone calls at a call center, using data extracted from Dataset 3. Our goal

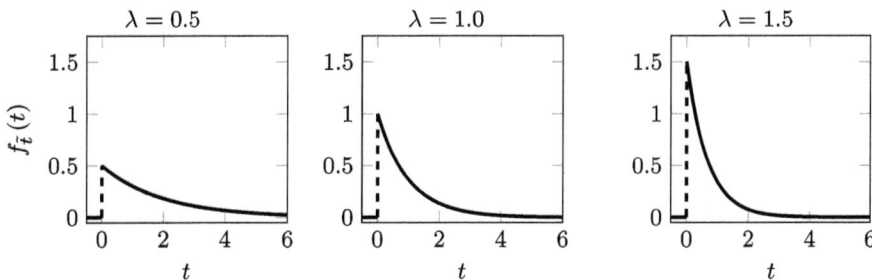

Figure 3.16 Exponential distribution. Probability density functions of exponential random variables with different parameter values.

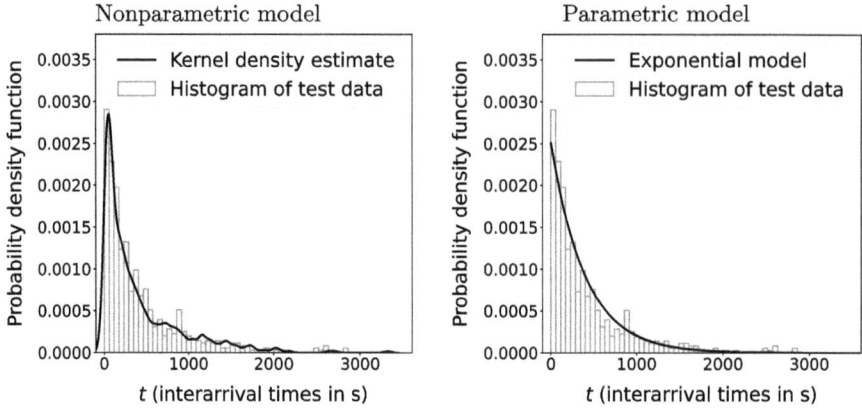

Figure 3.17 Parametric vs. nonparametric modeling of interarrival times at a call center.
The left graph shows a nonparametric pdf estimate, obtained by applying kernel density estimation with a Gaussian kernel (see Definition 3.25). The right column shows a parametric pdf estimate obtained via maximum-likelihood estimation based on an exponential model (applying Theorem 3.34, see Figure 3.22). Both estimates are superposed onto the histogram of the test set for comparison. The parametric model underfits short interarrival times, whereas the nonparametric model overfits noisy fluctuations in the training data.

is to estimate the distribution of interarrival times of calls arriving between 6 am and 7 am on weekdays. We model the interarrival time as a continuous random variable and estimate its pdf using a nonparametric and a parametric model. The nonparametric model is obtained by applying kernel density estimation with a Gaussian kernel (see Definition 3.25). The parametric model is obtained by fitting the exponential distribution in Definition 3.27 via maximum-likelihood estimation (applying Theorem 3.34, see Figure 3.22). Both models are fit to a training set containing data from January to June, and evaluated on a test set containing the remaining months (July to December).

Figure 3.17 shows the results. Both models provide a good approximation to the test data. It turns out that there is a large number of short interarrival times in the training data. The parametric exponential model is not able to fit this pattern, while remaining consistent with the rest of the interarrival times, so it underestimates the density in that region. This is an example of how parametric models may underfit the data, when their underlying assumptions do not hold. By contrast, the nonparametric estimator does not underfit the short interarrival times, but instead overfits spurious fluctuations in the training data, due to its greater flexibility.

An important property of the exponential distribution is that it is *memoryless*. Let $\tilde{t}$ be an exponential random variable. Conditioned on the event $\tilde{t} > t_0$, for any fixed $t_0 > 0$, the distribution of $t - t_0$ is also exponential (starting at t_0 instead of at zero). Imagine that $\tilde{t}$ represents the time you wait until someone answers the phone. If $\tilde{t}$ is memoryless, then no matter how long you have waited, the distribution of the time you have left is always the same. If you have ever tried calling the customer services of an airline to get a flight reimbursement, you know the feeling.

Lemma 3.29 (The exponential distribution is memoryless). *Let $\tilde{t}$ be a random variable distributed according to an exponential pdf with parameter $\lambda > 0$. We define the cdf of $\tilde{t}$ conditioned on the event $\tilde{t} > t_0$ as*

$$F_{\tilde{t}}(t \mid \tilde{t} > t_0) := \mathrm{P}(\tilde{t} \le t \mid \tilde{t} > t_0) \tag{3.130}$$

and the pdf conditioned on the same event as

$$f_{\tilde{t}}(t \mid \tilde{t} > t_0) := \frac{\mathrm{d} F_{\tilde{t}}(t \mid \tilde{t} > t_0)}{\mathrm{d}t}. \tag{3.131}$$

For any $t_0 \ge 0$, this conditional pdf is a copy of $f_{\tilde{t}}$ shifted to t_0,

$$f_{\tilde{t}}(t \mid \tilde{t} > t_0) = \lambda \exp(-\lambda(t - t_0)), \qquad \text{for } t > t_0, \tag{3.132}$$

and zero otherwise.

Proof By Definition 1.16 and Lemma 3.5, the conditional cdf of $\tilde{t}$ given $\tilde{t} \ge t_0$ is

$$F_{\tilde{t} \mid \tilde{t} > t_0}(t) = \mathrm{P}\left(\tilde{t} \le t \mid \tilde{t} > t_0\right) \tag{3.133}$$

$$= \frac{\mathrm{P}\left(t_0 < \tilde{t} \le t\right)}{\mathrm{P}\left(\tilde{t} > t_0\right)} \tag{3.134}$$

$$= \frac{F_{\tilde{t}}(t) - F_{\tilde{t}}(t_0)}{1 - F_{\tilde{t}}(t_0)} \tag{3.135}$$

$$= \frac{\exp\left(-\lambda t_0\right) - \exp\left(-\lambda t\right)}{\exp\left(-\lambda t_0\right)} \tag{3.136}$$

$$= 1 - \exp\left(-\lambda\left(t - t_0\right)\right) \tag{3.137}$$

for $t > t_0$. If $t \le t_0$, the conditional cdf equals zero because the conditional probability that $\tilde{t} \le t$ given $\tilde{t} > t_0$ is zero. Differentiating with respect to t yields the conditional pdf

$$f_{\tilde{t} \mid \tilde{t} > t_0}(t) = \lambda \exp\left(-\lambda\left(t - t_0\right)\right), \qquad \text{for } t > t_0, \tag{3.138}$$

and zero otherwise, which is an exponential pdf with parameter λ, shifted to begin at t_0. ∎

Figure 3.18 provides an intuitive, graphical explanation of the memoryless property. We truncate the exponential pdf at t_0 to reflect the information that values before t_0 are no

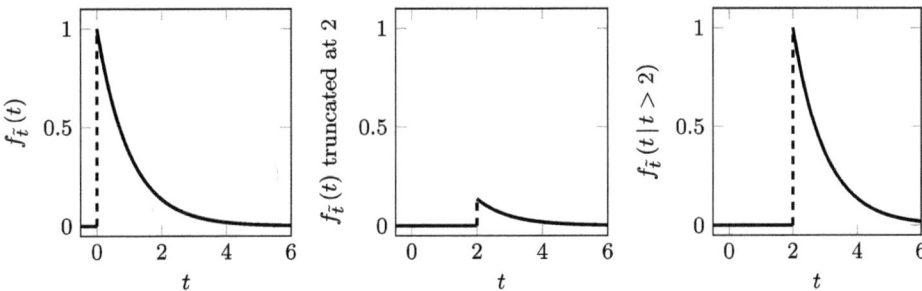

Figure 3.18 The exponential distribution is memoryless. Graphical explanation of why the exponential distribution is memoryless. Truncating the exponential pdf in the left plot at $t_0 := 2$ yields the exponential curve in the middle plot. Normalizing the curve, so that it is a valid density results in the shifted exponential pdf in the right plot, which coincides with the one derived in Lemma 3.29.

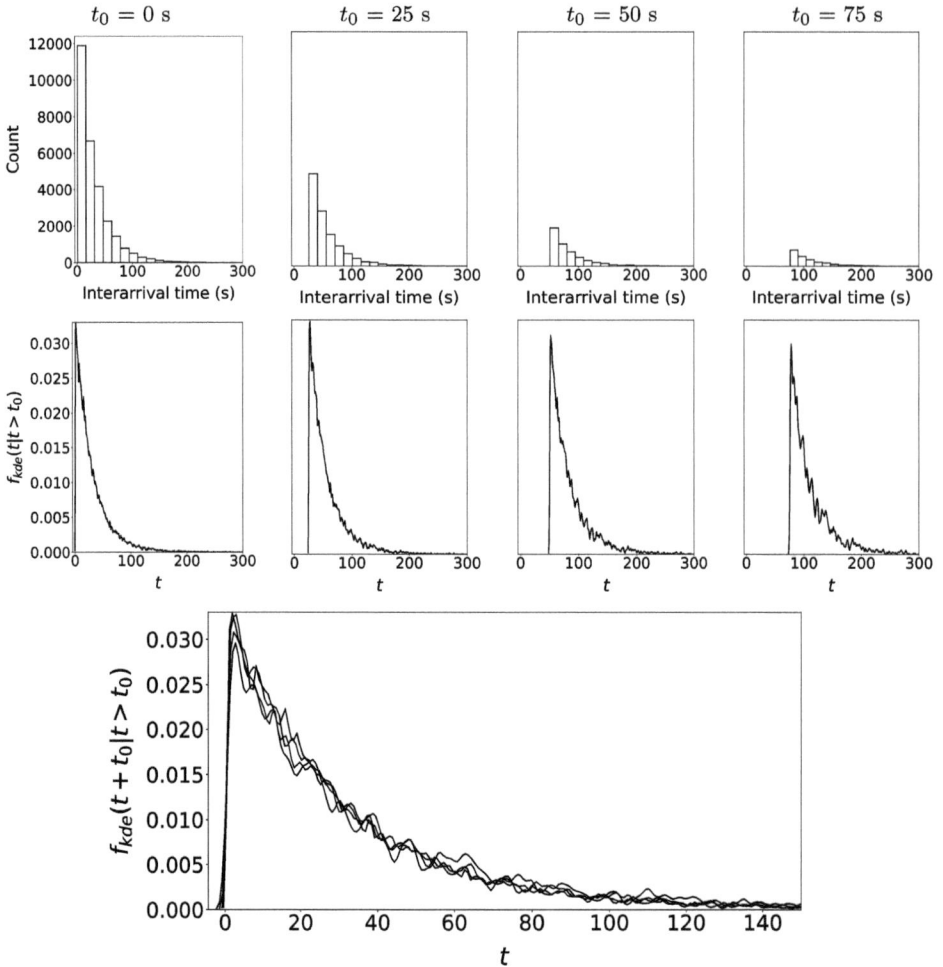

Figure 3.19 Memoryless property in real data. We consider calls arriving at a call center on weekdays from 9 am to 10 am, extracted from Dataset 3. The first row shows the histograms of interarrival times larger than t_0 for different values of t_0. The corresponding estimates of the conditional pdf obtained via kernel density estimation are shown in the second row. The plot at the bottom shows the superposition of the four conditional pdfs shifted to lie on top of each other. They are all very similar, which indicates that the distribution of the interarrival times is approximately memoryless.

longer possible. This results in an exponential curve starting at t_0. Renormalizing the curve, so that it is a valid pdf and its integral is equal to one, yields the conditional pdf derived in Lemma 3.29.

Figure 3.19 shows that the memoryless property arises in real data. We select calls from Dataset 3 that occurred between 9 am and 10 am during weekdays, and interpret the corresponding interarrival times as realizations of a random variable $\tilde{t}$. We then apply kernel density estimation with a Gaussian kernel (see Definition 3.25) to approximate the conditional pdf $f_{\tilde{t} \mid \tilde{t} > t_0}$ for different values of t_0 (this is achieved by only considering interarrival times greater than t_0). The resulting pdfs all resemble shifted copies of each other, indicating that the distribution is indeed approximately memoryless.

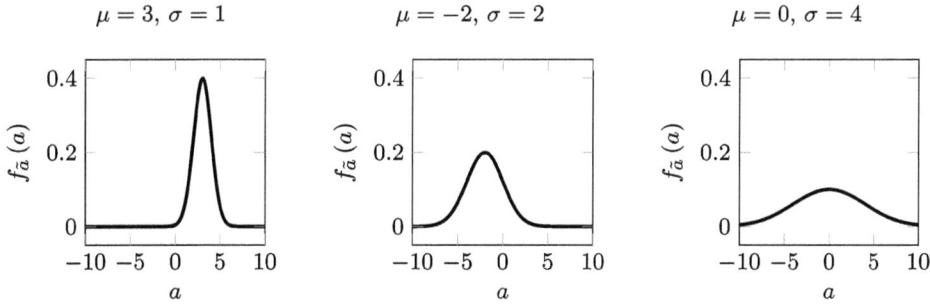

$$\mu = 3, \sigma = 1 \qquad\qquad \mu = -2, \sigma = 2 \qquad\qquad \mu = 0, \sigma = 4$$

Figure 3.20 Gaussian distribution. Probability density functions of Gaussian random variables with different means and standard deviations.

3.6.2 The Gaussian Distribution

The Gaussian or normal model is arguably the most popular parametric model in all of probability and statistics. It is used all over the place in the natural sciences and in engineering. The reason is that sums of independent quantities tend to have a Gaussian distribution. We describe this phenomenon, known as the central limit theorem, in more detail in Section 9.7.

Definition 3.30 (Gaussian distribution). *A Gaussian or normal random variable with mean μ and standard deviation $\sigma \geq 0$ has a pdf of the form*

$$f_{\tilde{a}}(a) = \frac{1}{\sqrt{2\pi}\sigma} \exp\left(-\frac{(a-\mu)^2}{2\sigma^2}\right). \tag{3.139}$$

The squared standard deviation σ^2 is called the variance. A Gaussian distribution with mean μ and variance σ^2 is often denoted by $\mathcal{N}(\mu, \sigma^2)$.

As explained in Sections 7.1 and 7.7, the mean and variance of a random variable represent the average of the random variable and its average squared deviation from the mean, respectively. However, here we interpret them as parameters that determine the shape of the parametric pdf. Figure 3.20 shows the pdfs of Gaussian random variables with different values of μ and σ^2. The bell-shaped pdf is centered at the mean μ. The standard-deviation parameter σ determines how concentrated the density is around μ.

If we shift and scale a Gaussian random variable, it remains Gaussian, with new mean and variance parameters.

Theorem 3.31 (Shifting and scaling a Gaussian random variable). *Let $\tilde{a}$ be a Gaussian random variable with mean μ and variance σ^2. The random variable*

$$\tilde{b} := \alpha\tilde{a} + \beta \tag{3.140}$$

is a Gaussian random variable with mean $\alpha\mu + \beta$ and variance $\alpha^2\sigma^2$.

Proof For simplicity, let us assume that $\alpha > 0$ (the same argument can be applied if $\alpha < 0$ with minor modifications). By Definition 3.3 and Theorem 3.14,

$$F_{\tilde{b}}(b) = \mathrm{P}(\alpha\tilde{a} + \beta \leq b) \tag{3.141}$$

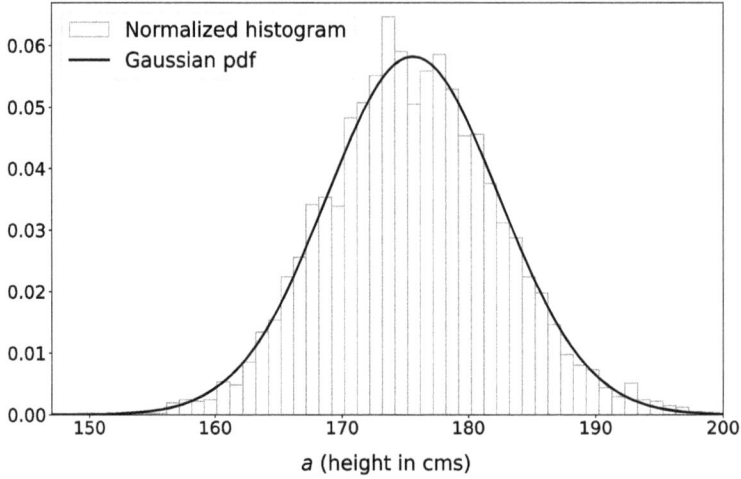

a (height in cms)

Figure 3.21 Gaussian parametric model for height in the US Army. The graph shows a Gaussian pdf estimate obtained via maximum-likelihood estimation (applying Theorem 3.35, see Figure 3.23) to fit the heights of 4,082 men in the United States Army, extracted from Dataset 5. The pdf is shown superimposed onto the histogram of the data used to fit the model.

$$= \mathrm{P}\left(\tilde{a} \le \frac{b-\beta}{\alpha}\right) \tag{3.142}$$

$$= \int_{-\infty}^{\frac{b-\beta}{\alpha}} \frac{1}{\sqrt{2\pi}\sigma} \exp\left(-\frac{(a-\mu)^2}{2\sigma^2}\right) \mathrm{d}a. \tag{3.143}$$

To obtain the pdf, we differentiate with respect to b, as dictated by Definition 3.13:

$$f_{\tilde{b}}(b) = \frac{1}{\sqrt{2\pi}\alpha\sigma} \exp\left(-\frac{\left(\frac{b-\beta}{\alpha}-\mu\right)^2}{2\sigma^2}\right) \tag{3.144}$$

$$= \frac{1}{\sqrt{2\pi}\alpha\sigma} \exp\left(-\frac{(b-\alpha\mu-\beta)^2}{2\alpha^2\sigma^2}\right). \tag{3.145}$$

∎

Gaussian random variables are often used to model continuous quantities that have approximately bell-shaped histograms, such as the height data in Figure 3.13. Figure 3.21 shows the result of fitting a Gaussian parametric model to these data via maximum-likelihood estimation (applying Theorem 3.35, see Figure 3.23). The fit is good, although the Gaussian model does not capture some fine-scale structure in the data. For example, the histogram of the data is not completely symmetric around the center of the fitted distribution.

3.7 Maximum-Likelihood Estimation

In this section, we explain how to fit continuous parametric models to data. Let f_θ be a nonnegative real-valued parametric pdf, which depends on a vector of parameters θ. Given a data point a, the pdf $f_\theta(a)$ evaluated at a is equal to the probability density at the data point, according to the parametric model. The higher the density, the more likely we are to observe the data point because by (3.56) the probability that the data point belongs to $[a - \epsilon, a]$

is approximately $f_\theta(a)\epsilon$ for small ϵ. It is therefore reasonable to choose the parameter θ to make the probability density as high as possible at the observed data. The probability density interpreted as a function of the model parameters is called the *likelihood*. This is analogous to our definition of likelihood for discrete models in Section 2.4, although in that case the likelihood is a probability, instead of a probability density.

In order to extend our reasoning to multiple data, x_1, x_2, ..., x_n, we assume that the observations are mutually independent and identically distributed (i.i.d.) according to the parametric model, as in Section 2.4. For continuous distributions, the i.i.d. assumption implies that the probability density of the data under the parametric model, and hence the likelihood, is $\prod_{i=1}^n f_\theta(x_i)$, as we explain in detail in Section 5.7 (see Definition 5.16). This product is often very small, so we typically consider its logarithm instead, to avoid numerical instabilities.

Definition 3.32 (Likelihood function for continuous models). *Let $f_\theta \colon \mathbb{R} \to \mathbb{R}^+$ be a parametric pdf dependent on a parameter vector θ, and $X := \{x_1, x_2, \ldots, x_n\}$ a real-valued dataset. The likelihood of the model given these data under i.i.d. assumptions is*

$$\mathcal{L}_X(\theta) := \prod_{i=1}^n f_\theta(x_i). \tag{3.146}$$

The log-likelihood function is the logarithm of the likelihood function,

$$\log \mathcal{L}_X(\theta) = \sum_{i=1}^n \log f_\theta(x_i). \tag{3.147}$$

Maximum-likelihood estimation selects the value of the parameters that maximizes the likelihood or the log-likelihood, and therefore the density of the observed data according to the parametric model, under i.i.d. assumptions. Maximizing the likelihood or the log-likelihood is equivalent because the logarithm is a strictly monotone function.

Definition 3.33 (Maximum-likelihood estimator). *Let $f_\theta \colon \mathbb{R} \to \mathbb{R}^+$ be a parametric model dependent on a parameter vector θ, $X := \{x_1, x_2, \ldots, x_n\}$ a real-valued dataset, and S the set of parameter values for which f_θ is a valid pdf. The maximum-likelihood estimator of θ is*

$$\theta_{\mathrm{ML}} := \arg\max_{\theta \in S} \mathcal{L}_X(\theta) \tag{3.148}$$

$$= \arg\max_{\theta \in S} \log \mathcal{L}_X(\theta). \tag{3.149}$$

Theorem 3.34 (Maximum-likelihood estimator for the exponential distribution). *Let $X := \{x_1, x_2, \ldots, x_n\}$ denote a set of real-valued, nonnegative data. The maximum-likelihood estimator of the parameter of the exponential distribution under i.i.d. assumptions equals*

$$\lambda_{\mathrm{ML}} = \frac{1}{\frac{1}{n}\sum_{i=1}^n x_i}. \tag{3.150}$$

Proof By Definition 3.32, the log-likelihood is

$$\log \mathcal{L}_X(\lambda) = \sum_{i=1}^{n} \log f_\lambda(x_i) \tag{3.151}$$

$$= \sum_{i=1}^{n} \log \lambda \exp(-\lambda x_i) \tag{3.152}$$

$$= n \log \lambda - \lambda \sum_{i=1}^{n} x_i. \tag{3.153}$$

To obtain the maximum-likelihood estimator, we maximize the log-likelihood, following Definition 3.33. The derivative and second derivatives of the log-likelihood function are

$$\frac{\mathrm{d} \log \mathcal{L}_X(\lambda)}{\mathrm{d}\lambda} = \frac{n}{\lambda} - \sum_{i=1}^{n} x_i, \tag{3.154}$$

$$\frac{\mathrm{d}^2 \log \mathcal{L}_X(\lambda)}{\mathrm{d}\lambda^2} = -\frac{n}{\lambda^2}. \tag{3.155}$$

The second derivative is negative for all positive values of λ, so the function is concave. Consequently, we can obtain the maximum by setting the first derivative equal to zero. This yields the equation,

$$\frac{n}{\lambda_{\mathrm{ML}}} = \sum_{i=1}^{n} x_i. \tag{3.156}$$

∎

Figure 3.22 shows the log-likelihood of an exponential model applied to the same call-center data (extracted from Dataset 3) used in Example 3.28. The figure also shows the corresponding maximum-likelihood fit and the fits produced by other values of the λ parameter. The fit achieved via maximum likelihood is better than the other two, but cannot account for the large number of short interarrival times because it violates the assumptions of the parametric model.

The following theorem derives the maximum-likelihood estimator corresponding to the Gaussian parametric model. The estimator of the mean parameter is the average of the data, which is also known as the sample mean, and is a popular estimator of the mean of a distribution (see Section 7.1.4). The estimator of the variance is the average of the squared deviation from the mean, which is very close to the sample variance, a popular estimator of the variance of a distribution, defined in Section 7.7.2.

Theorem 3.35 (Maximum-likelihood estimator for the Gaussian distribution)**.** *Let $X :=$ $\{x_1, x_2, \ldots, x_n\}$ denote a real-valued dataset. The maximum-likelihood estimators of the parameters of the Gaussian distribution under i.i.d. assumptions equal*

$$\mu_{\mathrm{ML}} = \frac{1}{n} \sum_{i=1}^{n} x_i, \tag{3.157}$$

$$\sigma_{\mathrm{ML}}^2 = \frac{1}{n} \sum_{i=1}^{n} (x_i - \mu_{\mathrm{ML}})^2. \tag{3.158}$$

Log-likelihood function

Maximum-likelihood estimate
$\lambda_{\mathrm{ML}} = 2.5 \cdot 10^{-3}$

$\lambda := 10^{-3}$

$\lambda := 4 \cdot 10^{-3}$

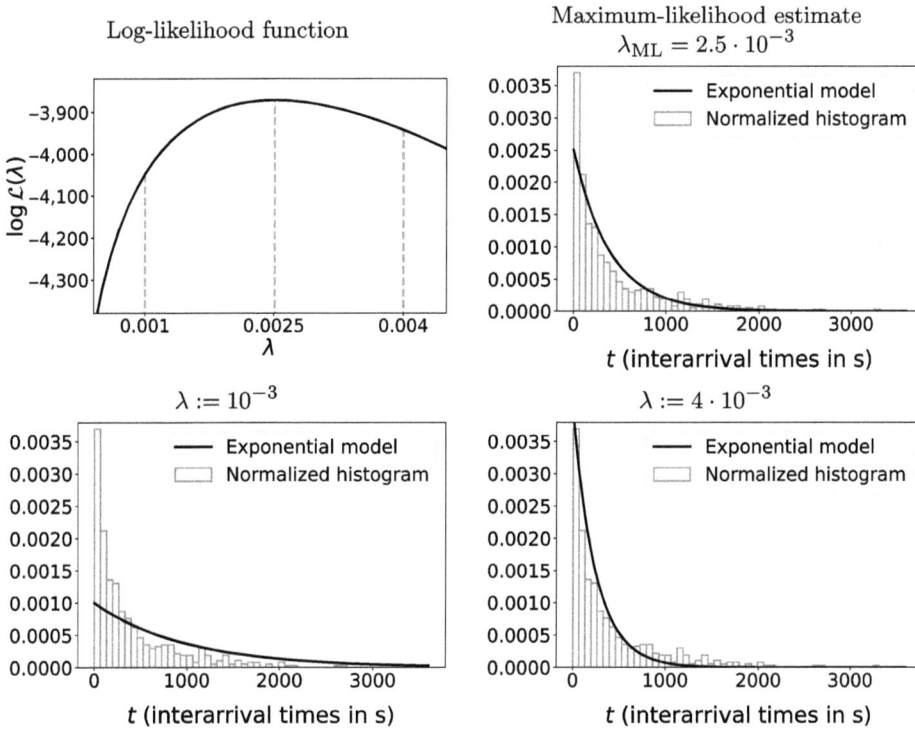

Figure 3.22 Exponential model for interarrival times at a call center. The top left graph shows the log-likelihood function of an exponential parametric model applied to the data described in Example 3.28. The top right graph shows the corresponding fit to the data obtained via maximum-likelihood estimation (applying Theorem 3.34), superimposed onto the histogram of the data used to fit the model. The parametric model is not able to fit the large number of calls with very short durations (leftmost bin of the histogram) because it is not consistent with the rest of the data under the assumptions of the exponential model. The two bottom graphs show the fits corresponding to two other choices of the parameter λ.

Proof By Definition 3.32, the likelihood function is

$$\mathcal{L}_X (\mu, \sigma) = \prod_{i=1}^{n} f_{\mu, \sigma} (x_i) \tag{3.159}$$

$$= \prod_{i=1}^{n} \frac{1}{\sqrt{2\pi}\sigma} \exp\left(-\frac{(x_i - \mu)^2}{2\sigma^2} \right), \tag{3.160}$$

and the log-likelihood is

$$\log \mathcal{L}_X (\mu, \sigma) = -\frac{n \log (2\pi)}{2} - n \log \sigma - \sum_{i=1}^{n} \frac{(x_i - \mu)^2}{2\sigma^2}. \tag{3.161}$$

By Definition 3.33, the maximum-likelihood estimators of the parameters μ and σ are

$$\{\mu_{\mathrm{ML}}, \sigma_{\mathrm{ML}}\} := \arg \max_{\{\mu, \sigma\}} \log \mathcal{L}_X (\mu, \sigma) \tag{3.162}$$

$$= \arg \max_{\{\mu, \sigma\}} -n \log \sigma - \sum_{i=1}^{n} \frac{(x_i - \mu)^2}{2\sigma^2}. \tag{3.163}$$

Log-likelihood function

Maximum-likelihood estimate
$\mu_{\mathrm{ML}} = 175.6$, $\sigma_{\mathrm{ML}} = 6.85$

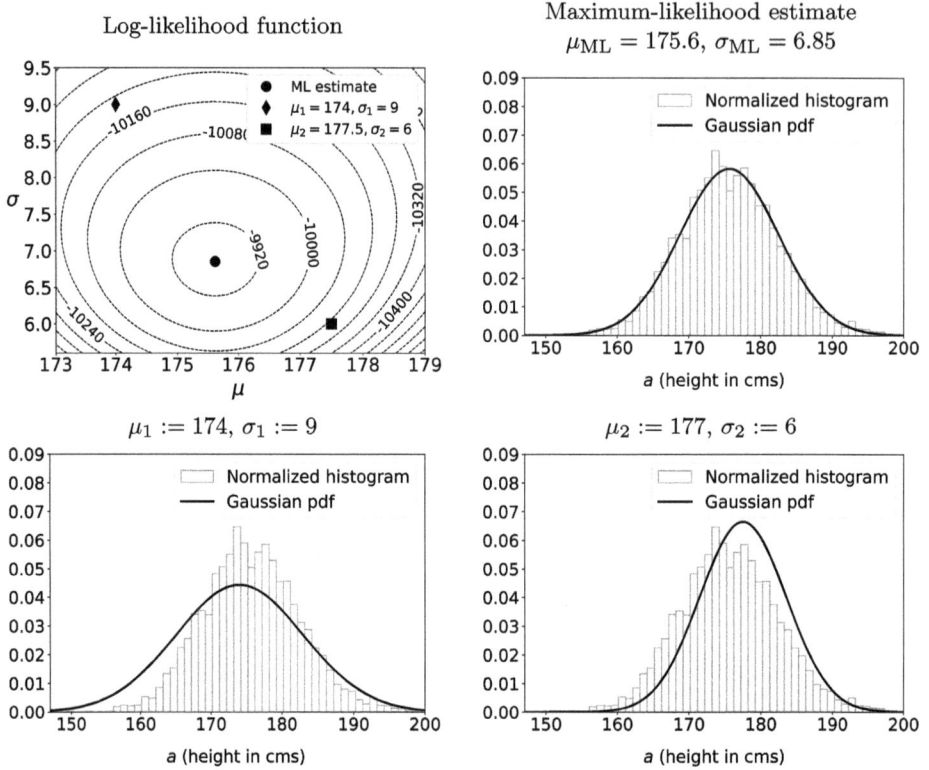

Figure 3.23 Gaussian model for height. The contour plot in the top left shows the log-likelihood function of a Gaussian parametric model applied to height data from 4,082 men in the United States army (see Figure 3.13). The graph in the top right shows the corresponding parametric fit obtained via maximum likelihood (applying Theorem 3.35), superimposed onto the histogram of the data used to fit the model. The two bottom graphs show the fits corresponding to two other choices of the model parameters.

The first and second partial derivative of the log-likelihood function with respect to μ equal

$$\frac{\partial \log \mathcal{L}_X(\mu, \sigma)}{\partial \mu} = \sum_{i=1}^{n} \frac{x_i - \mu}{\sigma^2}, \tag{3.164}$$

$$\frac{\partial^2 \log \mathcal{L}_X(\mu, \sigma)}{\partial \mu^2} = -\frac{n}{\sigma^2}. \tag{3.165}$$

For a fixed value of σ, the function is concave with respect to μ, so we can maximize it by setting the first partial derivative to zero. Regardless of the value of σ, the maximum is at $\mu_{\mathrm{ML}} = \frac{1}{n} \sum_{i=1}^{n} x_i$. We can therefore plug this value into the log likelihood and maximize with respect to σ. The derivative of the resulting function of σ is

$$\frac{\partial \log \mathcal{L}_X(\mu_{\mathrm{ML}}, \sigma)}{\partial \sigma} = -\frac{n}{\sigma} + \sum_{i=1}^{n} \frac{(x_i - \mu_{\mathrm{ML}})^2}{\sigma^3}. \tag{3.166}$$

The derivative is zero, if σ^2 is equal to $\sigma_{\mathrm{ML}}^2 = \frac{1}{n} \sum_{i=1}^{n} (x_i - \mu_{\mathrm{ML}})^2$. If σ^2 is between 0 and σ_{ML}, the derivative is positive; if it is larger, it is negative. The maximum is therefore at $\sigma^2 = \sigma_{\mathrm{ML}}^2$. ∎

Figure 3.23 shows the log-likelihood of a Gaussian model applied to the same height data from Dataset 5 used in Figure 3.13. The figure also shows the pdf corresponding to the maximum-likelihood parameter estimates and compares it to the pdfs corresponding to two other choices of model parameters, which provide a worse fit to the data.

3.8 Inverse-Transform Sampling

Simulation is a fundamental tool in probabilistic modeling. For example, it facilitates the approximation of complicated probabilities via the Monte Carlo method (see Section 1.7). The main strategy for generating simulated samples from a random variable decouples the process into two steps:

1 Generation of uniform samples in the unit interval $[0, 1]$.
2 Transformation of the uniform samples, so that they have the desired distribution.

Here we focus on the second step, assuming that we have access to a random-number generator, which produces independent samples following a uniform distribution in $[0, 1]$. Such generators are typically based on random physical phenomena or on algorithms that create a pseudorandom output with statistical properties that closely resemble those of random samples.

Consider the problem of transforming uniform random samples, so that they follow the distribution of a random variable $\tilde{a}$. The probability that the transformed uniform samples land in an interval $(x, y]$ (where $x < y$) should equal $P(x < \tilde{a} \leq y)$, for all possible choices of x and y. By Definition 3.17 and Theorem 3.14, the probability of obtaining a uniform sample in a subinterval of length ℓ is exactly ℓ (if the subinterval is within $[0, 1]$), as illustrated in the top graph of Figure 3.24. For example,

$$P(c < \tilde{u} \leq c + \ell) = \int_{u=c}^{c+\ell} f_{\tilde{u}}(u) \, du \qquad (3.167)$$

$$= \int_{u=c}^{c+\ell} du = \ell, \qquad (3.168)$$

as long as $c \geq 0$ and $c+\ell \leq 1$. Consequently, if we map all the uniform samples that fall in a subinterval of length $\ell := P(x < \tilde{a} \leq y)$ to $(x, y]$, then the probability that the transformed samples end up in $(x, y]$ is $P(x < \tilde{a} \leq y)$. The challenge is finding a transformation that achieves this for all possible values of x and y.

Recall that by Lemma 3.5, $P(x < \tilde{a} \leq y) = F_{\tilde{a}}(y) - F_{\tilde{a}}(x)$ (bottom left plot in Figure 3.24). If we map each point u in the unit interval to $F_{\tilde{a}}^{-1}(u)$, then $F_{\tilde{a}}(x)$ is mapped to x, $F_{\tilde{a}}(y)$ is mapped to y and the values in between are mapped to $(x, y]$ because the cdf (and therefore its inverse) is nondecreasing (bottom right plot in Figure 3.24). This is precisely what we want: The length of the interval mapping to $(x, y]$ is $F_{\tilde{a}}(y) - F_{\tilde{a}}(x) = P(x < \tilde{a} \leq y)$! This holds for any x and y, so the inverse cdf transforms the uniform samples so that they follow the desired distribution.

Definition 3.36 (Inverse-transform sampling). *Let $F_{\tilde{a}}$ be the cdf of a continuous distribution, where $\tilde{a}$ represents a random variable with a desired distribution. Given a random variable $\tilde{u}$, uniformly distributed in $[0, 1]$, we apply the following steps, in order to sample from $\tilde{a}$:*

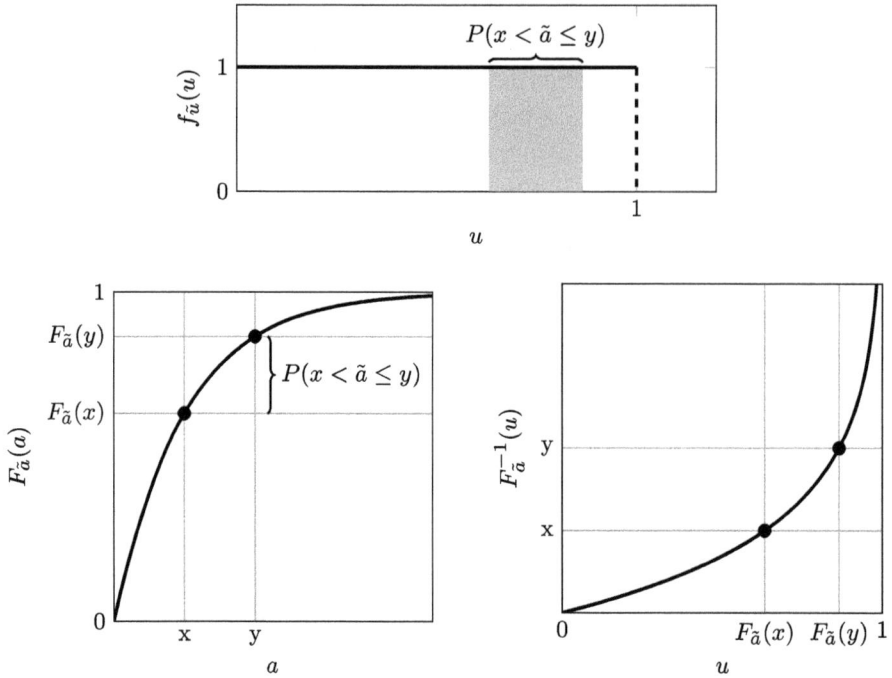

Figure 3.24 Inverse-transform sampling. Let $\tilde{u}$ be a uniform random variable in the unit interval $[0, 1]$. As illustrated in the top plot, the probability that $\tilde{u}$ belongs to an interval of length $P(x < \tilde{a} \leq y)$ is equal to $P(x < \tilde{a} \leq y)$. The bottom left plot shows that $P(x < \tilde{a} \leq y) = F_{\tilde{a}}(y) - F_{\tilde{a}}(x)$, as established in Lemma 3.5. The bottom right plot shows that for any $a \leq y$, the inverse cdf maps an interval of length $P(x < \tilde{a} \leq y)$ to $(x, y]$. Consequently, it transforms the uniform random variable so that it has the same distribution as $\tilde{a}$.

1 *Obtain a sample u of $\tilde{u}$.*
2 *Set $a := F_{\tilde{a}}^{-1}(u)$.*

The careful reader will point out that $F_{\tilde{a}}$ may not be invertible at every point. In such cases, we can just use the generalized inverse of the cdf defined in the proof of Theorem 3.22:

$$F_{\tilde{a}}^{-1}(u) := \min_{x} \{F_{\tilde{a}}(x) = u\}. \tag{3.169}$$

The following theorem provides a formal proof that inverse-transform sampling works.

Theorem 3.37 (Inverse-transform sampling works). *Let $F_{\tilde{a}}$ be the cdf of a continuous random variable $\tilde{a}$. Given a random variable $\tilde{u}$, uniformly distributed in $[0, 1]$, we define $\tilde{b} = F_{\tilde{a}}^{-1}(\tilde{u})$, where $F_{\tilde{a}}^{-1}$ is the inverse of the cdf, or the generalized inverse (3.169). Then, the cdf of b is $F_{\tilde{a}}$, so $\tilde{b}$ has the same distribution as $\tilde{a}$.*

Proof We just need to show that the cdf of $\tilde{b}$ is equal to $F_{\tilde{a}}$. As in the proof of Theorem 3.22, we use the fact that the events $\{F_{\tilde{a}}^{-1}(\tilde{u}) \leq y\}$ and $\{\tilde{u} \leq F_{\tilde{a}}(y)\}$ are equivalent because cdfs are nondecreasing. By Definitions 3.3 and 3.17, and Theorem 3.14,

$$F_{\tilde{b}}(y) := P(\tilde{b} \leq y) \tag{3.170}$$

$$= P\left(F_{\tilde{a}}^{-1}(\tilde{u}) \leq y\right) \tag{3.171}$$

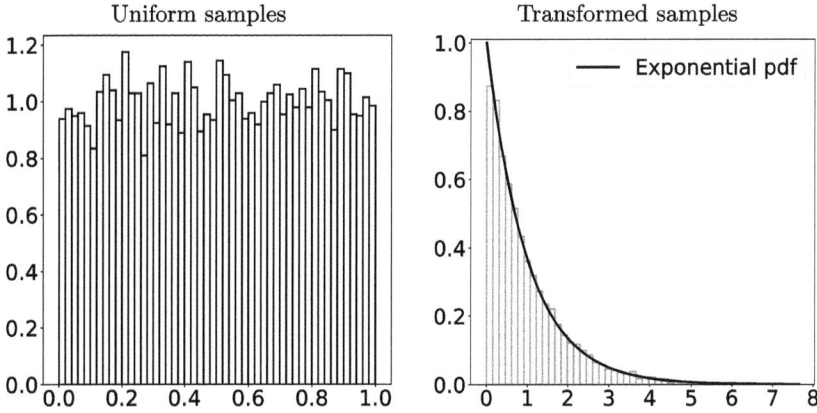

Figure 3.25 Exponential distribution simulated via inverse-transform sampling. The left graph shows the histogram of 10^4 realizations from a uniform distribution. The right graph shows the histogram of the transformed samples, after applying the inverse cdf (3.182) with $\lambda := 1$. The histogram closely approximates an exponential pdf with parameter equal to one, confirming that the transformed samples follow the desired exponential distribution.

$$= \mathrm{P}\left(\tilde{u} \leq F_{\tilde{a}}\left(y\right)\right) \tag{3.172}$$

$$= \int_{u=0}^{F_{\tilde{a}}(y)} f_{\tilde{u}}(u)\ \mathrm{d}u \tag{3.173}$$

$$= \int_{u=0}^{F_{\tilde{a}}(y)} \mathrm{d}u \tag{3.174}$$

$$= F_{\tilde{a}}\left(y\right). \tag{3.175}$$

∎

Example 3.38 (Sampling from an exponential distribution). In this example, we apply Definition 3.36 to simulate an exponential random variable $\tilde{a}$ with parameter λ. By Definitions 3.3 and 3.27, the cdf of the exponential distribution equals

$$F_{\tilde{a}}\left(x\right) := \mathrm{P}\left(\tilde{a} \leq x\right) \tag{3.176}$$

$$= \int_{-\infty}^{x} f_{\tilde{a}}(a)\ \mathrm{d}a \tag{3.177}$$

$$= \int_{0}^{x} \lambda \exp\left(-\lambda a\right)\ \mathrm{d}a \tag{3.178}$$

$$= 1 - \exp\left(-\lambda x\right) \tag{3.179}$$

because $-\exp\left(-\lambda a\right)$ is the antiderivative of $\lambda \exp\left(-\lambda a\right)$. The cdf $F_{\tilde{a}}$ shown in the bottom left graph in Figure 3.24 is actually an exponential cdf with $\lambda := 1$.

The exponential cdf is invertible. To compute the inverse $F_{\tilde{a}}^{-1}$ at u, we solve the equation

$$u = F_{\tilde{a}}\left(F_{\tilde{a}}^{-1}\left(u\right)\right) \tag{3.180}$$

$$= 1 - \exp\left(-\lambda F_{\tilde{a}}^{-1}\left(u\right)\right), \tag{3.181}$$

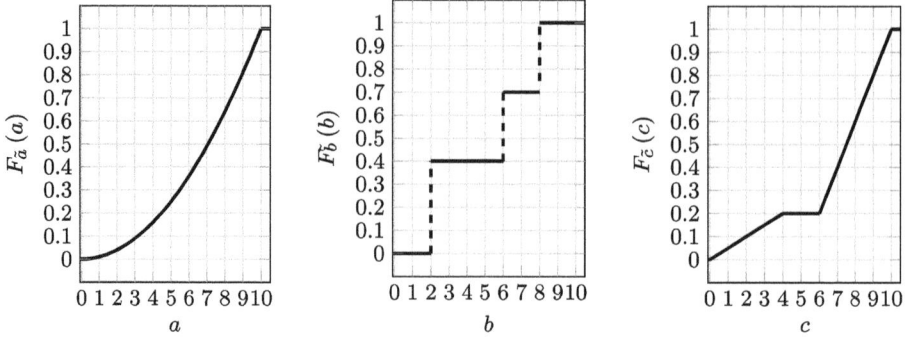

Figure 3.26 Cdfs of the random variables $\tilde{a}$, $\tilde{b}$ and $\tilde{c}$ in Exercise 3.4.

which yields

$$F_{\tilde{a}}^{-1}(u) = \frac{1}{\lambda} \log \left(\frac{1}{1-u} \right). \tag{3.182}$$

The bottom right graph in Figure 3.24 shows the inverse. Plugging a uniform random variable $\tilde{u}$ into it yields the random variable $\tilde{b} := F_{\tilde{a}}^{-1}(\tilde{u})$, which has the desired exponential cdf $F_{\tilde{a}}$ by Theorem 3.37, and hence is an exponential random variable with parameter λ. Figure 3.25 provides a numerical demonstration that applying (3.182) reshapes the uniform distribution to be exponential.

$\cdots\cdots\cdots\cdots\cdots\cdots\cdots\cdots\cdots\cdots\cdots\cdots\cdots\cdots\cdots\cdots\cdots\cdots\cdots\cdots\cdots$

Exercises

3.1 (Probability of individual points) Let $(\Omega, \mathcal{C}, \mathrm{P})$ be a probability space, and let $\tilde{a}$ be a random variable mapping Ω to an interval of the real line $\mathcal{I} \subseteq \mathbb{R}$. Prove that the probability measure can only assign nonzero probability to a countable number of events of the form $\tilde{a} = a$, $a \in \mathcal{I}$. You might find it useful to consider $S := \{s \in \mathbb{R} : \mathrm{P}(\tilde{a} = s) \neq 0\}$, the set of values that are assigned nonzero probability, and the following partition of S

$$A_n := \left\{ a \in S : \frac{1}{n-1} \geq \mathrm{P}(\tilde{a} = a) > \frac{1}{n} \right\}, \quad n = 2, 3, \ldots. \tag{3.183}$$

3.2 (Continuous cdf) Prove that the cdf $F_{\tilde{a}}$ of a random variable $\tilde{a}$ is continuous, if and only if the probability that it equals any single point $a \in \mathbb{R}$ is zero.

3.3 (Median of affine transformation) The median of a random variable $\tilde{a}$ is m. Express the median of the random variable $\tilde{b} := \alpha \tilde{a} + \beta$ as a function of m, where $\alpha > 0$ and $\beta \in \mathbb{R}$.

3.4 (Three cdfs) Figure 3.26 shows the cdfs of the random variables $\tilde{a}$, $\tilde{b}$, and $\tilde{c}$. The cdf of $\tilde{a}$ is quadratic, the cdf of $\tilde{b}$ is piecewise constant and the cdf of $\tilde{c}$ is piecewise linear.

    a What are the medians of $\tilde{a}$, $\tilde{b}$, and $\tilde{c}$?

    b Draw the pdf or the pmf of $\tilde{a}$, $\tilde{b}$, and $\tilde{c}$, annotating the horizontal and vertical axes of your plots.

3.5 (Cumulative distribution function) The random variable $\tilde{x}$ has the following cdf:

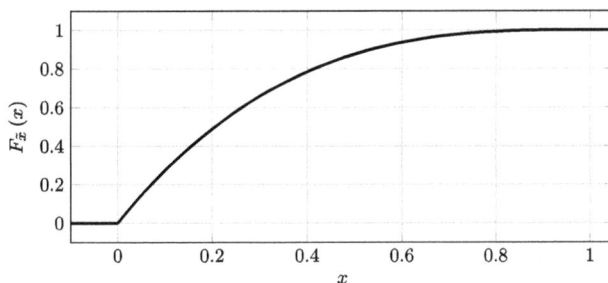

a What is the median of $\tilde{x}$?

b Is the probability density at 0.2 higher or lower than at 0.8? No need to calculate anything, just look at the cdf.

c If we know that $\tilde{x} \geq 0.2$, what is the conditional probability that $\tilde{x}$ is smaller than 0.4?

d You have access to the following three samples from a uniform distribution: 0.2, 0.4, 0.55. Use them to simulate three samples from $\tilde{x}$.

3.6 (Step pdf) Consider the following pdf of a random variable $\tilde{a}$:

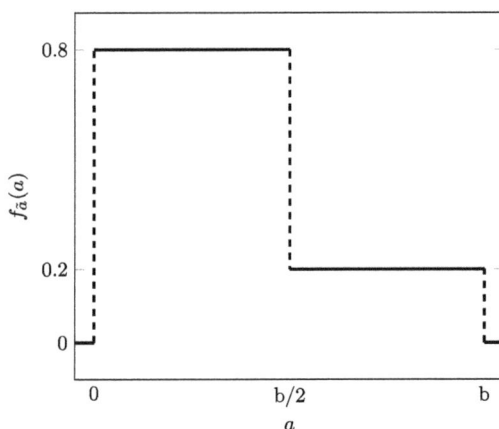

a What is the value of b?

b Derive the cdf of $\tilde{a}$ and use it to calculate its median.

3.7 (Pigs) A farmer weighs four pigs, the weights are: 150 kg, 110 kg, 180 kg, and 140 kg.

a Estimate the probability density of the pig weight using a normalized histogram with three bins: 80–120 kg, 120–160 kg, and 160–200 kg. Draw the histogram.

b If we model the pig weight as Gaussian and fit the parameters via maximum-likelihood estimation, what is the median of the corresponding parametric distribution?

3.8 (Scaled exponential distribution) If $\tilde{a}$ is exponential with parameter λ, what is the distribution of the random variable $\tilde{b} := \alpha \tilde{a}$, where α is a positive real constant?

3.9 (Gaussian pdf) Show that the integral of the pdf of a Gaussian random variable is equal to one for any value of the parameters μ and σ^2.

3.10 (Uniform distribution) Prove that if the random variable $\tilde{u}$ is uniformly distributed in the interval $[0, 1]$, then so is the random variable $\tilde{w} := 1 - \tilde{u}$.

3.11 (Nuclear power plant) A random variable $\tilde{t}$ with the following pdf

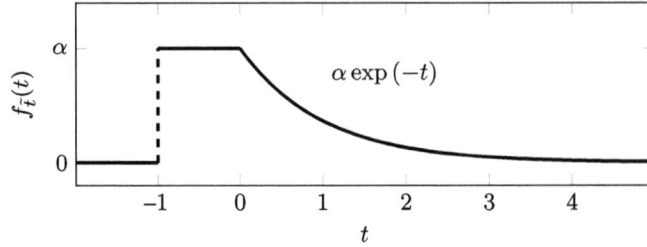

models the time when an accident occurs in a nuclear power plant. The pdf is constant during the time the station is built (between -1 and 0) and exponential with parameter 1 afterwards (from 0 to $+\infty$).

a Compute the value of the constant α.

b Compute the cdf of $\tilde{t}$ and plot it.

c Compute the pdf of $\tilde{t}$ conditioned on $\tilde{t} < 0$.

3.12 (Evaluating survival analysis) In survival analysis, the goal is to estimate the survival function $S_{\text{true}}(t) := P\left(\tilde{t} > t\right)$, which encodes the probability that an individual survives beyond any time t. The random variable $\tilde{t}$ represents the time of death. This exercise is inspired by a clever procedure proposed in Haider et al. (2020) to evaluate a survival-function estimate S_{est} without having access to the true survival function S_{true}.

a What is the distribution of the random variable $\tilde{w} := S_{\text{true}}(\tilde{t})$?

b Suggest a way of evaluating whether a survival-function estimate S_{est} is a good approximation to S_{true} using only S_{est} and a dataset of death times $t_1, \ldots, t_n$. Assume that the death times are i.i.d. samples from the same distribution as $\tilde{t}$, and n is large enough to ensure that empirical probabilities computed from the dataset are accurate.

3.13 (Feeding data into its own empirical cdf) Assume a dataset is sorted, so the data satisfy $x_1 \leq x_2 \leq \cdots \leq x_n$. If you apply the empirical cdf to each data point x_i, $1 \leq i \leq n$, to obtain a new data point $y_i := F_X(x_i)$, what are the new data equal to? Does your answer depend on the dataset?

3.14 (Uniform temperature) We model the temperature in a classroom as being uniformly distributed between 60°F and 80°F.

a Derive the cdf of the temperature and use it to compute the first quartile.

b Explain how to generate samples from our model, if we have access to samples from a uniform distribution between 0 and 1. Apply your approach to the following samples: 0.1, 0.8.

3.15 (Rounded-up measurements) You have access to the readings of a device that indicates whether a radioactive particle has decayed. However the readings are not continuous, you obtain a reading every second.

a A reasonable model for the time the particle takes to decay is that it is a random variable with pdf

$$f_{\tilde{t}}(t) := \begin{cases} \lambda \exp(-\lambda t), & \text{if } t \geq 0, \\ 0 & \text{otherwise,} \end{cases} \tag{3.184}$$

where λ is a fixed constant. The measurement device rounds up the time and outputs an integer number of seconds. If the time is 0.1, it outputs 1; if the time is 13.4, it outputs 14. Compute the pmf of the reading from the device. What kind of random variable is this?

b What is the pdf of the difference between the reading and the true time of decay, according to the model?

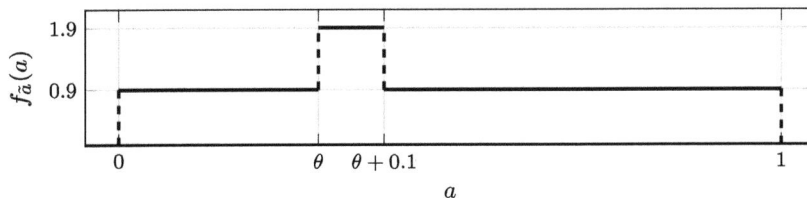

Figure 3.27 Pdf of the random variable $\tilde{a}$ in Exercise 3.18.

3.16 (Half life) The half life of a radioactive material quantifies how rapidly the material decays. It is the time that it takes for the material to be reduced to half because half of the radioactive particles have decayed. It is not immediately apparent why the time should be the same for any initial amount of material. In this exercise, you will show that this is the case (probabilistically) if the particles decay following an exponential distribution.

a Let $\tilde{t}$ be a random variable with a pdf of the form

$$f_{\tilde{t}}(t) := \begin{cases} \lambda \exp(-\lambda t), & \text{if } t \geq 0, \\ 0 & \text{otherwise,} \end{cases} \tag{3.185}$$

where λ is a fixed positive constant. We define the half life $t_{1/2}$ as the median of $\tilde{t}$. Compute $t_{1/2}$ in terms of λ. Then explain intuitively why this is a reasonable definition for the half life.

b Compute t such that $\mathrm{P}(t_{1/2} < \tilde{t} < t) = 1/4$, and express it as a function of $t_{1/2}$. Explain why the result is consistent with the intuitive meaning of half life.

c Compute $\mathrm{P}(\tilde{t} > kt_{1/2})$ for any integer k. Again, explain why the result is consistent with the intuitive meaning of half life.

3.17 (Earthquake) A geophysicist is trying to estimate the pdf of the interarrival times of earthquakes in a certain region. The available data (in years) are: 7.5, 10, 32.5.

a If she estimates the density using kernel density estimation with a rectangular kernel of width 10, what is the probability that the interarrival time is larger than 10?

b If she estimates the density by applying maximum-likelihood estimation with an exponential parametric model, what is the probability that the interarrival time is larger than 10?

c Briefly describe the advantages and disadvantages of the nonparametric and the parametric models.

3.18 (Uniform distribution with a bump) We consider the parametric pdf of a random variable $\tilde{a}$ depicted in Figure 3.27, where the parameter θ is between 0 and 0.9.

a Derive the cdf of $\tilde{a}$.

b Derive and sketch the log-likelihood function of the model for $0 \leq \theta \leq 0.9$ when the data equal 0.1, 0.4, 0.45, 0.7 (assume that the data are i.i.d.). Use the approximations $\log 0.9 = -0.05$ and $\log 1.9 = 0.27$.

3.19 (Planet) An astrophysicist determines that a good model for the pdf of the temperature in a newly discovered planet is

$$f_{\tilde{t}}(t) := \frac{\lambda \exp(-\lambda |t|)}{2}, \tag{3.186}$$

where t can be any real number.

a Compute the cdf of $\tilde{t}$.

b Compute the maximum-likelihood estimate of λ from the following data: 5, -50, -1, 100.

c What is the pdf of $\tilde{t}$ conditioned on the event $\tilde{t} > 0$?

3.20 (Triangular pdf) We are interested in fitting a model with a parametric pdf equal to

$$f_w(x) = \begin{cases} \frac{2x}{w^2}, & \text{for } 0 \le x \le w, \\ 0, & \text{otherwise,} \end{cases} \tag{3.187}$$

where the parameter w is nonnegative.

  a The observed values are 1.25, 0.4, 1.5, 1, and 1.2. What are the possible values of the parameter w?

  b Compute the likelihood function corresponding to these data and sketch it.

  c What is the maximum-likelihood estimator of w?

  d Assume that the data are indeed generated by the parametric model with $w := w_{\text{true}}$. Does the maximum-likelihood estimator systematically underestimate or overestimate the true parameter?

  e Generate a sample from a random variable that follows this parametric distribution with $w := 2$, using a uniform sample from the interval $[0, 1]$ equal to 0.64.

3.21 (Rat) We are interested in modeling the location of a rat that lives in a tunnel. The tunnel is 1 mile long. The first half mile is very dark. The second half is not as dark, but has more food. We decide to use a parametric model. The parametric pdf of the location is equal to 2α over $[0, 0.5]$, and to $2(1 - \alpha)$ over $[0.5, 1]$, where α is a parameter. Here is a plot of the pdf:

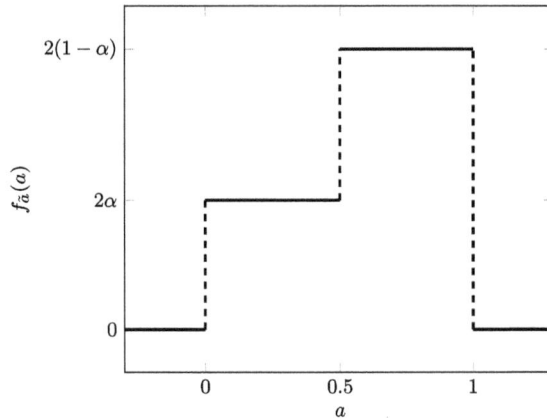

  a What two conditions must a function satisfy to be a valid pdf? For what values of α does this function satisfy them?

  b Compute the maximum-likelihood estimator of α, if we observe the following independent samples from the proposed model: 0.1, 0.8, 0.9, 0.7, 0.3.

  c What is the probability that the rat is in the first half of the tunnel, as a function of α? Based on your answer, suggest a way to estimate α based on the empirical-probability estimator. Compare it to the maximum-likelihood estimator.

  d Use kernel density estimation with a rectangular kernel of width 0.2 to approximate the probability density of the location. What is problematic about the estimate you obtain? How would you address this?

4

Multiple Discrete Variables

Overview

In this chapter, we explain how to model multiple uncertain discrete quantities. Section 4.1 shows that such quantities can be represented as random variables within the same probability space, and introduces the joint probability mass function, which characterizes their joint behavior. Section 4.2 explains how to obtain the distribution of individual quantities in models with multiple variables. Section 4.3 defines the conditional distribution of a random variable, which describes its behavior when the values of other variables are fixed. Sections 4.4 and 4.5 define independence and conditional independence for random variables. In Section 4.6, we discuss causal inference, where the goal is to identify causal effects between variables. Section 4.7 is our first encounter with the notorious curse of dimensionality, which is the reason why we need independence assumptions to make probabilistic models tractable. Sections 4.8 and 4.9 describe two popular models based on such assumptions: naive Bayes and Markov chains.

4.1 Multivariate Discrete Random Variables

In this section, we explain how to model the joint behavior of multiple uncertain discrete quantities. In Section 4.1.1, we show that such quantities can be represented as random variables within the same probability space. Section 4.1.2 introduces the joint probability mass function (pmf), which allows us to manipulate the random variables, without having to worry about the underlying probability space. In Section 4.1.3, we explain how to estimate the joint pmf from data.

When jointly modeling multiple random variables, we often group them as entries of a *random vector*:

$$\tilde{x} := \begin{bmatrix} \tilde{x}[1] \\ \tilde{x}[2] \\ \dots \\ \tilde{x}[d] \end{bmatrix}. \tag{4.1}$$

Here $\tilde{x}[i]$, $1 \leq i \leq d$, denotes the random variable that corresponds to the ith entry of the d-dimensional random vector $\tilde{x}$.

4.1.1 Mathematical Definition

In Section 2.1.1, we define discrete random variables as functions from a sample space to a discrete set. This mathematical framework enables us to characterize the joint behavior of several random variables very easily: We just define them *on the same probability space*. The outcome in the sample space simultaneously determines the value of all the random variables.

Example 4.1 (Rolling a die twice). In Example 2.1, we define a probability space representing two rolls of a six-sided die. Each outcome is encoded as a two-dimensional vector,

$$\omega := \begin{bmatrix} \omega_1 \\ \omega_2 \end{bmatrix}, \quad \omega_1, \omega_2 \in \{1, 2, 3, 4, 5, 6\}, \tag{4.2}$$

where ω_1 is the result of the first roll, and ω_2 the result of the second roll. The random variables

$$\tilde{a}(\omega) := \omega_1, \tag{4.3}$$

$$\tilde{b}(\omega) := \omega_2, \tag{4.4}$$

$$\tilde{c}(\omega) := \omega_1 + \omega_2, \tag{4.5}$$

represent the value of the first roll, the second roll, and the sum of the two rolls, respectively. If $\omega = \begin{bmatrix} 3 \\ 1 \end{bmatrix}$, then $\tilde{a}(\omega) = 3$, $\tilde{b}(\omega) = 1$, and $\tilde{c}(\omega) = 4$. If $\omega = \begin{bmatrix} 2 \\ 5 \end{bmatrix}$, then $\tilde{a}(\omega) = 2$, $\tilde{b}(\omega) = 5$, and $\tilde{c}(\omega) = 7$. We can therefore reason about the *joint distribution* of the random variables. For example, we can define the event that $\tilde{a}$ equals 3, and simultaneously $\tilde{c}$ is smaller than 6:

$$\{\tilde{a} = 3\} \cap \{\tilde{c} < 6\} := \{\omega \colon \tilde{a}(\omega) = 3 \text{ and } \tilde{c}(\omega) < 6\} \tag{4.6}$$

$$= \left\{ \begin{bmatrix} 3 \\ 1 \end{bmatrix}, \begin{bmatrix} 3 \\ 2 \end{bmatrix} \right\}. \tag{4.7}$$

. . . .

As we discuss in Section 2.1.2, the probability space used to define random variables is a mathematical abstraction. We only build it explicitly for simple pedagogical examples, like Example 4.1. Instead, we describe random variables using their associated probabilities.

Consider two discrete random variables $\tilde{a}$ and $\tilde{b}$ defined on the same probability space. By Definition 2.3, the probability measure of the probability space must assign a probability to the events $\tilde{a} = a$ and $\tilde{b} = b$, for every possible value of a and b. By the properties of probability spaces (see Definitions 1.7 and 1.15), the probability measure must also assign a probability to the intersection of the two events,

$$\left\{ \omega \in \Omega \colon \tilde{a}(\omega) = a, \ \tilde{b}(\omega) = b \right\}. \tag{4.8}$$

To simplify notation, we usually denote the intersection by $\{\tilde{a} = a, \tilde{b} = b\}$. In practice, we define and manipulate discrete random variables using the probabilities of such intersections, encoded by the joint probability mass function, as explained in the following section.

4.1.2 The Joint Probability Mass Function

The joint probability mass function (pmf) of two random variables encodes the probability that the random variables equal any pair of values in their respective ranges.

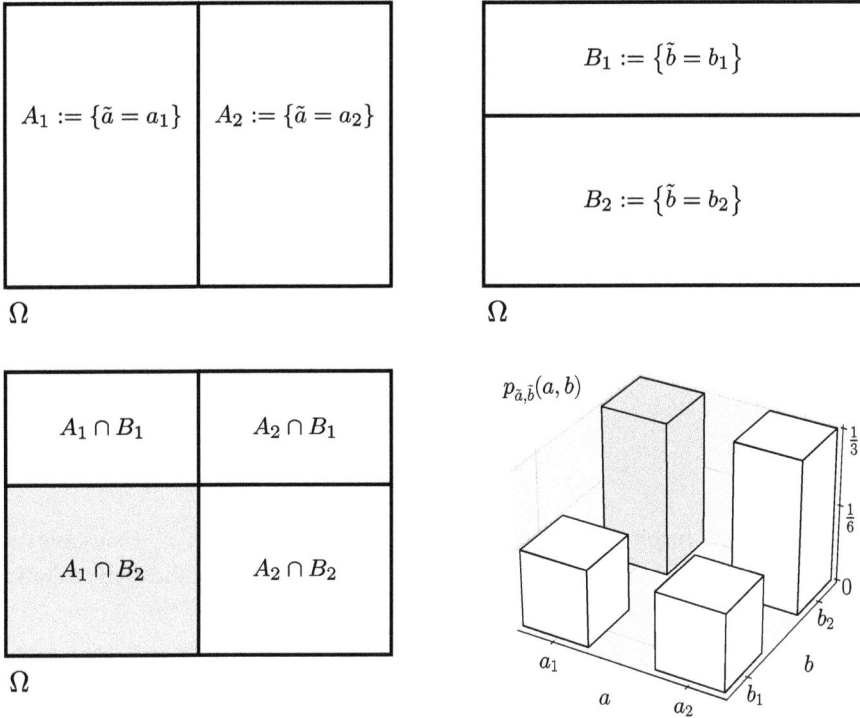

Figure 4.1 Joint probability mass function of two discrete random variables. The discrete random variables $\tilde{a}$ and $\tilde{b}$ are defined on the same probability space. The Venn diagrams in the top row show the two partitions of the sample space Ω induced by $\tilde{a}$ (left) and $\tilde{b}$ (right). For $i \in \{1, 2\}$, A_i and B_i denote the events containing the outcomes mapping to a_i and b_i, respectively. The bottom left Venn diagram shows the different intersections $A_i \cap B_j$, for $i, j \in \{1, 2\}$. These events contain outcomes ω such that $\tilde{a}(\omega) = a_i$ and $\tilde{b}(\omega) = b_j$. The joint pmf $p_{\tilde{a},\tilde{b}}(a_i, b_j)$, depicted in the bottom right plot, is defined as the probability of the event $A_i \cap B_j$, for $i, j = 1, 2$. For example, $p_{\tilde{a},\tilde{b}}(a_1, b_2)$ (shaded in gray) is equal to $P(A_1 \cap B_2)$, represented by the area of the corresponding event $A_1 \cap B_2$ in the Venn diagram (also shaded in gray).

Definition 4.2 (Joint probability mass function of two random variables). *Let $\tilde{a} : \Omega \to A$ and $\tilde{b} : \Omega \to B$ be discrete random variables with discrete ranges A and B, defined on the same probability space $(\Omega, \mathcal{C}, P)$. The joint pmf of $\tilde{a}$ and $\tilde{b}$ is*

$$p_{\tilde{a},\tilde{b}}(a, b) := P(\tilde{a} = a, \tilde{b} = b), \quad a \in A, b \in B. \tag{4.9}$$

In words, $p_{\tilde{a},\tilde{b}}(a, b)$ is the probability of $\tilde{a}$ and $\tilde{b}$ being equal to a and b at the same time. Figure 4.1 shows a simple example of a joint pmf and illustrates its connection to the underlying probability space.

We can generalize the definition of joint pmf to more than two random variables, or equivalently to the entries of a random vector.

Definition 4.3 (Joint probability mass function of a random vector). *Let $\tilde{x}$ be a vector with entries equal to d random variables $\tilde{x}[1] : \Omega \to R_1$, $\tilde{x}[2] : \Omega \to R_2$, ..., $\tilde{x}[d] : \Omega \to R_d$ defined on a probability space $(\Omega, \mathcal{C}, P)$, where R_i is the discrete range of $x[i]$ for $1 \leq i \leq d$. The joint pmf of $\tilde{x}$ is*

$$p_{\tilde{x}}(x) := \mathrm{P}\Big(\tilde{x}[1] = x[1], \tilde{x}[2] = x[2], \ldots, \tilde{x}[d] = x[d]\Big). \tag{4.10}$$

The joint pmf allows us to compute the probability that random variables or random vectors belong to any subset of their ranges. This means that we can fully characterize their behavior through the joint pmf, without having to refer to the underlying probability space.

Lemma 4.4. *Let $\tilde{a}$ and $\tilde{b}$ be discrete random variables with joint pmf $p_{\tilde{a},\tilde{b}}$. For any $S \subseteq A \times B$, where A and B denote the ranges of $\tilde{a}$ and $\tilde{b}$,*

$$\mathrm{P}\Big((\tilde{a},\tilde{b}) \in S\Big) = \sum_{(a,b)\in S} p_{\tilde{a},\tilde{b}}(a,b). \tag{4.11}$$

Let $\tilde{x}$ be a d-dimensional random vector with joint pmf $p_{\tilde{x}}$. For any $S \subseteq R_1 \times R_2 \times \cdots \times R_d$, where R_i is the range of $\tilde{x}[i]$ for $1 \leq i \leq d$,

$$\mathrm{P}(\tilde{x} \in S) = \sum_{x\in S} p_{\tilde{x}}(x). \tag{4.12}$$

Proof We prove the result for two variables, the general case for $d > 2$ variables follows by the same argument. The events $\{\tilde{a} = a, \tilde{b} = b\}$ are all disjoint for different values of a or b, so by Axiom 3 in Definition 1.9 and Definition 4.3,

$$\mathrm{P}\left((\tilde{a},\tilde{b}) \in S\right) = \mathrm{P}\left(\cup_{(a,b)\in S} \{\tilde{a} = a, \tilde{b} = b\}\right) \tag{4.13}$$

$$= \sum_{(a,b)\in S} \mathrm{P}\left(\tilde{a} = a, \tilde{b} = b\right) \tag{4.14}$$

$$= \sum_{(a,b)\in S} p_{\tilde{a},\tilde{b}}(a,b). \tag{4.15}$$

∎

It follows from Definition 4.3 that every joint pmf is nonnegative, and must sum to one. Conversely, any function that satisfies these two conditions can be interpreted as a joint pmf.

Theorem 4.5 (Properties of the joint pmf). *Let $\tilde{a}$ and $\tilde{b}$ be discrete random variables with joint pmf $p_{\tilde{a},\tilde{b}}$. The joint pmf is nonnegative and satisfies*

$$\sum_{a\in A}\sum_{b\in B} p_{\tilde{a},\tilde{b}}(a,b) = 1, \tag{4.16}$$

where A and B denote the ranges of $\tilde{a}$ and $\tilde{b}$, respectively.

Let $\tilde{x}$ be a d-dimensional random vector with joint pmf $p_{\tilde{x}}$. The joint pmf is nonnegative and satisfies

$$\sum_{x[1]\in R_1}\sum_{x[2]\in R_2} \cdots \sum_{x[d]\in R_d} p_{\tilde{x}}(x) = 1, \tag{4.17}$$

where R_i denotes the range of $\tilde{x}[i]$ for $1 \leq i \leq d$.

Let $R_1, R_2, \ldots, R_d$ be d discrete sets. Any nonnegative function $p : R_1 \times R_2 \times \cdots \times R_d \to [0, 1]$ satisfying

$$\sum_{x[1]\in R_1}\sum_{x[2]\in R_2} \cdots \sum_{x[d]\in R_d} p(x) = 1 \tag{4.18}$$

can be interpreted as the joint pmf of a random vector $\tilde{x}$ with range $R_1 \times R_2 \times \cdots \times R_d$.

Proof Joint pmfs are nonnegative because they represent probabilities. Since every outcome in the underlying space must be mapped to some pair $(a, b) \in A \times B$, by Lemma 4.4 and Axiom 2 in Definition 1.9,

$$\sum_{a \in A} \sum_{b \in B} p_{\tilde{a}, \tilde{b}}(a, b) = P\left(\cup_{a \in A, b \in B}\{\omega : \tilde{a}(\omega) = a, \tilde{b}(\omega) = b\}\right) \tag{4.19}$$

$$= P(\Omega) = 1, \tag{4.20}$$

where Ω denotes the sample space of the probability space in which the random variables are defined. In words, the sum is equal to the probability of the whole underlying sample space. The same argument establishes (4.17).

To prove that any nonnegative function $p : R_1 \times R_2 \times \cdots \times R_d \to [0, 1]$, which sums to one, is a valid joint pmf, we define a probability space where the sample space is $R_1 \times R_2 \times \cdots \times R_d$, the collection is the power set of the sample space, and each event S in the collection is assigned the probability

$$P(S) = \sum_{x \in S} p(x). \tag{4.21}$$

You can check that this yields a valid probability measure, where the random vector associated with the joint pmf p is the identity function. ∎

Example 4.6 (Simple example). We consider two random variables $\tilde{a}$ and $\tilde{b}$ with joint pmf,

$$p_{\tilde{a}, \tilde{b}}(1, 1) := 0.05, \quad p_{\tilde{a}, \tilde{b}}(1, 2) := 0.2, \quad p_{\tilde{a}, \tilde{b}}(1, 3) := 0.1, \tag{4.22}$$

$$p_{\tilde{a}, \tilde{b}}(2, 1) := 0.1, \quad p_{\tilde{a}, \tilde{b}}(2, 2) := 0.05, \quad p_{\tilde{a}, \tilde{b}}(2, 3) := 0.2, \tag{4.23}$$

$$p_{\tilde{a}, \tilde{b}}(3, 1) := 0.1, \quad p_{\tilde{a}, \tilde{b}}(3, 2) := 0.1, \quad p_{\tilde{a}, \tilde{b}}(3, 3) := 0.1. \tag{4.24}$$

The values add up to one and are nonnegative, so this is a valid joint pmf by Theorem 4.5. Figure 4.5 shows a 3D bar plot of the joint pmf on the upper left. To compute the probability that $\tilde{a}$ and $\tilde{b}$ belong to any subset of their joint range, we apply Lemma 4.4 and sum the joint pmf over the elements of the subset. For example,

$$P(\tilde{a} < 2, \tilde{b} > 1) = p_{\tilde{a}, \tilde{b}}(1, 2) + p_{\tilde{a}, \tilde{b}}(1, 3) \tag{4.25}$$

$$= 0.3. \tag{4.26}$$

4.1.3 The Empirical Joint Probability Mass Function

The joint pmf of multiple random variables encodes the probability that the random variables take any of their possible values at the same time. Therefore, a reasonable way to estimate the joint pmf is to leverage the empirical-probability estimator (see Definition 1.22) to approximate the corresponding probabilities. This is a generalization of the empirical pmf, defined in Section 2.2, to multiple variables.

Definition 4.7 (Empirical joint pmf). *Let $X := \{x_1, x_2, \ldots, x_n\}$ denote a dataset of d-dimensional vectors, where the jth entry takes values in a discrete set R_j for $1 \leq j \leq d$. The empirical joint pmf $p_X : R_1 \times R_2 \times \cdots \times R_d \to [0, 1]$ maps each d-dimensional vector $v \in R_1 \times R_2 \times \cdots \times R_d$ to the fraction of data points that equal v,*

Table 4.1 ***Empirical joint pmf of movie ratings.*** *The left table shows the number of users correspond-ing to all possible pairs of ratings for the movies Mission Impossible and Independence Day. These counts are normalized to yield the empirical joint pmf of the data on the right.*

Independence Day

Mission Impossible	1	2	3	4	5
1	2	3	5	1	0
2	3	12	18	11	5
3	5	14	37	41	17
4	6	15	20	47	19
5	0	0	4	12	17

Counts

Independence Day

Mission Impossible	1	2	3	4	5
1	0.6	1	1.6	0.3	0
2	1	3.8	5.7	3.5	1.6
3	1.6	4.5	11.8	13.1	5.4
4	1.9	4.8	6.4	15	6.1
5	0	0	1.3	3.8	5.4

Empirical joint pmf (%)

$$p_X(v) := \frac{1}{n}\sum_{i=1}^{n} 1\,(x_i = v)\,, \qquad (4.27)$$

where $1\,(x_i = v)$ *is an indicator function that is equal to one, if* $x_i = v$, *and to zero otherwise.*

The empirical joint pmf is nonnegative and sums to one (by the same argument as in the proof of Lemma 2.12), so we can interpret it as a valid joint pmf by Theorem 4.5. As usual, we should bear in mind that empirical probabilities may be noisy, if the data are limited.

Example 4.8 (Movie ratings). Dataset 8 contains ratings given by a group of users to popular movies. The ratings are integers between 1 and 5. We consider users that rated Inde-pendence Day and Mission Impossible, interpreting each pair of ratings as a joint realization of two random variables. Table 4.1 shows the empirical joint pmf of these random variables, computed following Definition 4.7. We can approximate different probabilities related to the ratings from the joint pmf estimate via Lemma 4.4. For example, the probability that both ratings are greater than or equal to 4 is

$$P(\text{both ratings} \geq 4) = \sum_{i=4}^{5}\sum_{j=4}^{5} p_X(i,j) \qquad (4.28)$$

$$= 0.303. \qquad (4.29)$$

. . . .

Example 4.9 (Precipitation in Oregon). We consider a dataset, extracted from Dataset 9, which contains hourly precipitation measurements from three Oregon weather stations in 2015: Coos Bay, Corvallis and John Day. We define a random vector of dimension three, where each entry is a Bernoulli random variable indicating whether there is precipitation in Coos Bay ($\tilde{x}[1] = 1$), Corvallis ($\tilde{x}[2] = 1$) or John Day ($\tilde{x}[3] = 1$). The empirical joint pmf of the data is shown in Figure 4.2. We see that 82.8% of the time there is no precipitation in any of the stations.

. . . .

Coos Bay

No   Yes

Corvallis   Corvallis

No   Yes   No   Yes

John Day   John Day   John Day   John Day

No   Yes   No   Yes   No   Yes   No   Yes

82.8 (7,253)   **1.8** (162)   **3.6** (318)   **0.4** (34)   **5.3** (462)   **0.3** (30)   **4.9** (432)   **0.8** (69)

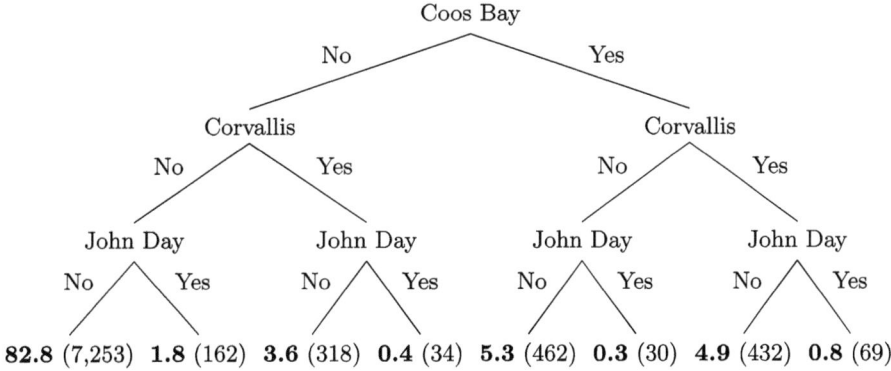

Figure 4.2 Empirical joint pmf of precipitation. The tree diagram shows the joint distribution of precipitation in the Oregon stations of Coos Bay, Corvallis and John Day in 2015, based on hourly measurements extracted from Dataset 9. The counts in parentheses are normalized to yield the empirical joint pmf (in bold). The joint pmf is expressed in percentage.

4.2 Marginal Distributions

In probabilistic models with multiple variables, it is often of interest to isolate the individual behavior of a single variable. This can be achieved by computing the pmf of the random variable, *marginalizing* out the rest of variables. In this context, the resulting pmf is called the *marginal* pmf of the random variable.

Theorem 4.10 (Marginal pmf). *Let $\tilde{a}$ and $\tilde{b}$ be discrete random variables with ranges A and B, respectively, and joint pmf $p_{\tilde{a},\tilde{b}}$. The marginal pmf of $\tilde{a}$ is obtained by summing the joint pmf over all the possible values of $\tilde{b}$,*

$$p_{\tilde{a}}(a) = \sum_{b \in B} p_{\tilde{a},\tilde{b}}(a,b). \tag{4.30}$$

Let $\tilde{x}$ be a random vector with joint pmf $p_{\tilde{x}}$. The marginal pmf of the ith entry $\tilde{x}[i]$ is obtained by summing the joint pmf over all the possible values of the other entries,

$$p_{\tilde{x}[i]}(a) = \sum_{b_1 \in R_1} \cdots \sum_{b_{i-1} \in R_{i-1}} \sum_{b_{i+1} \in R_{i+1}} \cdots \sum_{b_d \in R_d} p_{\tilde{x}}(b_1, \ldots, b_{i-1}, a, b_{i+1}, \ldots, b_d),$$

$$\tag{4.31}$$

where R_i denotes the range of the ith entry for $1 \leq i \leq d$.

Proof   We prove the bivariate case; the general case follows by the same argument. The event $\tilde{a} = a$ is the union of the disjoint events $\{\tilde{a} = a, \tilde{b} = b\}$ for $b \in B$. By Definition 2.2, Axiom 3 in Definition 1.9, and Definition 4.3,

$$p_{\tilde{a}}(a) := \mathrm{P}(\tilde{a} = a) \tag{4.32}$$

$$= \mathrm{P}(\cup_{b \in B} \{\tilde{a} = a, \tilde{b} = b\}) \tag{4.33}$$

$$= \sum_{b \in B} \mathrm{P}(\tilde{a} = a, \tilde{b} = b) \tag{4.34}$$

$$= \sum_{b \in B} p_{\tilde{a},\tilde{b}}(a,b). \tag{4.35}$$

∎

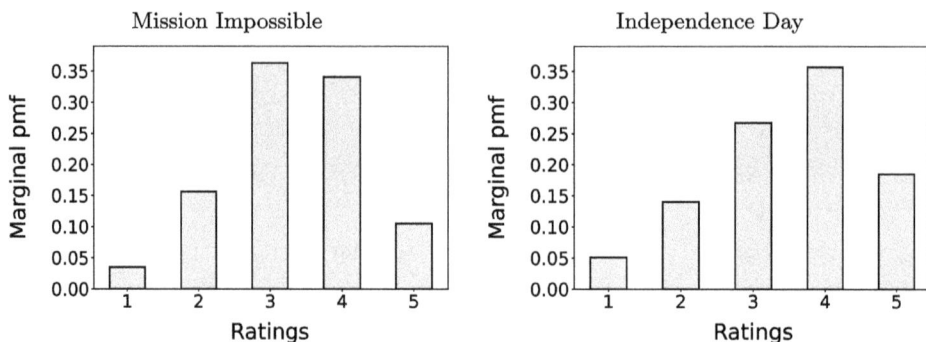

Figure 4.3 Marginal pmfs of movie ratings. Marginal pmf of the ratings of the movies Mission Impossible (left) and Independence Day (right), computed from the empirical joint pmf in Table 4.1 (see Example 4.8).

If we are interested in computing the joint pmf of several entries in a random vector, instead of just one, the marginalization process is essentially the same. Notation gets a bit complicated, so let us just consider an example with four entries. To compute the marginal pmf of the first and fourth entries, we marginalize out the second and third entries,

$$p_{\tilde{x}[1],\tilde{x}[4]}(a, d) = \mathrm{P}\left(\cup_{b \in R_2, c \in R_3} \{\tilde{x}[1] = a, \tilde{x}[2] = b, \tilde{x}[3] = c, \tilde{x}[4] = d\}\right) \quad (4.36)$$

$$= \sum_{b \in R_2} \sum_{c \in R_3} p_{\tilde{x}}(a, b, c, d). \quad (4.37)$$

Example 4.11 (Simple example: Marginal distribution). Consider the random variables $\tilde{a}$ and $\tilde{b}$ in Example 4.6. We compute the marginal pmf of $\tilde{a}$ by summing over b,

$$p_{\tilde{a}}(1) = 0.35, \qquad p_{\tilde{a}}(2) = 0.35, \qquad p_{\tilde{a}}(3) = 0.3. \quad (4.38)$$

The marginal pmf is displayed in the top right graph of Figure 4.5. The marginal pmf of $\tilde{b}$ is obtained by summing over a,

$$p_{\tilde{b}}(1) = 0.25, \qquad p_{\tilde{b}}(2) = 0.35, \qquad p_{\tilde{b}}(3) = 0.4. \quad (4.39)$$

Figure 4.3 shows the marginal pmfs of the movie ratings in Example 4.8, obtained by applying Theorem 4.10 to the joint pmf. It turns out that this group of users enjoyed Independence Day more than Mission Impossible.

Figure 4.4 shows the marginal distributions corresponding to the precipitation data in Example 4.9. By Theorem 4.10, to obtain the marginal joint pmfs of each pair of stations, we can sum the joint pmf over the remaining entry of the random vector. For example,

$$p_{\tilde{x}[1],\tilde{x}[2]}(a, b) = p_{\tilde{x}}(a, b, 0) + p_{\tilde{x}}(a, b, 1), \quad a, b \in \{0, 1\}. \quad (4.40)$$

Similarly, to obtain the marginal pmf of each station, we sum over the other two. For example,

$$p_{\tilde{x}[2]}(b) = \sum_{a=0}^{1} \sum_{c=0}^{1} p_{\tilde{x}}(a, b, c), \quad b \in \{0, 1\}. \quad (4.41)$$

From the data, we see that there is less precipitation in John Day than in Coos Bay or Corvallis.

	Corvallis	
Coos Bay	**No**	**Yes**
No	84.70	4.02
Yes	5.62	5.72

	John Day	
Coos Bay	**No**	**Yes**
No	86.43	2.24
Yes	10.21	1.13

	John Day	
Corvallis	**No**	**Yes**
No	88.07	2.19
Yes	8.56	1.18

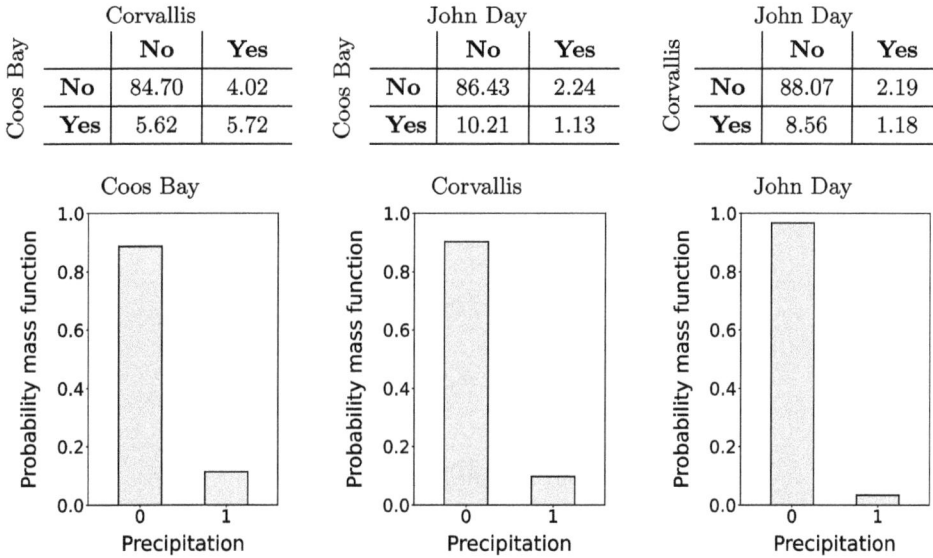

Figure 4.4 Marginal pmfs for precipitation data. The top row shows the marginal joint pmf of each pair of weather stations in Example 4.9, obtained from the empirical joint pmf in Figure 4.2. The graphs show the marginal pmf corresponding to each individual station.

In practice, we usually estimate marginal pmfs by directly applying Definition 2.11 to the data associated with the relevant variable, instead of first estimating the empirical joint pmf and then marginalizing. The following lemma shows that this is equivalent.

Lemma 4.12 (Marginal empirical pmf). *Let* (x_1, y_1), (x_2, y_2), ..., (x_n, y_n) *be* n *pairs of real-valued data. We denote by* $X := \{x_1, x_2, \ldots, x_n\}$ *and* $Y := \{y_1, y_2, \ldots, y_n\}$ *the datasets of first and second entries, respectively. By Definition 4.7, the joint empirical pmf of* X *and* Y *is*

$$p_{X,Y}(a, b) := \frac{1}{n} \sum_{i=1}^{n} 1\,(x_i = a, y_i = b), \qquad (4.42)$$

where $1\,(x_i = a, y_i = b)$ *is an indicator function that equals one, if both* $x_i = a$ *and* $y_i = b$, *and zero otherwise. By Definition 2.11, the empirical pmf of* X *is*

$$p_X(a) := \frac{1}{n} \sum_{i=1}^{n} 1\,(x_i = a), \qquad (4.43)$$

where $1\,(x_i = a)$ *is an indicator function that equals one, if* $x_i = a$ *and zero otherwise. For any* $a \in A$,

$$p_X(a) = \sum_{b \in B} p_{X,Y}(a, b), \qquad (4.44)$$

where A *and* B *denote the sets of distinct values in* X *and* Y, *respectively.*

Proof   Notice that $\sum_{b \in B} 1\left(x_i = a, y_i = b\right) = 1\left(x_i = a\right)$ because y_i must be equal to exactly one of the values in B. As a result,

$$\sum_{b \in B} p_{X,Y}\left(a, b\right) = \frac{1}{n} \sum_{i=1}^{n} \sum_{b \in B} 1\left(x_i = a, y_i = b\right) \tag{4.45}$$

$$= \frac{1}{n} \sum_{i=1}^{n} 1\left(x_i = a\right) \tag{4.46}$$

$$= p_X\left(a\right). \tag{4.47}$$

∎

4.3 Conditional Distributions

The conditional distribution of a random variable describes its behavior under the assumption that other random variables take fixed values. For discrete random variables, the distribution is typically specified using the conditional pmf.

Definition 4.13 (Conditional probability mass function). *Let $\tilde{a}$ and $\tilde{b}$ be discrete random variables. The conditional pmf of $\tilde{b}$ given $\tilde{a}$ is*

$$p_{\tilde{b} \mid \tilde{a}}\left(b \mid a\right) := \mathrm{P}\left(\tilde{b} = b \mid \tilde{a} = a\right) \tag{4.48}$$

$$= \frac{\mathrm{P}\left(\tilde{a} = a, \tilde{b} = b\right)}{\mathrm{P}\left(\tilde{a} = a\right)} \tag{4.49}$$

$$= \frac{p_{\tilde{a}, \tilde{b}}\left(a, b\right)}{p_{\tilde{a}}\left(a\right)}, \tag{4.50}$$

assuming $p_{\tilde{a}}\left(a\right) > 0$. Otherwise, the conditional pmf is undefined.

Let $\tilde{x}$ and $\tilde{y}$ be random vectors, each with multiple entries. The conditional pmf of $\tilde{y}$ given $\tilde{x}$ is

$$p_{\tilde{y} \mid \tilde{x}}\left(y \mid x\right) := \mathrm{P}\left(\tilde{y} = y \mid \tilde{x} = x\right) \tag{4.51}$$

$$= \frac{p_{\tilde{x}, \tilde{y}}\left(x, y\right)}{p_{\tilde{x}}\left(x\right)}, \tag{4.52}$$

assuming $p_{\tilde{x}}\left(x\right) > 0$. Otherwise, the conditional pmf is undefined.

The conditional pmf $p_{\tilde{b} \mid \tilde{a}}\left(\cdot \mid a\right)$ characterizes our uncertainty about $\tilde{b}$, conditioned on the event $\tilde{a} = a$. This object is a valid pmf of the random variable $\tilde{b}$, since it is nonnegative and

$$\sum_{b \in B} p_{\tilde{b} \mid \tilde{a}}\left(b \mid a\right) = \frac{\sum_{b \in B} p_{\tilde{a}, \tilde{b}}\left(a, b\right)}{p_{\tilde{a}}\left(a\right)} \tag{4.53}$$

$$= \frac{p_{\tilde{a}}\left(a\right)}{p_{\tilde{a}}\left(a\right)} = 1 \tag{4.54}$$

for any a such that $p_{\tilde{a}}\left(a\right) \neq 0$. However, the conditional pmf is *not* a pmf of the random variable we condition on (in this case $\tilde{a}$). Therefore, there is no reason for $\sum_{a \in A} p_{\tilde{b} \mid \tilde{a}}\left(b \mid a\right)$ to add up to one!

The following theorem provides a chain rule for discrete random variables and vectors, which follows directly from Definition 4.13 and the chain rule for probabilities of events (Theorem 1.17).

Theorem 4.14 (Chain rule for discrete random variables and vectors). *For any discrete random variables $\tilde{a}$ and $\tilde{b}$,*

$$p_{\tilde{a},\tilde{b}}(a, b) = p_{\tilde{a}}(a)\, p_{\tilde{b}\,|\,\tilde{a}}(b\,|\,a) \tag{4.55}$$

$$= p_{\tilde{b}}(b)\, p_{\tilde{a}\,|\,\tilde{b}}(a\,|\,b). \tag{4.56}$$

For any d-dimensional random vector $\tilde{x}$ with joint pmf $p_{\tilde{x}}$,

$$p_{\tilde{x}}(x) = p_{\tilde{x}[1]}(x[1])\prod_{i=2}^{d} p_{\tilde{x}[i]\,|\,\tilde{x}[1],\dots,\tilde{x}[i-1]}(x[i]\,|\,x[1],\dots,x[i-1]). \tag{4.57}$$

The order of indices in the random vector is completely arbitrary (any order works).

Example 4.15 (Simple example: Conditional distribution). In this example, we provide some intuition about the definition of the conditional pmf. Consider the random variables $\tilde{a}$ and $\tilde{b}$ in Example 4.6. To condition on $\tilde{a} = 2$, we could be tempted to just fix that value in the joint pmf, which yields

$$p_{\tilde{a},\tilde{b}}(2,1) = 0.1, \qquad p_{\tilde{a},\tilde{b}}(2,2) = 0.05, \qquad p_{\tilde{a},\tilde{b}}(2,3) = 0.2. \tag{4.58}$$

The problem is that this is not a valid pmf because its entries do not add up to one. However, we can ensure that they do, dividing each entry by their sum. The sum is equal to the marginal pmf of $\tilde{a}$ evaluated at 2, $p_{\tilde{a}}(2) = 0.35$, by Theorem 4.10. Consequently, our normalization is exactly equivalent to the formula in Definition 4.13. The resulting conditional pmf equals

$$p_{\tilde{b}\,|\,\tilde{a}}(1\,|\,2) = \frac{2}{7}, \qquad p_{\tilde{b}\,|\,\tilde{a}}(2\,|\,2) = \frac{1}{7}, \qquad p_{\tilde{b}\,|\,\tilde{a}}(3\,|\,2) = \frac{4}{7}. \tag{4.59}$$

The bottom row of Figure 4.5 shows the unnormalized and normalized values.

....

Example 4.16 (Voting data: Joint, marginal, and conditional distributions). In Example 1.25, we use conditional probabilities to analyze voting data, extracted from Dataset 1. Here we use the same data as a simple illustration of joint, marginal, and conditional distributions. Consider the Bernoulli random variables,

$$\tilde{b} = \begin{cases} 1 & \text{Yes on budget,} \\ 0 & \text{No on budget,} \end{cases} \tag{4.60}$$

$$\tilde{d} = \begin{cases} 1 & \text{Yes on duty-free exports,} \\ 0 & \text{No on duty-free exports.} \end{cases} \tag{4.61}$$

Table 4.2 shows the empirical joint pmf of the two random variables, computed from the data in Table 1.2, as well as the marginal and conditional pmfs of each random variable.

The conditional pmfs can be estimated in two different ways. We can approximate $p_{\tilde{b}\,|\,\tilde{d}}$ dividing the joint empirical pmf (see Definition 4.7) of $\tilde{b}$ and $\tilde{d}$ by the empirical pmf (see Definition 2.11) of $\tilde{d}$. Alternatively, for each element d of the range of $\tilde{d}$ (in this case, 0

Table 4.2 **Joint, marginal, and conditional pmfs.** *Joint, marginal, and conditional pmfs of the Bernoulli random variables $\tilde{b}$ and $\tilde{d}$ in Example 4.16.*

d

$p_{\tilde{b},\tilde{d}}$	0	1
0	0.35	0.05
1	0.22	0.38

b (labels the rows above)

$p_{\tilde{b}}$	$p_{\tilde{b}|\tilde{d}}(\cdot|0)$	$p_{\tilde{b}|\tilde{d}}(\cdot|1)$
0.40	0.61	0.12
0.60	0.39	0.88

$p_{\tilde{d}}$	0.57	0.43
$p_{\tilde{d}|\tilde{b}}(\cdot|0)$	0.87	0.13
$p_{\tilde{d}|\tilde{b}}(\cdot|1)$	0.37	0.63

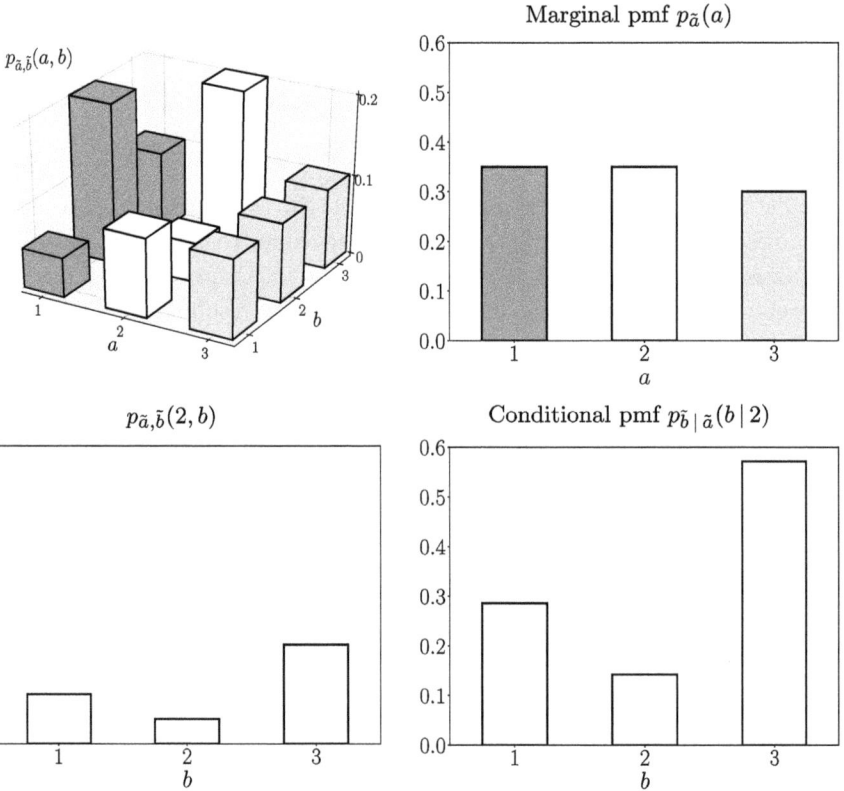

Figure 4.5 Joint, marginal and conditional pmfs. The joint pmf of the random variables in Example 4.6 is represented as a 3D bar plot in the top left graph. The top right plot shows the marginal pmf of $\tilde{a}$. Each entry is obtained by summing over the entries of the joint pmf represented with the same color in the 3D plot. The bottom left bar plot shows the joint pmf $p_{\tilde{a},\tilde{b}}(2,b)$, when we fix $\tilde{a} = 2$ (in white in the 3D bar plot). Dividing each entry by their total sum, which is equal to the marginal pmf of $\tilde{a}$ at 2, yields the conditional pmf of $\tilde{b}$ given $\tilde{a} = 2$, depicted by the bar plot on the bottom right.

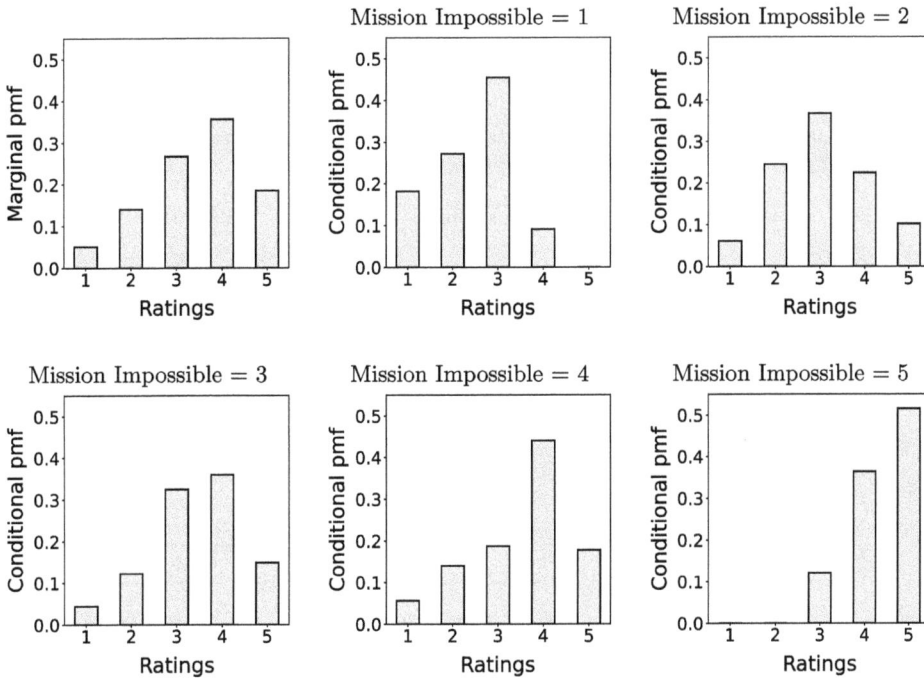

Figure 4.6 Conditional pmfs of movie ratings. The marginal pmf of the ratings for Independence Day is shown on the upper left. The remaining graphs show the conditional pmf of the ratings of Independence Day given different ratings for Mission Impossible (indicated at the top of each graph). Users who like Mission Impossible tend to rate Independence Day higher than users who don't like Mission Impossible.

and 1), we can compute the empirical pmf of $\tilde{b}$ only using the data points such that $\tilde{d} = d$. Exercise 4.2 shows that both approaches are equivalent.

. . . .

Figure 4.6 shows the conditional pmfs of the rating for Independence Day in Example 4.8 given the different ratings for Mission Impossible, and compares them to the marginal pmf. The rating for Mission Impossible provides a lot of information about the rating for Independence Day. Users who like Mission Impossible tend to give Independence Day higher ratings than users who don't like Mission Impossible.

Table 4.3 provides an example of conditional joint pmfs of two random variables given a third one. It shows the conditional joint pmfs of precipitation at each pair of weather stations in Example 4.9, given precipitation at the remaining station, and compares them to the marginal joint pmfs. These conditional joint pmfs are obtained from the joint pmf of the three values by applying Definition 4.13. For example, the conditional joint pmf of $\tilde{x}[1]$ and $\tilde{x}[2]$ given $\tilde{x}[3] = 1$ is

$$p_{\tilde{x}[1],\tilde{x}[2] \mid \tilde{x}[3]}(a,b \mid 1) = \frac{p_{\tilde{x}}(a,b,1)}{p_{\tilde{x}[3]}(1)}, \quad a,b \in \{0,1\}. \tag{4.62}$$

Figure 4.7 compares the marginal pmf of precipitation in Coos Bay (first row) with the conditional pmf given each of the other two stations (second row). It also shows the conditional pmf given both of the other stations (third row). To condition on two stations, we again apply Definition 4.13. For example, the conditional joint pmf of $\tilde{x}[1]$ (Coos Bay) given $\tilde{x}[2] = 0$ and $\tilde{x}[3] = 1$ is

Table 4.3 **Conditional joint pmfs for precipitation data.** *The left column shows the marginal joint pmfs of precipitation at different pairs of weather stations from Example 4.9. The center and right columns show the conditional joint pmfs of each pair given the precipitation status at the remaining station. Conditioning has a substantial effect on the distribution, especially if there is precipitation in the station we condition on. All pmfs are computed from the empirical joint pmf in Figure 4.2.*

	Corvallis				Corvallis				Corvallis	
Coos Bay	**0**	**1**		Coos Bay	**0**	**1**		Coos Bay	**0**	**1**
0	84.7	4.0		**0**	85.7	3.8		**0**	54.9	11.5
1	5.6	5.7		**1**	5.5	5.1		**1**	10.2	23.4

Marginal	**John Day = 0**	**John Day = 1**

	John Day				John Day				John Day	
Coos Bay	**0**	**1**		Coos Bay	**0**	**1**		Coos Bay	**0**	**1**
0	86.4	2.2		**0**	91.7	2.0		**0**	37.3	4.0
1	10.2	1.1		**1**	5.8	0.4		**1**	50.6	8.1

Marginal	**Corvallis = 0**	**Corvallis = 1**

	John Day				John Day				John Day	
Corvallis	**0**	**1**		Corvallis	**0**	**1**		Corvallis	**0**	**1**
0	88.1	2.2		**0**	93.4	2.1		**0**	46.5	3.0
1	8.6	1.2		**1**	4.1	0.4		**1**	43.5	7.0

Marginal	**Coos Bay = 0**	**Coos Bay = 1**

$$p_{\tilde{x}[1] \,|\, \tilde{x}[2], \tilde{x}[3]} \left(a \,|\, 0, 1 \right) = \frac{p_{\tilde{x}}\left(a, 0, 1 \right)}{p_{\tilde{x}[2], \tilde{x}[3]}(0, 1)}, \quad a \in \{0, 1\}. \tag{4.63}$$

The conditional pmfs indicate that the precipitation in Corvallis depends heavily on the precipitation in Coos Bay. The marginal probability of precipitation in Coos Bay is 11.3%. By contrast, the conditional probability is 58.7%, if there is precipitation in Corvallis, and just 6.2%, if there isn't. The dependence on John Day is also substantial, but not as strong. The conditional probability of precipitation in Coos Bay is 33.6%, if there is precipitation in John Day, and 10.6%, if there isn't.

4.4 Independence

When knowledge about a random variable $\tilde{a}$ does not affect our uncertainty about another random variable $\tilde{b}$, we say that $\tilde{a}$ and $\tilde{b}$ are *independent*. This can be expressed in terms of the conditional distributions of the random variables. If $\tilde{a}$ and $\tilde{b}$ are discrete, then for any a and b, we should have

$$p_{\tilde{b} \,|\, \tilde{a}}(b \,|\, a) = \mathrm{P}(\tilde{b} = b \,|\, \tilde{a} = a) \tag{4.64}$$

$$= \mathrm{P}(\tilde{b} = b) \tag{4.65}$$

$$= p_{\tilde{b}}(b), \tag{4.66}$$

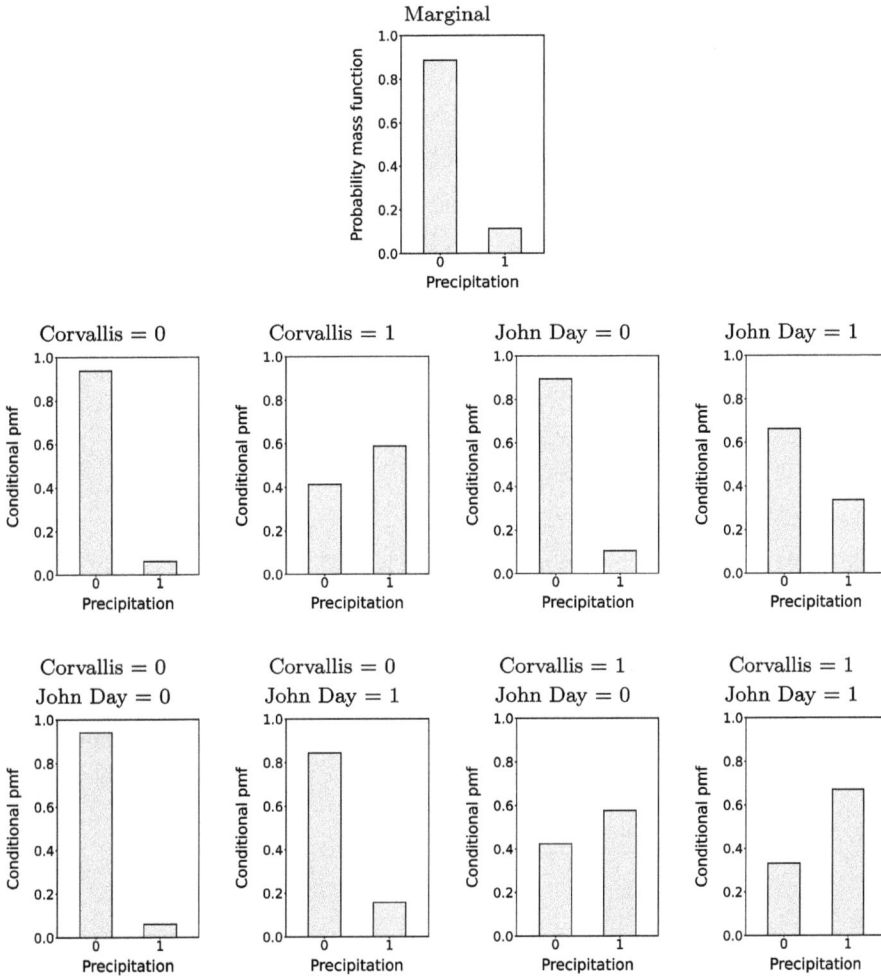

Figure 4.7 Conditional pmfs for precipitation data. The graph on the top shows the marginal pmf of precipitation in Coos Bay in Example 4.9. The second row shows the conditional pmf of precipitation in Coos Bay given different values of the random variable representing precipitation in Corvallis (first and second column), and also given different values of the random variable representing precipitation in John Day (third and fourth columns). The third row shows the pmf conditioned on both random variables representing Corvallis and John Day. All pmfs are computed from the empirical joint pmf in Figure 4.2.

assuming the conditional pmf is well defined (the pmf of $\tilde{b}$ is nonzero at b). Equivalently, by the chain rule (Theorem 4.14),

$$p_{\tilde{a},\tilde{b}}(a,b) = p_{\tilde{a}}(a)p_{\tilde{b}\,|\,\tilde{a}}(b\,|\,a) \tag{4.67}$$
$$= p_{\tilde{a}}(a)p_{\tilde{b}}(b). \tag{4.68}$$

This is consistent with the definition of independence for events (Definition 1.26), as it implies that the events $\tilde{a} = a$ and $\tilde{b} = b$ are independent.

Definition 4.17 (Independent discrete random variables). *Two discrete random variables* $\tilde{a}$ *and* $\tilde{b}$ *with respective ranges* A *and* B, *defined on the same probability space, are independent if and only if*

$$p_{\tilde{a},\tilde{b}}(a,b) = p_{\tilde{a}}(a)\,p_{\tilde{b}}(b) \quad for\ all\ a \in A, b \in B. \tag{4.69}$$

The definition can be extended to multiple random variables and, equivalently, to the entries of a random vector. In order to ensure that all possible conditional probabilities are the same as the marginal probabilities, we require that the joint pmf factorizes completely.

Definition 4.18 (Random vector with independent entries). *The* d *entries* $\tilde{x}[1]$, $\tilde{x}[2]$, ..., $\tilde{x}[d]$ *in a discrete random vector* $\tilde{x}$ *are independent if and only if*

$$p_{\tilde{x}}(x) = \prod_{i=1}^{d} p_{\tilde{x}[i]}(x[i]) \tag{4.70}$$

for all possible values of the entries.

Example 4.19 (Precipitation in Hawaii and Rhode Island). We consider hourly precipitation data measured at weather stations in Hilo (Hawaii) and Kingston (Rhode Island), extracted from Dataset 9. As in Example 4.9, we define a random vector of dimension two, where each entry is a Bernoulli random variable indicating whether there is precipitation in Hilo ($\tilde{x}[1] = 1$) and Kingston ($\tilde{x}[2] = 1$). We compute the empirical joint pmf of the random vector and use it to obtain the marginal and conditional pmfs of its entries. Figure 4.8 shows the pmfs. The marginal probability of precipitation at Hilo is 25.5%. The conditional probabilities

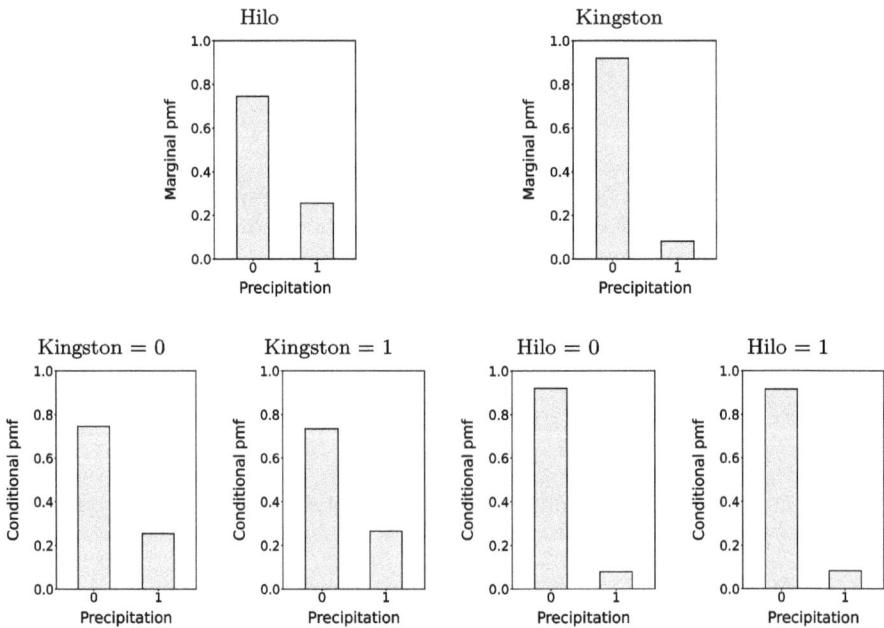

Figure 4.8 Precipitation in Hilo and Kingston is approximately independent. The graph on the top shows the marginal pmf of precipitation in Hilo (left) and Kingston (right) in Example 4.19. The second row shows the conditional pmf of precipitation in Hilo given the precipitation in Kingston (first and second column) and vice versa (third and fourth column). There is barely any change in the distributions when we condition, indicating that precipitation in both stations is approximately independent.

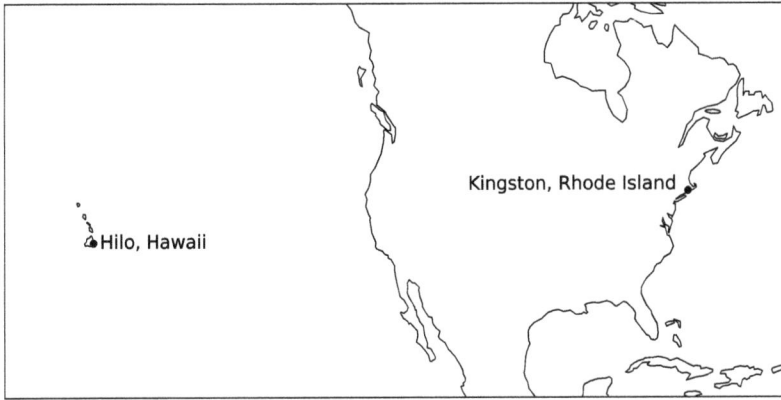

Figure 4.9 Weather stations in Kingston and Hilo. Location of the weather stations of Kingston (Rhode Island) and Hilo (Hawaii).

are very similar: 26.5%, if there is precipitation in Kingston, and 25.4%, if there isn't. The marginal probability of precipitation at Kingston is 8.1%. The conditional probabilities are again very similar: 8.4%, if there is precipitation in Hilo, and 8.0%, if there isn't. Strictly speaking, the marginal and conditional pmfs are not exactly the same, but the changes are very slight. Practically speaking, precipitation in both stations is approximately independent. This is not surprising given their geographical location, shown in Figure 4.9.

. . . .

Independence is a crucial modeling assumption, when we estimate probabilities from data. For instance, the pmf estimators in Sections 2.2 and 2.4 rely on the premise that the data are independent and identically distributed (i.i.d.), as captured by Definition 2.23. The following definition is equivalent.

Definition 4.20 (Independent identically distributed discrete random variables). *Let $\tilde{a}_1$, $\tilde{a}_2$, ..., $\tilde{a}_n$ be discrete random variables belonging to the same probability space. The random variables are identically distributed if their marginal pmfs are the same*

$$p_{\tilde{a}_1} = p_{\tilde{a}_2} = \cdots = p_{\tilde{a}_n} = p_{\tilde{a}}, \tag{4.71}$$

where $\tilde{a}$ is a random variable with the same distribution. The random variables are independent and identically distributed (i.i.d.), if their joint pmf is equal to the product of the marginal pmfs

$$p_{\tilde{a}_1, \tilde{a}_2, \ldots, \tilde{a}_n} (a_1, a_2, \ldots, a_n) = \prod_{i=1}^{d} p_{\tilde{a}} (a_i) \tag{4.72}$$

for all possible values of $a_1, a_2, \ldots, a_n$.

4.5 Conditional Independence

In order to motivate the definition of conditional independence, consider the conditional pmfs in Figure 4.7. Coos Bay is clearly dependent on John Day. The marginal probability of precipitation in Coos Bay is 11.3%. The conditional probability is 33.6%, if there is precipitation in John Day, and 10.6%, if there isn't.

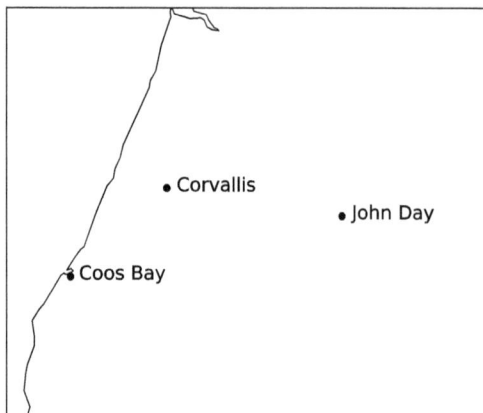

Figure 4.10 Weather stations in Oregon. Location of the weather stations of Coos Bay, Corvallis, and John Day in Oregon.

Now, let us condition on the precipitation in Corvallis. In that case, the precipitation in Coos Bay is much less dependent on the precipitation in John Day. Given precipitation in Corvallis, the conditional probability of precipitation in Coos Bay is 58.7%. Given precipitation in both Corvallis and John Day, the conditional probability is 67.0%. Given precipitation in Corvallis and no precipitation in John Day, the conditional probability is 57.6%. Similarly, if there is no precipitation in Corvallis, the conditional probability of precipitation in Coos Bay (6.2%) does not change very much when we also condition on the precipitation in John Day (15.6% if there is precipitation in John Day, and 6.0% if there isn't).

In summary, the dependence between Coos Bay and John Day is *greatly reduced when we condition on Corvallis*. This makes sense given their geographical locations; Corvallis is situated between them (see the map in Figure 4.10). Consequently, once we know whether there is precipitation in Corvallis, the precipitation in John Day is not as informative about the precipitation in Coos Bay. Conditional independence occurs when this effect is taken to the extreme.

Two random variables $\tilde{a}$ and $\tilde{b}$ are conditionally independent given a third random variable $\tilde{c}$, if there is no dependence between $\tilde{a}$ and $\tilde{b}$, *as long as the value of $\tilde{c}$ is known*. If $\tilde{a}$ and $\tilde{b}$ are discrete, for this to be the case, the conditional pmf $p_{\tilde{b} \mid \tilde{c}}$ of $\tilde{b}$ given $\tilde{c}$ should equal the conditional pmf $p_{\tilde{b} \mid \tilde{a}, \tilde{c}}$ of $\tilde{b}$ given both $\tilde{a}$ and $\tilde{c}$. Equivalently, by the chain rule (Theorem 4.14) applied to the conditional joint pmf $p_{\tilde{a}, \tilde{b} \mid \tilde{c}}$ of $\tilde{a}$ and $\tilde{b}$ given $\tilde{c}$,

$$p_{\tilde{a}, \tilde{b} \mid \tilde{c}} = p_{\tilde{a} \mid \tilde{c}} \, p_{\tilde{b} \mid \tilde{a}, \tilde{c}} \tag{4.73}$$

$$= p_{\tilde{a} \mid \tilde{c}} \, p_{\tilde{b} \mid \tilde{c}}. \tag{4.74}$$

This is consistent with the definition of conditional independence for events (Definition 1.31), as it implies that the events $\tilde{a} = a$ and $\tilde{b} = b$ are conditional independent given $\tilde{c} = c$, for any values of a, b, and c (such that $\mathrm{P}\left(\tilde{c} = c\right)$ is nonzero).

Definition 4.21 (Conditionally independent discrete random variables). *Two discrete random variables $\tilde{a}$ and $\tilde{b}$ with respective ranges A and B, defined on the same probability space, are conditionally independent given a random variable $\tilde{c}$ with range C, if and only if*

$$p_{\tilde{a}, \tilde{b} \mid \tilde{c}}\left(a, b \mid c\right) = p_{\tilde{a} \mid \tilde{c}}\left(a \mid c\right) p_{\tilde{b} \mid \tilde{c}}\left(b \mid c\right) \quad \textit{for all } a \in A, b \in B, c \in C. \tag{4.75}$$

The d entries $\tilde{x}[1]$, $\tilde{x}[2]$, ..., $\tilde{x}[d]$ in a discrete random vector $\tilde{x}$ are conditionally independent given a random variable $\tilde{a}$, if and only if

$$p_{\tilde{x}\,|\,\tilde{a}}\,(x\,|\,a) = \prod_{i=1}^{d} p_{\tilde{x}[i]\,|\,\tilde{a}}\,(x[i]\,|\,a) \tag{4.76}$$

for all possible values of x and a.

As discussed in Section 1.6, independence does *not* imply conditional independence or vice versa.

4.6 Causal Inference

Causal inference aims to identify causal effects from data, which is the ultimate goal in many data-science applications. In Section 4.6.1, we introduce the framework of potential outcomes, which allows us to define causal relationships formally using random variables. Section 4.6.2 describes Simpson's paradox, which illustrates the difficulty of performing causal inference in the presence of confounding factors. Section 4.6.3 explains how to neutralize confounding factors via randomization. Finally, in Section 4.6.4, we explain how to control for confounding factors, when randomization is not possible.

4.6.1 Potential Outcomes

Imagine that we are interested in evaluating the efficacy of a drug. Let $\tilde{t}$ be a Bernoulli random variable indicating whether a patient receives the drug ($\tilde{t} = 1$) or not ($\tilde{t} = 0$), and let $\tilde{y}$ be another Bernoulli random variable indicating whether the patient recovers ($\tilde{y} = 1$) or not ($\tilde{y} = 0$). If $\mathrm{P}(\tilde{y} = 1\,|\,\tilde{t} = 1)$ is larger than $\mathrm{P}(\tilde{y} = 1\,|\,\tilde{t} = 0)$, then treated patients recover more often than untreated patients. Can we conclude that the treatment is causing the recovery? No! To explain why, we introduce the concept of potential outcomes.[1]

In order to characterize the causal effect of a treatment, we consider two hypothetical scenarios: One where we treat all patients, and another where we treat no patients at all. The potential outcomes are random variables associated with each of these alternative scenarios. The potential outcome $\widetilde{po}_0$ indicates whether a patient would recover ($\widetilde{po}_0 = 1$) or not ($\widetilde{po}_0 = 0$) in the absence of treatment, *regardless of whether they are actually treated or not*. Similarly, the potential outcome $\widetilde{po}_1$ indicates whether a patient would recover ($\widetilde{po}_1 = 1$) or not ($\widetilde{po}_1 = 0$) if they were treated and is defined even if the patient is not treated.

In reality, both potential outcomes cannot be observed simultaneously: We cannot treat and not treat a patient at the same time! This is known as *the fundamental problem of causal inference* and is the main reason why causal inference is so challenging. As illustrated in Table 4.4, the observed outcome $\tilde{y}$ is equal to either $\widetilde{po}_0$ or $\widetilde{po}_1$, depending on the treatment $\tilde{t}$,

$$\tilde{y} := \begin{cases} \widetilde{po}_0 & \text{if} \quad \tilde{t} = 0, \\ \widetilde{po}_1 & \text{if} \quad \tilde{t} = 1. \end{cases} \tag{4.77}$$

---

[1] Notice that, here, outcome does not refer to an element in a sample space, as in Section 1.2. Instead, it is a random variable representing a quantity of interest that may (or may not) be affected by the treatment.

Table 4.4 ***Potential outcomes.*** *In order to characterize causal effects, we model the potential outcomes $\widetilde{po}_0$ and $\widetilde{po}_1$ of a treatment $\tilde{t}$ probabilistically. However, in practice, only one of the potential outcomes is observed. If $\tilde{t} = 0$ (indicated by a cross) we observe $\tilde{y} := \widetilde{po}_0$. If $\tilde{t} = 1$ (indicated by a check mark) we observe $\tilde{y} := \widetilde{po}_1$. In this example, the outcomes are either positive (smiley face) or negative (frownie face).*

Treatment	Observed outcome	Outcome if not treated	Outcome if treated
$\tilde{t}$	$\tilde{y}$	$\widetilde{po}_0$	$\widetilde{po}_1$
✗	☹	☹	?
✗	☺	☺	?
✓	☺	?	☺
✓	☹	?	☹
✓	☺	?	☺

If $\tilde{t} = 0$, we observe $\widetilde{po}_0$, but not $\widetilde{po}_1$. In that case we call $\widetilde{po}_1$ a *counterfactual* because it describes a situation that is counter to factual reality. Conversely, if $\tilde{t} = 1$, we only observe $\widetilde{po}_1$, and $\widetilde{po}_0$ is an unobserved counterfactual.

By modeling the potential outcomes, we can determine whether the treatment $\tilde{t}$ makes a difference or not. We consider that $\tilde{t}$ has a causal effect on $\tilde{y}$, if

$$P\left(\widetilde{po}_0 = 1\right) \neq P\left(\widetilde{po}_1 = 1\right), \tag{4.78}$$

because the probability of recovery is different depending on whether the patient takes the drug or not. Unfortunately, it is not always possible to compute these probabilities from the available data, especially in observational studies, where we have no control over which patients are selected to receive the treatment.

Example 4.22 (Drug efficacy: Observational study). We consider a fictitious observational study that evaluates a new drug. The treatment is represented by the random variable $\tilde{t}$. If $\tilde{t} = 1$, the patient is given the drug, so they belong to the *treatment group*. If $\tilde{t} = 0$, they are not treated, so they belong to the *control group*. The outcome of interest is recovery from a certain disease, represented by a random variable $\tilde{y}$, defined as in (4.77). The potential outcomes represent recovery with ($\widetilde{po}_1 = 1$) and without ($\widetilde{po}_0 = 1$) the drug.

It turns out that the efficacy of the drug depends on the age of the patients, but this is unknown to us. The random variable $\tilde{a}$ indicates whether a patient is old ($\tilde{a} = \text{old}$) or young ($\tilde{a} = \text{young}$). Half of the patients are young and half are old, so $p_{\tilde{a}}(\text{old}) = p_{\tilde{a}}(\text{young}) = 0.5$. For young patients, the recovery rate is relatively high (80%), as they are generally healthier, but the drug is completely useless:

$$p_{\widetilde{po}_0 \mid \tilde{a}}(1 \mid \text{young}) = p_{\widetilde{po}_1 \mid \tilde{a}}(1 \mid \text{young}) := 0.8. \tag{4.79}$$

For old patients, the recovery rate is lower and the drug is quite effective:

$$p_{\widetilde{po}_0 \mid \tilde{a}}(1 \mid \text{old}) := 0.2, \qquad p_{\widetilde{po}_1 \mid \tilde{a}}(1 \mid \text{old}) := 0.4. \tag{4.80}$$

Our goal is to determine the overall efficacy of the drug across all patients, which can be quantified by comparing $p_{\widetilde{po}_0}(1)$ and $p_{\widetilde{po}_1}(1)$. By Theorem 4.10 and the chain rule (Theorem 4.14),

$$p_{\widetilde{po}_0}(1) = p_{\tilde{a},\widetilde{po}_0}(\text{young}, 1) + p_{\tilde{a},\widetilde{po}_0}(\text{old}, 1) \tag{4.81}$$

$$= p_{\tilde{a}}(\text{young})p_{\widetilde{po}_0 \mid \tilde{a}}(1 \mid \text{young}) + p_{\tilde{a}}(\text{old})p_{\widetilde{po}_0 \mid \tilde{a}}(1 \mid \text{old}) \tag{4.82}$$

$$= 0.5 \cdot 0.8 + 0.5 \cdot 0.2 = 0.5. \tag{4.83}$$

Similarly,

$$p_{\widetilde{po}_1}(1) = p_{\tilde{a}}(\text{young})p_{\widetilde{po}_1 \mid \tilde{a}}(1 \mid \text{young}) + p_{\tilde{a}}(\text{old})p_{\widetilde{po}_1 \mid \tilde{a}}(1 \mid \text{old}) \tag{4.84}$$

$$= 0.5 \cdot 0.8 + 0.5 \cdot 0.4 = 0.6. \tag{4.85}$$

Notice that these probabilities do *not* depend on what patients are actually treated: they represent the recovery rate if nobody is treated ($p_{\widetilde{po}_0}(1)$) or if everyone is treated ($p_{\widetilde{po}_1}(1)$). The drug is quite effective, it increases the chance of recovery by 10%:

$$p_{\widetilde{po}_1}(1) - p_{\widetilde{po}_0}(1) = 0.1. \tag{4.86}$$

The key question is whether we can estimate the efficacy reliably from the data in our observational study. We only have access to the observed outcome $\tilde{y}$, which equals $\widetilde{po}_1$ for the patients in the treatment group, and $\widetilde{po}_0$ for the patients in the control group. This allows us to compute the conditional probabilities of the potential outcomes given the treatment, since by (4.77)

$$p_{\tilde{y} \mid \tilde{t}}(1 \mid 0) = p_{\widetilde{po}_0 \mid \tilde{t}}(1 \mid 0), \tag{4.87}$$

$$p_{\tilde{y} \mid \tilde{t}}(1 \mid 1) = p_{\widetilde{po}_1 \mid \tilde{t}}(1 \mid 1). \tag{4.88}$$

Unfortunately, these observed probabilities can be very different from $p_{\widetilde{po}_0}(1)$ and $p_{\widetilde{po}_1}(1)$. The reason is that they depend on the composition of the treatment and control groups. We define

$$\gamma_{\text{control}} := p_{\tilde{a} \mid \tilde{t}}(\text{young} \mid 0), \tag{4.89}$$

$$\gamma_{\text{treatment}} := p_{\tilde{a} \mid \tilde{t}}(\text{young} \mid 1) \tag{4.90}$$

to represent the conditional probability that a patient is young given that they are untreated, or treated, respectively. Equivalently, this is the fraction of the control and treatment groups that are young. We assume that the treatment $\tilde{t}$ and the potential outcomes (both $\widetilde{po}_0$ and $\widetilde{po}_1$) are conditionally independent given $\tilde{a}$ (see Definition 4.21). As we discuss in Example 4.26, this means that there is no systematic difference between the treated and untreated patients, once we condition on age. Under this assumption, by Theorem 4.10 (applied to the conditional joint pmf of $\widetilde{po}_0$ and $\tilde{a}$ given $\tilde{t}$) and the chain rule (Theorem 4.14),

$$p_{\widetilde{po}_0 \mid \tilde{t}}(1 \mid 0) \tag{4.91}$$

$$= p_{\widetilde{po}_0, \tilde{a} \mid \tilde{t}}(1, \text{young} \mid 0) + p_{\widetilde{po}_0, \tilde{a} \mid \tilde{t}}(1, \text{old} \mid 0) \tag{4.92}$$

$$= p_{\tilde{a} \mid \tilde{t}}(\text{young} \mid 0)p_{\widetilde{po}_0 \mid \tilde{a}, \tilde{t}}(1 \mid \text{young}, 0) + p_{\tilde{a} \mid \tilde{t}}(\text{old} \mid 0)p_{\widetilde{po}_0 \mid \tilde{a}, \tilde{t}}(1 \mid \text{old}, 0) \tag{4.93}$$

$$= p_{\tilde{a} \mid \tilde{t}}(\text{young} \mid 0)p_{\widetilde{po}_0 \mid \tilde{a}}(1 \mid \text{young}) + p_{\tilde{a} \mid \tilde{t}}(\text{old} \mid 0)p_{\widetilde{po}_0 \mid \tilde{a}}(1 \mid \text{old}) \tag{4.94}$$

$$= \gamma_{\text{control}}0.8 + (1 - \gamma_{\text{control}})0.2. \tag{4.95}$$

By the same argument,

$$p_{\widetilde{po}_1 \mid \tilde{t}}(1 \mid 1) \tag{4.96}$$

$$= p_{\tilde{a} \mid \tilde{t}}(\text{young} \mid 1) p_{\widetilde{po}_1 \mid \tilde{a}}(1 \mid \text{young}) + p_{\tilde{a} \mid \tilde{t}}(\text{old} \mid 1) p_{\widetilde{po}_1 \mid \tilde{a}}(1 \mid \text{old}) \tag{4.97}$$

$$= \gamma_{\text{treatment}} 0.8 + (1 - \gamma_{\text{treatment}}) 0.4. \tag{4.98}$$

The observed efficacy of the drug depends on γ_{control} and $\gamma_{\text{treatment}}$. Imagine that the treatment is optional and that the old patients are distrustful of new drugs, so that $\gamma_{\text{control}} := 0.4$ and $\gamma_{\text{treatment}} := 0.7$. Then the observed probabilities of recovery equal

$$p_{\tilde{y} \mid \tilde{t}}(1 \mid 0) = p_{\widetilde{po}_0 \mid \tilde{t}}(1 \mid 0) \tag{4.99}$$

$$= 0.4 \cdot 0.8 + 0.6 \cdot 0.2 = 0.44, \tag{4.100}$$

$$p_{\tilde{y} \mid \tilde{t}}(1 \mid 1) = p_{\widetilde{po}_1 \mid \tilde{t}}(1 \mid 1) \tag{4.101}$$

$$= 0.7 \cdot 0.8 + 0.3 \cdot 0.4 = 0.68. \tag{4.102}$$

From the observed data, it seems that the drug is more than twice as effective than it really is

$$p_{\tilde{y} \mid \tilde{t}}(1 \mid 1) - p_{\tilde{y} \mid \tilde{t}}(1 \mid 0) = 0.24 \tag{4.103}$$

The problem is that young patients are underrepresented in the control group and overrepresented in the treatment group. This boosts the apparent efficacy because those patients are more likely to recover *regardless of the treatment*. In this observational study, age acts as a *confounding factor* or *confounder*, which completely distorts the observed effect of the drug.

4.6.2 Simpson's Paradox

In Example 4.22, the age of the patients $\tilde{a}$ occludes the underlying causal relationship between recovery and treatment, acting as a confounder. In this section, we show that confounders are behind a notorious phenomenon in causal inference called Simpson's paradox.

While looking at data from NBA games in the 2014/2015 season (Dataset 10), I was puzzled to find that Courtney Lee from the Memphis Grizzlies had a better 3-point percentage (43.9%) than Stephen Curry (41.7%).[2] In case you don't follow basketball, Stephen Curry is considered the best 3-point shooter of all time. During the 2014/2015 season, he led the Golden State Warriors to an NBA championship and was declared the NBA Most Valuable Player, mostly due to his 3-point shooting prowess.

Let $\tilde{y}$ be a Bernoulli random variable representing whether a 3-point shot is made ($\tilde{y} = 1$) or not ($\tilde{y} = 0$). We are interested in determining the causal effect of the shooter being Lee or Curry. To this end, we interpret the player taking the shot as a *treatment* $\tilde{t}$. As you can see, in causal inference, the term treatment is often used in a figurative sense. According to the data, shown in Table 4.5,

$$0.439 = P\left(\tilde{y} = 1 \mid \tilde{t} = \text{Lee}\right) > P\left(\tilde{y} = 1 \mid \tilde{t} = \text{Curry}\right) = 0.417, \tag{4.104}$$

---

[2] Dataset 10 does not contain all games from the 2014/2015 season, so the percentages reported in Table 4.5 differ from the official percentages.

Table 4.5 **Who is the better shooter?** *The table reports the three-point shooting percentages of Stephen Curry and Courtney Lee in the 2014/2015 NBA season. Lee has a better overall percentage (highlighted in bold in the right column), but Curry has better percentages both for long threes (taken at a distance of more than 24 feet) and for short threes (highlighted in bold in the middle column). This apparent paradox is due to Curry shooting a larger proportion of long threes, which are more difficult to make for both players.*

	Stephen Curry	Courtney Lee
Short threes ($\leq$ 24 feet)	45/90 = **50.0%**	56/116 = 48.3%
Long threes (> 24 feet)	145/366 = **39.6%**	19/55 = 34.5%
Total	190/456 = 41.7%	75/171 = **43.9%**

which suggests that Lee shoots better than Curry! With all due respect to Lee, who was a fantastic player, this is very unexpected.

Let us dig a bit deeper into the data. The dataset indicates at what distance from the basket each shot was taken. We divide the 3-point shots in the dataset into short threes, taken at a distance 24 feet or less, and long threes, taken at a distance of more than 24 feet. For context, in the NBA the 3-point line is 23 feet and 9 inches away from the basket.[3] Surprisingly, for each type of shot, the pattern is exactly the opposite of what we observe in the aggregated data! Curry's rate is better for long threes, and also for short threes,

$$0.500 = P(\tilde{y} = 1 \mid \tilde{t} = \text{Curry}, \tilde{d} = \text{short}) > P(\tilde{y} = 1 \mid \tilde{t} = \text{Lee}, \tilde{d} = \text{short}) = 0.483,$$
$$0.396 = P(\tilde{y} = 1 \mid \tilde{t} = \text{Curry}, \tilde{d} = \text{long}) > P(\tilde{y} = 1 \mid \tilde{t} = \text{Lee}, \tilde{d} = \text{long}) = 0.345,$$

where $\tilde{d}$ is a random variable indicating whether the shot is long or short. This phenomenon, where a trend observed in subsets of the data (here, short threes and long threes) is reversed when the data are aggregated, is known as *Simpson's paradox*.

Simpson's paradox occurs due to the presence of a confounder. In our example, the confounder is the shot distance $\tilde{d}$. Lee takes many more short threes than Curry. The corresponding empirical conditional probabilities equal

$$0.678 = P(\tilde{d} = \text{short} \mid \tilde{t} = \text{Lee}) > P(\tilde{d} = \text{short} \mid \tilde{t} = \text{Curry}) = 0.197. \quad (4.105)$$

Even though he doesn't shoot short threes as well as Curry, this shot selection boosts Lee's overall 3-point rate because long threes are more difficult for both players. The following decomposition of the conditional probabilities of $\tilde{y}$ given $\tilde{t}$ reveals the effect of the confounder $\tilde{d}$. The decomposition follows from Theorem 4.10 (applied to the conditional joint pmf of $\tilde{d}$ and $\tilde{y}$ given $\tilde{t}$), and the chain rule (Theorem 4.14):

$$P\left(\tilde{y} = 1 \mid \tilde{t} = \text{Lee}\right) \qquad (4.106)$$
$$= P(\tilde{y} = 1, \tilde{d} = \text{short} \mid \tilde{t} = \text{Lee}) + P(\tilde{y} = 1, \tilde{d} = \text{long} \mid \tilde{t} = \text{Lee}) \qquad (4.107)$$

---

[3] At the corners it is closer, only 22 feet away.

$$= P(\tilde{d} = \text{short} \mid \tilde{t} = \text{Lee})P(\tilde{y} = 1 \mid \tilde{d} = \text{short}, \tilde{t} = \text{Lee})$$

$$+ P(\tilde{d} = \text{long} \mid \tilde{t} = \text{Lee})P(\tilde{y} = 1 \mid \tilde{d} = \text{long}, \tilde{t} = \text{Lee}) \tag{4.108}$$

$$= \mathbf{0.678} \cdot 0.483 + \mathbf{0.322} \cdot 0.345 = 0.439, \tag{4.109}$$

$$P\left(\tilde{y} = 1 \mid \tilde{t} = \text{Curry}\right) \tag{4.110}$$

$$= P(\tilde{y} = 1, \tilde{d} = \text{short} \mid \tilde{t} = \text{Curry}) + P(\tilde{y} = 1, \tilde{d} = \text{long} \mid \tilde{t} = \text{Curry}) \tag{4.111}$$

$$= P(\tilde{d} = \text{short} \mid \tilde{t} = \text{Curry})P(\tilde{y} = 1 \mid \tilde{d} = \text{short}, \tilde{t} = \text{Curry})$$

$$+ P(\tilde{d} = \text{long} \mid \tilde{t} = \text{Curry})P(\tilde{y} = 1 \mid \tilde{d} = \text{long}, \tilde{t} = \text{Curry}) \tag{4.112}$$

$$= \mathbf{0.197} \cdot 0.500 + \mathbf{0.803} \cdot 0.396 = 0.417. \tag{4.113}$$

Even though the conditional probability of $\tilde{y} = 1$ given $\tilde{d} = \text{short}$ and given $\tilde{d} = \text{long}$ are both larger for Curry, they are reweighted by the conditional probability of $\tilde{d}$ given $\tilde{t}$ (in bold) to yield a higher overall 3-point rate for Lee.

4.6.3 Randomized Experiments

Example 4.22 and Section 4.6.2 show that confounders can make it very challenging to perform causal inference from observational data. Fortunately, their effect can be neutralized by ensuring that the treatment is independent from the potential outcomes. In that case, we can easily estimate the probability of the potential outcomes from data.

Theorem 4.23 (Independence between potential outcomes and treatment enables causal inference). *Let $\tilde{y}$ be a random variable representing an observed outcome corresponding to two potential outcomes $\widetilde{\text{po}}_0$ and $\widetilde{\text{po}}_1$ associated with a treatment $\tilde{t}$,*

$$\tilde{y} := \begin{cases} \widetilde{\text{po}}_0 & \text{if} \quad \tilde{t} = 0, \\ \widetilde{\text{po}}_1 & \text{if} \quad \tilde{t} = 1. \end{cases} \tag{4.114}$$

If $\tilde{t}$ and $\widetilde{\text{po}}_0$ are independent, and $\tilde{t}$ and $\widetilde{\text{po}}_1$ are also independent, then

$$p_{\widetilde{\text{po}}_0}(1) = p_{\tilde{y} \mid \tilde{t}}(1 \mid 0), \tag{4.115}$$

$$p_{\widetilde{\text{po}}_1}(1) = p_{\tilde{y} \mid \tilde{t}}(1 \mid 1). \tag{4.116}$$

Proof By definition, $p_{\widetilde{\text{po}}_0 \mid \tilde{t}}(1 \mid 0) = p_{\tilde{y} \mid \tilde{t}}(1 \mid 0)$ and $p_{\widetilde{\text{po}}_1 \mid \tilde{t}}(1 \mid 1) = p_{\tilde{y} \mid \tilde{t}}(1 \mid 1)$. By the independence assumption (see Section 4.4), $p_{\widetilde{\text{po}}_0}(1) = p_{\widetilde{\text{po}}_0 \mid \tilde{t}}(1 \mid 0)$ and $p_{\widetilde{\text{po}}_1}(1) = p_{\widetilde{\text{po}}_1 \mid \tilde{t}}(1 \mid 1)$. ∎

The simplest and most effective way to render the treatment independent of the potential outcomes is to perform a *randomized experiment.*[4] The subjects in such an experiment are randomly divided into a treatment group and a control group. Each subject is assigned to one of the groups, independently from the remaining subjects, ensuring that the independence condition in Theorem 4.23 is satisfied.

In a randomized experiment, the members of the treatment and control groups are very unlikely to differ systematically, which neutralizes any possible confounders, even if we are not aware of their existence! In medicine, these experiments are known as *randomized*

---

[4] Here randomized experiment refers to an approach to gather data, as opposed to the virtual experiment in Section 1.1, which we use as a thought experiment to gain intuition about the properties of probability.

controlled trials. When possible, the trials are double blind, meaning that the participants and the personnel involved in the trials do not know what group each patient belongs to. This is necessary to avoid placebo effects, which could jeopardize the independence between the potential outcomes and the treatment.

Example 4.24 (Drug efficacy: Randomized controlled trial)**.** New drugs have to undergo strict randomized controlled trials in order to be officially approved. In this example, we analyze a randomized controlled trial to evaluate the drug in Example 4.22. Each participant in the trial is administered the drug independently at random with probability $1/2$. As a result, the treatment is independent from the potential outcomes. This might sound confusing at first. How can the treatment be independent of the outcome?

The answer is that randomization renders the *potential* outcomes independent from the treatment, but *not* the *observed* outcome. In our example, if the treatment is randomized then

$$p_{\widetilde{po}_1 \mid \tilde{t}}(1 \mid 0) = p_{\widetilde{po}_1 \mid \tilde{t}}(1 \mid 1) = p_{\widetilde{po}_1}(1) = 0.6. \tag{4.117}$$

This means that the probability that the drug works is the same (0.6) for patients in the treatment group and in the control group. Similarly,

$$p_{\widetilde{po}_0 \mid \tilde{t}}(1 \mid 0) = p_{\widetilde{po}_0 \mid \tilde{t}}(1 \mid 1) = p_{\widetilde{po}_0}(1) = 0.5. \tag{4.118}$$

However, the observed outcome $\tilde{y}$ is definitely not independent from the treatment. It is equal to the potential outcome $p_{\widetilde{po}_0}$ when $\tilde{t} = 0$, and to $p_{\widetilde{po}_1}$ when $\tilde{t} = 1$, so

$$p_{\tilde{y} \mid \tilde{t}}(1 \mid 0) = p_{\widetilde{po}_0 \mid \tilde{t}}(1 \mid 0) = 0.5, \tag{4.119}$$

$$p_{\tilde{y} \mid \tilde{t}}(1 \mid 1) = p_{\widetilde{po}_1 \mid \tilde{t}}(1 \mid 1) = 0.6. \tag{4.120}$$

This confirms that thanks to randomization, the conditional probabilities of the observed outcome given the treatment provide an accurate estimate of the corresponding potential outcomes, as established in Theorem 4.23.

Notice that, in contrast to the observational study in Example 4.22, in the randomized control trial, the age composition of the control and treatment groups is *the same*. Specifically, since subjects are randomly assigned independently from their age, the random variables $\tilde{a}$ and $\tilde{t}$ are independent, so that

$$\gamma_{\text{control}} := p_{\tilde{a} \mid \tilde{t}}(\text{young} \mid 0) = p_{\tilde{a} \mid \tilde{t}}(\text{young}) = 0.5, \tag{4.121}$$

$$\gamma_{\text{treatment}} := p_{\tilde{a} \mid \tilde{t}}(\text{young} \mid 1) = p_{\tilde{a} \mid \tilde{t}}(\text{young}) = 0.5. \tag{4.122}$$

This occurs automatically, even if we don't know that age is acting as a confounder. This showcases why randomization is so effective: The random assignment automatically balances the treatment and control groups with respect to all possible confounders, whether known or unknown!

Randomized controlled trials are essential in the development of vaccines, as illustrated by the following real-world example.

Example 4.25 (COVID-19 vaccine)**.** Vaccines protecting against COVID-19 played a crucial role in slowing the spread of the disease and reducing its severity. Randomized controlled trials were the main tool used to evaluate their efficacy and safety. In the case of the Pfizer vaccine, 43,448 patients were randomly divided into a treatment group of 21,720 patients, who received two doses of the vaccine, and a control group of 21,728, who received two placebo doses.

The number of COVID-19 cases with onset at least 7 days after the second dose were 8 for the treatment group (0.037%), and 162 for the control group (0.746%) (Polack et al., 2020). The randomized assignment guarantees that the only systematic difference between the two groups was the vaccine, which makes the results very convincing. In Example 10.33 and Figure 10.14, we apply hypothesis testing and confidence intervals to further analyze these data.

$\cdots\cdots\cdots\cdots\cdots\cdots\cdots\cdots\cdots\cdots\cdots\cdots\cdots\cdots\cdots\cdots\cdots\cdots\cdots\cdots\cdots\cdots\cdots\cdots\cdots$

4.6.4 Adjusting for Confounders

As explained in Section 4.6.3, randomization enables us to perform causal inference without having to worry about confounders. Unfortunately, randomized experiments have several important limitations. First, they are very expensive. Randomized controlled trials for new drugs typically cost millions of dollars. Second, randomization cannot always be applied due to ethical reasons. For example, it is unethical to administer treatments known to have negative effects, such as smoking or playing American football without a helmet. Third, in some situations, it is impossible to assign the treatment; all we have is observational data. This is definitely the case in our basketball example; choosing whether Curry or Lee takes each shot during a game is obviously not an option (they didn't even play in the same team!).

The following example shows that it is possible to *adjust* or *control* for the effect of a confounder as long as (1) we know about it, and (2) the treatment is conditionally independent from the potential outcomes given the confounder (see Section 4.5). This conditional-independence assumption guarantees that there are no systematic differences between untreated and treated subjects associated with each possible value of the confounder.

Example 4.26 (Drug efficacy: Adjusting for age). Imagine that the observational data in Example 4.22 is all we have. We cannot perform a randomized controlled trial. However, we do have access to the age of each patient. How can we use this information to improve our estimate of the drug efficacy?

By (4.82), we can decompose the probability of recovery for untreated patients $p_{\widetilde{po}_0}(1)$ as follows:

$$p_{\widetilde{po}_0}(1) = p_{\tilde{a}}(\text{young})p_{\widetilde{po}_0 \mid \tilde{a}}(1 \mid \text{young}) + p_{\tilde{a}}(\text{old})p_{\widetilde{po}_0 \mid \tilde{a}}(1 \mid \text{old}). \qquad (4.123)$$

If we are able to estimate the different terms of this equation from the data, we can combine them to approximate $p_{\widetilde{po}_0}(1)$. The marginal pmf $p_{\tilde{a}}$ is easily determined by computing what fraction of the patients are young or old. The key question is how to obtain $p_{\widetilde{po}_0 \mid \tilde{a}}(1 \mid \text{young})$ and $p_{\widetilde{po}_0 \mid \tilde{a}}(1 \mid \text{old})$. The observed rate of recovery for untreated young patients is

$$p_{\tilde{y} \mid \tilde{a}, \tilde{t}}(1 \mid \text{young}, 0) = p_{\widetilde{po}_0 \mid \tilde{a}, \tilde{t}}(1 \mid \text{young}, 0). \qquad (4.124)$$

In order for this to equal $p_{\widetilde{po}_0 \mid \tilde{a}}(1 \mid \text{young})$, $\widetilde{po}_0$ must be conditionally independent from $\tilde{t}$ given $\tilde{a} = \text{young}$ (see Definition 4.21), which ensures that

$$p_{\widetilde{po}_0 \mid \tilde{a}, \tilde{t}}(1 \mid \text{young}, 0) = p_{\widetilde{po}_0 \mid \tilde{a}}(1 \mid \text{young}). \qquad (4.125)$$

In that case, the observed rate of recovery for young patients $p_{\tilde{y}\,|\,\tilde{a},\tilde{t}}(1\,|\,\text{young}, 0)$ is equal to the true conditional rate $p_{\widetilde{\text{po}}_0\,|\,\tilde{a}}(1\,|\,\text{young}) = 0.8$!

Similarly, if we assume that $\widetilde{\text{po}}_0$ is conditionally independent from $\tilde{t}$ given $\tilde{a} = \text{old}$, the observed rate of recovery also reveals the conditional distribution of the potential outcome for untreated old patients,

$$p_{\tilde{y}\,|\,\tilde{a},\tilde{t}}(1\,|\,\text{old}, 0) = p_{\widetilde{\text{po}}_0\,|\,\tilde{a},\tilde{t}}(1\,|\,\text{old}, 0) = p_{\widetilde{\text{po}}_0\,|\,\tilde{a}}(1\,|\,\text{old}) = 0.2. \tag{4.126}$$

By the same argument, if $\widetilde{\text{po}}_1$ is also conditionally independent from $\tilde{t}$ given $\tilde{a}$,

$$p_{\tilde{y}\,|\,\tilde{a},\tilde{t}}(1\,|\,\text{young}, 1) = p_{\widetilde{\text{po}}_1\,|\,\tilde{a},\tilde{t}}(1\,|\,\text{young}, 1) = p_{\widetilde{\text{po}}_1\,|\,\tilde{a}}(1\,|\,\text{young}) = 0.8, \tag{4.127}$$

$$p_{\tilde{y}\,|\,\tilde{a},\tilde{t}}(1\,|\,\text{old}, 1) = p_{\widetilde{\text{po}}_1\,|\,\tilde{a},\tilde{t}}(1\,|\,\text{old}, 1) = p_{\widetilde{\text{po}}_1\,|\,\tilde{a}}(1\,|\,\text{old}) = 0.4. \tag{4.128}$$

Crucially, this reasoning is correct only when patient recovery is conditionally independent from the treatment given the age. In words, we are assuming that, within the young and old subpopulations, there are no additional confounders producing systematic differences between the control group and the treatment group. It is as if we had access to two randomized trials restricted to young and old patients.

Under the conditional-independence assumption, we can aggregate the conditional probability of recovery given age and treatment, and weight them by the marginal pmf of age to estimate the marginal pmfs of the potential outcomes:

$$p_{\widetilde{\text{po}}_0}(1) = \sum_{a \in \{\text{young},\text{old}\}} p_{\tilde{a}}(a) p_{\widetilde{\text{po}}_0\,|\,\tilde{a}}(1\,|\,a) \tag{4.129}$$

$$= \sum_{a \in \{\text{young},\text{old}\}} p_{\tilde{a}}(a) p_{\tilde{y}\,|\,\tilde{a},\tilde{t}}(1\,|\,a, 0) \tag{4.130}$$

$$= 0.5 \cdot 0.8 + 0.5 \cdot 0.2 = 0.5, \tag{4.131}$$

$$p_{\widetilde{\text{po}}_1}(1) = \sum_{a \in \{\text{young},\text{old}\}} p_{\tilde{a}}(a) p_{\widetilde{\text{po}}_1\,|\,\tilde{a}}(1\,|\,a) \tag{4.132}$$

$$= \sum_{a \in \{\text{young},\text{old}\}} p_{\tilde{a}}(a) p_{\tilde{y}\,|\,\tilde{a},\tilde{t}}(1\,|\,a, 1) \tag{4.133}$$

$$= 0.5 \cdot 0.8 + 0.5 \cdot 0.4 = 0.6. \tag{4.134}$$

Our correction yields an exact estimate of the probability of the potential outcomes. Unfortunately, in practice we often cannot completely rule out the presence of additional unknown confounders, which would violate the conditional-independence assumption.

· ·

The procedure derived in Example 4.26 is a standard approach to adjust for the effect of a confounder.

Theorem 4.27 (Adjusting for a confounder). *Let $\tilde{y}$ be a random variable that represents the observed outcome corresponding to two potential outcomes $\widetilde{\text{po}}_0$ and $\widetilde{\text{po}}_1$ associated with a treatment $\tilde{t}$,*

$$\tilde{y} := \begin{cases} \widetilde{\text{po}}_0 & \text{if} \quad \tilde{t} = 0, \\ \widetilde{\text{po}}_1 & \text{if} \quad \tilde{t} = 1, \end{cases}$$

and let $\tilde{c}$ be a confounder, represented by a discrete random variable with range C. If the treatment $\tilde{t}$ and the potential outcomes $\widetilde{\text{po}}_0$ and $\widetilde{\text{po}}_1$ are conditionally independent given $\tilde{c}$, then

$$p_{\widetilde{\text{po}}_0}(1) = \sum_{c \in C} p_{\tilde{c}}(c) p_{\tilde{y} \mid \tilde{c}, \tilde{t}}(1 \mid c, 0), \tag{4.135}$$

$$p_{\widetilde{\text{po}}_1}(1) = \sum_{c \in C} p_{\tilde{c}}(c) p_{\tilde{y} \mid \tilde{c}, \tilde{t}}(1 \mid c, 1). \tag{4.136}$$

Proof The proof follows the same logic as our derivations in Example 4.26. By the definition of the observed outcome and the conditional-independence assumption, we can estimate the conditional probability of each potential outcome given the confounder from the conditional pmf of the observed outcome given the confounder and the treatment:

$$p_{\widetilde{\text{po}}_0 \mid \tilde{c}}(1 \mid c) = p_{\widetilde{\text{po}}_0 \mid \tilde{c}, \tilde{t}}(1 \mid c, 0) \tag{4.137}$$

$$= p_{\tilde{y} \mid \tilde{c}, \tilde{t}}(1 \mid c, 0), \tag{4.138}$$

$$p_{\widetilde{\text{po}}_1 \mid \tilde{c}}(1 \mid c) = p_{\widetilde{\text{po}}_1 \mid \tilde{c}, \tilde{t}}(1 \mid c, 1) \tag{4.139}$$

$$= p_{\tilde{y} \mid \tilde{c}, \tilde{t}}(1 \mid c, 1). \tag{4.140}$$

The pmf of the potential outcome can then be reconstructed by applying Theorem 4.10 and the chain rule (Theorem 4.14),

$$p_{\widetilde{\text{po}}_0}(1) = \sum_{c \in C} p_{\tilde{c}, \widetilde{\text{po}}_0}(c, 1) \tag{4.141}$$

$$= \sum_{c \in C} p_{\tilde{c}}(c) p_{\widetilde{\text{po}}_0 \mid \tilde{c}}(1 \mid c) \tag{4.142}$$

$$= \sum_{c \in C} p_{\tilde{c}}(c) p_{\tilde{y} \mid \tilde{c}, \tilde{t}}(1 \mid c, 0), \tag{4.143}$$

$$p_{\widetilde{\text{po}}_1}(1) = \sum_{c \in C} p_{\tilde{c}}(c) p_{\widetilde{\text{po}}_1 \mid \tilde{c}}(1 \mid c) \tag{4.144}$$

$$= \sum_{c \in C} p_{\tilde{c}}(c) p_{\tilde{y} \mid \tilde{c}, \tilde{t}}(1 \mid c, 1). \tag{4.145}$$

∎

Example 4.28 (3-point shooting: Adjusting for shot distance). Let us apply Theorem 4.27 to the 3-point shooting example in Section 4.6.2, in order to estimate the probabilities of the potential outcomes associated with Curry and Lee. This yields an adjusted 3-point shooting rate, which suggests that Curry is the better shooter, as we had suspected all along,

$$p_{\widetilde{\text{po}}_{\text{Curry}}}(1) = \sum_{d \in \{\text{short}, \text{long}\}} p_{\tilde{d}}(d) p_{\tilde{y} \mid \tilde{d}, \tilde{t}}(1 \mid d, \text{Curry}) \tag{4.146}$$

$$= 0.329 \cdot 0.5 + 0.671 \cdot 0.396 \tag{4.147}$$

$$= 0.430, \tag{4.148}$$

$$p_{\widetilde{\text{po}}_{\text{Lee}}}(1) = \sum_{d \in \{\text{short}, \text{long}\}} p_{\tilde{d}}(d) p_{\tilde{y} \mid \tilde{d}, \tilde{t}}(1 \mid d, \text{Lee}) \tag{4.149}$$

$$= 0.329 \cdot 0.483 + 0.671 \cdot 0.345 \tag{4.150}$$

$$= 0.391, \tag{4.151}$$

where $p_{\bar{d}}$ is estimated from the data in Table 4.5 using the corresponding empirical pmf (see Definition 2.11).

The conditional-independence assumption in Theorem 4.27 is unlikely to be completely true for this setting. Even controlling for shot distance, Curry's shots were probably still more difficult than Lee's because opposing defenses were presumably much more focused on Curry than on Lee. Nonetheless, our adjusted shooting rate is definitively more informative about the abilities of the two players than the raw observed three-point rate! This often happens in applications of causal inference to observational data. In the absence of randomized experiments, it is very difficult to account for all possible confounders, but at least we should adjust for the ones we know about.

....

4.7 The Curse of Dimensionality

In the examples we consider in Sections 4.1–4.6, we model the joint distribution of at most three variables. In practice, probabilistic models often have many more. For instance, Dataset 9 contains measurements from 134 weather stations. Imagine that we want to build a model for precipitation, where each station is represented by a random variable. To simplify matters, we only model the presence or absence of precipitation, as in Examples 4.9 and 4.19, so each variable has just two possible values. If we try to estimate the joint pmf of these 134 Bernoulli random variables using empirical probabilities, we run into a problem. The joint pmf has $2^{134} > 10^{40}$ entries! In comparison, the number of available data (8,760 hourly measurements) is ridiculously small. In fact, as reported in Figure 4.11, 89% of the precipitation patterns in the data are observed just once, and only 1.9% of patterns are observed more than three times. As a result, the empirical probabilities that we compute are completely inaccurate: The vast majority are zero, and among the nonzero probabilities, most equal $1/n$ (where n is the number of data). This is highly problematic for any practical application of our model. For example, we cannot use it for forecasting future precipitation patterns, as we almost certainly will not have observed them previously.

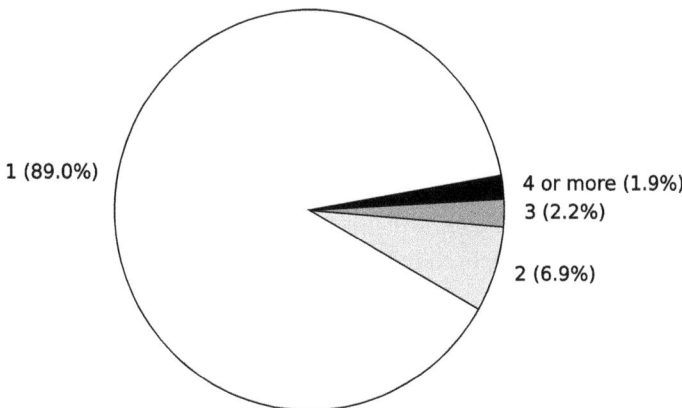

Figure 4.11 The curse of dimensionality. Dataset 9 contains 8,760 hourly precipitation measurements at 134 weather stations. The precipitation pattern associated with each measurement encodes in what stations there is precipitation. The pie chart reports how often precipitation patterns are repeated in the dataset. Most precipitation patterns occur just once (89.0%). Only 1.9% of patterns are observed more than three times.

Our precipitation example is not an exception, but rather the norm: Estimating the joint pmf of multiple random variables is usually *intractable* unless the number of variables is very small. This phenomenon is known as the *curse of dimensionality*. In order to ensure tractability, we need to make assumptions about the observed data and design simplified models with a number of parameters that does not explode exponentially as the number of variables increases. This is usually achieved via independence assumptions.

Let us assume that the precipitations at all weather stations are mutually independent in our precipitation example. Then the joint pmf is equal to the product of the marginal pmfs (see Definition 4.18), so we only need to estimate the 134 parameters associated with the Bernoulli random variables representing precipitation at each of the 134 stations. The bad news is that we are not modeling dependencies that could be important for our analysis. For instance, Figure 4.7 shows that there is a strong dependence between the precipitation in Coos Bay and Corvallis, which we may want to take into account. The key to effective probabilistic modeling is incorporating some crucial dependencies, while keeping the number of variables small enough to ensure that the model can be fit with the available data. The following sections describe two popular models based on this philosophy: naive Bayes and Markov chains.

4.8 Classification via Naive Bayes

In statistics and machine learning, classification is the problem of assigning a category or class to a data point. For example, we may want to identify what animal appears in a picture, or whether an email is spam or not. Assume that we have gathered a dataset of n examples (x_1, y_1), (x_2, y_2), ..., (x_n, y_n). Each example consists of a d-dimensional *feature* vector x_i (e.g. the pixels in a picture, or the text in an email) and its corresponding *label* $y_i \in \{1, 2, \dots, c\}$, indicating which of the c classes is associated with the example (e.g. *cat* in the case of the picture, or *spam* in the case of the email). The goal is to build a model from these data, which can be used to estimate the class of new examples from the corresponding features.

To perform classification via probabilistic modeling, we use the available data to learn the conditional probability of the class given the features. In more detail, we model the features as a d-dimensional random vector $\tilde{x}$ and the corresponding class label as a random variable $\tilde{y}$. We then estimate the conditional pmf $p_{\tilde{y} \mid \tilde{x}}$ of $\tilde{y}$ given $\tilde{x}$ from the data. The conditional pmf can be used to classify new examples by selecting the most likely class given the observed features. This approach is known as *maximum a posteriori* (MAP) estimation because the conditional distribution is the *posterior* distribution of the class once the features have been observed (this term is commonly used in Bayesian modeling, as explained in Section 6.7.1).

Definition 4.29 (MAP estimator from discrete features)**.** *Given a discrete random variable $\tilde{y}$ with range Y and a discrete random vector $\tilde{x}$, the maximum a posteriori (MAP) estimator of $\tilde{y}$ given $\tilde{x} = x$ is*

$$\text{MAP}(x) := \arg \max_{y \in Y} p_{\tilde{y} \mid \tilde{x}}(y \mid x). \tag{4.152}$$

The MAP estimator is optimal in the sense that it minimizes the probability of error, under the assumption that the estimated conditional pmf is correct.

Theorem 4.30 (MAP estimation is optimal). *Given a discrete random variable $\tilde{y}$ and a d-dimensional discrete random vector $\tilde{x}$, the MAP estimator of $\tilde{y}$ given $\tilde{x} = x$ minimizes the probability of error. Equivalently, for any estimator $h\colon \mathbb{R}^d \to \mathbb{R}$, the probability that the MAP estimator is correct is greater than or equal to the probability that h is correct:*

$$\mathrm{P}\left(\mathrm{MAP}(\tilde{x}) = \tilde{y}\right) \geq \mathrm{P}\left(h(\tilde{x}) = \tilde{y}\right). \tag{4.153}$$

Proof   To simplify notation, we denote the set of all possible values of the random vector $\tilde{x}$ by $\mathcal{X}$. By definition of the MAP estimator, for any $x \in \mathcal{X}$, $p_{\tilde{y}\,|\,\tilde{x}}(\mathrm{MAP}\,(x)\,|\,x) \geq p_{\tilde{y}\,|\,\tilde{x}}(h\,(x)\,|\,x)$. Therefore, by the law of total probability (Theorem 1.18),

$$\mathrm{P}\left(h(\tilde{x}) = \tilde{y}\right) = \sum_{x \in \mathcal{X}} \mathrm{P}\left(\tilde{x} = x\right) \mathrm{P}\left(\tilde{y} = h\,(x)\,|\,\tilde{x} = x\right) \tag{4.154}$$

$$= \sum_{x \in \mathcal{X}} p_{\tilde{x}}(x) p_{\tilde{y}\,|\,\tilde{x}}(h\,(x)\,|\,x) \tag{4.155}$$

$$\leq \sum_{x \in \mathcal{X}} p_{\tilde{x}}(x) p_{\tilde{y}\,|\,\tilde{x}}(\mathrm{MAP}\,(x)\,|\,x) \tag{4.156}$$

$$= \sum_{x \in \mathcal{X}} \mathrm{P}\left(\tilde{x} = x\right) \mathrm{P}\left(\tilde{y} = \mathrm{MAP}\,(x)\,|\,\tilde{x} = x\right) \tag{4.157}$$

$$= \mathrm{P}\left(\mathrm{MAP}(\tilde{x}) = \tilde{y}\right). \tag{4.158}$$

∎

Unfortunately, it is often intractable to compute the MAP estimator due to the curse of dimensionality. Let us illustrate this using the voting data in Dataset 1. The data consist of votes on 16 different issues by members of the United States House of Representatives, who are affiliated with the Democratic or Republican party. We consider the classification problem of determining the affiliation of a politician from their voting record.

We separate the dataset into a training set with 425 representatives and a test set with 10 representatives. We model the 16-dimensional voting record as a random vector $\tilde{x}$ with entries defined as follows:

$$\tilde{x}[j] := \begin{cases} 1 & \text{if the representative voted Yes on issue } j, \\ 0 & \text{otherwise.} \end{cases} \tag{4.159}$$

We model the affiliation as the Bernoulli random variable:

$$\tilde{y} = \begin{cases} R & \text{if the representative is a Republican,} \\ D & \text{if the representative is a Democrat.} \end{cases} \tag{4.160}$$

In order to determine the affiliation of a representative from their voting record x, we leverage the conditional pmf of $\tilde{y}$ given $\tilde{x}$: if $p_{\tilde{y}\,|\,\tilde{x}}(R\,|\,x) > p_{\tilde{y}\,|\,\tilde{x}}(D\,|\,x)$, we classify them as a Republican; if not, we classify them as a Democrat. The problem is that it is impossible to estimate the conditional pmf associated with every possible voting record. There are $2^{16} = 65,536$ possible values of x, and we only have 425 data points! After fitting the model with the available data, we are very likely to encounter politicians in the test set whose voting records have never been observed.

As mentioned in Section 4.7, a popular approach to address the curse of dimensionality is to make independence assumptions that simplify our probabilistic model. Here we cannot assume independence of the features $\tilde{x}$ and the corresponding class label $\tilde{y}$. Otherwise, we wouldn't be using the features at all! Instead, we assume that each entry of $\tilde{x}$ is conditionally independent from the rest of the entries given $\tilde{y}$. In that case, by Definitions 4.13 and 4.21, and Theorem 4.10,

$$p_{\tilde{y}\mid\tilde{x}}(y\mid x) := \frac{p_{\tilde{x},\tilde{y}}(x,y)}{p_{\tilde{x}}(x)} \tag{4.161}$$

$$= \frac{p_{\tilde{x},\tilde{y}}(x,y)}{\sum_{k=1}^{c} p_{\tilde{x},\tilde{y}}(x,k)} \tag{4.162}$$

$$= \frac{p_{\tilde{y}}(y) \prod_{j=1}^{d} p_{\tilde{x}[j]\mid\tilde{y}}(x[j]\mid y)}{\sum_{k=1}^{c} p_{\tilde{y}}(k) \prod_{j=1}^{d} p_{\tilde{x}[j]\mid\tilde{y}}(x[j]\mid k)}. \tag{4.163}$$

Under the conditional-independence assumption, computing the conditional pmf of the class given the features only requires approximating the conditional pmf of each separate feature given the class and the marginal pmf of the class. This dramatically reduces the number of parameters that we need to estimate from the data.

Example 1.34 shows that some of the features in our example are definitely not conditionally independent given political affiliation. We make the *naive* conditional-independence assumption anyway because it allows us to obtain an estimate tractably from the available data. This approach is consequently called *naive Bayes*, as (4.161) is analogous to the Bayes' rule for probabilities of events (see Theorem 1.20).

Definition 4.31 (Naive Bayes classifier). *Let* (x_1, y_1), (x_2, y_2), ..., (x_n, y_n) *denote a dataset with* n *examples, where* x_i *denotes a discrete vector with* d *features and* $y_i \in \{1, 2, \ldots, c\}$ *its corresponding class label (c is the number of classes) for* $1 \leq i \leq n$. *To classify a new example* x, *we:*

1 *Use empirical probabilities (see Exercise 4.2) to estimate the* c *conditional pmfs of the individual features given the class,* $p_{\tilde{x}[j]\mid\tilde{y}}(x[j]\mid k)$ *for* $j \in \{1, 2, \ldots, d\}$ *and* $k \in \{1, 2, \ldots, c\}$. *The random variables* $\tilde{x}[j]$ *and* $\tilde{y}$ *represent the* jth *feature and the class, respectively.*
2 *Compute the empirical pmf of the class* $\tilde{y}$, *following Definition 2.11.*
3 *Compute the MAP estimator of the class* $\tilde{y}$ *given the observed feature vector* $(\tilde{x} = x)$, *under the assumption that all entries of* $\tilde{x}$ *are conditionally independent given* $\tilde{y}$,

$$\mathrm{MAP}\,(x) := \arg \max_{y \in \{1,2,\ldots,c\}} p_{\tilde{y}\mid\tilde{x}}(y\mid x) \tag{4.164}$$

$$= \arg \max_{y \in \{1,2,\ldots,c\}} \frac{p_{\tilde{y}}(y) \prod_{j=1}^{d} p_{\tilde{x}[j]\mid\tilde{y}}(x[j]\mid y)}{\sum_{k=1}^{c} p_{\tilde{y}}(k) \prod_{j=1}^{d} p_{\tilde{x}[j]\mid\tilde{y}}(x[j]\mid k)}. \tag{4.165}$$

In the case of the voting data, the two class labels are R (Republican) and D (Democrat). In order to apply naive Bayes, we need to determine $p_{\tilde{x}[j]\mid\tilde{y}}(\cdot \mid R)$ and $p_{\tilde{x}[j]\mid\tilde{y}}(\cdot \mid D)$, for $1 \leq j \leq 16$, and also $p_{\tilde{y}}$. This requires estimating 33 parameters (there is only 1 parameter for each Bernoulli pmf), which is a dramatic reduction from the 65,535 parameters required to estimate the conditional pmf $\tilde{y}$ given $\tilde{x}$ without any assumptions. As a result, it is now tractable to fit the probabilistic model to the available data.

There are 263 Democrats and 162 Republicans, so by Definition 2.11, the entries of the empirical pmf of $\tilde{y}$ are $p_{\tilde{y}}(R) = 0.381$ and $p_{\tilde{y}}(D) = 0.619$. For each vote, we compute the fraction of Republicans who voted Yes to estimate $p_{\tilde{x}[j]\mid\tilde{y}}(1 \mid R)$ and the fraction of Democrats who voted Yes to approximate $p_{\tilde{x}[j]\mid\tilde{y}}(1 \mid D)$. Table 4.6 shows the resulting

Table 4.6 *Classification of political affiliation via naive Bayes.* The table shows the conditional probability of voting Yes on the 16 issues in Dataset 1 given political affiliation, i.e. depending on whether the representatives are Republicans (R) or Democrats (D). These are empirical probabilities estimated from the training set. We also show the voting record for two Democrats from the test set: Y indicates Yes, N indicates No and – indicates that they didn't vote. The first (Example D) has a voting pattern consistent with other Democrats in a majority of the issues. We highlight in bold the issues for which their vote is in agreement with Democrats in the training set (i.e. they voted Yes and $p_{\tilde{x}[j]\,|\,\tilde{y}}(1\,|\,D) > 0.5$, or they voted No and $p_{\tilde{x}[j]\,|\,\tilde{y}}(1\,|\,D) < 0.5$). By contrast, the Democrat who is wrongly classified (Misclassified D) has a voting pattern that tends to align with Republicans rather than with Democrats. We highlight in bold the issues for which their vote is in agreement with Republicans in the training set (i.e. they voted Yes and $p_{\tilde{x}[j]\,|\,\tilde{y}}(1\,|\,R) > 0.5$, or they voted No and $p_{\tilde{x}[j]\,|\,\tilde{y}}(1\,|\,R) < 0.5$).

j	1	2	3	4	5	6	7	8
$p_{\tilde{x}[j]\,|\,\tilde{y}}(1\,|\,R)$	0.19	0.50	0.14	0.99	0.95	0.90	0.24	0.15
$p_{\tilde{x}[j]\,|\,\tilde{y}}(1\,|\,D)$	0.61	0.50	0.89	0.05	0.22	0.47	0.78	0.83
Example D	N	–	**Y**	**N**	**N**	Y	**Y**	**Y**
Misclassified D	**Y**	Y	–	**Y**	**Y**	**Y**	N	N

j	9	10	11	12	13	14	15	16
$p_{\tilde{x}[j]\,|\,\tilde{y}}(1\,|\,R)$	0.11	0.55	0.14	0.87	0.86	0.98	0.09	0.66
$p_{\tilde{x}[j]\,|\,\tilde{y}}(1\,|\,D)$	0.76	0.47	0.51	0.15	0.29	0.35	0.64	0.94
Example D	**N**	Y	N	**N**	**N**	**N**	**Y**	–
Misclassified D	**Y**	**Y**	–	**Y**	**Y**	**Y**	N	N

probabilities. Finally, we classify each representative in the test set by computing the MAP estimator (4.164). The voting record of one of the test representatives is shown in Table 4.6 (row marked *Example D*). The representative did not vote for issues 2 and 16, so we omit the corresponding entries of $\tilde{x}$ in our calculation. The naive-Bayes estimate of the probability that the candidate is a Democrat is

$$p_{\tilde{y}\,|\,\tilde{x}}(D\,|\,x) = \frac{p_{\tilde{y}}(D) \prod_{i\in\{1,3,\ldots,15\}} p_{\tilde{x}[j]\,|\,\tilde{y}}(x[j]\,|\,D)}{p_{\tilde{y}}(D) \prod_{j\in\{1,3,\ldots,15\}} p_{\tilde{x}[j]\,|\,\tilde{y}}(x[j]\,|\,D) + p_{\tilde{y}}(R) \prod_{j\in\{1,3,\ldots,15\}} p_{\tilde{x}[j]\,|\,\tilde{y}}(x[j]\,|\,R)}$$

$$= 1 - 1.4 \, 10^{-8}. \tag{4.166}$$

We conclude correctly that the representative is a Democrat. Naive Bayes identifies the correct class for 9 of the 10 examples in the test set. The only error corresponds to a Democratic representative whose voting record is more aligned with the Republicans in the training set than with the Democrats (see row marked *Misclassified D* in Table 4.6).

4.9 Markov Chains

This section presents Markov chains, which are a popular model for data with temporal structure. Section 4.9.1 motivates and defines Markov chains. Section 4.9.2 shows that they can be represented in terms of a state vector and a transition matrix that governs the time evolution of the model. Section 4.9.3 explains how to analyze the asymptotic behavior of Markov chains and determine whether they converge to a stationary distribution.

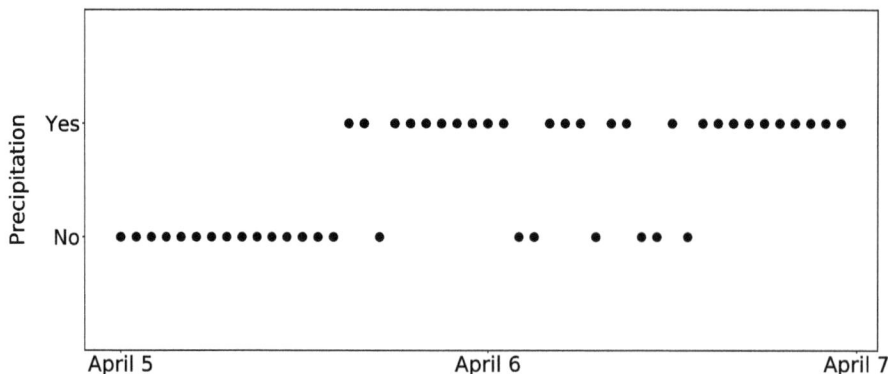

Figure 4.12 Precipitation in Coos Bay. Hourly precipitation in Coos Bay during two days in April 2015, extracted from Dataset 9. There is clear temporal dependence in the data: Changes from no precipitation to precipitation, and vice versa, are relatively rare.

4.9.1 Time-Homogeneous Markov Chains

Consider a *time series* of discrete measurements $x_1, x_2, \ldots, x_n$ representing a quantity of interest at different times. For example, x_i could indicate whether there is precipitation or not at time i in a certain location. Up to now, we have analyzed datasets assuming that they consist of independent, identically distributed samples from the same distribution (see Definition 2.23). However, this neglects crucial structure in time-series data. Specifically, temporal dependence between adjacent data points is not captured due to the independence assumption.

Figure 4.12 shows the hourly precipitation in Coos Bay (Oregon) over a couple of days, extracted from Dataset 9. There is obvious temporal dependence in the data: Precipitation occurs in stretches of multiple hours that are more or less continuous, so data with the same value tend to cluster together temporally. In order to account for such structure, we model the data as realizations from a sequence of dependent random variables $\tilde{a}_1, \tilde{a}_2, \ldots, \tilde{a}_n$.

Modeling a dependent sequence of random variables is challenging due to the curse of dimensionality (see Section 4.7). For simplicity, let us assume that the data are a binary sequence with two possible values, as in Figure 4.12. If we interpret these data as realizations of a sequence of Bernoulli random variables $\tilde{a}_1, \tilde{a}_2, \ldots, \tilde{a}_n$, then the joint pmf has 2^n entries, corresponding to all possible binary sequences of length n.[5] That is unfortunate, given that we only observe a single binary sequence with n data points! We need to make some assumptions to reduce the complexity of our model. Assuming the data are i.i.d. as in Definition 4.20 takes care of this problem: We just need to estimate a single pmf. However, this completely ignores temporal dependence. We need a model that is simple, so that we can fit it to our limited data but is also able to incorporate temporal dependence. Enter the Markov chain, which assumes that a sequence is completely characterized by memoryless one-step transitions.

---

[5] Strictly speaking, the degrees of freedom in the joint pmf are $2^n - 1$ because the entries of the joint pmf must sum to one.

Definition 4.32 (Markov chain). *Let $\tilde{a}_1$, $\tilde{a}_2$, ..., $\tilde{a}_n$ be discrete random variables belonging to the same probability space. The random variables form a Markov chain, if for any $2 \leq i \leq n - 1$, $\tilde{a}_{i+1}$ is conditionally independent of $\tilde{a}_1$, ..., $\tilde{a}_{i-1}$ given $\tilde{a}_i$, i.e.*

$$p_{\tilde{a}_{i+1} \mid \tilde{a}_1, \ldots, \tilde{a}_i} \left(a_{i+1} \mid a_1, a_2, \ldots, a_i\right) = p_{\tilde{a}_{i+1} \mid \tilde{a}_i} \left(a_{i+1} \mid a_i\right) \tag{4.167}$$

for any values of a_1, a_2, ..., a_n (see Section 4.5). Equivalently, by the chain rule (Theorem 4.14), the joint pmf of $\tilde{a}_1$, $\tilde{a}_2$, ..., $\tilde{a}_n$ equals

$$p_{\tilde{a}_1, \tilde{a}_2, \ldots, \tilde{a}_n} \left(a_1, a_2, \ldots, a_n\right) = p_{\tilde{a}_1}\left(a_1\right) \prod_{i=1}^{n-1} p_{\tilde{a}_{i+1} \mid \tilde{a}_1, \ldots, \tilde{a}_i} \left(a_{i+1} \mid a_1, a_2, \ldots, a_i\right) \tag{4.168}$$

$$= p_{\tilde{a}_1}\left(a_1\right) \prod_{i=1}^{n-1} p_{\tilde{a}_{i+1} \mid \tilde{a}_i} \left(a_{i+1} \mid a_i\right) \tag{4.169}$$

for any values of a_1, a_2, ..., a_n.

Assuming that our data are realizations from a Markov chain reduces the number of parameters enormously. In order to approximate the joint pmf in (4.169), we only need to estimate $p_{\tilde{a}_1}$ and the one-step conditional probabilities $p_{\tilde{a}_{i+1} \mid \tilde{a}_i}$, for $1 \leq i \leq n - 1$. If the random variables are Bernoulli, this reduces the number of parameters of the joint pmf from $2^n - 1$, to $2n - 1$, since we need to estimate $p_{\tilde{a}_1}$ and a conditional pmf $p_{\tilde{a}_{i+1} \mid \tilde{a}_i}(\cdot \mid a_i)$ for $1 \leq i \leq n - 1$ and $a_i \in \{0, 1\}$. Unfortunately, this is still problematic: We only have a single data point (x_i) from which to estimate each conditional pmf $p_{\tilde{a}_{i+1} \mid \tilde{a}_i}$! In order to make estimation tractable, an option is to assume that all of these conditional pmfs are the same. In that case, the Markov chain is said to be *time homogeneous*.

Definition 4.33 (Time-homogeneous Markov chain). *A Markov chain $\tilde{a}_1$, $\tilde{a}_2$, ..., $\tilde{a}_n$ is time homogeneous, if the conditional pmfs $p_{\tilde{a}_{i+1} \mid \tilde{a}_i} \left(a_{i+1} \mid a_i\right)$ are all equal to the same function p_{cond},*

$$p_{\tilde{a}_{i+1} \mid \tilde{a}_i} \left(a_{i+1} \mid a_i\right) = p_{\text{cond}} \left(a_{i+1} \mid a_i\right), \quad 1 \leq i \leq n - 1, \tag{4.170}$$

for $1 \leq i \leq n - 1$ and all possible values of a_1, a_2, ..., a_n. By (4.169), the joint pmf of $\tilde{a}_1$, $\tilde{a}_2$, ..., $\tilde{a}_n$ equals

$$p_{\tilde{a}_1, \tilde{a}_2, \ldots, \tilde{a}_n} \left(a_1, a_2, \ldots, a_n\right) = p_{\tilde{a}_1}\left(a_1\right) \prod_{i=1}^{n-1} p_{\text{cond}} \left(a_{i+1} \mid a_i\right) \tag{4.171}$$

for all possible values of a_1, a_2, ..., a_n.

When fitting a time-homogeneous Markov chain using data, we only need to estimate the one-step transition conditional pmf p_{cond} and the initial pmf $p_{\tilde{a}_1}$ to completely characterize the joint pmf. If the random variables are Bernoulli, this requires estimating three parameters, corresponding to $p_{\tilde{a}_1}$, $p_{\text{cond}}(\cdot \mid 0)$ and $p_{\text{cond}}(\cdot \mid 1)$. We have finally obtained a model that is tractable to fit using n data points. The following example applies it to the precipitation data in Figure 4.12.

Example 4.34 (Predicting hourly precipitation). We consider the problem of predicting hourly precipitation in Coos Bay, Oregon. We use binary hourly precipitation measurements from 2015, extracted from Dataset 9, as training data (a small segment is shown in Figure 4.12). We begin by interpreting the data as an i.i.d. sequence and compute the

Table 4.7 ***Precipitation statistics in Coos Bay.*** *The tables show the statistics of hourly precipitation in Coos Bay (Oregon) during 2015 represented as percentages. The top-left table shows the fraction of hours with and without precipitation. The top-right table shows the frequency of the different one-step transitions between precipitation and no precipitation. The bottom two tables show the frequency of the different possible two-step transitions.*

Marginal probabilities

No	88.7
Yes	11.3

One-step conditional probabilities

Hour $h+1$		No	Yes
	No	96.0	31.2
	Yes	4.0	68.8

(columns headed by Hour h*)*

Two-step conditional probabilities

No precipitation at hour $h-1$

Hour $h+1$		No	Yes
	No	97.1	49.4
	Yes	2.9	50.6

(columns headed by Hour h*)*

Precipitation at hour $h-1$

Hour $h+1$		No	Yes
	No	70.6	23.0
	Yes	29.4	77.0

(columns headed by Hour h*)*

corresponding empirical pmf, which is Bernoulli with a parameter equal to the frequency of precipitation in the training set (11.3%). Using this model, the best possible binary prediction is that it never rains. We test this prediction on hourly measurements from 2016. The resulting accuracy is 83.4% because in 2016 it rained 16.6% of the time.

In order to take into account the temporal structure of the data observed in Figure 4.12, we fit a two-state time-homogeneous Markov chain to the data. The top right corner of Table 4.7 shows the one-step transition conditional probabilities, estimated using the corresponding empirical conditional probabilities (see Definition 1.24). According to the Markov chain, when there is precipitation at time h, there is still precipitation at time $h+1$ with probability 0.688. By contrast, when there is no precipitation at time h, the probability of precipitation at time $h + 1$ plummets to 0.04. Based on this model, we predict precipitation, if there is precipitation the hour before, and no precipitation otherwise. This yields an improved accuracy of 87.3% for the 2016 test dataset.

A natural extension of the one-step Markov-chain model is to model two-step dependence. The bottom row of Table 4.7 shows two-step transition probabilities estimated using the corresponding empirical conditional probabilities. A quick analysis of the transition probabilities reveals that our data violate the conditional-independence assumption in Definition 4.32:

$$\text{P}(\text{Yes at } h + 1 \mid \text{Yes at } h) = 0.688 \tag{4.172}$$
$$\neq 0.770 = \text{P}(\text{Yes at } h + 1 \mid \text{Yes at } h, \text{Yes at } h - 1)$$
$$\neq 0.506 = \text{P}(\text{Yes at } h + 1 \mid \text{Yes at } h, \text{No at } h - 1).$$

However, the difference between these conditional probabilities is not very large. In fact, according to the two-step model, regardless of what happens at time $h - 1$, if there is precipitation at time h, then precipitation is more likely at time $h + 1$. Conversely, if there is no precipitation at h, then no precipitation is more likely at $h + 1$ (again, regardless of what happens at $h - 1$). Consequently, the binary predictions of the one-step and two-step models are the same.

This example illustrates two important points. First, conditional-independence assumptions almost never hold exactly. Second, there is often diminishing returns when modeling dependence between variables in a probabilistic model. In this case, incorporating one-step transitions improves performance substantially, but modeling the two-step transitions does not make a difference in our binary prediction, and hence in the accuracy of our model. Note, however, that this is a toy example. In practice, we would use a probabilistic estimate to forecast precipitation, not just a binary prediction, and we would evaluate the estimate using additional metrics beyond accuracy, as described in Section 12.9.

. . . .

4.9.2 The State Vector and the Transition Matrix

In this section, we explain how to define and manipulate time-homogeneous Markov chains. We restrict our attention to Markov chains that take values in a fixed finite set $S := \{s_1, s_2, \ldots, s_m\}$. For instance, the Markov chain in Example 4.34 has two states ($m = 2$): precipitation and no precipitation. We call the set of possible values S the *state space* of the Markov chain. For each element $s \in S$, if $\tilde{a}_i = s$ we say that the Markov chain is in state s at time i. We represent the marginal pmf of each entry $\tilde{a}_i$ in the Markov chain using an m-dimensional *state vector* π_i,

$$
\pi_i := \begin{bmatrix} p_{\tilde{a}_i}(s_1) \\ p_{\tilde{a}_i}(s_2) \\ \ldots \\ p_{\tilde{a}_i}(s_m) \end{bmatrix}, \qquad 1 \le i \le n, \tag{4.173}
$$

where n is the length of the Markov chain. Notice that $\sum_{j=1}^{m} \pi_i[j] = 1$ by Lemma 2.6.

The *transition matrix* T of a fine-state time-homogeneous Markov chains is an $m \times m$ matrix that contains all transition probabilities between states:

$$
T := \begin{bmatrix} p_{\text{cond}}(s_1 \mid s_1) & p_{\text{cond}}(s_1 \mid s_2) & \cdots & p_{\text{cond}}(s_1 \mid s_m) \\ p_{\text{cond}}(s_2 \mid s_1) & p_{\text{cond}}(s_2 \mid s_2) & \cdots & p_{\text{cond}}(s_2 \mid s_m) \\ \ldots & \ldots & \ldots & \ldots \\ p_{\text{cond}}(s_m \mid s_1) & p_{\text{cond}}(s_m \mid s_2) & \cdots & p_{\text{cond}}(s_m \mid s_m) \end{bmatrix}, \tag{4.174}
$$

where p_{cond} is defined as in Definition 4.33. Each column of the transition matrix contains a full conditional pmf, so its entries must sum to one by Lemma 2.6,

$$
\sum_{i=1}^{m} T_{ij} = 1, \qquad 1 \le j \le m. \tag{4.175}
$$

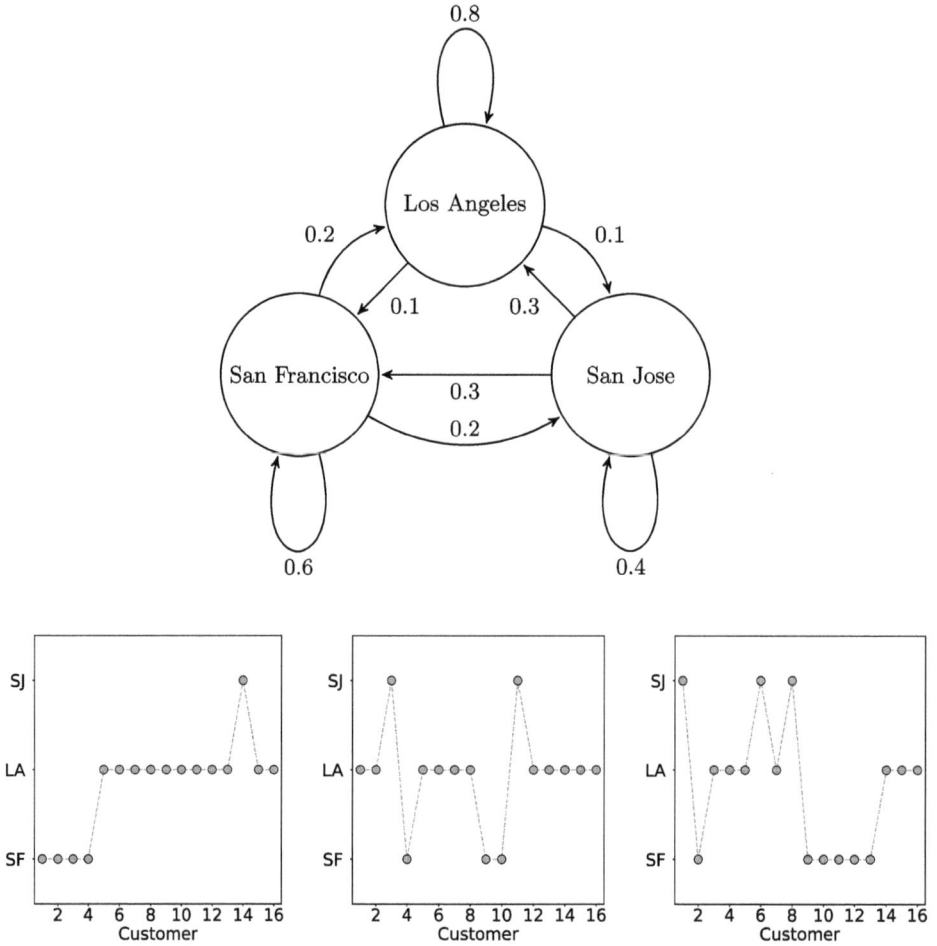

Figure 4.13 Finite-state time-homogeneous Markov chain modeling a rental car. The state diagram (top) describes the Markov chain in Example 4.35. Each arrow represents a transition between states, annotated by the corresponding transition probability. The graphs depict three realizations of the Markov chain.

The transition matrix for the Markov chain in Example 4.34 is

$$T := \begin{bmatrix} 0.960 & 0.312 \\ 0.040 & 0.688 \end{bmatrix}, \tag{4.176}$$

where the states s_1 and s_2 correspond to no precipitation and precipitation, respectively. The transition probabilities of finite-space time-homogeneous Markov chains can be visualized using a state diagram, which includes each state and the probability of every possible transition. Figure 4.13 shows the state diagram corresponding to the following example.

Example 4.35 (Car rental). A car-rental company hires you to model the location of their cars. The company operates in Los Angeles, San Francisco, and San Jose. Customers have the option to pick up and drop off the cars in different cities. You decide to model the location of the cars as a three-state time-homogeneous Markov chain $\tilde{a}_1, \tilde{a}_2, \ldots, \tilde{a}_n$. The random variable $\tilde{a}_1$ is the initial location and, for $i \geq 2$, $\tilde{a}_i$ represents the location after the $i-1$th customer. The transition probabilities, obtained from past data, are:

San Francisco	Los Angeles	San Jose	
0.6	0.1	0.3	San Francisco
0.2	0.8	0.3	Los Angeles
0.2	0.1	0.4	San Jose

If a car is in San Francisco, the probability that it moves to LA or San Jose is the same (0.2), whereas the probability that it stays in San Francisco is 0.6. If we assign state 1 to San Francisco, state 2 to Los Angeles and state 3 to San Jose, the transition matrix of the Markov chain is

$$T := \begin{bmatrix} 0.6 & 0.1 & 0.3 \\ 0.2 & 0.8 & 0.3 \\ 0.2 & 0.1 & 0.4 \end{bmatrix}. \tag{4.177}$$

The company allocates new cars evenly between the three cities, so the initial state vector is

$$\pi_1 := \begin{bmatrix} 1/3 \\ 1/3 \\ 1/3 \end{bmatrix}. \tag{4.178}$$

Figure 4.13 shows a state diagram of the Markov chain, as well as some realizations.

The initial state vector and the transition matrix completely characterize the joint pmf of the Markov chain. For example, we can compute the probability that the car starts in San Francisco and ends up in San Jose after the second customer:

$$p_{\tilde{a}_1, \tilde{a}_3}(1, 3) = \sum_{i=1}^{3} p_{\tilde{a}_1, \tilde{a}_2, \tilde{a}_3}(1, i, 3) \tag{4.179}$$

$$= \sum_{i=1}^{3} p_{\tilde{a}_1}(1) \, p_{\tilde{a}_2 \mid \tilde{a}_1}(i \mid 1) \, p_{\tilde{a}_3 \mid \tilde{a}_2}(3 \mid i) \tag{4.180}$$

$$= \pi_1[1] \sum_{i=1}^{3} T_{i1} T_{3i} \tag{4.181}$$

$$= \frac{0.6 \cdot 0.2 + 0.2 \cdot 0.1 + 0.2 \cdot 0.4}{3} = 7.33 \cdot 10^{-2}, \tag{4.182}$$

where we have combined Theorem 4.10, the chain rule (Theorem 4.14), and the conditional-independence assumption in Definition 4.32.

The following theorem establishes that we can obtain the state vector at time i by multiplying the transition matrix and the state vector at time $i - 1$.

Theorem 4.36 (State vector and transition matrix). *For an n-dimensional finite-state time-homogeneous Markov chain with transition matrix T and initial state vector π_1, the state vector at time i equals*

$$\pi_i = T \pi_{i-1} \tag{4.183}$$

$$= T^{i-1} \pi_1, \quad 2 \leq i \leq n, \tag{4.184}$$

where T^{i-1} means that we apply T $i-1$ times.

Proof Let $\tilde{a}_1, \tilde{a}_2, \ldots, \tilde{a}_n$ denote the Markov chain. Equation (4.183) follows from Theorem 4.10 and the chain rule (Theorem 4.14),

$$\pi_i := \begin{bmatrix} p_{\tilde{a}_i}(s_1) \\ p_{\tilde{a}_i}(s_2) \\ \cdots \\ p_{\tilde{a}_i}(s_m) \end{bmatrix} = \begin{bmatrix} \sum_{j=1}^m p_{\tilde{a}_{i-1},\tilde{a}_i}(s_j, s_1) \\ \sum_{j=1}^m p_{\tilde{a}_{i-1},\tilde{a}_i}(s_j, s_2) \\ \cdots \\ \sum_{j=1}^m p_{\tilde{a}_{i-1},\tilde{a}_i}(s_j, s_m) \end{bmatrix} = \begin{bmatrix} \sum_{j=1}^m p_{\tilde{a}_{i-1}}(s_j)\, p_{\tilde{a}_i \mid \tilde{a}_{i-1}}(s_1 \mid s_j) \\ \sum_{j=1}^m p_{\tilde{a}_{i-1}}(s_j)\, p_{\tilde{a}_i \mid \tilde{a}_{i-1}}(s_2 \mid s_j) \\ \cdots \\ \sum_{j=1}^m p_{\tilde{a}_{i-1}}(s_j)\, p_{\tilde{a}_i \mid \tilde{a}_{i-1}}(s_m \mid s_j) \end{bmatrix}$$

$$= \begin{bmatrix} p_{\tilde{a}_i \mid \tilde{a}_{i-1}}(s_1 \mid s_1) & p_{\tilde{a}_i \mid \tilde{a}_{i-1}}(s_1 \mid s_2) & \cdots & p_{\tilde{a}_i \mid \tilde{a}_{i-1}}(s_1 \mid s_m) \\ p_{\tilde{a}_i \mid \tilde{a}_{i-1}}(s_2 \mid s_1) & p_{\tilde{a}_i \mid \tilde{a}_{i-1}}(s_2 \mid s_2) & \cdots & p_{\tilde{a}_i \mid \tilde{a}_{i-1}}(s_2 \mid s_m) \\ & & \cdots & \\ p_{\tilde{a}_i \mid \tilde{a}_{i-1}}(s_m \mid s_1) & p_{\tilde{a}_i \mid \tilde{a}_{i-1}}(s_m \mid s_2) & \cdots & p_{\tilde{a}_i \mid \tilde{a}_{i-1}}(s_m \mid s_m) \end{bmatrix} \begin{bmatrix} p_{\tilde{a}_{i-1}}(s_1) \\ p_{\tilde{a}_{i-1}}(s_2) \\ \cdots \\ p_{\tilde{a}_{i-1}}(s_m) \end{bmatrix}$$

$$= T \pi_{i-1}. \tag{4.185}$$

Equation (4.184) is obtained by applying (4.183) $i-1$ times. ∎

Example 4.37 (Car rental: fifth customer). Let us leverage our Markov-chain model from Example 4.35 to estimate the probability that a car ends up in each of the three locations after the fifth customer. Applying Theorem 4.36, we obtain

$$\pi_6 = T^5 \pi_1 = \begin{bmatrix} 0.281 \\ 0.534 \\ 0.185 \end{bmatrix}. \tag{4.186}$$

According to the model, the car is more likely to end up in Los Angeles than in the other two locations.

. . . .

4.9.3 Stationary Distribution of a Markov Chain

In this section, we explain how to analyze the time evolution of finite-state time-homogeneous Markov chains. The following example shows that the eigendecomposition of the transition matrix can be very useful for this purpose.

Example 4.38 (Car rental: Asymptotic distribution). The rental company in Example 4.35 wants you to estimate the probability that the cars eventually end up in each of the different cities. These probabilities are encoded in the limit $\lim_{i \to \infty} \pi_i$ of the state vector π_i, as the time i tends to infinity. By Theorem 4.36, the limit can be expressed in terms of the transition matrix and the initial state vector

$$\lim_{i \to \infty} \pi_i = \lim_{i \to \infty} T^{i-1} \pi_1. \tag{4.187}$$

The term T^{i-1} makes it difficult to compute the limit. To address this, we leverage the eigendecomposition of the transition matrix T:

$$T = \underbrace{\begin{bmatrix} 0.27 & 0.37 & 0.37 \\ 0.55 & -0.50 & 0.13 \\ 0.18 & 0.13 & -0.50 \end{bmatrix}}_{Q} \underbrace{\begin{bmatrix} 1 & 0 & 0 \\ 0 & 0.57 & 0 \\ 0 & 0 & 0.23 \end{bmatrix}}_{\Lambda} \underbrace{\begin{bmatrix} 1 & 1 & 1 \\ 1.28 & -0.87 & 0.70 \\ 0.70 & 0.13 & -1.44 \end{bmatrix}}_{Q^{-1}}. \qquad (4.188)$$

The columns of Q contain the three eigenvectors of T, and the diagonal entries of Λ contain the corresponding eigenvalues. Let us to express T^{i-1} in terms of Q, Q^{-1} and the $i-1$th power of the diagonal matrix of eigenvalues Λ. Since $Q^{-1}Q$ is equal to the identity matrix,

$$T^{i-1} = (Q\Lambda Q^{-1})^{i-1} = Q\Lambda Q^{-1} Q\Lambda Q^{-1} \cdots Q\Lambda Q^{-1} \qquad (4.189)$$
$$= Q\Lambda^{i-1}Q^{-1}. \qquad (4.190)$$

This is very helpful because calculating the power of a diagonal matrix is easy: Λ^{i-1} is a diagonal matrix that contains the $i-1$th power of each eigenvalue. Combining (4.187) and (4.190) we obtain

$$\lim_{i\to\infty} \pi_i = \lim_{i\to\infty} (Q\Lambda Q^{-1})^{i-1}\pi_1 \qquad (4.191)$$

$$= \lim_{i\to\infty} Q \begin{bmatrix} 1^{i-1} & 0 & 0 \\ 0 & 0.57^{i-1} & 0 \\ 0 & 0 & 0.23^{i-1} \end{bmatrix} Q^{-1}\pi_1 \qquad (4.192)$$

$$= Q \begin{bmatrix} 1 & 0 & 0 \\ 0 & \lim_{i\to\infty} 0.57^{i-1} & 0 \\ 0 & 0 & \lim_{i\to\infty} 0.23^{i-1} \end{bmatrix} Q^{-1}\pi_1 \qquad (4.193)$$

$$= Q \begin{bmatrix} 1 & 0 & 0 \\ 0 & 0 & 0 \\ 0 & 0 & 0 \end{bmatrix} Q^{-1}\pi_1 \qquad (4.194)$$

$$= \begin{bmatrix} 0.27 \\ 0.55 \\ 0.18 \end{bmatrix} \sum_{j=1}^{3} \pi_1[j] \qquad (4.195)$$

$$= \begin{bmatrix} 0.27 \\ 0.55 \\ 0.18 \end{bmatrix}, \qquad (4.196)$$

where we have used the fact that the entries of the initial state vector sum to one by Lemma 2.6. The state vector always converges to the same pmf for all possible initial vectors! Specifically, it converges to the eigenvector corresponding to the eigenvalue that equals one because the terms depending on the other eigenvalues are multiplied by 0.57^{i-1} and 0.23^{i-1}, so they decrease exponentially. The numerical simulations in Figure 4.14 show that convergence is very fast. We conclude that no matter how the company initially allocates the cars, eventually, 27.3% end up in San Francisco, 54.5% in LA, and 18.2% in San Jose.

. . . .

The following theorem generalizes our derivation in Example 4.38, characterizing the relationship between the dynamics of a finite-state time-homogeneous Markov chain and the eigendecomposition of its transition matrix.

$$\pi_1 := \begin{bmatrix} 1/3 \\ 1/3 \\ 1/3 \end{bmatrix} \qquad \pi_1 := \begin{bmatrix} 1 \\ 0 \\ 0 \end{bmatrix} \qquad \pi_1 := \begin{bmatrix} 0.1 \\ 0.2 \\ 0.7 \end{bmatrix}$$

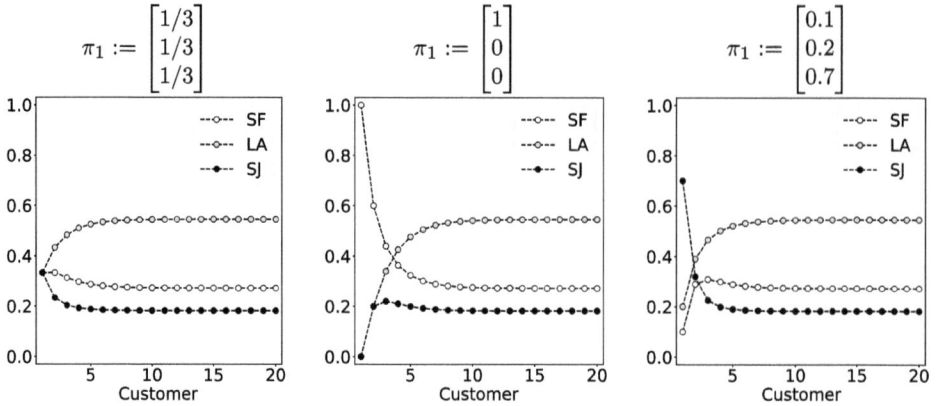

Figure 4.14 Convergence to a stationary distribution. The graphs show the evolution of the state vector of the Markov chain in Example 4.35 for different initial states. In all cases, the state vector quickly converges to the stationary distribution, corresponding to the eigenvector of the transition matrix with eigenvalue equal to one, as predicted by the analysis in Example 4.38.

Theorem 4.39. *Let $T \in \mathbb{R}^{m \times m}$ be the transition matrix of an m-state time-homogeneous Markov chain with eigendecomposition $T = Q\Lambda Q^{-1}$. The columns of Q are the m eigenvectors $q_1, \ldots, q_m$ and Λ is a diagonal matrix containing the corresponding eigenvalues $\lambda_1, \ldots, \lambda_m$. If we express the initial state vector $\pi_1 \in \mathbb{R}^m$ in terms of the eigenvectors,*

$$\pi_1 = \sum_{j=1}^{m} \alpha_j q_j, \tag{4.197}$$

where α_i is the ith entry of $\alpha := Q^{-1}\pi_1$, then for any $i \geq 1$ the state vector at time i equals

$$\pi_i = \sum_{j=1}^{m} \alpha_j \lambda_j^{i-1} q_j. \tag{4.198}$$

Proof Since q_j is the eigenvector of T associated with the eigenvalue λ_j, $T^{i-1}q_j = \lambda_j^{i-1}q_j$ for $1 \leq j \leq m$. Consequently, by Theorem 4.36,

$$\pi_i = T^{i-1}\pi_1 \tag{4.199}$$

$$= \sum_{j=1}^{m} \alpha_j T^{i-1} q_j \tag{4.200}$$

$$= \sum_{j=1}^{m} \alpha_j \lambda_j^{i-1} q_j. \tag{4.201}$$

∎

Consider a state vector π_* that is an eigenvector of the transition matrix of a Markov chain with eigenvalue equal to one. Then, by Theorem 4.36, once the state vector equals π_*, it will remain equal to π_* forever afterwards. Notice that this does *not* mean that the Markov chain is stuck at one of the states, but rather that from then on, all states will have fixed marginal probabilities (equal to the corresponding entry of π_*).

Definition 4.40 (Stationary distribution). *A state vector $\pi_* \in \mathbb{R}^m$ encodes a stationary distribution of a finite-state time-homogeneous Markov chain with transition matrix T, if $T\pi_* = \pi_*$.*

As illustrated by Example 4.38, if a finite-state time-homogeneous Markov chain has a single stationary distribution, and the remaining eigenvalues of the transition matrix have magnitudes smaller than one, then the state vector of the Markov chain converges to the stationary distribution *for any initial state vector*. The principle is the same as in the celebrated power method to compute eigenvalues. When we apply the transition matrix over and over, the component of the state vector associated with the stationary distribution is preserved because its corresponding eigenvalue equals one. By contrast, the other components shrink exponentially because their eigenvalues have magnitudes smaller than one. The following example shows that this occurs for the real precipitation data in Example 4.34.

Example 4.41 (Stationary distribution of precipitation). The eigendecomposition of the transition matrix in Example 4.34 is

$$T = \underbrace{\begin{bmatrix} 0.887 & -0.632 \\ 0.113 & -0.632 \end{bmatrix}}_{Q} \underbrace{\begin{bmatrix} 1 & 0 \\ 0 & 0.648 \end{bmatrix}}_{\Lambda} \underbrace{\begin{bmatrix} 1 & 1 \\ -0.179 & 1.40 \end{bmatrix}}_{Q^{-1}}. \tag{4.202}$$

We denote the two eigenvalues by λ_1 and λ_2, and the two corresponding eigenvectors (columns of Q) by q_1 and q_2. For any initial state vector π_1, let $\alpha := Q^{-1}\pi_1$ denote the coefficients of π_1 in the basis of eigenvectors. By Theorem 4.39,

$$\lim_{i \to \infty} \pi_i = \lim_{i \to \infty} \lambda_1^{i-1} \alpha_1 q_1 + \lambda_2^{i-1} \alpha_2 q_2 \tag{4.203}$$

$$= \alpha_1 q_1 \lim_{i \to \infty} 1 + \alpha_2 q_2 \lim_{i \to \infty} 0.648^{i-1} \tag{4.204}$$

$$= q_1 (\pi_1[1] + \pi_2[2]) \tag{4.205}$$

$$= q_1 = \begin{bmatrix} 0.887 \\ 0.113 \end{bmatrix}. \tag{4.206}$$

The stationary distribution to which the Markov chain converges is equal to the marginal empirical pmf reported in Table 4.7. Exercise 4.22 shows that this is not a coincidence.

. . . .

In some cases, the state vector of a finite-state time-homogeneous Markov chain may not converge to a stationary distribution, as illustrated by the following example.

Example 4.42 (Periodic Markov chain). We consider the three-state time-homogeneous Markov chain with the state diagram depicted in Figure 4.15. The transition matrix and corresponding eigendecomposition equal

$$T = \begin{bmatrix} 0 & 0.1 & 0 \\ 1 & 0 & 1 \\ 0 & 0.9 & 0 \end{bmatrix} \tag{4.207}$$

$$= \underbrace{\begin{bmatrix} 0.05 & 0.05 & 0.477 \\ 0.5 & -0.5 & 0 \\ 0.45 & 0.45 & -0.477 \end{bmatrix}}_{Q} \underbrace{\begin{bmatrix} 1 & 0 & 0 \\ 0 & -1 & 0 \\ 0 & 0 & 0 \end{bmatrix}}_{\Lambda} \underbrace{\begin{bmatrix} 1 & 1 & 1 \\ 1 & -1 & 1 \\ 1.89 & 0 & -0.21 \end{bmatrix}}_{Q^{-1}}. \tag{4.208}$$

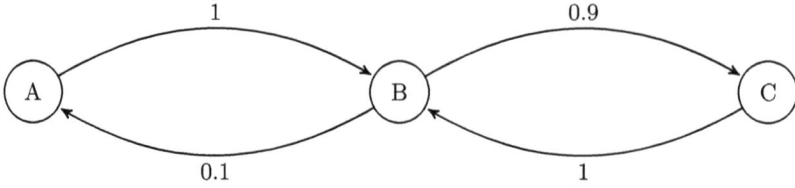

Figure 4.15 Periodic Markov chain. State diagram of a three-state time-homogeneous Markov chain with periodic dynamics. The chain alternates between states A or C, and state B.

The Markov chain has a stationary distribution corresponding to the eigenvector with eigenvalue equal to one,

$$\pi_* = \begin{bmatrix} 0.05 \\ 0.5 \\ 0.45 \end{bmatrix}. \tag{4.209}$$

If the Markov chain is initialized at this state vector, then the state vector will equal π_* indefinitely. However, let us assume an initial state vector where $\pi[2] = 0$, which means that the initial state is either A or C. In that case,

$$\pi_i = Q\Lambda^{i-1}Q^{-1}\pi_1 \tag{4.210}$$

$$= Q \begin{bmatrix} 1^{i-1} & 0 & 0 \\ 0 & (-1)^{i-1} & 0 \\ 0 & 0 & 0 \end{bmatrix} Q^{-1}\pi_1 \tag{4.211}$$

$$= \left(\begin{bmatrix} 0.05 \\ 0.5 \\ 0.45 \end{bmatrix} \begin{bmatrix} 1 & 1 & 1 \end{bmatrix} + (-1)^{i-1} \begin{bmatrix} 0.05 \\ -0.5 \\ 0.45 \end{bmatrix} \begin{bmatrix} 1 & -1 & 1 \end{bmatrix} \right) \pi_1 \tag{4.212}$$

$$= \begin{bmatrix} 0.05 \\ 0.5 \\ 0.45 \end{bmatrix} \sum_{j=1}^{3} \pi_1[j] + (-1)^{i-1} \begin{bmatrix} 0.05 \\ -0.5 \\ 0.45 \end{bmatrix} (\pi_1[1] - \pi_1[2] + \pi_1[3]) \tag{4.213}$$

$$= \begin{cases} \begin{bmatrix} 0.1 \\ 0 \\ 0.9 \end{bmatrix} & \text{if } i \text{ is odd,} \\[2em] \begin{bmatrix} 0 \\ 1 \\ 0 \end{bmatrix} & \text{if } i \text{ is even.} \end{cases} \tag{4.214}$$

The state vector oscillates between two possible values. This makes sense given the state diagram. If we begin at A or C at $i = 1$, the Markov chain is always in state B for even i, but for odd i, it moves to either A or C with probability 0.1 and 0.9, respectively.

· ·

Exercises

4.1 (Halloween) In Halloween Ellie and her brother Lou are offered a bowl with two chocolate bars. Lou grabs a random number of chocolate bars; he takes 0, 1, or 2 bars with probability 1/3. Ellie then grabs some bars out of the remaining ones, also with uniform probability (e.g. if two bars are remaining, she grabs 0, 1, or 2 with probability 1/3).

a Model the number of bars grabbed by Lou and the number of bars grabbed by Ellie as random variables, and compute their joint pmf.

b Compute the marginal pmf of the number of bars grabbed by Ellie.

c What is the conditional pmf of the number of bars grabbed by Lou, if we know that Ellie grabbed 1 bar?

4.2 (Empirical conditional pmf) Consider a dataset $\mathcal{D}$ formed by pairs of data points (x_1, y_1), $(x_2, y_2), \ldots, (x_n, y_n)$, where the first entry is in a discrete set A and the second in a discrete set B. We denote by $X := \{x_1, x_2, \ldots, x_n\}$ and $Y := \{y_1, y_2, \ldots, y_n\}$ the datasets containing the first and second entries, respectively. For any $a \in A$,

$$Y_a := \{y : (a, y) \in \mathcal{D}\} \tag{4.215}$$

is the dataset containing the values of the second entry, such that the data point (a, y) is in $\mathcal{D}$. If there are several data points with the same y values such that $(a, y) \in \mathcal{D}$, we include all of them in Y_a. We denote the total number of data points in Y_a by n_a. The empirical conditional pmf of the second entry given that the first equals $a \in A$ is

$$p_{Y \mid X}(b \mid a) := \frac{1}{n_a} \sum_{\{y \in Y_a\}} 1 \, (y = b), \qquad b \in B, \tag{4.216}$$

where $1 \, (y = b)$ is an indicator function that equals one, if $y = b$, and zero otherwise. Let $p_{X,Y}$ denote the empirical joint pmf of the two entries (see Definition 4.7), and p_X the empirical pmf of the first entry (see Definition 2.11). Show that for all $a \in A$ and $b \in B$,

$$p_{Y \mid X}(b \mid a) = \frac{p_{X,Y}(a, b)}{p_X(a)}. \tag{4.217}$$

4.3 (Noisy data) A signal $\tilde{x}$ is equal to -1 or 1 with probability $1/2$. When it is measured, the measurement $\tilde{y}$ equals $\tilde{x}$ with probability 0.9 and $-\tilde{x}$ with probability 0.1. When it is stored, the stored value $\tilde{z}$ equals $\tilde{y}$ with probability 0.9 and $-\tilde{y}$ with probability 0.1.

a If $\tilde{x} = 1$, what is the probability that $\tilde{z} = 1$?

b Are $\tilde{y}$ and $\tilde{z}$ conditionally independent given $\tilde{x}$?

c Are $\tilde{x}$ and $\tilde{z}$ conditionally independent given $\tilde{y}$?

4.4 (Fire alarm) We are interested in analyzing a newly installed fire alarm, which can ring if the battery malfunctions. We model the alarm as a Bernoulli random variable $\tilde{a}$ ($\tilde{a} = 1$ if the alarm rings, $\tilde{a} = 0$ if it doesn't), the occurrence of fire as another Bernoulli random variable $\tilde{x}$ ($\tilde{x} = 1$ if there is fire, $\tilde{x} = 0$ if there is no fire), and the behavior of the battery as a third Bernoulli random variable $\tilde{b}$ ($\tilde{b} = 1$ if the battery malfunctions, $\tilde{b} = 0$ if it is fine). After gathering some data, we determine that the probability of the alarm ringing depends on the fire alarm and the state of the battery in the following way:

	No fire	Fire
Battery is fine	0	0.8
Battery malfunctions	0.2	0.4

The table contains the probability of $\tilde{a} = 1$ conditioned on all possible values of $\tilde{x}$ and $\tilde{b}$. In all your calculations, assume that the occurrence of fire and the state of the battery are independent (i.e. $\tilde{x}$ and $\tilde{b}$ are independent). The probability of a fire is 0.2 and the probability of the battery malfunctioning is 0.5.

    a Compute the marginal pmf of $\tilde{a}$.

    b The alarm rings. Compute the conditional joint pmf of $\tilde{x}$ and $\tilde{b}$ given this event.

    c Are $\tilde{x}$ and $\tilde{b}$ conditionally independent given $\tilde{a}$? Justify your answer mathematically and intuitively.

4.5 (Volcano eruption) Three sensors on a volcano are checked each night to determine the risk of an eruption the next day. We represent the eruption by a Bernoulli random variable $\tilde{v}$ (equal to 1, if the volcano erupts, and to 0 otherwise) and the sensors by Bernoulli random variables $\tilde{s}_1$, $\tilde{s}_2$, and $\tilde{s}_3$ (equal to 1 if the sensor is activated and to 0 otherwise). The conditional probability that each sensor is activated given the state of the volcano is:

	$p_{\tilde{s}_1 \mid \tilde{v}}(1 \mid v)$	$p_{\tilde{s}_2 \mid \tilde{v}}(1 \mid v)$	$p_{\tilde{s}_3 \mid \tilde{v}}(1 \mid v)$
$v = 0$	0.2	0.2	0.2
$v = 1$	0.5	0.8	0.9

We model $\tilde{s}_1$, $\tilde{s}_2$, and $\tilde{s}_3$ as being conditionally independent given $\tilde{v}$. The probability of an eruption is 0.1.

    a What is the conditional probability of an eruption given that all sensors are activated (i.e. $\tilde{s}_1$, $\tilde{s}_2$, and $\tilde{s}_3$ are all equal to 1)?

    b Are $\tilde{s}_1$ and $\tilde{s}_2$ independent? Justify your answer mathematically and intuitively.

4.6 (Flu and COVID-19) An epidemiologist is studying the symptoms of the flu and COVID-19. She models the diseases and the symptoms as Bernoulli random variables: $\tilde{x}$ (flu), $\tilde{c}$ (COVID-19), $\tilde{s}$ (sore throat), and $\tilde{t}$ (high temperature). If the corresponding random variable equals 1, then the patient has the disease or the symptom. The random variables $\tilde{x}$ and $\tilde{c}$ are assumed to be independent. The probability that a patient has the flu is 0.1. The probability that a patient has COVID-19 is 0.2. The following tables contain the entries of $p_{\tilde{s},\tilde{t} \mid \tilde{c},\tilde{x}}(s,t \mid c,x)$ for all possible values of s, t, c, and x.

		$c = 0$				$c = 1$	
		$t = 0$	$t = 1$			$t = 0$	$t = 1$
$x = 0$	$s = 0$	0.8	0		$s = 0$	0	0.2
	$s = 1$	0.2	0		$s = 1$	0.4	0.4

		$t = 0$	$t = 1$			$t = 0$	$t = 1$
$x = 1$	$s = 0$	0	0.4		$s = 0$	0	0
	$s = 1$	0.2	0.4		$s = 1$	0	1

    a What is the probability that a patient has a high temperature?

    b What is the probability that a patient with a high temperature has COVID-19?

4.7 (Vibrations) During a period of high seismic activity, a group of scientists is trying to predict the occurrence of earthquakes by measuring vibrations in the ground. They model the occurrence of an earthquake as a random variable $\tilde{e}$ ($\tilde{e} = 1$ if there is an earthquake, and $\tilde{e} = 0$ if there isn't) and the vibrations as a random variable $\tilde{v}$ ($\tilde{v} = 0$ if there are no vibrations, $\tilde{v} = 1$ if there are small vibrations, and $\tilde{v} = 2$ if there are large vibrations). The joint pmf of $\tilde{e}$ and $\tilde{v}$ is:

Vibrations

$p_{\tilde{e},\tilde{v}}$	0	1	2
Earthquake 0	0.8	0.05	0
1	0	0.05	0.1

The sensor reading is modeled as a Bernoulli random variable $\tilde{s}$ that is conditionally independent of the earthquake given the vibrations. If there are no vibrations, the reading is always 0. If there are small vibrations, the reading is 1 with probability 0.5. If there are large vibrations, the reading is always 1.

a Derive the marginal pmf of $\tilde{s}$.

b What is the probability that there is an earthquake, if the sensor reading equals 1?

c Are the random variables $\tilde{s}$ and $\tilde{e}$ independent? Justify your answer mathematically, but also explain it intuitively.

4.8 (Footprints) A researcher uses a hidden camera to identify coyotes and wolves in Yellowstone. She then examines the footprints left by the animals to record their size and shape. These are the data:

Size

Coyote	Small	Medium	Large
Shape Oval	80	30	5
Circular	30	5	0

Size

Wolf	Small	Medium	Large
Oval	0	5	10
Circular	5	15	15

a Use the data to estimate the joint pmf of three random variables $\tilde{a}$, $\tilde{s}$, and $\tilde{h}$ representing the type of animal, the size of the footprint and the shape of the footprint, respectively.

b Her goal is to use these data to identify new footprints found in other parts of the park. According to the estimated joint pmf, what is the probability that a medium-sized circular-shaped footprint belongs to a wolf?

4.9 (Election) The made-up country Lalaland is composed of three states. Their presidential election is won by the candidate that wins more states. We model the result in each state as Bernoulli random variables $\tilde{s}_1$, $\tilde{s}_2$, and $\tilde{s}_3$ ($\tilde{s}_i = 1$ means that A wins state i). From the available data, we determine that the probability that candidate A wins state 1, 2, or 3 is equal to the same value: 0.6. In addition, we assume that $\tilde{s}_1$, $\tilde{s}_2$, and $\tilde{s}_3$ are mutually independent.

a What is the probability that candidate A wins the election?

b If B wins the election, what is the conditional probability that they won in state 2?

c Are $\tilde{s}_1$ and $\tilde{s}_2$ conditionally independent given the overall result of the election? Justify your answer mathematically and also explain it intuitively.

4.10 (Interview) A company is interviewing candidates for a data-scientist position. They estimate that the probability of a candidate being well qualified is 0.25. This is modeled by a random variable $\tilde{q}$ that equals 1 with probability 0.25, and -1 with probability 0.75. Candidates are interviewed separately by two interviewers. The decision of the interviewers are modeled as two random variables $\tilde{d}_1 = \tilde{e}_1\tilde{q}$ and $\tilde{d}_2 = \tilde{e}_2\tilde{q}$, where $\tilde{e}_1$ and $\tilde{e}_2$ are random variables representing whether the interviewers make a mistake or not. Each equals 1 with probability 0.8

(no mistake) and -1 with probability 0.2 (mistake). We assume that the interviews make mistakes independently, so that $\tilde{e}_1$, $\tilde{e}_2$, and $\tilde{q}$ are all mutually independent. The candidate is hired, if the decision of both interviewers is positive.

a What is the probability that the candidate is hired, if they are well qualified?
b What is the probability that the candidate is well qualified, if they are hired?
c Are $\tilde{d}_1$ and $\tilde{d}_2$ independent?
d Are $\tilde{d}_1$ and $\tilde{d}_2$ conditionally independent given $\tilde{q}$?

4.11 (Surgery) A doctor, called Ramiro, wants to evaluate two surgery procedures: A and B. There are two types of patients that receive the procedure, *mild* and *serious* cases. The truth is that procedure A is better. Mild cases recover with probability 0.9, if they receive A, and with probability 0.8, if they receive B. Serious cases recover with probability 0.5, if they receive A, and with probability 0.2, if they receive B.

a Ramiro gathers data from past surgeries. He determines that patients recover with probability 0.58, if they receive procedure A, and with probability 0.68, if they receive B. How is this possible? Justify your answer mathematically. (Hint: Start by computing what fraction of patients receiving each procedure are mild or serious cases.)
b Explain how Ramiro should analyze the data in order to obtain an accurate conclusion about the surgery procedures. Under what assumption does this work?
c Suggest how to design a follow-up study that would not require adjusting for confounding factors.

4.12 (Admissions) The following table shows the number of men and women that applied to the medical school and the art school at a university, and also the corresponding number of admitted applicants:

Medical school				Art school		
	Applicants	Admitted			Applicants	Admitted
Women	500	25		Women	500	50
Men	800	40		Men	200	25

a Estimate the conditional probability that an applicant is admitted if they are a man and if they are a woman.
b A group of rejected applicants are upset because they believe that they would have been accepted if *they were a woman*. In order to study whether the data supports this statement, we define the potential outcome $\widetilde{po}_{man}$ and $\widetilde{po}_{woman}$ to indicate whether an applicant is admitted, if they are a man ($\widetilde{po}_{man} = 1$) and if they are a woman ($\widetilde{po}_{woman} = 1$). The observed outcome $\tilde{y}$ is equal to $\widetilde{po}_{man}$, if the applicant is a man, and to $\widetilde{po}_{woman}$, if they are a woman. Assuming that the potential outcomes are conditionally independent of the sex of the applicant given the school they applied to, estimate $P(\widetilde{po}_{man} = 1)$ and $P(\widetilde{po}_{woman} = 1)$. Do the data support the thesis of the rejected applicants? (Hint: It may be useful to define random variables $\tilde{s}$ and $\tilde{x}$ to represent the sex of the applicant and the school they apply to, respectively.)

4.13 (Shop) A shop owner wants to determine whether customers are more likely to buy something, if music is playing. To test this, they decide to play music in the evening, and record whether customers of different ages make a purchase or not. Here are the data (each entry indicates number of customers):

Music		Young	Middle-aged	Old
Purchase	Yes	10	90	20
	No	10	10	20

No music		Young	Middle-aged	Old
Purchase	Yes	40	9	60
	No	40	1	40

a Estimate the conditional probability that a customer makes a purchase if (1) there is music and (2) there is no music.

b We define the potential outcomes $\widetilde{po}_1$ and $\widetilde{po}_0$ to indicate whether a customer makes a purchase when there is music ($\widetilde{po}_1 = 1$) or no music ($\widetilde{po}_0 = 1$). The observed outcome $\tilde{y}$ is equal to $\widetilde{po}_1$, if there is music, and to $\widetilde{po}_0$, if there is no music. Assuming that the potential outcomes are conditionally independent of the music given age, estimate the pmfs of $\widetilde{po}_1$ and $\widetilde{po}_0$. Under this assumption, does the presence of music increase or decrease purchases?

4.14 (Rackets) A tennis player is trying to decide between two racket types: A and B. Out of all the games she plays with model A, a fraction α takes place on clay and $1 - \alpha$ on grass. Out of all the games she plays with racket B, half take place on clay and half on grass. We define the potential outcomes $\widetilde{po}_A$ and $\widetilde{po}_B$ to indicate whether she wins a game, if she uses racket A ($\widetilde{po}_A = 1$), or if she uses racket B ($\widetilde{po}_B = 1$). The observed outcome $\tilde{y}$ is equal to $\widetilde{po}_A$, if the racket type $\tilde{r}$ is A, and to $\widetilde{po}_B$, if $\tilde{r} = B$. We assume that the potential outcomes are independent of the choice of racket given the surface (clay or grass).

a If α is equal to the fraction of total games that take place on clay, then the observed probability $P(\tilde{y} = 1 \mid \tilde{r} = A)$ equals $P(\widetilde{po}_A = 1)$. True or false? Justify your answer.

b In reality, racket A is better. Let $\tilde{s}$ be a random variable that represents the surface. We have:

$$P(\widetilde{po}_A = 1 \mid \tilde{s} = \text{clay}) := 0.8, \tag{4.218}$$
$$P(\widetilde{po}_B = 1 \mid \tilde{s} = \text{clay}) := 0.7, \tag{4.219}$$
$$P(\widetilde{po}_A = 1 \mid \tilde{s} = \text{grass}) := 0.2, \tag{4.220}$$
$$P(\widetilde{po}_B = 1 \mid \tilde{s} = \text{grass}) := 0.1. \tag{4.221}$$

For what values of α does Simpson's paradox occur, so that racket B seems better ($P(\tilde{y} = 1 \mid \tilde{r} = A) < P(\tilde{y} = 1 \mid \tilde{r} = B)$)?

4.15 (Missing data) A medical researcher is trying to determine the probability that a certain drug produces side effects. Unfortunately, for some patients, we don't know whether there were any side effects; the data are missing because they never returned to the doctor. Here is the dataset:

	Side effects	No side effects	Missing
Men	40	10	50
Women	30	60	10

a A common assumption when dealing with missing data is that the data are *missing at random*, meaning that each data point is missing with the same probability, independently from all the others. Do you think this holds here?

b What is the probability of a patient suffering side effects according to the observed data?

c Assuming that side effects are conditionally independent of the data being observed given the patient sex, estimate the true probability that a patient suffers side effects.

4.16 (Three players) A basketball team has three-star players called James, Kevin, and Kyrie, who are often injured. The following table shows the results of 10 games, and also what players were present (✓) or absent (✗) in each game.

Result	James	Kevin	Kyrie
Win	✓	✓	✓
Win	✓	✗	✓
Loss	✓	✗	✗
Win	✗	✓	✗
Win	✓	✓	✗
Loss	✓	✓	✗
Loss	✓	✗	✗
Win	✗	✓	✓
Loss	✗	✗	✓
Win	✓	✓	✓

a Can we estimate the probability that the team wins a game when none of the three players participate using the empirical-probability estimator?

b Apply naive Bayes to estimate the probability that the team wins a game when none of the three players participate.

4.17 (Spam detector) In order to build a spam classifier, we gather the following data. Each row corresponds to an email. The first column indicates whether the email is spam or not. The remaining columns indicate whether the email contains (✓) or not (✗) the word at the top.

	Miracle	Alternative	Medicine	Basketball
Spam	✓	✓	✓	✓
Not spam	✗	✗	✓	✓
Spam	✓	✗	✗	✗
Not spam	✗	✓	✗	✗
Spam	✓	✗	✗	✗
Not spam	✗	✓	✗	✓
Spam	✓	✗	✗	✗
Spam	✗	✓	✓	✗
Not spam	✓	✓	✗	✓
Not spam	✗	✗	✓	✓

We use a Bernoulli random variable $\tilde{y}$ to model whether the email is spam ($\tilde{y} = 1$) or not ($\tilde{y} = 0$), and a four-dimensional random vector $\tilde{x}$ to indicate whether the jth word is ($\tilde{x}[j] = 1$) or not ($\tilde{x}[j] = 0$) in the email for $j \in \{1, 2, 3, 4\}$.

a Your friend suggests that you just estimate the conditional pmf of $\tilde{y}$ given $\tilde{x}$ and then compute the resulting maximum a posteriori estimator. What is the problem with this approach?

b Apply naive Bayes to classify an email that reads *I hurt my foot playing basketball. Can you get me some medicine?*

c Apply naive Bayes to classify an email that reads *This alternative medicine is amazing, send us all your money!* Explain what shortcoming of the naive-Bayes classifier is illustrated by this example.

4.18 (Symbols) A linguist analyzes an inscription with three symbols:

$$*\square\square\square*\dagger\dagger\dagger*\dagger\dagger\dagger\square\dagger\dagger\dagger\square\dagger\dagger\dagger*\square*\square\square*\dagger\dagger\dagger\square\square*\square\dagger\dagger\dagger*$$

   a The linguist decides to model the sequence as a Markov chain. Estimate the transition matrix of the Markov chain. Do you think that the Markov assumption holds for this sequence? Explain why or why not.

   b Predict the next two symbols (after the last symbol of the observed sequence) using the most likely two-symbol sequence, according to the Markov-chain model.

4.19 (The Markov property) Let $\tilde{a}_1, \tilde{a}_2, \ldots, \tilde{a}_n$ be a Markov chain, where for any $1 < i < n$, $\tilde{a}_{i+1}$ is conditionally independent of $\tilde{a}_1, \ldots, \tilde{a}_{i-1}$ given $\tilde{a}_i$, i.e.

$$p_{\tilde{a}_{i+1} \mid \tilde{a}_1, \ldots, \tilde{a}_i} (a_{i+1} \mid a_1, a_2, \ldots, a_i) = p_{\tilde{a}_{i+1} \mid \tilde{a}_i} (a_{i+1} \mid a_i) \tag{4.222}$$

for any values of $a_1, a_2, \ldots, a_n$. Show that this implies that the future is conditionally independent from the past given the present:

$$p_{\tilde{a}_{i+1}, \tilde{a}_{i-1} \mid \tilde{a}_i} (a_{i+1}, a_{i-1} \mid a_i) = p_{\tilde{a}_{i+1} \mid \tilde{a}_i} (a_{i+1} \mid a_i) \, p_{\tilde{a}_{i-1} \mid \tilde{a}_i} (a_{i-1} \mid a_i) \tag{4.223}$$

for any $2 \leq i \leq n - 1$ and any values of a_{i-1}, a_i, a_{i+1}. (Hint: First, show

$$p_{\tilde{a}_{i+1} \mid \tilde{a}_{i-1}, \tilde{a}_i} (a_{i+1} \mid a_{i-1}, a_i) = p_{\tilde{a}_{i+1} \mid \tilde{a}_i} (a_{i+1} \mid a_i) \tag{4.224}$$

for any $2 \leq i \leq n - 1$ and any a_{i-1}, a_i, a_{i+1}.)

4.20 (Employment dynamics) A researcher is interested in modeling the employment dynamics of young people. She assumes that at age 18, everyone is a student and models the employment status at age $18 + i$ as a random variable $\tilde{x}_i$. She models the joint distribution of $\tilde{x}_0, \tilde{x}_1, \ldots$ as a Markov chain with transition matrix:

$$\begin{pmatrix} \text{Student} & \text{Employed} & \text{Unemployed} \\ 0.8 & 0 & 0 \\ 0.2 & 0.9 & 0.4 \\ 0 & 0.1 & 0.6 \end{pmatrix} \begin{matrix} \\ \text{Student} \\ \text{Employed} \\ \text{Unemployed} \end{matrix}$$

   a Let $\tilde{s}$ denote the number of years that a person remains a student after age 18. What is the pmf of $\tilde{s}$?

   b If we know that someone was employed at $i = 3$, what is the conditional probability that they were a student at $i = 1$ given this information?

   c Show that the state vector of any stationary distribution of the Markov chain must be of the form

$$\begin{bmatrix} 0 \\ \alpha \\ 1 - \alpha \end{bmatrix}, \tag{4.225}$$

where $0 \leq \alpha \leq 1$, and compute the value of α.

4.21 (Cellphones) A company that makes cellphones wants to model the sales of a new model they have just released. At the moment, 90% of the phones are in stock, 10% have been sold locally, and none have been exported. Based on past data, the company determines that each day a phone is sold with probability 0.2 and exported with probability 0.1. We define the following time-homogeneous Markov chain with three states to model this:

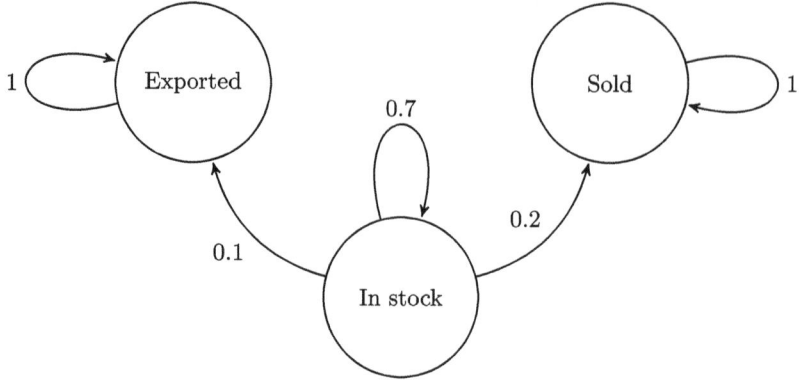

a What is the limit of the state vector π_i as $i \to \infty$?

b Simulate the Markov chain and plot the evolution of the state vector.

4.22 (Empirical state vector and transition matrix) Assume we have access to a sequence of observed data $X := \{x_1, \ldots, x_n\}$, which have m possible values $s_1, \ldots, s_m$. Let π_X be a state vector containing the empirical probabilities of each value

$$\pi_X[j] := p_X(s_j), \qquad 1 \le j \le m, \tag{4.226}$$

where p_X is the empirical pmf of the data, computed following Definition 2.11. We define the empirical transition matrix T_X as the matrix containing the empirical conditional probabilities (see Definition 1.24) of all possible transitions. For $1 \le i, j \le m$,

$$T_X[i, j] := \frac{1}{n_j} \sum_{l=2}^{n} 1\left(x_{l-1} = s_j, x_l = s_i\right). \tag{4.227}$$

The indicator function $1\left(x_{l-1} = s_j, x_l = s_i\right)$ is equal to one, if both $x_{l-1} = s_j$ and $x_l = s_i$, and to zero otherwise. The number of transitions starting at s_j is denoted by n_j. The entry $T_X[i, j]$ is the fraction of those transitions that end in s_i. Show that

$$T_X \pi_X \approx \pi_X, \tag{4.228}$$

so the empirical state vector is always approximately stationary with respect to the empirical transition matrix, which is what we observe in Example 4.41.

5

Multiple Continuous Variables

Overview

In this chapter, we explain how to jointly model multiple continuous quantities. Section 5.1 shows that such quantities can be represented as continuous random variables within the same probability space. We describe these random variables in terms of their joint cumulative distribution function or their joint probability density function (pdf), as discussed in Sections 5.2 and 5.3, respectively. Section 5.4 explains how to estimate the joint pdf from data, using a multidimensional generalization of kernel density estimation. Section 5.5 describes how to characterize the behavior of individual random variables in models with multiple variables. Section 5.6 defines the conditional distribution of continuous random variables given the value of other variables. Sections 5.7 and 5.8 discuss independence and conditional independence. In Section 5.9, we explain how to jointly simulate multiple continuous random variables. Finally, Section 5.10 introduces Gaussian random vectors, which are the most popular multidimensional parametric model for continuous data.

5.1 Joint Distribution of Continuous Random Variables

To model the joint behavior of multiple uncertain continuous quantities, we represent them as random variables belonging to a common probability space, following the same strategy as in Section 4.1.1. The only difference is that the random variables are continuous, instead of discrete. As explained in Section 3.1, we describe individual continuous random variables in terms of the probability that they belong to intervals of the real line. It is therefore natural to describe the joint distribution of multiple continuous random variables in terms of the probability that they belong to different intervals *at the same time*. Fortunately, these probabilities are guaranteed to exist if the random variables are defined in the same probability space, as explained in this section, so our strategy is mathematically sound.

Consider two continuous random variables $\tilde{a}$ and $\tilde{b}$ defined on the same probability space $\{P, C, \Omega\}$. We are interested in the event that $\tilde{a}$ is in an interval $A \subseteq \mathbb{R}$ and simultaneously $\tilde{b}$ is in another interval $B \subseteq \mathbb{R}$. Equivalently, the two-dimensional random vector formed by $\tilde{a}$ and $\tilde{b}$ is in the rectangle formed by the Cartesian product of the two intervals, $A \times B$. This event contains all outcomes ω of the sample space Ω such that $\tilde{a}(\omega)$ is in A and $\tilde{b}(\omega)$ is in B, as illustrated in Figure 5.1. Its probability equals

$$P\left(\begin{bmatrix} \tilde{a} \\ \tilde{b} \end{bmatrix} \in A \times B\right) := P(\{\omega : \tilde{a}(\omega) \in A \text{ and } \tilde{b}(\omega) \in B\}) \tag{5.1}$$

$$= P\left(A^{-1} \cap B^{-1}\right), \tag{5.2}$$

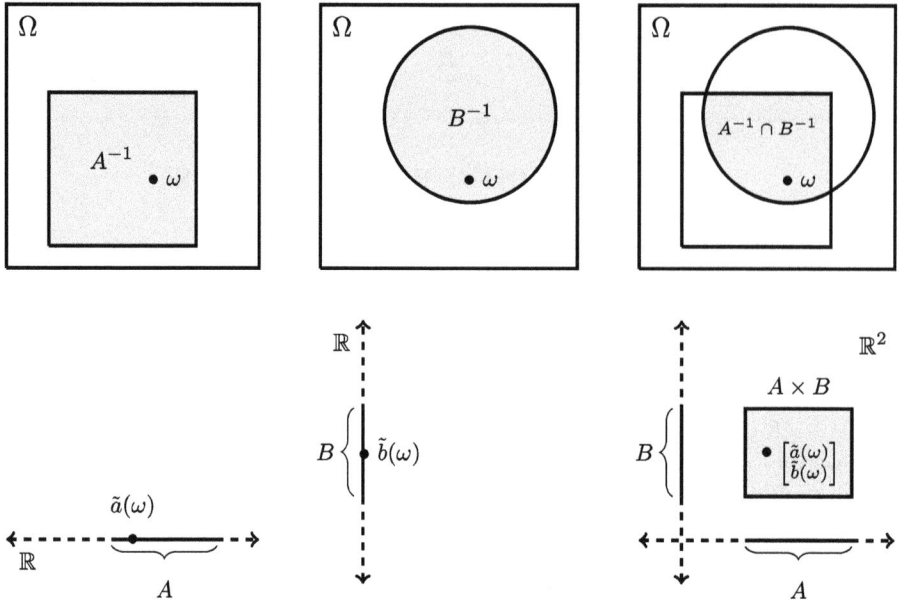

Figure 5.1 Joint distribution of continuous random variables. The continuous random variables $\tilde{a}$ and $\tilde{b}$ map outcomes in the sample space Ω to the real line $\mathbb{R}$. The left Venn diagram shows the event A^{-1} mapped by $\tilde{a}$ to the interval A, depicted on the horizontal real line. The middle Venn diagram shows the event B^{-1} mapped by $\tilde{b}$ to the interval B, depicted on the vertical real line. The right Venn diagram shows that outcomes mapped to A by $\tilde{a}$ and to B by $\tilde{b}$ are in the intersection $A^{-1} \cap B^{-1}$. These outcomes are mapped by the random vector $\begin{bmatrix} \tilde{a} \\ \tilde{b} \end{bmatrix}$ to the rectangle $A \times B$. The probability that the random vector belongs to $A \times B$ therefore equals $P(A^{-1} \cap B^{-1})$, represented by the area of the event $A^{-1} \cap B^{-1}$ in the Venn diagram.

where A^{-1} and B^{-1} are the events containing the outcomes that are mapped to A via $\tilde{a}$ and to B via $\tilde{b}$, respectively,

$$A^{-1} := \{\omega \colon \tilde{a}(\omega) \in A\}, \tag{5.3}$$

$$B^{-1} := \{\omega \colon \tilde{b}(\omega) \in B\}. \tag{5.4}$$

If $\tilde{a}$ and $\tilde{b}$ are continuous random variables satisfying Definition 3.1, then A^{-1} and B^{-1} belong to the collection $\mathcal{C}$ and are assigned probabilities by the probability measure P. By the properties of probability spaces (see Definitions 1.7 and 1.15), a probability must be assigned to the intersection of A^{-1} and B^{-1}, so we can rest assured that the probability of our event of interest is well defined. For the same reason, the probability that $\tilde{a}$ and $\tilde{b}$ simultaneously belong to any pair of Borel sets, formed by countable unions of intervals, is also well defined.

We can extend the same reasoning to more than two continuous random variables. Let $\tilde{x}$ be a random vector consisting of d continuous random variables $\tilde{x}[1]$, $\tilde{x}[2]$, ..., $\tilde{x}[d]$, defined on the same probability space $(\Omega, \mathcal{C}, \mathrm{P})$. For any d intervals $X_1, X_2, \ldots, X_d \subseteq \mathbb{R}$, the event that $\tilde{x}$ is in the hyperrectangle formed by the Cartesian product of the intervals, $X_1 \times X_2 \times \cdots \times X_d$, can be expressed as the intersection of d events:

$$\{\omega \colon \tilde{x}(\omega) \in X_1 \times X_2 \times \cdots \times X_d\} = \cap_{i=1}^{d} \{\omega \colon \tilde{x}[i](\omega) \in X_i\}. \tag{5.5}$$

These events are all in $\mathcal{C}$, as long $\tilde{x}[1], \tilde{x}[2], \ldots, \tilde{x}[d]$ are valid continuous random variables, so their intersection also belongs to $\mathcal{C}$ and is assigned a probability by the probability measure P. Similarly, the probability that $\tilde{x}$ belongs to any d-dimensional Borel set, defined as the Cartesian product of d Borel sets, is also well defined.

By Theorem 3.2, the probability that a continuous random variable belongs to any countable union of disjoint intervals is equal to the sum of the probabilities that it belongs to each individual interval. By the same logic, the probability that a d-dimensional continuous random vector belongs to any union of d-dimensional hyperrectangles (i.e. to any multidimensional Borel set) is equal to the sum of the probabilities that it belongs to each individual hyperrectangle. The reasoning is the same as in the proof of Theorem 3.2: The events mapped to the disjoint hyperrectangles by the random vector are all disjoint, so the probability of their union is equal to the sum of the individual probabilities. This means that we don't need to worry about the underlying probability space, as long as we have access to the probability that the random vector belongs to any d-dimensional hyperrectangle. As explained in Sections 5.2 and 5.3, we keep track of these probabilities using generalizations of the cumulative distribution function and the probability density function. This is analogous to how we define and manipulate discrete random vectors only via their joint pmf (see Section 4.1.2).

5.2 Joint Cumulative Distribution Function

The cumulative distribution function (cdf), introduced in Section 3.2, encodes the probability that a random variable is less than or equal to any real number. The joint cdf is a generalization of the cdf to two or more random variables.

Definition 5.1 (Joint cumulative distribution function). *Let $\tilde{a}$ and $\tilde{b}$ be random variables defined on the same probability space. The joint cdf of $\tilde{a}$ and $\tilde{b}$ is defined as*

$$F_{\tilde{a}, \tilde{b}}(a, b) := \mathrm{P}(\tilde{a} \leq a, \tilde{b} \leq b), \tag{5.6}$$

where $a, b \in \mathbb{R}$. In words, $F_{\tilde{a}, \tilde{b}}(a, b)$ is the probability of $\tilde{a}$ and $\tilde{b}$ being less than or equal to a and b, respectively, at the same time.

Let $\tilde{x}$ be a vector with entries equal to d continuous random variables, $\tilde{x}[i]$, $1 \leq i \leq d$, defined on the same probability space. The joint cdf of $\tilde{x}$ is

$$F_{\tilde{x}}(x) := \mathrm{P}\left(\tilde{x}[1] \leq x[1], \tilde{x}[2] \leq x[2], \ldots, \tilde{x}[d] \leq x[d]\right), \tag{5.7}$$

where $x \in \mathbb{R}^d$. In words, $F_{\tilde{x}}(x)$ is the probability that $\tilde{x}[i] \leq x[i]$, for all $i = 1, 2, \ldots, d$, at the same time.

The joint cdf has similar properties to the cdf (see Lemma 3.4). For simplicity, we state them for two variables, but analogous properties hold for the joint cdf of more than two random variables.

Lemma 5.2 (Properties of the joint cdf). *For any random variables $\tilde{a}$ and $\tilde{b}$ defined on the same probability space, the joint cdf $F_{\tilde{a}, \tilde{b}}$ of $\tilde{a}$ and $\tilde{b}$ satisfies*

$$\lim_{a \to -\infty} F_{\tilde{a},\tilde{b}}(a,b) = 0, \qquad for\ any\ b \in \mathbb{R}, \tag{5.8}$$

$$\lim_{b \to -\infty} F_{\tilde{a},\tilde{b}}(a,b) = 0, \qquad for\ any\ a \in \mathbb{R}, \tag{5.9}$$

$$\lim_{a \to \infty, b \to \infty} F_{\tilde{a},\tilde{b}}(a,b) = 1. \tag{5.10}$$

In addition, $F_{\tilde{a},\tilde{b}}$ is nondecreasing in each of its entries,

$$F_{\tilde{a},\tilde{b}}(a_1,b_1) \le F_{\tilde{a},\tilde{b}}(a_2,b_2), \quad if\ a_2 \ge a_1, b_2 \ge b_1. \tag{5.11}$$

Proof $F_{\tilde{a},\tilde{b}}(a,b)$ is the probability that both $\tilde{a} \le a$ and $\tilde{b} \le b$, so if we make either of them arbitrarily small, the probability becomes zero. This can be made mathematically rigorous using a similar argument to the proof of (3.16) in Lemma 3.4. Similarly, if we make a and b arbitrarily large, the probability becomes one.

If $a_2 \ge a_1$ and $b_2 \ge b_1$, then the event $\{\tilde{a} \le a_1\} \cap \{\tilde{b} \le b_1\}$ is a subset of the event $\{\tilde{a} \le a_2\} \cap \{\tilde{b} \le b_2\}$, so by Definition 5.1 and Lemma 1.12,

$$F_{\tilde{a},\tilde{b}}(a_1,b_1) = \mathrm{P}(\{\tilde{a} \le a_1\} \cap \{\tilde{b} \le b_1\}) \tag{5.12}$$

$$\le \mathrm{P}(\{\tilde{a} \le a_2\} \cap \{\tilde{b} \le b_2\}) \tag{5.13}$$

$$= F_{\tilde{a},\tilde{b}}(a_2,b_2). \tag{5.14}$$

∎

The following lemma shows how to compute probabilities that describe the joint behavior of two random variables using their joint cdf.

Lemma 5.3 (Probability of a rectangle). *Let $F_{\tilde{a},\tilde{b}}$ denote the cdf of two continuous random variables $\tilde{a}$ and $\tilde{b}$. For any $a_1 < a_2$ and $b_1 < b_2$,*

$$\mathrm{P}(a_1 < \tilde{a} \le a_2, b_1 < \tilde{b} \le b_2) \tag{5.15}$$

$$= F_{\tilde{a},\tilde{b}}(a_2,b_2) - F_{\tilde{a},\tilde{b}}(a_1,b_2) - F_{\tilde{a},\tilde{b}}(a_2,b_1) + F_{\tilde{a},\tilde{b}}(a_1,b_1). \tag{5.16}$$

Proof   For convenience, we define the 2D random vector

$$\tilde{x} := \begin{bmatrix} \tilde{a} \\ \tilde{b} \end{bmatrix}, \tag{5.17}$$

and the disjoint 2D sets

$$R := (a_1, a_2] \times (b_1, b_2], \tag{5.18}$$

$$S_1 := (-\infty, a_1] \times (b_1, b_2], \tag{5.19}$$

$$S_2 := (-\infty, a_1] \times (-\infty, b_1], \tag{5.20}$$

$$S_3 := (a_1, a_2] \times (-\infty, b_1], \tag{5.21}$$

which are depicted in Figure 5.2. Since the sets are disjoint, the probability that $\tilde{x}$ belongs to the union of any of these sets is equal to the sum of the probabilities that it belongs to each of the sets in the union. Formally, this follows from Axiom 3 in Definition 1.9 because the

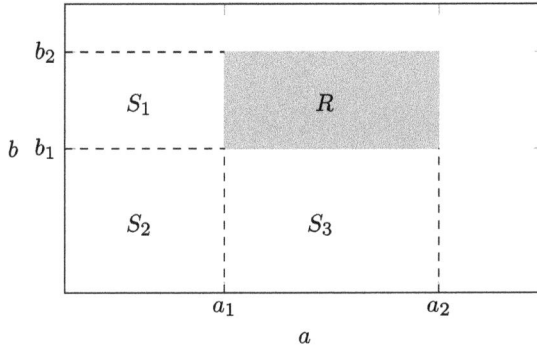

Figure 5.2 Probability of a rectangle. The graph shows the 2D sets defined in the proof of Lemma 5.3, which derives a formula for the probability that a 2D random vector belongs to the shaded rectangle, as a function of the joint cdf of the random vector.

events mapped to each of the 2D sets by $\tilde{x}$ are disjoint. We omit the details, which are very similar to the proof of Theorem 3.2. As a result, by Definition 5.1,

$$F_{\tilde{a},\tilde{b}}(a_2, b_2) := \mathrm{P}(\tilde{a} \leq a_2, \tilde{b} \leq b_2) \tag{5.22}$$

$$= \mathrm{P}(\tilde{x} \in S_1 \cup S_2 \cup S_3 \cup R) \tag{5.23}$$

$$= \mathrm{P}(\tilde{x} \in S_1) + \mathrm{P}(\tilde{x} \in S_2) + \mathrm{P}(\tilde{x} \in S_3) + \mathrm{P}(\tilde{x} \in R), \tag{5.24}$$

and by the same argument,

$$F_{\tilde{a},\tilde{b}}(a_1, b_1) = \mathrm{P}(\tilde{x} \in S_2), \tag{5.25}$$

$$F_{\tilde{a},\tilde{b}}(a_1, b_2) = \mathrm{P}(\tilde{x} \in S_1) + \mathrm{P}(\tilde{x} \in S_2), \tag{5.26}$$

$$F_{\tilde{a},\tilde{b}}(a_2, b_1) = \mathrm{P}(\tilde{x} \in S_2) + \mathrm{P}(\tilde{x} \in S_3). \tag{5.27}$$

Combining all of these equations yields the desired expression for the probability of R in terms of the joint cdf of $\tilde{a}$ and $\tilde{b}$,

$$\mathrm{P}(a_1 < \tilde{a} \leq a_2, b_1 < \tilde{b} \leq b_2) \tag{5.28}$$

$$= \mathrm{P}(R) \tag{5.29}$$

$$= F_{\tilde{a},\tilde{b}}(a_2, b_2) - \mathrm{P}(\tilde{x} \in S_1) - \mathrm{P}(\tilde{x} \in S_2) - \mathrm{P}(\tilde{x} \in S_3) \tag{5.30}$$

$$= F_{\tilde{a},\tilde{b}}(a_2, b_2) - F_{\tilde{a},\tilde{b}}(a_1, b_2) - F_{\tilde{a},\tilde{b}}(a_2, b_1) + F_{\tilde{a},\tilde{b}}(a_1, b_1). \tag{5.31}$$

∎

Armed with Lemma 5.3, we can express the probability that two random variables belong to any rectangle in $\mathbb{R}^2$ in terms of their joint cdf. The result can be generalized, using the same arguments, to yield an analogous (but more complicated) expression of the probability that a d-dimensional random vector belongs to any hyperrectangle in terms of its joint cdf. This implies that the joint cdf completely characterizes the behavior of a random vector. Indeed, we can decompose the probability that the random vector belongs to any countable union of hyperrectangles in $\mathbb{R}^d$ (i.e. any multidimensional Borel set) into the sum of the probabilities that it belongs to individual hyperrectangles, as explained in Section 5.1

(see also Theorem 3.2 for a proof in the one-dimensional case).[1] Consequently, we can use the joint cdf to express the probability that the random vector belongs to any Borel set. Unfortunately, it is cumbersome to manipulate random vectors via their joint cdf because the resulting expressions become very complicated. As a result, the joint cdf is useful mostly as a mathematical tool and is seldom estimated from data. Instead, we normally use probability densities to describe and manipulate multiple random variables, as described in Section 5.3.

5.3 Joint Probability Density Function

In Section 3.3, we define the probability density of a continuous random variable and show that it completely describes the behavior of the random variable. In this section, we introduce a multidimensional generalization of the probability density.

Let us consider two random variables $\tilde{a}$ and $\tilde{b}$ defined on the same probability space. We are interested in defining their joint probability density at a certain point $\begin{bmatrix} a \\ b \end{bmatrix}$. The probability that the two-dimensional random vector formed by the two random variables belongs to a small square of area ϵ^2 that includes $\begin{bmatrix} a \\ b \end{bmatrix}$ is

$$
\mathrm{P}\left(\begin{bmatrix} \tilde{a} \\ \tilde{b} \end{bmatrix} \in [a - \epsilon, a] \times [b - \epsilon, b] \right). \tag{5.32}
$$

To obtain the corresponding probability density $f_{\tilde{a},\tilde{b}}(a, b)$, we divide the probability by the area of the square and take the limit $\epsilon \to 0$:

$$
f_{\tilde{a},\tilde{b}}(a, b) := \lim_{\epsilon \to 0} \frac{\mathrm{P}\left(a - \epsilon < \tilde{a} \leq a, b - \epsilon < \tilde{b} \leq b \right)}{\epsilon^2}. \tag{5.33}
$$

The probability of the small square is approximately equal to $\epsilon^2 f_{\tilde{a},\tilde{b}}(a, b)$, for small enough ϵ, so $f_{\tilde{a},\tilde{b}}(a, b)$ is a measure of probability per unit area. As in the one-dimensional case, the probability density can be obtained by differentiating the corresponding joint cdf. By Lemma 5.3 and the definition of partial derivative,

$$
f_{\tilde{a},\tilde{b}}(a, b) \tag{5.34}
$$

$$
= \lim_{\epsilon \to 0} \frac{F_{\tilde{a},\tilde{b}}(a, b) - F_{\tilde{a},\tilde{b}}(a - \epsilon, b) - F_{\tilde{a},\tilde{b}}(a, b - \epsilon) + F_{\tilde{a},\tilde{b}}(a - \epsilon, b - \epsilon)}{\epsilon^2} \tag{5.35}
$$

$$
= \lim_{\epsilon \to 0} \frac{1}{\epsilon} \left(\lim_{\epsilon \to 0} \frac{F_{\tilde{a},\tilde{b}}(a, b) - F_{\tilde{a},\tilde{b}}(a - \epsilon, b)}{\epsilon} - \lim_{\epsilon \to 0} \frac{F_{\tilde{a},\tilde{b}}(a, b - \epsilon) - F_{\tilde{a},\tilde{b}}(a - \epsilon, b - \epsilon)}{\epsilon} \right)
$$

$$
= \lim_{\epsilon \to 0} \frac{1}{\epsilon} \left(\frac{\partial F_{\tilde{a},\tilde{b}}(a, b)}{\partial a} - \frac{\partial F_{\tilde{a},\tilde{b}}(a, b - \epsilon)}{\partial a} \right) = \frac{\partial^2 F_{\tilde{a},\tilde{b}}(a, b)}{\partial a\, \partial b}, \tag{5.36}
$$

assuming the joint cdf is differentiable. For d-dimensional random vectors, the probability density is the ratio between the probability of the random vectors belonging to a hypercube and the d-dimensional volume of the hypercube, as the volume goes to zero. As a result, it can be obtained by differentiating the joint cdf with respect to the d components of the random vector. The joint probability density function (pdf) of a d-dimensional continuous

---

[1] We only consider multidimensional Borel sets when describing random vectors, just as we only consider Borel sets when describing random variables, as mentioned at the end of Section 3.1, but this suffices for all practical purposes.

random vector encodes the probability density at any point of $\mathbb{R}^d$. For it to exist, the joint cdf needs to be differentiable.

Definition 5.4 (Joint probability density function). *If the joint cdf of two random variables* $\tilde{a}, \tilde{b}$ *is differentiable, then their joint pdf is*

$$f_{\tilde{a},\tilde{b}}(a,b) := \lim_{\epsilon \to 0} \frac{P\left(a - \epsilon < \tilde{a} \le a, b - \epsilon < \tilde{b} \le b\right)}{\epsilon^2} \tag{5.37}$$

$$= \frac{\partial^2 F_{\tilde{a},\tilde{b}}(a,b)}{\partial a \partial b}. \tag{5.38}$$

If the joint cdf of a d-dimensional random vector $\tilde{x}$ is differentiable, then its joint pdf is

$$f_{\tilde{x}}(x)$$
$$:= \lim_{\epsilon \to 0} \frac{P\left(x[1] - \epsilon < \tilde{x}[1] \le x[1], x[2] - \epsilon < \tilde{x}[2] \le x[2], \ldots, x[d] - \epsilon < \tilde{x}[d] \le x[d]\right)}{\epsilon^d}$$
$$= \frac{\partial^d F_{\tilde{x}}(x)}{\partial x[1] \partial x[2] \cdots \partial x[d]}. \tag{5.39}$$

The following theorem establishes that the joint pdf has very similar properties to the pdf (see Theorem 3.16). It is nonnegative, its integral equals one, and any function with these characteristics can be interpreted as a valid joint pdf. In addition, we can use the joint pdf to compute the probability that the corresponding random vector belongs to any multidimensional Borel set (i.e. to any countable union of hyperrectangles).

Theorem 5.5 (Properties of the joint pdf). *Let $\tilde{a}$ and $\tilde{b}$ be two continuous random variables with joint pdf $f_{\tilde{a},\tilde{b}}$. The joint pdf is nonnegative at every two-dimensional point of $\mathbb{R}^2$. In addition, for any two-dimensional Borel set $B \subseteq \mathbb{R}^2$,*

$$P\left(\begin{bmatrix} \tilde{a} \\ \tilde{b} \end{bmatrix} \in B\right) = \int_{\begin{bmatrix} a \\ b \end{bmatrix} \in B} f_{\tilde{a},\tilde{b}}(a,b) \, \mathrm{d}a \, \mathrm{d}b, \tag{5.40}$$

and

$$\int_{a=-\infty}^{\infty} \int_{b=-\infty}^{\infty} f_{\tilde{a},\tilde{b}}(a,b) \, \mathrm{d}a \, \mathrm{d}b = 1. \tag{5.41}$$

Let $\tilde{x}$ be a d-dimensional continuous random vector with joint pdf $f_{\tilde{x}}$. The joint pdf is nonnegative at every d-dimensional point of $\mathbb{R}^d$. In addition, for any d-dimensional Borel set $B \subseteq \mathbb{R}^d$,

$$P(\tilde{x} \in B) = \int_{x \in B} f_{\tilde{x}}(x) \, \mathrm{d}x \tag{5.42}$$

and

$$\int_{x \in \mathbb{R}^d} f_{\tilde{x}}(x) \, \mathrm{d}x = 1. \tag{5.43}$$

Conversely, any function $f : \mathbb{R}^d \to \mathbb{R}$ that is nonnegative and integrates to one over $\mathbb{R}^d$ can be interpreted as the joint pdf of a random vector $\tilde{x} : \Omega \to \mathbb{R}^d$ for some sample space Ω.

Proof   The joint pdf is nonnegative because it is obtained by differentiating the joint cdf, which is nondecreasing with respect to all of its entries, as established for two dimensions in Lemma 5.2.

Integrating the joint pdf over a Borel set B yields the probability that the corresponding random vector belongs to B by the same reasoning as in one dimension (see Section 3.3). We can partition B into disjoint hypersquares of volume ϵ^d (in two dimensions, squares of area ϵ^2). The probability that the random vector is in B is the sum of the probabilities that it is in any of the hyperrectangles, as explained in Section 5.1 (see also Theorem 3.2 for a proof in the one-dimensional case). As $\epsilon \to 0$, the probability converges to the value of the joint pdf in the hyperrectangle multiplied by ϵ^d. Consequently, the sum of these probabilities is a Riemann sum that converges to the integral of the joint pdf over B. This establishes (5.40) and (5.42). A proof based on the joint cdf, generalizing the proof of Theorem 3.14, is also possible.

The integral of the joint pdf over the whole space $\mathbb{R}^d$ equals one because every point in the sample space of the probability space associated with the random vector must map to some point in $\mathbb{R}^d$ (see the proof of Theorem 3.16). To prove that any nonnegative function f with this property is a valid joint pdf, we can apply the same argument as in the proof of Theorem 3.16. We select $\mathbb{R}^d$ as the sample space and the collection of Borel sets as the associated collection of events. The probability measure is defined by assigning the probability $\int_{x \in B} f(x) \, dx$ to each event B. In that case, the random vector associated with the joint pdf is the identity function.   ■

Example 5.6 (Triangle lake). A biologist is tracking an otter that lives in the triangular lake depicted in the left graph of Figure 5.3. She decides to model the location of the otter probabilistically, as a two-dimensional random vector with entries $\tilde{a}$ and $\tilde{b}$. The otter does not seem to prefer any specific region of the lake, so the biologist decides to model the location as being uniformly distributed, meaning that the joint pdf of $\tilde{a}$ and $\tilde{b}$ is equal to a constant c,

$$f_{\tilde{a},\tilde{b}}(a,b) = \begin{cases} c & \text{if } (a,b) \in \text{Lake}, \\ 0 & \text{otherwise}. \end{cases} \tag{5.44}$$

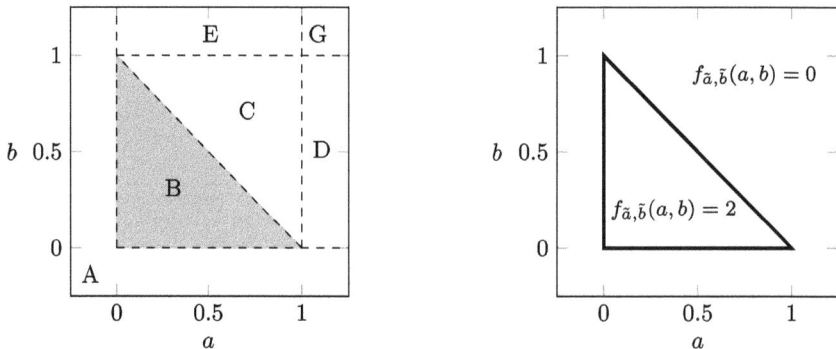

Figure 5.3 Triangle lake in Example 5.6. The left plot shows the lake shaded in gray. The different regions marked A to G are used to define the joint cdf. The right plot shows the corresponding joint pdf.

By Theorem 5.5, the integral of the joint pdf over $\mathbb{R}^2$ equals one. Consequently,

$$\int_{a=-\infty}^{\infty} \int_{b=-\infty}^{\infty} c \, da \, db = \int_{b=0}^{1} \int_{a=0}^{1-b} c \, da \, db \tag{5.45}$$

$$= c \int_{b=0}^{1} (1 - b) \, db \tag{5.46}$$

$$= \frac{c}{2} = 1, \tag{5.47}$$

so $c = 2$. The right graph of Figure 5.3 shows the joint pdf.

To compute the probability that the otter is in any subset of the lake, we integrate the joint pdf, applying (5.40). For example,

$$P(\{\tilde{a} \geq 0.6, \tilde{b} \leq 0.2\}) = \int_{b=0}^{0.2} \int_{a=0.6}^{1-b} 2 \, da \, db \tag{5.48}$$

$$= \int_{b=0}^{0.2} 2 \, (0.4 - b) \, db \tag{5.49}$$

$$= 2(0.08 - 0.02) \tag{5.50}$$

$$= 0.12. \tag{5.51}$$

Let us compute the joint cdf of $\tilde{a}$ and $\tilde{b}$. By Definition 5.1 $F_{\tilde{a}, \tilde{b}}(a, b)$ represents the probability that the otter is southwest of the point (a, b). We derive the joint cdf by dividing the range into the sets shown in the left graph of Figure 5.3 and applying (5.40). If $(a, b) \in A$, then $F_{\tilde{a}, \tilde{b}}(a, b) = 0$ because $P(\{\tilde{a} \leq a\} \cap \{\tilde{b} \leq b\}) = 0$. If $(a, b) \in B$,

$$F_{\tilde{a}, \tilde{b}}(a, b) = \int_{u=0}^{b} \int_{v=0}^{a} 2 \, dv \, du = 2ab. \tag{5.52}$$

If $(a, b) \in C$,

$$F_{\tilde{a}, \tilde{b}}(a, b) = \int_{u=0}^{1-a} \int_{v=0}^{a} 2 \, dv \, du + \int_{u=1-a}^{b} \int_{v=0}^{1-u} 2 \, dv \, du = 2a + 2b - b^2 - a^2 - 1.$$

Setting $a := 1$ in this expression, yields the value of the joint cdf at $(a, b) \in D$, since

$$F_{\tilde{a}, \tilde{b}}(a, b) = P(\tilde{a} \leq a, \tilde{b} \leq b) = P(\tilde{a} \leq 1, \tilde{b} \leq b) = 2b - b^2. \tag{5.53}$$

Exchanging the roles of a and b, we obtain $F_{\tilde{a}, \tilde{b}}(a, b) = 2a - a^2$ for $(a, b) \in E$ by the same reasoning. Finally, for $(a, b) \in G$ $F_{\tilde{a}, \tilde{b}}(a, b) = 1$ because $P(\tilde{a} \leq a, \tilde{b} \leq b) = 1$. Putting everything together,

$$F_{\tilde{a}, \tilde{b}}(a, b) = \begin{cases} 0 & \text{if } a < 0 \text{ or } b < 0, \\ 2ab, & \text{if } a \geq 0, b \geq 0, a + b \leq 1, \\ 2a + 2b - b^2 - a^2 - 1, & \text{if } a \leq 1, b \leq 1, a + b \geq 1, \\ 2b - b^2, & \text{if } a \geq 1, 0 \leq b \leq 1, \\ 2a - a^2, & \text{if } 0 \leq a \leq 1, b \geq 1, \\ 1, & \text{if } a \geq 1, b \geq 1. \end{cases} \tag{5.54}$$

This was rather painful! As illustrated by this example, the joint pdf is usually a more convenient way of describing the joint distribution of multiple continuous random variables than the joint cdf.

....

5.4 Multidimensional Density Estimation

In this section, we show how to obtain a nonparametric estimate of the joint pdf using a multidimensional extension of kernel density estimation, described for a single variable in Section 3.5.2. The idea is the same as in the univariate case: The density estimate is a superposition of shifted copies of a kernel function centered at the data points. The only difference is that the kernel function is multidimensional.

Definition 5.7 (Multidimensional kernel density estimation). *Consider a d-dimensional real-valued dataset $X := \{x_1, x_2, \ldots, x_n\}$, where $x_i \in \mathbb{R}^d$ for $1 \leq i \leq n$. The corresponding kernel density estimate at any point $a \in \mathbb{R}^d$ is*

$$f_{X,h}(a) := \frac{1}{n\,h^d} \sum_{i=1}^{n} K\left(\frac{a - x_i}{h}\right), \tag{5.55}$$

where $K: \mathbb{R}^d \to \mathbb{R}$ is a kernel function centered at the origin that satisfies

$$K(a) \geq 0 \quad for\ all\ a \in \mathbb{R}^d, \tag{5.56}$$

$$\int_{\mathbb{R}^d} K(a)\, \mathrm{d}a = 1. \tag{5.57}$$

The bandwidth h is a positive constant that governs the spread of the kernel function. Generalizations where h is a matrix that induces different spreads in different directions are possible.

A popular choice for the kernel in multidimensional kernel density estimation is the Gaussian function

$$K(a) := \frac{1}{(2\pi)^{\frac{d}{2}}} \exp\left(-\frac{||a||_2^2}{2}\right), \tag{5.58}$$

which is smooth and decays rapidly away from its center. Figure 5.4 shows a simple example, where we apply kernel density estimation with a Gaussian kernel to model three two-dimensional data points. The density estimate is visualized as a three-dimensional surface, as well as a contour plot.

As in the case of one-dimensional kernel density estimation, the bandwidth h has a strong influence on the density estimate. Increasing h dilates the kernel, so that the estimate takes into account more points. If the bandwidth is very small, individual samples have a large influence on the estimate. This makes it possible to reproduce irregular shapes more easily, but may also overfit spurious fluctuations in the observed data, especially if we don't have many observations. Increasing the bandwidth smooths out such fluctuations and yields more stable estimates. However, a value of h that is too large results in over-smoothing, and can eliminate meaningful structure from the estimate. The tradeoff is illustrated in Figure 5.5, where we show the contour plots of density estimates obtained using different bandwidths from hourly temperatures in Manhattan (Kansas) and Versailles (Kentucky), extracted from Dataset 9.

Unfortunately, kernel density estimation is only useful when the number of variables is small, due to the curse of dimensionality. As explained in Section 4.7, the possible values taken by multiple variables increases exponentially as the number of variables increases,

Scatterplot
x_1, x_2, x_3

Kernel density estimate
$$\frac{K(a-x_1)+K(a-x_2)+K(a-x_3)}{3}$$

Contour plot

$$\frac{K(a-x_1)}{3} \qquad\qquad \frac{K(a-x_2)}{3} \qquad\qquad \frac{K(a-x_3)}{3}$$

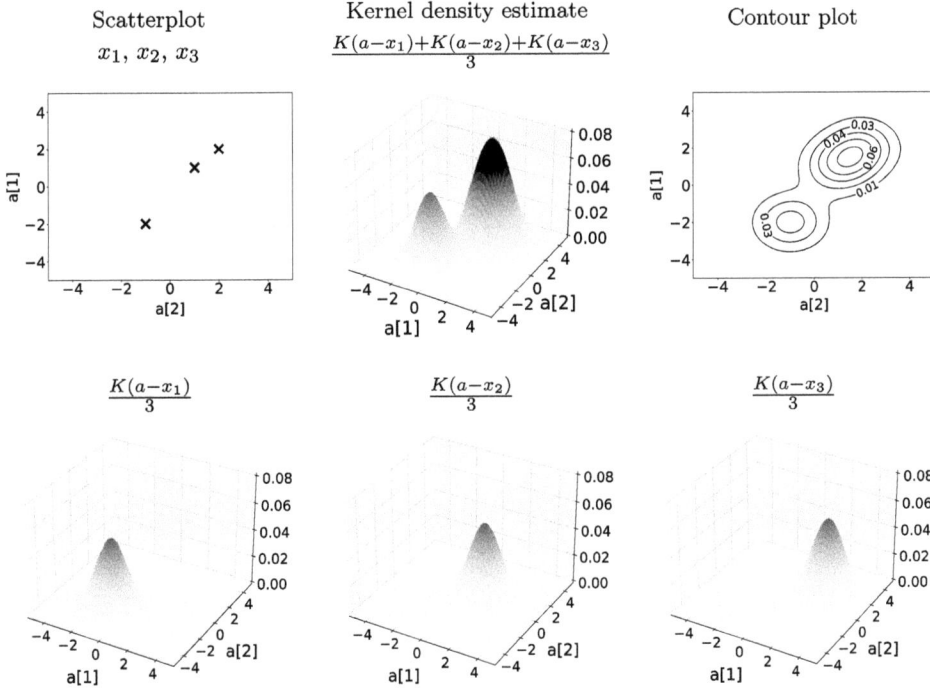

Figure 5.4 Kernel density estimation in 2D. We apply kernel density estimation to approximate the joint probability density corresponding to the three data points x_1, x_2, and x_3 shown in the top left graph. The density is estimated by summing shifted copies of a Gaussian kernel with bandwidth equal to one ($h := 1$ in Definition 5.7) centered at each data point. Each component is shown separately in the bottom row. The resulting joint pdf estimate is depicted in the graph at the center of the top row. The top right graph shows the corresponding contour plot.

which makes it impossible to estimate their joint pdf from data via nonparametric estimation, unless they are very few. To address this, we need to either make independence assumptions, or resort to parametric models such as the multidimensional Gaussian model, presented in Section 5.10.

5.5 Marginal Distributions

In probabilistic models consisting of multiple random variables, it is often useful to isolate the individual behavior of a single variable. In the case of discrete random variables, this is achieved by summing the joint pmf over the rest of the variables (see Section 4.2). In the case of continuous random variables, we can obtain the pdf of a single variable, which we call the *marginal* pdf, in a similar way: By integrating the joint pdf with respect to the remaining variables.

Theorem 5.8 (Marginal pdf). *Let $\tilde{a}$ and $\tilde{b}$ be continuous random variables with joint pdf $f_{\tilde{a},\tilde{b}}$. The marginal pdf of $\tilde{a}$ is*

$$f_{\tilde{a}}(a) = \int_{b=-\infty}^{\infty} f_{\tilde{a},\tilde{b}}(a,b) \ \mathrm{d}b. \tag{5.59}$$

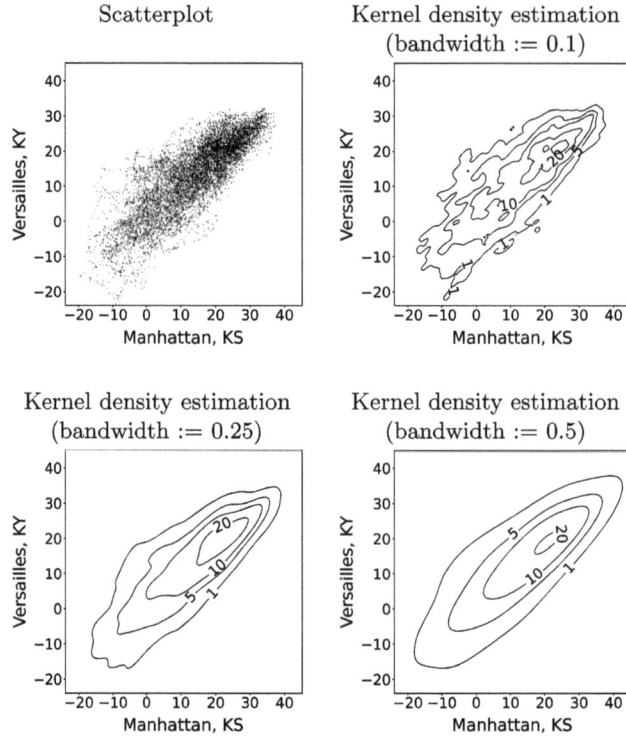

Figure 5.5 Joint pdf of temperature data. The top left graph shows a scatterplot of temperature data, measured hourly at weather stations in Manhattan (Kansas) and Versailles (Kentucky) in 2015. The remaining plots show the corresponding kernel density estimates obtained following Definition 5.7 using Gaussian kernels with different bandwidths.

Let $\tilde{x}$ be a d-dimensional random vector. The marginal pdf of $\tilde{x}[i]$ is

$$f_{\tilde{x}[i]}(a) = \int\limits_{b_1 \in \mathbb{R}} \cdots \int\limits_{b_{i-1} \in \mathbb{R}} \int\limits_{b_{i+1} \in \mathbb{R}} \tag{5.60}$$

$$\cdots \int\limits_{b_d \in \mathbb{R}} f_{\tilde{x}}\left(b[1], \ldots, b[i-1], a, b[i+1], \ldots, b[d]\right) \, \mathrm{d}b_1 \ldots \mathrm{d}b_{i-1} \, \mathrm{d}b_{i+1} \ldots \mathrm{d}b_d.$$

Proof We prove the bivariate case, the vector case follows by the same argument. By Definition 3.3 and (5.40), the marginal cdf of $\tilde{a}$ equals

$$F_{\tilde{a}}(a) = \mathrm{P}(\tilde{a} \le a) \tag{5.61}$$

$$= \int_{u=-\infty}^{a} \int_{b=-\infty}^{\infty} f_{\tilde{a},\tilde{b}}(u,b) \, \mathrm{d}b \, \mathrm{d}u. \tag{5.62}$$

Taking the derivative with respect to a yields the marginal pdf and completes the proof. ∎

Alternatively, we can estimate the marginal cdf of a random variable from the joint cdf by computing the limit when the remaining variables tend to infinity.

Lemma 5.9 (Marginal cumulative distribution function). *Let $\tilde{a}$ and $\tilde{b}$ be continuous random variables with joint cdf $F_{\tilde{a},\tilde{b}}$. The marginal cdf of $\tilde{a}$ is*

$$F_{\tilde{a}}(a) = \lim_{b \to \infty} F_{\tilde{a},\tilde{b}}(a, b). \tag{5.63}$$

Proof When $b \to \infty$, the limit of $F_{\tilde{a},\tilde{b}}(a, b)$ is the probability of $\tilde{a}$ being smaller than a, which is precisely the marginal cdf of $\tilde{a}$. This can be made mathematically rigorous using a similar argument to the proof of (3.16) in Lemma 3.4. ∎

If we are interested in computing the joint pdf of several entries in a random vector, instead of just one, we integrate the joint pdf of the vector with respect to the remaining variables. Notation gets a bit complicated, so let us consider an example with $d := 4$. To compute the marginal joint pdf of the first and fourth entries of random vector $\tilde{x}$, we integrate the joint pdf with respect to the second and third entries,

$$f_{\tilde{x}[1],\tilde{x}[4]}(a, d) = \int_{b=-\infty}^{\infty} \int_{c=-\infty}^{\infty} f_{\tilde{x}}(a, b, c, d) \, db \, dc. \tag{5.64}$$

Example 5.10 (Triangle lake: Marginal distribution). The biologist from Example 5.6 wants to compute the probability density of the horizontal coordinate $\tilde{a}$ of the otter. By Theorem 5.8, the marginal pdf of $\tilde{a}$ is obtained by integrating the joint pdf with respect to the vertical coordinate, which yields

$$f_{\tilde{a}}(a) = \int_{b=-\infty}^{\infty} f_{\tilde{a},\tilde{b}}(a, b) \, db \tag{5.65}$$

$$= \int_{b=0}^{1-a} 2 \, db \tag{5.66}$$

$$= 2(1 - a) \tag{5.67}$$

for $0 \leq a \leq 1$ and zero otherwise. The marginal pdf is shown in Figure 5.6.

Figure 5.7 shows the joint pdf and the marginal pdfs of the temperature in Manhattan (Kansas) and Versailles (Kentucky), obtained using data from Dataset 9. To estimate the marginal pdf of each variable, we can apply Theorem 5.8 and integrate their estimated joint pdf with respect to the other variable, but there is a simpler alternative: Applying one-dimensional kernel density estimation (see Section 3.5.2) to the temperature values at each station separately.

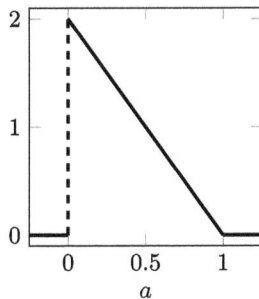

Figure 5.6 Marginal pdf. The plot shows the marginal pdf $f_{\tilde{a}}$ of the horizontal coordinate $\tilde{a}$ of the otter in Example 5.6.

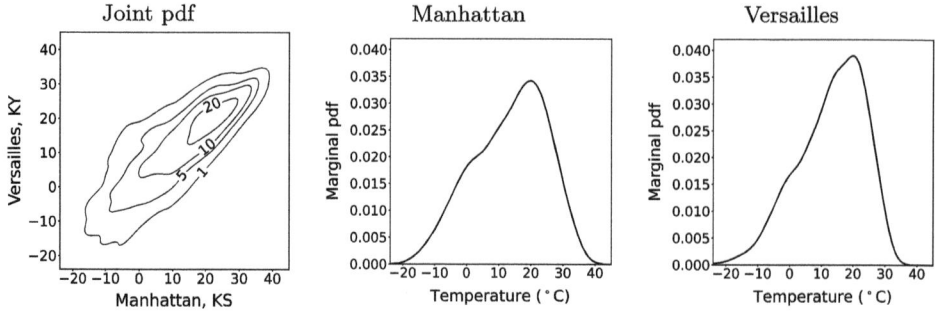

Figure 5.7 Marginal pdf of temperature data. The left graph shows a contour plot of the estimated joint pdf of the temperature in Manhattan (Kansas) and Versailles (Kentucky) in 2015. The center and right graphs show the estimated marginal densities of the temperatures in Manhattan and Versailles respectively, obtained by applying one-dimensional kernel density estimation to the data from each location, as described in Section 3.5.2.

5.6 Conditional Distributions

Conditional distributions allow us to update our uncertainty about certain variables in a model, when other variables are observed. For example, we may want to determine the conditional probability density of a random variable $\tilde{b}$ when another random variable $\tilde{a}$ equals a fixed value a. As explained in Section 3.3, the probability density of a random variable is the limit of the probability that it belongs to a small interval divided by the length of the interval, when the length tends to zero. Consequently, we could consider defining the conditional probability density of $\tilde{b}$ given $\tilde{a} = a$ at a point b as the following limit,

$$\lim_{\epsilon_1 \to 0} \frac{P\left(b - \epsilon_1 < \tilde{b} \leq b \,|\, \tilde{a} = a\right)}{\epsilon_1}. \tag{5.68}$$

This does not quite work. The conditional probability in the numerator is not well defined when $\tilde{a}$ is continuous because the probability of the event $\tilde{a} = a$ is zero, as explained in Section 3.1. Instead, we condition on the event that $\tilde{a}$ belongs to a small interval of length ϵ_2, which includes a. The resulting conditional density is

$$f_{\tilde{b}\,|\,a-\epsilon_2<\tilde{a}\leq a}(b) := \lim_{\epsilon_1 \to 0} \frac{P\left(b - \epsilon_1 < \tilde{b} \leq b \,|\, a - \epsilon_2 < \tilde{a} \leq a\right)}{\epsilon_1}. \tag{5.69}$$

To define a conditional density given $\tilde{a} = a$, we take the limit when $\epsilon_2 \to 0$, so that the interval collapses to a. The resulting density can be expressed as the ratio between the joint pdf of $\tilde{a}$ and $\tilde{b}$ and the marginal pdf of $\tilde{a}$, assuming they both exist. To simplify the derivation, we set ϵ_1 and ϵ_2 to equal a single value denoted by ϵ. By Definitions 1.16, 3.13, and 5.4,

$$f_{\tilde{b}\,|\,\tilde{a}}(b\,|\,a) := \lim_{\epsilon \to 0} f_{\tilde{b}\,|\,a-\epsilon<\tilde{a}\leq a}(b) \tag{5.70}$$

$$= \lim_{\epsilon \to 0} \frac{P\left(b - \epsilon < \tilde{b} \leq b \,|\, a - \epsilon < \tilde{a} \leq a\right)}{\epsilon} \tag{5.71}$$

$$= \lim_{\epsilon \to 0} \frac{1}{\epsilon} \frac{P\left(b - \epsilon < \tilde{b} \leq b, a - \epsilon < \tilde{a} \leq a\right)}{P\left(a - \epsilon < \tilde{a} \leq a\right)} \tag{5.72}$$

$$= \frac{\lim_{\epsilon \to 0} \dfrac{P\left(b - \epsilon < \tilde{b} \leq b, a - \epsilon < \tilde{a} \leq a\right)}{\epsilon^2}}{\lim_{\epsilon \to 0} \dfrac{P\left(a - \epsilon < \tilde{a} \leq a\right)}{\epsilon}} \tag{5.73}$$

$$= \frac{f_{\tilde{a},\tilde{b}}\left(a, b\right)}{f_{\tilde{a}}\left(a\right)}, \tag{5.74}$$

as long as $f_{\tilde{a}}\left(a\right) > 0$. Inspired by this reasoning, we define the conditional pdf of a continuous random variable $\tilde{b}$ given another continuous variable $\tilde{a}$ as the ratio between their joint pdf and the marginal pdf of $\tilde{a}$.

Definition 5.11 (Conditional pdf). *Let $\tilde{a}$ and $\tilde{b}$ be random variables with joint pdf $f_{\tilde{a},\tilde{b}}$. The conditional pdf of $\tilde{b}$ given $\tilde{a}$ is*

$$f_{\tilde{b}\,|\,\tilde{a}}\left(b\,|\,a\right) := \frac{f_{\tilde{a},\tilde{b}}\left(a, b\right)}{f_{\tilde{a}}\left(a\right)}, \quad \text{if } f_{\tilde{a}}\left(a\right) > 0, \tag{5.75}$$

and is undefined otherwise.

Let $\tilde{x}$ and $\tilde{y}$ be random vectors, each with multiple entries. The conditional pdf of $\tilde{y}$ given $\tilde{x}$ is

$$f_{\tilde{y}\,|\,\tilde{x}}\left(y\,|\,x\right) := \frac{f_{\tilde{x},\tilde{y}}\left(x, y\right)}{f_{\tilde{x}}\left(x\right)}, \tag{5.76}$$

assuming $f_{\tilde{x}}\left(x\right) > 0$. Otherwise, the conditional pdf is undefined.

The conditional pdf is a valid pdf, according to Theorem 3.16, because it is nonnegative by definition (the joint pdf and the marginal pdf are nonnegative), and its integral is equal to one. Indeed, for any a such that $f_{\tilde{a}}\left(a\right) > 0$, by Definition 5.11 and Theorem 5.8,

$$\int_{b=-\infty}^{\infty} f_{\tilde{b}\,|\,\tilde{a}}\left(b\,|\,a\right)\,\mathrm{d}b = \frac{\int_{b=-\infty}^{\infty} f_{\tilde{a},\tilde{b}}\left(a, b\right)\,\mathrm{d}b}{f_{\tilde{a}}\left(a\right)} \tag{5.77}$$

$$= \frac{f_{\tilde{a}}\left(a\right)}{f_{\tilde{a}}\left(a\right)} = 1. \tag{5.78}$$

We can also define the conditional pdf of some of the entries in a random vector given other entries. For an example with $d = 4$, the conditional pdf of the second and third entries given the first and fourth entries is

$$f_{\tilde{x}[2],\tilde{x}[3]\,|\,\tilde{x}[1],\tilde{x}[4]}\left(b, c\,|\,a, d\right) = \frac{f_{\tilde{x}}(a, b, c, d)}{f_{\tilde{x}[1],\tilde{x}[4]}(a, d)}. \tag{5.79}$$

All such conditional pdfs integrate to one, as long as we integrate them over the appropriate variables. For example, the integral with respect to b and c of the conditional pdf $f_{\tilde{x}[2],\tilde{x}[3]\,|\,\tilde{x}[1],\tilde{x}[4]}\left(b, c\,|\,a, d\right)$ in (5.79) is equal to one, but it does not make sense to integrate it with respect to a and d.

An immediate consequence of Definition 5.11 is the chain rule for continuous random variables, which is exactly the same as the chain rule for discrete random variables (Theorem 4.14), if we replace pmfs by pdfs.

Theorem 5.12 (Chain rule for continuous random variables). *For any continuous random variables* $\tilde{a}$ *and* $\tilde{b}$,

$$f_{\tilde{a},\tilde{b}}(a,b) = f_{\tilde{a}}(a)\, f_{\tilde{b}\,|\,\tilde{a}}(b\,|\,a) \tag{5.80}$$

$$= f_{\tilde{b}}(b)\, f_{\tilde{a}\,|\,\tilde{b}}(a\,|\,b), \tag{5.81}$$

as long as the marginal and conditional pdfs exist.

Similarly, the joint pdf of any d-dimensional random vector $\tilde{x}$ *can be decomposed as follows:*

$$f_{\tilde{x}}(x) = f_{\tilde{x}[1]}(x[1]) \prod_{i=2}^{d} f_{\tilde{x}[i]\,|\,\tilde{x}[1],\ldots,\tilde{x}[i-1]}(x[i]\,|\,x[1],\ldots,x[i-1]), \tag{5.82}$$

as long as the marginal and conditional pdfs exist. The order of indices in the random vector is completely arbitrary (any order works).

Example 5.13 (Triangle lake: Conditional distribution). The biologist from Example 5.6 wants to compute the probability density of the vertical coordinate $\tilde{b}$ of the otter given the horizontal coordinate $\tilde{a}$. By Definition 5.11, the conditional pdf is

$$f_{\tilde{b}\,|\,\tilde{a}}(b\,|\,a) = \frac{f_{\tilde{a},\tilde{b}}(a,b)}{f_{\tilde{a}}(a)} \tag{5.83}$$

$$= \frac{1}{1-a} \qquad 0 \le b \le 1-a, \tag{5.84}$$

where $f_{\tilde{a}}$ is calculated in Example 5.10. For fixed $\tilde{a} = a$, the density of probability for $\tilde{b}$ is constant between 0 and $1 - a$.

Figure 5.8 motivates the procedure to compute the conditional pdf intuitively, fixing $a :=$ 0.75. The top left graph shows the joint pdf. We are interested in the conditional pdf of $\tilde{b}$ given $\tilde{a} = 0.75$, which is the probability density along the vertical dashed line. The value of the joint pdf on that line, $f_{\tilde{a},\tilde{b}}(0.75, b)$, depicted in the bottom left graph, is *almost* a conditional pdf, but not quite, because its integral does not equal to one. By Theorem 5.8, integrating this function with respect to b yields the marginal pdf $f_{\tilde{a}}(0.75)$, illustrated in the top right graph. Consequently, we can divide $f_{\tilde{a},\tilde{b}}(0.75, b)$ by $f_{\tilde{a}}(0.75)$ to ensure that it integrates to one, obtaining a valid conditional pdf, shown in the bottom right graph. This yields exactly the same expression for the conditional pdf of $\tilde{b}$ given $\tilde{a} = 0.75$ as Definition 5.11.

. . . .

Example 5.14 (Conditional distribution of temperature). Figure 5.9 shows an estimate of the conditional pdf of the temperature in Versailles (Kentucky) given the temperature in Manhattan (Kansas), obtained using hourly measurements from Dataset 9. To compute the estimate, we divide the joint pdf, approximated via 2D kernel density estimation (see Figure 5.5), by the marginal pdf of the temperature in Manhattan, approximated via 1D kernel density estimation (see Figure 5.7). The results reveal that there is a strong dependence between the temperatures at the two locations. Each conditional pdf is approximately centered at the corresponding temperature in Manhattan.

Figure 5.10 shows estimates of the conditional joint pdf of the temperatures in Versailles and Corvallis (Oregon) given different temperatures in Manhattan, again using data from

Joint pdf $f_{\tilde{a},\tilde{b}}(a,b)$ Marginal pdf $f_{\tilde{a}}(a)$

$f_{\tilde{a},\tilde{b}}(0.75,b)$ $f_{\tilde{b}\mid\tilde{a}}(b\mid 0.75)$

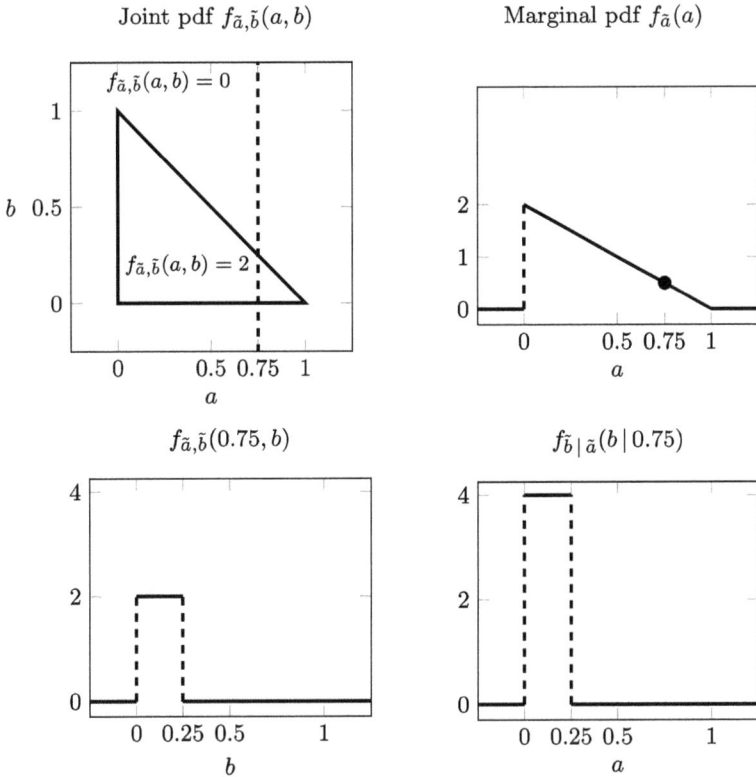

Figure 5.8 Conditional pdf. The top left diagram shows the joint pdf from Example 5.6. The dashed line corresponds to the values for which $a = 0.75$. The top right plot shows the marginal pdf of $f_{\tilde{a}}$. The value $f_{\tilde{a}}(0.75)$ of the marginal pdf at 0.75, indicated by the circular marker, is computed by integrating along the dashed line in the joint-pdf plot. The bottom left plot depicts the joint pdf restricted to that line, $f_{\tilde{a},\tilde{b}}(0.75,b)$. The bottom right plot shows the conditional pdf of $\tilde{b}$ given $\tilde{a} = 0.75$, which is the result of dividing $f_{\tilde{a},\tilde{b}}(0.75,b)$ by $f_{\tilde{a}}(0.75)$, so that it integrates to one.

Dataset 9. Let $\tilde{v}$, $\tilde{c}$, and $\tilde{m}$ be random variables representing the temperature in Versailles, Corvallis, and Manhattan, respectively. To obtain the estimates, we apply Definition 5.11

$$f_{\tilde{v},\tilde{c}\mid\tilde{m}}(v,c\mid t) = \frac{f_{\tilde{v},\tilde{c},\tilde{m}}(v,c,t)}{f_{\tilde{m}}(t)}, \tag{5.85}$$

estimating the joint pdf in the numerator and the marginal pdf in the denominator via kernel density estimation.

. . . .

5.7 Independence

As discussed in Section 4.4, when knowledge about a random variable $\tilde{a}$ does not affect our uncertainty about another random variable $\tilde{b}$, we say that $\tilde{a}$ and $\tilde{b}$ are *independent*. This can be expressed in terms of the conditional distributions of the random variables. If the variables are continuous, then we should have

$$P(\tilde{b} \in B \mid \tilde{a} = a) = P(\tilde{b} \in B) \tag{5.86}$$

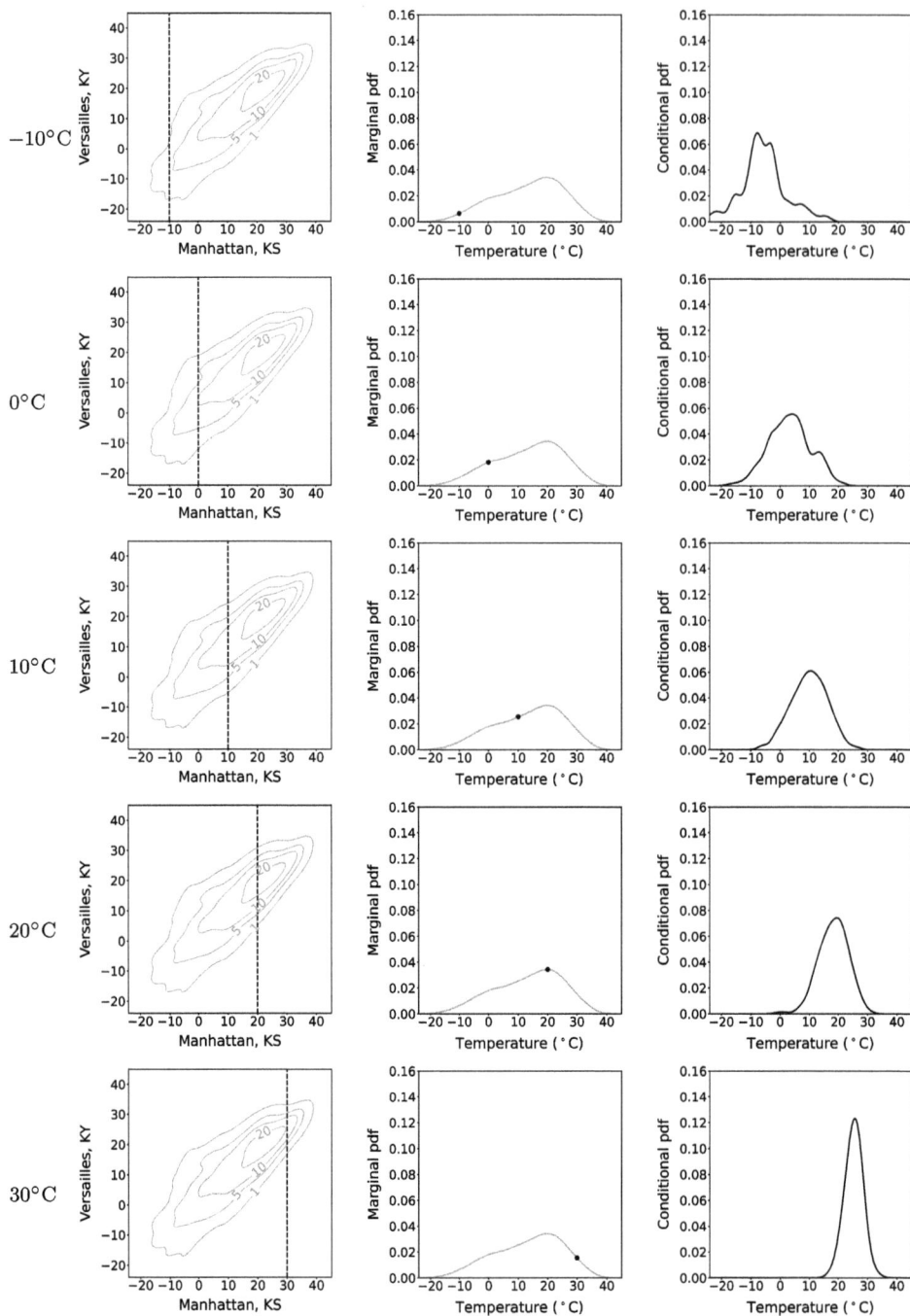

Figure 5.9 Conditional pdf of temperature data. The left column shows the contour plot of
the estimated joint pdf of the temperature in Manhattan (Kansas) and Versailles (Kentucky) in
2015. In each row, we condition on a different temperature t in Manhattan. The conditional pdf
of the temperature in Versailles given that the temperature in Manhattan equals t is shown in the
right column. It is obtained by dividing the joint pdf in the left column, restricted to the dashed
line, by the value of the marginal pdf of Manhattan at t (indicated by the black marker in the
middle column).

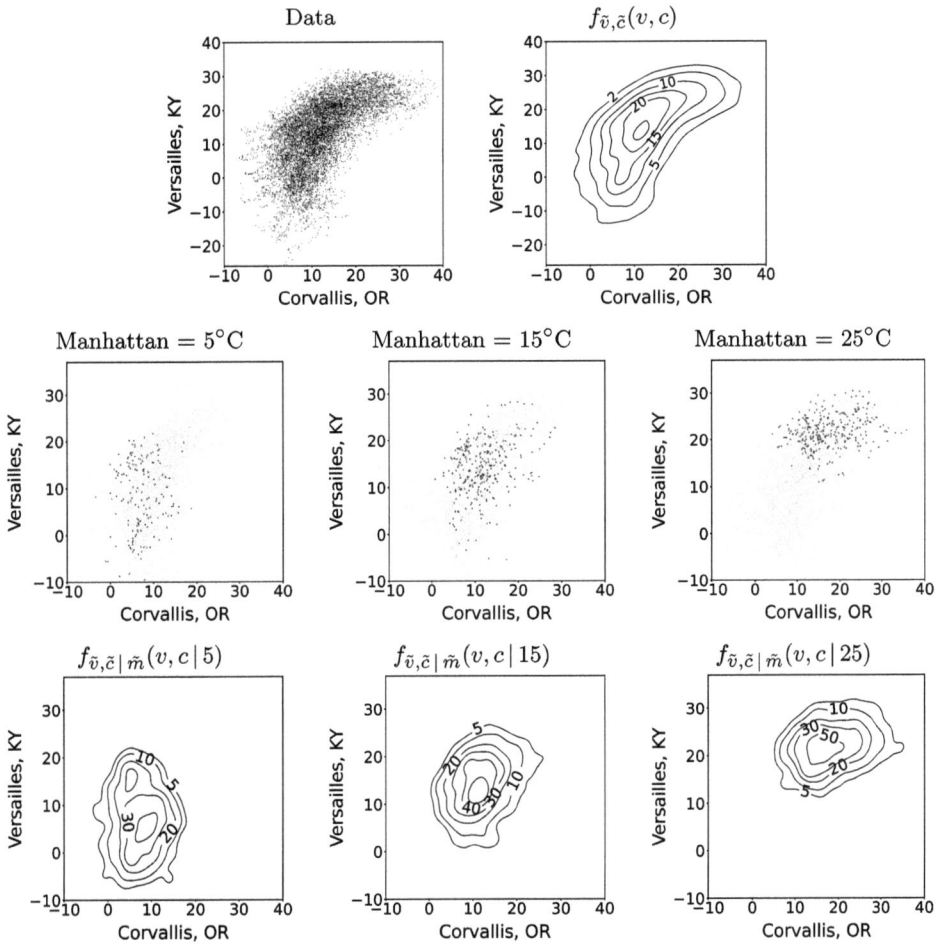

Figure 5.10 Conditional joint pdf of temperature data. The top row shows the scatterplot (left) and corresponding contour plot of the estimated joint pdf $f_{\tilde{v},\tilde{c}}$ (right) of the random variables $\tilde{v}$ and $\tilde{c}$, representing the temperature in Versailles (Kentucky) and Corvallis (Oregon), respectively. The second row highlights the data for which the temperature in Manhattan (Kansas) is close to t (within $1°C$), for different values of t indicated at the top of each column. The third row shows the conditional joint pdf $f_{\tilde{v},\tilde{c}\,|\,\tilde{m}}(v,c\,|\,t)$, where $\tilde{m}$ represents the temperature in Manhattan, estimated by applying (5.85).

for any Borel set B and any value of a, such that the conditional probability exists. For this to be the case, the conditional distribution of $\tilde{b}$ given $\tilde{a} = a$ needs to be the same as the marginal distribution of $\tilde{b}$. If the random variables have pdfs, we require $f_{\tilde{b}\,|\,\tilde{a}}(b\,|\,a) = f_{\tilde{b}}(b)$. By the chain rule (Theorem 5.12), this is equivalent to

$$f_{\tilde{a},\tilde{b}}(a,b) = f_{\tilde{a}}(a) f_{\tilde{b}\,|\,\tilde{a}}(b\,|\,a) \tag{5.87}$$

$$= f_{\tilde{a}}(a) f_{\tilde{b}}(b). \tag{5.88}$$

Definition 5.15 (Independent continuous random variables). *Two continuous random variables $\tilde{a}$ and $\tilde{b}$ with pdfs $f_{\tilde{a}}$ and $f_{\tilde{b}}$ are independent, if and only if*

$$f_{\tilde{a},\tilde{b}}(a,b) = f_{\tilde{a}}(a)\, f_{\tilde{b}}(b), \quad for\ all\ a,b \in \mathbb{R}. \tag{5.89}$$

The d entries $\tilde{x}[1]$, $\tilde{x}[2]$, ..., $\tilde{x}[d]$ in a continuous random vector $\tilde{x}$ are independent, if and only if

$$f_{\tilde{x}}(x) = \prod_{i=1}^{d} f_{\tilde{x}[i]}(x[i]), \quad for\ all\ x \in \mathbb{R}^d. \tag{5.90}$$

In Section 3.7, we define the likelihood $\mathcal{L}_X$ of a parametric model given a dataset X, as the product of the parametric pdf evaluated at each data point. This is equivalent to defining the likelihood as the joint pdf of the data (interpreted as a function of the parameters) under the assumption that all variables are independent, and have the same marginal distribution. We say that such random variables are independent and identically distributed, analogously to Definition 4.20.

Definition 5.16 (Independent identically distributed continuous random variables). *Let $\tilde{a}_1$, $\tilde{a}_2$, ..., $\tilde{a}_n$ be continuous random variables belonging to the same probability space. The random variables are identically distributed, if their marginal pdfs are the same*

$$f_{\tilde{a}_1} = f_{\tilde{a}_2} = \cdots = f_{\tilde{a}_n} = f_{\tilde{a}}, \tag{5.91}$$

where $\tilde{a}$ is a random variable with the same distribution. The random variables are independent and identically distributed (i.i.d.), if their joint pdf is equal to the product of the marginal pdfs

$$f_{\tilde{a}_1, \tilde{a}_2, ..., \tilde{a}_n}(a_1, a_2, ..., a_n) = \prod_{i=1}^{d} f_{\tilde{a}}(a_i) \tag{5.92}$$

for all possible values of a_1, a_2, ..., a_n.

In Figure 5.11, we analyze the joint distribution of the duration of a movie and its box-office revenue, using data from Dataset 11. The conditional pdfs of the box-office revenue given different durations (90, 100, and 110 minutes) are all quite similar to the marginal pdf, indicating that the two variables are approximately independent. Intuitively, independence implies that no matter where we *slice* the joint pdf of the two random variables (see the dashed lines in the bottom row of Figure 5.11), we always end up with the same conditional density. The example illustrates the difficulty of establishing that two quantities are completely independent in practice. Since we estimate the conditional densities from finite data, we cannot expect them to be exactly the same, even if the quantities are independent. In this case, it looks like longer movies may tend to have more revenue, but the effect is small.

5.8 Conditional Independence

Figure 5.12 shows that the temperatures in Versailles (Kentucky) and Corvallis (Oregon) are not independent. The distribution of temperatures in Versailles changes dramatically depending on the temperature in Corvallis. However, when we condition on the temperature in Manhattan (Kansas), there is much less dependence between the temperatures in

Comparison of marginal and conditional pdfs

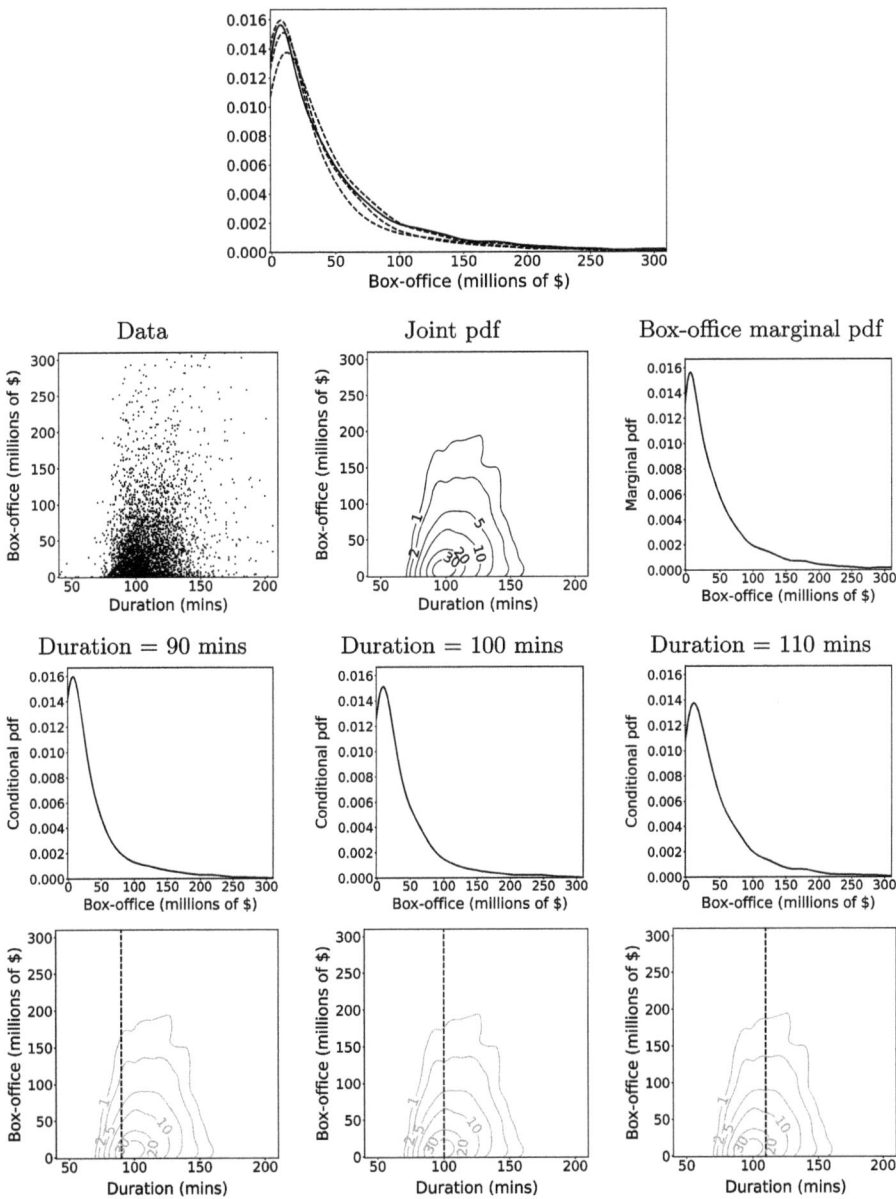

Figure 5.11 **Movie duration and box-office revenue are approximately independent.** The second row shows a scatterplot of movie box-office revenue and duration (left), the contour plot of the corresponding joint pdf (center), and the marginal pdf of the box-office revenue (right). The third row shows the estimated conditional pdf of box-office revenue conditioned on different durations. The dashed lines in the bottom row indicate the different slices of the joint pdf used to estimate each conditional pdf. The marginal and conditional pdfs, shown superposed in the top plot for comparison, are quite similar, which suggests that the quantities are approximately independent. The marginal and joint pdfs are estimated via kernel density estimation with Gaussian kernels, as described in Sections 3.5.2 and 5.4. The conditional pdfs are computed following Definition 5.11.

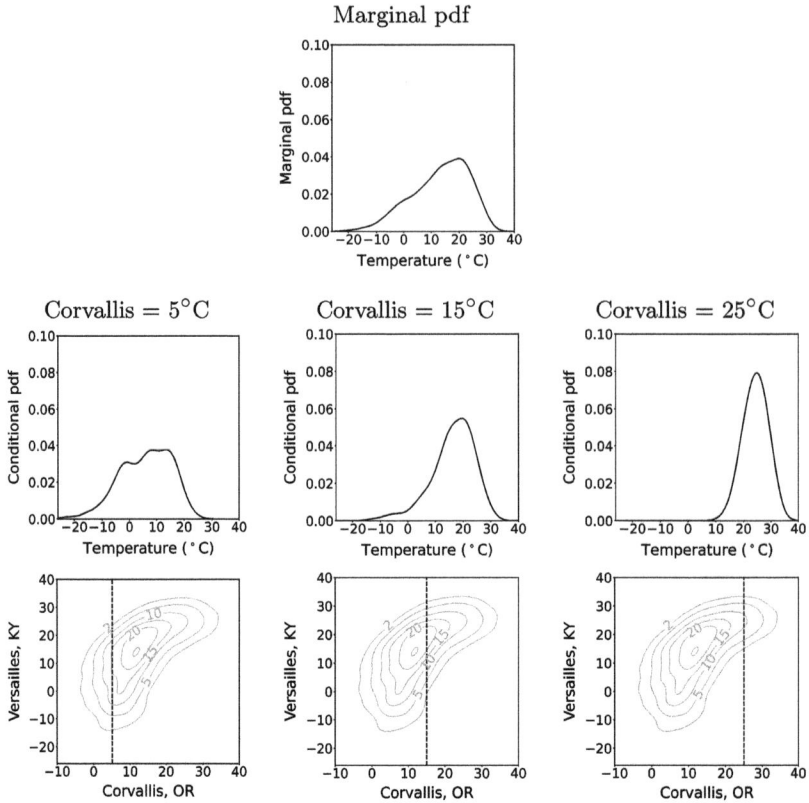

Figure 5.12 The temperatures in Kentucky and Oregon are not independent. The top plot shows the marginal pdf of the temperature in Versailles (Kentucky). The second row shows the conditional pdf of the temperature in Versailles given different temperatures in Corvallis (Oregon). The marginal and conditional pdfs are clearly very different. This is reflected in the shape of the estimated joint pdf of both temperatures, shown in the third row. The conditional pdfs correspond to normalized *slices* of the joint pdf, indicated by the dashed lines.

Versailles and Corvallis. Let $\tilde{v}$, $\tilde{c}$, and $\tilde{m}$ be random variables representing the temperature in Versailles, Corvallis, and Manhattan, respectively. Figure 5.13 shows that

$$f_{\tilde{v}\,|\,\tilde{m}}\left(v\,|\,15\right) \approx f_{\tilde{v}\,|\,\tilde{c},\tilde{m}}\left(v\,|\,5,15\right) \tag{5.93}$$

$$\approx f_{\tilde{v}\,|\,\tilde{c},\tilde{m}}\left(v\,|\,15,15\right) \tag{5.94}$$

and also

$$f_{\tilde{v}\,|\,\tilde{m}}\left(v\,|\,25\right) \approx f_{\tilde{v}\,|\,\tilde{c},\tilde{m}}\left(v\,|\,15,25\right) \tag{5.95}$$

$$\approx f_{\tilde{v}\,|\,\tilde{c},\tilde{m}}\left(v\,|\,25,25\right). \tag{5.96}$$

Once we condition on Manhattan, the conditional pdf of the temperature in Versailles does not change much if we condition (or not) on different temperatures in Corvallis. The temperature at Corvallis provides little information about the temperature in Versailles, *as long as we also know the temperature in Manhattan*. This makes sense given the geographic location of the weather stations, depicted in Figure 5.14: Manhattan is situated approximately between Corvallis and Versailles.

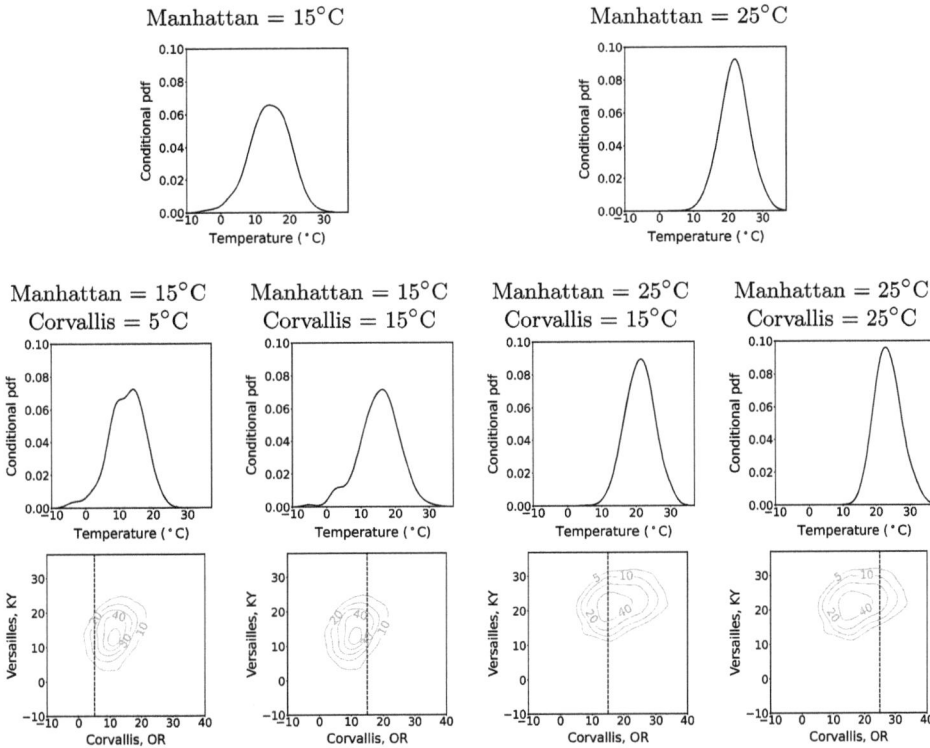

Figure 5.13 The temperatures in Kentucky and Oregon are approximately conditionally independent given the temperature in Kansas. The top row shows the conditional pdf of the temperature in Versailles (Kentucky) given different temperatures in Manhattan (Kansas). The second row shows the conditional pdf when we also condition on the temperature in Corvallis (Oregon). Conditioning on Corvallis barely changes the conditional pdfs (compare to Figure 5.12, where we do not condition on Manhattan). The third row shows the estimated conditional joint pdf of the temperature in Versailles and Corvallis, given the temperature in Manhattan. The conditional pdfs in the second row correspond to normalized *slices* of the conditional joint pdf, indicated by the dashed lines.

As discussed in Section 4.5, if our uncertainty about a random variable $\tilde{a}$ does not change when another random variable $\tilde{b}$ is revealed, *as long as the value of a third random variable $\tilde{c}$ is known*, then $\tilde{a}$ and $\tilde{b}$ are conditionally independent given $\tilde{c}$. If the variables are continuous, then we should have

$$P(\tilde{b} \in S \,|\, \tilde{a} = a, \tilde{c} = c) = P(\tilde{b} \in B \,|\, \tilde{c} = c) \tag{5.97}$$

for any Borel set B and any values of a and c, such that the conditional probabilities exist. For this to be the case, the conditional distribution of $\tilde{b}$ given $\tilde{a} = a$ and $\tilde{c} = c$ needs to be the same as the conditional distribution of $\tilde{b}$ given $\tilde{c} = c$. If the random variables have pdfs, we require $f_{\tilde{b} \,|\, \tilde{a}, \tilde{c}}(b \,|\, a, c) = f_{\tilde{b} \,|\, \tilde{c}}(b \,|\, c)$. By the chain rule (Theorem 5.12), this is equivalent to

$$f_{\tilde{a}, \tilde{b} \,|\, \tilde{c}}(a, b \,|\, c) = f_{\tilde{a} \,|\, \tilde{c}}(a \,|\, c) f_{\tilde{b} \,|\, \tilde{a}, \tilde{c}}(b \,|\, a, c) \tag{5.98}$$

$$= f_{\tilde{a} \,|\, \tilde{c}}(a \,|\, c) f_{\tilde{b} \,|\, \tilde{c}}(b \,|\, c). \tag{5.99}$$

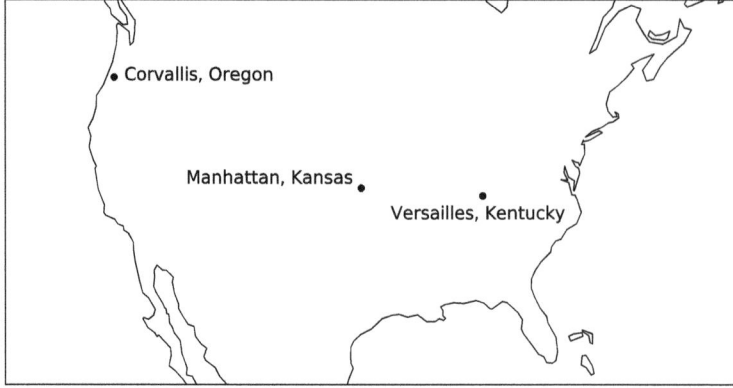

Figure 5.14 Weather stations in Oregon, Kansas, and Kentucky. Location of the weather stations of Corvallis (Oregon), Manhattan (Kansas) and Versailles (Kentucky). The geographic configuration explains why the temperature in Versailles and Corvallis are approximately conditionally independent given the temperature in Manhattan (see Figure 5.13).

Definition 5.17 (Conditionally independent continuous random variables). *Two continuous random variables $\tilde{a}$ and $\tilde{b}$ defined on the same probability space are conditionally independent given a random variable $\tilde{c}$, if and only if*

$$f_{\tilde{a},\tilde{b}\,|\,\tilde{c}}(a,b\,|\,c) = f_{\tilde{a}\,|\,\tilde{c}}(a\,|\,c)\,f_{\tilde{b}\,|\,\tilde{c}}(b\,|\,c), \quad for\ all\ a,b,c \in \mathbb{R}. \quad (5.100)$$

The d_1 entries $\tilde{x}[1]$, $\tilde{x}[2]$, ..., $\tilde{x}[d]$ in a continuous random vector $\tilde{x}$ are conditionally independent given a random variable $\tilde{a}$ if and only if

$$f_{\tilde{x}\,|\,\tilde{a}}(x\,|\,a) = \prod_{i=1}^{d} f_{\tilde{x}[i]\,|\,\tilde{a}}(x[i]\,|\,a), \quad for\ all\ x \in \mathbb{R}^d, a \in \mathbb{R}. \quad (5.101)$$

As discussed in Section 1.6, independence does *not* imply conditional independence or vice versa.

5.9 Jointly Simulating Multiple Random Variables

Inverse-transform sampling, presented in Section 3.8, allows us to simulate a single continuous random variable. In this section, we explain how to simulate the joint distribution of multiple continuous random variables, focusing on the random variables $\tilde{a}$ and $\tilde{b}$ in Example 5.6.

In order to simulate the random variables in Example 5.6, we could be tempted to simply generate samples from $\tilde{a}$ and $\tilde{b}$ separately via inverse-transform sampling. However, this does not work, as illustrated in the middle plot of Figure 5.15. The samples clearly have the wrong distribution. Many are outside of the triangle, and the ones inside are not uniformly distributed.

Our naive approach generates the right marginal distributions, but does not take into account the conditional distributions, which are consequently simulated incorrectly. To address this, we generate the random variables sequentially. First, we produce a sample a from the marginal distribution of $\tilde{a}$. Then, we produce a sample b from the *conditional distribution of $\tilde{b}$ given $\tilde{a} = a$*. This is achieved by applying inverse-transform sampling to

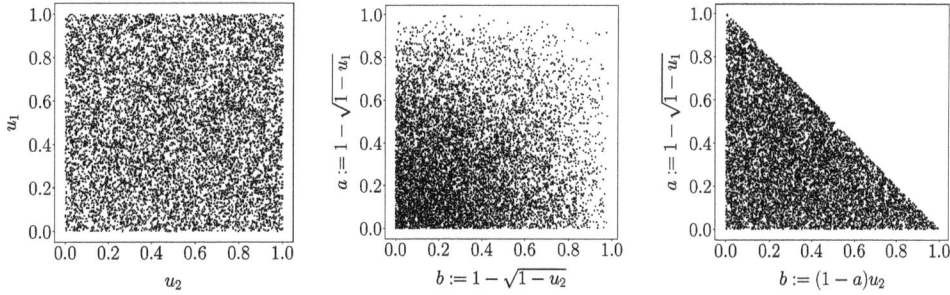

Figure 5.15 Simulating a triangular joint pdf. The left plot shows 10,000 samples, where the horizontal and vertical coordinates are independent and uniformly distributed in $[0, 1]$. The central plot shows the effect of transforming the vertical and horizontal coordinates via inverse-transform sampling, so that they are distributed according to the marginal distributions of $\tilde{a}$ and $\tilde{b}$ in Example 5.18. This does not reproduce the joint distribution of $\tilde{a}$ and $\tilde{b}$. The right plot shows the effect of transforming the vertical coordinate to simulate the marginal distribution of $\tilde{a}$, and the horizontal coordinate to simulate the conditional distribution of $\tilde{b}$ given that $\tilde{a}$ equals the vertical coordinate, as explained in Example 5.18. This does simulate the correct joint distribution of $\tilde{a}$ and $\tilde{b}$.

the conditional cdf of $\tilde{b}$ given $\tilde{a} = a$ (see Example 5.18 for the details). The resulting pair of values (a, b) have the correct joint distribution, as illustrated by the graph on the right of Figure 5.15. In conclusion, we can leverage inverse-transform sampling to jointly simulate random variables, as long as we apply it sequentially to ensure that the conditional distributions are correct.

Example 5.18 (Triangle lake: Sampling from the joint distribution). In order to sample from the joint pdf in Example 5.6, we first simulate $\tilde{a}$ via inverse-transform sampling. Integrating the marginal pdf of $\tilde{a}$ derived in Example 5.10 yields the marginal cdf of $\tilde{a}$, which equals $F_{\tilde{a}}(a) = 2a - a^2$ for $a \in [0, 1]$. Solving the equation

$$u = F_{\tilde{a}}\left(F_{\tilde{a}}^{-1}(u)\right) \tag{5.102}$$
$$= 2F_{\tilde{a}}^{-1}(u) - F_{\tilde{a}}^{-1}(u)^2 \tag{5.103}$$

yields the inverse cdf $F_{\tilde{a}}^{-1}(u) = 1 - \sqrt{1 - u}$. We generate a sample from the marginal distribution of $\tilde{a}$ by plugging a sample u_1 from the uniform distribution in $[0, 1]$ into $F_{\tilde{a}}^{-1}$:

$$a_{\text{samp}} := 1 - \sqrt{1 - u_1}. \tag{5.104}$$

We can apply the same approach to simulate the marginal distribution of $\tilde{b}$, which by symmetry is exactly the same as the marginal distribution of $\tilde{a}$. Using another sample u_2 from the uniform distribution in $[0, 1]$ this yields:

$$b_{\text{wrong}} := 1 - \sqrt{1 - u_2}. \tag{5.105}$$

As illustrated by the graph at the center of Figure 5.15, this doesn't work. The problem is that we are simulating the two random variables *independently*. To capture the dependence between the random variables, we instead sample from the conditional distribution of $\tilde{b}$ given

$\tilde{a} = a_{\text{samp}}$. As derived in Example 5.13, the conditional pdf of $\tilde{b}$ given $\tilde{a} = a_{\text{samp}}$ equals $(1 - a_{\text{samp}})^{-1}$ in $[0, 1 - a_{\text{samp}}]$. The conditional cdf is therefore

$$F_{\tilde{b}\,|\,\tilde{a}}(b\,|\,a_{\text{samp}}) := \mathrm{P}\,(\tilde{b} \leq b \,|\, \tilde{a} = a_{\text{samp}}) \tag{5.106}$$

$$= \int_{u=-\infty}^{b} f_{\tilde{b}\,|\,\tilde{a}}\,(u\,|\,a_{\text{samp}})\,\mathrm{d}u \tag{5.107}$$

$$= \frac{b}{1 - a_{\text{samp}}}. \tag{5.108}$$

Solving the equation

$$u = F_{\tilde{b}\,|\,\tilde{a}}\left(F_{\tilde{b}\,|\,\tilde{a}}^{-1}(u\,|\,a_{\text{samp}})\,|\,a_{\text{samp}}\right) \tag{5.109}$$

$$= \frac{F_{\tilde{b}\,|\,\tilde{a}}^{-1}(u\,|\,a_{\text{samp}})}{1 - a_{\text{samp}}} \tag{5.110}$$

yields the inverse conditional cdf $F_{\tilde{b}\,|\,\tilde{a}}^{-1}(u\,|\,a_{\text{samp}}) = (1 - a_{\text{samp}})u$. We obtain a sample from the conditional distribution by plugging the uniform sample u_2 into $F_{\tilde{b}\,|\,\tilde{a}}^{-1}(\cdot\,|\,a_{\text{samp}})$:

$$b_{\text{samp}} := (1 - a_{\text{samp}})u_2. \tag{5.111}$$

The resulting pair of values $(a_{\text{samp}}, b_{\text{samp}})$ correctly capture the dependence between $\tilde{a}$ and $\tilde{b}$, as shown in the graph on the right of Figure 5.15.

. . . .

5.10 Gaussian Random Vectors

As mentioned in Section 5.4, nonparametric density estimation is often intractable, unless the number of variables is small. This motivates the use of parametric models to estimate the joint pdf of continuous random vectors. Gaussian random vectors are a multidimensional generalization of Gaussian random variables, described in Section 3.6.2. They are perhaps the most popular parametric multidimensional models for continuous data. In Section 5.10.1, we derive the parametric distribution of Gaussian random vectors. In Section 5.10.2, we study the marginal and conditional distributions of the entries of these vectors, establishing that they are all Gaussian. In Section 5.10.3, we explain how to fit the multidimensional Gaussian distribution to data via maximum-likelihood estimation.

5.10.1 Definition

In this section, we explain how to derive a parametric model for multiple variables based on the Gaussian distribution. We begin with a simple approach, where we model each variable independently. Let $\tilde{x}$ denote a d-dimensional random vector. If we model each entry as a Gaussian random variable and assume that the entries are all independent, by Definitions 5.15 and 3.30, the joint pdf of the random vector equals

$$f_{\tilde{x}}(x) = \prod_{i=1}^{d} f_{\tilde{x}[i]}\left(\tilde{x}[i]\right) \tag{5.112}$$

$$= \prod_{i=1}^{d} \frac{1}{\sqrt{2\pi}\sigma_i} \exp\left(-\frac{(x[i] - \mu_i)^2}{2\sigma_i^2}\right) \tag{5.113}$$

$$= \frac{1}{(2\pi)^{\frac{d}{2}} \prod_{i=1}^{d} \sigma_i} \exp\left(-\frac{1}{2} \sum_{i=1}^{d} \frac{(x[i] - \mu_i)^2}{\sigma_i^2}\right), \tag{5.114}$$

where μ_i and σ_i denote the mean and standard-deviation parameters of each Gaussian for $1 \le i \le d$.

The parameters of our Gaussian model can be estimated from data by applying the maximum-likelihood estimator in Theorem 3.35 to each variable separately. Figure 5.16 shows an application to an anthropometric dataset. Unfortunately, the model does not provide a good fit to the data. The problem is that it cannot capture the dependence between the variables. To gain geometric intuition about why this is the case, let us consider the contour surfaces of our parametric Gaussian density, which are surfaces on which the joint pdf is equal to a constant c:

$$\left\{x \in \mathbb{R}^d : f_{\tilde{x}}(x) = c\right\} = \left\{x \in \mathbb{R}^d : \sum_{i=1}^{d} \frac{(x[i] - \mu_i)^2}{\sigma_i^2} = c'\right\}, \tag{5.115}$$

where $c' := -2\log\left(c\,(2\pi)^{\frac{d}{2}} \prod_{i=1}^{d} \sigma_i\right)$. The contour surfaces are concentric ellipsoids with axes that are aligned with the coordinate axes. The middle plot of Figure 5.16, shows the 2D contour ellipses for the model fit to the anthropometric data. The standard deviations σ_1 and σ_2 determine the vertical and horizontal spread of the ellipses, and the means μ_1 and μ_2 determine where they are centered. Looking at the data in the left scatterplot of Figure 5.16, it is clear that constraining the ellipsoid axes to lie along the coordinate axes is a major limitation. It is impossible to achieve a good fit to the data just by stretching the

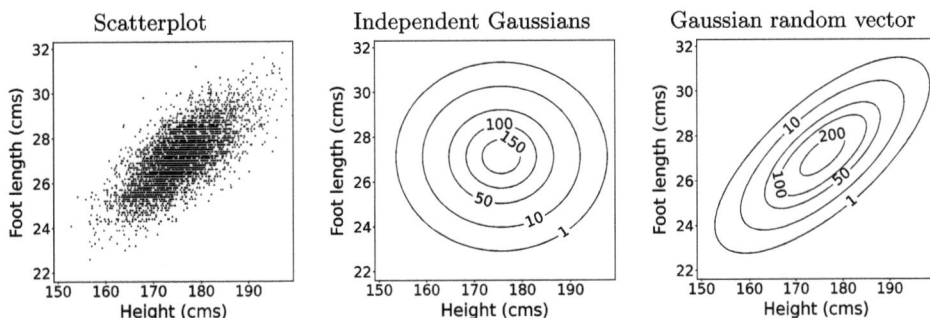

Figure 5.16 Gaussian parametric model of height and foot length. The left graph shows a scatterplot of the height and foot length of 4,082 men in the United States army, extracted from Dataset 5. The middle graph shows the contour lines of a Gaussian parametric model, where each entry is modeled independently. The model is not able to capture the dependence between height and foot length (taller people have longer feet). The right graph shows the density of a parametric Gaussian model that incorporates dependence between variables, achieving a much better fit to the data. The model parameters are selected via maximum-likelihood estimation, as described in Section 5.10.3.

ellipses horizontally or vertically because the data exhibit a diagonal arrangement, due to the dependence between the two variables.

In order to allow our parametric model to capture variable dependence, we incorporate *rotations* of the density with respect to the coordinate axes. This can be achieved by incorporating additional parameters in the form of d orthonormal vectors $u_1, u_2, \ldots, u_d$. The contour surfaces of an ellipsoid, whose axes are aligned with these vectors, equal

$$\left\{ x \in \mathbb{R}^d : \sum_{i=1}^{d} \frac{\left(u_i^T \left(x - \mu \right) \right)^2}{\sigma_i^2} = c' \right\} \tag{5.116}$$

for some constant c'. Here μ is the vector of mean parameters, such that $\mu[i] = \mu_i$, for $1 \leq i \leq d$. We can reformulate the equation for the ellipsoid as a quadratic function:

$$\sum_{i=1}^{d} \frac{\left(u_i^T \left(x - \mu \right) \right)^2}{\sigma_i^2} = \left(x - \mu \right)^T U \Lambda^{-1} U^T \left(x - \mu \right) \tag{5.117}$$

$$= \left(x - \mu \right)^T \Sigma^{-1} \left(x - \mu \right), \tag{5.118}$$

where

$$U := \begin{bmatrix} u_1 & u_2 & \cdots & u_d \end{bmatrix}, \qquad \Lambda := \begin{bmatrix} \sigma_1^2 & 0 & \cdots & 0 \\ 0 & \sigma_2^2 & \cdots & 0 \\ \cdots & \cdots & \cdots & \cdots \\ 0 & 0 & \cdots & \sigma_d^2 \end{bmatrix}, \qquad \Sigma := U \Lambda U^T.$$

By the spectral theorem (Theorem 11.19), any positive definite symmetric matrix Σ has an eigendecomposition of the form $U \Lambda U^T$ where U is orthogonal and Λ is diagonal. This means that we can use a matrix parameter Σ to encode the standard-deviation parameters σ_1, $\ldots, \sigma_d$ and the vector parameters $u_1, \ldots, u_d$. In order to obtain our desired flexible model, we replace the exponent in equation (5.114) by the quadratic function in (5.118):

$$f_{\tilde{x}} \left(x \right) = \frac{1}{\sqrt{\left(2\pi \right)^d \left| \Sigma \right|}} \exp \left(-\frac{1}{2} \left(x - \mu \right)^T \Sigma^{-1} \left(x - \mu \right) \right). \tag{5.119}$$

Including the determinant $\left| \Sigma \right|$ in the denominator of (5.119) ensures that the expression integrates to one. Σ needs to be positive definite, so that the density decays exponentially in every direction. If not, there are vectors v for which $v^T \Sigma v < 0$. In such directions, the density would explode to infinity (set $x := \mu + \alpha v$ where $\alpha \to \infty$). We call Σ the covariance-matrix parameter because it is the covariance matrix of the random vector (see Theorem 11.10). Notice that the joint pdf of our model no longer factorizes into a product of the marginals, so the entries of the random vector $\tilde{x}$ are not independent. As illustrated by the data in Figure 5.16, this is often important to fit multiple variables effectively.

In conclusion, we have derived a model that has exponentially decaying ellipsoidal contours, rotated with respect to the coordinate axes. Random vectors that have this parametric distribution are known as Gaussian random vectors.

Definition 5.19 (Gaussian random vector). *A Gaussian random vector $\tilde{x}$ of dimension d is a random vector with joint pdf*

$$f_{\tilde{x}}(x) = \frac{1}{\sqrt{(2\pi)^d |\Sigma|}} \exp\left(-\frac{1}{2}(x-\mu)^T \Sigma^{-1}(x-\mu)\right), \qquad (5.120)$$

where $|\Sigma|$ denotes the determinant of Σ. The joint pdf is parametrized by the mean vector $\mu \in \mathbb{R}^d$ and the covariance matrix $\Sigma \in \mathbb{R}^{d \times d}$. Σ must be invertible, symmetric, and positive definite (all its eigenvalues must be positive).

The following example shows how to derive the ellipsoidal contours of a Gaussian joint pdf. This requires reverse engineering our derivation of the covariance-matrix parameter Σ using its eigendecomposition.

Example 5.20 (Contour lines of two-dimensional Gaussian). We consider a two-dimensional Gaussian random vector where μ is the zero vector and

$$\Sigma = \begin{bmatrix} 0.5 & -0.3 \\ -0.3 & 0.5 \end{bmatrix}. \qquad (5.121)$$

Since μ is zero, the contour lines of the density correspond to the set of points where $x^T \Sigma^{-1} x$ is constant. These are ellipses satisfying the equation

$$x^T \Sigma^{-1} x = \frac{(u_1^T x)^2}{\lambda_1} + \frac{(u_2^T x)^2}{\lambda_2} = c' \qquad (5.122)$$

for different values of the constant c'. The eigenvectors of Σ are

$$u_1 = \begin{bmatrix} 1/\sqrt{2} \\ -1/\sqrt{2} \end{bmatrix}, \quad u_2 = \begin{bmatrix} 1/\sqrt{2} \\ 1/\sqrt{2} \end{bmatrix}, \qquad (5.123)$$

so the axes of the density contour lines are diagonally oriented. The spread in the direction of each eigenvector is proportional to the square root of the corresponding eigenvalue. The eigenvalues equal $\lambda_1 = 0.8$ and $\lambda_2 = 0.2$, so there is greater spread in the direction of u_1. Figure 5.17 shows a 3D plot of the joint pdf and a 2D plot of its ellipsoidal contour lines.

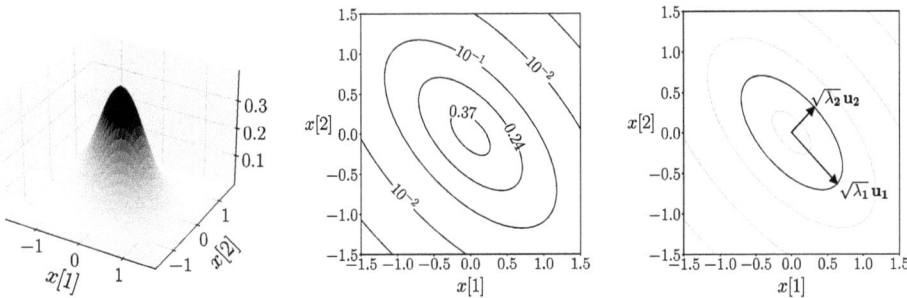

Figure 5.17 Probability density of a Gaussian vector. The left 3D plot depicts the joint pdf of the two-dimensional Gaussian random vector defined in Example 5.20. The middle plot shows the corresponding contour plot. The axes align with the eigenvectors of the covariance matrix. Their lengths are proportional to the square root of the corresponding eigenvalues. The right plot shows the contour line on which the density equals 0.24.

5.10.2 Marginal and Conditional Distributions

In this section, we study the marginal and conditional distributions of Gaussian random vectors. We begin with a simple example in two dimensions.

Example 5.21 (Marginal and conditional distributions in two dimensions). Let $\tilde{a}$ and $\tilde{b}$ be the two entries of a two-dimensional Gaussian random vector with zero mean and covariance matrix

$$\Sigma := \begin{bmatrix} 1 & \rho \\ \rho & 1 \end{bmatrix}. \tag{5.124}$$

As we explain in Example 8.8, ρ is the correlation coefficient of $\tilde{a}$ and $\tilde{b}$, but here we just interpret it as a model parameter determining the shape of the distribution. In order for Σ to be positive definite, we need $-1 < \rho < 1$ (this can be verified by checking its eigendecomposition, since a positive definite matrix has positive eigenvalues). The determinant of the covariance matrix is $1 - \rho^2$ and its inverse is

$$\Sigma^{-1} = \frac{1}{1 - \rho^2} \begin{bmatrix} 1 & -\rho \\ -\rho & 1 \end{bmatrix}, \tag{5.125}$$

so by Definition 5.19 the joint pdf of the random vector equals

$$f_{\tilde{a},\tilde{b}}(a,b) := \frac{1}{\sqrt{(2\pi)^2 |\Sigma|}} \exp\left(-\frac{1}{2} \begin{bmatrix} a \\ b \end{bmatrix}^T \Sigma^{-1} \begin{bmatrix} a \\ b \end{bmatrix} \right) \tag{5.126}$$

$$= \frac{1}{2\pi\sqrt{1-\rho^2}} \exp\left(-\frac{a^2 - 2\rho ab + b^2}{2(1-\rho^2)} \right) \tag{5.127}$$

$$= \frac{1}{2\pi\sqrt{1-\rho^2}} \exp\left(-\frac{(1-\rho^2)a^2 + (b - \rho a)^2}{2(1-\rho^2)} \right) \tag{5.128}$$

$$= \frac{1}{\sqrt{2\pi}} \exp\left(-\frac{a^2}{2} \right) \frac{1}{\sqrt{2\pi(1-\rho^2)}} \exp\left(-\frac{(b - \rho a)^2}{2(1-\rho^2)} \right). \tag{5.129}$$

These algebraic manipulations are often referred to as *completing the square*. Interpreted as a function of b, the term

$$\frac{1}{\sqrt{2\pi(1-\rho^2)}} \exp\left(-\frac{(b - \rho a)^2}{2(1-\rho^2)} \right) \tag{5.130}$$

is a Gaussian pdf with mean ρa and variance $1 - \rho^2$ by Definition 3.30, which means that it integrates to one with respect to b. By Theorem 5.8, the marginal pdf of $\tilde{a}$ therefore equals

$$f_{\tilde{a}}(a) = \int_{b=-\infty}^{\infty} f_{\tilde{a},\tilde{b}}(a,b) \, db \tag{5.131}$$

$$= \frac{1}{\sqrt{2\pi}} \exp\left(-\frac{a^2}{2} \right) \int_{b=-\infty}^{\infty} \frac{1}{\sqrt{2\pi(1-\rho^2)}} \exp\left(-\frac{(b - \rho a)^2}{2(1-\rho^2)} \right) \, db \tag{5.132}$$

$$= \frac{1}{\sqrt{2\pi}} \exp\left(-\frac{a^2}{2} \right). \tag{5.133}$$

The marginal distribution of $\tilde{a}$ is Gaussian with zero mean and unit variance, again by Definition 3.30. The exact same argument switching a and b implies that the marginal distribution of $\tilde{b}$ is also Gaussian with zero mean and unit variance.

By Definition 5.11, the Gaussian term with mean ρa and variance $1 - \rho^2$ is in fact the conditional pdf of $\tilde{b}$ given $\tilde{a} = a$:

$$f_{\tilde{b}|\tilde{a}}(b \mid a) := \frac{f_{\tilde{a},\tilde{b}}(a,b)}{f_{\tilde{a}}(a)} \tag{5.134}$$

$$= \frac{1}{\sqrt{2\pi(1-\rho^2)}} \exp\left(-\frac{(b-\rho a)^2}{2(1-\rho^2)}\right). \tag{5.135}$$

The correlation coefficient ρ completely determines the dependence between $\tilde{a}$ and $\tilde{b}$. When $\rho := 0$, the conditional distribution is Gaussian with zero mean and unit variance, so $f_{\tilde{b}|\tilde{a}}$ is equal to the marginal pdf of $\tilde{b}$, and the two variables are independent (see Section 5.7). When ρ approaches 1, the variables become highly dependent: The conditional pdf is centered at $\rho a \approx a$, and the variance $1 - \rho^2$ is very small, so the density is concentrated near a. Consequently, $\tilde{b}$ is close to $\tilde{a}$ with high probability. When ρ approaches -1, $\tilde{b}$ tends to be close to $-\tilde{a}$ instead, by the same argument. Figure 5.18 shows the joint pdf $f_{\tilde{a},\tilde{b}}$ and the corresponding conditional pdf $f_{\tilde{b}|\tilde{a}}$ for different values of ρ.

....

In Example 5.21, the marginal and conditional pdfs of the Gaussian random vector turn out to be Gaussian. This is no accident. The following theorem shows that the marginal and conditional distributions of the entries of any 2D Gaussian random vector are Gaussian.

Theorem 5.22 (Marginal and conditional distributions of a 2D Gaussian). *Let $\tilde{a}$ and $\tilde{b}$ be the two entries of a Gaussian random vector with mean $\mu \in \mathbb{R}^2$ and covariance matrix $\Sigma \in \mathbb{R}^{2\times 2}$, where Σ is positive definite and full rank. To ease notation, we express the mean and covariance in terms of the following parameters,*

$$\mu := \begin{bmatrix} \mu_{\tilde{a}} \\ \mu_{\tilde{b}} \end{bmatrix}, \tag{5.136}$$

$$\Sigma := \begin{bmatrix} \sigma_{\tilde{a}}^2 & \rho\sigma_{\tilde{a}}\sigma_{\tilde{b}} \\ \rho\sigma_{\tilde{a}}\sigma_{\tilde{b}} & \sigma_{\tilde{b}}^2 \end{bmatrix}, \tag{5.137}$$

where $-1 < \rho < 1$ (otherwise Σ is not positive definite, which can be verified checking its eigendecomposition).

The marginal distribution of $\tilde{a}$ is Gaussian with mean $\mu_{\tilde{a}}$ and standard deviation $\sigma_{\tilde{a}}$. The marginal distribution of $\tilde{b}$ is Gaussian with mean $\mu_{\tilde{b}}$ and standard deviation $\sigma_{\tilde{b}}$. The conditional distribution of $\tilde{b}$ given $\tilde{a} = a$ is Gaussian with mean

$$\mu_{\mathrm{cond}} := \mu_{\tilde{b}} + \frac{\rho\sigma_{\tilde{b}}(a - \mu_{\tilde{a}})}{\sigma_{\tilde{a}}} \tag{5.138}$$

and variance

$$\sigma_{\mathrm{cond}}^2 := (1 - \rho^2)\sigma_{\tilde{b}}^2. \tag{5.139}$$

Proof The proof follows the exact same reasoning as in Example 5.21. We define the *standardized* variables (see Section 8.2),

$$s(a) := \frac{a - \mu_{\tilde{a}}}{\sigma_{\tilde{a}}}, \tag{5.140}$$

$$s(b) := \frac{b - \mu_{\tilde{b}}}{\sigma_{\tilde{b}}}. \tag{5.141}$$

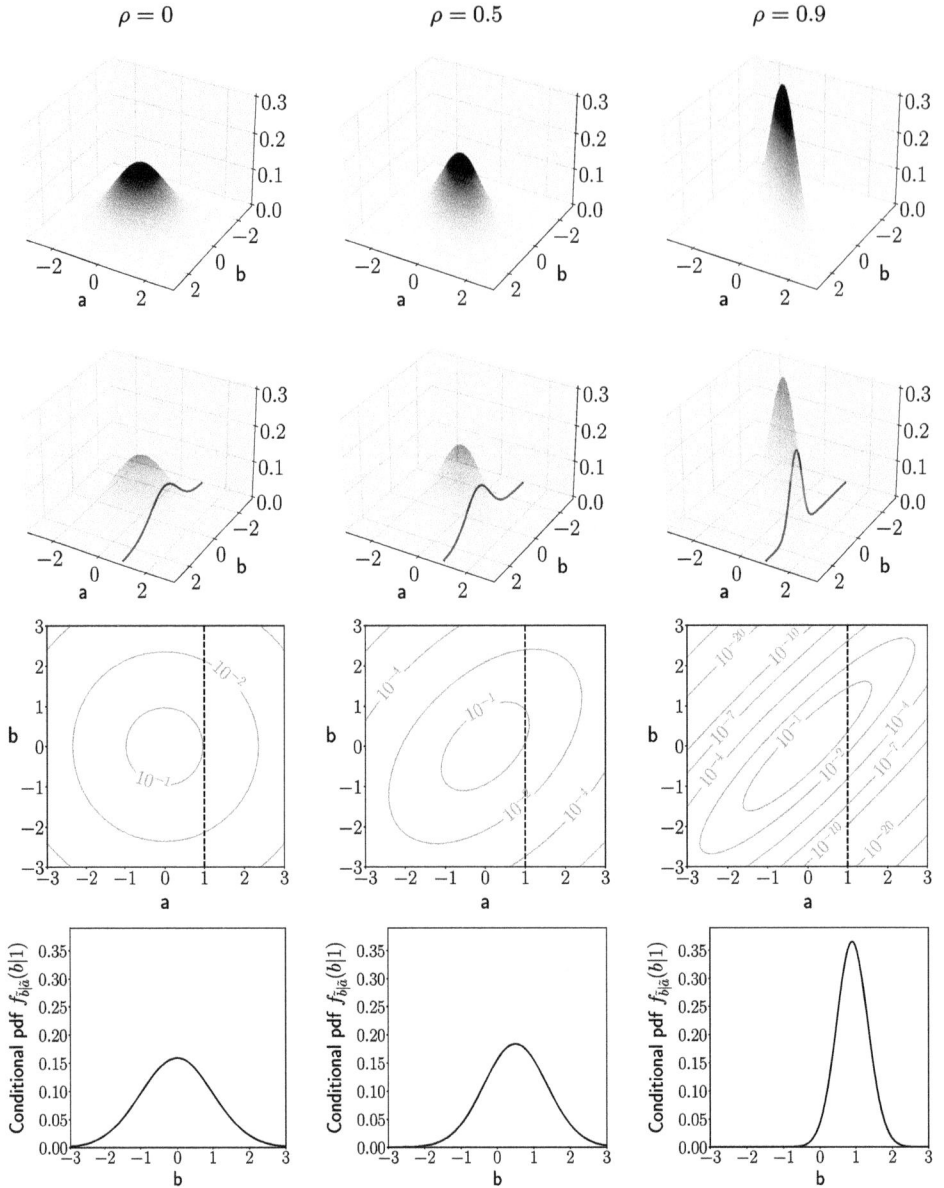

Figure 5.18 Conditional distributions of Gaussian random vectors. The first row shows the joint pdf of the Gaussian random vector in Example 5.21 for values of the correlation coefficient ρ indicated at the top of each column. In the second row we see that the slice of the density corresponding to $a = 1$ has a Gaussian shape. The third row shows the contour plots and a dashed line at $a = 1$. The fourth row shows the conditional pdf of $\tilde{b}$, obtained by normalizing the joint pdf restricted to the dashed line. As established in Example 5.21, the density is Gaussian with mean ρ and variance $1 - \rho^2$.

The determinant of Σ is $\sigma_{\tilde{a}}^2 \sigma_{\tilde{b}}^2 \left(1 - \rho^2\right)$ and its inverse equals

$$\Sigma^{-1} = \frac{1}{\sigma_{\tilde{a}}^2 \sigma_{\tilde{b}}^2 (1 - \rho^2)} \begin{bmatrix} \sigma_{\tilde{b}}^2 & -\rho\sigma_{\tilde{a}}\sigma_{\tilde{b}} \\ -\rho\sigma_{\tilde{a}}\sigma_{\tilde{b}} & \sigma_{\tilde{a}}^2 \end{bmatrix}, \tag{5.142}$$

so by Definition 5.19,

$$f_{\tilde{a},\tilde{b}}(a,b) := \frac{1}{\sqrt{(2\pi)^2 |\Sigma|}} \exp\left(-\frac{1}{2}\left(\begin{bmatrix} a \\ b \end{bmatrix} - \mu\right)^T \Sigma^{-1} \left(\begin{bmatrix} a \\ b \end{bmatrix} - \mu\right)\right) \tag{5.143}$$

$$= \frac{1}{2\pi\sigma_{\tilde{a}}\sigma_{\tilde{b}}\sqrt{1-\rho^2}} \exp\left(-\frac{s(a)^2 - 2\rho s(a)s(b) + s(b)^2}{2(1-\rho^2)}\right) \tag{5.144}$$

$$= \frac{1}{2\pi\sigma_{\tilde{a}}\sigma_{\tilde{b}}\sqrt{1-\rho^2}} \exp\left(-\frac{(1-\rho^2)s(a)^2 + (s(b) - \rho s(a))^2}{2(1-\rho^2)}\right) \tag{5.145}$$

$$= \frac{1}{\sqrt{2\pi}\sigma_{\tilde{a}}} \exp\left(-\frac{s(a)^2}{2}\right) \frac{1}{\sqrt{2\pi(1-\rho^2)}\sigma_{\tilde{b}}} \exp\left(-\frac{(s(b) - \rho s(a))^2}{2(1-\rho^2)}\right)$$

$$= \frac{1}{\sqrt{2\pi}\sigma_{\tilde{a}}} \exp\left(-\frac{(a-\mu_{\tilde{a}})^2}{2\sigma_{\tilde{a}}^2}\right) \frac{1}{\sqrt{2\pi(1-\rho^2)}\sigma_{\tilde{b}}} \exp\left(-\frac{(b-\mu_{\tilde{b}} - \rho\sigma_{\tilde{b}}s(a))^2}{2(1-\rho^2)\sigma_{\tilde{b}}^2}\right).$$

As in Example 5.21, we realize that the second term

$$\frac{1}{\sqrt{2\pi(1-\rho^2)}\sigma_{\tilde{b}}} \exp\left(-\frac{(b-\mu_{\tilde{b}} - \rho\sigma_{\tilde{b}}s(a))^2}{2(1-\rho^2)\sigma_{\tilde{b}}^2}\right) \tag{5.146}$$

is a Gaussian pdf with respect to b, so it integrates to one. By Theorem 5.8, the first term is the marginal pdf of $\tilde{a}$,

$$f_{\tilde{a}}(a) = \int_{b=-\infty}^{\infty} f_{\tilde{a},\tilde{b}}(a,b) \, db \tag{5.147}$$

$$= \frac{1}{\sqrt{2\pi}\sigma_{\tilde{a}}} \exp\left(-\frac{(a-\mu_{\tilde{a}})^2}{2\sigma_{\tilde{a}}^2}\right), \tag{5.148}$$

which is Gaussian with mean $\mu_{\tilde{a}}$ and variance $\sigma_{\tilde{a}}^2$ by Definition 3.30. By Definition 5.11, this implies that (5.146) is the conditional pdf of $\tilde{b}$. The marginal pdf of $\tilde{b}$ can be derived by switching the roles of a and b. ∎

Theorem 5.22 establishes that each component of a Gaussian 2D vector is Gaussian, and so are their conditional distributions. The correlation coefficient ρ governs the dependence between the two components. We elaborate further on the properties of the correlation coefficient in Chapter 8.

Our results for 2D Gaussian vectors generalize to higher dimensions. The marginal and conditional distributions of any subvector of a Gaussian random vector are Gaussian.

Theorem 5.23 (Marginal and conditional distributions of a multidimensional Gaussian). *Let $\tilde{z}$ denote a d-dimensional Gaussian random vector with mean $\mu \in \mathbb{R}^d$ and covariance matrix $\Sigma \in \mathbb{R}^{d \times d}$, which is positive definite and full rank. Assume that the subvectors $\tilde{x}$ and $\tilde{y}$ consist of the first m and $d - m$ entries, respectively, for any m such that $1 \leq m < d$,*

$$\tilde{z} := \begin{bmatrix} \tilde{x} \\ \tilde{y} \end{bmatrix}. \tag{5.149}$$

This is without loss of generality; we can reorder the entries of any random vector to place a subvector of interest at the beginning. We express μ and Σ as

$$\mu := \begin{bmatrix} \mu_{\tilde{x}} \\ \mu_{\tilde{y}} \end{bmatrix}, \tag{5.150}$$

$$\Sigma := \begin{bmatrix} \Sigma_{\tilde{x}} & \Sigma_{\tilde{x},\tilde{y}} \\ \Sigma_{\tilde{x},\tilde{y}}^T & \Sigma_{\tilde{y}} \end{bmatrix}, \tag{5.151}$$

where $\mu_{\tilde{x}} \in \mathbb{R}^m$, $\mu_{\tilde{y}} \in \mathbb{R}^{d-m}$, $\Sigma_{\tilde{x}} \in \mathbb{R}^{m \times m}$, $\Sigma_{\tilde{x},\tilde{y}} \in \mathbb{R}^{m \times (d-m)}$, $\Sigma_{\tilde{y}} \in \mathbb{R}^{(d-m) \times (d-m)}$.

Then, the marginal distribution of $\tilde{x}$ is Gaussian with mean $\mu_{\tilde{x}}$ and covariance matrix $\Sigma_{\tilde{x}}$. The marginal distribution of $\tilde{y}$ is Gaussian with mean $\mu_{\tilde{y}}$ and covariance matrix $\Sigma_{\tilde{y}}$. The conditional distribution of $\tilde{y}$ given $\tilde{x} = x$ is Gaussian with mean

$$\mu_{\mathrm{cond}} = \mu_{\tilde{y}} + \Sigma_{\tilde{x},\tilde{y}}^T \Sigma_{\tilde{x}}^{-1} (x - \mu_{\tilde{x}}) \tag{5.152}$$

and covariance matrix

$$\Sigma_{\mathrm{cond}} = \Sigma_{\tilde{y}} - \Sigma_{\tilde{x},\tilde{y}}^T \Sigma_{\tilde{x}}^{-1} \Sigma_{\tilde{x},\tilde{y}}. \tag{5.153}$$

Proof   The proof follows the same argument as the proof of Theorem 5.22. We factorize the joint pdf of $\tilde{z}$ into two terms,

$$f_{\tilde{z}}\left(\begin{bmatrix} x \\ y \end{bmatrix}\right) = \frac{1}{\sqrt{(2\pi)^m |\Sigma_{\tilde{x}}|}} \exp\left(-\frac{1}{2}(x - \mu_{\tilde{x}})^T \Sigma_{\tilde{x}}^{-1}(x - \mu_{\tilde{x}})\right)$$
$$\frac{1}{\sqrt{(2\pi)^{d-m} |\Sigma_{\mathrm{cond}}|}} \exp\left(-\frac{1}{2}(y - \mu_{\mathrm{cond}})^T \Sigma_{\mathrm{cond}}^{-1}(y - \mu_{\mathrm{cond}})\right). \tag{5.154}$$

This can be achieved through algebraic manipulations that rely on the Schur complement, which we omit as they are lengthy and not particularly insightful. We refer the reader to Jordan (2009) for a detailed derivation. By Definition 5.19, the term on the second line of (5.154) is the joint pdf of a Gaussian random vector, which integrates to one. The term on the first line is therefore the marginal pdf $f_{\tilde{x}}$ of $\tilde{x}$ (see Section 5.5). The marginal pdf of $\tilde{y}$ can be derived by switching the roles of $\tilde{x}$ and $\tilde{y}$. By Definition 5.11, dividing the joint pdf by $f_{\tilde{x}}$ establishes that the Gaussian pdf on the second line is the conditional pdf of $\tilde{y}$ given $\tilde{x} = x$. ∎

5.10.3 Maximum-Likelihood Estimation

In order to fit a multidimensional Gaussian distribution to a d-dimensional dataset $X := \{x_1, \ldots, x_n\}$, we apply the maximum-likelihood method described in Section 3.7. This requires maximizing the likelihood of the data with respect to the mean and covariance parameters. Assuming i.i.d. data sampled from a Gaussian distribution with unknown parameters μ and Σ, by Definitions 3.32 and 5.19 the likelihood is

$$\mathcal{L}(\mu, \Sigma) := \prod_{i=1}^{n} f_{\mu,\Sigma}(x_i) \tag{5.155}$$

$$= \prod_{i=1}^{n} \frac{1}{\sqrt{(2\pi)^d |\Sigma|}} \exp\left(-\frac{1}{2}(x_i - \mu)^T \Sigma^{-1}(x_i - \mu)\right), \tag{5.156}$$

and the log-likelihood is

$$\log \mathcal{L}(\mu, \Sigma) := -n \log \sqrt{(2\pi)^d |\Sigma|} - \sum_{i=1}^{n} \frac{1}{2} (x_i - \mu)^T \Sigma^{-1} (x_i - \mu). \qquad (5.157)$$

The following theorem provides the maximum-likelihood estimator of the parameters of the Gaussian random vector. The mean parameter is obtained by averaging the data and is therefore equal to the sample mean (see Definition 11.4). The covariance matrix is obtained by averaging the outer product of each data point with itself, after subtracting the mean, which is essentially equal to the sample covariance matrix (see Definition 11.14).[2]

Theorem 5.24 (Maximum-likelihood estimation for Gaussian random vectors). *Let* $X := \{x_1, \ldots, x_n\}$ *be a dataset containing* n *data points in* $\mathbb{R}^d$. *The maximum-likelihood estimators of the parameters of a Gaussian random vector given these data are*

$$\mu_{\text{ML}} = \frac{1}{n} \sum_{i=1}^{n} x_i, \qquad (5.158)$$

$$\Sigma_{\text{ML}} = \frac{1}{n} \sum_{i=1}^{n} (x_i - \mu_{\text{ML}})(x_i - \mu_{\text{ML}})^T. \qquad (5.159)$$

Proof For fixed Σ, the gradient and Hessian of the log likelihood with respect to μ equal

$$\nabla_\mu \log \mathcal{L}(\mu, \Sigma) = \Sigma^{-1} \sum_{i=1}^{n} (x_i - \mu), \qquad (5.160)$$

$$\nabla_\mu^2 \log \mathcal{L}(\mu, \Sigma) = -\Sigma^{-1}. \qquad (5.161)$$

Since Σ is positive definite, Σ^{-1} is also positive definite (the eigenvalues of Σ^{-1} are just the inverses of the eigenvalues of Σ). This means that the log likelihood is concave with respect to μ and can be maximized by setting the gradient to zero, which yields (5.158). Maximizing the log likelihood with respect to the covariance matrix is more complicated. We plug in μ_{ML}, since it doesn't depend on the value of Σ, and compute the gradient of the log likelihood with respect to Σ^{-1},

$$\nabla_{\Sigma^{-1}} \log \mathcal{L}(\mu_{\text{ML}}, \Sigma) = \frac{n}{2} \Sigma - \frac{1}{2} \sum_{i=1}^{n} (x_i - \mu_{\text{ML}}) (x_i - \mu_{\text{ML}})^T. \qquad (5.162)$$

Setting this expression to zero, we obtain (5.159). Deriving (5.162) requires lengthy matrix computations, which we omit. We refer the reader to Jordan (2009) for the gory details. ∎

Figure 5.19 shows the result of applying the Gaussian parametric model to anthropometric data, extracted from Dataset 5. The model is fit via maximum-likelihood estimation,

---

[2] There is a minor discrepancy in the scaling factor, which is set to $n - 1$ in Definition 11.14 so that the estimator is unbiased, but this doesn't make any difference in practice, unless n is very small.

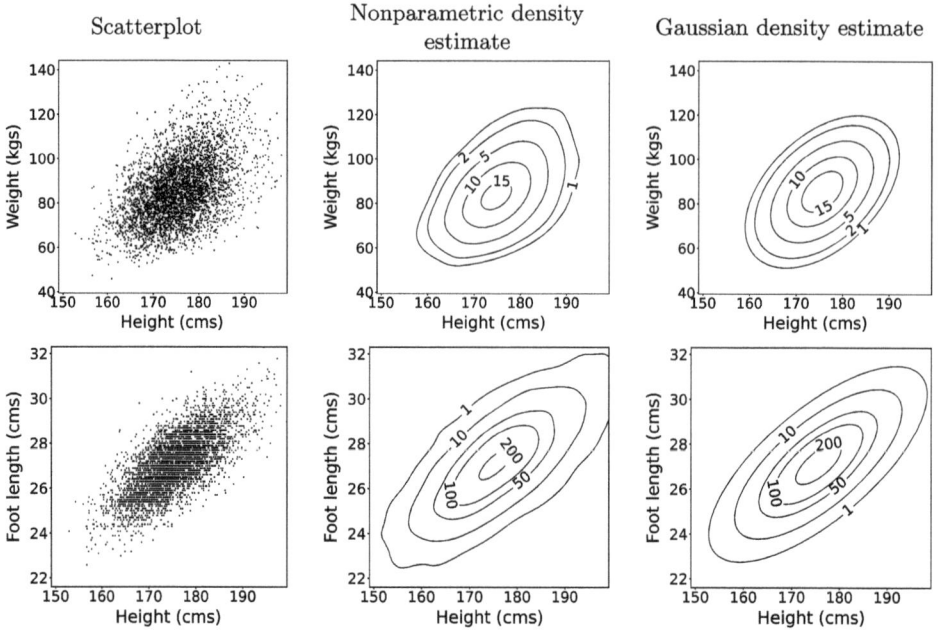

Figure 5.19 Gaussian parametric modeling of anthropometric data. The first column shows the scatterplot of the height and weight (first row) and the height and foot length (second row) of 4,082 men in the United States Army, extracted from Dataset 5. The second column shows the contour lines of the corresponding nonparametric estimate of the joint density obtained via kernel density estimation with a Gaussian kernel, following Definition 5.7. The third column shows the density of a parametric Gaussian model fit via maximum-likelihood estimation.

applying Theorem 5.24. The resulting estimated densities are similar to the density estimates produced by kernel density estimation, following Definition 5.7.

Exercises

5.1 (Identities) Let $\tilde{x}$, $\tilde{y}$, and $\tilde{z}$ be continuous random variables defined in the same probability space. Do the following identities hold? Justify your answer only if you think they do.

$$\int_{y=-\infty}^{\infty} f_{\tilde{x}\mid\tilde{y},\tilde{z}}\left(x\mid y,z\right)\,\mathrm{d}y = f_{\tilde{x}\mid\tilde{z}}\left(x\mid z\right), \tag{5.163}$$

$$\int_{x=-\infty}^{\infty} f_{\tilde{x}\mid\tilde{y},\tilde{z}}\left(x\mid y,z\right)\,\mathrm{d}x = 1, \tag{5.164}$$

$$\int_{y=-\infty}^{\infty} f_{\tilde{x}\mid\tilde{y},\tilde{z}}\left(x\mid y,z\right)\,\mathrm{d}y = 1, \tag{5.165}$$

$$\int_{x=-\infty}^{\infty} f_{\tilde{x},\tilde{y}\mid\tilde{z}}\left(x,y\mid z\right)\,\mathrm{d}x = f_{\tilde{y}\mid\tilde{z}}\left(y\mid z\right), \tag{5.166}$$

$$\int_{z=-\infty}^{\infty} f_{\tilde{x},\tilde{y}\mid\tilde{z}}\left(x,y\mid z\right)\,\mathrm{d}z = f_{\tilde{y}\mid\tilde{z}}\left(y\mid z\right). \tag{5.167}$$

5.2 (Cross) Assume that the joint pdf of $\tilde{x}$ and $\tilde{y}$ is uniform in the shaded region:

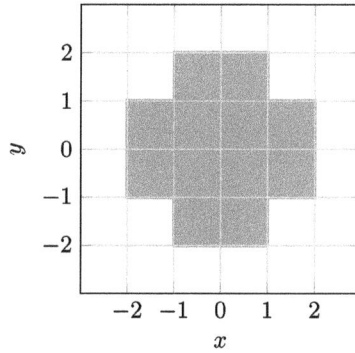

a Compute the marginal pdf of $\tilde{x}$.

b Compute the conditional pdf of $\tilde{y}$ given $\tilde{x}$. Are $\tilde{x}$ and $\tilde{y}$ independent?

5.3 (Two random variables) The random variables $\tilde{a}$ and $\tilde{b}$ are defined in the same probability space. The random variable $\tilde{a}$ is uniformly distributed in $[0, 2]$. If $0 \leq \tilde{a} \leq 1$, $\tilde{b}$ is uniformly distributed in $[0, 1]$. If $1 \leq \tilde{a} \leq 2$, $\tilde{b}$ is uniformly distributed in $[0, 2]$.

a Derive the conditional pdf of $\tilde{a}$ given $\tilde{b}$.

b Explain how to sample from the joint distribution of $\tilde{a}$ and $\tilde{b}$ using pairs of independent samples (u_1, u_2) uniformly distributed in $[0, 1]$. Apply the procedure to the pair $u_1 := 0.7$, $u_2 := 0.2$.

5.4 (Piecewise-constant density) Figure 5.20 shows the piecewise-constant joint pdf $f_{\tilde{a}, \tilde{b}}(a, b)$ of two random variables $\tilde{a}$ and $\tilde{b}$. Inside each square, the density is equal to the constant indicated at the center. Out of the squares, the density is zero.

a Derive the marginal pdf of $\tilde{a}$.

b Derive the marginal cdf of $\tilde{b}$.

c What is the conditional probability that $\tilde{b} < 0.5$ given that $\tilde{a}$ is equal to 1.6?

d Obtain joint samples of $\tilde{a}$ and $\tilde{b}$ given two independent samples $u_1 := 0.1$ and $u_2 := 0.7$ from a uniform distribution in the unit interval.

5.5 (Samples and independence) Figure 5.21 shows scatterplots of samples drawn independently at random from three different joint pdfs $f_{\tilde{a}, \tilde{b}}(a, b)$. For each of the plots, explain whether you think that $\tilde{a}$ and $\tilde{b}$ are independent and why.

5.6 (Independence of continuous random variables) Let $\tilde{x}$ and $\tilde{y}$ be two continuous random variables defined in the same probability space. Prove that the following statements hold, or provide a counterexample.

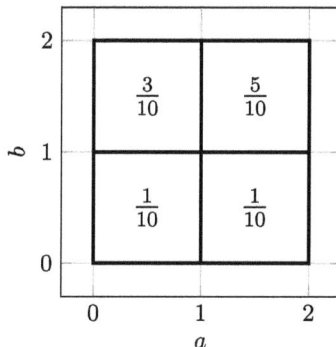

Figure 5.20 Joint pdf of the random variables $\tilde{a}$ and $\tilde{b}$ in Exercise 5.4.

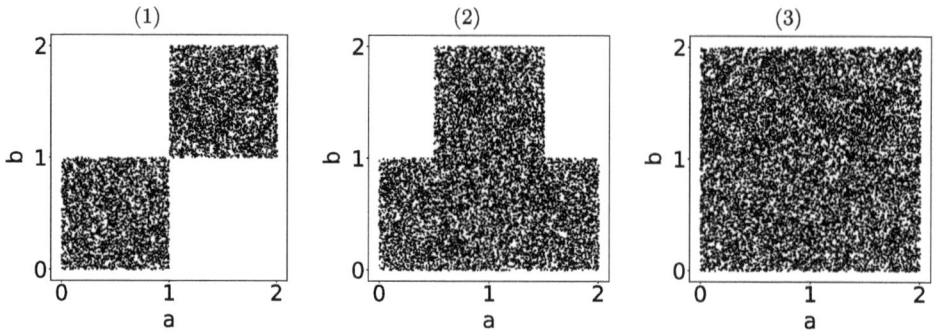

Figure 5.21 Plots for Exercise 5.5.

   a If the conditional pdf of $\tilde{y}$ given $\tilde{x}$ satisfies

$$f_{\tilde{y}\,|\,\tilde{x}}(y\,|\,x_1) = f_{\tilde{y}\,|\,\tilde{x}}(y\,|\,x_2) \tag{5.168}$$

for all y and all x_1 and x_2, then $\tilde{x}$ and $\tilde{y}$ are independent.
   b If the joint pdf of $\tilde{x}$ and $\tilde{y}$ is nonzero in the gray region:

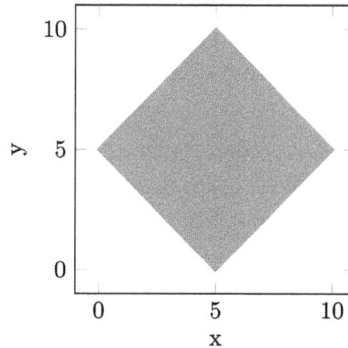

then $\tilde{x}$ and $\tilde{y}$ cannot be independent.
   c If for two fixed values x_1 and x_2, $f_{\tilde{x},\tilde{y}}(x_1,y) \neq f_{\tilde{x},\tilde{y}}(x_2,y)$ for all y such that either $f_{\tilde{x},\tilde{y}}(x_1,y)$ or $f_{\tilde{x},\tilde{y}}(x_2,y)$ are nonzero, then $\tilde{x}$ and $\tilde{y}$ cannot be independent.

5.7   (Regression to the mean) Regression[3] to the mean is a phenomenon often observed in data analysis: When an extreme value of a certain quantity is measured, the following measurements tend to be less extreme and closer to the average value of the quantity (i.e. its mean, see Chapter 7). After unusually high test scores, students tend to perform worse. After a disappointing performance, athletes tend to do better. There is often a strong temptation to provide intuitive causal interpretations to such occurrences: *The student didn't study as much after doing well in the midterm, the athlete was especially motivated after their poor result in the previous race.* However, here we show that the phenomenon arises for independent measurements and has a simple probabilistic explanation.

   a Let $\tilde{a}$ and $\tilde{b}$ be two independent random variables with the same distribution. We observe that $\tilde{a}$ is equal to the 95th percentile of its distribution (see Definition 3.9). What is the probability that $\tilde{b}$ is smaller than or equal to $\tilde{a}$ conditioned on this event? What if $\tilde{a}$ is equal to the xth percentile of its distribution?
   b How is the previous question relevant to regression to the mean?
   c If $\tilde{a}$ and $\tilde{b}$ are independent, why does the value of $\tilde{a}$ seem to have an effect on $\tilde{b}$?

---

[3] Here regression does not refer to estimation, as in Section 7.8.3 and Chapter 12, but instead means *reversion*.

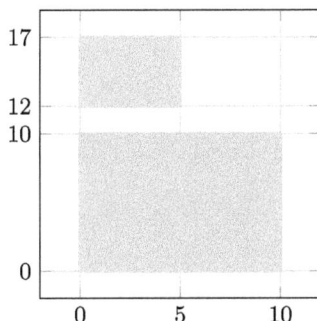

Figure 5.22 The two ponds in Exercise 5.8.

5.8 (Frog) A frog lives in a garden where there are two ponds, see Figure 5.22. It spends 1/4 of its time in the large pond and the rest in the small pond. When it is in either of the ponds, we model its position as uniformly distributed.

   a What is the joint pdf of the position of the frog in the diagram?
   b What is the marginal pdf of the horizontal coordinate of the frog's position? Sketch the pdf.
   c If we know that the horizontal coordinate of the frog is 3, what is the conditional pdf of its vertical coordinate given this information? Sketch the pdf.
   d Is the vertical coordinate of the frog independent from the horizontal coordinate?
   e Is the vertical coordinate of the frog conditionally independent from the horizontal coordinate given the event *the frog is in the small pond*?

5.9 (Sonar) An ocean scientist, called Laure, is trying to determine the depth of the sea at a certain location. She knows that it must be between 5 and 10 km, but that is all she knows. To capture this uncertainty, Laure models the depth as uniformly distributed between 5 and 10 km. She uses two separate sonar measurements to estimate the depth. She models the measurements as two random variables $\tilde{s}_1$ and $\tilde{s}_2$. If the depth is equal to x, then each sonar measurement is uniformly distributed between $x - 0.25$ and $x + 0.25$. The two measurements are conditionally independent given the depth.

   a Compute and sketch the pdf of the first sonar measurement $\tilde{s}_1$.
   b Compute the conditional pdf of the depth conditioned on the measurements being equal to 7 km and 7.1 km.
   c Compute the joint pdf of the two sonar measurements $\tilde{s}_1$ and $\tilde{s}_2$. Are the two measurements independent? Justify your answer mathematically and explain it intuitively.

5.10 (Rufus) Naroa's dog Rufus lives in her garden, which is the shaded area in Figure 5.23. Naroa models his position within the garden as uniformly distributed (i.e. the probability density of his position is the same at every point of the garden).

   a If $\tilde{x}$ is the position of Rufus on the x axis in Figure 5.23 and $\tilde{y}$ his position on the y axis, what is the value of the joint pdf of $\tilde{x}$ and $\tilde{y}$ on all of the x-y plane?
   b Derive the pdf of $\tilde{y}$ and sketch it.
   c Derive the conditional pdf of $\tilde{x}$ given $\tilde{y}$ and sketch it.
   d Are $\tilde{x}$ and $\tilde{y}$ independent? Justify your answer.
   e Given two independent samples from a uniform random variable equal to 0.1 and 0.3, generate a joint sample from $\tilde{x}$ and $\tilde{y}$.

5.11 (Spider on a wall) A spider lives on a wall of Aria's living room. There is a painting on the wall, behind which the spider likes to hide. Figure 5.24 shows a diagram of the wall; it is 10 feet high and 10 feet wide.

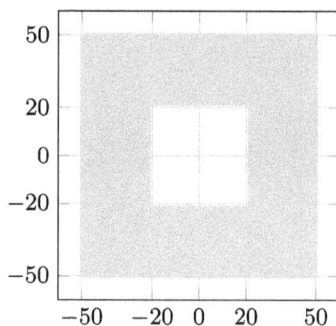

Figure 5.23 Naroa's garden (in gray) from Exercise 5.10.

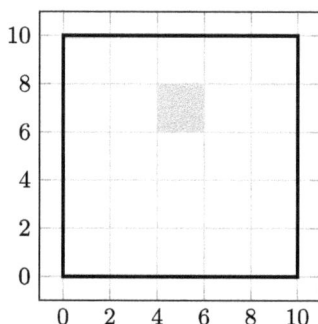

Figure 5.24 The wall (white with black edges) and painting (gray) in Exercise 5.11.

After observing the spider for a while, Aria determines that (1) it spends twice the time behind the painting than on the rest of the wall, (2) it never crawls on the painting or leaves the wall, (3) if it is not behind the painting then it is equally likely to be anywhere on the wall. Since she cannot see it behind the painting, she assumes that when it is there, it is also equally likely to be at any spot.

a Model the position of the spider as a bivariate random variable and derive its pdf.
b Compute the pdf of the height at which the spider is located and sketch it.
c Compute the conditional cdf of the height at which the spider is located, given that you can see it (i.e. it's not under the painting) and sketch it.

5.12 (Simulating a constant density) Explain how to simulate a random two-dimensional vector $\tilde{x}$ with a joint pdf $f_{\tilde{x}}(x)$ that is uniformly distributed in the shaded region in Figure 5.25. Assume that you have access to independent samples from a uniform distribution in $[0, 1]$. Explain your method, justifying why it works. Then implement it and plot a scatterplot of 1,000 samples to verify that they are indeed uniformly distributed within the shaded region.

5.13 (Multivariate Gaussian pdf) In this exercise, we show that the multivariate Gaussian pdf is a valid joint pdf. It is obviously nonnegative, so we just need to prove that the integral of the joint pdf over the corresponding domain equals one.

a Prove that the integral of a d-dimensional multivariate Gaussian pdf over $\mathbb{R}^d$ equals one, if the covariance matrix is diagonal.
b Based on the geometric interpretation of the multivariate Gaussian pdf with non-diagonal covariance matrix Σ in Section 5.10.1, argue why the integral of a d-dimensional multivariate Gaussian pdf over $\mathbb{R}^d$ should equal one, and then prove it mathematically.

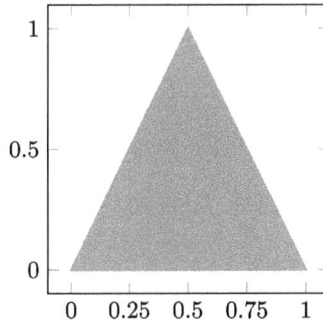

Figure 5.25 Region that we aim to sample from uniformly in Exercise 5.12.

5.14 (Exotic fruit) A biologist, called Ani, is studying a newly discovered plant that produces edible fruit. She has measured 10 specimens to obtain the following data:

Fruit weight (kg)	1.5	2.3	0.8	1.2	2.0	1.2	0.7	2.7	2.3	0.6
Stem height (cm)	20	15	18	19	17	22	21	14	17	22
Stem radius (mm)	8	12	6	10	17	12	9	14	10	8
Number of leaves	110	94	152	123	78	60	111	83	85	90
Median leaf length (cm)	12	21	9	14	19	15	7	29	22	15

Ani wants to use these data to model the conditional pdf of the fruit weight given the other four features.

a Her first thought is to estimate the conditional pdf by selecting a relevant subset of the data, and then applying kernel density estimation to the weight values within the subset. What is the problem with this approach?

b Model all the features as jointly Gaussian and use the model to compute the conditional pdf of the fruit weight for a plant that has a stem height of 15 cm, a stem radius of 20 mm, 120 leaves, and median leaf length of 8 cm.

5.15 (Gaussian Bayesian model) Eliseo is trying to use a noisy sensor reading to determine the temperature on top of a mountain. He expects the temperature to be close to 0°C. He decides to encode the uncertainty about the sensor reading using a Bayesian model (see Section 6.7). He models the reading as a random variable $\tilde{y}$ that has a Gaussian distribution when conditioned on a parameter $\tilde{\mu}$, which is also modeled as a random variable. The mean parameter of the Gaussian distribution is equal to $\tilde{\mu}$, and the variance parameter is equal to one. The random variable $\tilde{\mu}$ is Gaussian with mean zero and variance σ^2.

a Show that $\begin{bmatrix} \tilde{\mu} \\ \tilde{y} \end{bmatrix}$ is a Gaussian random vector with zero mean and derive its covariance-matrix parameter. The following formula for the inverse of a 2x2 matrix might be useful:

$$\begin{bmatrix} a & b \\ c & d \end{bmatrix}^{-1} = \frac{1}{ad - bc} \begin{bmatrix} d & -b \\ -c & a \end{bmatrix}. \tag{5.169}$$

b Derive the posterior distribution of $\tilde{\mu}$ given $\tilde{y} = y$.

c Sketch the prior and posterior distribution of $\tilde{\mu}$ when $y = -1$ and $\sigma^2 = 0.01$. If he chooses this prior, is Eliseo confident that the temperature is close to 0°? Is the posterior very different from the prior?

d Sketch the prior and posterior distribution of $\tilde{\mu}$ when $y = -1$ and $\sigma^2 = 10$. If he chooses this prior, is Eliseo confident that the temperature is close to 0°? Is the posterior very different from the prior?

6

Discrete and Continuous Variables

Overview

Probabilistic models usually include both discrete and continuous quantities. In Section 6.1, we explain how to jointly model the behavior of such quantities by representing them as random variables belonging to the same probability space. Section 6.2 defines the conditional distribution of a continuous random variable given a discrete variable and introduces mixture models. Section 6.3 defines the conditional distribution of a discrete random variable given a continuous random variable. Section 6.4 discusses independence and conditional independence between discrete and continuous variables. Sections 6.5 and 6.6 describe Gaussian discriminant analysis and Gaussian mixture models, two popular probabilistic models for classification and clustering, respectively. Finally, Section 6.7 presents the Bayesian framework for parametric modeling.

6.1 Joint Distribution of Discrete and Continuous Variables

In Section 4.1.1 we model the joint behavior of multiple uncertain discrete quantities by representing them as random variables in a common probability space. In Section 5.1, we model multiple uncertain continuous quantities in the same way. In this section, we show that this approach can also be used to jointly model discrete and continuous quantities.

Let $\tilde{d}\colon \Omega \to R_{\tilde{d}}$ be a discrete random variable with range $R_{\tilde{d}}$, and $\tilde{c}\colon \Omega \to \mathbb{R}$ a continuous random variable. The variables are defined in the same probability space $\{\Omega, \mathcal{C}, \mathrm{P}\}$, so each outcome $\omega \in \Omega$ is mapped to a value $\tilde{d}(\omega) \in R_{\tilde{d}}$ by $\tilde{d}$, and to a real number $\tilde{c}(\omega)$ by $\tilde{c}$. In order to characterize the joint behavior of the random variables, we consider the event that $\tilde{d}$ is equal to a certain value $d \in R_{\tilde{d}}$ and $\tilde{c}$ is in a Borel set C (i.e. an interval or a countable union of intervals in $\mathbb{R}$) *at the same time*. The event containing the outcomes mapped to d by $\tilde{d}$ is

$$D := \left\{\omega\colon \tilde{d}(\omega) = d\right\}. \tag{6.1}$$

The event containing the outcomes mapped to C by $\tilde{c}$ is

$$C^{-1} := \left\{\omega\colon \tilde{c}(\omega) \in C\right\}. \tag{6.2}$$

The event that $\tilde{d} = d$ and $\tilde{c} \in C$ at the same time is the intersection of D and C^{-1}, as illustrated in Figure 6.1. D and C^{-1} belong to the collection $\mathcal{C}$ of the probability space, as long as $\tilde{d}$ is a valid discrete random variable (see Definition 2.3), and $\tilde{c}$ is a valid continuous random variable (see Definition 3.1). The probability $\mathrm{P}(D \cap C^{-1})$, which we denote by

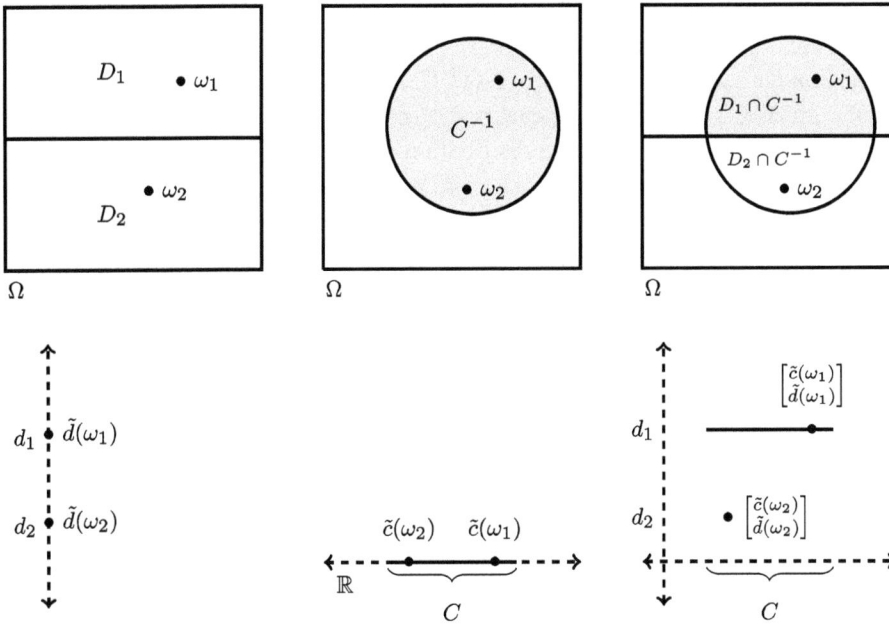

Figure 6.1 Joint distribution of discrete and continuous random variables. The discrete random variable $\tilde{d}$ and the continuous random variable $\tilde{c}$ map outcomes in the sample space Ω to a discrete set $\mathbb{R}_{\tilde{d}} := \{d_1, d_2\}$ and the real line $\mathbb{R}$, respectively. The top left Venn diagram shows the partition $\{D_1, D_2\}$ of Ω associated with $\tilde{d}$. The outcomes $\omega_1 \in D_1$ and $\omega_2 \in D_2$ are mapped to $\tilde{d}(\omega_1) = d_1$ and $\tilde{d}(\omega_2) = d_2$. The top middle Venn diagram shows the event of outcomes C^{-1} mapped by $\tilde{c}$ to the interval C, depicted on the horizontal real line. Both ω_1 and ω_2 are mapped to C by $\tilde{c}$, so they belong to C^{-1}. The top right Venn diagram shows that the outcomes mapped to d_1 by $\tilde{d}$ and simultaneously to C by $\tilde{c}$ are in the intersection $D_1 \cap C^{-1}$, which includes ω_1, but not ω_2. In $\mathbb{R}^2$, these outcomes are mapped to the Cartesian product $C \times d_1$, depicted as a horizontal segment with vertical coordinate d_1. The probability that the vector $\begin{bmatrix} \tilde{c} \\ \tilde{d} \end{bmatrix}$ belongs to $C \times d_1$ is $\mathrm{P}(D_1 \cap C^{-1})$, which equals the area of $D_1 \cap C^{-1}$ in the Venn diagram.

$\mathrm{P}(\tilde{d} = d, \tilde{c} \in C)$, is therefore well defined, by the properties of probability spaces (see Definitions 1.7 and 1.15).

We usually describe the behavior of multiple discrete random variables using their joint pmf (see Section 4.1.2), and the behavior of multiple continuous random variables using their joint pdf (see Section 5.3). The challenge when we consider models with discrete and continuous variables is that neither of these objects is well defined.[1] The joint cdf is well defined, but unfortunately it is also very annoying to manipulate (see Example 5.6). Instead, we typically use marginal and conditional distributions to characterize the joint distribution of discrete and continuous random variables, as explained in the following sections.

---

[1]   It is possible to define a probability density for discrete random variables using Dirac deltas, but this requires leveraging the machinery of generalized functions.

6.2 Conditional Distribution of Continuous Variables Given Discrete Variables

Consider a continuous random variable $\tilde{c}\colon \Omega \to \mathbb{R}$ and a discrete random variable $\tilde{d}\colon \Omega \to R_{\tilde{d}}$ defined on the same probability space $\{\Omega, \mathcal{C}, \mathrm{P}\}$. To characterize the conditional distribution of $\tilde{c}$ given $\tilde{d}$, we consider the events $\tilde{c} \le c$ and $\tilde{d} = d$, for any $c \in \mathbb{R}$ and $d \in R_{\tilde{d}}$. We define the conditional cumulative distribution function (cdf) of $\tilde{c}$ given $\tilde{d} = d$ to equal the conditional probability of $\tilde{c} \le c$ given $\tilde{d} = d$, analogously to Definition 3.3. If the conditional cdf is differentiable, then we interpret its derivative as a conditional probability density function (pdf), which we denote by $f_{\tilde{c} \mid \tilde{d}}(c \mid d)$. The conditional pdf is the ratio between the conditional probability that $\tilde{c}$ belongs to a small interval of length ϵ given $\tilde{d} = d$ and ϵ, when $\epsilon \to 0$ (see Section 3.3),

$$f_{\tilde{c} \mid \tilde{d}}(c \mid d) := \frac{\mathrm{d}F_{\tilde{c} \mid \tilde{d}}(c \mid d)}{\mathrm{d}c} \tag{6.3}$$

$$= \lim_{\epsilon \to 0} \frac{F_{\tilde{c} \mid \tilde{d}}(c \mid d) - F_{\tilde{c} \mid \tilde{d}}(c - \epsilon \mid d)}{\epsilon} \tag{6.4}$$

$$= \lim_{\epsilon \to 0} \frac{\mathrm{P}(c - \epsilon < \tilde{c} \le c \mid \tilde{d} = d)}{\epsilon}. \tag{6.5}$$

Definition 6.1 (Conditional cdf and pdf of a continuous random variable given a discrete random variable). *Let $\tilde{c}$ and $\tilde{d}$ be a continuous and a discrete random variable defined on the same probability space. For d such that $p_{\tilde{d}}(d) \neq 0$, the conditional cdf of $\tilde{c}$ given $\tilde{d}$ is*

$$F_{\tilde{c} \mid \tilde{d}}(c \mid d) := \mathrm{P}\left(\tilde{c} \le c \mid \tilde{d} = d\right). \tag{6.6}$$

If $F_{\tilde{c} \mid \tilde{d}}$ is differentiable, the conditional pdf of $\tilde{c}$ given $\tilde{d}$ is

$$f_{\tilde{c} \mid \tilde{d}}(c \mid d) := \frac{\mathrm{d}F_{\tilde{c} \mid \tilde{d}}(c \mid d)}{\mathrm{d}c}. \tag{6.7}$$

The following theorem provides a formula to compute the marginal pdf of a continuous random variable from its conditional pdf given a discrete random variable.

Theorem 6.2. *Let $F_{\tilde{c} \mid \tilde{d}}$ and $f_{\tilde{c} \mid \tilde{d}}$ be the conditional cdf and pdf of a continuous random variable $\tilde{c}$ given a discrete random variable $\tilde{d}$, with range $R_{\tilde{d}}$. Then, for any $c \in \mathbb{R}$,*

$$F_{\tilde{c}}(c) = \sum_{d \in R_{\tilde{d}}} p_{\tilde{d}}(d) \, F_{\tilde{c} \mid \tilde{d}}(c \mid d), \tag{6.8}$$

$$f_{\tilde{c}}(c) = \sum_{d \in R_{\tilde{d}}} p_{\tilde{d}}(d) \, f_{\tilde{c} \mid \tilde{d}}(c \mid d). \tag{6.9}$$

Proof The events $\tilde{d} = d$, for $d \in R_{\tilde{d}}$, are a partition of the whole probability space (one of them must happen and they are all disjoint), so by Definition 3.3 and the law of total probability (Theorem 1.18),

$$F_{\tilde{c}}(c) = \mathrm{P}\left(\tilde{c} \le c\right) \tag{6.10}$$

$$= \sum_{d \in R_{\tilde{d}}} \mathrm{P}\left(\tilde{d} = d\right) \mathrm{P}\left(\tilde{c} \le c \mid \tilde{d} = d\right) \tag{6.11}$$

$$= \sum_{d \in R_{\tilde{d}}} p_{\tilde{d}}(d) \, F_{\tilde{c} \mid \tilde{d}}(c \mid d). \tag{6.12}$$

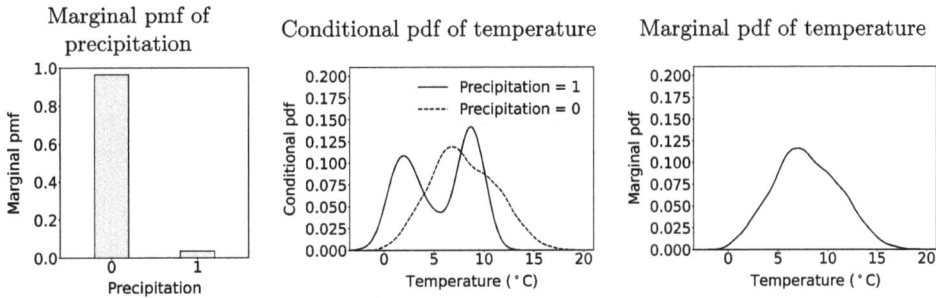

Figure 6.2 Joint distribution of precipitation and temperature in Mauna Loa. The left graph shows the marginal pmf of precipitation on the Mauna Loa volcano (Hawaii); there is precipitation only about 3.6% of the time. The middle graph shows the conditional pdf of the temperature on the volcano depending on whether it rains or snows (precipitation = 1) or not (precipitation = 0), obtained via kernel density estimation. The right graph shows the marginal pdf of temperature, which is equal to the sum of the conditional pdfs weighted by the entries of the pmf, as established in Theorem 6.2.

Following Definition 3.13, we differentiate the cdf to obtain the pdf, which yields (6.9) by Definition 6.1. ∎

Figure 6.2 illustrates Theorem 6.2, using weather data from the Mauna Loa volcano (Hawaii), extracted from Dataset 9. The figure shows the marginal pmf of precipitation (left), the conditional pdf of temperature given precipitation or no precipitation (center), and the marginal pdf of temperature (right). The marginal pdf is equal to the sum of the conditional pdfs weighted by the entries of the pmf. The weighted sum is close to the conditional pdf given no precipitation because the probability of precipitation is low (less than 4%).

In *mixture models*, a continuous quantity is modeled using a parametric distribution, whose parameters depend on the value of a discrete quantity. If a Gaussian is used as the continuous distribution, the distribution is referred to as a Gaussian mixture. In Sections 6.5 and 6.6, we describe applications of Gaussian mixture models to the tasks of classification and clustering, respectively.

Example 6.3 (Gaussian mixture model for height). In Figure 3.23, we show the result of fitting a Gaussian parametric model to height data from 4,082 men in the United States army, extracted from Dataset 5. If we instead fit the Gaussian model to the whole population, which also includes 1,986 women, then the fit is not as good, as shown in the lower right plot of Figure 6.3. However, the height of the women is well approximated by a Gaussian distribution, as long as it is modeled separately from the men (top center graph in Figure 6.3). Motivated by this, we model height as a continuous random variable $\tilde{h}$ that is Gaussian when conditioned on a discrete Bernoulli random variable $\tilde{s}$, representing sex.

We fit the conditional Gaussian pdfs of $\tilde{h}$ given $\tilde{s}$ via maximum-likelihood estimation, applying Theorem 3.35. For women, the mean is $\mu_{\text{women}} := 163$ cm and the standard deviation is $\sigma_{\text{women}} := 6.4$ cm. For men, the mean is $\mu_{\text{men}} := 176$ cm and the standard deviation is $\sigma_{\text{men}} := 6.9$ cm. By Definition 2.11, the entries of the empirical pmf of $\tilde{s}$ are $\alpha_{\text{women}} := 0.33$ and $\alpha_{\text{men}} := 0.67$ because one third of the individuals are women. The top row of Figure 6.3 shows the marginal pmf of $\tilde{s}$ and the conditional pdfs of $\tilde{h}$.

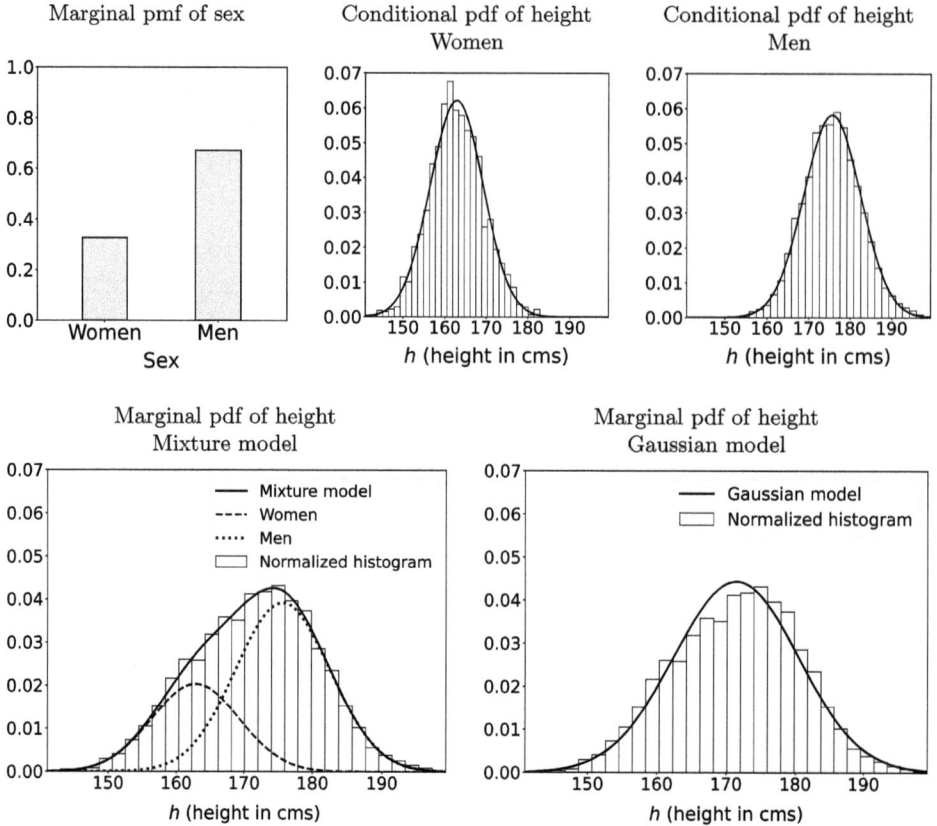

Figure 6.3 Gaussian mixture model for height. The top left graph shows the sex distribution in the population considered in Example 6.3. The height of each sex is modeled separately using a Gaussian distribution with different parameters (middle and right graphs in the top row). The bottom left graph shows the resulting marginal pdf of height in the whole population, obtained by applying Theorem 6.2. The resulting Gaussian mixture provides a better model than the Gaussian distribution on the bottom right, obtained by fitting the data from all individuals regardless of sex.

By (6.9) the marginal pdf of $\tilde{h}$ is

$$f_{\tilde{h}}(h)$$

$$= p_{\tilde{s}}(\text{woman}) f_{\tilde{h} \mid \tilde{s}}(h \mid \text{woman}) + p_{\tilde{s}}(\text{man}) f_{\tilde{h} \mid \tilde{s}}(h \mid \text{woman}) \tag{6.13}$$

$$= \frac{\alpha_{\text{women}}}{\sqrt{2\pi}\sigma_{\text{women}}} \exp\left(-\frac{1}{2}\left(\frac{h - \mu_{\text{women}}}{\sigma_{\text{women}}}\right)^2\right) + \frac{\alpha_{\text{men}}}{\sqrt{2\pi}\sigma_{\text{men}}} \exp\left(-\frac{1}{2}\left(\frac{h - \mu_{\text{men}}}{\sigma_{\text{men}}}\right)^2\right)$$

$$= \frac{0.33}{6.4\sqrt{2\pi}} \exp\left(-\frac{1}{2}\left(\frac{h - 163}{6.4}\right)^2\right) + \frac{0.67}{6.9\sqrt{2\pi}} \exp\left(-\frac{1}{2}\left(\frac{h - 176}{6.9}\right)^2\right). \tag{6.14}$$

The Gaussian pdfs are *mixed* via weighted averaging to produce the marginal pdf of the height, which is *not* Gaussian. The bottom row of Figure 6.3 shows that the resulting marginal pdf (left) achieves a better approximation to the data than fitting a single Gaussian model to the whole dataset (right).

6.3 Conditional Distribution of Discrete Variables Given Continuous Variables

In Section 6.2, we explain how to characterize the conditional distribution of a continuous variable given a discrete variable. In this section, we consider the conditional distribution of a discrete variable given a continuous variable. Recall that the probability that a continuous random variable $\tilde{c}$ is equal to any single specific value c equals zero (see Section 3.1). Consequently, it is not immediately obvious how to define the conditional pmf of a discrete random variable $\tilde{d}$ given $\tilde{c} = c$. This is the same issue we encounter in Section 5.6, when defining the conditional pdf of a continuous random variable given another continuous random variable. To address it, we mimic the approach in Section 5.6. We define the conditional pmf of $\tilde{d}$ given that $\tilde{c}$ is in a small interval that includes c, and then take the limit when the length of the interval shrinks to zero.

Definition 6.4 (Conditional pmf of a discrete random variable given a continuous random variable). *Let $\tilde{c}$ and $\tilde{d}$ be a continuous and a discrete random variable defined on the same probability space. For any d in the range of $\tilde{d}$, the conditional pmf of $\tilde{d}$ given $\tilde{c}$ is*

$$p_{\tilde{d}\,|\,\tilde{c}}\,(d\,|\,c) := \lim_{\epsilon \to 0} \mathrm{P}\left(\tilde{d} = d\,|\,c - \epsilon < \tilde{c} \le c\right), \tag{6.15}$$

assuming that $f_{\tilde{c}}(c) > 0$.

Analogously to Theorem 6.2, if we have access to the conditional pmf $p_{\tilde{d}\,|\,\tilde{c}}$ of a discrete random variable $\tilde{d}$ given a continuous random variable $\tilde{c}$, we can obtain the marginal pmf of $\tilde{d}$ by integrating $p_{\tilde{d}\,|\,\tilde{c}}$ weighted by the marginal pdf of $\tilde{c}$.

Theorem 6.5. *Let $p_{\tilde{d}\,|\,\tilde{c}}$ be the conditional pmf of a discrete random variable $\tilde{d}$ given a continuous random variable $\tilde{c}$ with pdf $f_{\tilde{c}}$. Then, for any d in the range of $\tilde{d}$, the marginal pmf of $\tilde{d}$ equals*

$$p_{\tilde{d}}\,(d) = \int_{c=-\infty}^{\infty} f_{\tilde{c}}\,(c)\, p_{\tilde{d}\,|\,\tilde{c}}\,(d\,|\,c)\ \mathrm{d}c. \tag{6.16}$$

Proof   Consider a discretized grid $\{\ldots, c_{-1}, c_0, c_1, \ldots\}$ with step size ϵ covering the whole real line. We express the marginal pmf of the discrete variable, as

$$p_{\tilde{d}}\,(d) = \sum_{i=-\infty}^{\infty} \mathrm{P}\left(\tilde{d} = d, c_i - \epsilon < \tilde{c} \le c_i\right) \tag{6.17}$$

by the law of total probability (Theorem 1.18). In the limit when $\epsilon \to 0$, the sum becomes a Riemann integral

$$\lim_{\epsilon \to 0} \sum_{i=-\infty}^{\infty} \frac{\mathrm{P}\left(\tilde{d} = d, c_i - \epsilon < \tilde{c} \le c_i\right)}{\epsilon} \epsilon = \int_{c=-\infty}^{\infty} \lim_{\epsilon \to 0} \frac{\mathrm{P}\left(\tilde{d} = d, c - \epsilon < \tilde{c} \le c\right)}{\epsilon} \mathrm{d}c.$$

Consequently, by the chain rule (Theorem 1.17), Definitions 3.13 and 6.4,

$$p_{\tilde{d}}\,(d) = \int_{c=-\infty}^{\infty} \lim_{\epsilon \to 0} \frac{\mathrm{P}\left(\tilde{d} = d, c - \epsilon < \tilde{c} \le c\right)}{\epsilon} \mathrm{d}c \tag{6.18}$$

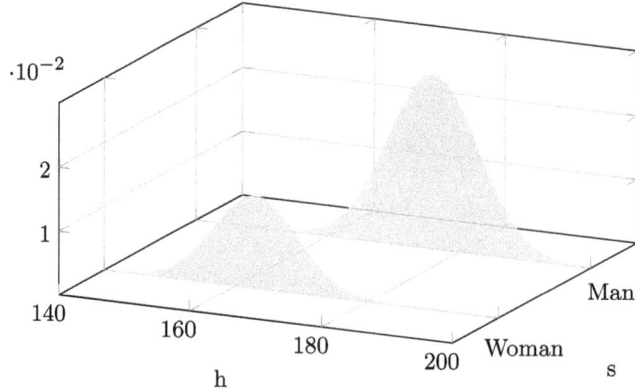

Figure 6.4 Chain rule for Gaussian mixture model. The graph shows the function $p_{\tilde{s}}(s) f_{\tilde{h}\,|\,\tilde{s}}(h\,|\,s) = f_{\tilde{h}}(h) p_{\tilde{s}\,|\,\tilde{h}}(s\,|\,h)$ for the variables $\tilde{s}$ and $\tilde{h}$ in Example 6.3.

$$= \int_{c=-\infty}^{\infty} \lim_{\epsilon \to 0} \frac{\mathrm{P}\left(c - \epsilon < \tilde{c} \leq c\right)}{\epsilon} \cdot \mathrm{P}\left(\tilde{d} = d \,|\, c - \epsilon < \tilde{c} \leq c\right) \mathrm{d}c \qquad (6.19)$$

$$= \int_{c=-\infty}^{\infty} f_{\tilde{c}}(c)\, p_{\tilde{d}\,|\,\tilde{c}}(d\,|\,c)\ \mathrm{d}c. \qquad (6.20)$$

$\blacksquare$

The following theorem establishes a chain rule for jointly distributed discrete and continuous random variables.

Theorem 6.6 (Chain rule for discrete and continuous random variables). *Let $\tilde{c}$ and $\tilde{d}$ be a continuous and a discrete random variable defined on the same probability space. If $\tilde{c}$ has a pdf $f_{\tilde{c}}$ and a conditional pdf $f_{\tilde{c}\,|\,\tilde{d}}$,*

$$p_{\tilde{d}}(d)\, f_{\tilde{c}\,|\,\tilde{d}}(c\,|\,d) = f_{\tilde{c}}(c)\, p_{\tilde{d}\,|\,\tilde{c}}(d\,|\,c), \qquad (6.21)$$

where $p_{\tilde{d}}$ and $p_{\tilde{d}\,|\,\tilde{c}}$ are the marginal pmf of $\tilde{d}$ and the conditional pmf of $\tilde{d}$ given $\tilde{c}$, respectively.

Proof   We express the conditional pdf of $\tilde{c}$ given $\tilde{d} = d$ as the limit (6.5), multiply it by the marginal pmf of $\tilde{d}$, and apply the chain rule (Theorem 1.17) twice within the limit, to obtain

$$p_{\tilde{d}}(d)\, f_{\tilde{c}\,|\,\tilde{d}}(c\,|\,d) = \lim_{\epsilon \to 0} \mathrm{P}(\tilde{d} = d) \frac{\mathrm{P}\left(c - \epsilon < \tilde{c} \leq c \,|\, \tilde{d} = d\right)}{\epsilon} \qquad (6.22)$$

$$= \lim_{\epsilon \to 0} \frac{\mathrm{P}\left(\tilde{d} = d, c - \epsilon < \tilde{c} \leq c\right)}{\epsilon} \qquad (6.23)$$

$$= \lim_{\epsilon \to 0} \frac{\mathrm{P}\left(c - \epsilon < \tilde{c} \leq c\right)}{\epsilon} \mathrm{P}\left(\tilde{d} = d \,|\, c - \epsilon < \tilde{c} \leq c\right) \qquad (6.24)$$

$$= f_{\tilde{c}}(c)\, p_{\tilde{d}\,|\,\tilde{c}}(d\,|\,c), \qquad (6.25)$$

where the last step follows from Definitions 3.13 and 6.4. $\blacksquare$

By Theorem 4.14, for two discrete random variables $\tilde{d}_1$ and $\tilde{d}_2$ with joint pmf $p_{\tilde{d}_1,\tilde{d}_2}$, and any d_1 and d_2 in their respective ranges,

$$p_{\tilde{d}_1}(d_1)\, p_{\tilde{d}_2 \mid \tilde{d}_1}(d_2 \mid d_1) = p_{\tilde{d}_1, \tilde{d}_2}(d_1, d_2)\,. \tag{6.26}$$

By Theorem 5.12, for two continuous random variables $\tilde{c}_1$ and $\tilde{c}_2$ with joint pdf $f_{\tilde{c}_1, \tilde{c}_2}$, and any real values c_1 and c_2,

$$f_{\tilde{c}_1}(c_1)\, f_{\tilde{c}_2 \mid \tilde{c}_1}(c_2 \mid c_1) = f_{\tilde{c}_1, \tilde{c}_2}(c_1, c_2)\,. \tag{6.27}$$

Consequently, the function that appears on both sides of the chain-rule equality in Theorem 6.6,

$$p_{\tilde{d}}(d)\, f_{\tilde{c} \mid \tilde{d}}(c \mid d) = f_{\tilde{c}}(c)\, p_{\tilde{d} \mid \tilde{c}}(d \mid c)\,, \tag{6.28}$$

is analogous to the joint pmf of two discrete random vectors, and the joint pdf of two continuous random variables. Figure 6.4 plots this function for the Gaussian mixture model in Example 6.3.

Theorem 6.6 yields a version of Bayes' rule for discrete and continuous random variables. Assume that we know the marginal distribution of the discrete variable $\tilde{d}$, and the conditional distribution of the continuous variable $\tilde{c}$ given $\tilde{d}$. Then, by Theorems 6.6 and 6.2, the conditional pmf of $\tilde{d}$ given $\tilde{c}$ equals

$$p_{\tilde{d} \mid \tilde{c}}(d \mid c) = \frac{p_{\tilde{d}}(d)\, f_{\tilde{c} \mid \tilde{d}}(c \mid d)}{f_{\tilde{c}}(c)} \tag{6.29}$$

$$= \frac{p_{\tilde{d}}(d)\, f_{\tilde{c} \mid \tilde{d}}(c \mid d)}{\sum_{d \in R_{\tilde{d}}} p_{\tilde{d}}(d)\, f_{\tilde{c} \mid \tilde{d}}(c \mid d)}\,, \tag{6.30}$$

where $R_{\tilde{d}}$ is the range of $\tilde{d}$. This is useful to perform classification, as we explain in Section 6.5. The left plot of Figure 6.5 shows the entries of the conditional pmf of precipitation given temperature at the Mauna Loa weather station, computed by applying (6.30) using the marginal pmf and the conditional pdfs shown in Figure 6.2.

In the case of the Gaussian mixture model from Example 6.3, (6.30) allows us to compute the conditional probability of an individual being male given that their height is h:

$$p_{\tilde{s} \mid \tilde{h}}(\mathrm{man} \mid h) \tag{6.31}$$

$$= \frac{p_{\tilde{s}}(\mathrm{man})\, f_{\tilde{h} \mid \tilde{s}}(h \mid \mathrm{man})}{p_{\tilde{s}}(\mathrm{woman})\, f_{\tilde{h} \mid \tilde{s}}(h \mid \mathrm{woman}) + p_{\tilde{s}}(\mathrm{man})\, f_{\tilde{h} \mid \tilde{s}}(h \mid \mathrm{man})} \tag{6.32}$$

$$= \frac{\frac{\alpha_{\mathrm{men}}}{\sqrt{2\pi}\sigma_{\mathrm{men}}} \exp\left(-\frac{1}{2}\left(\frac{h - \mu_{\mathrm{men}}}{\sigma_{\mathrm{men}}}\right)^2\right)}{\frac{\alpha_{\mathrm{women}}}{\sqrt{2\pi}\sigma_{\mathrm{women}}} \exp\left(-\frac{1}{2}\left(\frac{h - \mu_{\mathrm{women}}}{\sigma_{\mathrm{women}}}\right)^2\right) + \frac{\alpha_{\mathrm{men}}}{\sqrt{2\pi}\sigma_{\mathrm{men}}} \exp\left(-\frac{1}{2}\left(\frac{h - \mu_{\mathrm{men}}}{\sigma_{\mathrm{men}}}\right)^2\right)} \tag{6.33}$$

$$= \frac{1}{1 + \frac{\alpha_{\mathrm{women}}}{\alpha_{\mathrm{men}}} \frac{\sigma_{\mathrm{men}}}{\sigma_{\mathrm{women}}} \exp\left(\frac{1}{2}\left(\frac{h - \mu_{\mathrm{men}}}{\sigma_{\mathrm{men}}}\right)^2 - \frac{1}{2}\left(\frac{h - \mu_{\mathrm{women}}}{\sigma_{\mathrm{women}}}\right)^2\right)} \tag{6.34}$$

$$= \frac{1}{1 + 0.53 \exp\left(0.98 + 0.28h - 0.0017h^2\right)}\,. \tag{6.35}$$

The right plot of Figure 6.5 shows the entries of the conditional pmf as a function of height. At small heights, the probability of woman is very close to one and the probability of man is close to zero. As we increase the height, the conditional probability of man increases toward

Precipitation conditioned on
temperature

Height conditioned on sex

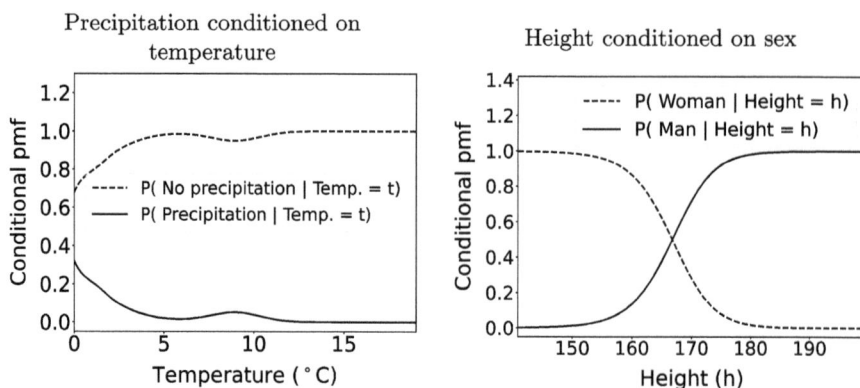

Figure 6.5 Conditional distribution of discrete variables given continuous variables. The left graph shows the conditional pmf of precipitation at the Mauna Loa volcano (Hawaii) given the temperature, computed via (6.30) using the pmf and conditional pdfs from Figure 6.2. The probability of precipitation is higher at low temperatures. The right graph shows the conditional pmf of sex given height, derived in (6.35). The probability of woman is close to one for small heights, and then undergoes a sharp transition, becoming almost zero for large heights.

one and the conditional probability of woman decreases toward zero, with a quick transition between 160 and 170 cm. The curve looks like a logistic function, which is commonly used to build classification models, as explained in Section 12.5. The following theorem establishes that when the variances of the conditional Gaussian distributions are the same, the entries of the conditional pmf are in fact logistic functions of the temperature.

Theorem 6.7 (Logistic function). *Let $\tilde{d}$ be a Bernoulli random variable with parameter α and let $\tilde{c}$ be a continuous random variable defined on the same probability space. If the conditional distributions of $\tilde{c}$ given $\tilde{d} = 0$ and $\tilde{d} = 1$ are both Gaussian with mean parameters μ_0 and μ_1, respectively, and the same standard-deviation parameter σ, then the conditional pmf of $\tilde{d}$ given $\tilde{c} = c$ equals*

$$p_{\tilde{d}\,|\,\tilde{c}}(1\,|\,c) = \frac{1}{1 + \eta \exp(-\beta c)}, \tag{6.36}$$

which is a logistic function of c where

$$\eta := \frac{1-\alpha}{\alpha} \exp\left(\frac{1}{2\sigma^2}\left(\mu_1^2 - \mu_0^2\right)\right), \tag{6.37}$$

$$\beta := \frac{\mu_1 - \mu_0}{\sigma^2}. \tag{6.38}$$

Proof　By the assumptions, applying (6.30) yields

$$p_{\tilde{d}\,|\,\tilde{c}}(1\,|\,c) = \frac{p_{\tilde{d}}(1)\,f_{\tilde{c}\,|\,\tilde{d}}(c\,|\,1)}{p_{\tilde{d}}(0)\,f_{\tilde{c}\,|\,\tilde{d}}(c\,|\,0) + p_{\tilde{d}}(1)\,f_{\tilde{c}\,|\,\tilde{d}}(c\,|\,1)} \tag{6.39}$$

$$= \frac{\frac{\alpha}{\sigma\sqrt{2\pi}} \exp\left(-\frac{1}{2}\left(\frac{c-\mu_1}{\sigma}\right)^2\right)}{\frac{\alpha}{\sigma\sqrt{2\pi}} \exp\left(-\frac{1}{2}\left(\frac{c-\mu_1}{\sigma}\right)^2\right) + \frac{1-\alpha}{\sigma\sqrt{2\pi}} \exp\left(-\frac{1}{2}\left(\frac{c-\mu_0}{\sigma}\right)^2\right)} \tag{6.40}$$

$$= \frac{1}{1 + \frac{1-\alpha}{\alpha} \exp\left(\frac{1}{2}\left(\frac{c-\mu_1}{\sigma}\right)^2 - \frac{1}{2}\left(\frac{c-\mu_0}{\sigma}\right)^2\right)} \tag{6.41}$$

$$= \frac{1}{1 + \frac{1-\alpha}{\alpha} \exp\left(\frac{\mu_0-\mu_1}{\sigma^2}c + \frac{1}{2\sigma^2}\left(\mu_1^2 - \mu_0^2\right)\right)}. \tag{6.42}$$

■

6.4 Independence

Two random variables are independent, if the conditional distribution of one given the other is the same as the marginal distribution, as explained in Section 4.4 for discrete variables and Section 5.7 for continuous variables. When one variable is discrete and the other continuous, we can define independence between them in the same way.

Definition 6.8 (Independence). *A discrete random variable $\tilde{d}$ and a continuous random variable $\tilde{c}$ with pdf $f_{\tilde{c}}$, defined on the same probability space, are independent, if and only if*

$$p_{\tilde{d}\,|\,\tilde{c}}(d\,|\,c) = p_{\tilde{d}}(d), \tag{6.43}$$

$$f_{\tilde{c}\,|\,\tilde{d}}(c\,|\,d) = f_{\tilde{c}}(c) \tag{6.44}$$

for all possible values of c and d.

Notice that (6.43) implies (6.44) and vice versa. If (6.43) holds, then by the chain rule (Theorem 6.6),

$$f_{\tilde{c}\,|\,\tilde{d}}(c\,|\,d) = \frac{f_{\tilde{c}}(c)\,p_{\tilde{d}\,|\,\tilde{c}}(d\,|\,c)}{p_{\tilde{d}}(d)} \tag{6.45}$$

$$= \frac{f_{\tilde{c}}(c)\,p_{\tilde{d}}(d)}{p_{\tilde{d}}(d)} \tag{6.46}$$

$$= f_{\tilde{c}}(c). \tag{6.47}$$

Figure 6.6 shows the conditional and marginal pdfs and pmfs of height and handedness, estimated using data extracted from Dataset 5. Handedness indicates whether the person is left- or right-handed. The conditional and marginal distributions are very similar, which suggests that height and handedness are independent.

Two random variables are conditionally independent given a third one, if the conditional distribution of the first variable given the third does not change when we also condition on the second variable, as explained in Section 4.5 for discrete variables and Section 5.8 for continuous variables. The following definition extends this notion to situations where the first variable is discrete and the second continuous. The third value can be either discrete or continuous.

Definition 6.9 (Conditionally independent discrete and continuous random variables). *A discrete random variable $\tilde{d}$ and a continuous random variable $\tilde{c}$, defined on the same probability space, are conditionally independent given a random variable $\tilde{a}$, if and only if*

$$p_{\tilde{d}\,|\,\tilde{c},\tilde{a}}(d\,|\,c,a) = p_{\tilde{d}\,|\,\tilde{a}}(d\,|\,a), \tag{6.48}$$

$$f_{\tilde{c}\,|\,\tilde{d},\tilde{a}}(c\,|\,d,a) = f_{\tilde{c}\,|\,\tilde{a}}(c\,|\,a) \tag{6.49}$$

for all possible values of a, c and d.

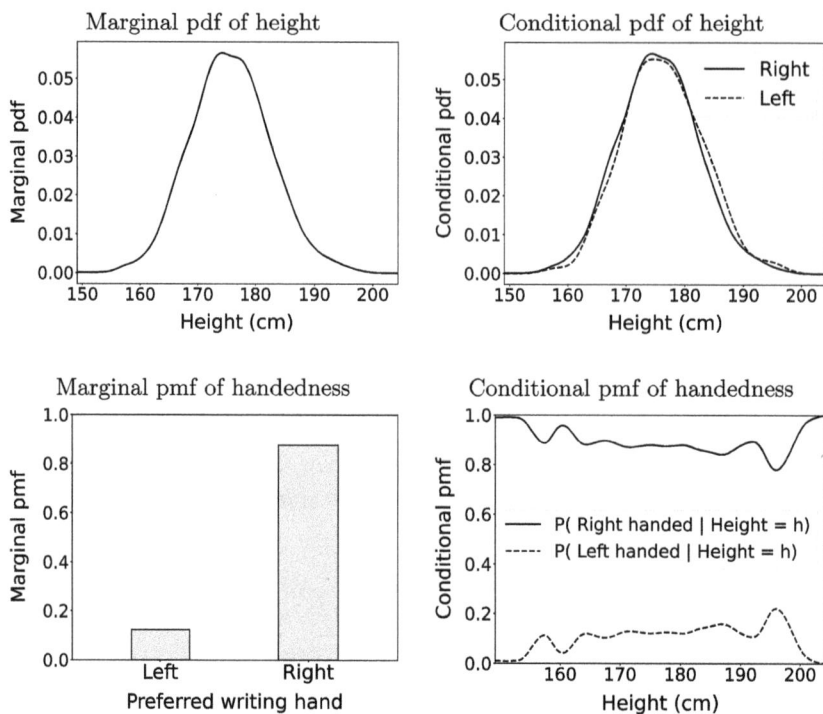

Figure 6.6 Joint distribution of height and handedness. The top left graph shows the marginal pdf of the height in Dataset 5. The top right graph shows the conditional pdf of height given handedness. The conditional pdfs and the marginal pdf are very similar, suggesting that height and handedness are independent. All pdfs are estimated via kernel density estimation with a Gaussian kernel, following Definition 3.25. The bottom left plot shows the empirical pmf of handedness (see Definition 2.11); about 11.5% of the individuals are left-handed. The bottom right plot shows the conditional pmf of handedness given height, obtained by applying Theorem 6.6. The conditional pmf is approximately constant and equal to the marginal, except at small and large heights, where the number of data is small.

6.5 Classification via Gaussian Discriminant Analysis

In Section 4.8, we define the task of classification, where the goal is to separate data into predefined classes. Here, we present an approach to perform classification from continuous features, based on Gaussian mixture models. As a motivating application, we consider the automatic diagnosis of Alzheimer's disease, a neurodegenerative disease that causes dementia.

We assume that we have available n examples (x_1, y_1), (x_2, y_2), ..., (x_n, y_n). Each example consists of a d-dimensional vector x_i of features and its corresponding label, denoted by $y_i \in \{1, 2, \ldots, c\}$, where c is the number of classes. In our application of interest, there are two classes: healthy individuals and Alzheimer's patients. We consider two features consisting of the volumes of the hippocampus and entorhinal cortex, which are regions of the brain that shrink due to Alzheimer's. The volumes are measured using magnetic resonance imaging and normalized by the intracranial volume of the head (to take into account that some people have larger heads than others). Our goal is to determine whether a person has Alzheimer's from these features.

In order to tackle the classification task, we interpret each example as a sample from the joint distribution of a d-dimensional random vector $\tilde{x}$ representing the features, and a random variable $\tilde{y}$ representing the corresponding class label. We apply *maximum a posteriori* (MAP) estimation, as in Section 4.8, to classify each new example based on the associated feature vector x. The MAP estimator is obtained by maximizing the conditional pmf of $\tilde{y}$ given $\tilde{x} = x$, analogously to the discrete case (see Definition 4.29).

Definition 6.10 (MAP estimator from continuous features). *Given a discrete random variable $\tilde{y}$ with range Y and a continuous random vector $\tilde{x}$, the maximum a posteriori (MAP) estimator of $\tilde{y}$ given $\tilde{x} = x$ is*

$$\mathrm{MAP}(x) := \arg\max_{y \in Y} p_{\tilde{y} \mid \tilde{x}}(y \mid x). \tag{6.50}$$

Just as in classification from discrete features (see Theorem 4.30), the MAP estimator is optimal in the sense that it minimizes the probability of error.

Theorem 6.11 (MAP estimation is optimal). *Given a discrete random variable $\tilde{y}$ and a d-dimensional continuous random vector $\tilde{x}$, the MAP estimator of $\tilde{y}$ given $\tilde{x} = x$, minimizes the probability of error. Equivalently, the probability that the MAP estimator is correct is greater than or equal to the probability that any estimator $h \colon \mathbb{R}^d \to \mathbb{R}$ is correct:*

$$\mathrm{P}\left(\mathrm{MAP}(\tilde{x}) = \tilde{y}\right) \geq \mathrm{P}\left(h(\tilde{x}) = \tilde{y}\right). \tag{6.51}$$

Proof By the same argument as in the proof of Theorem 6.5, we can express the probability that $h(\tilde{x}) = \tilde{y}$, for a fixed h, as follows:

$$\mathrm{P}\left(h(\tilde{x}) = \tilde{y}\right) = \int_{x[1]=-\infty}^{\infty} \cdots \int_{x[d]=-\infty}^{\infty} f_{\tilde{x}}(x) \mathrm{P}\left(\tilde{y} = h\left(x\right) \mid \tilde{x} = x\right) \, \mathrm{d}x[1] \ldots \mathrm{d}x[d]$$

$$= \int_{x[1]=-\infty}^{\infty} \cdots \int_{x[d]=-\infty}^{\infty} f_{\tilde{x}}(x) p_{\tilde{y} \mid \tilde{x}}(h\left(x\right) \mid x) \, \mathrm{d}x[1] \ldots \mathrm{d}x[d]. \tag{6.52}$$

For each fixed x, $p_{\tilde{y} \mid \tilde{x}}(\mathrm{MAP}\left(x\right) \mid x) \geq p_{\tilde{y} \mid \tilde{x}}(h\left(x\right) \mid x)$, by definition of the MAP estimator. Consequently,

$$\mathrm{P}\left(h(\tilde{x}) = \tilde{y}\right) \leq \int_{x[1]=-\infty}^{\infty} \cdots \int_{x[d]=-\infty}^{\infty} f_{\tilde{x}}(x) p_{\tilde{y} \mid \tilde{x}}(\mathrm{MAP}\left(x\right) \mid x) \, \mathrm{d}x[1] \ldots \mathrm{d}x[d]$$

$$\tag{6.53}$$

$$= \int_{x[1]=-\infty}^{\infty} \cdots \int_{x[d]=-\infty}^{\infty} f_{\tilde{x}}(x) \mathrm{P}\left(\tilde{y} = \mathrm{MAP}\left(x\right) \mid \tilde{x} = x\right) \, \mathrm{d}x[1] \ldots \mathrm{d}x[d]$$

$$= \mathrm{P}\left(\mathrm{MAP}(\tilde{x}) = \tilde{y}\right). \tag{6.54}$$

∎

Unfortunately, the MAP estimator is intractable to compute, due to the curse of dimensionality, unless the features are very few. Otherwise, the number of possible values of the feature vector $\tilde{x}$ scales exponentially with its dimension d, as explained in Section 4.7, which makes it impossible to estimate all the possible conditional pmfs of $\tilde{y}$ given $\tilde{x}$ from the training data. In Section 4.8, we circumvent the curse of dimensionality by making naive conditional independence assumptions that render MAP estimation tractable. Here we take a different route: We assume that the distribution of the features $\tilde{x}$ follows a parametric model, with parameters that depend on the class $\tilde{y}$. Specifically, we model $\tilde{x}$ as a Gaussian mixture model where the discrete random variable is $\tilde{y}$. We fit the parametric model to the training data and use the resulting distributions to estimate the conditional pmf $p_{\tilde{y}\mid\tilde{x}}$ of the class given the features. This classification method is known as *Gaussian discriminant analysis*.

Definition 6.12 (Gaussian discriminant analysis). *Let* (x_1, y_1), (x_2, y_2), ..., (x_n, y_n) *be a dataset of* n *examples, where* x_i *denotes a* d-*dimensional continuous vector and* $y_i \in \{1, 2, \ldots, c\}$ *its corresponding class (c is the number of classes). To perform Gaussian discriminant analysis, we fit a Gaussian mixture model to the data.*

The pmf of $\tilde{y}$ *is estimated using the corresponding empirical pmf (see Definition 2.11). The resulting estimate for* $p_{\tilde{y}}(k)$ *is*

$$\alpha_k := \frac{n_k}{n}, \quad k \in \{1, 2, \ldots, c\}, \tag{6.55}$$

where n_k *is the number of examples in class* k.

Let us denote the examples belonging to class k *by* $(x_1^{[k]}, y_1^{[k]})$, $(x_2^{[k]}, y_2^{[k]})$, ..., $(x_{n_k}^{[k]}, y_{n_k}^{[k]})$. *We model the conditional pdf of* $\tilde{x}$ *given* $\tilde{y} = k$ *as a Gaussian random vector, obtaining the corresponding mean and covariance-matrix parameters via maximum-likelihood estimation (Theorem 5.24),*

$$\mu_k = \frac{1}{n_k} \sum_{i=1}^{n_k} x_i^{[k]}, \tag{6.56}$$

$$\Sigma_k = \frac{1}{n_k} \sum_{i=1}^{n_k} (x_i^{[k]} - \mu_k)(x_i^{[k]} - \mu_k)^T. \tag{6.57}$$

By (6.30) and Definition 5.19, the probability of class y *given* $\tilde{x} = x$ *equals*

$$p_{\tilde{y}\mid\tilde{x}}(y \mid x) = \frac{p_{\tilde{y}}(y) f_{\tilde{x}\mid\tilde{y}}(x \mid y)}{\sum_{k=1}^{c} p_{\tilde{y}}(k) p_{\tilde{x}\mid\tilde{y}}(x \mid k)} \tag{6.58}$$

$$= \frac{\frac{\alpha_y}{\sqrt{(2\pi)^d |\Sigma_y|}} \exp\left(-\frac{1}{2}(x - \mu_y)^T \Sigma_y^{-1}(x - \mu_y)\right)}{\sum_{k=1}^{c} \frac{\alpha_k}{\sqrt{(2\pi)^d |\Sigma_k|}} \exp\left(-\frac{1}{2}(x - \mu_k)^T \Sigma_k^{-1}(x - \mu_k)\right)} \tag{6.59}$$

for $1 \leq y \leq c$. *To classify a new example with features* x_{new}, *we compute the MAP estimator of* $\tilde{y}$ *given* $\tilde{x} = x_{\text{new}}$ *following Definition 6.10*

$$\text{MAP}(x) := \arg \max_{y \in \{1,2,\ldots,c\}} p_{\tilde{y}\mid\tilde{x}}(y \mid x_{\text{new}}). \tag{6.60}$$

In order to fit a Gaussian mixture model, we need to estimate a d-dimensional mean and a $d \times d$ covariance matrix for each class (as well as the marginal pmf of the class). The number of parameters scales *quadratically* with the number of features d, instead of exponentially, which makes it tractable to fit the model, as long as the number of features is not extremely large.

The left column in Figure 6.7 shows the result of applying Gaussian discriminant analysis to perform diagnosis of Alzheimer's disease using real-world data extracted from Dataset 12. The top graph shows the contour lines of the two Gaussian conditional distributions of the features given each class. The middle graph depicts the estimated conditional probability of Alzheimer's given the features. If we diagnose Alzheimer's when the conditional probability is above 0.5, the resulting accuracy (the fraction of correct diagnoses) is 75.8%. For comparison, a naive classifier that never diagnoses Alzheimer's attains an accuracy of 72.9% because the fraction of healthy subjects in the dataset is 72.9%. Note, however, that accuracy is not necessarily the most important metric in healthcare applications and should be complemented with other measures of performance. Section 12.9 provides an in-depth discussion of how to evaluate classification models.

A critical consideration in real-world applications is how a classification model *generalizes* to held-out test data, not included in the training set used to fit the model. To evaluate generalization, we use a test dataset consisting of a completely different cohort of individuals. The estimated conditional probabilities of Alzheimer's for the test data are depicted in the bottom graph of Figure 6.7. Comparing the training and test data reveals a challenge that often arises in practice: There is a systematic difference between the distribution of the features in the two datasets. This is referred to as *domain shift* in the machine-learning literature. In this case, our method is robust to the shift: The model yields an accuracy of 81.5% on the test data. In comparison, the fraction of healthy subjects in the test population is 78.4%, so we again achieve an improvement of around 3% over the naive baseline.

Gaussian discriminant analysis is also known as *quadratic discriminant analysis*. To see why, let us consider the values of the feature vector x for which the conditional probability of two classes a and b are equal,

$$\frac{p_{\tilde{y}\,|\,\tilde{x}}(a\,|\,x)}{p_{\tilde{y}\,|\,\tilde{x}}(b\,|\,x)} = 1. \tag{6.61}$$

The resulting hypersurface separates the feature space between a region where a is more likely and another region where b is more likely, so we call it a *decision boundary*. By (6.59), we can express the boundary equation (6.61) in terms of the model parameters:

$$\frac{\alpha_a\sqrt{|\Sigma_b|}}{\alpha_b\sqrt{|\Sigma_a|}}\exp\left(\frac{1}{2}\left(x-\mu_b\right)^T\Sigma_b^{-1}\left(x-\mu_b\right) - \frac{1}{2}\left(x-\mu_a\right)^T\Sigma_a^{-1}\left(x-\mu_a\right)\right) = 1.$$

Taking logarithms, we obtain a quadratic hypersurface:

$$\frac{1}{2}\left(x-\mu_b\right)^T\Sigma_b^{-1}\left(x-\mu_b\right) - \frac{1}{2}\left(x-\mu_a\right)^T\Sigma_a^{-1}\left(x-\mu_a\right) + \log\left(\frac{\alpha_a\sqrt{|\Sigma_b|}}{\alpha_b\sqrt{|\Sigma_a|}}\right) = 0. \tag{6.62}$$

The middle left plot of Figure 6.7 shows this quadratic decision boundary for the Alzheimer's example, where the boundary is a one-dimensional curve because $d := 2$.

Gaussian discriminant analysis can be adapted to yield a linear decision boundary by imposing the constraint that the covariance-matrix parameters $\Sigma_1, \ldots, \Sigma_c$ in Definition 6.12 are all equal to the same matrix Σ. The matrix is estimated via maximum-likelihood estimation using all the data,

$$\Sigma := \frac{1}{n}\sum_{i=1}^{n}(x_i - \mu_{y_i})(x_i - \mu_{y_i})^T, \tag{6.63}$$

where each data point is centered using the mean parameter corresponding to its respective class. Setting both covariance matrices equal to Σ in (6.62) yields

$$\beta^T x + \xi = 0, \tag{6.64}$$

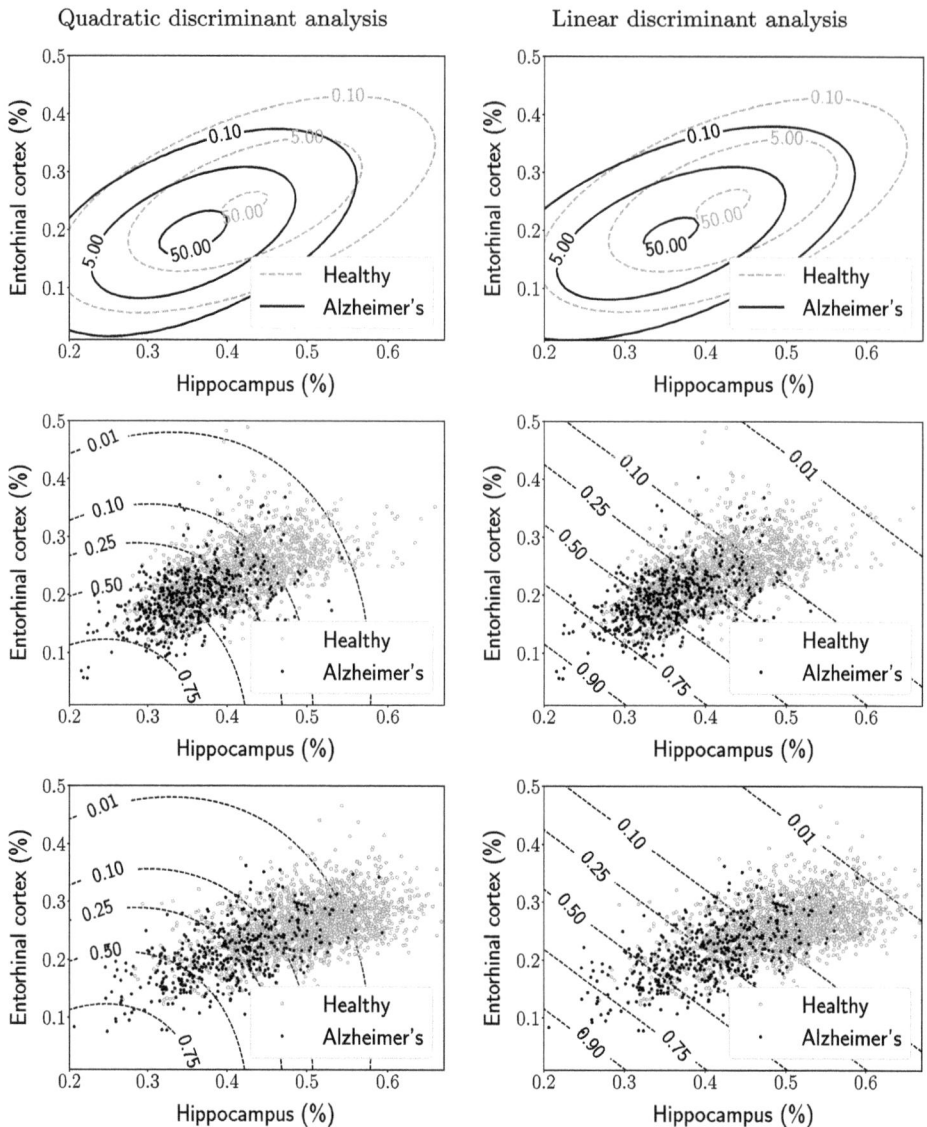

Figure 6.7 Diagnosis of Alzheimer's disease via Gaussian discriminant analysis. The top row shows the contour lines of Gaussian distributions obtained by fitting a training set of normalized volumes of the hippocampus and the entorhinal cortex, extracted from 1,926 magnetic-resonance scans in the Alzheimer's Disease Neuroimaging Initiative dataset (ADNI, 2020). The data are separated into two groups: healthy individuals (gray) and patients with Alzheimer's disease (black). The middle row shows the probability of Alzheimer's computed via (6.30), superposed onto the normalized volumes in the training set. The bottom row shows the same probability superposed onto the normalized volumes in a test set, extracted from 2,046 scans in the National Alzheimer's Coordinating Center dataset (NACC, 2020). The left column corresponds to quadratic discriminant analysis, where each Gaussian distribution has a different covariance matrix. The right column corresponds to linear discriminant analysis, where the Gaussians have the same covariance matrix.

where

$$\beta := \Sigma^{-1} \left(\mu_a - \mu_b \right), \tag{6.65}$$

$$\xi := \frac{1}{2} \mu_b^T \Sigma^{-1} \mu_b - \frac{1}{2} \mu_a^T \Sigma^{-1} \mu_a + \log \frac{\alpha_a}{\alpha_b}, \tag{6.66}$$

which is indeed a linear function of the feature vector x, so the decision boundary is a hyperplane. This version of Gaussian discriminant analysis is called *linear discriminant analysis*.

The right column of Figure 6.7 shows the result of applying linear discriminant analysis to our Alzheimer's example. The top plot shows the contour lines of the two Gaussian conditional distributions, which are identical shifted copies of one another. The contour lines of the conditional probability of Alzheimer's given the data are shown in the middle and bottom plots, superposed on the training and test data, respectively. The decision boundaries are indeed linear. The accuracy is 75.9% for the training data and 81.5% for the test data, essentially the same as that of quadratic discriminant analysis. For these features, it may be difficult to outperform a simple linear classifier. In order to improve performance, we would need to incorporate additional features into our model.

6.6 Clustering via Gaussian Mixture Models

6.6.1 Latent Variables

In Example 6.3, we model height data using a mixture model, which separates the data into two groups, according to sex. In this section, we explain how to fit such a model *without knowing what group each data point belongs to*. To this end, we define a *latent* or *hidden* discrete variable representing the group and infer its value from the data. This is called *unsupervised* learning, as opposed to the *supervised* setting of Section 6.5, where the group associated with each data point is known. An important application of unsupervised learning is *clustering*, which is a fundamental task in exploratory data analysis. The goal is to separate the data into groups or clusters of similar points. This can be achieved by fitting a mixture model with a discrete latent variable, and then using the estimated latent variable to identify the clusters.

We focus our discussion on Gaussian mixture models, which are the most popular mixture model used for clustering, but the same ideas can be adapted to other parametric distributions. Our objective is to separate a dataset of d-dimensional feature vectors $x_1, x_2, \ldots, x_n$ into m clusters, where the number of clusters m is fixed beforehand. We interpret the data as realizations from a continuous random variable $\tilde{x}$ that depends on a latent discrete random variable $\tilde{k}$ with range $\{1, 2, \ldots, m\}$. If $\tilde{k} = k$, then $\tilde{x}$ belongs to the kth cluster, and its conditional distribution is Gaussian with parameters that depend on k. After fitting the mixture model, the data can be clustered by computing the conditional probability that each data point belongs to each of the clusters.

6.6.2 Fitting Gaussian Mixture Models with Latent Variables

In this section, we explain how to fit a Gaussian mixture model dependent on a latent variable $\tilde{k}$ to data, represented by a random vector $\tilde{x}$. Our goal is to estimate the pmf of $\tilde{k}$ and the parameters of the conditional Gaussian distributions of $\tilde{x}$ given the different possible values of $\tilde{k}$. As explained in Section 3.7, a reasonable approach to fit the parameters of a parametric model is to maximize its likelihood, which equals its probability density at the observed data, usually under an i.i.d. assumption. To ease notation, let us denote the parameters of the Gaussian mixture model by

$$\theta := \left\{ \{\alpha_k\}_{k=1}^m, \{\mu_k\}_{k=1}^m, \{\Sigma_k\}_{k=1}^m \right\}. \tag{6.67}$$

The parameter $\alpha_k \in [0,1]$ represents the probability that $\tilde{k} = k$, so it must satisfy $\sum_{k=1}^m \alpha_k = 1$. The parameters $\mu_k \in \mathbb{R}^d$ and $\Sigma_k \in \mathbb{R}^{d \times d}$ are the mean and covariance-matrix parameters of $\tilde{x}$ given $\tilde{k} = k$, respectively, so Σ_k must be symmetric and positive definite (see Definition 5.19). Analogously to Definition 3.32, we define the conditional likelihood of the ith data point given $\tilde{k} = k$ as

$$\mathcal{L}_{i,k}(\mu_k, \Sigma_k) := f_{\tilde{x} \mid \tilde{k}}(x_i \mid k) \tag{6.68}$$

$$= \frac{1}{\sqrt{(2\pi)^d |\Sigma_k|}} \exp\left(-\frac{1}{2} (x_i - \mu_k)^T \Sigma_k^{-1} (x_i - \mu_k) \right). \tag{6.69}$$

If the data are i.i.d., by Definition 3.32 and Theorem 6.2, the likelihood of the whole dataset equals

$$\mathcal{L}(\theta) := \prod_{i=1}^n f_{\tilde{x}}(x_i) \tag{6.70}$$

$$= \prod_{i=1}^n \sum_{k=1}^m p_{\tilde{k}}(k) f_{\tilde{x} \mid \tilde{k}}(x_i \mid k) \tag{6.71}$$

$$= \prod_{i=1}^n \sum_{k=1}^m \alpha_k \mathcal{L}_{i,k}(\mu_k, \Sigma_k), \tag{6.72}$$

and the corresponding log-likelihood is

$$\log \mathcal{L}(\theta) = \sum_{i=1}^n \log \sum_{k=1}^m \alpha_k \mathcal{L}_{i,k}(\mu_k, \Sigma_k). \tag{6.73}$$

Up to now, whenever we have encountered a log-likelihood function, it has been relatively straightforward to derive the parameter values that maximize it (see Sections 2.4, 3.7, or 5.10.3). This is not the case for the log-likelihood in (6.73). In particular, it is not concave and has several local maxima. Our strategy to fit Gaussian mixture models is to maximize the log-likelihood iteratively, until we reach a local maximum. This can be achieved in different ways, but by far the most popular is the *expectation-maximization* algorithm. The algorithm jointly fits the parameters and an auxiliary quantity $\gamma_{i,k}$, which we call *membership probabilities*.

The kth membership probability $\gamma_{i,k}$ associated with the ith data point x_i is the conditional probability that x_i belongs to the kth cluster given $\tilde{x} = x_i$. By (6.30),

$$\gamma_{i,k} := p_{\tilde{k} \mid \tilde{x}}(k \mid x_i) \tag{6.74}$$

$$= \frac{p_{\tilde{k}}(k) f_{\tilde{x}_i \mid \tilde{k}}(x_i \mid k)}{\sum_{l=1}^{m} p_{\tilde{k}}(l) f_{\tilde{x}_i \mid \tilde{k}}(x_i \mid l)} \tag{6.75}$$

$$= \frac{\alpha_k \mathcal{L}_{i,k}(\mu_k, \Sigma_k)}{\sum_{l=1}^{m} \alpha_l \mathcal{L}_{i,l}(\mu_l, \Sigma_l)}. \tag{6.76}$$

The membership probabilities are useful for clustering. Once we fit the mixture model, we assign each data point i to the cluster with the highest membership probability, as it is the most likely according to the model,

$$\hat{k}_i := \arg\max_k p_{\tilde{k} \mid \tilde{x}}(k \mid x_i) \tag{6.77}$$

$$= \arg\max_k \gamma_{i,k}. \tag{6.78}$$

The sum of $\gamma_{i,k}$ over all data points can be interpreted as the *effective cardinality* (number of points) of the kth cluster,

$$n_k := \sum_{i=1}^{n} \gamma_{i,k}. \tag{6.79}$$

The expectation-maximization algorithm alternates between updating the membership probabilities in the *expectation* step, and the parameters α_k, μ_k, and Σ_k for each cluster k in the *maximization* step. It can be shown that each update of the expectation-maximization algorithm increases the log-likelihood of the model (we refer the reader to Section 9.4 in Bishop (2006) for a proof).

Section 6.6.3 provides a formal justification of the updates in the expectation-maximization algorithm, but first let us motivate them intuitively. Imagine that we know what cluster each data point corresponds to. Then we should set α_k equal to the fraction of data points in cluster k, and μ_k and Σ_k equal to the sample mean and sample covariance matrix of the data points in that cluster (since that is the appropriate maximum-likelihood estimator by Theorem 5.24). Of course, we don't know the cluster assignments, but if we have an estimate of the membership probabilities, we can use them as a *soft assignment*. We estimate α_k as the ratio between the effective cardinality and the total number of points, and μ_k and Σ_k as the sample mean and sample covariance matrix *weighted by the membership probabilities*, so that the contribution of each point depends on the probability that it belongs to the cluster. This yields the updates (6.82), (6.83), and (6.84) in Definition 6.13.

Definition 6.13 (Expectation maximization for Gaussian mixture models). *Let x_1, x_2, ..., x_n be n d-dimensional real-valued vectors. To fit a Gaussian mixture models with m clusters to these data, we first initialize the parameters $\{\alpha_k\}_{k=1}^{m}$, $\{\mu_k\}_{k=1}^{m}$ and $\{\Sigma_k\}_{k=1}^{m}$, ensuring that $\alpha_k \in [0, 1]$, $\sum_{k=1}^{m} \alpha_k = 1$, $\mu_k \in \mathbb{R}^d$ and every $\Sigma_k \in \mathbb{R}^{d \times d}$ is symmetric and positive definite. This can be done by separating the points into m groups (either randomly or using some other clustering approach, such as k-means), and applying (6.82), (6.83), and (6.84). We then repeat the following steps until the log-likelihood converges:*

1 Update the membership probabilities $\gamma_{i,k}$ for $1 \leq i \leq n$, $1 \leq k \leq m$, and also the corresponding effective cluster cardinality n_k:

$$\gamma_{i,k} := \frac{\alpha_k \mathcal{L}_{i,k}(\mu_k, \Sigma_k)}{\sum_{l=1}^{m} \alpha_l \mathcal{L}_{i,l}(\mu_l, \Sigma_l)}, \tag{6.80}$$

$$n_k := \sum_{i=1}^{n} \gamma_{i,k}. \tag{6.81}$$

2 Update the parameters of each Gaussian conditional distribution. For $1 \leq k \leq m$, set

$$\alpha_k := \frac{n_k}{n}, \tag{6.82}$$

$$\mu_k := \frac{1}{n_k} \sum_{i=1}^{n} \gamma_{i,k} x_i, \tag{6.83}$$

$$\Sigma_k := \frac{1}{n_k} \sum_{i=1}^{n} \gamma_{i,k} \left(x_i - \mu_k\right) \left(x_i - \mu_k\right)^T. \tag{6.84}$$

After convergence, we assign a cluster $\hat{k}_i \in \{1, \ldots, m\}$ to the ith data point for $1 \leq i \leq n$ by setting

$$\hat{k}_i := \arg\max_k \gamma_{i,k}. \tag{6.85}$$

Example 6.14 (Clustering according to height). In this example, we fit a Gaussian mixture model with two clusters to the height data in Example 6.3, without using the fact that we know the sex of each individual. Figure 6.8 shows the convergence of the log-likelihood over the iterations of the expectation-maximization algorithm, as well as the convergence of the parameters in the mixture model. Figure 6.9 depicts the corresponding mixture distribution

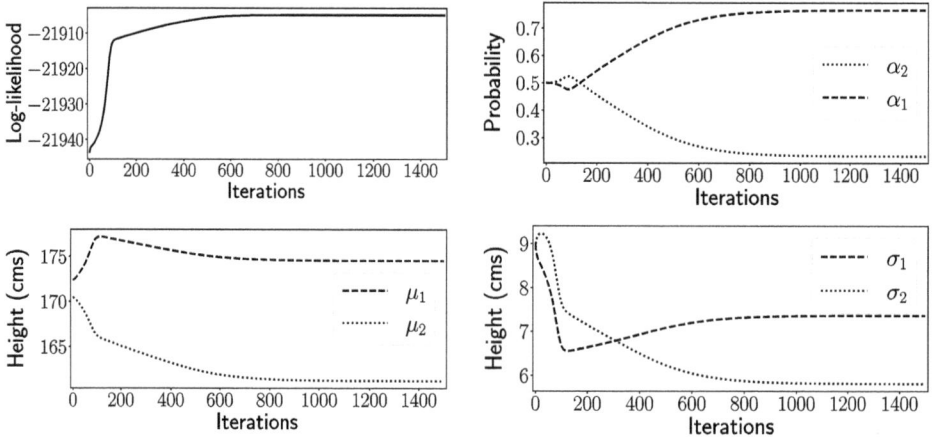

Figure 6.8 Convergence of the expectation-maximization algorithm to fit a Gaussian mixture model. The top left graph shows the evolution of the log-likelihood (6.73) when we apply expectation maximization to fit a Gaussian mixture model with two clusters to the height data in Example 6.3. The log-likelihood increases monotonically and eventually converges. The remaining plots show the convergence of the parameters α_1 and α_2 (top right), μ_1 and μ_2 (bottom left), and σ_1 and σ_2 (bottom right).

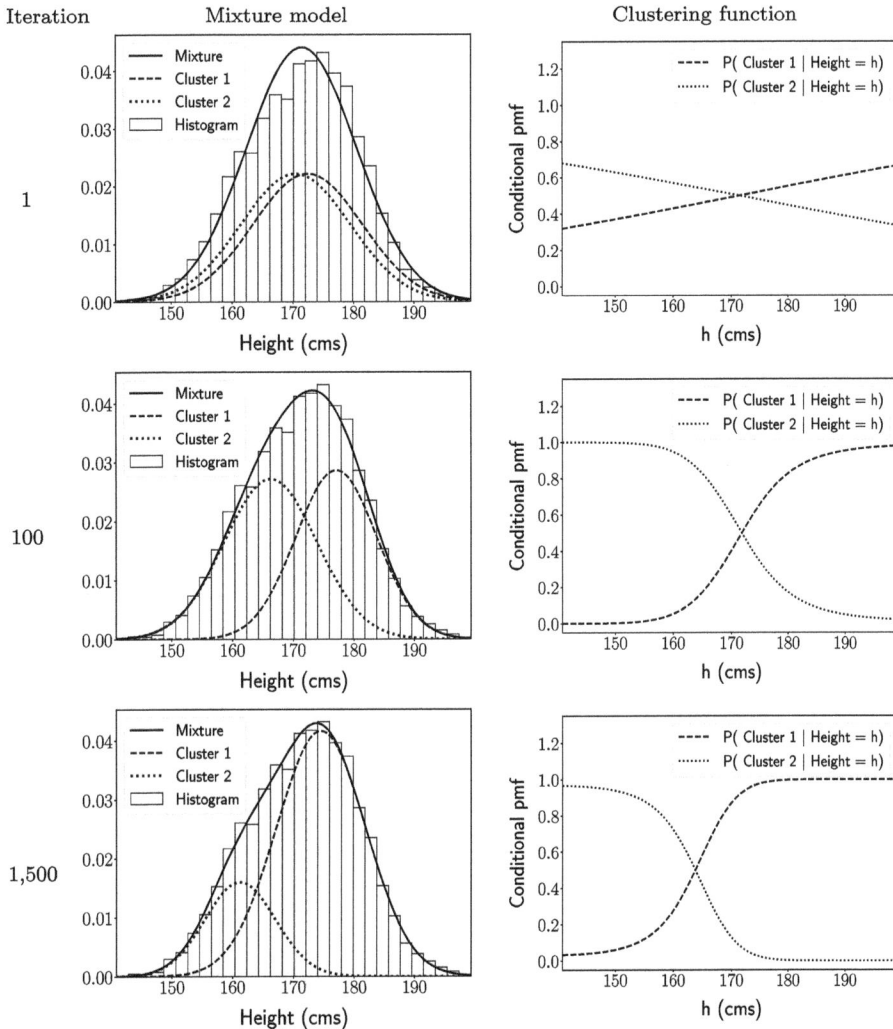

Figure 6.9 Gaussian mixture model for height. The plots show the results of fitting a Gaussian mixture model with two clusters to the height data in Example 6.3. The left column shows the probability density of the model (solid line) compared to the data (represented as a histogram), as well as the contributions of each cluster to the density at different iterations of the expectation-maximization algorithm. The fit clearly improves as the iterations proceed. The right column shows the conditional probabilities of each cluster given the data at the same iterations. After convergence, the conditional pmf is very similar the conditional pmf of sex given height in the right plot of Figure 6.5, even though here we do not have access to sex information.

and the conditional probabilities of each cluster given the data for some of the iterations. The model parameters eventually converge to provide a good fit to the data.

The two clusters identified by the model make a lot of sense. Even though the model has no access to sex information, the mean (175 cm) and standard deviation (7.4 cm) of the first cluster are close to the mean (176 cm) and standard deviation (6.9 cm) of the men. Likewise, the mean (161 cm) and standard deviation (5.8 cm) of the second cluster are close

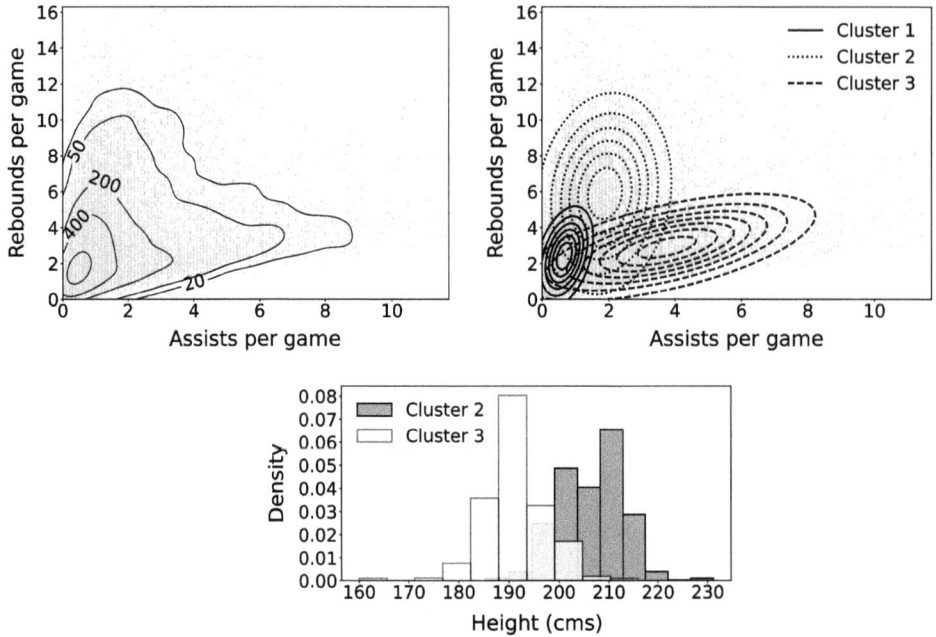

Figure 6.10 Gaussian mixture model for basketball players. The top left plot shows a scatter-plot of the assists and rebounds per game of NBA players between 1996 and 2019, as well as the corresponding probability density estimated via multidimensional kernel density estimation (see Definition 5.7). The top right plot shows the Gaussian densities associated with each of the three clusters identified by a Gaussian mixture model (fit to the data via expectation maximization, following Definition 6.13). The bottom plot shows histograms of the heights of players assigned to clusters 2 and 3. The players in cluster 2 are taller than those in cluster 3, which suggests that the clustering algorithm automatically separates the players into guards, who tend to be shorter, and centers and forwards, who tend to be taller.

to the mean (163 cm) and standard deviation (6.4 cm) of the women. The Gaussian mixture model is able to automatically identify the two subgroups in the population.

⋯⋯⋯

Example 6.15 (Clustering basketball players). In this example, we apply a Gaussian mixture model to analyze basketball statistics. The top left plot of Figure 6.10 shows a scatterplot of the assists and rebounds per game of NBA players between 1996 and 2019, extracted from Dataset 13. We apply a Gaussian mixture model with three clusters to the data, obtaining the conditional Gaussian distributions shown in the top right plot of Figure 6.10. Cluster 1 corresponds to players for which the number of assists and rebounds per game are very small. The two other clusters are more interesting.

Cluster 2 consists of players that catch a lot of rebounds, but do not assist very much. By contrast, cluster 3 consists of players who assist a lot, but do not catch many rebounds. The model is able to automatically group players with similar responsibilities. This is confirmed by the height distributions of the players assigned to each cluster, shown at the bottom of Figure 6.10 (note that we are not using this information to produce the clusters). Players in cluster 2 are taller, as is typical of centers and forwards, whose role is to catch rebounds rather than pass the ball. Players in cluster 3 are shorter, as is typical of guards, who usually give more assists, and catch less rebounds.

⋯⋯⋯

6.6.3 Derivation of the Expectation-Maximization Updates

Our strategy to fit Gaussian mixture models is to try to maximize the log-likelihood $\log \mathcal{L}(\theta)$ defined in (6.73) by finding parameter values for which its gradient is zero. Unfortunately, setting the gradient to zero does not yield equations that we can solve in closed form (in contrast to the models in Sections 2.4, 3.7, or 5.10.3). In order to make some progress, recall that we already know how to maximize the Gaussian conditional log-likelihood $\mathcal{L}_{i,k}(\mu_k, \Sigma_k)$, $1 \leq k \leq m$, $1 \leq i \leq n$, with respect to μ_k and Σ_k (see Section 5.10.3). Inspired by this, we express the log-likelihood as a function of the conditional log-likehood:

$$\nabla_{\mu_k,\Sigma_k} \log \mathcal{L}(\theta) = \sum_{i=1}^{n} \frac{\alpha_k \nabla_{\mu_k,\Sigma_k} \mathcal{L}_{i,k}(\mu_k, \Sigma_k)}{\sum_{l=1}^{m} \alpha_l \mathcal{L}_{i,l}(\mu_l, \Sigma_l)} \tag{6.86}$$

$$= \sum_{i=1}^{n} \frac{\alpha_k \mathcal{L}_{i,k}(\mu_k, \Sigma_k)}{\sum_{l=1}^{m} \alpha_l \mathcal{L}_{i,l}(\mu_l, \Sigma_l)} \nabla_{\mu_k,\Sigma_k} \log \mathcal{L}_{i,k}(\mu_k, \Sigma_k) \tag{6.87}$$

$$= \sum_{i=1}^{n} \gamma_{i,k} \nabla_{\mu_k,\Sigma_k} \log \mathcal{L}_{i,k}(\mu_k, \Sigma_k), \tag{6.88}$$

where we have used the fact that

$$\nabla_{\mu_k,\Sigma_k} \log \mathcal{L}_{i,k}(\mu_k, \Sigma_k) = \mathcal{L}_{i,k}(\mu_k, \Sigma_k)^{-1} \nabla_{\mu_k,\Sigma_k} \mathcal{L}_{i,k}(\mu_k, \Sigma_k). \tag{6.89}$$

If all the membership probabilities $\gamma_{i,k}$ were constant with respect to the model parameters, then it is not too difficult to maximize this expression with respect to μ_k and Σ_k, as we explain in the following paragraph. However, this is not entirely correct, as $\gamma_{i,k}$ is a function of μ_k, Σ_k and the rest of model parameters. The key idea of the expectation-maximization algorithm is to ignore this and update the model parameters assuming that each $\gamma_{i,k}$ is constant.

Under the assumption that $\gamma_{i,k}$ is fixed, the gradient of the log-likelihood with respect to μ_k and Σ_k can be obtained following the same derivations as in the proof of Theorem 5.24. Setting the gradient with respect to the mean parameter,

$$\sum_{i=1}^{n} \gamma_{i,k} \nabla_{\mu_k} \log \mathcal{L}_{i,k}(\mu_k, \Sigma_k) = \sum_{i=1}^{n} \gamma_{i,k} \Sigma_k^{-1} (x_i - \mu_k) \tag{6.90}$$

$$= \Sigma_k^{-1} \sum_{i=1}^{n} \gamma_{i,k} (x_i - \mu_k), \tag{6.91}$$

to zero yields

$$\mu_k^* = \frac{\sum_{i=1}^{n} \gamma_{i,k} x_i}{\sum_{i=1}^{n} \gamma_{i,k}} \tag{6.92}$$

$$= \frac{1}{n_k} \sum_{i=1}^{n} \gamma_{i,k} x_i. \tag{6.93}$$

Plugging this value into the conditional log-likelihood, and setting the gradient with respect to the inverse of the covariance-matrix parameter,

$$\sum_{i=1}^{n} \gamma_{i,k} \nabla_{\Sigma_k^{-1}} \log \mathcal{L}_{i,k}(\mu_k, \Sigma_k) = \sum_{i=1}^{n} \frac{\gamma_{i,k}}{2} \left(\Sigma_k - (x_i - \mu_k^*)(x_i - \mu_k^*)^T \right), \quad (6.94)$$

to zero yields

$$\Sigma_k^* = \frac{\sum_{i=1}^{n} \gamma_{i,k} (x_i - \mu_k^*)(x_i - \mu_k^*)^T}{\sum_{i=1}^{n} \gamma_{i,k}} \quad (6.95)$$

$$= \frac{1}{n_k} \sum_{i=1}^{n} \gamma_{i,k} (x_i - \mu_k^*)(x_i - \mu_k^*)^T. \quad (6.96)$$

In order to determine the update for α_k, $1 \le k \le m$, we need to enforce the constraint that $\sum_{k=1}^{m} \alpha_k = 1$. This is achieved by setting $\alpha_m := 1 - \sum_{k=1}^{m-1} \alpha_k$, so that the log-likelihood equals

$$\log \mathcal{L}(\theta) = \sum_{i=1}^{n} \log \left(\sum_{k=1}^{m-1} \alpha_k \mathcal{L}_{i,k}(\mu_k, \Sigma_k) + \left(1 - \sum_{k=1}^{m-1} \alpha_k \right) \mathcal{L}_{i,m}(\mu_m, \Sigma_m) \right). \quad (6.97)$$

The partial derivative of the log-likelihood with respect to α_k is

$$\frac{\partial \log \mathcal{L}(\theta)}{\partial \alpha_k} = \sum_{i=1}^{n} \frac{\mathcal{L}_{i,k}(\mu_k, \Sigma_k) - \mathcal{L}_{i,m}(\mu_m, \Sigma_m)}{\sum_{l=1}^{m} \alpha_l \mathcal{L}_{i,l}(\mu_l, \Sigma_l)}. \quad (6.98)$$

Multiplying the partial derivative by α_k yields

$$\alpha_k \frac{\partial \log \mathcal{L}(\theta)}{\partial \alpha_k} = \sum_{i=1}^{n} \frac{\alpha_k \mathcal{L}_{i,k}(\mu_k, \Sigma_k) - \alpha_k \mathcal{L}_{i,m}(\mu_m, \Sigma_m)}{\sum_{l=1}^{m} \alpha_l \mathcal{L}_{i,l}(\mu_l, \Sigma_l)} \quad (6.99)$$

$$= \sum_{i=1}^{n} \left(\gamma_{i,k} - \frac{\alpha_k \mathcal{L}_{i,m}(\mu_m, \Sigma_m)}{\sum_{l=1}^{m} \alpha_l \mathcal{L}_{i,l}(\mu_l, \Sigma_l)} \right) \quad (6.100)$$

$$= n_k - \frac{\alpha_k}{\alpha_m} \sum_{i=1}^{n} \frac{\alpha_m \mathcal{L}_{i,m}(\mu_m, \Sigma_m)}{\sum_{l=1}^{m} \alpha_l \mathcal{L}_{i,l}(\mu_l, \Sigma_l)} \quad (6.101)$$

$$= n_k - \frac{\alpha_k}{\alpha_m} \sum_{i=1}^{n} \gamma_{i,m} \quad (6.102)$$

$$= n_k - \frac{\alpha_k}{\alpha_m} n_m. \quad (6.103)$$

We now assume that $\gamma_{i,k}$ is constant for $1 \le i \le n$, $1 \le k \le m$, and therefore so is n_k (but remember that this is not true in general!), and find the values of α_k for which the partial derivatives vanish. We sum (6.103) over k and set it equal to zero:

$$\sum_{k=1}^{m} n_k - \sum_{k=1}^{m} \frac{\alpha_k}{\alpha_m} n_m = n - \frac{n_m}{\alpha_m} = 0, \quad (6.104)$$

where we have used the fact that $\sum_{k=1}^{m} n_k = n$ and $\sum_{k=1}^{m} \alpha_k = 1$. We conclude that α_m should equal $\frac{n_m}{n}$. Plugging this into (6.103) yields $n_k - \alpha_k n$, which equals zero, if α_k equals

$$\alpha_k^* := \frac{n_k}{n}, \quad 1 \le k \le m. \quad (6.105)$$

We check that indeed $\sum_{k=1}^{m} \alpha_k^* = 1$.

To summarize, we have derived parameter values for which the gradient of the log-likelihood vanishes, as long as $\gamma_{i,k}$ is constant for all $1 \leq i \leq n$, $1 \leq k \leq m$. Of course, there is a catch. As we already mentioned, $\gamma_{i,k}$ depends on the parameters! In order for the gradient of the log-likelihood to actually vanish, we also need the following equation to hold

$$\gamma_{i,k} = \frac{\alpha_k^* \mathcal{L}_{i,k}(\mu_k^*, \Sigma_k^*)}{\sum_{l=1}^{m} \alpha_l^* \mathcal{L}_{i,l}(\mu_l^*, \Sigma_l^*)}. \tag{6.106}$$

In order to resolve this chicken-and-egg problem, we take a pragmatic approach. We alternate between (1) updating μ_k, Σ_k and α_k using (6.93), (6.96), and (6.105), and (2) updating $\gamma_{i,k}$ via (6.106). Our hope is that the alternating scheme will eventually converge to parameter values that satisfy all the equations. Step 1 is known as the *maximization* step because it maximizes the log-likelihood with respect to the parameters. Step 2 is known as the *expectation* step because the log-likelihood function can be interpreted as the expected log-likelihood with respect to the membership probabilities. Putting everything together, we obtain the algorithm in Definition 6.13.

Figures 6.8 and 6.9 show the expectation-maximization updates converging to a local maximum of the log-likelihood. Even after convergence, we are not guaranteed to reach a global maximum because the function is nonconvex. However, we must remember that our ultimate goal is obtaining a *useful model*, and Gaussian mixture models often uncover useful clusters in real data, as illustrated by Examples 6.14 and 6.15.

6.7 Bayesian Parametric Modeling

6.7.1 The Bayesian Framework

In traditional parametric modeling, described in Sections 2.3 and 3.6, we model data using predefined distributions that depend on a small number of parameters. These parameters are deterministic quantities, chosen so that the parametric model approximates the data as accurately as possible (often via maximum-likelihood estimation). By contrast, Bayesian parametric modeling represents the parameters as *random variables*. This allows us to quantify our uncertainty about these parameters in a principled way.

Let $\tilde{\theta}$ be a random variable, or random vector, representing the parameters in a Bayesian parametric model, and let $\tilde{x}$ be a random vector representing the data. Fitting the model to the data requires making two decisions. We need to choose:

1 The marginal pmf $p_{\tilde{\theta}}$ or marginal pdf $f_{\tilde{\theta}}$ of $\tilde{\theta}$. We call this the *prior* distribution of the parameters because it captures our uncertainty about them *before* seeing the data.
2 The conditional pmf $p_{\tilde{x}|\tilde{\theta}}$ or the conditional pdf $f_{\tilde{x}|\tilde{\theta}}$ of the data $\tilde{x}$ given the parameters. The conditional distribution is called the *likelihood* of the data. For a fixed value of the parameters, this is the same as the likelihood defined in Sections 2.4 and 3.7; it represents the probability or the probability density of the data according to the parametric model. The crucial difference is that in traditional parametric modeling, the likelihood has no probabilistic meaning, it is just interpreted as a function of the parameters. In Bayesian modeling, it is a conditional pmf or conditional pdf.

The main goal in Bayesian parametric modeling is to compute the conditional distribution of the parameters $\tilde{\theta}$ given the observed data x, encoded by the conditional pmf $p_{\tilde{\theta}|\tilde{x}}$ or the conditional pdf $f_{\tilde{\theta}\,|\,\tilde{x}}$ evaluated at $\tilde{x} = x$. This is called the *posterior* distribution of the parameters because it represents our uncertainty about them *after* seeing the data.

Example 6.16 (Single coin flip). We consider the problem of modeling a single coin flip, using a parametric Bernoulli model (see Definition 2.15). The Bernoulli parameter represents the probability of heads. In traditional parametric modeling, the parameter is a deterministic variable θ restricted to the unit interval $[0, 1]$. By Definition 2.24, the likelihood is the Bernoulli pmf p_θ evaluated at the data point x, which represents the coin flip, interpreted as a function of θ:

$$\mathcal{L}(\theta) := p_\theta(x) = \begin{cases} \theta, & \text{if } x = \text{heads}, \\ 1 - \theta, & \text{if } x = \text{tails}. \end{cases} \tag{6.107}$$

By Definition 2.25, the maximum-likelihood estimator is

$$\theta_{\mathrm{ML}} := \arg\max_{0 \le \theta \le 1} \mathcal{L}(\theta) = \begin{cases} \arg\max_{0 \le \theta \le 1} \theta = 1, & \text{if } x = \text{heads}, \\ \arg\max_{0 \le \theta \le 1} 1 - \theta = 0, & \text{if } x = \text{tails}. \end{cases} \tag{6.108}$$

Under the traditional framework, there is no uncertainty regarding the parameter: The estimate equals either one (if x is heads) or zero (if x is tails).

In order to quantify our uncertainty about the Bernoulli parameter, we switch to a Bayesian viewpoint, and model it as a random variable $\tilde{\theta}$. We consider two possible marginal pdfs for $\tilde{\theta}$: A uniform pdf in $[0, 1]$ and a triangular pdf of the form

$$f_{\tilde{\theta}}(\theta) = 2\theta \quad \text{for } \theta \in [0, 1]. \tag{6.109}$$

The two prior distributions encode different assumptions about the coin flip. According to the uniform prior, all values of the parameter are equally possible. By contrast, the triangular prior assigns higher density to parameter values near one, indicating that we expect the coin to be more likely to land on heads.

Conditioned on $\tilde{\theta} = \theta$, we model the coin flip $\tilde{x}$ as a Bernoulli random variable with parameter θ, as in the traditional parametric approach. The likelihood is the conditional pmf of $\tilde{x}$ given θ:

$$p_{\tilde{x}\,|\,\tilde{\theta}}(x \,|\, \theta) = \begin{cases} \theta, & \text{if } x = \text{heads}, \\ 1 - \theta, & \text{if } x = \text{tails}. \end{cases} \tag{6.110}$$

This is exactly the same as the likelihood (6.107) of the traditional parametric model, but it has *a very different interpretation*: It is a conditional probability mass function, and not a function of a deterministic parameter.

We observe that the coin lands on tails. The posterior pdf of the parameter given this data point depends on the prior that we choose. For the uniform prior, by Theorems 6.6 and 6.5, it equals

$$f_{\tilde{\theta}\,|\,\tilde{x}}(\theta \,|\, \text{tails}) = \frac{f_{\tilde{\theta}}(\theta)\, p_{\tilde{x}\,|\,\tilde{\theta}}(\text{tails} \,|\, \theta)}{p_{\tilde{x}}(\text{tails})} \tag{6.111}$$

$$= \frac{1 - \theta}{\int_{u=-\infty}^{\infty} f_{\tilde{\theta}}(u)\, p_{\tilde{x}\,|\,\tilde{\theta}}(\text{tails} \,|\, u)\, \mathrm{d}u} \tag{6.112}$$

$$= \frac{1 - \theta}{\int_{u=0}^{1} 1 - u \ du} \qquad (6.113)$$

$$= 2(1 - \theta). \qquad (6.114)$$

The posterior pdf is skewed toward zero, indicating that the probability of tails is higher than the probability of heads. For the triangular prior, by the same argument, the posterior pdf equals

$$f_{\tilde{\theta} \mid \tilde{x}} (\theta \mid \text{tails}) = \frac{f_{\tilde{\theta}} (\theta) \, p_{\tilde{x} \mid \tilde{\theta}} (\text{tails} \mid \theta)}{p_{\tilde{x}} (\text{tails})} \qquad (6.115)$$

$$= \frac{2\theta (1 - \theta)}{\int_{u=-\infty}^{\infty} f_{\tilde{\theta}} (u) \, p_{\tilde{x} \mid \tilde{\theta}} (\text{tails} \mid u) \ du} \qquad (6.116)$$

$$= \frac{2\theta (1 - \theta)}{\int_{u=0}^{1} 2u (1 - u) \ du} \qquad (6.117)$$

$$= 6\theta (1 - \theta). \qquad (6.118)$$

The posterior pdf is centered, correcting our initial assumption that heads is more likely. Figure 6.11 shows the prior and posterior pdfs for the two priors.

Example 6.16 just considers a single data point. If more data are available, we need to model the joint distribution of the random variables representing the data. A common assumption is that the data are conditionally independent given the parameters. This is analogous to the i.i.d. assumption in Sections 2.4 and 3.7, where the data are assumed to be independent and identically distributed according to the chosen parametric distribution. Let $\tilde{\theta}$ represent the parameters in a Bayesian model, and let $\tilde{x}$ be an n-dimensional random

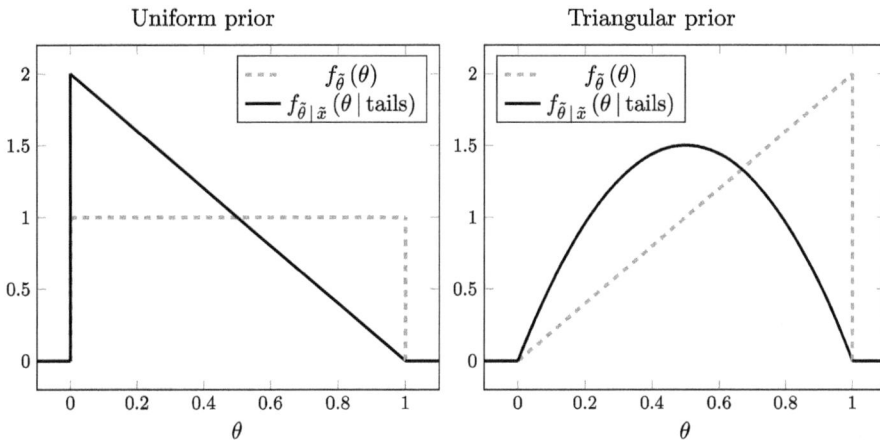

Figure 6.11 Prior and posterior distributions for the coin flip in Example 6.16. The prior distribution on the left is uniform, so it does not make any assumptions about the fairness of the coin. The prior pdf on the right encodes the assumption that the coin may be more prone to land on heads. The corresponding posterior pdfs reflect our updated uncertainty about the coin-flip Bernoulli parameter after we observe that the coin lands on tails.

vector representing the data. If we assume conditional independence, the likelihood of the data equals

$$p_{\tilde{x} \mid \tilde{\theta}}(x \mid \theta) = \prod_{i=1}^{n} p_{\tilde{x}[i] \mid \tilde{\theta}}(x[i] \mid \theta) \tag{6.119}$$

if the data are discrete, or

$$f_{\tilde{x} \mid \tilde{\theta}}(x \mid \theta) = \prod_{i=1}^{n} f_{\tilde{x}[i] \mid \tilde{\theta}}(x[i] \mid \theta) \tag{6.120}$$

if the data are continuous.

6.7.2 The Beta Distribution and Conjugate Priors

In discrete parametric models, the parameters often encode a probability. This is the case for the Bernoulli, geometric, and binomial parametric distributions defined in Section 2.3. The beta distribution is a useful model for parameters representing probabilities because its associated pdf is restricted to the unit interval (just like a probability). The shape of the pdf depends on two parameters a and b, as illustrated in Figure 6.12. The priors and posterior distributions in Example 6.16 are all beta distributions.

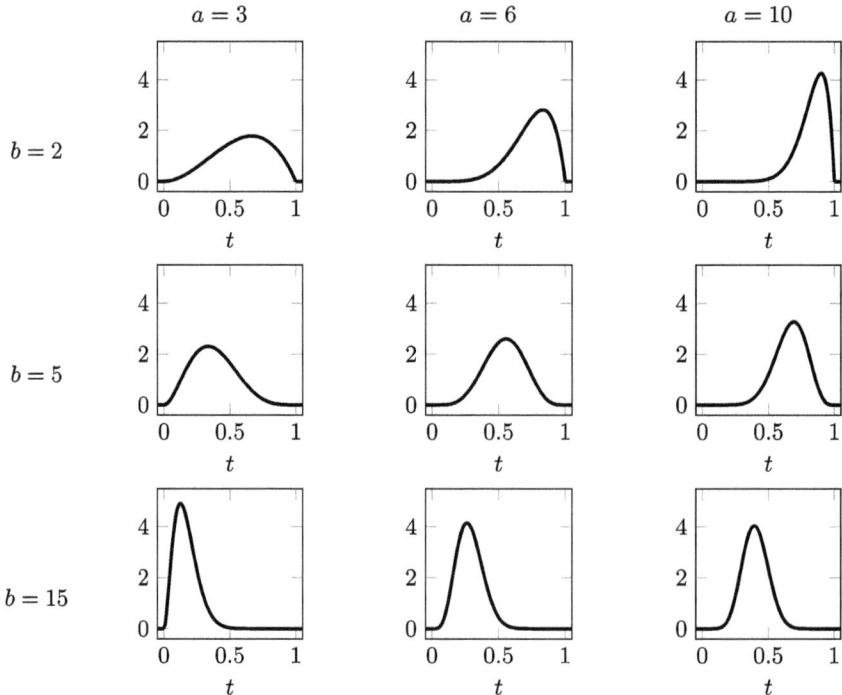

Figure 6.12 Beta distribution. The probability density of the beta distribution has at most a single maximum whose location is determined by the value of the a and b parameters. When a is larger than b, the maximum is located toward the right of the unit interval. When we increase b, it shifts to the left. Increasing both a and b results in a pdf that is more concentrated around the maximum.

Definition 6.17 (Beta distribution). *The pdf of a beta random variable $\tilde{t}$ with parameters a and b is*

$$f_{\tilde{t}}(t) := \frac{t^{a-1}(1-t)^{b-1}}{\beta(a,b)}, \qquad if \quad 0 \leq t \leq 1, \tag{6.121}$$

and zero otherwise, where a and b are assumed to be positive real numbers and

$$\beta(a,b) := \int_{u} u^{a-1}(1-u)^{b-1} \, \mathrm{d}u \tag{6.122}$$

is a special function called the beta function or Euler integral of the first kind, which does not have a closed form, but can be computed numerically.

To illustrate the use of the beta distribution in Bayesian parametric modeling, let us consider the problem of estimating the posterior distribution of a Bernoulli parameter $\tilde{\theta}$ from n data points. We model the data as an n-dimensional random vector $\tilde{x}$, with entries that are conditionally independent given $\tilde{\theta}$. Under these assumptions, the likelihood of an observed data vector $x \in \mathbb{R}^n$ is

$$p_{\tilde{x}|\tilde{\theta}}(x \mid \theta) = \prod_{i=1}^{n} p_{\tilde{x}[i]|\tilde{\theta}}(x[i] \mid \theta) \tag{6.123}$$

$$= \theta^{n_1}(1-\theta)^{n-n_1}, \tag{6.124}$$

where n_1 indicates the number of ones in x. Notice that n_1 can also be interpreted as the realization of a binomial random variable $\tilde{y}$ with parameters n and θ (see Example 2.16). In that case, the likelihood equals

$$p_{\tilde{y}|\tilde{\theta}}(n_1 \mid \theta) = \binom{n}{n_1} \theta^{n_1}(1-\theta)^{n-n_1}. \tag{6.125}$$

This is completely equivalent to (6.124). The only difference is the fixed scaling factor $\binom{n}{n_1}$, which does not depend on θ.

As established in the following theorem, for both likelihoods, if we choose a beta prior for the Bernoulli parameter, then the posterior is also a beta distribution. When this occurs, we say that the distribution is a *conjugate prior* to the likelihood.

Theorem 6.18 (The beta distribution is a conjugate prior to the Bernoulli and binomial likelihoods). *Let the continuous random variable $\tilde{\theta}$ represent a parameter restricted to the unit interval. We assume that the prior distribution of $\tilde{\theta}$ is a beta distribution with parameters a and b, and that we have available n data points. Conditioned on $\tilde{\theta} = \theta$, the data are modeled as either (1) n conditionally independent random variables following a Bernoulli distribution with parameter θ, or (2) a binomial random variable with parameters n and θ. In both cases, if we observe n_0 zeros and n_1 ones, the posterior distribution of $\tilde{\theta}$ is a beta distribution with parameters $a + n_1$ and $b + n_0$.*

Proof We prove the result for assumption (1). The proof for assumption (2) is essentially the same (we need to incorporate the scaling factor $\binom{n}{n_1}$, but it just cancels out). The likelihood of the observed data x, interpreted as a realization of an n-dimensional random vector $\tilde{x}$, is given by (6.124). Consequently, by Theorems 6.6 and 6.5 and Definition 6.17,

$$f_{\tilde{\theta}\mid\tilde{x}}\left(\theta\mid x\right) = \frac{f_{\tilde{\theta}}\left(\theta\right)p_{\tilde{x}\mid\tilde{\theta}}\left(x\mid\theta\right)}{p_{\tilde{x}}\left(x\right)} \tag{6.126}$$

$$= \frac{f_{\tilde{\theta}}\left(\theta\right)p_{\tilde{x}\mid\tilde{\theta}}\left(x\mid\theta\right)}{\int_{u=0}^{1}f_{\tilde{\theta}}\left(u\right)p_{\tilde{x}\mid\tilde{\theta}}\left(x\mid u\right)\,du} \tag{6.127}$$

$$= \frac{\theta^{a-1}\left(1-\theta\right)^{b-1}\theta^{n_1}\left(1-\theta\right)^{n_0}}{\int_{u=0}^{1}u^{a-1}\left(1-u\right)^{b-1}u^{n_1}\left(1-u\right)^{n_0}\,du} \tag{6.128}$$

$$= \frac{\theta^{a+n_1-1}\left(1-\theta\right)^{b+n_0-1}}{\int_{u=0}^{1}u^{a+n_1-1}\left(1-u\right)^{b+n_0-1}\,du}, \tag{6.129}$$

which is the pdf of a beta random variable with parameters $a + n_1$ and $b + n_0$ by Definition 6.17. ∎

The Bayesian update to the beta prior in Theorem 6.18 is very intuitive. We sum the number of observed ones n_1 to the a parameter, and the number of observed zeros n_0 to the b parameter. When we observe more ones, the a parameter becomes larger than the b parameter, which skews the pdf of $\tilde{\theta}$ toward one, as depicted in Figure 6.12. Conversely, if we observe more zeros, the b parameter becomes larger, which skews the pdf of $\tilde{\theta}$ toward zero. The more data we observe, the larger both parameters become, which results in a more concentrated pdf, and hence reduced uncertainty about the parameter $\tilde{\theta}$. In the following example, we apply the theorem to analyze poll data from the 2020 US presidential election.

Example 6.19 (Poll in Pennsylvania). In this example, we explain how to predict the result of an election from poll data via Bayesian modeling. We focus on a poll carried out in Pennsylvania before the 2020 US election. The poll reports that 281 people intend to vote for Trump and 300 for Biden. Our goal is to estimate the probability that Trump or Biden wins in Pennsylvania from these data. This example was written before the 2020 election, where Pennsylvania turned out to be a pivotal state.

The data are from a real poll (Ipsos, 2020), where we ignore other candidates and undecided voters for simplicity. We assume that the $n := 581$ people in the poll are chosen independently and uniformly at random with replacement from the population of Pennsylvania. This is an idealized assumption. In reality, polls are more likely to reach certain subpopulations.

We begin by fitting a traditional parametric model to the data. Under the assumption that the people in the poll are chosen uniformly at random from the population, the probability that each of them votes for Trump is equal to the fraction of Trump voters in Pennsylvania, which we represent by a parameter θ. If the poll participants are chosen independently, then the data can be modeled as i.i.d. Bernoulli random variables (where 1 indicates a Trump vote, and 0 a Biden vote) with parameter θ. By Example 2.26, the maximum-likelihood estimate of θ is the fraction of Trump voters in the poll

$$\theta_{\mathrm{ML}} = \frac{281}{581} = 0.484. \tag{6.130}$$

The estimator indicates that Biden winning the election is more consistent with the data, since the parameter is smaller than 0.5. However, it says nothing about the *probability* that Biden (or Trump) wins, i.e. the probability that θ is greater than 0.5. In traditional parametric modeling, θ is a deterministic parameter, so this statement does not even make sense! Example 9.47 shows that confidence intervals are not helpful for this purpose either. If we want to quantify how likely it is for either candidate to win, then we must resort to a Bayesian approach.

From a Bayesian viewpoint, the fraction of Trump voters is modeled as a random variable $\tilde{\theta}$. This allows us to consider the conditional probability that Trump wins ($\tilde{\theta} > 0.5$) given that the data, modeled as a random vector $\tilde{x}$, are equal to the observed value x. Under the assumption that the poll participants are chosen independently and uniformly at random with replacement from the population, the conditional distribution of the number of Trump voters in the poll given $\tilde{\theta} = \theta$ is binomial with parameters n and θ. If we choose the prior of $\tilde{\theta}$ to be a beta distribution with parameters a and b, the posterior pdf is a beta distribution with parameters $a + 281$ and $b + 300$ by Theorem 6.18. By Theorem 3.14 and Definition 6.17, our probability of interest is therefore equal to

$$P\left(\tilde{\theta} > 0.5\right) = \int_{\theta=0.5}^{1} f_{\theta \mid \tilde{x}}(\theta \mid x) \, \mathrm{d}\theta \tag{6.131}$$

$$= \frac{1}{\beta\left(a + 281, b + 300\right)} \int_{\theta=0.5}^{1} \theta^{a+281-1} \left(1 - \theta\right)^{b+300-1} \, \mathrm{d}\theta. \tag{6.132}$$

Figure 6.13 shows the prior and posterior pdfs, and the corresponding probability estimates, for values of a and b that encode different assumptions. In the left column, the priors are favorable to Biden; in the center, they are neutral; and in the right column, they are favorable to Trump. Our confidence in the prior is reflected by the shape of the prior pdf. In the top row, we are not very confident in our prior assumption, so the prior pdf is very spread out, and the posterior pdf is mainly driven by the data. Conversely, in the center and especially the bottom row, the prior pdf is more concentrated and has a stronger effect on the posterior pdf. As a result, the estimated probability that Trump wins depends heavily on the chosen prior. Overall, our analysis is moderately optimistic for Biden. In most scenarios, the model predicts that he wins Pennsylvania. However, the probability that Trump wins is not negligible for any of the neutral priors, and even for the low-confidence prior that favors Biden (20%). In the actual election, Biden ended up winning Pennsylvania by a very narrow margin.

. . . .

6.7.3 How Not to Predict an Election

In this section, we study the effect of independence assumptions in Bayesian probabilistic modeling. As a motivating example, we consider the United States presidential election. In the United States, the president is elected by the Electoral College, which is formed by electors determined by the result in each state (and Washington D.C.). If we want to predict who wins the election, we need a probabilistic model that includes every state. Let us represent the result in each state as a Bernoulli random variable. The joint pmf of these 51 Bernoulli variables has $2^{51} > 10^{15}$ entries, so it is impossible to estimate. This is a manifestation of the curse of dimensionality, described in Section 4.7. As explained in Sections 4.8 and 4.9, independence assumptions can be leveraged to obtain a tractable model. However, such independence assumptions can have a dramatic effect on the resulting prediction. To illustrate this, we study a fictitious election.

In our fictitious election, there are two candidates, a Republican and a Democrat. The number of states is 51. Each state contributes a single elector to the Electoral College. The result of the vote in state i is modeled as a Bernoulli random variable $\tilde{s}_i$, which equals one, if the Republican candidate wins, and zero otherwise. The result of the whole election

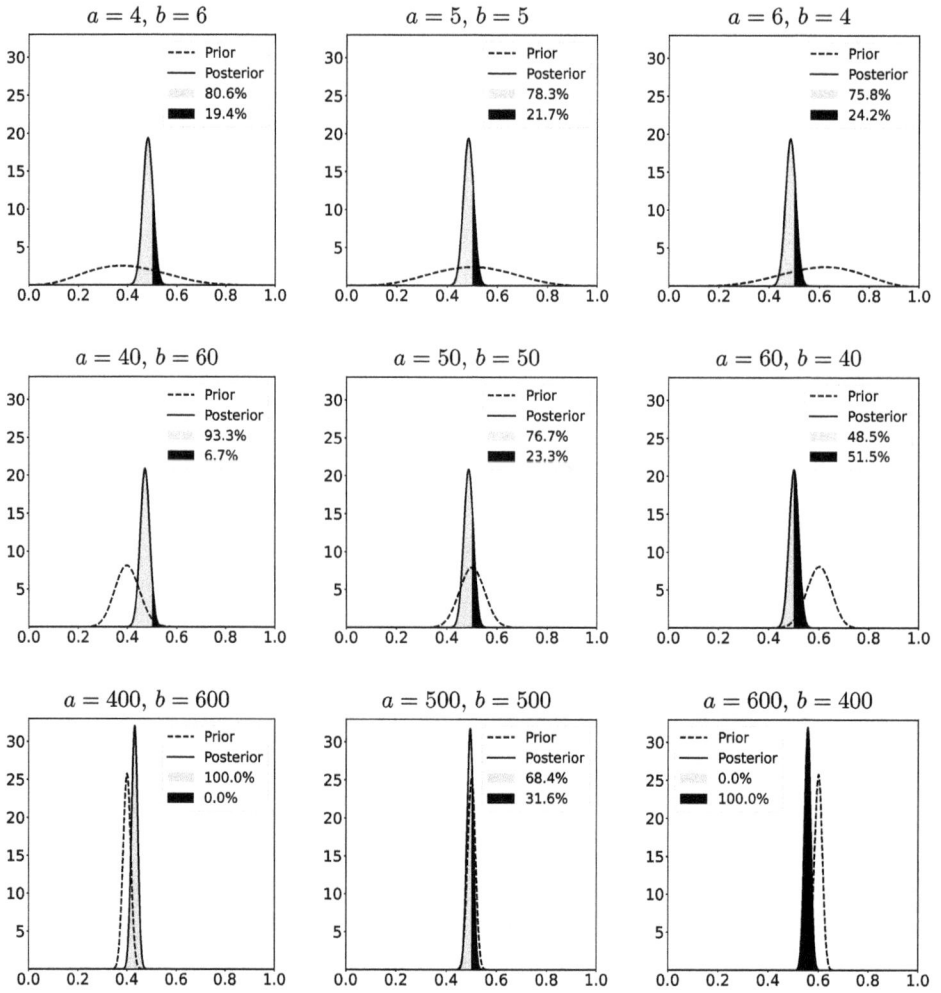

Figure 6.13 Bayesian analysis of poll data for the 2020 US presidential election. The graphs show the prior (dashed line) and posterior (solid line) pdf of the parameter $\tilde{\theta}$ in Example 6.19, which represents the fraction of Trump voters in Pennsylvania. The area of the shaded region under the posterior pdf is the probability that Biden (gray) or Trump (black) win Pennsylvania according to the model. The priors in the left column favor Biden, the ones in the center column are neutral, and the ones in the right column favor Trump. Our confidence in the prior is reflected in the prior pdf. In the top row, it is very spread out, so the posterior pdf is mainly influenced by the data, yielding similar conclusions for all three columns (Biden is likely to win, but not overwhelmingly). By contrast, in the center and especially the bottom row, the prior pdf is more concentrated and has a greater influence on the posterior pdf.

is represented by a random variable $\tilde{e}$, which equals one if the Republican candidate wins more than 25 states:

$$\tilde{e} := \begin{cases} 1 & \text{if } \sum_{i=1}^{51} \tilde{s}_i > 25, \\ 0 & \text{otherwise.} \end{cases} \tag{6.133}$$

The Republican candidate has more support among rural voters, whereas the Democrat has more support among urban voters. Consequently, the relative turnouts of rural and urban voters in each state are crucial. We take a Bayesian perspective and model these quantities as random variables. The rural turnout $\tilde{r}_i$, $1 \leq i \leq 51$, indicates what fraction of the voters that actually vote on election day are rural. We assume that the conditional probability that the Republican candidate wins state i, given that the turnout $\tilde{r}_i$ equals r, where $0 \leq r \leq 1$, is

$$p_{\tilde{s}_i \mid \tilde{r}_i}(1 \mid r) := 0.6r + 0.1(1 - r). \tag{6.134}$$

If the rural turnout is 100%, the Republican candidate wins the state with probability 0.6. If it is 0%, they win with probability 0.1.

We model the rural turnout in each state as a uniform random variable in $[0, 1]$. As a result, by Theorem 6.5 the marginal probability of the Republican candidate winning state i is

$$p_{\tilde{s}_i}(1) = \int_{r=0}^{1} f_{\tilde{r}_i}(r) p_{\tilde{s}_i \mid \tilde{r}_i}(1 \mid r) \, \mathrm{d}r \tag{6.135}$$

$$= \int_{r=0}^{1} (0.6r + 0.1(1 - r)) \, \mathrm{d}r \tag{6.136}$$

$$= 0.35. \tag{6.137}$$

In order to predict the result of the whole election, we need to compute the probability that the total number of states won by the Republican candidate,

$$\tilde{t} := \sum_{i=1}^{51} \tilde{s}_i, \tag{6.138}$$

is greater than 25. This requires modeling the joint distribution of the rural turnouts $\tilde{r}_1, \ldots, \tilde{r}_{51}$. As discussed earlier, we cannot possibly estimate the joint pmf of 51 random variables without making independence assumptions. We consider two options: (1) the rural turnouts are all mutually independent; (2) the rural turnouts are highly dependent. It turns out (no pun intended) that this is a critical decision.

Assumption 1: State Turnouts Are Independent

If the turnouts $\tilde{r}_i$ are all mutually independent for $1 \leq i \leq 51$, then so are the results in each state $\tilde{s}_i$. Under this assumption, the number of states won by the Republican candidate $\tilde{t}$ is a binomial random variable with parameters $n := 51$ and $\theta := 0.35$. By Definition 2.17 and Theorem 2.5, the Republican wins the election with probability

$$p_{\tilde{e}}(1) = \mathrm{P}\left(\tilde{t} > 25\right) \tag{6.139}$$

$$= \sum_{i=26}^{51} p_{\tilde{t}}(i) \tag{6.140}$$

$$= \sum_{i=26}^{51} \binom{51}{i} 0.35^i (1 - 0.35)^{51-i} \tag{6.141}$$

$$= 0.014. \tag{6.142}$$

Under the independence assumption, it is very unlikely for the rural turnout to be simultaneously high in enough states for the Republican candidate to win the election.

Assumption 2: State Turnouts Are Highly Dependent

In the real US presidential election, turnouts in different states are dependent, especially for states that are close geographically or have similar demographics. We consider an extreme case, where the rural turnout is the same in all states and is therefore represented by a single random variable $\tilde{r}$. We model the distribution of $\tilde{r}$ as uniform in $[0, 1]$, so the marginal distribution of the rural turnout is exactly the same as under Assumption 1 and the probability that the Republican wins each individual state is still $p_{\tilde{s}_i}(1) = 0.35$, $1 \leq i \leq 51$.

In order to model the result of the whole election, we need to characterize the conditional joint distribution of $\tilde{s}_1, \ldots, \tilde{s}_{51}$ given $\tilde{r}$. To obtain a tractable model, we assume that these random variables are conditionally independent given the turnout $\tilde{r}$. This implies that the conditional distribution of $\tilde{t}$ given $\tilde{r} = r$ is binomial with parameters $n := 51$ and $\theta := 0.6r + 0.1(1 - r)$. Consequently, by Theorems 6.5 and 2.5, and Definition 2.17, the probability of the Republican candidate winning the election is

$$p_{\tilde{e}}(1) = \int_{r=0}^{1} f_{\tilde{r}}(r) p_{\tilde{e} \mid \tilde{r}}(1 \mid r) \, dr \tag{6.143}$$

$$= \int_{r=0}^{1} f_{\tilde{r}}(r) P\left(\tilde{t} > 25 \mid \tilde{r} = r\right) \, dr \tag{6.144}$$

$$= \int_{r=0}^{1} \sum_{i=26}^{51} \binom{51}{i} p_{\tilde{t} \mid \tilde{r}}(i \mid r) \, dr \tag{6.145}$$

$$= \int_{r=0}^{1} \sum_{i=26}^{51} \binom{51}{i} (0.5r + 0.1)^i (1 - (0.5r + 0.1))^{51-i} \, dr \tag{6.146}$$

$$= 0.204. \tag{6.147}$$

The integral that appears in this expression does not have a closed form (although it can be computed using tabulated beta functions or Euler integrals). In practice, most Bayesian models result in expressions that cannot be solved by hand. Instead, the Monte Carlo method, described in Section 1.7, is applied to compute the probabilities of interest. In our case, this involves the following steps:

1 We generate a sample r of the rural turnout $\tilde{r}$ by simulating a uniform distribution in the unit interval.
2 We generate a sample s_i of the outcome in each state $\tilde{s}_i$, $1 \leq i \leq 51$, by sampling from independent Bernoulli random variables with parameter $0.6r + 0.1(1 - r)$.
3 We check whether $\sum_{i=1}^{51} s_i > 25$, in which case the Republican candidate wins the election.

Our final estimate of $p_{\tilde{e}}(1)$ is the fraction of simulations in which the Republican candidate wins the election. This produces an accurate estimate, as long as we simulate enough elections.

If the turnout is the same everywhere, when the rural turnout is high, it is high for all states simultaneously, which tips the election in favor of the Republican candidate. As a result, the probability of the Republican candidate winning (1 in 5) is much higher than under the independent-turnout assumption (1 in 66). The difference between the two models is evident in the respective pmfs of the number of states won by the Republican candidate, which are shown in Figure 6.14. Ignored dependencies across states is one of the reasons why

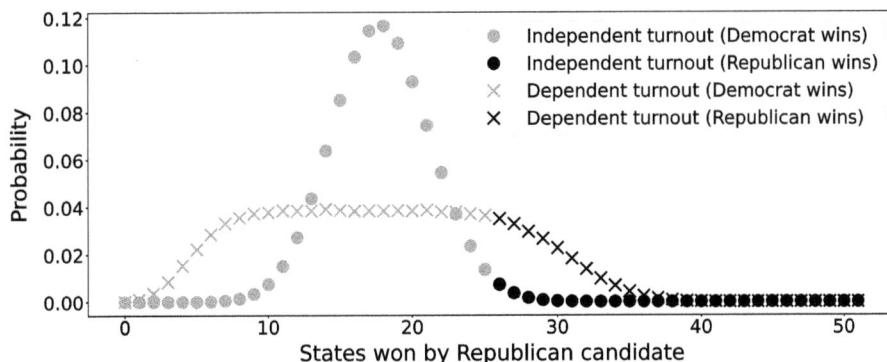

Figure 6.14 Independence assumptions can have a dramatic influence on probabilistic models. The plot shows the pmf of the states won by the Republican candidate in the fictitious election from Section 6.7.3, under two different assumptions about the rural turnouts of different states: (1) mutual independence (circles) and (2) complete dependence (crosses). The probability that the Republican candidate wins more than 25 states, and therefore the whole election, correspond to the sum of the entries highlighted in black. The probability is much lower under the independence assumption.

most models underestimated the probability of Trump winning in 2016. Section 9.7.3 provides another example of the consequences of overly optimistic independence assumptions, inspired by the 2008 financial crisis.

Exercises

6.1 (Shared car) Three friends called Carlos, Dani, and Felix share a car. We model the distance traveled by the car each time it is driven as a random variable $\tilde{d}$. This random variable is uniformly distributed between 0 and x km, where x is a constant that depends on the driver. If Carlos drives, $x := 10$. If Dani drives, $x := 20$. If Felix drives, $x := 30$. We assume that the three friends take the car with the same probability.

   a What is the pdf of $\tilde{d}$?

   b If the last time it was used the car was driven for 15 km, what is the probability that it was driven by Dani?

6.2 (Buckets) A toddler called Azul has two buckets. One is large and can fit 2 kg of sand. The other is small and can fit 1 kg of sand. Azul chooses to play with the big bucket with probability 1/4; otherwise, she plays with the small one. We model her choice of bucket as a Bernoulli random variable $\tilde{b}$, equal to 0, if the bucket is small, and 1, if it is big. Azul grabs an amount of sand $\tilde{s}_1$ that is uniformly distributed between 0 and the capacity of the bucket.

   a What is the probability that Azul grabs less than 1 kg of sand?

   b Azul grabs another amount of sand $\tilde{s}_2$ that has the same distribution as $\tilde{s}_1$ and is conditionally independent from $\tilde{s}_1$ given $\tilde{b}$. If $\tilde{s}_1 = 1/2$ and $\tilde{s}_2 = 3/4$, what is the probability that she chose the large bucket?

   c Now assume that conditioned on $\tilde{s}_1 = s_1$, $\tilde{s}_2$ is uniformly distributed between 0 and s_1. If $\tilde{s}_1 = 1/2$ and $\tilde{s}_2 = 3/4$, what is the probability that she chose the large bucket?

6.3 (Balls) Aarush and Ishan are playing a game. They have two balls that look identical, but one is hollow and weighs much less than the other one, which is solid. Ishan throws one of the balls and Aarush has to guess which one it is. Aarush estimates that when Ishan throws the hollow ball, the ball lands at a distance that is uniformly distributed between 5 and 10 m. By contrast, when he throws the solid ball, the distance is uniformly distributed between 1 and 6 m. He knows that Ishan likes to throw the hollow ball, so he estimates that he will choose that ball with probability 3/4.

    a What is the pdf of the distance traveled by the ball?

    b Ishan throws once and the ball lands at 3 m. If he throws again, what is the conditional pdf of the distance for the second throw given this information, if we assume that both throws are conditionally independent given the type of ball? Derive the conditional pdf mathematically and explain intuitively why the result makes sense.

6.4   (Computer defect) We model the time (in years) until a computer breaks down as a random variable $\tilde{t}$. The time depends on whether the computer has a defect or not, which is modeled by a random variable $\tilde{d}$. If the computer has a defect ($\tilde{d} = 1$), then $\tilde{t}$ is an exponential random variable with parameter 2. If it does not ($\tilde{d} = 0$), then $\tilde{t}$ is an exponential random variable with parameter 1. The probability that the computer has a defect is 0.1.

    a What is the probability that the computer lasts for more than 2 years?

    b If the computer lasts for 2 years, what is the conditional probability that it has a defect, given this information?

6.5   (Wolf) We model the location of a wolf in a national park as a random variable $\tilde{x}$. When it snows, the wolf stays away from certain areas of the park. We define a random variable $\tilde{s}$ to represent whether it snows ($\tilde{s} = 1$) or not ($\tilde{s} = 0$). We assume that it snows one fourth of the time, so $P(\tilde{s} = 1) = 0.25$. The conditional pdfs of $\tilde{x}$ given $\tilde{s}$ are depicted in the graphs underneath. Each density is constant in the corresponding shaded area, and zero elsewhere.

$$f_{\tilde{x}\,|\,\tilde{s}}(x\,|\,0) \qquad\qquad\qquad f_{\tilde{x}\,|\,\tilde{s}}(x\,|\,1)$$

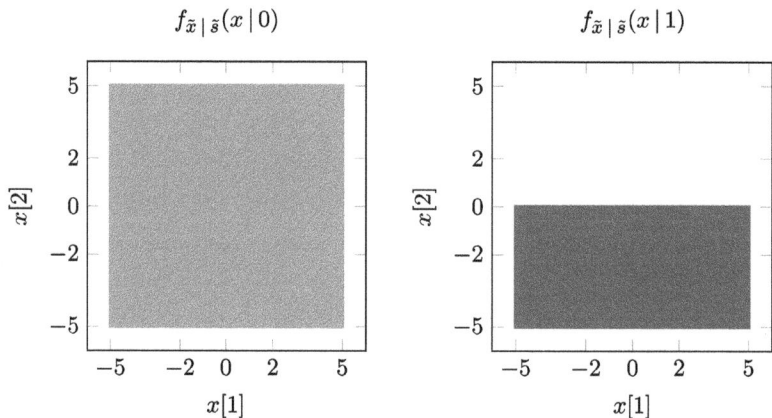

We observe that the position of the wolf is (-1,-1). What is the conditional probability that it is snowing?

6.6   (Potatoes) Your aunt from Idaho asks you to analyze the production of her potato farm. Using data gathered over 45 years, you determine that the yearly production depends mainly on two factors: The weather and the presence of a beetle, which ruins the plants. You model the weather using a random variable $\tilde{w}$, and the presence of the beetle using a random variable $\tilde{b}$. $\tilde{w} = 1$ means good weather and $\tilde{w} = 0$ means bad weather. $\tilde{b} = 1$ means that the beetle is present and $\tilde{b} = 0$ that it is absent. Out of the 45 years, the following table shows how many years had good/bad weather, and in how many the beetle was present.

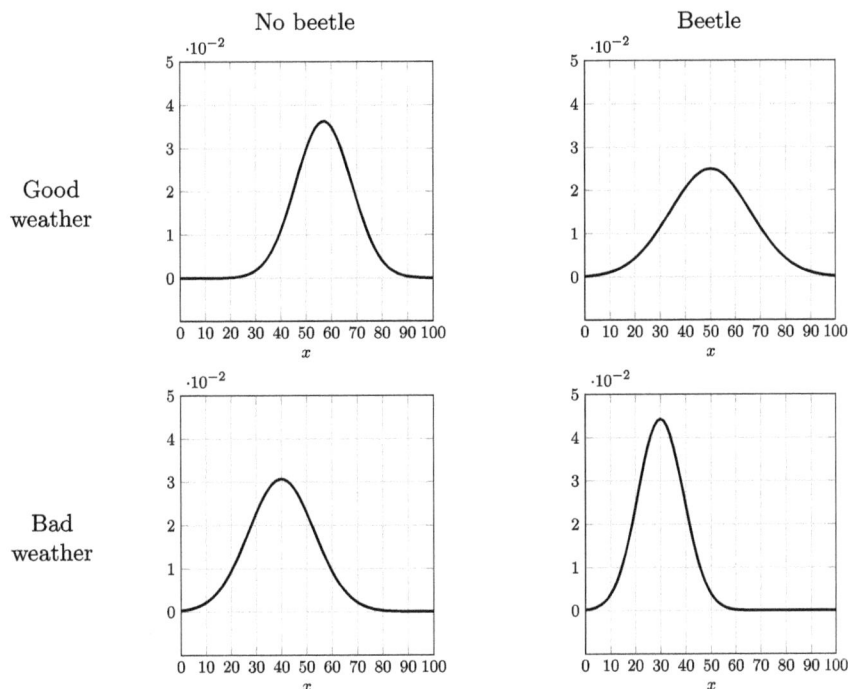

Figure 6.15 Estimated conditional pdf of the potato production in tons given the weather and the presence of the beetle in Exercise 6.6.

	No beetle	Beetle
Good weather	5	10
Bad weather	10	20

You model the potato production during 1 year as a continuous random variable $\tilde{x}$. Figure 6.15 shows the estimated conditional pdf of $\tilde{x}$ given $\tilde{b}$ and $\tilde{w}$, obtained by fitting a Gaussian distribution to the data.

a Is the distribution of $\tilde{x}$ Gaussian?

b Estimate the marginal pmfs of $\tilde{w}$ and $\tilde{b}$.

c According to your model, are $\tilde{w}$ and $\tilde{b}$ independent?

d Fifty years ago, the potato production was 40 tons. There is no data about the beetle, but you find out online that the weather that year was good. Given this information, determine whether it is more likely that the beetle was present or absent.

6.7 (Halloween parade) The city of New York hires you to estimate whether it will rain during the Halloween parade. Based on past data, you determine that the chance of rain is 20%. You model this using a random variable $\tilde{r}$ with pmf

$$p_{\tilde{r}}(1) = 0.2, \qquad p_{\tilde{r}}(0) = 0.8,$$

where $\tilde{r} = 1$ means that it rains and $\tilde{r} = 0$ that it doesn't. Your first idea is to be lazy, and just use the forecast of a certain website, modeled as a random variable $\tilde{w}$. Analyzing data from previous forecasts, you determine that

$$P(\tilde{w} = 1 \mid \tilde{r} = 1) = 0.8, \qquad P(\tilde{w} = 0 \mid \tilde{r} = 0) = 0.75.$$

a What is the probability that the website is wrong?

b Unsatisfied with the accuracy of the website, you look at the data used for the forecast (they are available online). Surprisingly, the relative humidity of the air is not used, so you decide to incorporate it in your prediction in the form of a random variable $\tilde{h}$. Is it more reasonable to assume that $\tilde{h}$ and $\tilde{w}$ are independent or that they are conditionally independent given $\tilde{r}$? Explain why.

c You assume that $\tilde{h}$ and $\tilde{w}$ are conditionally independent given $\tilde{r}$. More research establishes that conditioned on $\tilde{r} = 1$, $\tilde{h}$ is uniformly distributed between 0.5 and 0.7, whereas conditioned on $\tilde{r} = 0$, $\tilde{h}$ is uniformly distributed between 0.1 and 0.6. Compute the conditional pmf of $\tilde{r}$ given $\tilde{w}$ and $\tilde{h}$. Use the distribution to determine whether you would predict rain for every possible value of $\tilde{w}$ and $\tilde{h}$.

d What is the probability that you make a mistake, under your assumptions?

6.8 (Fish) Two species of fish, which we call a and b, live in a river. A biologist called Roma wants to identify fish found in the river. There are roughly the same number of fish a as of fish b, so a reasonable assumption is that the probability that a fish is from species a or b is the same. Roma models the river as a unit interval and the position of the fish as a random variable $\tilde{x}$. She estimates the conditional pdfs of $\tilde{x}$ given the species $\tilde{s}$ of the fish and determines that they equal $2 - 2x$ (given $\tilde{s} = a$) and $2x$ (given $\tilde{s} = b$) when $0 \leq x \leq 1$, and zero otherwise. The pdfs are shown in the following two graphs.

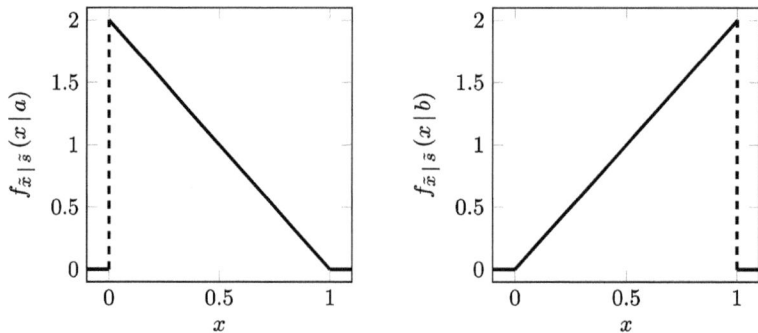

a Compute the pdf of $\tilde{x}$.

b If Roma finds a fish at position 0.25, what is the probability that it belongs to species b?

c Compute the conditional cdf of $\tilde{x}$ given the species of the fish.

d Apply inverse transform sampling to simulate a sample from the joint distribution of the species and the position of the fish, using the following two independent samples obtained from a uniform distribution: 0.8, 0.64.

6.9 (Samples and conditional independence) Figure 6.16 shows scatterplots of samples drawn independently at random from three different joint distributions of a discrete random variable $\tilde{c}$ and two continuous random variables $\tilde{a}$ and $\tilde{b}$. The joint distribution of $\tilde{a}$ and $\tilde{b}$ is exactly the same as in the corresponding plot of Exercise 5.5. In the graphs a, b, and c are the sampled values of $\tilde{a}$, $\tilde{b}$, and $\tilde{c}$, respectively. For each of the plots, explain whether you think that $\tilde{a}$ and $\tilde{b}$ are conditionally independent given $\tilde{c}$ and why.

6.10 (Chad) You find your coworker Chad really annoying. He often works from home, but when he is in the office and you walk by his desk, he insists on showing you pictures of his pet iguana. You would like to be able to predict when he is in the office, in order to avoid him as much as possible. Another of Chad's annoying habits is to crank up the AC, so you decide to use the temperature in the office to predict his presence. After a month you have gathered the following data.

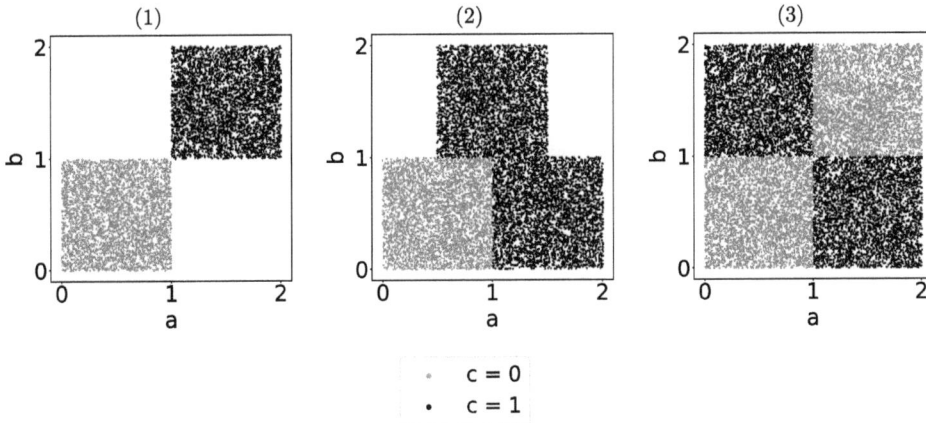

Figure 6.16 Plots for Exercise 6.9.

Chad	61	65	59	61	61	65	61	63	63	59
No Chad	68	70	68	64	64	–	–	–	–	–

Temperature (^{o} F)

a You model the temperature using a random variable $\tilde{t}$. Use kernel density estimation, with a rectangular kernel of width 2, to estimate the conditional pdf of $\tilde{t}$ given the presence or absence of Chad. Sketch the conditional pdfs.

b We model the presence of Chad using a parameter c ($c = 1$ means he is present, $c = 0$ that he is absent). If the temperature is $68°$, does a maximum-likelihood estimate of c predict that Chad is in the office?

c Under our probabilistic model, what is the probability that Chad is in the office?

d Now we take a Bayesian approach and model the presence or absence of Chad using a random variable $\tilde{c}$, which is equal to 1 if he is there and 0 if he is not. Estimate the prior pmf of $\tilde{c}$ from the data.

e If the temperature is $64°$, use the posterior distribution of $\tilde{c}$ to compute the probability that Chad is in the office.

f What problem do we run into if the temperature is $57°$? Explain how using parametric estimation could address this issue.

6.11 (Mixture model with fixed mean) Let $\tilde{d}$ be a Bernoulli random variable with parameter θ and let $\tilde{c}$ be a continuous random variable defined on the same probability space. The conditional distributions of $\tilde{c}$ given $\tilde{d} = 0$ and $\tilde{d} = 1$ are both Gaussian with zero mean and standard-deviation parameters σ_0 and σ_1, respectively. Derive the conditional pmf of $\tilde{d}$ given $\tilde{c}$. Plot $p_{\tilde{d} \mid \tilde{c}}(1 \mid c)$ as a function of c for $\theta := 0.5$, $\sigma_0 := 1$, and $\sigma_1 := 0.5$.

6.12 (K-means and Gaussian mixture models) In k-means clustering, the data are clustered into k clusters by repeating the following two steps until convergence: (1) Each data point is assigned the cluster whose center is closest (in ℓ_2 norm). (2) The center of each cluster is updated by averaging the entries of the data assigned to it. This method, called Lloyd's algorithm, is similar to the expectation-maximization method for fitting Gaussian mixture models. What are the main differences?

6.13 (Alternative model for coin flip) In this exercise, we consider another prior for the coin flip in Example 6.16. We model the parameter of the Bernoulli random variable $\tilde{\theta}$ as a uniform random variable in the interval $[0.5, 1]$.

    a Provide a possible brief justification for the prior and compute the probability that the result of the coin flip is heads or tails under this model.

    b Compute the posterior pdf of $\tilde{\theta}$ if we observe tails, and if we observe heads. Sketch the conditional pdfs.

    c You observe 100 coin flips and they all turn out to be tails. Do you think you should reconsider your prior? If so, why?

6.14 (Two coin flips) In this exercise, we again model a coin using a Bayesian approach, as in Example 6.16. The probability of heads is represented by a single random variable $\tilde{\theta}$, which is uniformly distributed in $[0, 1]$. We flip the coin twice. Given $\tilde{\theta} = \theta$, the coin flips are modeled as Bernoulli random variables $\tilde{x}_1$ and $\tilde{x}_2$ with parameter θ. For example, $\tilde{x}_1 = 1$ and $\tilde{x}_2 = 0$ indicate that the first flip is heads and the second tails. We assume that $\tilde{x}_1$ and $\tilde{x}_2$ are conditionally independent given $\tilde{\theta}$.

    a Compute the joint pmf of $\tilde{x}_1$ and $\tilde{x}_2$.

    b We observe that the two coin flips are heads. What is the conditional probability that $\tilde{\theta} < 1/2$, given this event?

6.15 (PT Cruiser) Idoia has an old PT Cruiser. She is interested in modeling the probability that it breaks down and cannot be used anymore. She takes a Bayesian perspective and models the probability that the car breaks down each time she uses it as a random variable $\tilde{\theta}$, which is uniformly distributed in the unit interval. More precisely, given $\tilde{\theta} = \theta$, the event that the car breaks down has probability θ and is conditionally independent of whatever happened before. If the car breaks down the tth time she uses it, compute the posterior pdf of $\tilde{\theta}$. How does the shape change as a function of t?

7

Averaging

Overview

Averaging is a fundamental operation in probability and statistics. In Section 7.1, we define an averaging procedure for random variables, known as the mean. Section 7.2 shows that the mean is linear, and Section 7.3 shows that the mean of the product of independent random variables is equal to the product of their means. In Section 7.4, we derive the mean of several popular distributions. In Section 7.5, we explain that the mean can be distorted by extreme values, unlike the median. Sections 7.6 defines the mean square, which represents the average squared value of a random variable. Section 7.7 defines the variance, which quantifies the average variation of a random variable. In Section 7.8, we define the conditional mean, discussing its connection to regression. Finally, in Section 7.9, we discuss the estimation of causal effects using conditional averages.

7.1 The Mean

In this section, we define an averaging operation for random quantities, known as the mean. Sections 7.1.1 and 7.1.2 define the mean for discrete and continuous random variables, respectively. In Section 7.1.3, we explain how to compute the mean of quantities that depend on both discrete and continuous random variables. Finally, Section 7.1.4 describes how to estimate the mean from data.

7.1.1 Discrete Random Variables

Intuitively, the mean of a discrete random variable $\tilde{a}$ is the value we obtain, when we average many realizations or samples of $\tilde{a}$. Let $x_1, \ldots, x_n$ be n such samples, which consequently belong to the range A of $\tilde{a}$. Their arithmetic average equals

$$\frac{1}{n} \sum_{i=1}^{n} x_i = \sum_{a \in A} a \cdot \frac{\text{number of samples equal to } a}{n} \tag{7.1}$$

because every sample must equal some element of A. According to our intuitive definition of probability (1.1), we can interpret the probability $\mathrm{P}\left(\tilde{a} = a\right)$ as the number of times that the random variable $\tilde{a}$ is equal to a, when we repeat an experiment representing the uncertain phenomenon modeled by the underlying probability space many times. If we assume that the samples in (7.1) are the result of such repetitions and n is very large, then

$$\mathrm{P}(\tilde{a} = a) \approx \frac{\text{number of samples equal to } a}{n}. \tag{7.2}$$

Plugging (7.2) into (7.1) yields a formula for the average of many samples from the distribution of $\tilde{a}$, in terms of the entries of its pmf:

$$\frac{1}{n}\sum_{i=1}^{n} x_i \approx \sum_{a \in A} a\mathrm{P}(\tilde{a} = a) \tag{7.3}$$

$$= \sum_{a \in A} a p_{\tilde{a}}(a). \tag{7.4}$$

This averaging procedure, known as the *expectation* operator, maps the random variable $\tilde{a}$ to a single number, which we call the *mean* or *expected value* of $\tilde{a}$.

Definition 7.1 (Mean of a discrete random variable). *The mean, expected value or first moment of a discrete random variable $\tilde{a}$ with range A and pmf $p_{\tilde{a}}$ is*

$$\mathbb{E}\left[\tilde{a}\right] := \sum_{a \in A} a p_{\tilde{a}}\left(a\right) \tag{7.5}$$

if the sum converges.

It is possible for a random variable to have an infinite mean, if the sum in Definition 7.1 tends to infinity, or to have no mean at all, when the sum is not well defined. We provide examples of such random variables and discuss the implications in Section 9.6.

Lemma 7.2 (Mean of a Bernoulli random variable). *The mean of a Bernoulli random variable $\tilde{a}$ with parameter θ is $\mathbb{E}[\tilde{a}] = \theta$.*

Proof By Definitions 7.1 and 2.15,

$$\mathbb{E}\left[\tilde{a}\right] = 0 \cdot p_{\tilde{a}}\left(0\right) + 1 \cdot p_{\tilde{a}}\left(1\right) \tag{7.6}$$

$$= \theta. \tag{7.7}$$

■

Example 7.3 (Expected goals in soccer game). In Example 2.7, we consider a discrete random variable $\tilde{g}$ that represents the goal difference in a soccer game between Barcelona and Atlético de Madrid. The pmf of $\tilde{g}$ equals $p_{\tilde{g}}(-2) := 0.1$, $p_{\tilde{g}}(-1) := 0.2$, $p_{\tilde{g}}(0) := 0.3$, $p_{\tilde{g}}(1) := 0.25$, $p_{\tilde{g}}(2) := 0.1$, $p_{\tilde{g}}(3) := 0.05$. By Definition 7.1, the mean of $\tilde{g}$ is

$$\mathbb{E}[\tilde{g}] := \sum_{g=-2}^{3} g p_{\tilde{g}}\left(g\right) \tag{7.8}$$

$$= -2 \cdot 0.1 - 1 \cdot 0.2 + 0 \cdot 0.3 + 1 \cdot 0.25 + 2 \cdot 0.1 + 3 \cdot 0.05 \tag{7.9}$$

$$= 0.2. \tag{7.10}$$

In probabilistic modeling, we are often interested in the behavior of functions of random variables. Imagine that we want to compute the mean of $h(\tilde{a})$, where h is a deterministic function and $\tilde{a}$ is a discrete random variable. Consider applying h to n samples $x_1, \ldots, x_n$ generated according to the distribution of $\tilde{a}$ for large n, and then averaging the transformed samples. If we interpret the samples as in our intuitive derivation of the mean in (7.4), the arithmetic average of the transformed samples approximately equals

$$\frac{1}{n}\sum_{i=1}^{n} h(x_i) = \sum_{a \in A} h(a) \cdot \frac{\text{number of samples equal to } a}{n} \tag{7.11}$$

$$\approx \sum_{a \in A} h(a)p_{\tilde{a}}(a), \tag{7.12}$$

where A is the range of $\tilde{a}$. This motivates the following generalization of Definition 7.1.

Definition 7.4 (Mean of a function of a discrete random variable). *Let $\tilde{a}$ be a discrete random variable with range A and pmf $p_{\tilde{a}}$ and let h be a deterministic function. Then,*

$$\mathbb{E}\left[h(\tilde{a})\right] := \sum_{a \in A} h(a)p_{\tilde{a}}(a). \tag{7.13}$$

The mean of a function of a discrete random variable can also be computed by first deriving the pmf of the transformed variable and then applying Definition 7.1. Exercise 7.2 proves that this yields the same result.

Example 7.5 (Expected points in soccer game). In Example 2.9, we derive the distribution of points $\tilde{x} := h(\tilde{g})$ corresponding to the goal difference $\tilde{g}$ from Example 7.3. A win is worth three points, a draw one point, and a loss zero points, so

$$h(g) := \begin{cases} 0 & \text{if } g < 0, \\ 1 & \text{if } g = 0, \\ 3 & \text{if } g > 0. \end{cases} \tag{7.14}$$

By Definition 7.4, we can compute the mean of the points without deriving the pmf of $\tilde{x}$:

$$\mathbb{E}[\tilde{x}] = \mathbb{E}\left[h(\tilde{g})\right] \tag{7.15}$$

$$= \sum_{g=-2}^{3} h(g)p_{\tilde{g}}(g) \tag{7.16}$$

$$= 0 \cdot 0.1 + 0 \cdot 0.2 + 1 \cdot 0.3 + 3 \cdot 0.25 + 3 \cdot 0.1 + 3 \cdot 0.05 \tag{7.17}$$

$$= 1.5. \tag{7.18}$$

We can extend our intuitive definition of the mean to functions of multiple discrete random variables. For instance, let us consider a bivariate function $h \colon \mathbb{R}^2 \to \mathbb{R}$ applied to n pairs of samples $(x_1, y_1), (x_2, y_2), \ldots, (x_n, y_n)$. We assume that each pair of samples has the same joint distribution as the random variables $\tilde{a}$ and $\tilde{b}$, with respective ranges A and B. Then, under the assumptions of our intuitive derivation in (7.4), for large n we have

$$\frac{1}{n}\sum_{i=1}^{n} h(x_i, y_i) = \sum_{a \in A}\sum_{b \in B} h(a, b) \cdot \frac{\text{number of pairs } (x, y) \text{ such that } x = a \text{ and } y = b}{n}$$

$$\approx \sum_{a \in A}\sum_{b \in B} h(a, b)p_{\tilde{a}, \tilde{b}}(a, b). \tag{7.19}$$

The same logic applies to more than two random variables and to the entries of a d-dimensional random vector.

Definition 7.6 (Mean of a function of multiple discrete random variables). *If $\tilde{a}$ and $\tilde{b}$ are discrete random variables with ranges A and B, the mean or expected value of a function $h(\tilde{a}, \tilde{b})$ of $\tilde{a}$ and $\tilde{b}$, $h \colon \mathbb{R}^2 \to \mathbb{R}$, is*

Table 7.1 **Cats and dogs.** *Joint pmf* $p_{\tilde{c}, \tilde{d}}(c, d)$ *of the random variables $\tilde{c}$ and $\tilde{d}$ in Example 7.7.*

| | | \multicolumn{4}{c}{Cats (c)} |
		0	**1**	**2**	**3**
Dogs (d)	**0**	0.35	0.15	0.1	0.05
	1	0.2	0.05	0.03	0
	2	0.05	0.02	0	0

$$\mathbb{E}[h(\tilde{a}, \tilde{b})] := \sum_{a \in A} \sum_{b \in B} h(a, b)\, p_{\tilde{a}, \tilde{b}}(a, b) \tag{7.20}$$

if the sum converges.

Let $\tilde{x}$ *be a d-dimensional discrete random vector, where the ith entry $\tilde{x}[i]$, $1 \le i \le d$, is a random variable with range R_i. The mean or expected value of a function $h(\tilde{x})$ of $\tilde{x}$, $h : \mathbb{R}^d \to \mathbb{R}$, is*

$$\mathbb{E}[h(\tilde{x})] := \sum_{x[1] \in R_1} \sum_{x[2] \in R_2} \cdots \sum_{x[d] \in R_d} h(x)\, p_{\tilde{x}}(x) \tag{7.21}$$

if the sum converges.

Example 7.7 (Cats and dogs). A producer of pet food wants to compute the expected total number of cats and dogs per household in a certain city. Table 7.1 shows the joint pmf of the cats and dogs, represented by the random variables $\tilde{c}$ and $\tilde{d}$. The expected number of pets is the mean of the sum $\tilde{c} + \tilde{d}$. By Definition 7.6, it equals

$$\mathbb{E}[\tilde{c} + \tilde{d}] = \sum_{c=0}^{3} \sum_{d=0}^{2} (c + d) p_{\tilde{c}, \tilde{d}}(c, d) \tag{7.22}$$

$$= 0.15 + 2 \cdot 0.1 + 3 \cdot 0.05 + 0.2 + 2 \cdot 0.05 + 3 \cdot 0.03 + 2 \cdot 0.05 + 3 \cdot 0.02$$

$$= 1.05. \tag{7.23}$$

7.1.2 Continuous Random Variables

Just as in the discrete case, we motivate the definition of the mean of a continuous random variable $\tilde{a}$ by analyzing the average of many samples of $\tilde{a}$. Let $x_1, \ldots, x_n$ be n real-valued numbers representing such samples. We cannot reason about the probability that the samples equal specific values because that probability is always zero (see Section 3.1). Instead, we partition the real line into intervals $(a_m - \epsilon, a_m]$ of length ϵ, where m is an integer and $a_m := m\epsilon$, so that the intervals are disjoint and cover the whole real line.

Since each sample follows the distribution of $\tilde{a}$, the probability that it belongs to the interval $(a_m - \epsilon, a_m]$ is $\mathrm{P}(a_m - \epsilon < \tilde{a} \le a_m)$. For small ϵ, this probability is close to $f_{\tilde{a}}(a_m)\epsilon$, where $f_{\tilde{a}}$ is the pdf of $\tilde{a}$, by (3.56). Consequently, if we interpret the samples according to our intuitive definition of probability (1.1), for very small ϵ and very large n,

$$\frac{1}{n} \sum_{i=1}^{n} x_i \approx \sum_{m \in \mathbb{Z}} \frac{a_m \cdot \text{number of samples between } a_m - \epsilon \text{ and } a_m}{n} \tag{7.24}$$

$$\approx \sum_{m \in \mathbb{Z}} a_m \mathrm{P}\left(a_m - \epsilon < \tilde{a} \le a_m\right) \tag{7.25}$$

$$\approx \sum_{m \in \mathbb{Z}} a_m f_{\tilde{a}}(a_m)\epsilon. \tag{7.26}$$

If we take the limit $\epsilon \to 0$, where the length of each interval tends to zero, then the Riemann sum (7.26) converges to the integral $\int_{a \in \mathbb{R}} a f_{\tilde{a}}(a) \, \mathrm{d}a$. This motivates our definition of the mean of a continuous random variable. A similar argument can be used to motivate our definition of the mean of a function of a continuous random variable.

Definition 7.8 (Mean of a continuous random variable). *The mean, expected value or first moment of a continuous random variable $\tilde{a}$ is defined as*

$$\mathbb{E}\left[\tilde{a}\right] := \int_{a=-\infty}^{\infty} a f_{\tilde{a}}(a) \, \mathrm{d}a \tag{7.27}$$

if the integral converges.

Let $\tilde{a}$ be a continuous random variable with pdf $f_{\tilde{a}}$ and let $h \colon \mathbb{R} \to \mathbb{R}$ be a deterministic function, such that $h(\tilde{a})$ is a valid random variable. Then,

$$\mathbb{E}\left[h(\tilde{a})\right] := \int_{a \in \mathbb{R}} h(a) f_{\tilde{a}}(a) \, \mathrm{d}a \tag{7.28}$$

if the integral converges.

As in the discrete case, it is possible for a continuous random variable to have an infinite mean, if the integral in Definition 7.8 tends to infinity. Similarly, a continuous random variable does not have a mean if the integral is not well defined. We provide examples and discuss the implications in Section 9.6.

The mean of a uniform random variable is the midpoint of the interval where its density is nonzero, so it is equal to its median. For most distributions, this is not the case. The mean and median of a random variable can be very different, as we discuss in Section 7.5.

Lemma 7.9 (Mean of a uniform random variable). *The mean of random variable $\tilde{u}$ that is uniformly distributed in the interval $[a, b]$, $b > a$, equals*

$$\mathbb{E}[\tilde{u}] = \frac{a+b}{2}. \tag{7.29}$$

Proof By Definitions 7.8 and 3.17,

$$\mathbb{E}\left[\tilde{u}\right] = \int_{u=-\infty}^{\infty} u f_{\tilde{a}}(u) \, \mathrm{d}u \tag{7.30}$$

$$= \int_{u=a}^{b} \frac{u}{b-a} \, \mathrm{d}u \tag{7.31}$$

$$= \frac{b^2 - a^2}{2(b-a)} \tag{7.32}$$

$$= \frac{a+b}{2}. \tag{7.33}$$

∎

Example 7.10 (Expected power). In Example 3.21, we explain how to derive the pdf of the power $\tilde{w}$ dissipated in a resistor, as a function of the pdf of the current $\tilde{c}$ passing through it. However, if we are only interested in the mean power, we can compute it directly from the pdf $f_{\tilde{c}}$ of $\tilde{c}$ by applying (7.28) because the power is a deterministic function of the current: $\tilde{w} = r\tilde{c}^2$, where r denotes the resistance of the resistor. For instance, if the current is uniformly distributed between -1 and 1 amperes,

$$\mathbb{E}[\tilde{w}] = \mathbb{E}\left[r\tilde{c}^2\right] \tag{7.34}$$

$$= \int_{c=-1}^{1} \frac{rc^2}{2} \, dc \tag{7.35}$$

$$= \frac{r}{3}. \tag{7.36}$$

We now generalize Definition 7.8 to define the mean of a function of multiple continuous random variables. The definition can be justified following a similar argument to (7.26).

Definition 7.11 (Mean of a function of multiple continuous random variables). *If $\tilde{a}$ and $\tilde{b}$ are continuous random variables with joint pdf $f_{\tilde{a},\tilde{b}}$, the mean or expected value of a function $h(\tilde{a}, \tilde{b})$, $h \colon \mathbb{R}^2 \to \mathbb{R}$, of $\tilde{a}$ and $\tilde{b}$ is*

$$\mathbb{E}[h(\tilde{a}, \tilde{b})] := \int_{a=-\infty}^{\infty} \int_{b=-\infty}^{\infty} h(a, b) \, f_{\tilde{a},\tilde{b}}(a, b) \, db \, da, \tag{7.37}$$

if the integral converges.

If $\tilde{x}$ is a d-dimensional continuous random vector, the mean or expected value of a function $h(\tilde{x})$, $h \colon \mathbb{R}^d \to \mathbb{R}$, of $\tilde{x}$ is

$$\mathbb{E}[h(\tilde{x})] := \int_{x[1] \in \mathbb{R}} \int_{x[2] \in \mathbb{R}} \cdots \int_{x[d] \in \mathbb{R}} h(x) \, f_{\tilde{x}}(x) \, dx \tag{7.38}$$

if the integral converges.

Example 7.12 (Sugar). Anselm puts his hand in a bag that has 1 kg of sugar and grabs an amount of sugar that is uniformly distributed between 0 and 1 kg. While transferring the sugar to another bag, he spills an amount that is uniformly distributed between 0 and the amount that he grabbed. What is the expected amount of spilled sugar?

We represent the grabbed sugar by a random variable $\tilde{g}$ and the spilled sugar by a random variable $\tilde{s}$. Setting $h(\tilde{g}, \tilde{s}) := \tilde{s}$ in Definition 7.11, by our assumptions and the chain rule for continuous random variables (Theorem 5.12),

$$\mathbb{E}[\tilde{s}] = \int_g \int_s s f_{\tilde{g},\tilde{s}}(g, s) \, ds \, dg \tag{7.39}$$

$$= \int_g \int_s s f_{\tilde{g}}(g) f_{\tilde{s}\,|\,\tilde{g}}(s\,|\,g) \, ds \, dg \tag{7.40}$$

$$= \int_{g=0}^{1} \int_{s=0}^{g} \frac{s}{g} \, ds \, dg \tag{7.41}$$

$$= \int_{g=0}^{1} \frac{g}{2} \, dg \tag{7.42}$$

$$= \frac{1}{4}. \tag{7.43}$$

7.1.3 Discrete and Continuous Random Variables

To compute the mean of a quantity that depends on a discrete and continuous random variable, we use their marginal and conditional distributions. The definition can be justified following a similar argument to our motivation of Definitions 7.6 and 7.8.

Definition 7.13 (Mean of a discrete and a continuous random variable). *If $\tilde{c}$ is a continuous random variable and $\tilde{d}$ a discrete random variable with range D defined on the same probability space, the mean or expected value of a function $h(\tilde{c}, \tilde{d})$ of $\tilde{c}$ and $\tilde{d}$ is*

$$\mathbb{E}\left[h(\tilde{c}, \tilde{d})\right] := \int_{c=-\infty}^{\infty} \sum_{d \in D} h(c, d) f_{\tilde{c}}(c) p_{\tilde{d}|\tilde{c}}(d \mid c) \, dc \qquad (7.44)$$

$$= \sum_{d \in D} \int_{c=-\infty}^{\infty} h(c, d) p_{\tilde{d}}(d) f_{\tilde{c}|\tilde{d}}(c \mid d) \, dc \qquad (7.45)$$

if the sum and integral converge. The equality follows from Theorem 6.6.

Example 7.14 (Mean of a Bayesian coin flip). In Example 6.16, we model the result of a coin flip $\tilde{x}$ as a Bernoulli random variable ($\tilde{x} = 1$ is heads and $\tilde{x} = 0$ is tails) with a random parameter $\tilde{\theta}$. Let us compute the mean of $\tilde{x}$ under the two different priors for $\tilde{\theta}$ described in Example 6.16. If $\tilde{\theta}$ is uniformly distributed in $[0, 1]$, by Definition 7.13,

$$\mathbb{E}[\tilde{x}] = \int_{\theta=-\infty}^{\infty} \sum_{x=0}^{1} x f_{\tilde{\theta}}(\theta) p_{\tilde{x}|\tilde{\theta}}(x \mid \theta) \, d\theta \qquad (7.46)$$

$$= \int_{0}^{1} \theta \, d\theta \qquad (7.47)$$

$$= \frac{1}{2}. \qquad (7.48)$$

If $\tilde{\theta}$ has a triangular pdf $f_{\tilde{\theta}}(\theta) = 2\theta$, $\theta \in [0, 1]$, then

$$\mathbb{E}[\tilde{x}] = \int_{\theta=-\infty}^{\infty} \sum_{x=0}^{1} x f_{\tilde{\theta}}(\theta) p_{\tilde{x}|\tilde{\theta}}(x \mid \theta) \, d\theta \qquad (7.49)$$

$$= \int_{0}^{1} 2\theta^2 \, d\theta \qquad (7.50)$$

$$= \frac{2}{3}. \qquad (7.51)$$

7.1.4 The Sample Mean

As explained in Sections 7.1.1, 7.1.2, and 7.1.3, the mean of a random variable can be interpreted intuitively as an average of samples that follow the same distribution as the random variable. Consequently, if we want to approximate the mean of a random variable using data, it makes sense to estimate it via averaging. This estimator is known as the *sample mean* because it is a mean computed from samples.

Definition 7.15 (Sample mean). *Let $X := \{x_1, x_2, \ldots, x_n\}$ denote a real-valued dataset. The sample mean is the arithmetic average*

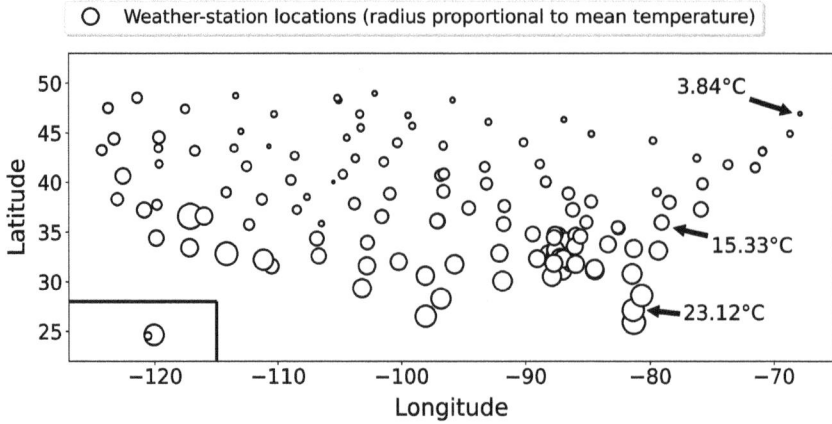

Figure 7.1 Mean temperature in the United States. The graph provides a visualization of the mean temperature at 134 weather stations in the United States in 2015. The bottom left corner shows the two stations in Hawaii (one is in Mauna Loa at high altitude, so its mean temperature is low). The radius of each circular marker is proportional to the mean temperature. The mean temperature increases from north to south. The mean temperatures at Durham (North Carolina), Limestone (Maine) and Sebring (Florida) are included for reference.

$$m(X) := \frac{1}{n} \sum_{i=1}^{n} x_i. \tag{7.52}$$

Note that although the mathematical formula for the mean changes depending on whether the random variables involved are discrete (Definition 7.1) or continuous (Definition 7.8), *the sample mean is always the same*. We just average the data. In Section 9.5, we show that the sample mean provides an accurate estimate of the mean for distributions with finite variance, as long as it is computed from independent samples.

Example 7.16 (Mean temperature). Figure 7.1 shows the sample means of hourly temperatures measured at 134 weather stations in the United States in 2015, extracted from Dataset 9. The means provide an effective summary of the data, revealing that it is colder at higher latitudes and warmer at lower latitudes.
....

7.2 Linearity of Expectation

The mean is a sum or an integral weighted by probabilities or densities of probability, respectively. As a result, it is linear, as illustrated by the following example.

Example 7.17 (Cost of a latte). The owner of a café wants to estimate the mean cost of a latte. She models the price of one kilogram of coffee as a random variable $\tilde{c}$ with a mean of 2.5 dollars, and the price of a gallon of milk as a random variable $\tilde{m}$ with a mean of 3.5 dollars. A latte in her café has 0.02 kg of coffee and 0.1 gallons of milk, so by Definition 7.11 and Theorem 5.8,

$$\mathbb{E}[\tilde{\ell}] = \mathbb{E}[0.02\tilde{c} + 0.1\tilde{m}] \tag{7.53}$$

$$= \int_{c \in \mathbb{R}} \int_{m \in \mathbb{R}} (0.02c + 0.1m) f_{\tilde{c}, \tilde{m}}(c, m) \, \mathrm{d}c \, \mathrm{d}m \tag{7.54}$$

$$= 0.02 \int_{c \in \mathbb{R}} c \int_{m \in \mathbb{R}} f_{\tilde{c},\tilde{m}}(c,m) \; \mathrm{d}m \; \mathrm{d}c + 0.1 \int_{m \in \mathbb{R}} m \int_{c \in \mathbb{R}} f_{\tilde{c},\tilde{m}}(c,m) \; \mathrm{d}c \; \mathrm{d}m$$

$$= 0.02 \int_{c \in \mathbb{R}} c f_{\tilde{c}}(c) \; \mathrm{d}c + 0.1 \int_{m \in \mathbb{R}} m f_{\tilde{m}}(m) \; \mathrm{d}m \tag{7.55}$$

$$= 0.02 \, \mathbb{E}[\tilde{c}] + 0.1 \, \mathbb{E}[\tilde{m}] \tag{7.56}$$

$$= 0.4. \tag{7.57}$$

The expected cost is 40 cents.

In Example 7.17, we show that $\mathbb{E}[0.02\tilde{c} + 0.1\tilde{m}] = 0.02 \, \mathbb{E}[\tilde{c}] + 0.1 \, \mathbb{E}[\tilde{m}]$, i.e. that the mean of the linear combination of the random variables is equal to the linear combination of their respective means. This is a fundamental property called *linearity of expectation*, which we use all the time in probabilistic modeling.

Theorem 7.18 (Linearity of expectation). *For any constants* $c_1, c_2 \in \mathbb{R}$, *any functions* $h_1, h_2 \colon \mathbb{R}^2 \to \mathbb{R}$ *and any continuous or discrete random variables* $\tilde{a}$ *and* $\tilde{b}$,

$$\mathbb{E}\left[c_1 \, h_1(\tilde{a}, \tilde{b}) + c_2 \, h_2(\tilde{a}, \tilde{b}) \right] = c_1 \, \mathbb{E}\left[h_1(\tilde{a}, \tilde{b}) \right] + c_2 \, \mathbb{E}\left[h_2(\tilde{a}, \tilde{b}) \right]. \tag{7.58}$$

Proof The theorem follows immediately from the linearity of sums and integrals, just as in Example 7.17. ■

Linearity of expectation makes it possible to compute the mean of linear functions of random variables, without having to derive their joint pdf or pmf, which is usually much more complicated. We illustrate this by deriving the mean of a binomial random variable.

Lemma 7.19 (Mean of a binomial random variable). *The mean of a binomial random variable* $\tilde{a}$ *with parameters* n *and* θ *equals* $\mathbb{E}[\tilde{a}] = n\theta$.

Proof The binomial random variable $\tilde{a}$ can be represented as a function of n independent Bernoulli random variables $\tilde{b}_1, \ldots, \tilde{b}_n$ with parameter θ. Specifically, $\tilde{a}$ is equal to the number of Bernoulli variables that equal one, or equivalently to their sum. To prove this formally, we can apply the exact same reasoning as in Example 2.16, representing each coin flip as a Bernoulli random variable, which equals one if the coin flip is heads, so that their sum is equal to the number of heads. By linearity of expectation and Lemma 7.2,

$$\mathbb{E}[\tilde{a}] = \mathbb{E}\left[\sum_{k=1}^{n} \tilde{b}_k \right] \tag{7.59}$$

$$= \sum_{k=1}^{n} \mathbb{E}[\tilde{b}_k] \tag{7.60}$$

$$= n\theta. \tag{7.61}$$

■

7.3 Independent Random Variables

If two random variables are independent, then the mean of their product equals the product of their means.

Theorem 7.20 (Mean of the product of independent random variables). *If $\tilde{a}$ and $\tilde{b}$ are independent random variables defined on the same probability space, and $g, h \colon \mathbb{R} \to \mathbb{R}$ are deterministic real-valued functions, then*

$$\mathbb{E}\left[g\left(\tilde{a}\right) h(\tilde{b})\right] = \mathbb{E}\left[g\left(\tilde{a}\right)\right] \mathbb{E}[h(\tilde{b})], \tag{7.62}$$

as long as all the expected values in the expression are well defined.

Proof If $\tilde{a}$ and $\tilde{b}$ are continuous, then the joint pdf factorizes into the product of the marginals (see Definition 5.15), so by Definition 7.11,

$$\mathbb{E}\left[g\left(\tilde{a}\right) h(\tilde{b})\right] = \int_{a=-\infty}^{\infty} \int_{b=-\infty}^{\infty} g\left(a\right) h\left(b\right) f_{\tilde{a},\tilde{b}}\left(a,b\right) \, \mathrm{d}a \, \mathrm{d}b \tag{7.63}$$

$$= \int_{a=-\infty}^{\infty} \int_{b=-\infty}^{\infty} g\left(a\right) h\left(b\right) f_{\tilde{a}}\left(a\right) f_{\tilde{b}}\left(b\right) \, \mathrm{d}a \, \mathrm{d}b \tag{7.64}$$

$$= \mathbb{E}\left[g\left(\tilde{a}\right)\right] \mathbb{E}[h(\tilde{b})]. \tag{7.65}$$

The proof is the same for discrete random variables. In that case, the joint pmf factorizes into the product of the marginal pmfs (see Definition 4.17). If $\tilde{a}$ is discrete and $\tilde{b}$ is continuous, then, by independence, the conditional pdf of $\tilde{b}$ given $\tilde{a}$ is equal to the marginal pdf of $\tilde{b}$ (see Definition 6.8). Consequently, by Definition 7.13,

$$\mathbb{E}\left[g\left(\tilde{a}\right) h(\tilde{b})\right] = \sum_{a \in A} \int_{b=-\infty}^{\infty} g\left(a\right) h\left(b\right) p_{\tilde{a}}\left(a\right) f_{\tilde{b}\,|\,\tilde{a}}\left(b\,|\,a\right) \, \mathrm{d}b \tag{7.66}$$

$$= \sum_{a \in A} g\left(a\right) p_{\tilde{a}}\left(a\right) \int_{b=-\infty}^{\infty} h\left(b\right) f_{\tilde{b}}\left(b\right) \, \mathrm{d}b \tag{7.67}$$

$$= \mathbb{E}\left[g\left(\tilde{a}\right)\right] \mathbb{E}[h(\tilde{b})]. \tag{7.68}$$

■

The following example shows that the independence assumption is crucial in Theorem 7.20.

Example 7.21 (Restaurant). The owner of a restaurant wants to estimate their expected revenue per night. Looking at past data, she determines that the mean number of customers per night is 50 and the mean amount spent per customer is 40 dollars. She concludes that the mean revenue must be 2,000 dollars. However, looking at the actual data, she realizes that this is not the case! The reason is that the number of customers and the money spent by each customer are not independent.

It turns out that a good model for the customers in the restaurant is that each night is busy or calm with probability 1/2. In busy nights, there are 80 customers and each spends 60 dollars (perhaps motivated by the lively atmosphere). In calm nights, there are 20 customers, and each spends 20 dollars. Consequently, the mean of the random variable $\tilde{c}$, representing the number of customers, is 50, and the mean of the random variable $\tilde{a}$, representing the amount spent per customer, is 40 dollars. By Definition 7.6, the mean revenue equals

$$\mathbb{E}[\tilde{c}\tilde{a}] = \sum_{c \in \{20,80\}} \sum_{a \in \{20,60\}} c\,b\,p_{\tilde{c},\tilde{a}}\left(c,a\right) \tag{7.69}$$

$$= \frac{20 \cdot 20}{2} + \frac{80 \cdot 60}{2} \tag{7.70}$$

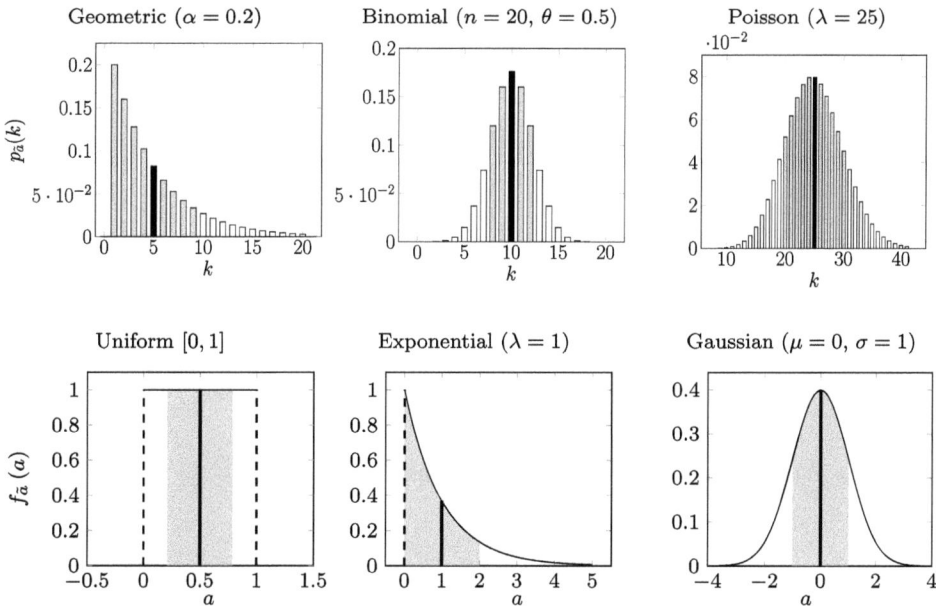

Figure 7.2 Mean and standard deviation of popular distributions. The graphs show the pmfs of several discrete distributions (top row) and the pdfs of several continuous distributions (bottom row). The mean of the random variable is indicated in black. Values within one standard deviation of the mean are shaded in gray. The means and variances of these distributions are derived in Sections 7.4 and 7.7.3, except for the mean and variance of the binomial, derived in Lemmas 7.19 and 9.12, and the mean and variance of the uniform, derived in Lemmas 7.9 and 7.35.

$$= 2600 \neq 2000 = \mathbb{E}[\tilde{c}]\mathbb{E}[\tilde{a}] \tag{7.71}$$

since the joint pmf entries equal $p_{\tilde{c},\tilde{a}}(20, 20) = 0.5$, $p_{\tilde{c},\tilde{a}}(80, 60) = 0.5$, and $p_{\tilde{c},\tilde{a}}(c, a) = 0$ for all other values of a and c. The mean of the product is different from the product of the means, but this does not contradict Theorem 7.20 because the random variables are dependent (when there are more customers, each customer spends more).

7.4 Mean of Parametric Distributions

In this section, we derive the mean of different parametric distributions as a function of their parameters. Figure 7.2 shows a visualization of the means superposed onto the corresponding pmf or pdf.

In the case of a geometric random variable with parameter α, the mean is $1/\alpha$. Consequently, the expected number of times that we need to flip a fair coin until it lands on heads is two.

Lemma 7.22 (Mean of a geometric random variable). *The mean of a geometric random variable $\tilde{a}$ with parameter α equals $\mathbb{E}[\tilde{a}] = 1/\alpha$.*

Proof We apply the geometric series identity $\sum_{k=1}^{\infty} kr^{k-1} = (1 - r)^{-2}$, which holds for $0 < r < 1$. Setting $r = 1 - \alpha$, by Definitions 7.1 and 2.20,

$$\mathbb{E}[\tilde{a}] = \sum_{a=1}^{\infty} a\, p_{\tilde{a}}(a) \tag{7.72}$$

$$= \alpha \sum_{a=1}^{\infty} a \, (1-\alpha)^{a-1} \tag{7.73}$$

$$= \frac{1}{\alpha}. \tag{7.74}$$

∎

The mean of a Poisson random variable is equal to its parameter λ. This justifies our interpretation of λ in Example 2.21, as the expected number of earthquakes per year.

Lemma 7.23 (Mean of a Poisson random variable). *The mean of a Poisson random variable $\tilde{a}$ with parameter λ equals*

$$\mathbb{E}[\tilde{a}] = \lambda. \tag{7.75}$$

Proof   The Taylor series expansion of the exponential function is

$$\sum_{k=0}^{\infty} \frac{\lambda^k}{k!} = \exp(\lambda), \tag{7.76}$$

so, by Definitions 7.1 and 2.22 and the change of variable $m := a - 1$,

$$\mathbb{E}[\tilde{a}] = \sum_{a=0}^{\infty} a \, p_{\tilde{a}}(a) \tag{7.77}$$

$$= \sum_{a=1}^{\infty} \frac{\lambda^a \exp(-\lambda)}{(a-1)!} \tag{7.78}$$

$$= \lambda \exp(-\lambda) \sum_{m=0}^{\infty} \frac{\lambda^m}{m!} \tag{7.79}$$

$$= \lambda. \tag{7.80}$$

∎

The mean of an exponential random variable is the inverse of its parameter λ.

Lemma 7.24 (Mean of an exponential random variable). *The mean of an exponential random variable $\tilde{a}$ with parameter λ equals $1/\lambda$.*

Proof   Applying integration by parts, by Definitions 7.8 and 3.27,

$$\mathbb{E}[\tilde{a}] = \int_{a=-\infty}^{\infty} a f_{\tilde{a}}(a) \, \mathrm{d}a \tag{7.81}$$

$$= \int_{a=0}^{\infty} a \lambda \exp(-\lambda a) \, \mathrm{d}a \tag{7.82}$$

$$= -a \exp(-\lambda a)]_0^{\infty} + \frac{1}{\lambda} \int_0^{\infty} \lambda \exp(-\lambda a) \, \mathrm{d}a \tag{7.83}$$

$$= \frac{1}{\lambda} \tag{7.84}$$

since the integral of the pdf over the positive real line equals one by Theorem 3.16.   ∎

Reassuringly, the mean of a Gaussian random variable is equal to its mean parameter μ.

Lemma 7.25 (Mean of a Gaussian random variable). *The mean of a Gaussian random variable $\tilde{a}$ with mean parameter μ and standard-deviation parameter σ equals μ.*

Proof By Definitions 7.8 and 3.30, applying the change of variables $t := (a - \mu)/\sigma$,

$$\mathbb{E}\left[\tilde{a}\right] = \int_{a=-\infty}^{\infty} a f_{\tilde{a}}(a)\, \mathrm{d}a \tag{7.85}$$

$$= \int_{a=-\infty}^{\infty} \frac{a}{\sqrt{2\pi}\sigma} \exp\left(-\frac{(a-\mu)^2}{2\sigma^2}\right) \mathrm{d}a \tag{7.86}$$

$$= \frac{\sigma}{\sqrt{2\pi}} \int_{t=-\infty}^{\infty} t \exp\left(-\frac{t^2}{2}\right) \mathrm{d}t + \frac{\mu}{\sqrt{2\pi}} \int_{t=-\infty}^{\infty} \exp\left(-\frac{t^2}{2}\right) \mathrm{d}t \tag{7.87}$$

$$= -\frac{\sigma}{\sqrt{2\pi}} \exp\left(-\frac{t^2}{2}\right)\Bigg]_{-\infty}^{\infty} + \frac{\mu}{\sqrt{2\pi}} \int_{t=-\infty}^{\infty} \exp\left(-\frac{t^2}{2}\right) \mathrm{d}t \tag{7.88}$$

$$= \mu. \tag{7.89}$$

The last step holds because $\exp\left(-t^2/2\right)/\sqrt{2\pi}$ is the pdf of a Gaussian random variable with mean zero and unit variance, so its integral over the real line equals one by Theorem 3.16. ∎

The means of all the parametric distributions that we consider in this section are very simple functions of the corresponding parameters: They are equal to the parameter or to its inverse. This suggests a simple approach to estimate the parameters from data: Expressing the parameter as a function of the mean, and then replacing the mean with the sample mean computed from the data. For example, the parameter of a Bernoulli random variable is equal to its mean, so we estimate it using the sample mean. For the Poisson and Gaussian distributions, the parameter estimate is also equal to the sample mean. For the geometric and exponential distributions, the parameter estimate is the inverse of the sample mean, by the same reasoning. This estimation strategy is called the *method of moments*. For the aforementioned distributions, it yields the same estimator as maximum-likelihood estimation (see Table 7.2).

Table 7.2 *Mean of parametric models and the corresponding maximum-likelihood estimator. The table reports the maximum-likelihood estimator and the mean of several popular parametric distributions, as a function of their parameters. For these distributions, maximum-likelihood estimation is equivalent to the method of moments, which produces a parameter estimate by expressing the parameter as a function of the mean and plugging in the sample mean $m(X)$ instead. The means are derived in Section 7.4, except for the mean of the binomial, derived in Lemma 7.19.*

Distribution	Parameter	Maximum-likelihood estimator	Mean
Bernoulli	θ	Example 2.26: $m(X)$	θ
Geometric	α	Exercise 2.13: $m(X)^{-1}$	α^{-1}
Poisson	λ	Example 2.28: $m(X)$	λ
Exponential	λ	Theorem 3.34: $m(X)^{-1}$	λ^{-1}
Gaussian	μ	Theorem 3.35: $m(X)$	μ

7.5 Sensitivity of the Mean to Extreme Values

The mean is often interpreted as representing a *typical* value taken by a random variable. However, it can be severely distorted by a small subset of extreme values or outliers. In such cases, the median is a good alternative, due to its robustness to outliers.

Example 7.26 (Mean vs. median). Consider a random variable $\tilde{a}$ that is uniformly distributed in $[-4.5, 4.5]$, but can also belong to a small interval of extreme values $[x - 0.5, x + 0.5]$, where x is a large constant. Figure 7.3 depicts the pdf. By Definition 7.8 the mean of $\tilde{a}$ equals

$$\mathbb{E}\left[\tilde{a}\right] = \int_{a=-4.5}^{4.5} a f_{\tilde{a}}\left(a\right) \, \mathrm{d}a + \int_{a=x-0.5}^{x+0.5} a f_{\tilde{a}}\left(a\right) \, \mathrm{d}a \tag{7.90}$$

$$= \frac{1}{10} \frac{(x+0.5)^2 - (x-0.5)^2}{2} \tag{7.91}$$

$$= \frac{x}{10}. \tag{7.92}$$

The mean scales linearly with x. It is completely driven by the extreme values, even though by Theorem 3.14 their total probability is only

$$\mathrm{P}\left(x - 0.5 \leq \tilde{a} \leq x + 0.5\right) = \int_{a=x-0.5}^{x+0.5} f_{\tilde{a}}\left(a\right) \, \mathrm{d}a \tag{7.93}$$

$$= \frac{1}{10}. \tag{7.94}$$

By contrast, the median ignores these outliers. By Definition 3.3 and Theorem 3.14, the cdf of $\tilde{a}$ between -4.5 and 4.5 is equal to

$$F_{\tilde{a}}\left(q\right) = \int_{-4.5}^{q} f_{\tilde{a}}\left(a\right) \, \mathrm{d}a \tag{7.95}$$

$$= \frac{q + 4.5}{10}. \tag{7.96}$$

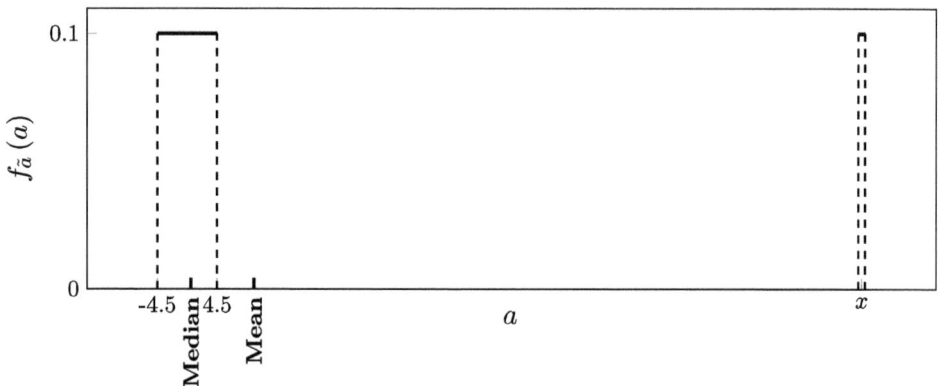

Figure 7.3 The mean can be far from the median. Distribution of the random variable in Example 7.26. The mean is $x/10$, so for large x it is very different to the median, which always equals 0.5.

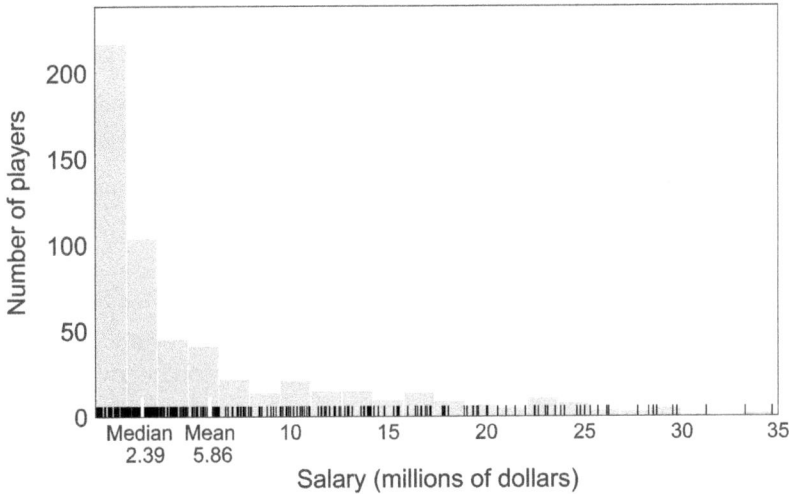

Figure 7.4 Salaries of NBA players in the 2017/2018 season. The plot shows a rug plot and a histogram of the salaries, as well as their sample mean and their median (computed following Definition 3.12).

By Definition 3.10, the median is the value of q for which this expression equals $1/2$. Consequently, the median is 0.5, and does not depend at all on x, so it is robust to the outliers.

· ·

Example 7.27 (NBA salaries). Figure 7.4 shows a real-world example, where the mean is distorted by extreme values. The data are salaries of NBA players in the 2017/2018 season (Dataset 14). The sample mean is not a good characterization of a *typical* salary because it is inflated by the massive salaries of the best-paid players. The sample mean of the salaries (5.86 million) is more than twice as large as the median salary (2.39 million). In fact, less than one third of the players (32.1%) earn more than the sample mean.

· ·

7.6 The Mean Square

The *mean square* provides a measure of the *magnitude* of a random variable. In mathematics, squaring is often used to quantify the magnitude of different objects because it avoids cancellations between positive and negative quantities. The magnitude $|a|$ of a scalar a is the square root of its square $\sqrt{a^2}$. The length or Euclidean norm $||v||_2$ of a vector v is the square root of the sum of its squared entries $\sqrt{\sum_i v[i]^2}$. Motivated by this, we use the mean of the square of a random variable to quantify its average magnitude or energy. The mean square aggregates all the squared values of the random variable, preventing cancellations when the random variable takes negative values.

Definition 7.28 (Mean square). *The mean square or second moment of a random variable $\tilde{a}$ is the mean of $\tilde{a}^2$: $\mathbb{E}\left[\tilde{a}^2\right]$.*

The mean square can be applied to quantify the difference between two random variables. In particular, we can use it as a metric to evaluate estimators. This metric is called the mean squared error.

Definition 7.29 (Mean squared error). *The mean squared error (MSE) between an estimator $\tilde{e}$ and a random variable $\tilde{a}$ is $\mathbb{E}[(\tilde{a} - \tilde{e})^2]$.*

An interesting property of the mean of a random variable is that it is the constant that best approximates the random variable in terms of MSE.

Theorem 7.30 (Constant minimum MSE estimator). *For any random variable $\tilde{a}$ with mean $\mathbb{E}[\tilde{a}]$, the best constant estimator in terms of MSE is $\mathbb{E}[\tilde{a}]$,*

$$\mathbb{E}[\tilde{a}] = \arg\min_{c \in \mathbb{R}} \mathbb{E}\left[(c - \tilde{a})^2\right]. \tag{7.97}$$

Proof Let $\text{MSE}(c) := \mathbb{E}\left[(c - \tilde{a})^2\right] = c^2 - 2c\mathbb{E}[\tilde{a}] + \mathbb{E}[\tilde{a}^2]$ denote the mean squared error as a function of the constant estimate c. The first and second derivatives of the MSE with respect to c equal

$$\frac{d\,\text{MSE}(c)}{dc} = 2(c - \mathbb{E}[\tilde{a}]), \tag{7.98}$$

$$\frac{d^2\,\text{MSE}(c)}{dc^2} = 2. \tag{7.99}$$

The error is strictly convex because the second derivative is positive. Therefore, its minimum is attained at $c = \mathbb{E}[\tilde{a}]$, where the first derivative is zero.   ∎

When performing estimation of uncertain quantities from data, we typically approximate the mean squared error using the average of the squared errors, known as the *residual sum of squares*. Minimizing this cost function is known as *least-squares* estimation. The following lemma is the finite-data counterpart to Theorem 7.30; it establishes that the sample mean is the best constant least-squares estimator.

Theorem 7.31 (Constant least-squares estimator). *Let $X := \{x_1, x_2, \ldots, x_n\}$ denote a real-valued dataset. The best constant estimator, in terms of the residual sum of squares, is the sample mean:*

$$m(X) = \arg\min_{c} \sum_{i=1}^{n} (x_i - c)^2. \tag{7.100}$$

Proof We follow the same steps as in the proof of Theorem 7.30. As a function of the constant estimate c the residual sum of squares equals

$$\text{RSS}(c) := \sum_{i=1}^{n} (x_i - c)^2 \tag{7.101}$$

$$= nc^2 - 2c\sum_{i=1}^{n} x_i + \sum_{i=1}^{n} x_i^2. \tag{7.102}$$

Its derivatives are

$$\frac{d\,\text{RSS}(c)}{dc} = 2\left(nc - \sum_{i=1}^{n} x_i\right) \tag{7.103}$$

$$= 2n\left(c - m(X)\right), \tag{7.104}$$

$$\frac{d^2\,\text{RSS}(c)}{dc^2} = 2n. \tag{7.105}$$

The error is strictly convex, so the minimum is at $c = m(X)$, where the first derivative is zero. ∎

7.7 The Variance

7.7.1 Definition

The mean square of the difference between a random variable and its mean is called the *variance* of the random variable. It quantifies the variation of the random variable around its mean. The square root of this quantity can be interpreted as an average deviation from the mean; it is called the *standard deviation* of the random variable.

Definition 7.32 (Variance and standard deviation). *The variance of a random variable $\tilde{a}$ is the mean square deviation from the mean,*

$$\text{Var}\left[\tilde{a}\right] := \mathbb{E}\left[\left(\tilde{a} - \mathbb{E}\left[\tilde{a}\right]\right)^2\right]. \tag{7.106}$$

The variance is sometimes referred to as the second central moment of $\tilde{a}$. The standard deviation $\sigma_{\tilde{a}}$ of $\tilde{a}$ is the square root of the variance,

$$\sigma_{\tilde{a}} := \sqrt{\text{Var}\left[\tilde{a}\right]}. \tag{7.107}$$

Just as in the case of the mean, it is possible for a random variable to have infinite variance, or to not have a well-defined variance. This is the case, for instance, if the random variable does not have a finite mean.

The following lemma provides a convenient way to compute the variance: Subtracting the squared mean from the mean square.

Lemma 7.33. *For any random variable $\tilde{a}$ with finite variance,*

$$\text{Var}\left[\tilde{a}\right] = \mathbb{E}\left[\tilde{a}^2\right] - \mathbb{E}\left[\tilde{a}\right]^2. \tag{7.108}$$

Proof By Definition 7.32 and linearity of expectation (Theorem 7.18),

$$\text{Var}\left[\tilde{a}\right] := \mathbb{E}\left[\left(\tilde{a} - \mathbb{E}\left[\tilde{a}\right]\right)^2\right] \tag{7.109}$$

$$= \mathbb{E}\left[\tilde{a}^2 - 2\tilde{a}\mathbb{E}\left[\tilde{a}\right] + \mathbb{E}\left[\tilde{a}\right]\right]^2 \tag{7.110}$$

$$= \mathbb{E}\left[\tilde{a}^2\right] - 2\mathbb{E}\left[\tilde{a}\right]\mathbb{E}\left[\tilde{a}\right] + \mathbb{E}\left[\tilde{a}\right]^2 \tag{7.111}$$

$$= \mathbb{E}\left[\tilde{a}^2\right] - \mathbb{E}\left[\tilde{a}\right]^2. \tag{7.112}$$

∎

Lemma 7.34 (Variance of a Bernoulli random variable). *The variance of a Bernoulli random variable $\tilde{a}$ with parameter θ equals $\text{Var}[\tilde{a}] = \theta(1 - \theta)$.*

Proof By Definitions 7.4 and 2.15, the mean square equals

$$\mathbb{E}\left[\tilde{a}^2\right] = 0 \cdot p_{\tilde{a}}\left(0\right) + 1 \cdot p_{\tilde{a}}\left(1\right) \tag{7.113}$$

$$= \theta. \tag{7.114}$$

The result then follows from Lemmas 7.33 and 7.2. ∎

The variance of a Bernoulli random variable is maximized when $\theta = 0.5$ (you can verify this by taking derivatives with respect to θ). This makes sense, as that value of θ maximizes the variability of the samples.

The following lemma derives the variance of a uniform random variable. The standard deviation turns out to be slightly more than one fourth ($1/\sqrt{12} \approx 0.289$) of the length of the interval.

Lemma 7.35 (Variance of a uniform random variable). *The variance of a random variable $\tilde{u}$ that is uniformly distributed in the interval $[a, b]$, $b > a$, equals*

$$\text{Var}[\tilde{u}] = \frac{(b-a)^2}{12}. \tag{7.115}$$

Proof   By Definitions 7.8 and 3.17, the mean square equals

$$\mathbb{E}\left[\tilde{u}^2\right] = \int_{u=-\infty}^{\infty} u^2 f_{\tilde{a}}(u)\, du \tag{7.116}$$

$$= \int_{u=a}^{b} \frac{u^2}{b-a}\, du \tag{7.117}$$

$$= \frac{b^3 - a^3}{3(b-a)}. \tag{7.118}$$

Subtracting the squared mean, derived in Lemma 7.9, from the mean square yields the result by Lemma 7.33,

$$\text{Var}[\tilde{u}] = \mathbb{E}\left[\tilde{u}^2\right] - \mathbb{E}\left[\tilde{u}\right]^2 \tag{7.119}$$

$$= \frac{b^3 - a^3}{3(b-a)} - \left(\frac{a+b}{2}\right)^2 \tag{7.120}$$

$$= \frac{(b-a)^2}{12}. \tag{7.121}$$

∎

Unlike the mean, the variance is not linear. The following lemma shows what happens to the variance of a random variable, when we scale and shift it.

Lemma 7.36 (Variance of a scaled and shifted random variable). *Let $\tilde{a}$ be a random variable with finite variance. For any constants c_1 and c_2,*

$$\text{Var}\left[c_1\tilde{a} + c_2\right] = c_1^2\, \text{Var}\left[\tilde{a}\right]. \tag{7.122}$$

Proof   By linearity of expectation (Theorem 7.18) and Definition 7.32,

$$\text{Var}\left[c_1\tilde{a} + c_2\right] = \mathbb{E}\left[\left(c_1\tilde{a} + c_2 - \mathbb{E}\left[c_1\tilde{a} + c_2\right]\right)^2\right] \tag{7.123}$$

$$= \mathbb{E}\left[\left(c_1\tilde{a} + c_2 - c_1\mathbb{E}\left[\tilde{a}\right] - c_2\right)^2\right] \tag{7.124}$$

$$= c_1^2\, \mathbb{E}\left[\left(\tilde{a} - \mathbb{E}\left[\tilde{a}\right]\right)^2\right] \tag{7.125}$$

$$= c_1^2\, \text{Var}\left[\tilde{a}\right]. \tag{7.126}$$

∎

The result is very intuitive. If we shift the random variable by a constant value, it follows from linearity of expectation that the mean is also shifted by the same value. Consequently, the variance is not affected because the variance only measures the deviation from the mean. If we multiply a random variable by a constant, the standard deviation is scaled by the same factor.

7.7.2 The Sample Variance

In order to estimate the variance from data, we compute the average squared deviation from the sample mean. The resulting estimator is called the sample variance.

Definition 7.37 (The sample variance and sample standard deviation). *Let* $X := \{x_1, x_2, \ldots, x_n\}$ *denote a real-valued dataset. The sample variance is the average squared deviation from the sample mean* $m(X)$,

$$v(X) := \frac{1}{n-1} \sum_{i=1}^{n} (x_i - m(X))^2. \tag{7.127}$$

The square root of the sample variance is called the sample standard deviation.

In Definition 7.37 we divide by $n-1$ when computing the average, instead of by n. This is to ensure that the estimator is unbiased, when we compute it from independent samples, as established in Theorem 9.8. In practice, it does not make much of a difference whether we divide by n or $n-1$, unless n is very small.

Example 7.38 (Variance of temperature). Figure 7.5 shows the sample variance of the temperature at three weather stations in the United States in 2015, extracted from Dataset 9. The mean temperature at the three stations is very similar, but their probability densities look very different. This is captured by the respective variances. A higher variance indicates that the pdf is more spread out because the temperature is farther from its mean on average.

Figure 7.6 shows the sample standard deviation of the temperature in weather stations all over the United States, complementing the map of the mean temperatures in Figure 7.1. We

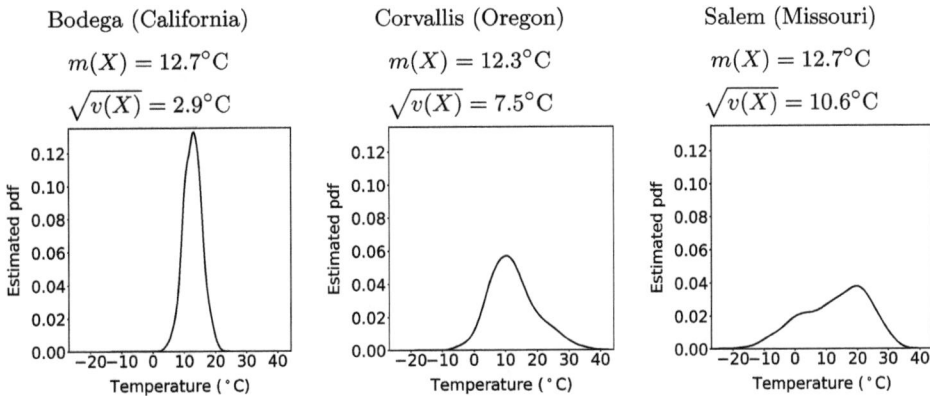

Figure 7.5 Distributions with similar mean but different variances. The plots show the pdfs of the temperature in Bodega (California), Corvallis (Oregon), and Salem (Missouri) in 2015, estimated via kernel density estimation following Definition 3.25. The mean temperatures at the three stations are almost the same, but their standard deviations are very different.

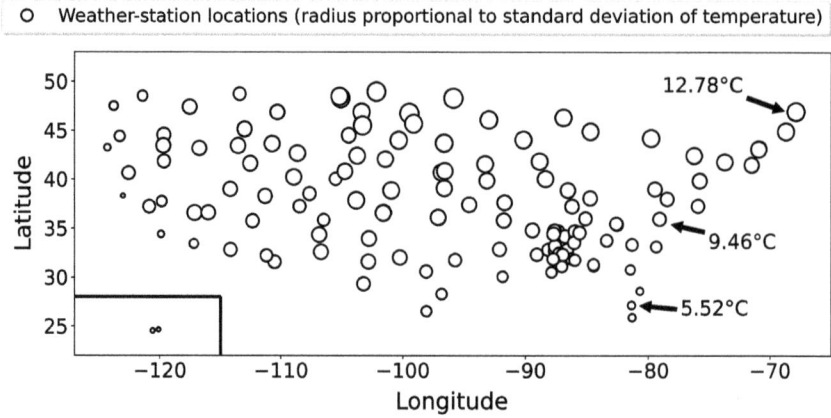

○ Weather-station locations (radius proportional to standard deviation of temperature)

Figure 7.6 Standard deviation of the temperature in the United States. The graph shows the standard deviation of the temperature at 134 weather stations in the United States in 2015. The bottom left corner shows the two stations in Hawaii. The radius of each circular marker is proportional to the standard deviation. The standard deviation is lowest at stations situated on the coast, especially on the West Coast, Hawaii, and Florida. The standard deviations of the temperatures at Durham (North Carolina), Limestone (Maine), and Sebring (Florida) are included for reference.

observe that coastal locations have very small variance, and that the variance is highest in the north and the northeast.

· ·

7.7.3 Variance of Parametric Distributions

In this section, we derive the variance of several popular parametric distributions. Figure 7.2 provides a visual summary of these results by displaying the range of values that fall within one standard deviation of the mean for each of the distributions.

Lemma 7.39 (Variance of a geometric random variable). *The variance of a geometric random variable $\tilde{a}$ with parameter α equals*

$$\text{Var}[\tilde{a}] = \frac{1 - \alpha}{\alpha^2}. \tag{7.128}$$

Proof To derive the mean square, we apply the geometric series identity $\sum_{k=1}^{\infty} k^2 r^k = r(1 + r)(1 - r)^{-3}$, which holds for $0 < r < 1$. Setting $r := 1 - \alpha$, by Definitions 7.4 and 2.20,

$$\mathbb{E}\left[\tilde{a}^2\right] = \sum_{a=1}^{\infty} a^2 \, p_{\tilde{a}}\left(a\right) \tag{7.129}$$

$$= \sum_{a=1}^{\infty} a^2 \, \alpha \left(1 - \alpha\right)^{a-1} \tag{7.130}$$

$$= \frac{\alpha}{1 - \alpha} \sum_{a=1}^{\infty} a^2 \left(1 - \alpha\right)^{a} \tag{7.131}$$

$$= \frac{2 - \alpha}{\alpha^2}. \tag{7.132}$$

The result then follows from Lemmas 7.33 and 7.22. ∎

An interesting property of the Poisson distribution is that the mean is equal to the variance.

Lemma 7.40 (Variance of a Poisson random variable). *The variance of a Poisson random variable $\tilde{a}$ with parameter λ equals λ.*

Proof By Definitions 7.4 and 2.22, applying the changes of variable $k := a - 2$ and $m := a - 1$, the mean square equals

$$\mathbb{E}\left[\tilde{a}^2\right] = \sum_{a=1}^{\infty} a^2 \, p_{\tilde{a}}\left(a\right) \tag{7.133}$$

$$= \sum_{a=1}^{\infty} \frac{a\lambda^a \exp\left(-\lambda\right)}{(a-1)!} \tag{7.134}$$

$$= \exp\left(-\lambda\right) \left(\sum_{a=2}^{\infty} \frac{(a-1)\lambda^a}{(a-1)!} + \sum_{a=1}^{\infty} \frac{\lambda^a}{(a-1)!} \right) \tag{7.135}$$

$$= \exp\left(-\lambda\right) \left(\lambda^2 \sum_{k=0}^{\infty} \frac{\lambda^k}{k!} + \lambda \sum_{m=0}^{\infty} \frac{\lambda^m}{m!} \right) \tag{7.136}$$

$$= \lambda^2 + \lambda, \tag{7.137}$$

where the last step follows from (7.76). The result then follows from Lemmas 7.33 and 7.23. ∎

In the case of the exponential random variable, the standard deviation is equal to the mean.

Lemma 7.41 (Variance of an exponential random variable). *The variance of an exponential random variable $\tilde{a}$ with parameter λ equals $1/\lambda^2$.*

Proof By Definitions 7.8 and 3.27, applying integration by parts,

$$\mathbb{E}\left[\tilde{a}^2\right] = \int_{a=-\infty}^{\infty} a^2 f_{\tilde{a}}\left(a\right) \, \mathrm{d}a \tag{7.138}$$

$$= \int_{a=0}^{\infty} a^2 \lambda \exp\left(-\lambda a\right) \, \mathrm{d}a \tag{7.139}$$

$$= -a^2 \exp\left(-\lambda a\right)\big]_0^{\infty} + \frac{2}{\lambda} \int_0^{\infty} a\lambda \exp\left(-\lambda a\right) \, \mathrm{d}a \tag{7.140}$$

$$= \frac{2}{\lambda^2}. \tag{7.141}$$

The result then follows from Lemmas 7.33 and 7.24. ∎

Example 7.42 (Variance of call center data). According to Lemma 7.41, quantities modeled as exponential random variables should have a standard deviation that is similar to their mean. In Example 3.28, we fit an exponential random variable to data consisting of inter-arrival times of calls at a call center (Dataset 3). For calls arriving at the bank call center on weekdays from 9 am to 10 am, the sample mean equals 30.8 and the sample standard deviation is 33.6, which is indeed quite close.

. . . .

The variance of a Gaussian random variable is equal to its variance parameter. It therefore makes sense that the maximum-likelihood estimator of this parameter is almost equal to the sample variance (compare Theorem 3.35 and Definition 7.37). The only difference is that the sum is normalized by n instead of $n - 1$, which does not make much difference, as long as n is not very small.

Lemma 7.43 (Variance of a Gaussian random variable). *The variance of a Gaussian random variable $\tilde{a}$ with mean parameter μ and standard-deviation parameter σ equals σ^2.*

Proof   By Definitions 7.8 and 3.30, and the change of variables $t := (a - \mu)/\sigma$, the mean square equals

$$\mathbb{E}\left[\tilde{a}^2\right] = \int_{a=-\infty}^{\infty} a^2 f_{\tilde{a}}(a)\, \mathrm{d}a \tag{7.142}$$

$$= \int_{a=-\infty}^{\infty} \frac{a^2}{\sqrt{2\pi}\sigma} \exp\left(-\frac{(a-\mu)^2}{2\sigma^2}\right) \mathrm{d}a \tag{7.143}$$

$$= \frac{\sigma^2}{\sqrt{2\pi}} \int_{t=-\infty}^{\infty} t^2 \exp\left(-\frac{t^2}{2}\right) \mathrm{d}t + \frac{2\mu\sigma}{\sqrt{2\pi}} \int_{t=-\infty}^{\infty} t \exp\left(-\frac{t^2}{2}\right) \mathrm{d}t$$

$$+ \mu^2 \int_{t=-\infty}^{\infty} \frac{\exp\left(-t^2/2\right)}{\sqrt{2\pi}}\, \mathrm{d}t. \tag{7.144}$$

We compute the first term in (7.144) via derivation by parts.

$$\frac{\sigma^2}{\sqrt{2\pi}} \int_{t=-\infty}^{\infty} t^2 \exp\left(-\frac{t^2}{2}\right) \mathrm{d}t = \frac{-\sigma^2 t}{\sqrt{2\pi}} \exp\left(-\frac{t^2}{2}\right)\Bigg]_{-\infty}^{\infty} + \sigma^2 \int_{t=-\infty}^{\infty} \frac{\exp\left(-t^2/2\right)}{\sqrt{2\pi}}\, \mathrm{d}t$$

$$= \sigma^2. \tag{7.145}$$

The last step holds because $\exp\left(-t^2/2\right)/\sqrt{2\pi}$ is the pdf of a Gaussian random variable with mean zero and unit variance, so its integral over the real line equals one by Theorem 3.16. The second term in (7.144) is zero since

$$\int_{t=-\infty}^{\infty} t \exp\left(-\frac{t^2}{2}\right) \mathrm{d}t = -\exp\left(-\frac{t^2}{2}\right)\Bigg]_{-\infty}^{\infty} = 0. \tag{7.146}$$

The third term in (7.144) is μ^2, since the integral of $\exp\left(-t^2/2\right)/\sqrt{2\pi}$ equals one. We conclude that

$$\mathbb{E}\left[\tilde{a}^2\right] = \sigma^2 + \mu^2, \tag{7.147}$$

so the result then follows from Lemmas 7.33 and 7.25.   ∎

7.8 The Conditional Mean

The conditional mean of a random variable represents how its average value varies depending on the value of other random variables. In Section 7.8.1, we define the conditional mean formally and explain how to estimate it from data. Section 7.8.2 explains how to analyze the probabilistic behavior of the conditional mean, motivated by iterated expectation, an identity that allows us to compute the mean from the conditional mean. Section 7.8.3 introduces the problem of regression and shows that the conditional mean is an optimal estimator in terms of mean squared error.

7.8.1 The Conditional Mean Function

The conditional mean function of a random variable $\tilde{b}$ given another random variable $\tilde{a}$ is equal to the mean of $\tilde{b}$ conditioned on the event $\tilde{a} = a$. It can be computed using the conditional distribution of $\tilde{b}$ given $\tilde{a}$.

Definition 7.44 (The conditional mean function). *Let $\tilde{a}$ be a random variable or a random vector, and let $\tilde{b}$ be another random variable belonging to the same probability space. The conditional mean function $\mu_{\tilde{b} \mid \tilde{a}}(a)$ is the mean of $\tilde{b}$ computed according to the conditional distribution of $\tilde{b}$ given $\tilde{a} = a$. If $\tilde{b}$ is a discrete random variable, then*

$$\mu_{\tilde{b} \mid \tilde{a}}(a) := \sum_{b \in B} b\, p_{\tilde{b} \mid \tilde{a}}(b \mid a), \tag{7.148}$$

where B is the range of $\tilde{b}$ conditioned on $\tilde{a} = a$. If $\tilde{b}$ is a continuous random variable with conditional pdf $f_{\tilde{b} \mid \tilde{a}}$, then

$$\mu_{\tilde{b} \mid \tilde{a}}(a) := \int_{b=-\infty}^{\infty} b\, f_{\tilde{b} \mid \tilde{a}}(b \mid a)\; \mathrm{d}b. \tag{7.149}$$

The conditional mean function $\mu_{\tilde{b} \mid \tilde{a}}(a)$ is only defined for values of a such that the conditional distribution of $\tilde{b}$ given $\tilde{a} = a$ exists.

In addition, for any deterministic function $h \colon \mathbb{R} \to \mathbb{R}$, we define the conditional mean function of $h(\tilde{a}, \tilde{b})$ given $\tilde{a} = a$ as

$$\mu_{h(\tilde{a},\tilde{b}) \mid \tilde{a}}(a) := \sum_{b \in B} h(a, b)\, p_{\tilde{b} \mid \tilde{a}}(b \mid a) \tag{7.150}$$

if $\tilde{b}$ is discrete, and

$$\mu_{h(\tilde{a},\tilde{b}) \mid \tilde{a}}(a) := \int_{b=-\infty}^{\infty} h(a, b)\, f_{\tilde{b} \mid \tilde{a}}(b \mid a)\; \mathrm{d}b \tag{7.151}$$

if $\tilde{b}$ is continuous, assuming the conditional distribution of $\tilde{b}$ given $\tilde{a} = a$ exists.

The conditional mean $\mu_{\tilde{b} \mid \tilde{a}}(a)$ is often denoted by $\mathbb{E}[\tilde{b} \mid \tilde{a} = a]$ in the literature. We have chosen our notation to emphasize that $\mu_{\tilde{b} \mid \tilde{a}}(a)$ is a *deterministic function* of a.

Example 7.45 (Simple example: Conditional mean function). Consider the random variables $\tilde{a}$ and $\tilde{b}$ in Example 4.6. In order to compute the conditional mean function of $\tilde{b}$ given $\tilde{a} = a$, we first need to derive the conditional pmf of $\tilde{b}$ given $\tilde{a}$. Following Definition 4.13, we divide the joint pmf by the marginal pmf of $\tilde{a}$ (computed in Example 4.11) to obtain

$$p_{\tilde{b} \mid \tilde{a}}(1 \mid 1) = \frac{1}{7}, \qquad p_{\tilde{b} \mid \tilde{a}}(2 \mid 1) = \frac{4}{7}, \qquad p_{\tilde{b} \mid \tilde{a}}(3 \mid 1) = \frac{2}{7}, \tag{7.152}$$

$$p_{\tilde{b} \mid \tilde{a}}(1 \mid 2) = \frac{2}{7}, \qquad p_{\tilde{b} \mid \tilde{a}}(2 \mid 2) = \frac{1}{7}, \qquad p_{\tilde{b} \mid \tilde{a}}(3 \mid 2) = \frac{4}{7}, \tag{7.153}$$

$$p_{\tilde{b} \mid \tilde{a}}(1 \mid 3) = \frac{1}{3}, \qquad p_{\tilde{b} \mid \tilde{a}}(2 \mid 3) = \frac{1}{3}, \qquad p_{\tilde{b} \mid \tilde{a}}(3 \mid 3) = \frac{1}{3}. \tag{7.154}$$

By Definition 7.44, the conditional mean function of $\tilde{b}$ given $\tilde{a}$ is

$$\mu_{\tilde{b} \mid \tilde{a}}(1) = \sum_{b \in B} b\, p_{\tilde{b} \mid \tilde{a}}(b \mid 1) = \frac{15}{7}, \tag{7.155}$$

Conditional mean function   Pdf of conditional mean

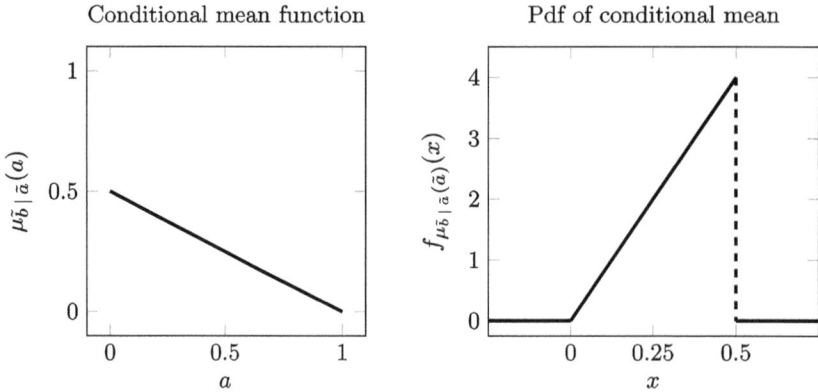

Figure 7.7 Conditional mean of the vertical coordinate in the triangle lake. The black line in the left plot is the conditional mean function of the vertical coordinate of the otter in Example 5.6 given the horizontal coordinate, as derived in Example 7.46. The pdf of the conditional mean, computed in Example 7.53, is shown on the right.

$$\mu_{\tilde{b}\,|\,\tilde{a}}(2) = \sum_{b \in B} b\, p_{\tilde{b}\,|\,\tilde{a}}(b\,|\,2) = \frac{16}{7}, \tag{7.156}$$

$$\mu_{\tilde{b}\,|\,\tilde{a}}(3) = \sum_{b \in B} b\, p_{\tilde{b}\,|\,\tilde{a}}(b\,|\,3) = 2. \tag{7.157}$$

The conditional mean function $\mu_{\tilde{b}\,|\,\tilde{a}}$ is a deterministic function of its input, defined for the three values of a (1, 2, and 3) such that the conditional distribution of $\tilde{b}$ given $\tilde{a} = a$ exists.

. . . .

Example 7.46 (Triangle lake: Conditional mean function). In Example 5.6, the position of an otter is modeled as being uniformly distributed in a triangular lake. In Example 5.13, we derive the conditional pdf of the vertical coordinate $\tilde{b}$ given the horizontal coordinate $\tilde{a}$,

$$f_{\tilde{b}\,|\,\tilde{a}}(b\,|\,a) = \frac{1}{1-a}, \qquad 0 \le b \le 1-a, \tag{7.158}$$

which exists for $a \in [0, 1]$. By Definition 7.44, the conditional mean function of $\tilde{b}$ given $\tilde{a} = a$ is equal to

$$\mu_{\tilde{b}\,|\,\tilde{a}}(a) = \int_{b \in \mathbb{R}} b f_{\tilde{b}\,|\,\tilde{a}}(b\,|\,a)\ \mathrm{d}b \tag{7.159}$$

$$= \int_{b=0}^{1-a} \frac{b}{1-a}\ \mathrm{d}b \tag{7.160}$$

$$= \frac{(1-a)^2}{2(1-a)} \tag{7.161}$$

$$= \frac{1-a}{2} \tag{7.162}$$

for $a \in [0, 1]$. The conditional mean function is depicted in the left plot of Figure 7.7.

. . . .

In order to estimate the conditional mean function from real data, we use the sample conditional mean.

Rating for Independence Day   Temperature in Versailles (Kansas)

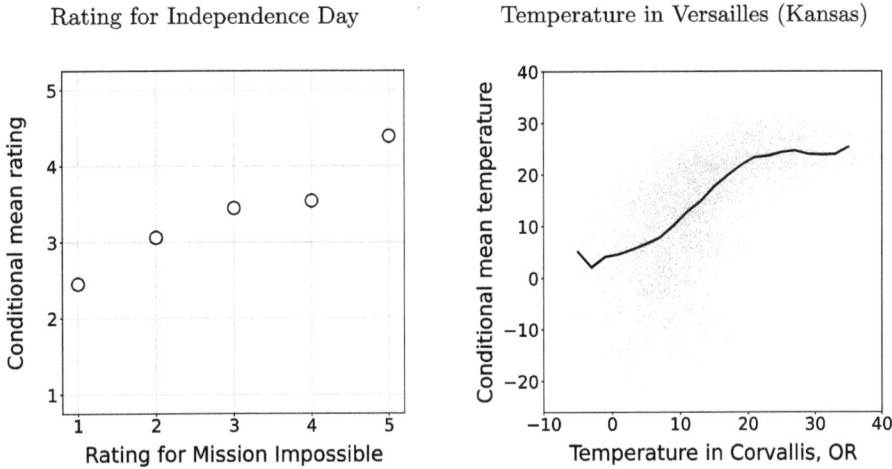

Figure 7.8 Estimating the conditional mean function from data. The left plot shows the sample conditional mean function of the rating for Independence Day given the rating for Mission Impossible, computed following Definition 7.47. The right plot shows the sample conditional mean function of the temperature in Versailles (Kansas) given the temperature in Corvallis (Oregon), computed following Definition 7.49 with $\epsilon := 1°C$.

Definition 7.47 (Sample conditional mean function of discrete data). *Consider a dataset $\mathcal{D}$ formed by pairs (x_1, y_1), (x_2, y_2), ..., (x_n, y_n), where x_i belongs to a discrete set A for $1 \leq i \leq n$. We interpret the samples as realizations from a pair of random variables $\tilde{a}$ and $\tilde{b}$ belonging to the same probability space. For any $a \in A$, let*

$$Y_a := \{y : (a, y) \in \mathcal{D}\} \tag{7.163}$$

denote the values of the second entry, such that the corresponding first entry is equal to a. If there are several data points with the same y value, such that $(a, y) \in \mathcal{D}$, we include all of them in Y_a. The sample conditional mean function of $\tilde{b}$ given $\tilde{a} = a$ is the arithmetic average of the elements of Y_a,

$$\widehat{m}_{\tilde{b} \mid \tilde{a}}(a) := \frac{1}{n_a} \sum_{y \in Y_a} y, \tag{7.164}$$

where n_a is the number of data points in Y_a.

Example 7.48 (Movie ratings: Sample conditional mean function). The left plot in Figure 7.8 shows the sample conditional mean function of the rating for the movie Independence Day given the rating for the movie Mission Impossible. The function is computed following Definition 7.47 using the data in Example 4.8, extracted from Dataset 8. The conditional mean function is monotonically increasing: The higher the rating for Mission Impossible, the higher the conditional mean of the rating for Independence Day.
. . . .

When approximating the conditional mean of a random variable $\tilde{b}$ given the value of a *continuous* random variable $\tilde{a}$, we encounter a challenge. The sample conditional mean given $\tilde{a} = a$ is the average of the data points such that $\tilde{a}$ equals a (the dataset Y_a defined in (7.163)). The problem is that, for continuous random variables, a takes uncountably infinite possible values, but the available data are finite. Consequently, Y_a is empty for most values of a! To overcome this issue, we have two options. The first is to approximate the conditional pdf or pmf of $\tilde{b}$ (as in Example 5.14) and use it to compute the conditional mean. The second is to average over data points such that $\tilde{a}$ is close to a, instead of exactly equal to a, assuming that the conditional mean function is smooth. This smoothness assumption is governed by a parameter ϵ, which has an analogous role to the histogram bin width or the kernel bandwidth in density estimation (see Section 3.5).

Definition 7.49 (Sample conditional mean of continuous data). *Consider a dataset $\mathcal{D}$ formed by pairs (x_1, y_1), (x_2, y_2), ..., (x_n, y_n), where x_i and y_i, $1 \leq i \leq n$, are real values. We interpret the samples as realizations from a pair of random variables $\tilde{a}$ and $\tilde{b}$ belonging to the same probability space. For any $a \in \mathbb{R}$ and a fixed positive constant ϵ, let*

$$Y_{a,\epsilon} := \{y \colon (x, y) \in \mathcal{D} \text{ for } |x - a| \leq \epsilon\} \tag{7.165}$$

denote the values taken by the second variable when the first one is close to a (at a distance no greater than ϵ). The sample conditional mean function of $\tilde{b}$ given $\tilde{a} = a$ is the average of $Y_{a,\epsilon}$,

$$\widehat{m}_{\tilde{b} \mid \tilde{a}}(a) := \frac{1}{n_a} \sum_{y \in Y_{a,\epsilon}} y, \tag{7.166}$$

where n_a is the number of data points in $Y_{a,\epsilon}$.

Example 7.50 (Sample conditional mean temperature). The right plot in Figure 7.8 shows the sample conditional mean function of the temperature in Versailles (Kansas) given the temperature in Corvallis (Oregon), computed using hourly temperature measurements in 2015 extracted from Dataset 9. We set the ϵ parameter in Definition 7.49 to equal $1°$. Between $0°$ and $20°$, the average temperature in Versailles grows proportionally to the temperature in Corvallis. Beyond $20°$, the sample conditional mean is essentially constant. Below $0°$, the sample conditional mean is noisy because of the small number of data in that region.

....

7.8.2 The Conditional Mean and Iterated Expectation

In some situations, the conditional mean function $\mu_{\tilde{b} \mid \tilde{a}}$ of a random variable $\tilde{b}$ given another random variable $\tilde{a}$ is much easier to derive than the mean of $\tilde{b}$ (for instance, in Example 7.57). It is therefore useful to be able to compute the mean of $\tilde{b}$ from $\mu_{\tilde{b} \mid \tilde{a}}$. This can be achieved via *iterated expectation*.

To motivate iterated expectation, we begin with an intuitive derivation. Consider a dataset $\mathcal{D}$ formed by pairs (x_1, y_1), (x_2, y_2), ..., (x_n, y_n), which we assume to be samples from the joint distribution of two random variables $\tilde{a}$ and $\tilde{b}$. As explained in Section 7.1, the mean of a random variable $\tilde{b}$ is defined to approximate the value we obtain, when we average many samples with the same distribution as $\tilde{b}$, i.e.

$$\mathbb{E}[\tilde{b}] \approx \frac{1}{n} \sum_{i=1}^{n} y_i. \tag{7.167}$$

Analogously, we interpret the conditional mean function of $\tilde{b}$ given $\tilde{a} = a$, as the average of the second entry y_i in the dataset, if we only consider data points whose first entry equals a:

$$\mu_{\tilde{b} \mid \tilde{a}}(a) \approx \frac{1}{n_a} \sum_{y \in Y_a} y, \tag{7.168}$$

where Y_a contains the values of y such that (a, y) belongs to the dataset $\mathcal{D}$, as in Definition 7.47, and n_a is the number of samples in Y_a.

Now, let us express the mean of $\tilde{b}$ in terms of the conditional mean function. If $\tilde{a}$ is discrete with range A, each y_i (for $1 \leq i \leq n$) belongs to exactly one of the datasets Y_a (specifically to Y_{x_i}), so summing $y_1, \ldots, y_n$ is equivalent to summing all the samples in all of the datasets Y_a for $a \in A$. Consequently, by (7.167),

$$\mathbb{E}[\tilde{b}] \approx \frac{1}{n} \sum_{i=1}^{n} y_i \tag{7.169}$$

$$= \frac{1}{n} \sum_{a \in A} \sum_{y \in Y_a} y. \tag{7.170}$$

According to our intuitive definition of probability in (1.1), the probability $p_{\tilde{a}}(a) := \mathrm{P}(\tilde{a} = a)$ that $\tilde{a}$ equals a can be approximated by the fraction n_a/n of first-entry values equal to a, when n is large. Therefore, by (7.170) and (7.168),

$$\mathbb{E}[\tilde{b}] \approx \sum_{a \in A} \frac{n_A}{n} \frac{1}{n_A} \sum_{y \in Y_a} y \tag{7.171}$$

$$\approx \sum_{a \in A} \frac{n_A}{n} \mu_{\tilde{b} \mid \tilde{a}}(a) \tag{7.172}$$

$$\approx \sum_{a \in A} p_{\tilde{a}}(a) \mu_{\tilde{b} \mid \tilde{a}}(a). \tag{7.173}$$

The resulting expression is the sum of the conditional mean function at every possible value of $\tilde{a}$, weighted by the corresponding probability.

If $\tilde{a}$ is continuous, a similar argument (involving a grid and limits, as in the intuitive derivation of the mean of a continuous random variable in Section 7.1.2), yields an expression for the mean of $\tilde{b}$ as a weighted integral of the conditional mean function,

$$\mathbb{E}[\tilde{b}] \approx \int_{a=-\infty}^{\infty} f_{\tilde{a}}(a) \mu_{\tilde{b} \mid \tilde{a}}(a) \, \mathrm{d}a. \tag{7.174}$$

The conditional mean function $\mu_{\tilde{b} \mid \tilde{a}}(a)$ is a deterministic function of a, as explained in Section 7.8.1. Consequently, setting $h := \mu_{\tilde{b} \mid \tilde{a}}$ in Definitions 7.4 and 7.8, we can interpret (7.173) and (7.174) as the mean of the random variable $\mu_{\tilde{b} \mid \tilde{a}}(\tilde{a})$, obtained by plugging $\tilde{a}$ into the conditional mean function. This random variable is known as the conditional mean of $\tilde{b}$ given $\tilde{a}$. It describes our uncertainty about the conditional average of $\tilde{b}$ given $\tilde{a}$, when $\tilde{a}$ is unknown.

Definition 7.51 (The conditional mean). *Let $\tilde{a}$ be a random variable or a random vector, and let $\tilde{b}$ be another random variable belonging to the same probability space. The conditional mean of $\tilde{b}$ given $\tilde{a}$ is the random variable $\mu_{\tilde{b} \mid \tilde{a}}(\tilde{a})$ obtained by plugging $\tilde{a}$ into the conditional mean function of $\tilde{b}$ given $\tilde{a}$ (see Definition 7.44).*

A common notation for $\mu_{\tilde{b}\,|\,\tilde{a}}(\tilde{a})$ in the literature is $\mathbb{E}[\tilde{b}\,|\,\tilde{a}]$. We have chosen our notation to avoid confusion: Expected values or means are *deterministic* quantities, but the conditional mean is a *random variable*. Consequently, we describe its behavior using its pmf if it is discrete, or its cdf or pdf if it is continuous.

Example 7.52 (Simple example: Conditional mean). In Example 7.45, the conditional mean function only takes three different values: 15/7, 16/7, and 2, depending on whether $\tilde{a}$ is equal to 1, 2, or 3, respectively. It is therefore straightforward to compute the pmf of $\mu_{\tilde{b}\,|\,\tilde{a}}(\tilde{a})$ from the marginal pmf of $\tilde{a}$ derived in Example 4.11:

$$p_{\mu_{\tilde{b}\,|\,\tilde{a}}(\tilde{a})}\left(\frac{15}{7}\right) = \mathrm{P}\left(\mu_{\tilde{b}\,|\,\tilde{a}}(\tilde{a}) = \frac{15}{7}\right) \tag{7.175}$$

$$= \mathrm{P}\left(\tilde{a} = 1\right) \tag{7.176}$$

$$= 0.35. \tag{7.177}$$

Similarly,

$$p_{\mu_{\tilde{b}\,|\,\tilde{a}}(\tilde{a})}\left(\frac{16}{7}\right) = 0.35, \tag{7.178}$$

$$p_{\mu_{\tilde{b}\,|\,\tilde{a}}(\tilde{a})}(2) = 0.3. \tag{7.179}$$

Example 7.53 (Triangle lake: Conditional mean). To obtain the conditional mean of the otter's vertical coordinate $\tilde{b}$ given the horizontal coordinate $\tilde{a}$, we plug $\tilde{a}$ into the conditional mean function derived in Example 7.46:

$$\mu_{\tilde{b}\,|\,\tilde{a}}(\tilde{a}) = \frac{1-\tilde{a}}{2}. \tag{7.180}$$

This is a continuous random variable. We derive its cdf using the marginal pdf of $\tilde{a}$ derived in Example 5.10. By Definition 3.3 and Theorem 3.14, for $x \in [0, 0.5]$,

$$F_{\mu_{\tilde{b}\,|\,\tilde{a}}(\tilde{a})}(x) = \mathrm{P}\left(\mu_{\tilde{b}\,|\,\tilde{a}}(\tilde{a}) \le x\right) \tag{7.181}$$

$$= \mathrm{P}\left(\frac{1-\tilde{a}}{2} \le x\right) \tag{7.182}$$

$$= \mathrm{P}\left(\tilde{a} \ge 1 - 2x\right) \tag{7.183}$$

$$= \int_{1-2x}^{1} p_{\tilde{a}}(t)\,\mathrm{d}t \tag{7.184}$$

$$= \int_{1-2x}^{1} 2(1-t)\,\mathrm{d}t \tag{7.185}$$

$$= 4x^2. \tag{7.186}$$

For $x < 0$, $F_{\mu_{\tilde{b}\,|\,\tilde{a}}(\tilde{a})}(x) = 0$, and for $x > 0.5$, $F_{\mu_{\tilde{b}\,|\,\tilde{a}}(\tilde{a})}(x) = 1$, by the same argument. To obtain the pdf, we differentiate with respect to x, as dictated by Definition 3.13:

$$f_{\mu_{\tilde{b}\,|\,\tilde{a}}(\tilde{a})}(x) = 8x \tag{7.187}$$

for $x \in [0, 0.5]$ and zero otherwise. The pdf is depicted in the right plot of Figure 7.7.

The identity derived intuitively in (7.173) and (7.174) is known as iterated expectation because we are taking the expected value of a conditional expected value. The following theorem confirms that this works, proving formally that the mean of a random variable is indeed equal to the mean of its conditional mean.

Theorem 7.54 (Iterated expectation). *For any random variables $\tilde{a}$ and $\tilde{b}$ belonging to the same probability space,*

$$\mathbb{E}\left[\mu_{\tilde{b}\mid\tilde{a}}(\tilde{a})\right] = \mathbb{E}\left[\tilde{b}\right]. \tag{7.188}$$

Similarly, for any function $h\colon \mathbb{R}^2 \to \mathbb{R}$,

$$\mathbb{E}[\mu_{h(\tilde{a},\tilde{b})\mid\tilde{a}}(\tilde{a})] = \mathbb{E}\left[h(\tilde{a},\tilde{b})\right]. \tag{7.189}$$

In both cases, we assume that the means exist and are finite.

Proof   We prove the result for continuous random variables. The proof for discrete random variables, and for functions that depend on both continuous and discrete random variables, is essentially the same. By Definition 7.51, the chain rule (Theorem 5.12) and Definition 7.11,

$$\mathbb{E}[\mu_{h(\tilde{a},\tilde{b})\mid\tilde{a}}(\tilde{a})] = \int_{a=-\infty}^{\infty} f_{\tilde{a}}(a)\, \mu_{h(\tilde{a},\tilde{b})\mid\tilde{a}}(a)\, \mathrm{d}a \tag{7.190}$$

$$= \int_{a=-\infty}^{\infty} \int_{b=-\infty}^{\infty} f_{\tilde{a}}(a)\, f_{\tilde{b}\mid\tilde{a}}(b\mid a)\, h(a,b)\, \mathrm{d}b\, \mathrm{d}a \tag{7.191}$$

$$= \int_{a=-\infty}^{\infty} \int_{b=-\infty}^{\infty} f_{\tilde{a},\tilde{b}}(a,b)\, h(a,b)\, \mathrm{d}b\, \mathrm{d}a \tag{7.192}$$

$$= \mathbb{E}[h(\tilde{a},\tilde{b})]. \tag{7.193}$$

Setting $h(\tilde{a},\tilde{b}) := \tilde{b}$ establishes (7.188). ∎

Example 7.55 (Simple example: Iterated expectation). By Definition 7.1 and the pmf derived in Example 7.52, the mean of the conditional mean $\mu_{\tilde{b}\mid\tilde{a}}(\tilde{a})$ equals

$$\mathbb{E}\left[\mu_{\tilde{b}\mid\tilde{a}}(\tilde{a})\right] = \sum_{x\in\{2,15/7,16/7\}} x p_{\mu_{\tilde{b}\mid\tilde{a}}(\tilde{a})}(x) \tag{7.194}$$

$$= 2\cdot 0.3 + \frac{15}{7}\cdot 0.35 + \frac{16}{7}\cdot 0.35 \tag{7.195}$$

$$= 2.15. \tag{7.196}$$

We verify that this indeed equals the mean of $\tilde{b}$, using the marginal pmf derived in Example 4.11,

$$\mathbb{E}[\tilde{b}] = \sum_{b=1}^{3} b p_{\tilde{b}}(b) \tag{7.197}$$

$$= 1\cdot 0.25 + 2\cdot 0.35 + 3\cdot 0.4 \tag{7.198}$$

$$= 2.15 = \mathbb{E}\left[\mu_{\tilde{b}\mid\tilde{a}}(\tilde{a})\right]. \tag{7.199}$$

Example 7.56 (Triangle lake: iterated expectation). By Definition 7.8 and the pdf derived in Example 7.53, the mean of the conditional mean $\mu_{\tilde{b}\mid\tilde{a}}(\tilde{a})$ is

$$\mathbb{E}\left[\mu_{\tilde{b}\,|\,\tilde{a}}(\tilde{a})\right] = \int_{x=-\infty}^{\infty} x f_{\mu_{\tilde{b}\,|\,\tilde{a}}(\tilde{a})}(x) \,\mathrm{d}x \tag{7.200}$$

$$= \int_{x=0}^{\frac{1}{2}} 8x^2 \,\mathrm{d}x = \frac{1}{3}. \tag{7.201}$$

The marginal pdf of $\tilde{b}$ (see Example 5.10) equals

$$f_{\tilde{b}}(b) = \int_{a=-\infty}^{\infty} f_{\tilde{a},\tilde{b}}(a,b) \,\mathrm{d}a = 2(1-b) \tag{7.202}$$

for $0 \le b \le 1$ and zero otherwise. We confirm that the mean of $\tilde{b}$ is equal to the mean of the conditional mean,

$$\mathbb{E}[\tilde{b}] = \int_{b=-\infty}^{\infty} b f_{\tilde{b}}(b) \,\mathrm{d}b \tag{7.203}$$

$$= \int_{b=0}^{1} 2(1-b)b \,\mathrm{d}b \tag{7.204}$$

$$= \frac{1}{3} = \mathbb{E}\left[\mu_{\tilde{b}\,|\,\tilde{a}}(\tilde{a})\right]. \tag{7.205}$$

Iterated expectation allows us to compute means very easily, when we have access to the conditional distribution of our quantity of interest.

Example 7.57 (Computer). We want to model the time until a computer breaks down. We know that the time depends on how often the computer is turned off, and whether the owner is careful. We model the time $\tilde{t}$ as an exponential random variable with a parameter $\tilde{\lambda}$ that is itself a random variable. The random parameter is defined as

$$\tilde{\lambda} := \frac{1}{\tilde{o} + \tilde{c}}. \tag{7.206}$$

The random variable $\tilde{o}$ represents the fraction of time the computer is off; it is uniformly distributed in $[0, 1]$. The random variable $\tilde{c}$ represents how careful the owner is; it is also uniformly distributed in $[0, 1]$. In order to compute the mean of $\tilde{t}$, we can derive the corresponding pdf, but this is pretty complicated. Fortunately, iterated expectation comes to the rescue. Conditioned on $\tilde{o} = o$ and $\tilde{c} = c$, $\tilde{t}$ is exponential with parameter $\lambda := \frac{1}{o+c}$, so by Lemma 7.24 the conditional mean function equals

$$\mu_{\tilde{t}\,|\,\tilde{o},\tilde{c}}(o, c) = \frac{1}{\lambda} \tag{7.207}$$

$$= o + c. \tag{7.208}$$

By Lemma 7.9, the means of $\tilde{o}$ and $\tilde{c}$ equal 0.5, so by iterated expectation (Theorem 7.54) and linearity of expectation (Theorem 7.18),

$$\mathbb{E}\left[\tilde{t}\right] = \mathbb{E}\left[\mu_{\tilde{t}\,|\,\tilde{o},\tilde{c}}(o, c)\right] \tag{7.209}$$

$$= \mathbb{E}\left[\tilde{o} + \tilde{c}\right] \tag{7.210}$$

$$= 0.5 + 0.5 \tag{7.211}$$

$$= 1. \tag{7.212}$$

Example 7.58 (Mean of a mixture model). In Example 6.3, we use a Gaussian mixture model to model height in a population. In the model, height is represented by a continuous random variable $\tilde{h}$ and sex as a discrete random variable $\tilde{s}$. The sample mean of the height equals 163 cm for the women ($\tilde{s} = 0$) and 176 cm for the men ($\tilde{s} = 1$). This coincides with our estimate of the mean parameters of the corresponding Gaussian distributions because the maximum-likelihood estimator of the mean parameter is the sample mean by Theorem 3.35. The conditional mean function therefore equals

$$\mu_{\tilde{h} \mid \tilde{s}}(0) = 163, \tag{7.213}$$

$$\mu_{\tilde{h} \mid \tilde{s}}(1) = 176. \tag{7.214}$$

The sex $\tilde{s}$ is Bernoulli with parameter 0.67. Iterated expectation allows us to compute the mean of $\tilde{h}$ without having to derive its marginal pdf

$$\mathbb{E}[\tilde{h}] = \mathbb{E}[\mu_{\tilde{h} \mid \tilde{s}}(\tilde{s})] \tag{7.215}$$

$$= p_{\tilde{s}}(0)\mu_{\tilde{h} \mid \tilde{s}}(0) + p_{\tilde{s}}(1)\mu_{\tilde{h} \mid \tilde{s}}(1) \tag{7.216}$$

$$= 171.7 \text{ cm}. \tag{7.217}$$

7.8.3 Regression via Conditional Averaging

Regression is the problem of estimating a certain quantity of interest, called the *response*, from observed variables called *features*. From a probabilistic viewpoint, the goal is to approximate a random variable $\tilde{y}$, representing the response, as a deterministic function of a random vector $\tilde{x}$, representing the features. In this section, we study the regression problem under the assumption that we know the joint distribution of the features and the response.

A popular metric to evaluate a regression estimator is the mean squared error (MSE) between the estimator and the response (see Definition 7.29). By Theorem 7.30, the mean is the best constant estimator of a random variable in terms of MSE. Consequently, a reasonable estimator of the response $\tilde{y}$ given that the features equal a certain value, $\tilde{x} = x$, is the conditional mean function $\mu_{\tilde{y} \mid \tilde{x}}(x)$ evaluated at x. This estimator is optimal, in the sense that no other estimator can have smaller MSE.

Theorem 7.59 (Minimum MSE estimator). *Let $\tilde{x}$ and $\tilde{y}$ be a random vector and a random variable, representing the features and response in a regression problem. Among all estimators of the response $\tilde{y}$ that only depend on the features $\tilde{x}$, the conditional mean $\mu_{\tilde{y} \mid \tilde{x}}(\tilde{x})$ achieves the minimum MSE,*

$$\mu_{\tilde{y} \mid \tilde{x}}(\tilde{x}) = \arg\min_{h(\tilde{x})} \mathbb{E}\left[(\tilde{y} - h(\tilde{x}))^2\right]. \tag{7.218}$$

Because of this, the conditional mean is often referred to as the minimum mean-squared-error (MMSE) estimator.

Proof Let $h(\tilde{x})$ be an arbitrary estimator of $\tilde{y}$. Our goal is to show that this estimator cannot achieve a smaller MSE than the conditional mean. By linearity of expectation (Theorem 7.18),

$$\mathbb{E}\left[(\tilde{y} - h(\tilde{x}))^2\right] \tag{7.219}$$

$$= \mathbb{E}\left[\left(\tilde{y} - \mu_{\tilde{y}\,|\,\tilde{x}}(\tilde{x}) + \mu_{\tilde{y}\,|\,\tilde{x}}(\tilde{x}) - h(\tilde{x})\right)^2\right] \tag{7.220}$$

$$= \mathbb{E}\left[(\tilde{y} - \mu_{\tilde{y}\,|\,\tilde{x}}(\tilde{x}))^2\right] + \mathbb{E}\left[(\mu_{\tilde{y}\,|\,\tilde{x}}(\tilde{x}) - h(\tilde{x}))^2\right]$$
$$+ 2\mathbb{E}\left[(\tilde{y} - \mu_{\tilde{y}\,|\,\tilde{x}}(\tilde{x}))(\mu_{\tilde{y}\,|\,\tilde{x}}(\tilde{x}) - h(\tilde{x}))\right].$$

The term

$$\mathbb{E}\left[(\tilde{y} - \mu_{\tilde{y}\,|\,\tilde{x}}(\tilde{x}))(\mu_{\tilde{y}\,|\,\tilde{x}}(\tilde{x}) - h(\tilde{x}))\right] \tag{7.221}$$

$$= \mathbb{E}\left[\tilde{y}\mu_{\tilde{y}\,|\,\tilde{x}}(\tilde{x})\right] - \mathbb{E}[\mu_{\tilde{y}\,|\,\tilde{x}}(\tilde{x})^2] + \mathbb{E}\left[\mu_{\tilde{y}\,|\,\tilde{x}}(\tilde{x})h(\tilde{x})\right] - \mathbb{E}\left[\tilde{y}h(\tilde{x})\right] \tag{7.222}$$

equals zero, as we prove in what follows, which implies

$$\mathbb{E}\left[(\tilde{y} - h(\tilde{x}))^2\right] = \mathbb{E}\left[(\tilde{y} - \mu_{\tilde{y}\,|\,\tilde{x}}(\tilde{x}))^2\right] + \mathbb{E}\left[(\mu_{\tilde{y}\,|\,\tilde{x}}(\tilde{x}) - h(\tilde{x}))^2\right] \tag{7.223}$$

$$\geq \mathbb{E}\left[(\tilde{y} - \mu_{\tilde{y}\,|\,\tilde{x}}(\tilde{x}))^2\right] \tag{7.224}$$

because the second term in (7.223) is nonnegative (it is an integral or sum of a nonnegative quantity).

All that remains is to prove that (7.222) is zero. Assuming that $\tilde{x}$ is continuous[1], by Definition 7.44,

$$\mu_{\tilde{y}h(\tilde{x})\,|\,\tilde{x}}(x) = \int_{y=-\infty}^{\infty} yh(x)f_{\tilde{y}\,|\,\tilde{x}}(y\,|\,x)\,\mathrm{d}y \tag{7.225}$$

$$= h(x)\int_{y=-\infty}^{\infty} yf_{\tilde{y}\,|\,\tilde{x}}(y\,|\,x)\,\mathrm{d}y \tag{7.226}$$

$$= \mu_{\tilde{y}\,|\,\tilde{x}}(x)h(x). \tag{7.227}$$

Consequently, by iterated expectation (Theorem 7.54),

$$\mathbb{E}\left[\tilde{y}h(\tilde{x})\right] = \mathbb{E}\left[\mu_{\tilde{y}h(\tilde{x})\,|\,\tilde{x}}(\tilde{x})\right] \tag{7.228}$$

$$= \mathbb{E}\left[\mu_{\tilde{y}\,|\,\tilde{x}}(\tilde{x})h(\tilde{x})\right], \tag{7.229}$$

so both terms cancel out in (7.222). Setting $h := \mu_{\tilde{y}\,|\,\tilde{x}}$ in (7.229) implies

$$\mathbb{E}\left[\tilde{y}\mu_{\tilde{y}\,|\,\tilde{x}}(\tilde{x})\right] = \mathbb{E}[\mu_{\tilde{y}\,|\,\tilde{x}}(\tilde{x})^2], \tag{7.230}$$

so those two terms also cancel out, and the proof is complete. ∎

Example 7.60 (Cats and dogs: MMSE estimator). Example 7.7 models the number of cats and dogs in a city, represented by the random variables $\tilde{c}$ and $\tilde{d}$, respectively. We consider the problem of estimating the number of cats, if we know the number of dogs. The conditional pmf of $\tilde{c}$ given $\tilde{d}$, computed following Definition 4.13, is shown in Table 7.3. By Definition 7.44, the conditional mean function equals

---

[1] The same argument holds if $\tilde{x}$ is discrete, replacing the integral by a sum and the conditional pdf by the conditional pmf.

Table 7.3 **Conditional pmf of cats given dogs.**
Conditional pmf $p_{\tilde{c}|\tilde{d}}(c|d)$ *of* $\tilde{c}$ *given* $\tilde{d}$ *in Example 7.7.*

		0	1	2	3
Cats (c)					
Dogs (d)	**0**	0.54	0.23	0.15	0.08
	1	0.71	0.18	0.11	0
	2	0.71	0.29	0	0

$$\mu_{\tilde{c}|\tilde{d}}(0) = \sum_{c=0}^{3} c\, p_{\tilde{c}|\tilde{d}}(c\,|\,0) = 0.77, \qquad (7.231)$$

$$\mu_{\tilde{c}|\tilde{d}}(1) = \sum_{c=0}^{3} c\, p_{\tilde{c}|\tilde{d}}(c\,|\,1) = 0.39, \qquad (7.232)$$

$$\mu_{\tilde{c}|\tilde{d}}(2) = \sum_{c=0}^{3} c\, p_{\tilde{c}|\tilde{d}}(c\,|\,2) = 0.29. \qquad (7.233)$$

By Theorem 7.59, this is the optimal estimator of $\tilde{c}$ given $\tilde{d}$ in terms of MSE. The MSE equals

$$\mathbb{E}\left[\left(\tilde{c} - \mu_{\tilde{c}|\tilde{d}}(\tilde{d})\right)^2\right] = \sum_{c=0}^{3}\sum_{d=0}^{2} p_{\tilde{c},\tilde{d}}(c,d)\left(c - \mu_{\tilde{c}|\tilde{d}}(d)\right)^2 \qquad (7.234)$$

$$= 0.756. \qquad (7.235)$$

Notice that our estimator is optimal in terms of MSE, but it is abysmal in terms of probability of error: Nobody can own 0.77, 0.4, or 0.29 cats, so we are always wrong! By Theorem 4.30, the best estimate in terms of probability of error is the MAP estimator of $\tilde{c}$ given $\tilde{d} = d$,

$$\text{MAP}(0) = \arg\max_{c} p_{\tilde{c}|\tilde{d}}(c\,|\,0) = 0, \qquad (7.236)$$

$$\text{MAP}(1) = \arg\max_{c} p_{\tilde{c}|\tilde{d}}(c\,|\,1) = 0, \qquad (7.237)$$

$$\text{MAP}(2) = \arg\max_{c} p_{\tilde{c}|\tilde{d}}(c\,|\,2) = 0. \qquad (7.238)$$

The MAP estimator equals zero, no matter what value of $\tilde{d}$ is observed. Its MSE is

$$\mathbb{E}\left[\left(\tilde{c} - \text{MAP}(\tilde{d})\right)^2\right] = \sum_{c=0}^{3}\sum_{d=0}^{2} p_{\tilde{c},\tilde{d}}(c,d)c^2 \qquad (7.239)$$

$$= 1.19, \qquad (7.240)$$

which is substantially larger than the MSE of the MMSE estimator.

....

Theorem 7.59 makes regression sound easy, but this could not be farther from the truth! The conditional mean is indeed an optimal estimator, but it is *intractable to compute* unless the number of features is very small. The reason is that the number of possible conditional distributions of the response given the features explodes exponentially. For instance, consider the regression problem of estimating the rating for the movie Independence Day from the rating for Mission Impossible, in Example 7.48. The MMSE estimator can be easily approximated via the sample conditional mean function, shown in the left plot of Figure 7.8, because there is only one feature. However, what if we want to estimate the rating for Independence Day from the ratings of *100 other movies*? Then, to approximate the minimum MSE estimator, we need to compute the sample conditional mean given all the possible ratings of the 100 movies. However, there are $5^{100} > 10^{68}$ possibilities, so this is impossible! When we encounter a new user, they are likely to have a unique set of ratings that we have not observed before. This is a manifestation of the curse of dimensionality, described in Section 4.7. In Chapter 12 we describe several methods that address this challenge and enable us to perform regression in practice.

7.9 The Average Treatment Effect

In Section 4.6, we explain how to estimate the causal effect of a treatment on a binary outcome with two possible values (e.g. whether a patient recovers or not). In this section, we study outcomes with multiple possible values.[2] As a motivating example, consider the question: *Does capitalizing the title of YouTube videos increase the number of views?* Our goal is to determine whether the treatment (capitalization of the title) has a causal effect on the number of views, which is our outcome of interest.

Applying the causal-inference framework presented in Section 4.6.1, we represent the treatment as a random variable $\tilde{t}$. In our motivating example, $\tilde{t} = 1$ indicates that the title is set to all caps and $\tilde{t} = 0$ indicates that the title is set to proper case. In order to study the causal effect of the treatment, we define its associated potential outcomes. In our example, the potential outcome $\widetilde{po}_0$ represents the number of views in a hypothetical situation where all titles are in proper case. Conversely, the potential outcome $\widetilde{po}_1$ represents the number of views, if all titles are in all caps. The *average treatment effect* (ATE) is the difference between the means of the two potential outcomes. It quantifies the mean causal effect of the treatment on the outcome of interest.

Definition 7.61 (Average treatment effect). *Given two potential outcomes* $\widetilde{po}_0$ *and* $\widetilde{po}_1$ *associated with a treatment* $\tilde{t}$, *the average treatment effect is*

$$\mathrm{ATE} := \mathbb{E}\left[\widetilde{po}_1\right] - \mathbb{E}\left[\widetilde{po}_0\right]. \tag{7.241}$$

The challenge in estimating the ATE is that our observations of the potential outcomes are incomplete, due to the fundamental problem of causal inference (see Section 4.6.1). The observed outcome $\tilde{y}$ is equal to $\widetilde{po}_0$, only if $\tilde{t} = 0$, and to $\widetilde{po}_1$, only if $\tilde{t} = 1$:

$$\tilde{y} := \begin{cases} \widetilde{po}_0 & \text{if} \quad \tilde{t} = 0, \\ \widetilde{po}_1 & \text{if} \quad \tilde{t} = 1. \end{cases} \tag{7.242}$$

In order to determine the ATE, we need to estimate the mean of the potential outcomes $\mathbb{E}\left[\widetilde{po}_0\right]$ and $\mathbb{E}\left[\widetilde{po}_1\right]$ from the observed outcome $\tilde{y}$. An option is to use the conditional means of the observed outcome given the treatment, $\mu_{\tilde{y}\,|\,\tilde{t}}(0)$ and $\mu_{\tilde{y}\,|\,\tilde{t}}(1)$. However, this is *not*

---

[2] Recall that in causal inference, the term *outcome* does not refer to an element in a sample space, as in Section 1.2. Instead, it is a random variable representing a quantity of interest that may (or may not) be affected by the treatment.

necessarily a good estimate. By Definition 7.44 and (7.242), the conditional mean function given $\tilde{t} = 1$ equals

$$\mu_{\tilde{y}\,|\,\tilde{t}}(1) = \sum_{y=0}^{\infty} y\, p_{\tilde{y}\,|\,\tilde{t}}(y\,|\,1) \tag{7.243}$$

$$= \sum_{y=0}^{\infty} y\, p_{\widetilde{po}_1\,|\,\tilde{t}}(y\,|\,1), \tag{7.244}$$

where we assume that the views are nonnegative integers. By contrast, by Definition 7.1, the mean of the corresponding potential outcome $\widetilde{po}_1$ equals

$$\mathbb{E}\left[\widetilde{po}_1\right] = \sum_{y=0}^{\infty} y\, p_{\widetilde{po}_1}(y). \tag{7.245}$$

These values are the same only if the treatment and the potential outcome are independent. In that case, by Definition 4.17, the conditional pmf $p_{\widetilde{po}_1\,|\,\tilde{t}}(\cdot\,|\,1)$ is the same as the marginal pmf $p_{\widetilde{po}_1}$. In our YouTube example, this means that the distribution of views for the videos with capitalized titles ($\tilde{t} = 1$) is the same as the distribution of views for all videos. If that is the case, then we can use the conditional mean to compute the correct ATE.

Theorem 7.62 (Estimation of the average treatment effect). *Let $\tilde{y}$ be a random variable defined as in (7.242), where $\widetilde{po}_0$ and $\widetilde{po}_1$ denote the two potential outcomes associated with a treatment $\tilde{t}$. If $\tilde{t}$ and $\widetilde{po}_0$ are independent, and $\tilde{t}$ and $\widetilde{po}_1$ are also independent, then the average treatment effect, introduced in Definition 7.61, equals*

$$\text{ATE} = \mu_{\tilde{y}\,|\,\tilde{t}}(1) - \mu_{\tilde{y}\,|\,\tilde{t}}(0). \tag{7.246}$$

Proof We assume that the potential outcomes are continuous, the same argument holds for discrete outcomes, replacing the integrals by sums and the pdfs by pmfs. If $\tilde{t} = 1$, then $\tilde{y} = \widetilde{po}_1$, so $\mu_{\tilde{y}\,|\,\tilde{t}}(1) = \mu_{\widetilde{po}_1\,|\,\tilde{t}}(1)$. By the independence assumption and Definition 5.15, the conditional pdf of $\widetilde{po}_1$ given $\tilde{t}$ is equal to the marginal pdf, so by Definitions 7.44 and 7.8,

$$\mu_{\widetilde{po}_1\,|\,\tilde{t}}(1) = \int_{y=-\infty}^{\infty} y f_{\widetilde{po}_1\,|\,\tilde{t}}(y\,|\,1)\,dy \tag{7.247}$$

$$= \int_{y=-\infty}^{\infty} y f_{\widetilde{po}_1}(y)\,dy \tag{7.248}$$

$$= \mathbb{E}\left[\widetilde{po}_1\right]. \tag{7.249}$$

By the same argument, $\mu_{\tilde{y}\,|\,\tilde{t}}(0) = \mathbb{E}\left[\widetilde{po}_0\right]$. Plugging these identities into Definition 7.61 completes the proof. ∎

An important consequence of Theorem 7.62 is that the ATE can be estimated reliably from randomized experiments, such as randomized controlled trials, which guarantee independence between the treatment and the potential outcomes, as discussed in Section 4.6.3.

Example 7.63 (All caps titles in YouTube videos). In order to evaluate the effect of capitalization on YouTube videos, we randomized the title capitalization of 46 videos on probability and statistics for data science. Each time a video was posted, the title was set to all caps with probability 1/2 and to proper case with probability 1/2. The sample mean of the number of

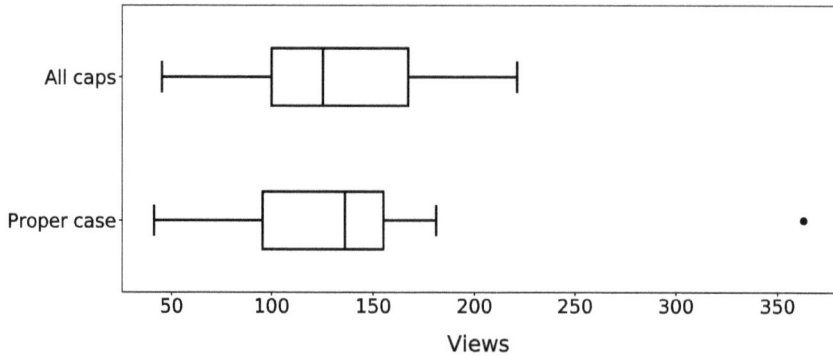

Figure 7.9 Effect of title capitalization in YouTube videos. The graph compares the distribution of the number of views of videos with all-caps and proper-case titles in Example 7.63, using box plots (see Section 3.2.3). The distributions are very similar. Since capitalization was randomized, this is strong evidence that it has essentially no causal effect on the number of views.

. . . .

views for the 19 videos with all-caps titles was 133. The sample mean of the number of views for the 27 videos with proper-case titles was 132. Due to the treatment randomization, we can apply Theorem 7.62 and estimate the ATE as the difference between these conditional averages, which is very small ($133 - 132 = 1$). This suggests that capitalization makes very little difference in the number of views. Figure 7.9 shows the distribution of the number of views for both types of title.

. . . .

In observational studies, where we have no control over the treatment, we should not trust ATE estimates computed using the difference between observed conditional means. The reason is that confounders can completely distort the estimate. This is the same phenomenon described in Example 4.22 and Section 4.6.2.

Example 7.64 (Private lessons). Our goal in this example is to study the causal effect of private math lessons on the performance of Portuguese high-school students, using data extracted from Dataset 15. The observed outcome of interest is the grade obtained in a math class, represented by a random variable $\tilde{y}$. The private lessons are represented by a random variable $\tilde{t}$, since we interpret them as a treatment. $\tilde{t} = 1$ indicates that the student received private lessons, so they belong to the treatment group. $\tilde{t} = 0$ indicates that they did not receive private lessons, so they belong to the control group.

The conditional mean function of $\tilde{y}$ given $\tilde{t}$, computed following Definition 7.47, equals

$$\mu_{\tilde{y}\,|\,\tilde{t}}(1) = 10.92, \qquad \mu_{\tilde{y}\,|\,\tilde{t}}(0) = 9.99. \tag{7.250}$$

We define the *observed* ATE as the difference between these two values:

$$\text{observed ATE} := \mu_{\tilde{y}\,|\,\tilde{t}}(1) - \mu_{\tilde{y}\,|\,\tilde{t}}(0) = 0.93. \tag{7.251}$$

According to the observed ATE, private lessons increase a student grade by one point (out of 20) on average, which is quite substantial. In a randomized study, the observed ATE is equal to the true ATE by Theorem 7.62, so we could be tempted to conclude that private lessons are effective. However, our data are *not* randomized, so we should determine whether there are any systematic differences between the treatment and control groups, which might distort the observed ATE.

The data also report what students had previously failed the course, which is a possible confounding factor. We represent it using a random variable $\tilde{c}$: $\tilde{c} = 1$ indicates that the student previously failed the course, and $\tilde{c} = 0$ that they didn't. The conditional mean function of the grade $\tilde{y}$ given $\tilde{c}$ and $\tilde{t}$ for students that had previously failed, computed following Definition 7.47, is

$$\mu_{\tilde{y}|\tilde{c},\tilde{t}}(1,1) = 8.95, \qquad \mu_{\tilde{y}|\tilde{c},\tilde{t}}(1,0) = 6.66. \tag{7.252}$$

For these students, it looks like the private lessons may be helpful: The conditional average grade is more than two points higher given $\tilde{t} = 1$. The conditional mean function for students who had not failed is

$$\mu_{\tilde{y}|\tilde{c},\tilde{t}}(0,1) = 11.20, \qquad \mu_{\tilde{y}|\tilde{c},\tilde{t}}(0,0) = 11.31. \tag{7.253}$$

These students have better grades than the *previously failed* group, whether they receive private lessons or not. Those that receive private lessons have a slightly worse grade on average than those who don't.

A key consideration when analyzing a possible confounding factor is whether it is independent from the treatment. Our confounder $\tilde{c}$ is definitely not independent from the treatment because

$$p_{\tilde{c}|\tilde{t}}(1\,|\,1) = 0.122 \quad \text{and} \quad p_{\tilde{c}|\tilde{t}}(1\,|\,0) = 0.285 \tag{7.254}$$

are very different, which contradicts Definition 4.17. Students in the control group are more than twice as likely to have failed than those in the treatment group. This is very problematic for our analysis. Students who previously failed have lower grades than the rest of the students. Consequently, the treatment group may have a higher average grade just because it contains less such students, *not because the private lessons are useful*.

To analyze the effect of the confounder quantitatively, we derive an expression of the conditional mean function of the observed grades $\tilde{y}$ given $\tilde{t}$ as a function of the conditional mean function of $\tilde{y}$ given $\tilde{t}$ and also $\tilde{c}$. By Definition 7.44,

$$\mu_{\tilde{y}|\tilde{t}}(t) = \int_{y=-\infty}^{\infty} y f_{\tilde{y}|\tilde{t}}(y\,|\,t) \, \mathrm{d}y \tag{7.255}$$

$$= \sum_{c=0}^{1} p_{\tilde{c}|\tilde{t}}(c\,|\,t) \int_{y=-\infty}^{\infty} y f_{\tilde{y}|\tilde{c},\tilde{t}}(y\,|\,c,t) \, \mathrm{d}y \tag{7.256}$$

$$= \sum_{c=0}^{1} p_{\tilde{c}|\tilde{t}}(c\,|\,t) \mu_{\tilde{y}|\tilde{c},\tilde{t}}(c,t), \tag{7.257}$$

where we have used the fact that, by the chain rule for discrete and continuous random variables (Theorem 6.6),

$$f_{\tilde{y}|\tilde{t}}(y\,|\,t) = \sum_{c=0}^{1} p_{\tilde{c}|\tilde{t}}(c\,|\,t) f_{\tilde{y}|\tilde{c},\tilde{t}}(y\,|\,c,t). \tag{7.258}$$

We can use (7.257) to separate the contributions of the *previously failed* and *not previously failed* groups to the observed conditional mean:

$$\mu_{\tilde{y}\,|\,\tilde{t}}(1) = p_{\tilde{c}\,|\,\tilde{t}}(0\,|\,1)\mu_{\tilde{y}\,|\,\tilde{c},\tilde{t}}(0,1) + p_{\tilde{c}\,|\,\tilde{t}}(1\,|\,1)\mu_{\tilde{y}\,|\,\tilde{c},\tilde{t}}(1,1) \tag{7.259}$$

$$= 0.878 \cdot 11.20 + 0.122 \cdot 8.95 \tag{7.260}$$

$$= \underset{\underset{\tilde{c}=0}{\uparrow}}{9.83} + \underset{\underset{\tilde{c}=1}{\uparrow}}{1.09} = 10.92. \tag{7.261}$$

Similarly,

$$\mu_{\tilde{y}\,|\,\tilde{t}}(0) = p_{\tilde{c}\,|\,\tilde{t}}(0\,|\,0)\mu_{\tilde{y}\,|\,\tilde{c},\tilde{t}}(0,0) + p_{\tilde{c}\,|\,\tilde{t}}(1\,|\,0)\mu_{\tilde{y}\,|\,\tilde{c},\tilde{t}}(1,0) \tag{7.262}$$

$$= 0.715 \cdot 11.31 + 0.285 \cdot 6.66 \tag{7.263}$$

$$= \underset{\underset{\tilde{c}=0}{\uparrow}}{8.09} + \underset{\underset{\tilde{c}=1}{\uparrow}}{1.90} = 9.99. \tag{7.264}$$

It turns out that the contribution of the *not previously failed* group drives the observed ATE up, *even though private lessons seem to make no difference for that group.* By contrast, the contribution of the *previously failed* group drives the observed ATE down, even though private lessons do seem to result in higher grades for that group. As we had feared, the observed ATE is severely distorted by the dependence between the confounder and the treatment, just like in Simpson's paradox (see Section 4.6.2).

. . . .

In Section 4.6.4, we explain how to adjust for confounders in the case of binary outcomes. The same procedure can be used to control for confounders when estimating the ATE. Crucially, we again require that the potential outcomes be conditionally independent from the treatment given the confounder. In that case, the conditional means are not distorted by additional confounders, and we can correct the ATE by reweighting the conditional means of the outcome given the treatment and the confounder.

Theorem 7.65 (Adjusting the ATE to control for confounders). *Let $\tilde{y}$ be a random variable following the definition in* (7.242), *where $\widetilde{po}_0$ and $\widetilde{po}_1$ denote the two potential outcomes associated with a treatment $\tilde{t}$, and let $\tilde{c}$ be a confounder, represented by a discrete random variable with range C. If the treatment $\tilde{t}$ and the potential outcomes $\widetilde{po}_0$ and $\widetilde{po}_1$ are conditionally independent given $\tilde{c}$, then the average treatment effect, introduced in Definition 7.61, equals*

$$\text{ATE} = \sum_{c \in C} p_{\tilde{c}}(c)\mu_{\tilde{y}\,|\,\tilde{c},\tilde{t}}(c,1) - \sum_{c \in C} p_{\tilde{c}}(c)\mu_{\tilde{y}\,|\,\tilde{c},\tilde{t}}(c,0). \tag{7.265}$$

Proof If $\tilde{t} = 1$, then $\tilde{y} = \widetilde{po}_1$, so $\mu_{\tilde{y}\,|\,\tilde{c},\tilde{t}}(c,1) = \mu_{\widetilde{po}_1\,|\,\tilde{c},\tilde{t}}(c,1)$. By the conditional independence assumption and Definition 7.44,

$$\mu_{\tilde{y}\,|\,\tilde{c},\tilde{t}}(c,1) = \mu_{\widetilde{po}_1\,|\,\tilde{c},\tilde{t}}(c,1) = \int_x x f_{\widetilde{po}_1\,|\,\tilde{c},\tilde{t}}(x\,|\,c,1)\ \mathrm{d}x \tag{7.266}$$

$$= \int_x x f_{\widetilde{po}_1\,|\,\tilde{c}}(x\,|\,c)\ \mathrm{d}x \tag{7.267}$$

$$= \mu_{\widetilde{po}_1\,|\,\tilde{c}}(c). \tag{7.268}$$

By the same argument,

$$\mu_{\tilde{y}\,|\,\tilde{c},\tilde{t}}(c,0) = \mu_{\widetilde{po}_0\,|\,\tilde{c}}(c). \tag{7.269}$$

We have assumed that the potential outcomes are continuous, the same argument holds for discrete outcomes replacing the integrals by sums and the pdfs by pmfs.

The result then follows from aggregating the conditional means given the confounder and applying iterated expectation (Theorem 7.54),

$$\sum_{c \in C} p_{\tilde{c}}(c) \mu_{\tilde{y} \mid \tilde{c}, \tilde{t}}(c, 1) - \sum_{c \in C} p_{\tilde{c}}(c) \mu_{\tilde{y} \mid \tilde{c}, \tilde{t}}(c, 0) \tag{7.270}$$

$$= \sum_{c \in C} p_{\tilde{c}}(c) \mu_{\widetilde{po}_1 \mid \tilde{c}}(c) - \sum_{c \in C} p_{\tilde{c}}(c) \mu_{\widetilde{po}_0 \mid \tilde{c}}(c) \tag{7.271}$$

$$= \mathbb{E}\left[\mu_{\widetilde{po}_1 \mid \tilde{c}}(\tilde{c})\right] - \mathbb{E}\left[\mu_{\widetilde{po}_0 \mid \tilde{c}}(\tilde{c})\right] \tag{7.272}$$

$$= \mathbb{E}\left[\widetilde{po}_1\right] - \mathbb{E}\left[\widetilde{po}_0\right]. \tag{7.273}$$

∎

Example 7.66 (Private lessons: Adjusting for previous failure). In Example 7.64 the observed ATE is distorted by a confounder. The fractions of students, which previously failed the course, are very different in the treatment and control groups, see (7.254). In order to correct for this, we apply Theorem 7.65. From the data, the marginal probability of previous failure (estimated as the fraction of students that previously failed, following Definition 2.11) equals 0.21. The adjusted ATE is obtained by using this probability to reweight the conditional means given the treatment and the confounder, reported in (7.252) and (7.253),

$$\text{adjusted ATE} := \sum_{c=0}^{1} p_{\tilde{c}}(c) \mu_{\tilde{y} \mid \tilde{c}, \tilde{t}}(c, 1) - \sum_{c=0}^{1} p_{\tilde{c}}(c) \mu_{\tilde{y} \mid \tilde{c}, \tilde{t}}(c, 0) \tag{7.274}$$

$$= (0.79 \cdot 11.20 + 0.21 \cdot 8.95) - (0.79 \cdot 11.31 + 0.21 \cdot 6.66) \tag{7.275}$$

$$= 0.39. \tag{7.276}$$

After the correction, the ATE is reduced to less than half, from 0.93 to 0.39.

Should we conclude that private lessons are not very useful? Definitely not! The conditional independence assumption in Theorem 7.65 implies that within the *not previously failed* group and the *previously failed* group, the students that take private lessons are not systematically different from the students that do not take private lessons. It is difficult to believe that this is the case: Students often take private lessons *because they need them*. If the students receiving private lessons are systematically weaker, then our estimated ATE underestimates the true ATE.

A counterargument is that students often take private lessons *because they can afford them*. It is therefore possible that students taking private lessons have more resources at their disposal, and would do better anyway. In that case, our estimated ATE overestimates the true ATE. In conclusion, more analysis is needed in order to control for additional confounders, such as academic ability and socioeconomic status. This example reflects the shortcomings of observational studies. Randomizing the treatment automatically controls for all possible confounders, but it is often problematic to implement in practice. Imagine the reaction of some parents, if their kid were randomly assigned to a control group without private lessons...

Exercises

7.1 (Race) The organizers of a race with three different distances (5, 10, and 20 km) ask the participants to declare what distance they plan to run. However, the participants can then choose to change their mind. Table 7.4 shows the joint pmf of two random variables representing the distance declared by the participants ($\tilde{d}$) and the actual distance they ran ($\tilde{a}$), respectively.

   a Compute the mean of the difference $\tilde{d} - \tilde{a}$ between the declared and actual distance.
   b Compute the variance of the actual distance $\tilde{a}$.

Table 7.4 *Table for Exercise 7.1.*

Actual distance

$p_{\tilde{d},\tilde{a}}$	5 km	10 km	20 km
5 km	0.2	0.1	0
10 km	0.1	0.2	0
20 km	0.1	0.2	0.1

Declared distance

7.2 (Mean of a function) Let $\tilde{a}$ be a discrete random variable with range A and pmf $p_{\tilde{a}}$, and let $\tilde{b} := h(\tilde{a})$ be a random variable with range B, obtained by applying a deterministic function $h \colon \mathbb{R} \to \mathbb{R}$ to $\tilde{a}$. Prove that our definition of $\mathbb{E}[h(\tilde{a})]$ is sound, i.e. that

$$\mathbb{E}[\tilde{b}] := \sum_{b \in B} b\, p_{\tilde{b}}(b) \tag{7.277}$$

$$= \sum_{a \in A} h(a) p_{\tilde{a}}(a). \tag{7.278}$$

7.3 (Presents) In a class with n children, the teacher asks the parents to leave a present for each kid under the Christmas tree in the classroom. The day after, each child picks a present at random. What is the expected number of children that end up getting the present bought by their own parents?

7.4 (Cats and dogs) Compute the standard deviation of the total number of animals (cats and dogs) per household in Example 7.7.

7.5 (Mean squared difference) Let $\tilde{a}$ and $\tilde{b}$ be two independent random variables with the same mean μ and standard deviation σ^2. What is their mean squared difference $\mathbb{E}\left[(\tilde{a} - \tilde{b})^2\right]$ equal to?

7.6 (Basketball player) Each time a basketball player attempts a shot, it is a 2-point shot with probability α and a 3-point shot with probability $1 - \alpha$, where $0 \le \alpha \le 1$. The probability that she makes the 2-point shot is 0.5. The probability that she makes the 3-point shot is 0.4. Compute the standard deviation of the points she scores per shot, as a function of α. What are the maximum and minimum values of the standard deviation, and at what values of α are they attained?

7.7 (Computer defect) Recall the model in Exercise 6.4 for the time $\tilde{t}$ a computer lasts until it breaks down, depending on the presence of a defect $\tilde{d}$.

   a Is the conditional expectation of $\tilde{t}$ given $\tilde{d}$ a discrete or a continuous random variable? What is its pmf or pdf?
   b What is the variance of $\tilde{t}$?
   c A company buys 100 of these computers. If the time until they break down is distributed as explained in Exercise 6.4, and they are all independent, what is the mean and variance of the number of computers that break down during the first year?

7.8 (Hen breeds) Figure 7.10 shows the conditional pmf of the eggs per week laid by two different breeds of hen (bantam and Sussex) given the weather (cold or warm).

a Martin lives in Florida, where it is always warm. If he buys four bantam and three Sussex hens, what is the mean of the total number of eggs that they will produce? Do *not* assume independence between the different hens.

b Adela wants to buy a single hen. Let α be the probability of warm weather where she lives. For what values of α would a Sussex hen produce more eggs than a bantam hen on average?

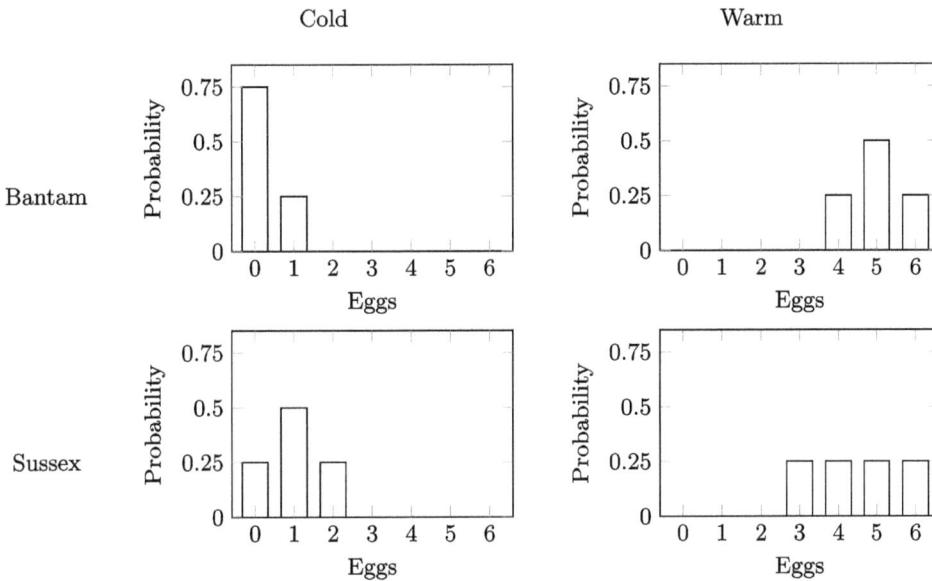

Figure 7.10 Graphs for Exercise 7.8.

7.9 (Hotel questionnaire) A hotel wants to determine average customer satisfaction using a questionnaire that asks customers for a discrete score between 1 (worst) and 5 (best). Twenty percent of the customers are retired. The average score given by the retired customers is 2, and they answer the questionnaire with probability 0.9. The average score given by the rest of the customers is 4, and they answer the questionnaire with probability 0.25. We define a Bernoulli random variable $\tilde{r}$ to represent whether a customer is retired ($\tilde{r} = 1$) or not ($\tilde{r} = 0$), another Bernoulli random variable $\tilde{q}$ to represent whether they answer the questionnaire ($\tilde{q} = 1$) or not ($\tilde{q} = 0$), and a discrete random variable $\tilde{s}$ to represent the score. We assume that $\tilde{q}$ and $\tilde{s}$ are conditionally independent given $\tilde{r}$.

a What is the mean of the score $\tilde{s}$?
b What is the mean of the *observed* score, i.e. the conditional mean of $\tilde{s}$ given $\tilde{q} = 1$?

7.10 (Another basketball player) From past statistics, we determine that the probability that Dimitra makes a 2-point shot, a 3-point shot, and a free throw is 0.5, 0.3, and 0.8 respectively. We model the number of 2-point shots attempted by Dimitra as being uniformly distributed between 0 and 4 (i.e. she attempts 0, 1, 2, 3, or 4 shots with the same probability). The distributions of the attempted three-point shots and free throws are the same. What is the mean number of points scored by Dimitra?

7.11 (Life expectancy) During the Middle Ages, life expectancy was much shorter than in modern times. According to some calculations, the mean age of death was 37.5 years. To some extent, this was driven by the high infant mortality: It has been estimated that the probability of dying during the first year of age was 0.25. If we assign an age of death equal to zero to anyone dying in their first year, what was the mean age of death of the people who survived the first year?

7.12 (Buckets) What is the variance of the amount of sand $\tilde{s}_1$ first grabbed by the toddler in Exercise 6.2?

7.13 (Potatoes) Recall the model for weather, beetle presence, and potato production in Exercise 6.6.

    a This year the weather is good, but the beetle is present. What is the best estimate for the potato production in terms of mean squared error according to the model?

    b Compute the conditional mean of the potato production, according to the model, if the weather is bad.

7.14 (Wolf) Recall the model for the location $\tilde{x}$ of the wolf in Exercise 6.5.

    a What is the mean of $\tilde{x}[2]$?

    b What is the minimum mean-squared-error estimator of $\tilde{x}[1]$ given $\tilde{x}[2]$?

7.15 (Law of conditional variance) We define the conditional variance function $\nu_{\tilde{b}\,|\,\tilde{a}}(a)$ of a random variable $\tilde{b}$ given another random variable $\tilde{a}$ as the variance of $\tilde{b}$, conditioned on the event $\tilde{a} = a$. Analogously to the conditional mean, we define the conditional variance $\nu_{\tilde{b}\,|\,\tilde{a}}(\tilde{a})$ of $\tilde{b}$ given $\tilde{a}$ as the random variable obtained by plugging $\tilde{a}$ into the conditional variance function.

    a Compute the conditional variance function $\nu_{\tilde{b}\,|\,\tilde{a}}$ for the random variables $\tilde{a}$ and $\tilde{b}$ in Example 4.6.

    b Compute the pmf of the conditional variance $\nu_{\tilde{b}\,|\,\tilde{a}}(\tilde{a})$ for the random variables $\tilde{a}$ and $\tilde{b}$ in Example 4.6.

    c Prove the law of conditional variance:

$$\mathrm{Var}[\tilde{b}] = \mathbb{E}\left[\nu_{\tilde{b}\,|\,\tilde{a}}(\tilde{a})\right] + \mathrm{Var}\left[\mu_{\tilde{b}\,|\,\tilde{a}}(\tilde{a})\right]. \tag{7.279}$$

    d Verify that the law of conditional variance holds for the random variables $\tilde{a}$ and $\tilde{b}$ in Example 4.6.

7.16 (Faulty database) An entry in a database can be modeled as a Gaussian random variable $\tilde{a}$ with zero mean and variance σ^2. Due to a programming error, for a certain fixed constant x, the entry is replaced by $-x$ with probability 0.1 and x with probability 0.1, independently of the content of the entry. Otherwise, the entry remains the same. We model the resulting entry as a random variable $\tilde{b}$.

    a Compute the variance of $\tilde{b}$.

    b Derive the minimum mean-squared error estimator of $\tilde{a}$ given $\tilde{b} = b$, for all possible values of b.

7.17 (Coffee and life expectancy) An observational study to analyze the effect of drinking coffee on life expectancy reports the following data:

Number of people				Average life span (years)		
	Coffee	No coffee			Coffee	No coffee
Smokers	80	20		Smokers	68	70
Nonsmokers	40	60		Nonsmokers	82	80

a What is the observed average treatment effect (ATE) of drinking coffee on life expectancy?

b Estimate the ATE of drinking coffee on life expectancy, adjusting for smoking. Assume that the potential outcomes associated with the life span of the study participants are conditionally independent from the treatment (drinking coffee or not) given their smoking status.

7.18 (Weight-loss supplement) In order to evaluate the effectiveness of a weight-loss supplement, we record the weight gain/loss in kilograms after 1 month of a treatment group that takes the supplement and a control group that doesn't. In bold, we highlight the participants who report that they exercise daily.
Supplement: 0.5, **–4.6**, **–7.5**, 1.1, **–3.4**, **–1.4**, **3.2**, –1.9, **–3.6**, 2.4, 1.2, **4.2**, **–2.5**, **–3.5**, 4.7
No supplement: 1.6, **–6.7**, 2.3, **–4.5**, **3.4**, –0.6, 2.3, 1.5, 1.7, 2.6, 0.5, 1.2, **–3.6**, 2.1, –1.2, 0.4

a Compute the observed average treatment effect (ATE) given by the difference in average weight gain/loss between the treatment and control groups.

b Estimate the ATE making any adjustments that you think are reasonable. Explain under what assumptions your estimate approximates the true ATE.

7.19 (Chocolate bar) A company hires a data scientist with dubious ethics to market a chocolate bar that is supposed to help reduce cholesterol. The data scientist designs a study, where half of the participants are men and half are women. The treatment group in the study eats one chocolate bar daily for a week. In reality, the average treatment effect (ATE) of doing this is +10 mg/dL for men, and the same for women. If they don't take the bar, the men in the study have an average cholesterol level of 140 mg/dL and the women have an average cholesterol level of 120 mg/dL. The data scientist assigns each man to the treatment group with probability α and to the control group with probability $1 - \alpha$, in order to manipulate the observed ATE. The women are assigned to either group with probability 1/2.

a Derive the observed ATE as a function of α.

b Evaluate your expression for α equal to 0.05, 0.5, and 0.95.

8

Correlation

Overview

Correlation quantifies to what extent two uncertain quantities are linearly related, on average. Section 8.1 defines the correlation coefficient between random variables that are normalized to have zero mean and unit variance. Section 8.2 extends the definition to all random variables and introduces the covariance, a related measure of linear dependence. In Section 8.3, we describe how to estimate the correlation coefficient and the covariance from data. In Section 8.4, we introduce linear regression and study its connection with correlation. Section 8.5 establishes several important properties of the correlation coefficient. In Section 8.6, we discuss the differences between uncorrelation and independence. Section 8.7 provides geometric intuition about correlation, based on the insight that the covariance can be interpreted as an inner product between random variables. Finally, Section 8.8 explains why correlation between two quantities does not necessarily imply a causal relationship.

8.1 Correlation between Standardized Quantities

The correlation between two random variables quantifies the average *linear* dependence between them. In this section, we assume that the random variables are *standardized*, which means that their mean is zero and their variance is equal to one (we extend our definitions to nonstandardized variables in Section 8.2).

In order to quantify the linear dependence between two standardized random variables $\tilde{a}$ and $\tilde{b}$, we approximate $\tilde{b}$ using a linear function of $\tilde{a}$, obtained multiplying $\tilde{a}$ by a constant scaling factor. Intuitively, the better the linear approximation, the more related the two variables are. We evaluate the approximation using the mean squared error (MSE), introduced in Definition 7.29. As established in the following theorem, the scaling factor that minimizes the MSE is the mean of the product between the two variables. We call this quantity the *correlation coefficient* of $\tilde{a}$ and $\tilde{b}$.

Theorem 8.1 (Linear dependence between standardized random variables). *Let $\tilde{a}$ and $\tilde{b}$ be two random variables with zero mean and unit variance, belonging to the same probability space. The correlation coefficient between $\tilde{a}$ and $\tilde{b}$, defined as*

$$\rho_{\tilde{a},\tilde{b}} := \mathbb{E}[\tilde{a}\tilde{b}], \tag{8.1}$$

is the scaling factor that achieves the best linear approximation in terms of MSE,

$$\rho_{\tilde{a},\tilde{b}} = \arg\min_{\beta} \mathbb{E}\left[(\tilde{b} - \beta\tilde{a})^2\right]. \tag{8.2}$$

Consequently, the linear minimum MSE estimator of $\tilde{b}$ given $\tilde{a}$ is $\rho_{\tilde{a},\tilde{b}}\,\tilde{a}$. The corresponding MSE equals $1 - \rho_{\tilde{a},\tilde{b}}^2$.

Proof By linearity of expectation (Theorem 7.18),

$$\mathrm{MSE}(\beta) := \mathbb{E}\left[(\tilde{b} - \beta\tilde{a})^2\right] = \mathbb{E}\left[\tilde{b}^2 - 2\beta\tilde{a}\tilde{b} + \beta^2\tilde{a}^2\right] \tag{8.3}$$

$$= \mathbb{E}[\tilde{b}^2] + \beta^2\mathbb{E}[\tilde{a}^2] - 2\beta\mathbb{E}[\tilde{a}\tilde{b}] \tag{8.4}$$

$$= 1 + \beta^2 - 2\beta\mathbb{E}[\tilde{a}\tilde{b}]. \tag{8.5}$$

Taking derivatives reveals that the quadratic function is convex:

$$\frac{\mathrm{d}\,\mathrm{MSE}(\beta)}{\mathrm{d}\beta} = 2\left(\beta - \mathbb{E}[\tilde{a}\tilde{b}]\right), \tag{8.6}$$

$$\frac{\mathrm{d}^2\,\mathrm{MSE}(\beta)}{\mathrm{d}\beta^2} = 2, \tag{8.7}$$

so we can set the first derivative to zero to find the minimum. As a result, the optimal scaling factor is $\rho_{\tilde{a},\tilde{b}} := \mathbb{E}[\tilde{a}\tilde{b}]$.

The MSE corresponding to the optimal linear estimator $\rho_{\tilde{a},\tilde{b}}\,\tilde{a}$ equals

$$\mathbb{E}\left[(\tilde{b} - \rho_{\tilde{a},\tilde{b}}\,\tilde{a})^2\right] = \mathbb{E}[\tilde{b}^2] + \rho_{\tilde{a},\tilde{b}}^2\mathbb{E}[\tilde{a}^2] - 2\rho_{\tilde{a},\tilde{b}}\mathbb{E}[\tilde{a}\tilde{b}] \tag{8.8}$$

$$= 1 - \rho_{\tilde{a},\tilde{b}}^2 \tag{8.9}$$

by linearity of expectation (Theorem 7.18), and the assumption that $\tilde{a}$ and $\tilde{b}$ have zero mean and unit variance, which implies that their mean square is one by Lemma 7.33. ∎

By Theorem 8.1, the error we incur when we approximate the standardized random variable $\tilde{b}$ as a linear function of the variable $\tilde{a}$ is equal to $1 - \rho_{\tilde{a},\tilde{b}}{}^2$. As we prove in Section 8.5.1, the correlation coefficient $\rho_{\tilde{a},\tilde{b}}$ is bounded between -1 and 1. When it equals zero, the linear estimator does not depend on $\tilde{a}$ at all and the MSE attains its maximum value. In that case, we say that the variables are *uncorrelated*, meaning that there is no linear dependence between $\tilde{a}$ and $\tilde{b}$. When $\rho_{\tilde{a},\tilde{b}}$ is nonzero, the accuracy of the linear approximation increases proportionally to the square of the correlation coefficient. In fact, the approximation is perfect when $\rho_{\tilde{a},\tilde{b}}$ is 1 or -1! The correlation coefficient therefore quantifies to what extent we can approximate $\tilde{b}$ as a linear function of $\tilde{a}$.

Figure 8.1 shows samples from different pairs of standardized random variables $\tilde{a}$ and $\tilde{b}$ with varying correlation coefficients. We decompose the samples into two components, corresponding to the best linear estimator of $\tilde{b}$ given $\tilde{a}$ and the residual of the linear estimator:

$$\tilde{b} = \underbrace{\rho_{\tilde{a},\tilde{b}}\tilde{a}}_{\text{Best linear estimator given } \tilde{a}} + \underbrace{\tilde{b} - \rho_{\tilde{a},\tilde{b}}\tilde{a}}_{\text{Residual}}. \tag{8.10}$$

The magnitude of the correlation coefficient $\rho_{\tilde{a},\tilde{b}}$ governs the relative magnitude of the two components. When $|\rho_{\tilde{a},\tilde{b}}|$ is close to one, the residual is small, and $\tilde{b}$ is well approximated by the linear estimator. When $|\rho_{\tilde{a},\tilde{b}}|$ is small, the residual dominates, so there is little linear dependence between the variables. When $\rho_{\tilde{a},\tilde{b}}$ is zero, the variables are uncorrelated, and $\tilde{b}$ is equal to the residual. If $\rho_{\tilde{a},\tilde{b}}$ is positive, the linear estimator is proportional to $\tilde{a}$, so $\tilde{a}$ and $\tilde{b}$ are said to be *positively correlated*. If $\rho_{\tilde{a},\tilde{b}}$ is negative, then the linear approximation is proportional to $-\tilde{a}$. In that case, the variables are said to be *negatively correlated*.

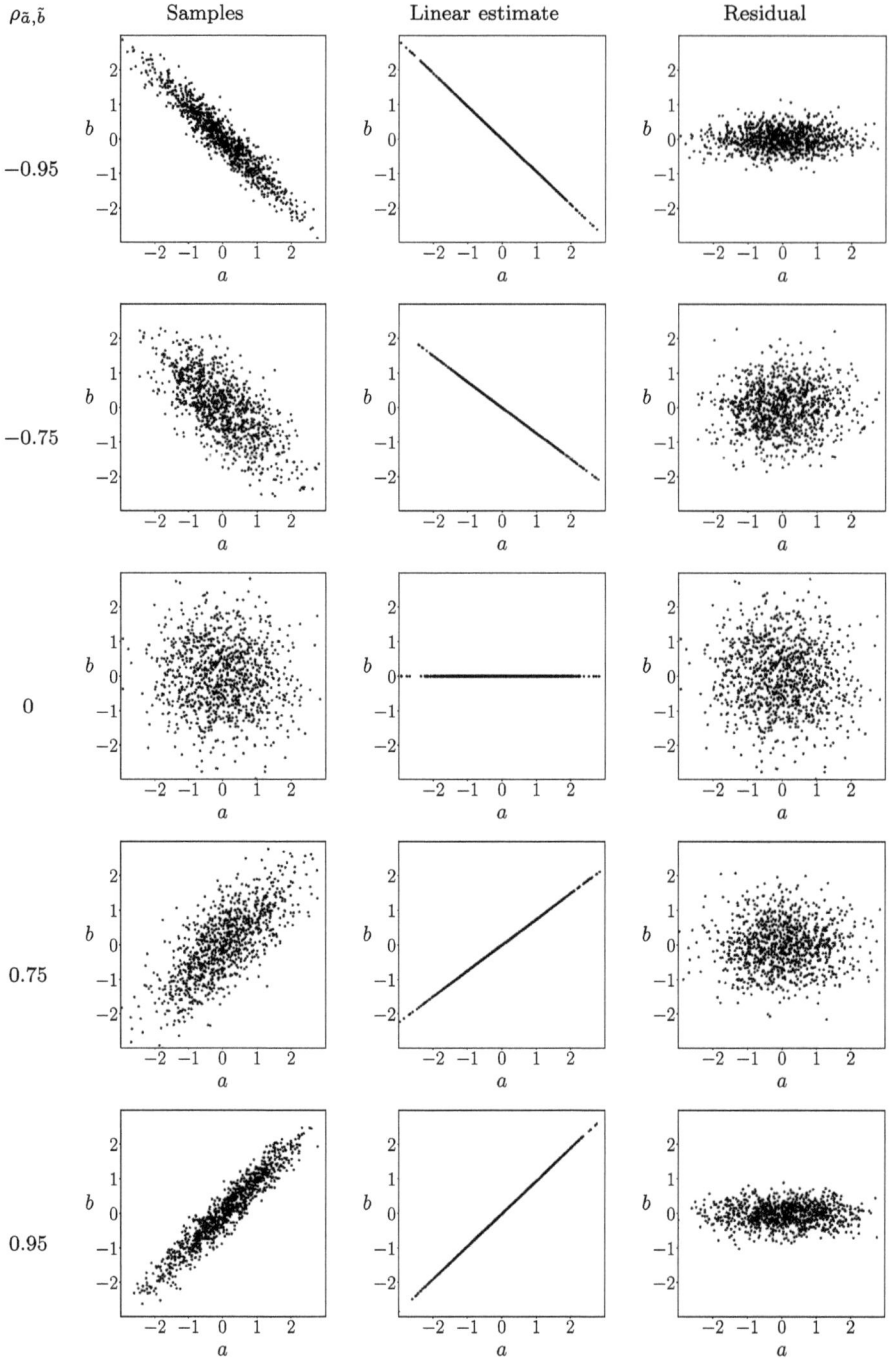

Figure 8.1 The correlation coefficient quantifies linear dependence between random variables. The left column shows scatterplots of 1,000 i.i.d. samples from two Gaussian random variables $\tilde{a}$ and $\tilde{b}$ with zero mean, unit variance and different correlation coefficients. The central column shows the best linear approximation $\rho_{\tilde{a},\tilde{b}}a$ of $\tilde{b}$ given $\tilde{a} = a$ for these samples. The right column shows the corresponding residual $\tilde{b} - \rho_{\tilde{a},\tilde{b}}\tilde{a}$. The sign of $\rho_{\tilde{a},\tilde{b}}$ indicates whether $\tilde{b}$ is proportional to $\tilde{a}$ ($\rho_{\tilde{a},\tilde{b}} > 0$) or $-\tilde{a}$ ($\rho_{\tilde{a},\tilde{b}} < 0$) on average. The magnitude of $\rho_{\tilde{a},\tilde{b}}$ determines the accuracy of the linear approximation.

8.2 Correlation and Covariance

Section 8.1 defines the correlation coefficient between standardized random variables with zero mean and unit variance. In this section, we consider random variables with arbitrary means and variances. To ease notation, let us denote the means of two random variables $\tilde{a}$ and $\tilde{b}$ by $\mu_{\tilde{a}}$ and $\mu_{\tilde{b}}$, and their standard deviations by $\sigma_{\tilde{a}}$ and $\sigma_{\tilde{b}}$. A useful first step is to *standardize* the random variables by centering and normalizing them. The resulting variables have zero mean and unit variance.

Definition 8.2 (Standardized variable). *Given a random variable $\tilde{a}$ with mean $\mu_{\tilde{a}}$ and standard deviation $\sigma_{\tilde{a}}$, the standardized version $s(\tilde{a})$ of $\tilde{a}$ is obtained by subtracting the mean and dividing the result by the standard deviation,*

$$s(\tilde{a}) := \frac{\tilde{a} - \mu_{\tilde{a}}}{\sigma_{\tilde{a}}}. \tag{8.11}$$

Lemma 8.3 (Mean and variance of a standardized variable). *The standardized variable $s(\tilde{a})$, corresponding to a random variable $\tilde{a}$ with mean $\mu_{\tilde{a}}$ and standard deviation $\sigma_{\tilde{a}}$, has zero mean and unit variance.*

Proof By linearity of expectation (Theorem 7.18) and Definition 7.32,

$$\mathbb{E}\left[s(\tilde{a})\right] = \mathbb{E}\left[\frac{\tilde{a} - \mu_{\tilde{a}}}{\sigma_{\tilde{a}}}\right] \tag{8.12}$$

$$= \frac{\mathbb{E}\left[\tilde{a}\right] - \mu_{\tilde{a}}}{\sigma_{\tilde{a}}} = 0, \tag{8.13}$$

$$\mathrm{Var}\left[s(\tilde{a})\right] = \mathbb{E}\left[s(\tilde{a})^2\right] \tag{8.14}$$

$$= \mathbb{E}\left[\frac{(\tilde{a} - \mu_{\tilde{a}})^2}{\sigma_{\tilde{a}}^2}\right] \tag{8.15}$$

$$= \frac{\mathbb{E}\left[(\tilde{a} - \mu_{\tilde{a}})^2\right]}{\sigma_{\tilde{a}}^2} = 1. \tag{8.16}$$

∎

Since the standardized random variables $s(\tilde{a})$ and $s(\tilde{b})$ have zero mean and unit variance, the correlation coefficient $\rho_{s(\tilde{a}),s(\tilde{b})}$ quantifies to what extent $s(\tilde{a})$ and $s(\tilde{b})$ are linearly related, as explained in Section 8.1. By Theorem 8.1, the best linear approximation of $s(\tilde{b})$ given $s(\tilde{a})$ is $\rho_{s(\tilde{a}),s(\tilde{b})}s(\tilde{a})$. Since $\tilde{b} = \sigma_{\tilde{b}}s(\tilde{b}) + \mu_{\tilde{b}}$, this provides an affine approximation of $\tilde{b}$ given $\tilde{a}$, :

$$\tilde{b} \approx \sigma_{\tilde{b}}\rho_{s(\tilde{a}),s(\tilde{b})}s(\tilde{a}) + \mu_{\tilde{b}} \tag{8.17}$$

$$= \frac{\sigma_{\tilde{b}}\rho_{s(\tilde{a}),s(\tilde{b})}(\tilde{a} - \mu_{\tilde{a}})}{\sigma_{\tilde{a}}} + \mu_{\tilde{b}}. \tag{8.18}$$

In Section 8.4, we show that this is the best possible affine approximation in terms of mean squared error. The correlation coefficient $\rho_{s(\tilde{a}),s(\tilde{b})}$ therefore determines to what extent $\tilde{b}$ can be approximated using an affine function of $\tilde{a}$ (or vice versa). Motivated by this, we define the correlation coefficient between $\tilde{a}$ and $\tilde{b}$ as the correlation coefficient between their standardized counterparts $s(\tilde{a})$ and $s(\tilde{b})$. This yields a measure of linear dependence that is invariant to positive scaling and shifting. For any positive scaling factors β_1 and β_2, and any additive constants α_1 and α_2, the standardized counterparts of $\beta_1\tilde{a} + \alpha_1$ and $\beta_2\tilde{b} + \alpha_2$ are $s(\tilde{a})$ and $s(\tilde{b})$, respectively, so the correlation coefficient remains the same.

In terms of $\tilde{a}$ and $\tilde{b}$ and their means and variances, the correlation coefficient equals

$$\rho_{\tilde{a},\tilde{b}} := \rho_{s(\tilde{a}),s(\tilde{b})} \tag{8.19}$$

$$= \mathbb{E}\left[s(\tilde{a})s(\tilde{b})\right] \tag{8.20}$$

$$= \mathbb{E}\left[\frac{\tilde{a} - \mu_{\tilde{a}}}{\sigma_{\tilde{a}}} \cdot \frac{\tilde{b} - \mu_{\tilde{b}}}{\sigma_{\tilde{b}}}\right] \tag{8.21}$$

$$= \frac{\mathbb{E}\left[(\tilde{a} - \mu_{\tilde{a}})(\tilde{b} - \mu_{\tilde{b}})\right]}{\sigma_{\tilde{a}}\,\sigma_{\tilde{b}}}, \tag{8.22}$$

where the last step follows from linearity of expectation (Theorem 7.18). The numerator is the mean of the product between $\tilde{a}$ and $\tilde{b}$, after centering the variables. This quantity is known as the covariance between $\tilde{a}$ and $\tilde{b}$.

Definition 8.4 (Covariance). *Let $\tilde{a}$ and $\tilde{b}$ be two random variables with finite mean belonging to the same probability space. The covariance of $\tilde{a}$ and $\tilde{b}$ is*

$$\mathrm{Cov}[\tilde{a}, \tilde{b}] := \mathbb{E}\left[(\tilde{a} - \mu_{\tilde{a}})(\tilde{b} - \mu_{\tilde{b}})\right], \tag{8.23}$$

where $\mu_{\tilde{a}}$ and $\mu_{\tilde{b}}$ denote the means of $\tilde{a}$ and $\tilde{b}$.

The following lemma shows that the covariance can be obtained by subtracting the product of the means from the mean of the product.

Lemma 8.5. *The covariance of two random variables $\tilde{a}$ and $\tilde{b}$ with means $\mu_{\tilde{a}}$ and $\mu_{\tilde{b}}$ equals*

$$\mathrm{Cov}[\tilde{a}, \tilde{b}] := \mathbb{E}[\tilde{a}\tilde{b}] - \mu_{\tilde{a}}\mu_{\tilde{b}}. \tag{8.24}$$

Proof By linearity of expectation (Theorem 7.18),

$$\mathbb{E}\left[(\tilde{a} - \mu_{\tilde{a}})(\tilde{b} - \mu_{\tilde{b}})\right] = \mathbb{E}[\tilde{a}\tilde{b}] - \mathbb{E}[\tilde{a}]\mu_{\tilde{b}} - \mu_{\tilde{a}}\mathbb{E}[\tilde{b}] + \mu_{\tilde{a}}\mu_{\tilde{b}} \tag{8.25}$$

$$= \mathbb{E}[\tilde{a}\tilde{b}] - \mu_{\tilde{a}}\mu_{\tilde{b}}. \tag{8.26}$$

∎

The correlation coefficient between two random variables is usually defined as the covariance normalized by the standard deviations of the variables. This coincides with our initial definition (8.22).

Definition 8.6 (Correlation coefficient). *Let $\tilde{a}$ and $\tilde{b}$ be two random variables with finite mean and variance belonging to the same probability space. The correlation coefficient of $\tilde{a}$ and $\tilde{b}$ is*

$$\rho_{\tilde{a},\tilde{b}} := \frac{\mathrm{Cov}[\tilde{a}, \tilde{b}]}{\sigma_{\tilde{a}}\sigma_{\tilde{b}}}, \tag{8.27}$$

where $\sigma_{\tilde{a}}$ and $\sigma_{\tilde{b}}$ are the standard deviations of $\tilde{a}$ and $\tilde{b}$, respectively.

When the covariance, or equivalently the correlation coefficient, between two variables is positive, we say that the variables are positively correlated. If it is negative, we say they are negatively correlated. If the covariance is zero, then the variables are uncorrelated, which means that there is no linear dependence between them.

Example 8.7 (Correlation between cats and dogs). Example 7.7 models the number of cats and dogs in a city, represented by the random variables $\tilde{c}$ and $\tilde{d}$, respectively. In order to summarize the dependence between the two quantities, we evaluate the correlation between $\tilde{c}$ and $\tilde{d}$. By Definitions 7.4 and 7.1,

$$\mathbb{E}[\tilde{c}\,\tilde{d}] := \sum_{c=0}^{3}\sum_{d=0}^{2} c\,d\,p_{\tilde{c},\tilde{d}}(c,d) \tag{8.28}$$

$$= 1 \cdot 0.05 + 2(0.03 + 0.02) \tag{8.29}$$

$$= 0.15, \tag{8.30}$$

$$\mathbb{E}[\tilde{c}] := \sum_{c=0}^{3}\sum_{d=0}^{2} c\,p_{\tilde{c},\tilde{d}}(c,d) = 0.63, \tag{8.31}$$

$$\mathbb{E}[\tilde{d}] := \sum_{c=0}^{3}\sum_{d=0}^{2} d\,p_{\tilde{c},\tilde{d}}(c,d) = 0.42, \tag{8.32}$$

so, by Lemma 8.5, the covariance equals

$$\text{Cov}[\tilde{c}, \tilde{d}] = \mathbb{E}[\tilde{c}\,\tilde{d}] - \mathbb{E}[\tilde{c}]\mathbb{E}[\tilde{d}] \tag{8.33}$$

$$= -0.115. \tag{8.34}$$

The number of cats and the number of dogs are negatively correlated. This indicates that, on average, people with more dogs have less cats (and vice versa). By Lemma 7.33 the variances of $\tilde{c}$ and $\tilde{d}$ equal

$$\text{Var}[\tilde{c}] = \mathbb{E}[\tilde{c}^2] - \mathbb{E}[\tilde{c}]^2 = 0.793, \tag{8.35}$$

$$\text{Var}[\tilde{d}] = \mathbb{E}[\tilde{d}^2] - \mathbb{E}[\tilde{d}]^2 = 0.384. \tag{8.36}$$

To obtain the correlation coefficient, we divide the covariance by the product of the standard deviations following Definition 8.6,

$$\rho_{\tilde{c},\tilde{d}} := \frac{\text{Cov}[\tilde{c}, \tilde{d}]}{\sqrt{\text{Var}[\tilde{c}]\text{Var}[\tilde{d}]}} \tag{8.37}$$

$$= -0.208. \tag{8.38}$$

This provides a normalized quantification of the linear dependence between the number of cats and the number of dogs.

....

Example 8.8 (Dependence between Gaussian random variables). In Example 5.21, we study a bivariate Gaussian random vector with zero mean and covariance matrix

$$\Sigma := \begin{bmatrix} 1 & \rho \\ \rho & 1 \end{bmatrix}, \tag{8.39}$$

where $-1 < \rho < 1$ is a parameter. The entries of the vector are denoted by $\tilde{a}$ and $\tilde{b}$. By Theorem 5.22 and Lemmas 7.25 and 7.43, the means of $\tilde{a}$ and $\tilde{b}$ are zero, and their variances both equal one. Let us derive the correlation coefficient of $\tilde{a}$ and $\tilde{b}$. Since the variables have unit variance, the correlation coefficient is equal to the covariance. By Definition 7.44, the conditional mean function of $\tilde{a}\tilde{b}$ given $\tilde{a}$ is

$$\mu_{\tilde{a}\tilde{b}\,|\,\tilde{a}}(a) = \int_{b=-\infty}^{\infty} ab\, f_{\tilde{b}\,|\,\tilde{a}}(b\,|\,a)\,\mathrm{d}b \tag{8.40}$$

$$= a\,\mu_{\tilde{b}\,|\,\tilde{a}}(a) \tag{8.41}$$

$$= \rho a^2 \tag{8.42}$$

because the conditional distribution of $\tilde{b}$ given $\tilde{a} = a$ is Gaussian with mean ρa, as established in Example 5.21, so $\mu_{\tilde{b}\,|\,\tilde{a}}(a) = \rho a$ by Lemma 7.25. Consequently, by Definitions 8.6 and 8.4 and iterated expectation (Theorem 7.54),

$$\rho_{\tilde{a},\tilde{b}} = \frac{\mathrm{Cov}[\tilde{a},\tilde{b}]}{\sigma_{\tilde{a}}\sigma_{\tilde{b}}} \tag{8.43}$$

$$= \mathbb{E}\left[\tilde{a}\tilde{b}\right] \tag{8.44}$$

$$= \mathbb{E}\left[\mu_{\tilde{a}\tilde{b}\,|\,\tilde{a}}(\tilde{a})\right] \tag{8.45}$$

$$= \mathbb{E}\left[\rho\tilde{a}^2\right] \tag{8.46}$$

$$= \rho\mathbb{E}\left[\tilde{a}^2\right] \tag{8.47}$$

$$= \rho. \tag{8.48}$$

The correlation coefficient is equal to the parameter ρ. Example 5.21 shows that ρ completely determines the dependence between $\tilde{a}$ and $\tilde{b}$ (see Figure 5.18).

8.3 Estimating Correlation from Data

The covariance of two random variables is the mean of their product after centering. It is therefore natural to use averaging in order to estimate the covariance from data. As in the definition of the sample variance, we average dividing by $n - 1$ (where n is the number of data points), in order to ensure that the estimate is unbiased (see Exercise 9.4).

Definition 8.9 (Sample covariance). *Given a dataset formed by n real-valued pairs (x_1, y_1), (x_2, y_2), ..., (x_n, y_n), $1 \le i \le n$, let $X := \{x_1, x_2, \ldots, x_n\}$ and $Y := \{y_1, y_2, \ldots, y_n\}$. The sample covariance of the data equals*

$$c(X, Y) := \frac{1}{n-1}\sum_{i=1}^{n}(x_i - m(X))(y_i - m(Y)), \tag{8.49}$$

where $m(X)$ and $m(Y)$ denote the sample means of X and Y (see Definition 7.15).

In order to estimate the correlation coefficient from data, we apply Definition 8.6, replacing the covariance and variances by the sample covariance and sample variances, respectively.

Definition 8.10 (Sample correlation coefficient). *Given a dataset formed by n real-valued pairs (x_1, y_1), (x_2, y_2), ..., (x_n, y_n), $1 \le i \le n$, let $X := \{x_1, x_2, \ldots, x_n\}$ and $Y := \{y_1, y_2, \ldots, y_n\}$. The sample correlation coefficient of the data equals*

$$\rho_{X,Y} := \frac{c(X, Y)}{\sqrt{v(X)v(Y)}}, \tag{8.50}$$

where $c(X, Y)$ is the sample covariance of X and Y, and $v(X)$ and $v(Y)$ are the sample variances of X and Y (see Definition 7.37), respectively.

As explained in Section 8.2, the correlation coefficient of two random variables $\tilde{a}$ and $\tilde{b}$ can be interpreted as the optimal scaling factor, when approximating the standardized variable $s(\tilde{b})$ as a linear function of the standardized variable $s(\tilde{a})$. A similar interpretation holds for the sample correlation coefficient. To show this, we first need to define what it means to standardize data.

Definition 8.11 (Standardized data). *Let* $X := \{x_1, x_2, \ldots, x_n\}$ *denote a real-valued dataset. The corresponding standardized data are obtained by subtracting the sample mean* $m(X)$ *and dividing by the sample standard deviation* $\sqrt{v(X)}$,

$$s(x_i) := \frac{x_i - m(X)}{\sqrt{v(X)}}, \quad 1 \le i \le n. \tag{8.51}$$

The following theorem is the finite-data counterpart to Theorem 8.1.

Theorem 8.12 (Linear dependence in standardized data). *Given a dataset formed by* n *pairs* (x_1, y_1), (x_2, y_2), $\ldots$, (x_n, y_n), $1 \le i \le n$, *let* $s(x_i)$ *and* $s(y_i)$ *denote the corresponding standardized data, for* $1 \le i \le n$. *The optimal linear estimator of the second standardized variable given the first, in terms of the residual sum of squares, is obtained by scaling the first standardized variable using the sample correlation coefficient* $\rho_{X,Y}$:

$$\rho_{X,Y} = \arg\min_{\beta} \sum_{i=1}^{n} (s(y_i) - \beta s(x_i))^2. \tag{8.52}$$

Proof The sample mean of a standardized dataset is zero and its sample variance is one (see Exercise 8.7). Consequently,

$$\sum_{i=1}^{n} s(x_i)^2 = \sum_{i=1}^{n} s(y_i)^2 = n - 1, \tag{8.53}$$

and the sample correlation coefficient of the standardized variables is equal to their sample covariance,

$$\rho_{X,Y} = \frac{1}{n-1} \sum_{i=1}^{n} s(x_i)s(y_i). \tag{8.54}$$

This allows us to express the residual sum of squares of any linear approximation as a simple function of the corresponding scaling factor β:

$$\text{RSS}(\beta) := \sum_{i=1}^{n} (s(y_i) - \beta s(x_i))^2 \tag{8.55}$$

$$= \sum_{i=1}^{n} s(y_i)^2 + \beta^2 \sum_{i=1}^{n} s(x_i)^2 - 2\beta \sum_{i=1}^{n} s(x_i)s(y_i) \tag{8.56}$$

$$= (n-1)\left(1 + \beta^2 - 2\beta\rho_{X,Y}\right). \tag{8.57}$$

The derivatives of the error with respect to β are

$$\frac{d\,\text{RSS}(\beta)}{d\beta} = 2(n-1)(\beta - \rho_{X,Y}), \tag{8.58}$$

$$\frac{d^2\,\text{RSS}(\beta)}{d\beta^2} = 2(n-1), \tag{8.59}$$

so the error is strictly convex, and we can minimize it by setting the first derivative to zero. We conclude that the optimal scaling is equal to the sample correlation coefficient $\rho_{X,Y}$. ∎

Example 8.13 (Height of NBA players). Height is obviously very important in basketball. Here we study how to it relates to the offensive productivity of players in the NBA between

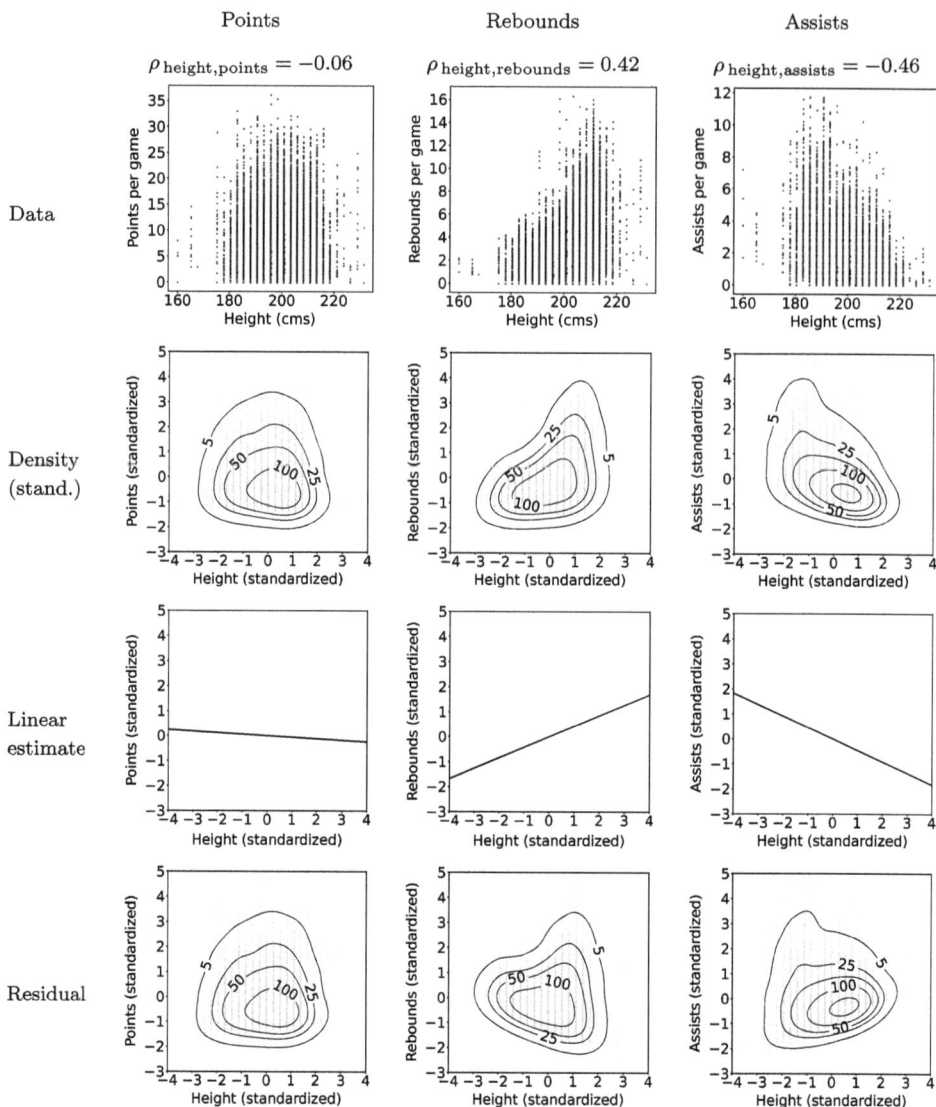

Figure 8.2 Linear dependence with height. The top row shows scatterplots of the points (left), rebounds (center), and assists (right) per game of NBA players against their height. The second row shows the corresponding bivariate probability densities estimated via kernel density estimation (see Definition 5.7), standardized using the sample means and variances. The third row shows the best linear approximation of each quantity given height. The fourth row shows the corresponding residuals, obtained by subtracting the linear estimate from each data point. The correlation coefficients indicate that rebounds are positively correlated with height, and assists are negatively correlated with height. Points are also negatively correlated with height, but the linear relationship is much weaker.

1996 and 2019, using Dataset 13. Our data consist of the height of each player and their three main offensive stats: Points, rebounds, and assists per game.

The sample correlation coefficient is -0.06 for height and points, 0.42 for height and rebounds, and -0.46 for height and assists. On average, taller players capture more rebounds, give less assists, and score slightly less than shorter players. Figure 8.2 shows scatterplots of the data, together with the corresponding densities after standardizing. The bottom two rows depict the decomposition of each standardized stat into its linear estimate given the standardized height and the corresponding residual. For each pair (h_i, y_i), $1 \le i \le n$, where h_i is the height and y_i is points, rebounds, or assists,

$$s(y_i) = \underbrace{\rho_{H,Y} s(h_i)}_{\text{Linear estimate given } s(h_i)} + \underbrace{y_i - \rho_{H,Y} s(h_i)}_{\text{Residual}}. \tag{8.60}$$

H is the set of all heights, and Y the set of all values of the selected stat.

. . . .

8.4 Simple Linear Regression

As we explain in Section 7.8.3, regression is the problem of estimating a quantity of interest, called the response, as a function of the observed features. In this section, we consider simple linear regression, where there is a single feature and the regression estimator is constrained to be affine. The general regression problem with multiple input variables is discussed in detail in Chapter 12.

Let us model the response and the feature as random variables. Our goal is to approximate the response $\tilde{b}$ using a linear transformation of the feature $\tilde{a}$. Theorem 8.1 establishes that $\rho_{\tilde{a},\tilde{b}} s(\tilde{a})$ is the linear minimum MSE estimator of $s(\tilde{b})$, where $s(\tilde{a})$ and $s(\tilde{b})$ are standardized variables obtained from $\tilde{a}$ and $\tilde{b}$ following Definition 8.2. Since

$$\tilde{b} = \sigma_{\tilde{b}} s(\tilde{b}) + \mu_{\tilde{b}}, \tag{8.61}$$

a reasonable estimator for $\tilde{b}$ is

$$\sigma_{\tilde{b}} \rho_{\tilde{a},\tilde{b}} s(\tilde{a}) + \mu_{\tilde{b}}, \tag{8.62}$$

as derived in (8.18). The transformation is illustrated in Figure 8.3. Strictly speaking, the estimator is affine because it consists of a linear term and an additive constant. However, it is commonly referred to as a linear estimator of $\tilde{b}$ given $\tilde{a}$. It turns out that this is the best possible affine estimator in terms of mean squared error (MSE), so it is known as the linear MMSE (minimum MSE) estimator.

Theorem 8.14 (Linear MMSE estimator)**.** *Let $\tilde{a}$ and $\tilde{b}$ be two random variables with means $\mu_{\tilde{a}}$ and $\mu_{\tilde{b}}$, variances $\sigma_{\tilde{a}}^2$ and $\sigma_{\tilde{b}}^2$, and correlation coefficient $\rho_{\tilde{a},\tilde{b}}$. The linear minimum MSE estimator of $\tilde{b}$ given $\tilde{a} = a$ is*

$$\ell_{\mathrm{MMSE}}(a) := \beta_{\mathrm{MMSE}} a + \alpha_{\mathrm{MMSE}} \tag{8.63}$$

$$= \sigma_{\tilde{b}} \, \rho_{\tilde{a},\tilde{b}} \left(\frac{a - \mu_{\tilde{a}}}{\sigma_{\tilde{a}}} \right) + \mu_{\tilde{b}}. \tag{8.64}$$

This estimator yields the best affine estimator of $\tilde{b}$ given $\tilde{a}$, in terms of MSE:

$$\beta_{\mathrm{MMSE}}, \alpha_{\mathrm{MMSE}} = \arg\min_{\beta,\alpha} \mathbb{E}\left[(\tilde{b} - \beta\tilde{a} - \alpha)^2 \right]. \tag{8.65}$$

Proof We denote the MSE as a function of β and α by

$$\mathrm{MSE}(\beta, \alpha) := \mathbb{E}\left[(\tilde{b} - \beta\tilde{a} - \alpha)^2 \right]. \tag{8.66}$$

$$\mu_{\tilde{a}} := 6, \ \sigma_{\tilde{a}} := 0.5$$
$$\mu_{\tilde{b}} := 4, \ \sigma_{\tilde{b}} := 2$$

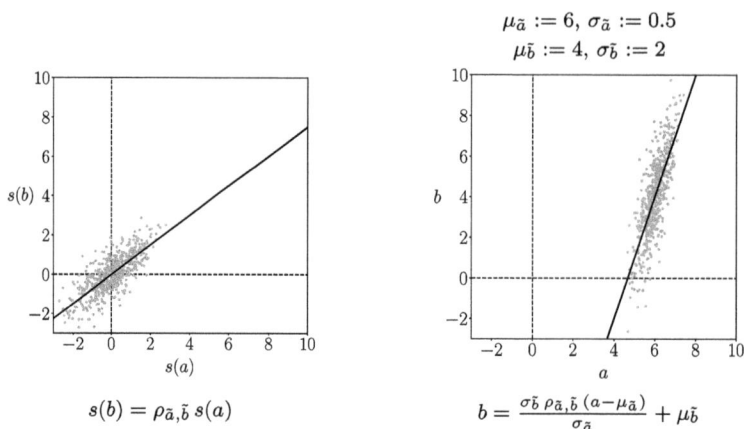

$$s(b) = \rho_{\tilde{a},\tilde{b}} \, s(a)$$

$$b = \frac{\sigma_{\tilde{b}} \, \rho_{\tilde{a},\tilde{b}} \, (a-\mu_{\tilde{a}})}{\sigma_{\tilde{a}}} + \mu_{\tilde{b}}$$

Figure 8.3 Affine estimation of nonstandardized random variables. The left plot shows 500 samples (gray dots) from two standardized random variables $s(\tilde{a})$ and $s(\tilde{b})$, and the corresponding linear minimum mean-squared-error estimate of $s(\tilde{b})$ given $s(\tilde{a})$ (black line). The right plot shows the original data before standardization (gray dots), and the corresponding affine estimate. The affine estimate (black line) is obtained by shifting the linear estimate for the standardized variables according to the means and standard deviations, as described by the equation underneath the graph.

If we fix $\beta \in \mathbb{R}$, the optimal value of α, denoted by $\alpha^*(\beta)$, for that particular β is the best constant estimate of the random variable $\tilde{b} - \beta\tilde{a}$, which equals

$$\alpha^*(\beta) := \arg\min_{\alpha} \mathbb{E}\left[(\tilde{b} - \beta\tilde{a} - \alpha)^2\right] \tag{8.67}$$

$$= \mathbb{E}\left[\tilde{b} - \beta\tilde{a}\right] \tag{8.68}$$

$$= \mu_{\tilde{b}} - \beta\mu_{\tilde{a}} \tag{8.69}$$

by Theorem 7.30 and linearity of expectation (Theorem 7.18). Consequently, for any β and any α $\mathrm{MSE}(\beta, \alpha) \geq \mathrm{MSE}(\beta, \alpha^*(\beta))$. To obtain β_{MMSE}, we fix α to $\alpha^*(\beta)$ and minimize the resulting MSE,

$$\beta_{\mathrm{MMSE}} = \arg\min_{\beta} \mathrm{MSE}(\beta, \alpha^*(\beta)). \tag{8.70}$$

By linearity of expectation (Theorem 7.18) and Definitions 7.32 and 8.4,

$$\mathrm{MSE}(\beta, \alpha^*(\beta)) = \mathbb{E}\left[(\tilde{b} - \beta\tilde{a} - \mu_{\tilde{b}} + \beta\mu_{\tilde{a}})^2\right] \tag{8.71}$$

$$= \mathbb{E}\left[(\tilde{b} - \mu_{\tilde{b}})^2\right] + \beta^2 \mathbb{E}\left[(\tilde{a} - \mu_{\tilde{a}})^2\right] - 2\beta\mathbb{E}\left[(\tilde{a} - \mu_{\tilde{a}})(\tilde{b} - \mu_{\tilde{b}})\right]$$

$$= \sigma_{\tilde{b}}^2 + \sigma_{\tilde{a}}^2 \beta^2 - 2\mathrm{Cov}[\tilde{a}, \tilde{b}]\beta. \tag{8.72}$$

The derivatives with respect to β equal

$$\frac{d\,\mathrm{MSE}(\beta, \alpha^*(\beta))}{d\beta} = 2\left(\sigma_{\tilde{a}}^2 \beta - \mathrm{Cov}[\tilde{a}, \tilde{b}]\right), \tag{8.73}$$

$$\frac{d^2\,\mathrm{MSE}(\beta, \alpha^*(\beta))}{d\beta^2} = 2\sigma_{\tilde{a}}^2 \geq 0. \tag{8.74}$$

The function is strictly convex, as long as the variance of $\tilde{a}$ is nonzero, so we can obtain β_{MMSE} by setting the derivative to zero,

$$\beta_{\mathrm{MMSE}} = \frac{\mathrm{Cov}[\tilde{a}, \tilde{b}]}{\sigma_{\tilde{a}}^2} \tag{8.75}$$

$$= \frac{\rho_{\tilde{a}, \tilde{b}} \, \sigma_{\tilde{b}}}{\sigma_{\tilde{a}}}. \tag{8.76}$$

The corresponding value of $\alpha^*(\beta)$ is

$$\alpha_{\mathrm{MMSE}} = \alpha^*(\beta_{\mathrm{MMSE}}) \tag{8.77}$$

$$= \mu_{\tilde{b}} - \beta_{\mathrm{MMSE}}\mu_{\tilde{a}} \tag{8.78}$$

$$= \mu_{\tilde{b}} - \frac{\rho_{\tilde{a}, \tilde{b}} \, \sigma_{\tilde{b}} \, \mu_{\tilde{a}}}{\sigma_{\tilde{a}}}. \tag{8.79}$$

Finally, we verify that

$$\ell_{\mathrm{MMSE}}(a) := \beta_{\mathrm{MMSE}}a + \alpha_{\mathrm{MMSE}} \tag{8.80}$$

$$= \sigma_{\tilde{b}} \, \rho_{\tilde{a}, \tilde{b}} \left(\frac{a - \mu_{\tilde{a}}}{\sigma_{\tilde{a}}} \right) + \mu_{\tilde{b}}. \tag{8.81}$$

■

Example 8.15 (Cats and dogs: Linear MMSE estimation). In Example 8.7, we compute the means, variances, and covariances of the number of cats and dogs from the joint pmf in Example 7.7. Here we use these statistics to obtain a linear estimator of the number of cats given the number of dogs. By Theorem 8.14, the linear MMSE estimator is

$$\ell_{\mathrm{MMSE}}(d) := \sigma_{\tilde{c}} \, \rho_{\tilde{d}, \tilde{c}} \left(\frac{d - \mu_{\tilde{d}}}{\sigma_{\tilde{d}}} \right) + \mu_{\tilde{c}} \tag{8.82}$$

$$= -0.3d + 0.755, \tag{8.83}$$

which yields the estimates

$$\ell_{\mathrm{MMSE}}(0) = 0.755, \tag{8.84}$$

$$\ell_{\mathrm{MMSE}}(1) = 0.455, \tag{8.85}$$

$$\ell_{\mathrm{MMSE}}(2) = 0.155. \tag{8.86}$$

The corresponding MSE equals

$$\mathbb{E}\left[\left(\tilde{c} - \ell_{\mathrm{MMSE}}(\tilde{d}) \right)^2 \right] = \sum_{c=0}^{3} \sum_{d=0}^{2} p_{\tilde{c}, \tilde{d}}(c, d) \, (c + 0.3d - 0.756)^2 \tag{8.87}$$

$$= 0.759. \tag{8.88}$$

The MSE is only slightly larger than the optimal MSE (0.756) achieved by the conditional-mean estimator derived in Example 7.60.

⋯⋯⋯⋯⋯⋯⋯⋯⋯⋯⋯⋯⋯⋯⋯⋯⋯⋯⋯⋯⋯⋯⋯⋯⋯⋯⋯⋯⋯⋯

As illustrated by Example 8.15, the MSE of the linear MMSE estimator is always lower bounded by the optimal MSE achieved by the nonlinear conditional-mean estimator. For Gaussian random variables, linear estimation attains this lower bound because the conditional mean function is linear.

Theorem 8.16 (Linear estimation is optimal for Gaussian random variables). *Let $\tilde{a}$ and $\tilde{b}$ be random variables representing the response and the feature in a simple regression problem. We assume that $\tilde{a}$ and $\tilde{b}$ are jointly Gaussian, meaning that*

$$\begin{bmatrix} \tilde{a} \\ \tilde{b} \end{bmatrix} \tag{8.89}$$

is a Gaussian random vector with mean parameter

$$\mu := \begin{bmatrix} \mu_{\tilde{a}} \\ \mu_{\tilde{b}} \end{bmatrix} \tag{8.90}$$

and covariance-matrix parameter

$$\Sigma := \begin{bmatrix} \sigma_{\tilde{a}}^2 & \rho\sigma_{\tilde{a}}\sigma_{\tilde{b}} \\ \rho\sigma_{\tilde{a}}\sigma_{\tilde{b}} & \sigma_{\tilde{b}}^2 \end{bmatrix}, \tag{8.91}$$

where $\sigma_{\tilde{a}} > 0$, $\sigma_{\tilde{b}} > 0$, and $-1 < \rho < 1$ to ensure that Σ is full rank. The parametrization of the covariance matrix is without loss of generality; any positive definite symmetric matrix can be written like this. The minimum MSE (MMSE) estimator of $\tilde{b}$ given $\tilde{a} = a$ is the linear estimator

$$\ell(a) = \frac{\rho\sigma_{\tilde{b}}(a - \mu_{\tilde{a}})}{\sigma_{\tilde{a}}} + \mu_{\tilde{b}}. \tag{8.92}$$

Proof By Theorem 5.22, the conditional mean function of $\tilde{b}$ given $\tilde{a}$ is

$$\mu_{\tilde{b}\,|\,\tilde{a}}(a) = \frac{\rho\sigma_{\tilde{b}}(a - \mu_{\tilde{a}})}{\sigma_{\tilde{a}}} + \mu_{\tilde{b}}. \tag{8.93}$$

This is an affine function of $\tilde{a}$, so it must equal the linear MMSE estimator of $\tilde{b}$ given $\tilde{a}$ because the conditional mean is the optimal estimator in terms of MSE by Theorem 7.59.

As a bonus, we obtain an expression for the correlation coefficient of the variables in terms of the covariance-matrix parameters. Since (8.93) must equal the expression of the optimal affine estimator in Theorem 8.14, the correlation coefficient between $\tilde{a}$ and $\tilde{b}$ is equal to the parameter ρ. ∎

In practice, we perform regression using data. Assume that we have available n pairs of a feature and a corresponding response: (x_1, y_1), (x_2, y_2), ..., (x_n, y_n). We can approximate the linear minimum MSE estimator in Theorem 8.14 by replacing the means with sample means, the variances with sample variances and the correlation coefficient with the sample correlation coefficient. This yields

$$\beta_{\text{MMSE}} \approx \rho_{X,Y} \sqrt{\frac{v(Y)}{v(X)}}, \tag{8.94}$$

$$\alpha_{\text{MMSE}} \approx m(Y) - \rho_{X,Y} \sqrt{\frac{v(Y)}{v(X)}} m(X), \tag{8.95}$$

where $m(X)$ and $m(Y)$ are the corresponding sample means (see Definition 7.15), $v(X)$ and $v(Y)$ the corresponding sample variances (see Definition 7.37), and $\rho_{X,Y}$ the sample correlation coefficient of the feature and the response (see Definition 8.10). $X := \{x_1, x_2, \ldots, x_n\}$ contains the feature values, and $Y := \{y_1, y_2, \ldots, y_n\}$ the responses.

Alternatively, we can fit an affine model to the data directly, approximating the response of each data point y_i by an affine function of the feature $\ell(x_i) := \beta x_i + \alpha$. If we select the affine function to minimize the residual sum of squares,

$$\text{RSS}\,(\beta, \alpha) := \sum_{i=1}^{n} (y_i - \beta x_i - \alpha)^2, \tag{8.96}$$

the resulting estimator is known as the *ordinary least-squares* estimator. It turns out that these two approaches are equivalent, as established in the following theorem.

Theorem 8.17 (Simple linear regression via ordinary least squares). *Given a dataset formed by n real-valued pairs of a single feature and the corresponding response* (x_1, y_1), (x_2, y_2), $\dots$, (x_n, y_n), $1 \le i \le n$, *let* $X := \{x_1, x_2, \dots, x_n\}$ *and* $Y := \{y_1, y_2, \dots, y_n\}$. *The ordinary least-squares (OLS) estimator of the response given the feature x is*

$$\ell_{\text{OLS}}(x) := \beta_{\text{OLS}}\, x + \alpha_{\text{OLS}} \tag{8.97}$$

$$= \sqrt{v(Y)}\rho_{X,Y}\left(\frac{x - m(X)}{\sqrt{v(X)}}\right) + m(Y), \tag{8.98}$$

where $m(X)$ *and* $m(Y)$ *are the sample means of X and Y,* $v(X)$ *and* $v(Y)$ *the corresponding sample variances, and* $\rho_{X,Y}$ *the sample correlation coefficient of the feature and the response.*

The OLS estimator is the optimal affine estimator, in the sense that it minimizes the sum of squared errors,

$$\beta_{\text{OLS}}, \alpha_{\text{OLS}} = \arg\min_{\beta, \alpha} \sum_{i=1}^{n} (y_i - \beta x_i - \alpha)^2. \tag{8.99}$$

Proof We follow a similar argument to the proof of Theorem 8.14. We denote the residual sum of squares, as a function of the linear coefficient β and the additive constant α, by

$$\text{RSS}(\beta, \alpha) := \sum_{i=1}^{n} (y_i - \beta x_i - \alpha)^2. \tag{8.100}$$

We denote by $\alpha^*(\beta)$ the optimal value of α for a fixed value of β. Equivalently, $\alpha^*(\beta)$ is the best constant estimate of $y_i - \beta x_i$, $1 \le i \le n$, in terms of the sum of squared errors. By Theorem 7.31,

$$\alpha^*(\beta) := \arg\min_{\alpha} \sum_{i=1}^{n} (y_i - \beta x_i - \alpha)^2 \tag{8.101}$$

$$= \frac{1}{n} \sum_{i=1}^{n} (y_i - \beta x_i) \tag{8.102}$$

$$= m(Y) - \beta m(X). \tag{8.103}$$

Consequently, for any β and any α $\text{RSS}(\beta, \alpha) \ge \text{RSS}(\beta, \alpha^*(\beta))$. To obtain β_{OLS}, we fix α to $\alpha^*(\beta)$ and minimize the sum of squared errors,

$$\beta_{\text{OLS}} = \arg\min_{\beta} \text{RSS}(\beta, \alpha^*(\beta)). \tag{8.104}$$

By Definitions 7.37, and 8.9,

$$\text{RSS}(\beta, \alpha^*(\beta)) \tag{8.105}$$

$$= \sum_{i=1}^{n} (y_i - \beta x_i - m(Y) + \beta m(X))^2 \tag{8.106}$$

$$= \sum_{i=1}^{n} (y_i - m(Y))^2 + \beta^2 \sum_{i=1}^{n} (x_i - m(X))^2 - 2\beta \sum_{i=1}^{n} (y_i - m(Y))(x_i - m(X))$$

$$= (n-1) \left(v(Y) + \beta^2 v(X) - 2\beta c(X, Y) \right). \tag{8.107}$$

The derivatives with respect to β equal

$$\frac{\text{d} \, \text{RSS}(\beta, \alpha^*(\beta))}{\text{d}\beta} = 2(n-1) \left(v(X)\beta - c(X, Y) \right), \tag{8.108}$$

$$\frac{\text{d}^2 \, \text{RSS}(\beta, \alpha^*(\beta))}{\text{d}\beta^2} = 2(n-1)v(X) \geq 0. \tag{8.109}$$

The function is strictly convex, as long as the sample variance of the feature is nonzero, so we can obtain β_{OLS} by setting the derivative to zero,

$$\beta_{\text{OLS}} = \frac{c(X, Y)}{v(X)} \tag{8.110}$$

$$= \rho_{X,Y} \sqrt{\frac{v(Y)}{v(X)}}. \tag{8.111}$$

The corresponding value of $\alpha^*(\beta)$ is

$$\alpha_{\text{OLS}} = \alpha^*(\beta_{\text{OLS}}) \tag{8.112}$$

$$= m(Y) - \rho_{X,Y} \sqrt{\frac{v(Y)}{v(X)}} m(X). \tag{8.113}$$

■

Example 8.18 (Comparing linear and nonlinear regression). We consider the problem of estimating points, rebounds, and assists of NBA players as a function of their height, using the same data as in Example 8.13. We consider the affine OLS estimator, defined in Theorem 8.17, and also a nonlinear estimator based on the sample conditional mean (see Definition 7.47).

The results are shown in Figure 8.4. The conditional mean is more flexible than the OLS estimator. As a result, it is much noisier in regions where we don't have a lot of data. However, this flexibility also means that it can capture nonlinear structure. For example, the general trend is for assists to be inversely correlated to height because shorter players tend to be responsible for creating scoring opportunities for teammates. This is captured by the OLS estimator. However, the conditional mean of assists per game is roughly constant for players under 190 cm. This makes sense: If we only consider short players, then height is no longer indicative of their role (all of them are guards). Consequently, height is not inversely correlated with assists for such players. This pattern is captured by the nonlinear conditional-mean estimator, but not by the OLS estimator.

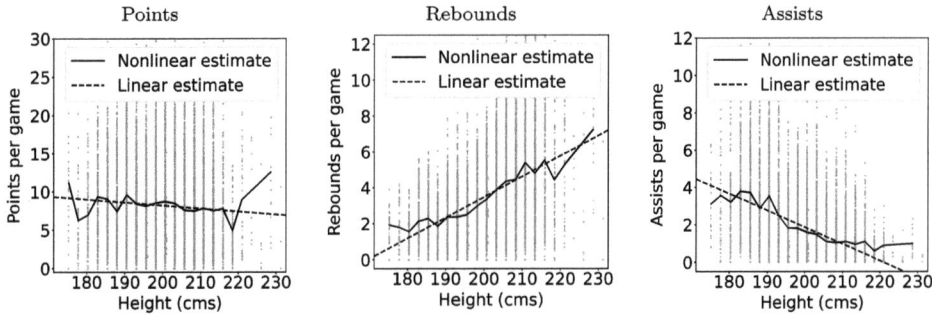

Figure 8.4 Comparison of linear and nonlinear regression. Estimates of the points (left), rebounds (center), and assists (right) per game of NBA players between 1996 and 2019, based on their height. The graphs show the affine ordinary least-squares (OLS) estimate (dashed line) and the sample conditional mean (solid line), superposed on a scatterplot of the data. The OLS estimator captures the overall relationship between each response (points, rebounds, or assists) and the height, but cannot reflect nonlinear structure, in contrast to the conditional mean. For example, assists are approximately constant for players shorter than 190 cm, but this is ignored by the OLS estimator.

8.5 Properties of the Correlation Coefficient

In this section, we establish three important properties of the correlation coefficient. Analogous properties hold for the sample correlation coefficient, as we show in Exercises 8.8 and 8.9. Section 8.5.1 proves that the correlation coefficient $\rho_{\tilde{a},\tilde{b}}$ between two random variables $\tilde{a}$ and $\tilde{b}$ is always bounded between -1 and 1. Section 8.5.2 establishes that, when $\rho_{\tilde{a},\tilde{b}}$ equals -1 or 1, there is an exact linear relationship between $\tilde{a}$ and $\tilde{b}$. Section 8.5.3 shows that the squared correlation coefficient is equal to the fraction of variance in $\tilde{b}$ that can be explained linearly in terms of $\tilde{a}$ (and vice versa). These three properties are all consequences of the following theorem, which provides a formula for the MSE of the linear MMSE estimator, as a function of the correlation coefficient.

Theorem 8.19 (Mean squared error of the linear MMSE estimator). *Let $\tilde{a}$ and $\tilde{b}$ be two random variables with finite variance, belonging to the same probability space. Let $\ell_{\mathrm{MMSE}}(\tilde{a})$ be the linear MMSE estimator of $\tilde{b}$ given $\tilde{a}$, as defined in Theorem 8.14. The MSE of this estimator equals*

$$\mathbb{E}\left[(\tilde{b} - \ell_{\mathrm{MMSE}}(\tilde{a}))^2\right] = \left(1 - \rho_{\tilde{a},\tilde{b}}^2\right)\sigma_{\tilde{b}}^2, \tag{8.114}$$

where $\sigma_{\tilde{b}}^2$ is the variance of $\tilde{b}$ and $\rho_{\tilde{a},\tilde{b}}$ is the correlation coefficient of $\tilde{a}$ and $\tilde{b}$.

Proof Let $s(\tilde{a})$ and $s(\tilde{b})$ denote the standardized counterparts of $\tilde{a}$ and $\tilde{b}$. By Theorem 8.14, linearity of expectation (Theorem 7.18), Lemma 8.3, and Definition 8.6,

$$\mathbb{E}\left[(\ell_{\mathrm{MMSE}}(\tilde{a}) - \tilde{b})^2\right] = \mathbb{E}\left[(\rho_{\tilde{a},\tilde{b}}\,\sigma_{\tilde{b}}s(\tilde{a}) + \mu_{\tilde{b}} - \tilde{b})^2\right] \tag{8.115}$$

$$= \sigma_{\tilde{b}}^2\,\mathbb{E}\left[(\rho_{\tilde{a},\tilde{b}}s(\tilde{a}) - s(\tilde{b}))^2\right] \tag{8.116}$$

$$= \sigma_{\tilde{b}}^2\left(\rho_{\tilde{a},\tilde{b}}^2\mathbb{E}[s(\tilde{a})^2] + \mathbb{E}[s(\tilde{b})^2] - 2\rho_{\tilde{a},\tilde{b}}\mathbb{E}\left[s(\tilde{a})s(\tilde{b})\right]\right) \tag{8.117}$$

$$= \sigma_{\tilde{b}}^2\left(1 - \rho_{\tilde{a},\tilde{b}}^2\right). \tag{8.118}$$

∎

8.5.1 The Correlation Coefficient Is Bounded

The following lemma establishes that the magnitude of the correlation coefficient cannot be greater than one.

Theorem 8.20. *The coefficient $\rho_{\tilde{a},\tilde{b}}$ of any random variables $\tilde{a}$ and $\tilde{b}$ with nonzero, bounded variance satisfies*

$$-1 \le \rho_{\tilde{a},\tilde{b}} \le 1. \tag{8.119}$$

Proof By Theorem 8.19

$$\sigma_{\tilde{b}}^2 \left(1 - \rho_{\tilde{a},\tilde{b}}^2\right) = \mathbb{E}\left[(\tilde{b} - \ell_{\mathrm{MMSE}}(\tilde{a}))^2\right] \ge 0, \tag{8.120}$$

where $\ell_{\mathrm{MMSE}}(\tilde{a})$ is the linear MMSE estimator of $\tilde{b}$ given $\tilde{a}$. The inequality follows from the fact that the expected value of a nonnegative quantity is nonnegative, as it is just a sum or integral of nonnegative values. Since $\sigma_{\tilde{b}}^2 > 0$, (8.120) implies $\rho_{\tilde{a},\tilde{b}}^2 \le 1$, which in turns means that $-1 \le \rho_{\tilde{a},\tilde{b}} \le 1$. $\blacksquare$

8.5.2 Complete Linear Dependence

If the correlation coefficient between two random variables equals -1 or 1, then the random variables can be expressed *exactly* as an affine function of each other. The reason is that the residual of the linear MMSE estimator has zero variance.

Theorem 8.21 (Complete linear dependence). *Let $\tilde{a}$ and $\tilde{b}$ be two random variables with nonzero, bounded variance belonging to the same probability space. If $\rho_{\tilde{a},\tilde{b}} = 1$ or $\rho_{\tilde{a},\tilde{b}} = -1$, then $\tilde{b}$ is an affine function of $\tilde{a}$ with probability one. The affine function is the linear MMSE estimator of $\tilde{b}$ given $\tilde{a}$, denoted by $\ell_{\mathrm{MMSE}}(\tilde{a})$:*

$$\mathrm{P}(\tilde{b} = \ell_{\mathrm{MMSE}}(\tilde{a})) = 1. \tag{8.121}$$

Proof By Theorem 8.19, if $|\rho_{\tilde{a},\tilde{b}}| = 1$, the MSE of the linear MMSE estimator is zero,

$$\mathbb{E}\left[(\ell_{\mathrm{MMSE}}(\tilde{a}) - \tilde{b})^2\right] = \left(1 - \rho_{\tilde{a},\tilde{b}}^2\right)\sigma_{\tilde{b}}^2 = 0, \tag{8.122}$$

which implies (8.121) by Corollary 9.20. $\blacksquare$

The scatterplots in Figure 8.5 illustrate Theorem 8.21. As the correlation coefficient approaches one, the data concentrate on a straight line, indicating that the random variables are almost completely linearly dependent.

8.5.3 Explained Variance

In this section, we provide an interpretation of the correlation coefficient in terms of explained variance. We will need the following result, which expresses the variance of a sum of random variables in terms of their individual variances and the covariance between them.

Theorem 8.22 (Variance of the sum of two random variables). *The variance of the sum of two random variables $\tilde{a}$ and $\tilde{b}$ with finite variance equals*

$$\mathrm{Var}[\tilde{a} + \tilde{b}] = \mathrm{Var}[\tilde{a}] + \mathrm{Var}[\tilde{b}] + 2\,\mathrm{Cov}[\tilde{a}, \tilde{b}]. \tag{8.123}$$

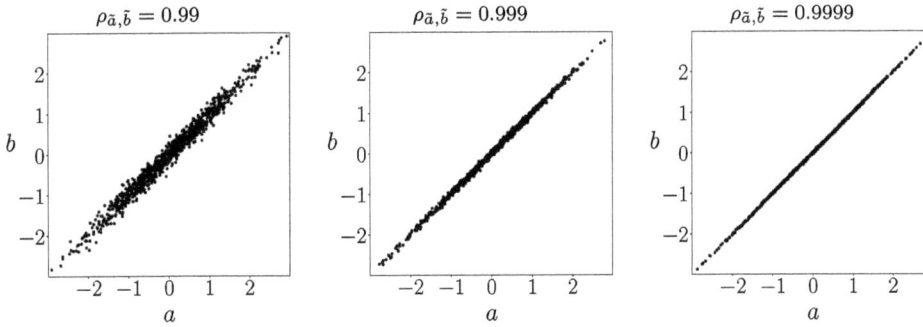

Figure 8.5 Complete linear dependence. The plots show 1,000 i.i.d. samples from two Gaussian random variables $\tilde{a}$ and $\tilde{b}$ with zero mean, unit variance, and different correlation coefficients. As the correlation coefficient approaches one, the samples lie increasingly close to the diagonal, implying an almost exact linear relationship between $\tilde{b}$ and $\tilde{a}$, as predicted by Theorem 8.21.

Proof The result follows from linearity of expectation (Theorem 7.18) and Definitions 7.32 and 8.4:

$$\text{Var}[\tilde{a} + \tilde{b}] = \mathbb{E}\left[(\tilde{a} + \tilde{b} - \mathbb{E}[\tilde{a} + \tilde{b}])^2\right] \tag{8.124}$$

$$= \mathbb{E}\left[(\tilde{a} - \mathbb{E}[\tilde{a}])^2\right] + \mathbb{E}\left[(\tilde{b} - \mathbb{E}[\tilde{b}])^2\right] + 2\mathbb{E}\left[(\tilde{a} - \mathbb{E}[\tilde{a}])(\tilde{b} - \mathbb{E}[\tilde{b}])\right]$$

$$= \text{Var}[\tilde{a}] + \text{Var}[\tilde{b}] + 2\,\text{Cov}[\tilde{a}, \tilde{b}]. \tag{8.125}$$

∎

By Theorem 8.22, if two random variables are positively correlated, their fluctuations reinforce each other, so that the variance of their sum is larger than the sum of their variances. Conversely, if they are negatively correlated, the fluctuations cancel out, and the variance of their sum is smaller than the sum of their variances. An immediate corollary of the theorem is that the variance of the sum of uncorrelated variables is exactly equal to the sum of their variances.

Corollary 8.23. *If two random variables $\tilde{a}$ and $\tilde{b}$ with finite variance are uncorrelated, then*

$$\text{Var}[\tilde{a} + \tilde{b}] = \text{Var}[\tilde{a}] + \text{Var}[\tilde{b}]. \tag{8.126}$$

Consider the decomposition of $\tilde{b}$ into the sum of the linear MMSE estimator $\ell_{\text{MMSE}}(\tilde{a})$ and the corresponding residual

$$\tilde{b} = \underbrace{\ell_{\text{MMSE}}(\tilde{a})}_{\text{Linear MMSE estimator}} + \underbrace{\tilde{b} - \ell_{\text{MMSE}}(\tilde{a})}_{\text{Residual}}. \tag{8.127}$$

The following lemma shows that the residual is uncorrelated with $\tilde{a}$. This makes sense: If they were correlated, then we could use an affine function of $\tilde{a}$ to approximate the residual and improve the linear MMSE estimator. However, this is impossible because the linear MMSE estimator is the optimal affine estimator by Theorem 8.14.

Lemma 8.24 (Uncorrelated residual). *Let $\tilde{a}$ and $\tilde{b}$ be two random variables with bounded variance belonging to the same probability space. The residual of the linear MMSE estimator $\ell_{\mathrm{MMSE}}(\tilde{a})$ of $\tilde{b}$ given $\tilde{a}$, defined in Theorem 8.14, has zero mean and is uncorrelated with $\tilde{a}$.*

Proof By Theorem 8.14 and linearity of expectation (Theorem 7.18),

$$\mathbb{E}\left[\tilde{b} - \ell_{\mathrm{MMSE}}(\tilde{a})\right] = \mathbb{E}\left[\tilde{b} - \sigma_{\tilde{b}}\,\rho_{\tilde{a},\tilde{b}}\left(\frac{\tilde{a} - \mu_{\tilde{a}}}{\sigma_{\tilde{a}}}\right) - \mu_{\tilde{b}}\right] \tag{8.128}$$

$$= \mu_{\tilde{b}} - \mu_{\tilde{b}} - \sigma_{\tilde{b}}\,\rho_{\tilde{a},\tilde{b}}\left(\frac{\mu_{\tilde{a}} - \mu_{\tilde{a}}}{\sigma_{\tilde{a}}}\right) = 0. \tag{8.129}$$

Consequently, by Definition 8.4 and linearity of expectation,

$$\mathrm{Cov}\left[\tilde{a}, \tilde{b} - \ell_{\mathrm{MMSE}}(\tilde{a})\right] = \mathbb{E}\left[(\tilde{a} - \mu_{\tilde{a}})\left(\tilde{b} - \sigma_{\tilde{b}}\,\rho_{\tilde{a},\tilde{b}}\left(\frac{\tilde{a} - \mu_{\tilde{a}}}{\sigma_{\tilde{a}}}\right) - \mu_{\tilde{b}}\right)\right] \tag{8.130}$$

$$= \sigma_{\tilde{a}}\,\sigma_{\tilde{b}}\mathbb{E}\left[s(\tilde{a})(s(\tilde{b}) - \rho_{\tilde{a},\tilde{b}}s(\tilde{a}))\right] \tag{8.131}$$

$$= \sigma_{\tilde{a}}\,\sigma_{\tilde{b}}\left(\mathbb{E}[s(\tilde{a})s(\tilde{b})] - \rho_{\tilde{a},\tilde{b}}\mathbb{E}[s(\tilde{a})^2]\right) \tag{8.132}$$

$$= \sigma_{\tilde{a}}\,\sigma_{\tilde{b}}(\rho_{\tilde{a},\tilde{b}} - \rho_{\tilde{a},\tilde{b}}) \tag{8.133}$$

$$= 0 \tag{8.134}$$

since $\mathbb{E}[s(\tilde{a})^2] = \mathrm{Var}\,[s(\tilde{a})] = 1$ by Lemma 8.3 and $\mathbb{E}[s(\tilde{a})s(\tilde{b})] = \rho_{\tilde{a},\tilde{b}}$, where $s(\tilde{a})$ and $s(\tilde{b})$ denote the standardized counterparts of $\tilde{a}$ and $\tilde{b}$. ∎

Lemma 8.24 establishes that the two components in (8.127) are uncorrelated. Therefore, the variance of $\tilde{b}$ is equal to the sum of their variances by Corollary 8.23. This yields a variance decomposition that is directly tied to the correlation coefficient. Figure 8.6 illustrates the decomposition.

Theorem 8.25 (Variance decomposition). *Let $\tilde{a}$ and $\tilde{b}$ be two random variables with bounded variance belonging to the same probability space. The variance of $\tilde{b}$ can be decomposed into the sum of the variance of the linear MMSE estimator $\ell_{\mathrm{MMSE}}(\tilde{a})$ of $\tilde{b}$ given $\tilde{a}$ and the variance of the residual $\tilde{b} - \ell_{\mathrm{MMSE}}(\tilde{a})$,*

$$\mathrm{Var}\,[\tilde{b}] = \mathrm{Var}\,[\ell_{\mathrm{MMSE}}(\tilde{a})] + \mathrm{Var}\,[\tilde{b} - \ell_{\mathrm{MMSE}}(\tilde{a})]. \tag{8.135}$$

The fraction of variance corresponding to the linear estimator is equal to the squared correlation coefficient $\rho_{\tilde{a},\tilde{b}}^2$ of $\tilde{a}$ and $\tilde{b}$:

$$\rho_{\tilde{a},\tilde{b}}^2 = \frac{\mathrm{Var}\,[\ell_{\mathrm{MMSE}}(\tilde{a})]}{\mathrm{Var}[\tilde{b}]}, \tag{8.136}$$

$$1 - \rho_{\tilde{a},\tilde{b}}^2 = \frac{\mathrm{Var}\,[\tilde{b} - \ell_{\mathrm{MMSE}}(\tilde{a})]}{\mathrm{Var}[\tilde{b}]}. \tag{8.137}$$

Proof By Lemma 8.24, the residual and $\tilde{a}$ are uncorrelated. This implies that the residual is also uncorrelated with any affine function of $\tilde{a}$ (see Exercise 8.2), and in particular with $\ell_{\mathrm{MMSE}}(\tilde{a})$. The variance decomposition (8.135) then follows from Corollary 8.23.

$\rho_{\tilde{a},\tilde{b}}$ / R^2 $\tilde{b}$ Linear estimate $\rho_{\tilde{a},\tilde{b}}\tilde{a}$ Residual $\tilde{b} - \rho_{\tilde{a},\tilde{b}}\tilde{a}$

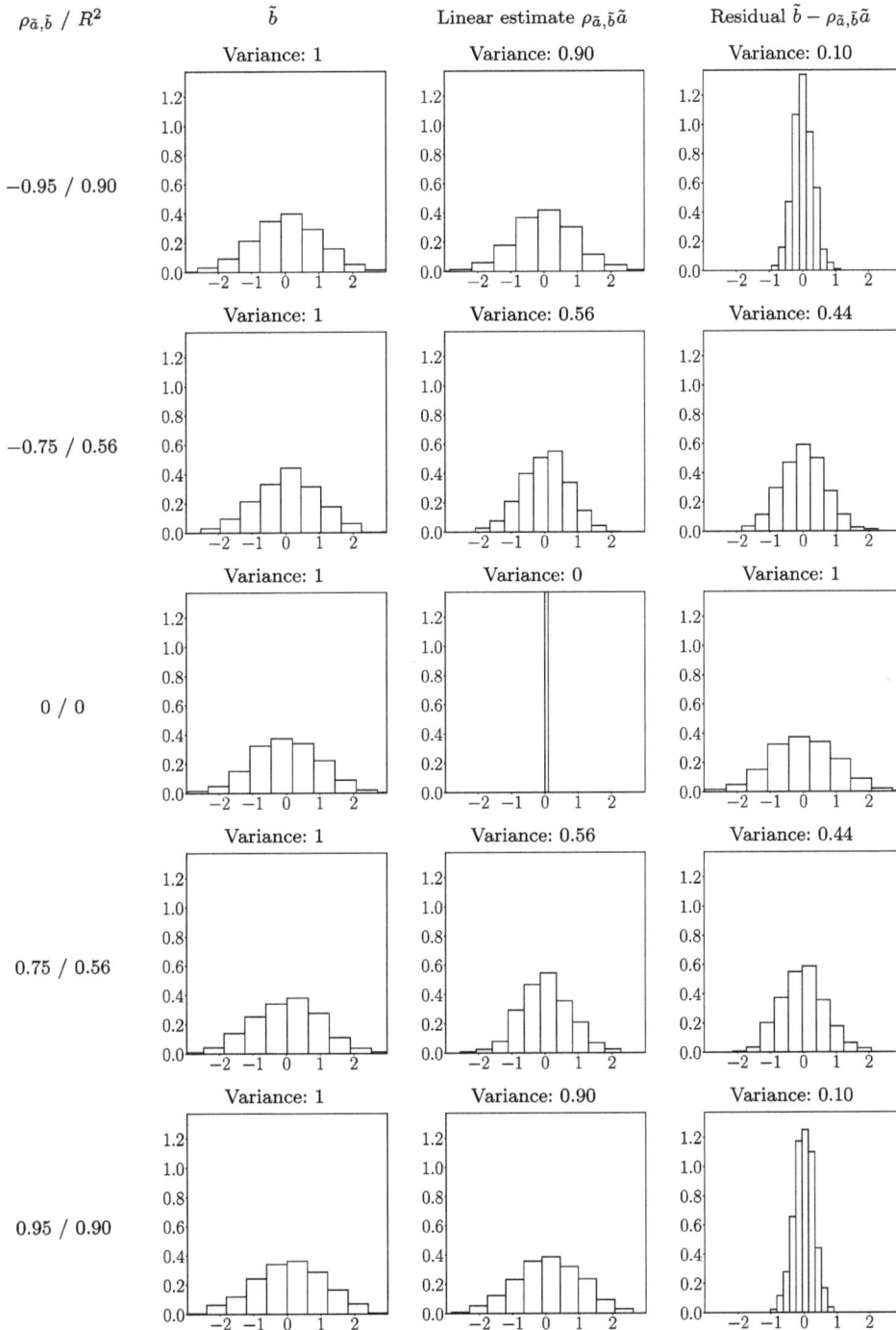

Figure 8.6 Decomposition of variance and coefficient of determination. The figure depicts the different terms in the variance decomposition of Theorem 8.25 for the two Gaussian random variables $\tilde{a}$ and $\tilde{b}$ in Figure 8.1. The left column shows the histogram of $\tilde{b}$. The middle column shows the histogram of the linear MMSE estimator of $\tilde{b}$ given $\tilde{a}$, $\rho_{\tilde{a},\tilde{b}}\tilde{a}$. The right column shows the histogram of the residual $\tilde{b} - \rho_{\tilde{a},\tilde{b}}\tilde{a}$. As established in Theorem 8.25, the fraction of variance explained by the linear estimator is equal to the coefficient of determination R^2.

By (8.129) the mean of the residual is zero, so its variance is equal to the MSE of the linear MMSE estimator, derived in Theorem 8.19:

$$\text{Var}[\tilde{b} - \ell_{\text{MMSE}}(\tilde{a})] = \mathbb{E}\left[(\tilde{b} - \ell_{\text{MMSE}}(\tilde{a}))^2\right] \tag{8.138}$$

$$= (1 - \rho_{\tilde{a},\tilde{b}}^2)\text{Var}\left[\tilde{b}\right]. \tag{8.139}$$

Consequently, by (8.139),

$$\text{Var}\left[\ell_{\text{MMSE}}(\tilde{a})\right] = \text{Var}[\tilde{b}] - \text{Var}[\tilde{b} - \ell_{\text{MMSE}}(\tilde{a})] \tag{8.140}$$

$$= \rho_{\tilde{a},\tilde{b}}^2 \text{Var}\left[\tilde{b}\right]. \tag{8.141}$$

∎

The variance decomposition in Theorem 8.25 provides an intuitive way to evaluate the linear MMSE estimator: Computing the fraction of the response variance that it *explains*. This metric is known as the *coefficient of determination* R^2.

Definition 8.26 (Coefficient of determination in simple linear regression). *Let $\tilde{a}$ and $\tilde{b}$ be random variables representing the feature and the response of a regression problem, and let ℓ_{MMSE} be the linear minimum MSE estimator of $\tilde{b}$ given $\tilde{a}$. The coefficient of determination is the ratio between the variance of the linear MMSE estimator and the variance of the response,*

$$R^2 := \frac{\text{Var}\left[\ell_{\text{MMSE}}(\tilde{a})\right]}{\text{Var}[\tilde{b}]}. \tag{8.142}$$

By Theorem 8.25, the coefficient of determination equals the square of the correlation coefficient, and can also be expressed in terms of the MSE of the linear MMSE. By Theorem 8.20, it is bounded between zero and one.

Theorem 8.27 (Properties of the coefficient of determination). *Let $\tilde{a}$ and $\tilde{b}$ be random variables representing the feature and the response of a regression problem, and let ℓ_{MMSE} be the linear minimum MSE estimator of $\tilde{b}$ given $\tilde{a}$. The coefficient of determination equals*

$$R^2 = \rho_{\tilde{a},\tilde{b}}^2 \tag{8.143}$$

$$= 1 - \frac{\text{MSE}}{\text{Var}[\tilde{b}]}, \qquad \text{MSE} := \mathbb{E}\left[(\tilde{b} - \ell_{\text{MMSE}}(\tilde{a}))^2\right], \tag{8.144}$$

where $\rho_{\tilde{a},\tilde{b}}$ is the correlation coefficient between $\tilde{a}$ and $\tilde{b}$. In addition, R^2 is bounded between zero and one:

$$0 \leq R^2 \leq 1. \tag{8.145}$$

Proof By (8.136) in Theorem 8.25,

$$R^2 := \frac{\text{Var}[\ell_{\text{MMSE}}(\tilde{a})]}{\text{Var}[\tilde{b}]} = \rho_{\tilde{a},\tilde{b}}^2. \tag{8.146}$$

Consequently, by Theorem 8.20, $0 \leq R^2 \leq 1$.

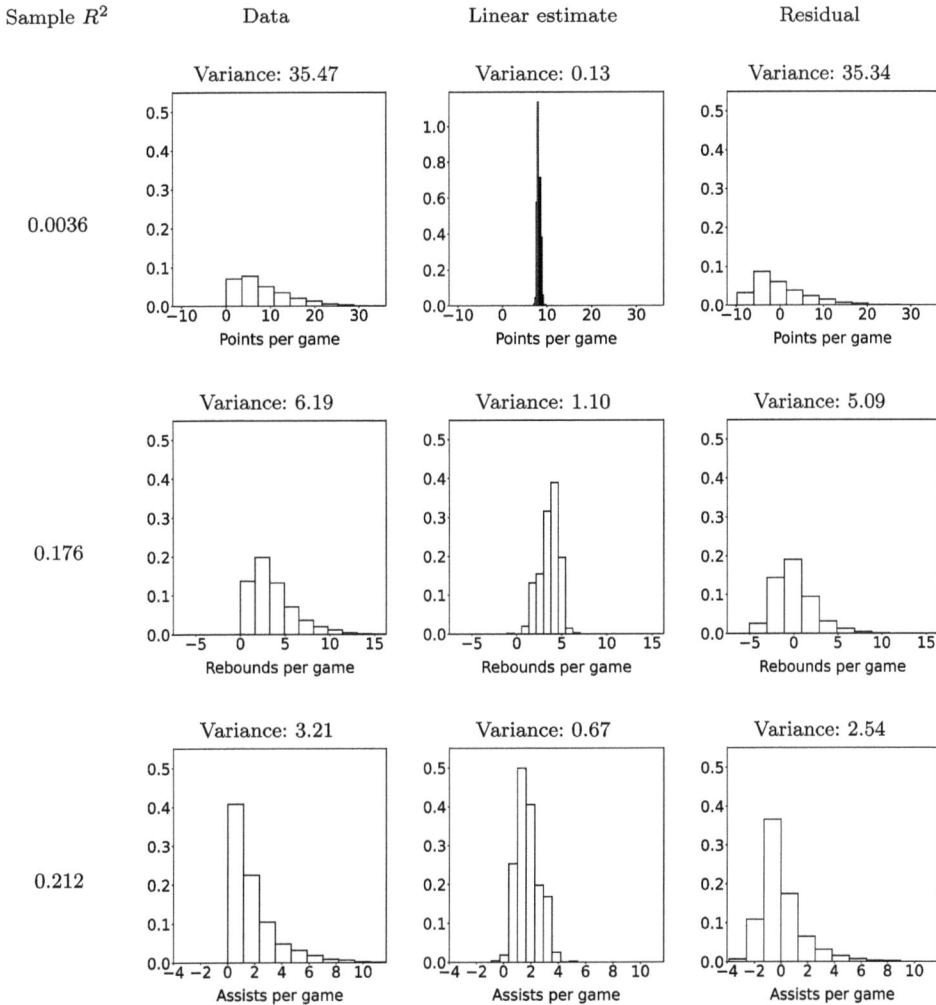

Figure 8.7 Decomposition of sample variance and sample coefficient of determination for NBA data. The figure depicts the different terms in the variance decomposition of Theorem 8.25 estimated from the NBA data in Figure 8.2. The left column shows the histogram of the points (top), rebounds (center) and assists (bottom) per game. The middle column shows the histogram of the ordinary-least-squares estimates of each stat given height. The right column shows the histogram of the corresponding residual. As established in Exercise 8.9, the fraction of sample variance explained by the linear estimator is equal to the sample coefficient of determination.

By (8.135) in Theorem 8.25,

$$R^2 = \frac{\mathrm{Var}\left[\tilde{b}\right] - \mathrm{Var}\left[\tilde{b} - \ell_{\mathrm{MMSE}}(\tilde{a})\right]}{\mathrm{Var}[\tilde{b}]} \tag{8.147}$$

$$= 1 - \frac{\mathrm{Var}[\tilde{b} - \ell_{\mathrm{MMSE}}(\tilde{a})]}{\mathrm{Var}[\tilde{b}]} \tag{8.148}$$

$$= 1 - \frac{\mathbb{E}\left[(\tilde{b} - \ell_{\mathrm{MMSE}}(\tilde{a}))^2\right]}{\mathrm{Var}[\tilde{b}]}, \tag{8.149}$$

where the last equality holds because the mean of the residual $\tilde{b} - \ell_{\mathrm{MMSE}}(\tilde{a})$ is zero by (8.129).   ∎

As illustrated by Figure 8.6, R^2 provides a normalized evaluation of the linear MMSE estimator. When R^2 is close to one, the estimator explains most of the variance in the response because the feature and the response are highly correlated, so the MSE is close to zero. Conversely, when R^2 is close to zero, the estimator explains almost no variance because the feature and the response are almost uncorrelated.

Exercise 8.9 establishes a decomposition of the sample variance that is analogous to the variance decomposition in Theorem 8.25 and proves that the sample coefficient of determination is equal to the squared sample correlation coefficient. Figure 8.7 illustrates the decomposition of sample variance using the same NBA data as in Figure 8.2.

8.6 Uncorrelation and Independence

If two random variables are independent, they are also uncorrelated. This makes sense, since lack of dependence necessarily implies lack of linear dependence.

Lemma 8.28 (Independence implies uncorrelation). *If two random variables are independent, then they are uncorrelated.*

Proof By Lemma 8.5 and Theorem 7.20, if two random variables $\tilde{a}$ and $\tilde{b}$ are independent,

$$\mathrm{Cov}[\tilde{a}, \tilde{b}] = \mathbb{E}[\tilde{a}\tilde{b}] - \mathbb{E}[\tilde{a}]\,\mathbb{E}[\tilde{b}] \tag{8.150}$$

$$= \mathbb{E}[\tilde{a}]\,\mathbb{E}[\tilde{b}] - \mathbb{E}[\tilde{a}]\,\mathbb{E}[\tilde{b}] \tag{8.151}$$

$$= 0. \tag{8.152}$$

∎

In the case of Gaussian random variables, uncorrelation implies independence. This establishes that the dependence between Gaussian random variables is purely linear.

Theorem 8.29 (Uncorrelation implies independence for Gaussian random variables). *Let $\tilde{a}$ and $\tilde{b}$ be the two entries of a Gaussian random vector. If $\tilde{a}$ and $\tilde{b}$ are uncorrelated, then they are also independent.*

Proof Without loss of generality, we parametrize the covariance matrix of the Gaussian random vector as

$$\Sigma := \begin{bmatrix} \sigma_{\tilde{a}}^2 & \rho\sigma_{\tilde{a}}\sigma_{\tilde{b}} \\ \rho\sigma_{\tilde{a}}\sigma_{\tilde{b}} & \sigma_{\tilde{b}}^2 \end{bmatrix}. \tag{8.153}$$

In the proof of Theorem 8.16, we show that the correlation coefficient of $\tilde{a}$ and $\tilde{b}$ is equal to ρ. If the variables are uncorrelated, then ρ is zero, so the covariance-matrix parameter is diagonal and its inverse equals

$$\Sigma^{-1} = \begin{bmatrix} \frac{1}{\sigma_{\tilde{a}}^2} & 0 \\ 0 & \frac{1}{\sigma_{\tilde{b}}^2} \end{bmatrix}. \tag{8.154}$$

In terms of the standardized variables

$$s(a) := \frac{a - \mu_{\tilde{a}}}{\sigma_{\tilde{a}}}, \qquad s(b) := \frac{b - \mu_{\tilde{b}}}{\sigma_{\tilde{b}}}, \tag{8.155}$$

we have

$$f_{\tilde{a},\tilde{b}}(a, b) = \frac{1}{2\pi \sqrt{|\Sigma|}} \exp\left(-\frac{1}{2} \begin{bmatrix} a - \mu_{\tilde{a}} \\ b - \mu_{\tilde{b}} \end{bmatrix}^T \Sigma^{-1} \begin{bmatrix} a - \mu_{\tilde{a}} \\ b - \mu_{\tilde{b}} \end{bmatrix}\right) \tag{8.156}$$

$$= \frac{1}{2\pi \sigma_{\tilde{a}} \sigma_{\tilde{b}}} \exp\left(-\frac{s(a)^2 + s(b)^2}{2}\right) \tag{8.157}$$

$$= \frac{1}{\sqrt{2\pi}\sigma_{\tilde{a}}} \exp\left(-\frac{s(a)^2}{2}\right) \frac{1}{\sqrt{2\pi}\sigma_{\tilde{b}}} \exp\left(-\frac{s(b)^2}{2}\right) \tag{8.158}$$

$$= f_{\tilde{a}}(a) f_{\tilde{b}}(b). \tag{8.159}$$

The joint pdf factors into the product of the marginal pdfs, so the two random variables are independent by Definition 5.15. ∎

Figure 8.8 shows a visualization of the joint pdf of two uncorrelated Gaussian random variables, and also some of their conditional pdfs. The variables are independent, so all the conditional pdfs are the same. For non-Gaussian random variables, uncorrelation does *not* necessarily imply independence, as demonstrated by the following two examples.

Example 8.30 (Uncorrelation does not imply independence). Let $\tilde{a}$ and $\tilde{b}$ be two random variables with the following joint probability mass function, depicted in Figure 8.9,

$$p_{\tilde{a},\tilde{b}}(0, 0) = \frac{1}{2}, \qquad p_{\tilde{a},\tilde{b}}(1, -1) = \frac{1}{4}, \qquad p_{\tilde{a},\tilde{b}}(1, 1) = \frac{1}{4}. \tag{8.160}$$

By Definition 7.4 the mean of $\tilde{b}$ is zero:

$$\mathbb{E}[\tilde{b}] = \sum_{a=0}^{1} \sum_{b=-1}^{1} b \, p_{\tilde{a},\tilde{b}}(a, b) \tag{8.161}$$

$$= 0 \cdot \frac{1}{2} - 1 \cdot \frac{1}{4} + 1 \cdot \frac{1}{4} = 0. \tag{8.162}$$

By Lemma 8.5,

$$\mathrm{Cov}[\tilde{a}, \tilde{b}] = \mathbb{E}[\tilde{a}\tilde{b}] - \mathbb{E}[\tilde{a}]\mathbb{E}[\tilde{b}] \tag{8.163}$$

$$= \mathbb{E}[\tilde{a}\tilde{b}] \tag{8.164}$$

$$= \sum_{a=0}^{1} \sum_{b=-1}^{1} ab \, p_{\tilde{a},\tilde{b}}(a, b) \tag{8.165}$$

$$= 0 \cdot \frac{1}{2} - 1 \cdot \frac{1}{4} + 1 \cdot \frac{1}{4} = 0, \tag{8.166}$$

so the random variables are uncorrelated. However, they are not independent. By Definition 4.13 and Theorem 4.10, the conditional pmf of $\tilde{b}$ given $\tilde{a} = 0$ is

$$p_{\tilde{b}|\tilde{a}}(b \,|\, 0) = \frac{p_{\tilde{a},\tilde{b}}(0, b)}{\sum_{b \in \{-1,0,1\}} p_{\tilde{a},\tilde{b}}(0, b)} = \begin{cases} 1 & \text{if } b = 0, \\ 0 & \text{otherwise.} \end{cases} \tag{8.167}$$

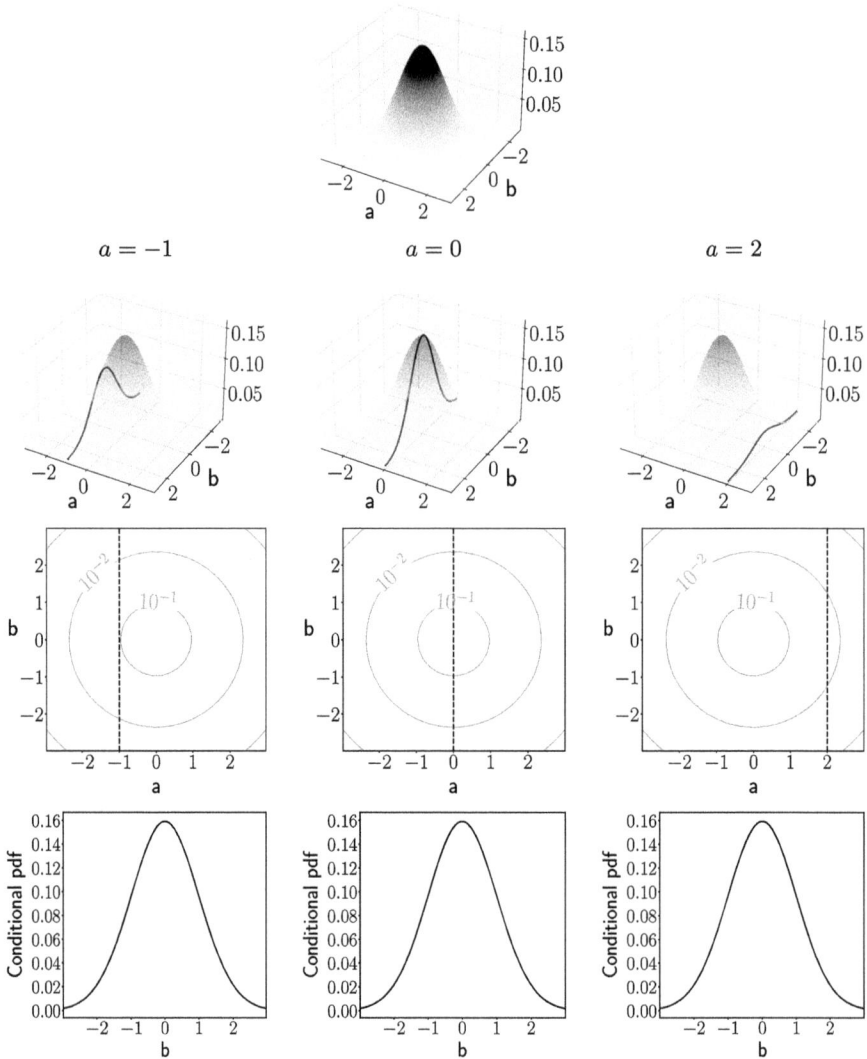

Figure 8.8 Uncorrelated Gaussian random variables are independent. The graph at the top depicts the joint pdf of two uncorrelated Gaussian random variables. The second row shows slices of the density when one of the variables is fixed to a, for different values of a. The third row shows the position of the slices on the contour plot of the joint pdf. The fourth row shows the corresponding conditional pdf of $\tilde{b}$ given $\tilde{a} = a$. All the conditional distributions are the same because a and b are independent, as established in Theorem 8.29.

This is very different from the conditional pmf of $\tilde{b}$ given $\tilde{a} = 1$, which equals

$$p_{\tilde{b}\,|\,\tilde{a}}(b\,|\,1) = \frac{p_{\tilde{a},\tilde{b}}(1,b)}{\sum_{b\in\{-1,0,1\}} p_{\tilde{a},\tilde{b}}(1,b)} = \begin{cases} 0.5 & \text{if } b = -1, \\ 0.5 & \text{if } b = 1, \\ 0 & \text{otherwise.} \end{cases} \qquad (8.168)$$

Consequently, the random variables are dependent (see Section 4.4).

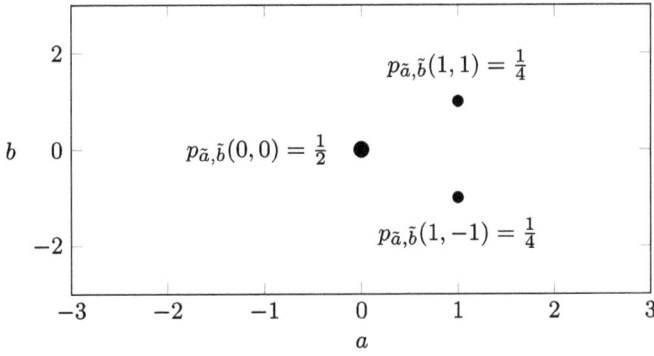

Figure 8.9 Uncorrelation does not imply independence. The plot depicts the joint probability mass function of the random variables in Example 8.30. The radius of each circular marker is proportional to the value of the joint pmf at that location. The random variables are uncorrelated, but not independent.

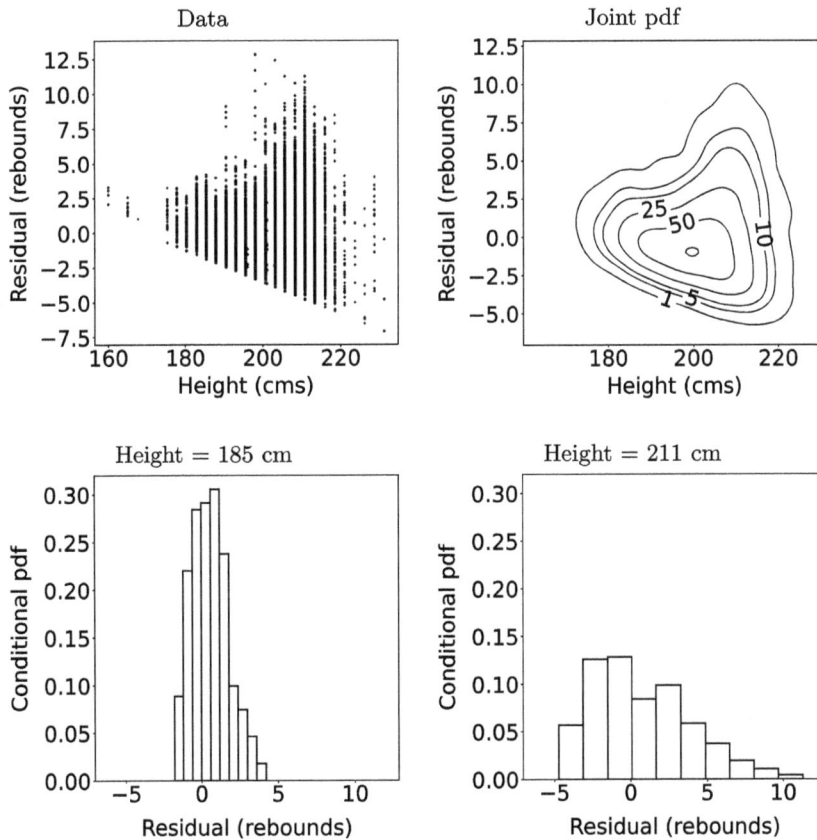

Figure 8.10 Uncorrelated residuals are not necessarily independent. The top row shows a scatterplot and a kernel density estimate (see Definition 5.7) of the residuals of the OLS estimator of rebounds given height in Example 8.18. The bottom row shows histograms of the residual conditioned on two different height values. The conditional distributions of the residual given the two heights are very different, indicating that the residual and height are not independent, even though they are uncorrelated.

Example 8.31 (Uncorrelated OLS residual). When we perform simple linear regression via ordinary least squares (OLS), the residual is uncorrelated with the feature. Consider a dataset of real-valued pairs (x_1, y_1), (x_2, y_2), ..., (x_n, y_n), where we estimate the response y_i for $1 \leq i \leq n$ using the OLS estimator $\ell_{\text{OLS}}(x_i)$ of y_i given the corresponding feature value x_i (see Theorem 8.17). The sample correlation coefficient between the features and the corresponding residuals

$$r_i := y_i - \ell_{\text{OLS}}(x_i), \qquad 1 \leq i \leq n, \tag{8.169}$$

is zero. Exercise 8.9 provides a formal proof. Intuitively, it makes sense for the feature and the residual to be uncorrelated. Otherwise, we would be able to compute a nonzero affine estimate of the residual and use it to improve the OLS estimator. However, this is impossible because Theorem 8.17 establishes that the OLS estimator is the optimal affine estimator.

Although they are always uncorrelated, the residual and the feature are not necessarily independent because there can exist nonlinear dependence between the feature and the response. Figure 8.10 shows the residual of the OLS estimator of rebounds given height in Example 8.18. As expected, the sample correlation coefficient of the residual and height equals zero, so the residual is uncorrelated with height. However, the two quantities are definitely not independent. The bottom row of the figure shows estimates of the conditional pdf of the residual given two different height values. For each height h, we approximate the corresponding conditional pdf using a histogram of the residuals corresponding to data points for which the height equals h, following Definition 3.23. The two conditional distributions are very different.

....

8.7 A Geometric Analysis of Correlation

In this section, we study correlation from a geometric perspective. We interpret zero-mean random variables as vectors in a vector space, where the covariance is an inner product, as explained in Section 8.7.1. Section 8.7.2 establishes that, in this vector space, the standard deviation can be interpreted as the *length* of a random variable, and the correlation coefficient as the cosine of the *angle* between random variables. Section 8.7.3 builds upon these insights to provide geometric intuition about simple linear regression, establishing an equivalence between mean-squared-error minimization and orthogonal projections.

8.7.1 The Inner-Product Space of Zero-Mean Random Variables

Random variables belonging to a common probability space can be interpreted as vectors. We often think of vectors as a list of numbers with fixed length, but the mathematical definition of vector is much more general.

Definition 8.32 (Vector space). *A vector space is a set $\mathcal{V}$ of objects that admit two operations, a vector sum that is commutative and associative, and a multiplication between scalars and vectors that is associative. The operations are distributive with respect to each other. In order for $\mathcal{V}$ to be a valid vector space, the following conditions must hold:*

- *For any vector $v \in \mathcal{V}$ and any scalar $\beta \in \mathbb{R}$, the scalar multiple βv is a vector belonging to $\mathcal{V}$.*

- *For any vectors $v_1, v_2 \in \mathcal{V}$, the vector sum $v_1 + v_2$ is a vector belonging to $\mathcal{V}$.*
- *There exists a zero vector 0, such that $v + 0 = v$ for any vector $v \in \mathcal{V}$.*
- *For any vector $v \in \mathcal{V}$, there exists an additive inverse $-v \in \mathcal{V}$ such that $v + (-v) = 0$.*

Theorem 8.33 (Random variables form a vector space). *The random variables belonging to a probability space form a vector space.*

Proof The two first conditions in Definition 8.32 hold because multiplying a random variable by a real scalar yields another random variable in the same probability space, and so does summing two random variables. This is not completely obvious in the case of continuous random variables but can be proved formally by showing that the resulting random variable is measurable.

In order to define the zero vector and the additive inverse in Definition 8.32, we first need to establish what it means for two random variables to be *equal*. Two random variables in the same probability space are considered equal, if they are equal with probability one, i.e. $\tilde{a} = \tilde{b}$ means

$$\mathrm{P}\left(\tilde{a} = \tilde{b}\right) = 1. \tag{8.170}$$

We define the zero vector as a random variable $\tilde{0}$ that is equal to zero with probability one. Then, for any random variable $\tilde{a}$, $\tilde{a} + \tilde{0} = \tilde{a}$. We define the additive inverse of a random variable $\tilde{a}$ as $-\tilde{a}$, since $\tilde{a} - \tilde{a} = 0$ with probability one, so $\tilde{a} + (-\tilde{a}) = \tilde{0}$. We conclude that random variables associated with the same probability space form a vector space. ∎

If we restrict our attention to zero-mean random variables, the covariance can be interpreted as an inner product between random variables. Let us recall the properties of inner products between vectors.

Definition 8.34 (Inner product). *An inner product on a vector space $\mathcal{V}$ is an operation $\langle \cdot, \cdot \rangle$ that maps each pair of vectors to a real number and is:*

- *Symmetric: for any vectors $v_1, v_2 \in \mathcal{V}$,*

$$\langle v_1, v_2 \rangle = \langle v_2, v_1 \rangle. \tag{8.171}$$

- *Linear: for any scalar $\beta \in \mathbb{R}$ and any vectors $v_1, v_2, v_3 \in \mathcal{V}$,*

$$\langle \beta\, v_1, v_2 \rangle = \beta \langle v_1, v_2 \rangle, \tag{8.172}$$
$$\langle v_1 + v_2, v_3 \rangle = \langle v_1, v_3 \rangle + \langle v_2, v_3 \rangle. \tag{8.173}$$

- *Positive semidefinite: $\langle v, v \rangle$ is nonnegative for any vector $v \in \mathcal{V}$ and if $\langle v, v \rangle = 0$, then $v = 0$, i.e. v must be the zero vector.*

The following theorem establishes that the covariance is an inner product for zero-mean random variables. We require the mean of the random variables to be zero, so that the operation is positive semidefinite. By Definition 8.4, the covariance of zero-mean random variables is the mean of their product, so its interpretation as an inner product is very natural.

Theorem 8.35 (Covariance as an inner product). *The covariance is a valid inner product between zero-mean random variables.*

Proof We verify that the covariance satisfies Definition 8.34. The covariance between two random variables $\tilde{a}$ and $\tilde{b}$ is symmetric

$$\text{Cov}[\tilde{a}, \tilde{b}] := \mathbb{E}\left[(\tilde{a} - \mathbb{E}[\tilde{a}])(\tilde{b} - \mathbb{E}[\tilde{b}])\right] \tag{8.174}$$

$$= \mathbb{E}\left[(\tilde{b} - \mathbb{E}[\tilde{b}])(\tilde{a} - \mathbb{E}[\tilde{a}])\right] \tag{8.175}$$

$$= \text{Cov}[\tilde{b}, \tilde{a}]. \tag{8.176}$$

It is also linear by linearity of expectation (Theorem 7.18). For any $\beta \in \mathbb{R}$ and any random variables $\tilde{a}$, $\tilde{b}$, and $\tilde{c}$ (with finite covariances), by Definition 8.4 and Lemma 8.5,

$$\text{Cov}[\beta\tilde{a}, \tilde{b}] := \mathbb{E}\left[(\beta\tilde{a} - \mathbb{E}[\beta\tilde{a}])(\tilde{b} - \mathbb{E}[\tilde{b}])\right] \tag{8.177}$$

$$= \beta\mathbb{E}\left[(\tilde{a} - \mathbb{E}[\tilde{a}])(\tilde{b} - \mathbb{E}[\tilde{b}])\right] \tag{8.178}$$

$$= \beta\text{Cov}[\tilde{a}, \tilde{b}], \tag{8.179}$$

$$\text{Cov}[\tilde{a} + \tilde{b}, \tilde{c}] := \mathbb{E}[(\tilde{a} + \tilde{b})\tilde{c}] - \mathbb{E}[\tilde{a} + \tilde{b}]\mathbb{E}[\tilde{c}] \tag{8.180}$$

$$= \mathbb{E}[\tilde{a}\tilde{c}] - \mathbb{E}[\tilde{a}]\mathbb{E}[\tilde{c}] + \mathbb{E}[\tilde{b}\tilde{c}] - \mathbb{E}[\tilde{b}]\mathbb{E}[\tilde{c}] \tag{8.181}$$

$$= \text{Cov}[\tilde{a}, \tilde{c}] + \text{Cov}[\tilde{b}, \tilde{c}]. \tag{8.182}$$

The covariance is positive semidefinite, as long as for any zero-mean random variable $\tilde{a}$, $\mathbb{E}[\tilde{a}^2] = 0$ implies $\text{P}(\tilde{a} = 0) = 1$. This follows from Chebyshev's inequality, as we establish in Corollary 9.20. ∎

8.7.2 The Geometry of Zero-Mean Random Variables

The inner product associated with a vector space can be used to define a norm. For any vector v in the vector space,

$$||v|| := \sqrt{\langle v, v \rangle}. \tag{8.183}$$

Geometrically, $||v||$ can be thought of as the *length* of the vector. The norm associated with the covariance inner product in Theorem 8.35 is the standard deviation. For a zero-mean random variable $\tilde{a}$, by Definitions 8.4 and 7.32,

$$||\tilde{a}|| := \sqrt{\text{Cov}[\tilde{a}, \tilde{a}]} \tag{8.184}$$

$$= \sqrt{\mathbb{E}[\tilde{a}^2]} \tag{8.185}$$

$$= \sqrt{\text{Var}[\tilde{a}]}. \tag{8.186}$$

Therefore, we can interpret the standard deviation of $\tilde{a}$ as its length, as depicted by the left diagram in Figure 8.11.

A natural measure of the similarity between two vectors is the cosine of the angle between them, which equals their inner product normalized by the product of their norms. In the case of the covariance inner product in Theorem 8.35, we have

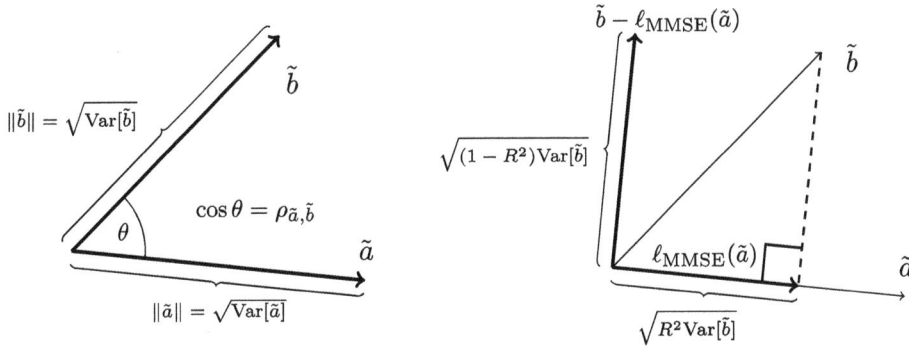

Figure 8.11 Geometric interpretation of correlation for zero-mean random variables. The left diagram illustrates the geometric interpretation of correlation for zero-mean random variables, described in Section 8.7. The lengths of the vectors representing the random variables are equal to their standard deviations. The cosine of the angle between the vectors is equal to the correlation coefficient. The right diagram depicts the connection between linear regression and orthogonal projections, described in Section 8.7.3. The linear MMSE estimator of the response $\tilde{b}$ given the feature $\tilde{a}$ is the projection of $\tilde{b}$ onto the line spanned by $\tilde{a}$. The estimator and the residual are orthogonal, and hence uncorrelated. The squared length of $\tilde{b}$ (its variance) can be decomposed into the sum of the squared lengths (variances) of the estimator and the residual. The ratio between the length of the estimate and of the response is equal to the coefficient of determination R^2.

$$\cos \theta = \frac{\langle \tilde{a}, \tilde{b} \rangle}{||\tilde{a}|| \, ||\tilde{b}||} \tag{8.187}$$

$$= \frac{\text{Cov}[\tilde{a}, \tilde{b}]}{\sqrt{\text{Var}[\tilde{a}]\text{Var}[\tilde{b}]}} \tag{8.188}$$

$$= \rho_{\tilde{a},\tilde{b}}, \tag{8.189}$$

where θ denotes the angle between $\tilde{a}$ and $\tilde{b}$ (see the left diagram in Figure 8.11). The cosine is exactly equal to the correlation coefficient! This analogy provides a geometric perspective on correlation. If two random variables are positively correlated, then the cosine of their angle is positive, so the corresponding vectors point in the same direction. Conversely, if the variables are negatively correlated, then the cosine is negative, and the vectors point in opposite directions. If the variables are uncorrelated, then the cosine is zero, which means that the vectors are orthogonal.

The analogy between the cosine and the correlation coefficient implies some of the properties of the correlation coefficient derived in Section 8.5. Since it is the cosine of an angle, the correlation coefficient is between -1 and 1, as established in Theorem 8.20. When the correlation coefficient is exactly equal to 1 or -1, the two vectors are collinear, so each random variable can be expressed as a linear scaling of the other, as established in Theorem 8.21.

8.7.3 Simple Linear Regression via Orthogonal Projection

In this section, we show how the interpretation of zero-mean random variables as vectors can be leveraged to gain geometric intuition about the simple linear regression problem discussed in Section 8.4, and to rederive the linear MMSE estimator.

Our goal is to estimate a zero-mean random variable $\tilde{b}$ by scaling another zero-mean random variable $\tilde{a}$. Minimizing the mean squared error of the estimate is equivalent to minimizing the squared norm induced by the covariance inner product. Indeed, since $\tilde{a}$ and $\tilde{b}$ have mean zero, then $\tilde{b} - \beta\tilde{a}$ also has zero mean for any β by linearity of expectation (Theorem 7.18), so

$$\mathbb{E}\left[(\tilde{b} - \beta\tilde{a})^2\right] = \text{Var}\left[\tilde{b} - \beta\tilde{a}\right] \tag{8.190}$$

$$= \|\tilde{b} - \beta\tilde{a}\|^2. \tag{8.191}$$

Interpreting $\tilde{a}$ and $\tilde{b}$ as vectors, we want the closest point to $\tilde{b}$ that is collinear with $\tilde{a}$. In linear algebra, this is called the *projection* of $\tilde{b}$ onto $\tilde{a}$ (or rather the projection onto the subspace spanned by $\tilde{a}$). Geometrically, the distance between $\tilde{b}$ and $\beta\tilde{a}$ is minimized when the residual $\tilde{b} - \beta\tilde{a}$ is orthogonal to $\tilde{a}$ (see the right diagram in Figure 8.11). Consequently, the projection $\beta_{\text{proj}}\tilde{a}$ satisfies

$$\langle \tilde{a}, \tilde{b} - \beta_{\text{proj}}\tilde{a} \rangle = 0, \tag{8.192}$$

which implies $\beta_{\text{proj}} \|\tilde{a}\|^2 = \langle \tilde{a}, \tilde{b} \rangle$, so

$$\beta_{\text{proj}} = \frac{\langle \tilde{a}, \tilde{b} \rangle}{\|\tilde{a}\|^2} \tag{8.193}$$

$$= \frac{\text{Cov}\left[\tilde{a}, \tilde{b}\right]}{\text{Var}\left[\tilde{a}\right]} \tag{8.194}$$

$$= \rho_{\tilde{a},\tilde{b}}\sqrt{\frac{\text{Var}[\tilde{b}]}{\text{Var}[\tilde{a}]}} = \beta_{\text{MMSE}}. \tag{8.195}$$

We obtain the linear coefficient of the linear MMSE estimator $\ell_{\text{MMSE}}(\tilde{a})$ defined in Theorem 8.14.

Finally, we can also rederive the decomposition of variance in Theorem 8.25 from this geometric perspective, as depicted by the right diagram in Figure 8.11. The orthogonality between $\tilde{a}$ and the residual $\tilde{a} - \ell_{\text{MMSE}}(\tilde{a})$ implies that the two random variables are uncorrelated, which is consistent with Lemma 8.24. It also implies that $\ell_{\text{MMSE}}(\tilde{a})$ and the residual are orthogonal (since $\ell_{\text{MMSE}}(\tilde{a})$ lies in the same direction as $\tilde{a}$), so by the Pythagorean theorem,

$$\text{Var}[\tilde{b}] = \|\tilde{b}\|^2 = \|\ell_{\text{MMSE}}(\tilde{a})\|^2 + \|\tilde{b} - \ell_{\text{MMSE}}(\tilde{a})\|^2 \tag{8.196}$$

$$= \text{Var}\left[\ell_{\text{MMSE}}(\tilde{a})\right] + \text{Var}\left[\tilde{b} - \ell_{\text{MMSE}}(\tilde{a})\right], \tag{8.197}$$

which is the desired decomposition of variance. By (8.195) the squared length of the projection equals

$$\|\ell_{\text{MMSE}}(\tilde{a})\|^2 = \|\beta_{\text{MMSE}}\tilde{a}\|^2 \tag{8.198}$$

$$= \beta_{\text{MMSE}}^2 \|\tilde{a}\|^2 \tag{8.199}$$

$$= \frac{\rho_{\tilde{a},\tilde{b}}^2 \text{Var}[\tilde{b}]}{\text{Var}[\tilde{a}]} \text{Var}[\tilde{a}] \tag{8.200}$$

$$= \rho_{\tilde{a},\tilde{b}}^2 \text{Var}[\tilde{b}], \tag{8.201}$$

so the fraction of the squared length of $\tilde{b}$ covered by the linear estimator is the coefficient of determination $R^2 = \rho_{\tilde{a},\tilde{b}}^2$, as established in Theorem 8.27.

8.8 Correlation (Usually) Does Not Imply Causation

Great care must be taken when interpreting correlation in terms of causal effects. For example, unemployment in Spain is negatively correlated with temperature, as shown in the left graph of Figure 8.12. Unemployment is lower, on average, when the temperatures are higher. However, this does *not* mean that higher temperatures *cause* the unemployment to decrease. If the temperature in Spain suddenly rose by several degrees, it seems unlikely that unemployment would automatically decrease as a result.

The goal of causal inference is to determine whether a variable, usually known as the *treatment*, has a causal effect on another variable, called the *outcome*. The potential outcome $\widetilde{po}_t$ represents the distribution of the outcome of interest in a hypothetical scenario where the treatment $\tilde{t}$ equals t. In Sections 4.6.1 and 7.9, we study causal effects of binary treatments. Here, we consider discrete or continuous treatments with multiple possible values. In our example, the treatment $\tilde{t}$ is the temperature, which is typically modeled as continuous. The corresponding potential outcome $\widetilde{po}_t$ represents unemployment in a hypothetical situation where the temperature in Spain is equal to t.

When trying to characterize causal effects from data, we encounter the fundamental problem of causal inference, mentioned in Section 4.6.1. Our measurements of the potential outcome $\widetilde{po}_t$ are extremely incomplete. As illustrated in Figure 8.13, for each data point, the observed outcome is equal to one specific potential outcome corresponding to the observed treatment:

$$\tilde{y} := \widetilde{po}_t \quad \text{if} \quad \tilde{t} = t. \tag{8.202}$$

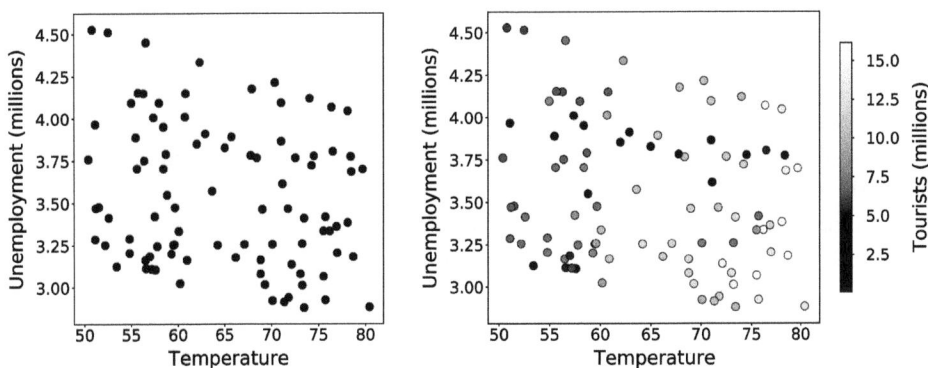

Figure 8.12 Unemployment and temperature. The scatterplot on the left shows monthly measurements of the number of unemployed people and the temperature in Spain (in °F) between 2015 and 2022. The correlation coefficient is -0.21. The scatterplot on the right shows the number of tourists visiting in Spain (indicated by the color of each marker). When the temperature is high, there are more tourists, which explains the negative correlation between unemployment and temperature. We exclude April and May 2020 from our analysis because there were no tourists due to the COVID-19 pandemic.

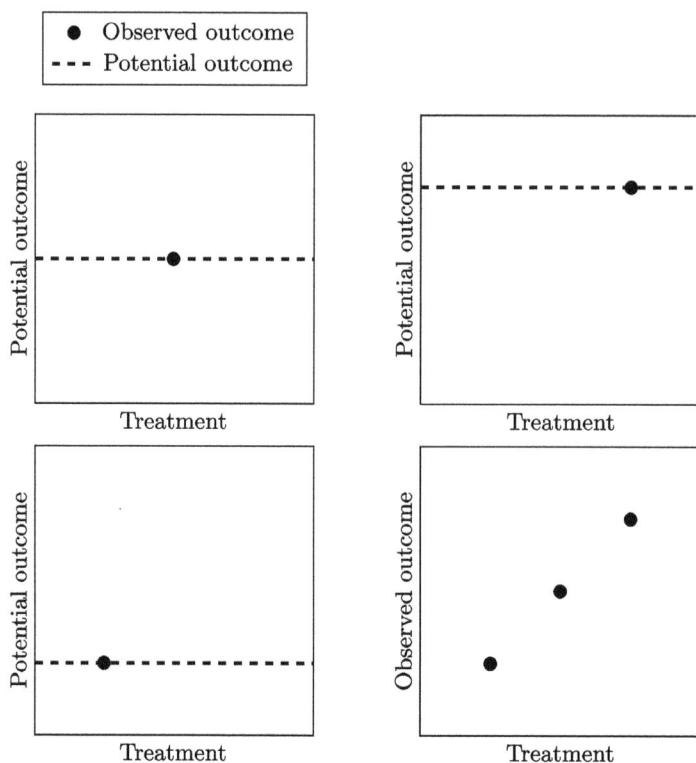

Figure 8.13 Correlation does not imply causation. The top two diagrams and the bottom left diagram provide an example depiction of the causal relationship between a treatment and a potential outcome. Each diagram shows a sample of the observed potential outcome corresponding to the observed treatment, indicated by a circular marker. The dashed lines represent the unobserved counterfactual potential outcome, which would have been observed for all other values of the treatment. In every case, the potential outcome is constant with respect to the treatment because the treatment has no causal effect. However, the treatment is proportional to the potential outcome, which induces a spurious positive correlation between the observed outcome and the treatment, as shown in the bottom right plot.

The potential outcomes associated with all other values of the treatment are unobserved *counterfactuals*. In our example, if the temperature is 60°, then we only get to see $\widetilde{\mathrm{po}}_{60}$. We don't know what would have happened if the temperature had been equal to another value, such as 50° or 70°.

A common assumption when analyzing nonbinary treatments is that the average causal dependence between the treatment and the outcome is *linear*. More precisely,

$$\mathbb{E}\left[\widetilde{\mathrm{po}}_t\right] = \beta_{\mathrm{treat}}t, \tag{8.203}$$

where β_{treat} is a coefficient that quantifies the causal effect. In our example, if $\beta_{\mathrm{treat}} = -1$, this would mean that increasing temperatures by 1° decreases unemployment on average by one unit (one million people).

Under the linear assumption, a reasonable way to estimate the causal effect is to compute the covariance between the observed outcome $\tilde{y}$ and the treatment $\tilde{t}$. By Theorem 8.14 (setting $\tilde{b} := \tilde{y}$ and $\tilde{a} := \tilde{t}$) and Definition 8.6, the linear coefficient of the MMSE estimator of $\tilde{y}$ given $\tilde{t}$ is

$$\beta_{\text{MMSE}} = \frac{\sigma_{\tilde{y}} \, \rho_{\tilde{t}, \tilde{y}}}{\sigma_{\tilde{t}}} \tag{8.204}$$

$$= \text{Cov} \left[\tilde{y}, \tilde{t} \right], \tag{8.205}$$

assuming the treatment is standardized, so $\sigma_{\tilde{t}} = 1$. As a result, if the dependence between the outcome and the response is indeed linear, then

$$\tilde{y} \approx \beta_{\text{MMSE}} \tilde{t} = \text{Cov} \left[\tilde{y}, \tilde{t} \right] \tilde{t}, \tag{8.206}$$

which motivates estimating β_{treat} using $\text{Cov} \left[\tilde{y}, \tilde{t} \right]$. The following theorem shows that this works, but only as long as the treatment and the potential outcome are *independent*, which can be ensured by randomizing the treatment (see Section 4.6.3). This is an extension of Theorems 4.23 and 7.62 to nonbinary treatments.

Theorem 8.36 (Correlation can imply causation). *Let $\widetilde{\text{po}}_t$ denote the potential outcome associated with a discrete or continuous treatment $\tilde{t}$, which is standardized, so $\mathbb{E}[\tilde{t}] = 0$ and $\mathbb{E}[\tilde{t}^2] = 1$. The observed outcome $\tilde{y}$ is equal to $\widetilde{\text{po}}_t$ when $\tilde{t} = t$. We assume that the treatment has a linear effect on the mean of $\widetilde{\text{po}}_t$: $\mathbb{E}\left[\widetilde{\text{po}}_t\right] = \beta_{\text{treat}} t$, for some real constant β_{treat}. If $\widetilde{\text{po}}_t$ and $\tilde{t}$ are independent for all t, then*

$$\text{Cov} \left[\tilde{y}, \tilde{t} \right] = \beta_{\text{treat}}. \tag{8.207}$$

Proof Without loss of generality, we assume that all variables are continuous. In order to derive the covariance of $\tilde{y}$ and $\tilde{t}$, we first compute the conditional mean function of their product given $\tilde{t} = t$, and then apply iterated expectation. Under the independence assumption $f_{\widetilde{\text{po}}_t | \tilde{t}} = f_{\widetilde{\text{po}}_t}$ (see Section 5.7), so by Definition 7.44 the conditional mean function is

$$\mu_{\tilde{y}\tilde{t} | \tilde{t}}(t) = \int_{y=-\infty}^{\infty} yt f_{\tilde{y} | \tilde{t}}(y \,|\, t) \, \mathrm{d}y \tag{8.208}$$

$$= \int_{y=-\infty}^{\infty} yt f_{\widetilde{\text{po}}_t | \tilde{t}}(y \,|\, t) \, \mathrm{d}y \tag{8.209}$$

$$= t \int_{y=-\infty}^{\infty} y f_{\widetilde{\text{po}}_t}(y) \, \mathrm{d}y \tag{8.210}$$

$$= t \mathbb{E}\left[\widetilde{\text{po}}_t\right] \tag{8.211}$$

$$= \beta_{\text{treat}} t^2. \tag{8.212}$$

By Definition 8.4, iterated expectation (Theorem 7.54) and the assumption that $\mathbb{E}[\tilde{t}] = 0$ and $\mathbb{E}[\tilde{t}^2] = 1$,

$$\text{Cov} \left[\tilde{y}, \tilde{t} \right] = \mathbb{E}\left[\tilde{y}\tilde{t}\right] = \mathbb{E}\left[\mu_{\tilde{y}\tilde{t} | \tilde{t}}(t)\right] \tag{8.213}$$

$$= \mathbb{E}\left[\beta_{\text{treat}} \tilde{t}^2\right] \tag{8.214}$$

$$= \beta_{\text{treat}} \mathbb{E}\left[\tilde{t}^2\right] \tag{8.215}$$

$$= \beta_{\text{treat}}. \tag{8.216}$$

■

Randomizing the treatment guarantees that the potential outcomes and the treatment are independent, but in many cases, randomization is not possible. This includes our motivating example: Setting the temperature in Spain to a random value and measuring the corresponding unemployment is not an option! Consequently, we cannot guarantee that the potential outcome associated with unemployment is independent from the temperature, so the observed correlation does not necessarily imply a causal linear effect.

In fact, the main reason why temperature and unemployment in Spain are correlated is that more tourists visit Spain when the temperatures are higher, as shown in the right graph in Figure 8.12. The number of tourists does have a causal effect on unemployment. If one million more tourists were suddenly to arrive in Spain, then unemployment would surely decrease as a consequence. The number of tourists is a confounding factor or confounder, which induces a spurious correlation between temperature and unemployment. A similar effect occurs in Figure 8.13: The potential outcome is constant with respect to the treatment for each observation, but there is a spurious correlation between the observed outcome and the treatment. This is caused by the choice of treatment, which is proportional to the potential outcome.

In order to study the effect of confounders on the correlation between a treatment and an outcome of interest, we consider a situation where the potential outcome depends linearly on the treatment, and also on a confounder. More precisely, we define the potential outcome $\widetilde{\text{po}}_t$ as a linear combination of the treatment, a confounder $\tilde{c}$, and an additional variable $\tilde{z}$ that captures sources of randomness unrelated to the treatment and the confounder:

$$\widetilde{\text{po}}_t := \beta_{\text{treat}}t + \beta_{\text{conf}}\tilde{c} + \tilde{z}. \tag{8.217}$$

Assuming that all random variables are centered to have zero mean, the mean causal effect is linear $\mathbb{E}\left[\widetilde{\text{po}}_t\right] = \beta_{\text{treat}}t$, as in Theorem 8.36. However, the treatment $\tilde{t}$ and the potential outcome $\widetilde{\text{po}}_t$ are not independent, as required by the theorem, unless the treatment is independent from the confounder $\tilde{c}$. The following theorem shows that, as a result, the covariance between the treatment and the observed response does not necessarily reflect the causal effect of the treatment.

Theorem 8.37 (Correlation does not imply causation because of confounders). *Let $\widetilde{\text{po}}_t$ denote a potential outcome associated with a certain treatment $\tilde{t}$, given by*

$$\widetilde{\text{po}}_t := \beta_{\text{treat}}t + \beta_{\text{conf}}\tilde{c} + \tilde{z}, \tag{8.218}$$

where β_{treat} and β_{conf} are real-valued constants, and $\tilde{c}$ and $\tilde{z}$ are random variables. The observed outcome $\tilde{y}$ equals $\widetilde{\text{po}}_t$, if the treatment $\tilde{t}$ equals t, so

$$\tilde{y} = \beta_{\text{treat}}\tilde{t} + \beta_{\text{conf}}\tilde{c} + \tilde{z}. \tag{8.219}$$

We assume that the random variables $\tilde{t}$ and $\tilde{c}$ are standardized to have zero mean and unit variance. If $\tilde{z}$ is independent from $\tilde{t}$ and $\tilde{c}$, then the covariance between the observed outcome and the treatment is dictated by the confounder. It equals

$$\text{Cov}\left[\tilde{y}, \tilde{t}\right] = \beta_{\text{treat}} + \beta_{\text{conf}}\sigma_{\tilde{t},\tilde{c}}, \tag{8.220}$$

where $\sigma_{\tilde{t},\tilde{c}}$ is the covariance between $\tilde{t}$ and the confounder $\tilde{c}$.

Proof Under our assumptions, by Definition 8.4 and linearity of expectation (Theorem 7.18),

$$\text{Cov}\left[\tilde{y}, \tilde{t}\right] = \mathbb{E}\left[\tilde{y}\tilde{t}\right] \tag{8.221}$$

$$= \mathbb{E}\left[\beta_{\text{treat}}\tilde{t}^2 + \beta_{\text{conf}}\tilde{c}\tilde{t} + \tilde{z}\tilde{t}\right] \tag{8.222}$$

$$= \beta_{\text{treat}}\mathbb{E}\left[\tilde{t}^2\right] + \beta_{\text{conf}}\mathbb{E}\left[\tilde{c}\tilde{t}\right] + \mathbb{E}\left[\tilde{z}\right]\mathbb{E}\left[\tilde{t}\right] \tag{8.223}$$

$$= \beta_{\text{treat}} + \beta_{\text{conf}}\sigma_{\tilde{t},\tilde{c}}. \tag{8.224}$$

∎

Theorem 8.37 does not contradict Theorem 8.36. If we randomize the treatment, then this renders it independent from $\tilde{c}$, so $\sigma_{\tilde{t},\tilde{c}} = 0$, and the covariance is indeed equal to β_{treat}.

Example 8.38 (Guinea-pig rescue: Spurious correlation). In this example we illustrate the consequences of Theorem 8.37 in a fictional scenario, where a zoologist called Imara runs a rescue center for underweight guinea pigs. She wants to find out if a nutritional supplement can help them gain weight. She decides to test the supplement by feeding different quantities to the guinea pigs for a week, and then checking their weight gain. We model the amount of supplement provided to each guinea pig as a treatment random variable $\tilde{t}$. It turns out that the supplement has no causal effect on weight gain, but food intake does. To capture this, we model the potential outcome, representing weight gain, as in (8.217), setting $\beta_{\text{treat}} := 0$ and $\beta_{\text{conf}} := 1$:

$$\widetilde{\text{po}}_t := \tilde{c} + \tilde{z}, \tag{8.225}$$

where the confounder $\tilde{c}$ represents the food intake, and $\tilde{z}$ captures factors affecting weight gain that are unrelated to the supplement and the food intake.

Being familiar with Theorem 8.36, Imara assumes that the causal effect is linear, and estimates it by computing the covariance between the treatment $\tilde{t}$ and the observed outcome $\tilde{y}$, $\mathbb{E}\left[\tilde{y}\tilde{t}\right]$ (after standardizing the treatment). She finds that the covariance is positive, which suggests that the weight gain is proportional to the supplement eaten by the pigs. To double-check, she asks her friend Christina to repeat the experiment the following week while she's on a trip to Ohio. Puzzlingly, the covariance turns out to be negative, which suggests that the supplement is making the pigs lose weight! Imara and Christina consult their friend Dimitris, who advises them to repeat the experiment once again, randomizing the treatment to ensure that the guarantees in Theorem 8.36 are valid. Making each guinea pig eat a random amount of supplement is hard, but Imara and Christina manage to do it. This time the covariance is zero, indicating that the supplement is useless.

The seemingly contradictory results of these experiments are explained by Theorem 8.37. In each of the three feeding scenarios, the covariance $\sigma_{\tilde{t},\tilde{c}}$ between the treatment and the confounder was different, which induced a different covariance between the observed outcome and the treatment.

When Imara fed the pigs, she mixed the supplement with the food. Consequently, the pigs that ate more food also ate more supplement, so the covariance between the treatment and the confounder was positive: $\sigma_{\tilde{t},\tilde{c}} := 0.8$. This induced a spurious positive correlation between the observed outcome and the treatment: The pigs that ate more supplement gained more weight. The top left graph in Figure 8.14 shows simulated samples of the supplement and food intake, under the assumption that they are jointly Gaussian. The top right graph

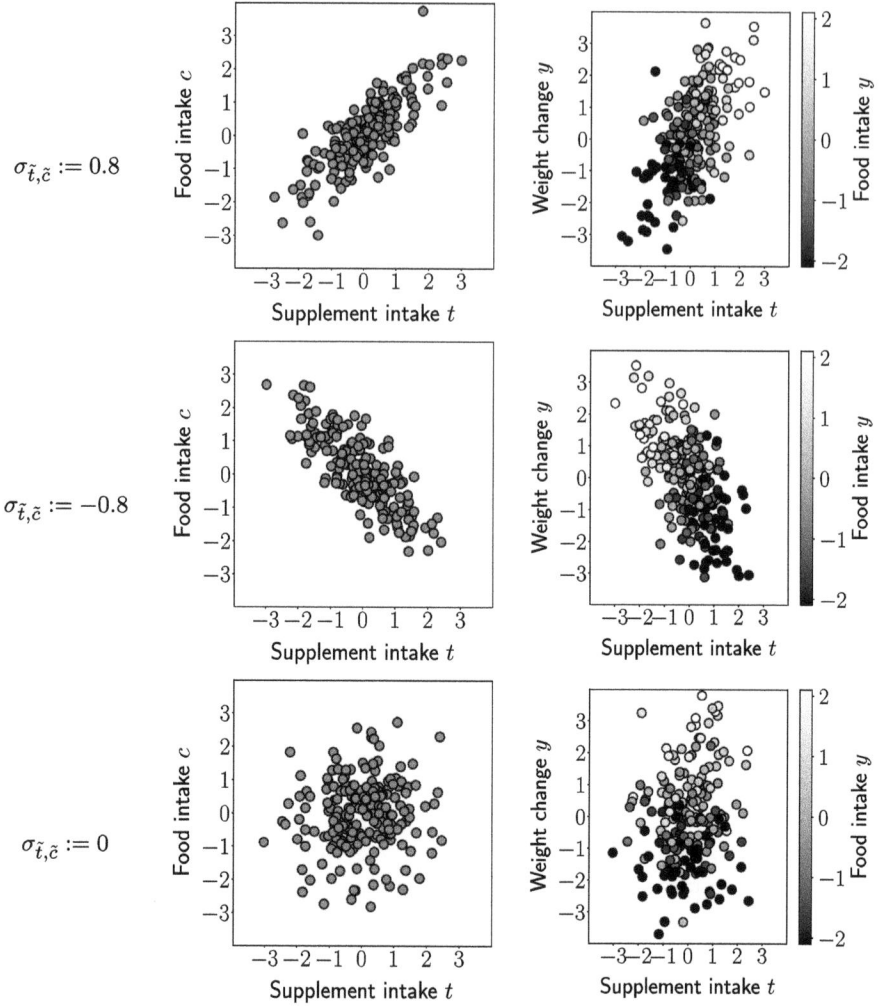

Figure 8.14 Unobserved confounders distort the observed correlation. The left column shows scatterplots of 200 simulated samples, corresponding to the supplement intake $\tilde{t}$ and the food intake $\tilde{c}$ of 200 guinea pigs in Example 8.38, assuming that $\tilde{t}$ and $\tilde{c}$ are jointly Gaussian. The right column shows scatterplots of the corresponding samples of $\tilde{t}$ and the weight gain $\tilde{y}$, assuming that in (8.225) $\tilde{z}$ is a Gaussian random variable with zero mean and unit variance. The marker representing each sample is color-coded to indicate the corresponding value of $\tilde{c}$. In each row, the covariance $\sigma_{\tilde{t},\tilde{c}}$ between $\tilde{t}$ and $\tilde{c}$ is different. This covariance governs the correlation between $\tilde{t}$ and $\tilde{y}$, as established in Theorem 8.37.

shows a scatterplot of the weight change and food intake, assuming that the variance of $\tilde{z}$ in (8.225) is one. The marker corresponding to each sample is color-coded to indicate the corresponding value of the food intake. We observe that the pigs that gained more weight, indeed ate more food (as well as more supplement), confirming that the positive correlation between the treatment and the observed outcome is due to the confounder.

When Christina fed the pigs, she gave them the supplement after their meal. Consequently, the pigs that ate more food were full by the time the supplement was offered, so they ate less supplement, resulting in a negative covariance between the treatment and the confounder:

$\sigma_{\tilde{t},\tilde{c}} := -0.8$. In this case, the confounder induced a negative spurious correlation between the observed outcome and the treatment: The pigs that ate more supplement gained less weight (because they ate less food!). The middle row in Figure 8.14 shows scatterplots of the treatment and the confounder, and of the treatment and the observed outcome.

Finally, randomization of the supplement eaten by each pig rendered the treatment independent from the confounder, and hence uncorrelated: $\sigma_{\tilde{t},\tilde{c}} := 0$. This ensured that the potential outcome was independent from the treatment, satisfying the requirement in Theorem 8.36. Consequently, there was no spurious correlation, and the observed covariance correctly revealed that the treatment had no causal effect. The bottom row in Figure 8.14 shows scatterplots of the treatment and the confounder, and of the treatment and the observed outcome.

....

In conclusion, correlation computed from data cannot be interpreted causally, unless the treatment is randomized. Otherwise, unobserved confounders may completely distort it. In Section 12.1.4, we discuss how to adjust for known confounders using linear regression.

Exercises

8.1 (Three random variables) If a random variable $\tilde{w}$ is positively correlated with another random variable $\tilde{y}$, and $\tilde{y}$ is positively correlated with a third random variable $\tilde{z}$, can $\tilde{w}$ and $\tilde{z}$ be negatively correlated? Provide an example of three such variables, or prove that it is impossible.

8.2 (Properties of covariance) Prove the following properties of the covariance. For any random variables $\tilde{a}$ and $\tilde{b}$ with finite variance:

a For any $\alpha, \beta \in \mathbb{R}$,

$$\mathrm{Cov}[\beta\tilde{a} + \alpha, \tilde{b}] = \beta\mathrm{Cov}[\tilde{a}, \tilde{b}], \tag{8.226}$$

in particular, if $\tilde{a}$ and $\tilde{b}$ are uncorrelated, then so are $\beta\tilde{a} + \alpha$ and $\tilde{b}$.

b

$$-\sqrt{\mathrm{Var}[\tilde{a}]\mathrm{Var}\left[\tilde{b}\right]} \leq \mathrm{Cov}[\tilde{a}, \tilde{b}] \leq \sqrt{\mathrm{Var}[\tilde{a}]\mathrm{Var}\left[\tilde{b}\right]}. \tag{8.227}$$

c

$$\mathrm{Cov}[\tilde{a} + \tilde{b}, \tilde{a} - \tilde{b}] = \mathrm{Var}[\tilde{a}] - \mathrm{Var}[\tilde{b}]. \tag{8.228}$$

8.3 (Dunning–Kruger effect) The Dunning–Kruger effect is a well-known cognitive bias, which causes people who are less competent to overestimate their ability relative to more competent people. Recent studies suggest that this effect can arise due to statistical artifacts. Here, we illustrate how easily this can occur, if one is not careful.

Consider a study that measures true competence, represented by a random variable $\tilde{t}$, and also self-evaluated competence, represented by another random variable $\tilde{s}$. We assume that both $\tilde{t}$ and $\tilde{s}$ are standardized, so they have zero mean and unit variance. The investigators are interested in the correlation between the true competence $\tilde{t}$ and the difference $\tilde{d} := \tilde{s} - \tilde{t}$ between the self-evaluated and true competence. Their hypothesis is that $\tilde{t}$ and $\tilde{d}$ should be negatively correlated because less competent people overestimate their competence. However, it turns out that self-assessment for their task of interest is impossible. As a result, the true competence $\tilde{t}$ and the self-evaluated competence $\tilde{s}$ are completely uncorrelated, so the data are useless (we could equivalently replace $\tilde{s}$ by random noise). Compute the correlation coefficient between $\tilde{t}$ and $\tilde{d}$

under these assumptions to show that these useless data actually support the hypothesis of the researchers!

8.4   (Cross) Are the random variables $\tilde{x}$ and $\tilde{y}$ in Exercise 5.2 correlated?

8.5   (Rufus) Are the random variables $\tilde{x}$ and $\tilde{y}$ in Exercise 5.10 correlated?

8.6   (Bernoulli random variables) Does uncorrelation imply independence for Bernoulli random variables?

8.7   (Sample mean and variance of standardized data) Given a dataset formed by n pairs (x_1, y_1), (x_2, y_2), ..., (x_n, y_n), $1 \leq i \leq n$, let $s(x_i)$ and $s(y_i)$ denote the corresponding standardized data, for $1 \leq i \leq n$ (see Definition 8.11). We define the standardized datasets as $S_X := \{s(x_1), s(x_2), \ldots, s(x_n)\}$ and $S_Y := \{s(y_1), s(y_2), \ldots, s(y_n)\}$. Show that the sample mean of the standardized data is zero,

$$m(S_X) = m(S_Y) = 0, \tag{8.229}$$

the sample variance is one,

$$v(S_X) = v(S_Y) = 1, \tag{8.230}$$

and the sample covariance between S_X and S_Y is equal to the sample correlation coefficient of the original data.

8.8   (Sample correlation coefficient) In this example, we study the sample correlation coefficient $\rho_{X,Y}$ from Definition 8.10. We consider a dataset (x_1, y_1), (x_2, y_2), ..., (x_n, y_n), and define $X := \{x_1, x_2, \ldots, x_n\}$ and $Y := \{y_1, y_2, \ldots, y_n\}$. In addition, we denote the ordinary least-squares (OLS) estimator of y_i given x_i by $\ell_{\text{OLS}}(x_i)$ and the corresponding residual by

$$r_i := y_i - \ell_{\text{OLS}}(x_i), \qquad 1 \leq i \leq n. \tag{8.231}$$

a   Prove that

$$\frac{1}{n-1} \sum_{i=1}^{n} r_i^2 = \left(1 - \rho_{X,Y}^2\right) v(Y), \tag{8.232}$$

where $v(Y)$ is the sample variance of Y.

b   Prove that the sample correlation coefficient satisfies the same bounds as the correlation coefficient,

$$-1 \leq \rho_{X,Y} \leq 1, \tag{8.233}$$

as long as $v(Y)$ is not zero.

c   Prove that if $\rho_{X,Y} = \pm 1$, then $y_i = \beta x_i + \alpha$, $1 \leq i \leq n$, for some constants $\alpha, \beta \in \mathbb{R}$.

8.9   (Decomposition of sample variance and sample coefficient of determination) For a dataset (x_1, y_1), (x_2, y_2), ..., (x_n, y_n), we denote the ordinary least squares (OLS) estimator of y_i given x_i by $\ell_{\text{OLS}}(x_i)$ and the corresponding residual by $r_i := y_i - \ell_{\text{OLS}}(x_i)$, $1 \leq i \leq n$. Our goal is to derive a decomposition of the sample variance, and to characterize the sample coefficient of determination.

a   Prove that the first entry is uncorrelated with the residual, i.e. the sample covariance between the feature and the residual is zero,

$$c(X, R) = 0, \tag{8.234}$$

where $X := \{x_1, x_2, \ldots, x_n\}$ and $R := \{r_1, r_2, \ldots, r_n\}$.

b   Show that the sample variance of the sum $M := \{a_1 + b_1, a_2 + b_2, \ldots, a_n + b_n\}$ of the pairs (a_1, b_1), (a_2, b_2), ..., (a_n, b_n) equals

$$v(M) = v(A) + v(B) + 2c(A, B), \tag{8.235}$$

where $v(A)$ and $v(B)$ are the sample variances of $A := \{a_1, a_2, \ldots, a_n\}$ and $B := \{b_1, b_2, \ldots, b_n\}$, and $c(A, B)$ is the sample covariance of A and B.

c Prove that for any constants $\beta, \alpha \in \mathbb{R}$, and any $A := \{a_1, a_2, \ldots, a_n\}$ and $B := \{b_1, b_2, \ldots, b_n\}$, the sample covariance between

$$A_{\beta,\alpha} := \{\beta a_1 + \alpha, \beta a_2 + \alpha, \ldots, \beta a_n + \alpha\} \tag{8.236}$$

and B equals

$$c(A_{\beta,\alpha}, B) = \beta c(A, B). \tag{8.237}$$

Use the result to derive a decomposition of the sample variance $v(Y)$ of the response $Y := \{y_1, y_2, \ldots, y_n\}$, as the sum of the sample variance of the OLS estimates

$$L := \{\ell_{\text{OLS}}(x_1), \ell_{\text{OLS}}(x_2), \ldots, \ell_{\text{OLS}}(x_n)\} \tag{8.238}$$

and the residuals R,

$$v(Y) = v(L) + v(R). \tag{8.239}$$

d Conclude that the sample coefficient of determination is equal to the squared sample correlation coefficient

$$R^2 := \frac{v(L)}{v(Y)} = \rho_{X,Y}^2. \tag{8.240}$$

8.10 (Noisy measurement) We are interested in measuring a certain quantity modeled by a random variable $\tilde{x}$ with zero mean and unit variance. Our available measurements $\tilde{y} := \tilde{x} + \tilde{z}$ are corrupted by additive random noise, modeled as a random variable $\tilde{z}$ with zero mean and variance σ^2, which is independent from $\tilde{x}$.

a What is the best linear estimator of $\tilde{x}$ given $\tilde{y} = y$?
b What is the corresponding mean squared error?
c What happens to the estimator and the error when $\sigma \to 0$? Explain why this makes sense.
d What happens to the estimator and the error when $\sigma \to \infty$? Explain why this makes sense.

8.11 (Affine function) Two random variables $\tilde{a}$ and $\tilde{b}$ have means $\mu_{\tilde{a}} := 2$ and $\mu_{\tilde{b}} := -2$ and variances $\sigma_{\tilde{a}}^2 := 4$ and $\sigma_{\tilde{b}}^2 := 9$ respectively. The correlation coefficient between $\tilde{a}$ and $\tilde{b}$ is 0.25. What is the linear minimum mean-squared-error estimator of $\tilde{c} := 2\tilde{b} + 3$ given $\tilde{a} = a$?

8.12 (Averaging noisy data) We want to estimate a signal represented by a zero-mean random variable $\tilde{x}$ with unit variance. We have access to n measurements $\tilde{y}_1, \tilde{y}_2, \ldots, \tilde{y}_n$, where $\tilde{y}_i := \tilde{x} + \tilde{z}_i$ for $1 \leq i \leq n$. Each $\tilde{z}_i$ is a zero-mean random variable with variance σ^2. The random variables $\tilde{x}$, $\tilde{z}_1, \tilde{z}_2, \ldots, \tilde{z}_n$ are all mutually independent. We decide to estimate $\tilde{x}$ by scaling the sum of all measurements: The estimator is $\alpha \sum_{i=1}^n \tilde{y}_i$ for some $\alpha \in \mathbb{R}$.

a What value of α minimizes the mean squared error?
b What does the estimator tend to when $\sigma^2 \to 0$ and $\sigma^2 \to \infty$?
c What is the mean squared error of the estimator? How does it scale with n?

8.13 (Interference) We model a signal of interest, which is known to be nonnegative, as a random variable $\tilde{a}$ with mean μ and variance σ^2. The signal cannot be observed directly. Instead, we have access to a measurement, modeled as a random variable $\tilde{y}$. The measurement $\tilde{y}$ equals $\tilde{w}\tilde{a}$, where $\tilde{w}$ is an interfering signal that is equal to -1 with probability 1/2 and 1 with probability 1/2. We assume that $\tilde{w}$ and $\tilde{a}$ are independent.

a What is the linear minimum mean-squared-error estimator (MMSE) estimator of $\tilde{a}$ given $\tilde{y} = y$?

b What is the mean-squared error (MSE) of the linear MMSE estimator?

c Propose a nonlinear estimator that achieves lower MSE than the linear MMSE estimator.

8.14 (Exam) The following table shows the number of hours that a group of students dedicated to study for an exam and their resulting grades.

Hours	5	6	2	6	5	5	3	3	6	8
Grade	85	90	30	80	90	80	60	40	90	85

a Approximate the minimum MSE estimator of the grade given the number of hours from the data and sketch it on a graph. What is the main advantage and disadvantage of this estimator?

b Approximate the linear minimum MSE estimator of the grade given the number of hours from the data, and sketch it superposed onto the minimum MSE estimator. What is the main advantage and disadvantage of this estimator?

8.15 (Ice cream sales) Alma is hired as a data scientist for a company that sells ice cream. In the first meeting she attends, the marketing department announces that there is a high correlation between the company's advertising expenses and sales, which suggests that the ads are paying off. Alma decides to investigate. The sample covariance matrix of advertising expenses, sales, and temperature, which contains their sample variances and sample covariances (see Definition 11.14), is:

	Advertising	Sales	Temperature
Advertising	900	960	540
Sales	960	1600	720
Temperature	540	720	400

Compute the sample correlation coefficient between advertising expenses and sales. Should Alma agree with the marketing department?

8.16 (Nonconfounding variable) In this exercise, we analyze the influence of the nonconfounding random variable $\tilde{z}$ in (8.217), which captures sources of randomness unrelated to the treatment and the confounder in Theorem 8.37.

a Express the correlation coefficient $\rho_{\tilde{y},\tilde{t}}$ between the observed outcome $\tilde{y}$ and the treatment $\tilde{t}$ in Theorem 8.37, as a function of the standard deviation $\sigma_{\tilde{z}}$ of the random variable $\tilde{z}$.

b What is the limit of $\rho_{\tilde{y},\tilde{t}}$ as $\sigma_{\tilde{z}} \to \infty$? Explain intuitively why this is the case.

c What is $\rho_{\tilde{y},\tilde{t}}$ in the three scenarios of Example 8.38, if $\sigma_{\tilde{z}} := 1$ as in Figure 8.14?

9

Estimation of Population Parameters

Overview

Estimation of parameters describing a population is a fundamental challenge in many scientific disciplines. Section 9.1 introduces random sampling, a simple approach that yields accurate estimates from limited data. Sections 9.2 and 9.3 define the bias and standard error, which measure the average error and the standard deviation of an estimator. In Section 9.4, we derive deviation bounds, which characterize the probabilistic behavior of a random variable just based on its mean and variance. In Section 9.5, we prove the celebrated law of large numbers and use it to derive theoretical guarantees for estimators based on averaging. Section 9.6 provides a word of caution, describing several situations in which the law of large numbers does not hold. Section 9.7 discusses another fundamental mathematical phenomenon: The central limit theorem, according to which averages of independent quantities tend to have Gaussian distributions. In Section 9.8, we introduce confidence intervals, which quantify the uncertainty associated with parameter estimates computed from data. Finally, in Section 9.9, we introduce the bootstrap, a popular computational technique to estimate standard errors and build confidence intervals.

9.1 Random Sampling

Imagine that we want to estimate the mean (i.e. the arithmetic average) of the weights of all the pigeons in New York. Statistics that describe entire populations, such as the mean weight of New York pigeons, are called *population parameters*. Estimating population parameters is a fundamental problem in many applications, such as economics, healthcare, sociology, and political science.

In theory, we could compute the mean weight of New York pigeons exactly by catching every single pigeon, weighing it, and then averaging all the weights. However, there are more than one million pigeons in the city,[1] so this approach would not be possible in practice. As in our pigeon example, measuring a whole population is often intractable or too costly. Consequently, population parameters are usually estimated from a subset or *sample* of the population. In our motivating example, we would catch a few pigeons and average their weights to estimate the mean. A crucial question is how to choose what individuals to measure. A simple strategy is to just pick individuals at random from the population. As illustrated by the following two examples, this strategy, known as *random sampling*, often provides very accurate estimates from surprisingly small amounts of data.

---

[1] According to a quick Internet search. There seems to be considerable debate about the exact number.

Example 9.1 (Estimating the mean height). In this example, we estimate the mean height of a population via random sampling in a controlled scenario, where we know the true population mean. We use 4,082 men in the United States army as the complete ground-truth population. The data are extracted from Dataset 5, as in Figures 3.6 and 3.13. We define the population mean

$$\mu_{\text{pop}} := \frac{1}{N} \sum_{i=1}^{N} h_i \tag{9.1}$$

$$= 175.6, \tag{9.2}$$

as the arithmetic average of all the heights h_1, h_2, ..., h_N of the $N := 4,082$ individuals in the population. We perform random sampling by selecting $n := 400$ individuals independently and uniformly at random with replacement and measuring their height. More precisely, we repeatedly pick a random index k from $\{1, 2, \ldots, N\}$ and set $x_j := h_k$ for $1 \leq j \leq n$. Figure 9.1 shows three different datasets of 400 samples obtained in this way.

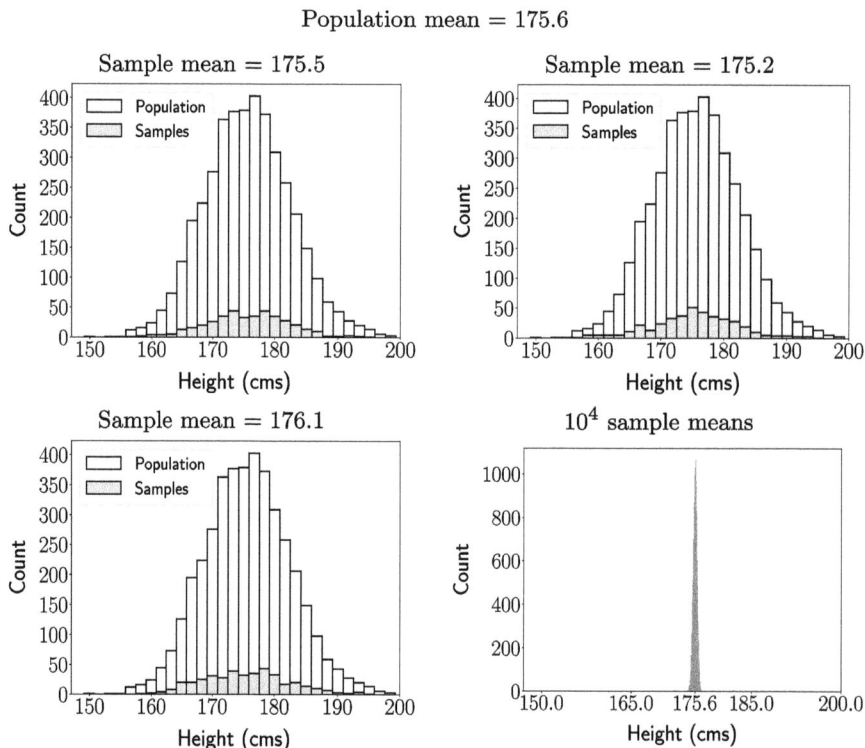

Figure 9.1 Estimating the mean height in a population via random sampling. In the top row and the bottom left, the white histogram represents the heights of a population of 4,082 individuals, extracted from Dataset 5 (see Figures 3.6 and 3.13). The gray histograms represent the values of 400 random samples measured uniformly at random with replacement from the population. The sample mean of the random samples is indicated at the top of each graph. The plot on the bottom right shows a histogram of the sample means of 10,000 sets of 400 random samples obtained in the same way. The histogram concentrates tightly around the population mean, which equals 175.6 cm.

Our estimator for the population mean is the sample mean of the resulting n random samples $x_1, \ldots, x_n$,

$$m := \frac{1}{n} \sum_{j=1}^{n} x_j, \qquad (9.3)$$

computed following Definition 7.15. Notice that the value of the sample mean changes depending on the selected samples. In Figure 9.1, each of the sampled datasets yields a different estimate. However, the three estimates are very close to the population mean. The figure also includes a histogram of sample means computed by repeating the sampling process 10,000 times. The histogram is tightly concentrated around the population mean, indicating that the estimator is very likely to be accurate.

. . . .

Example 9.2 (Prevalence of COVID-19). During the COVID-19 pandemic, determining the population prevalence, meaning what fraction of the population had the disease, was a key challenge. In this example, we consider the problem of estimating the population prevalence of COVID-19 in New York in a hypothetical scenario where 5% of New Yorkers have the virus. Our strategy is based on random sampling: We test 1,000 individuals chosen independently and uniformly at random with replacement from the entire population of 8.8 million people. Our estimate of the prevalence is simply the proportion of positive tests. For the sake of simplicity, we assume that the tests are perfect; if a person has the disease, they test positive. Figure 9.2 shows the result of simulating this procedure multiple times. The prevalence estimates are extremely accurate.

. . . .

In the remainder of this chapter, we study estimators of population parameters obtained via random sampling. Our analysis has to be probabilistic because the measurements are random: Repeating them would yield different data, and therefore a different estimate of the population parameter. To account for this, we model the measurements and the corresponding estimators as random variables. This enables us to encode our assumptions about the sampling process more formally.

Definition 9.3 (Probabilistic model of random samples). *Let $a_1, a_2, \ldots, a_N$ denote the values of a quantity of interest in a population of size N. The indices $\tilde{k}_1, \tilde{k}_2, \ldots, \tilde{k}_n$ are said to be drawn independently and uniformly at random with replacement, if they are independent random variables such that*

$$\mathrm{P}\left(\tilde{k}_j = i\right) = \frac{1}{N}, \qquad 1 \leq i \leq N, \ 1 \leq j \leq n. \qquad (9.4)$$

The corresponding dataset of n random samples $\tilde{x}_1, \tilde{x}_2, \ldots, \tilde{x}_n$, where

$$\tilde{x}_j = a_{\tilde{k}_j}, \qquad 1 \leq j \leq n, \qquad (9.5)$$

is said to be obtained independently and uniformly at random with replacement.

In Chapters 1–8, we define estimators of different quantities of interest, such as the sample mean (Definition 7.15) or the sample variance (Definition 7.37), as deterministic functions of the available data. Here we interpret the data as random variables, following Definition 9.3, so the corresponding estimators are also random variables. For instance, if we apply

Population proportion = 0.05

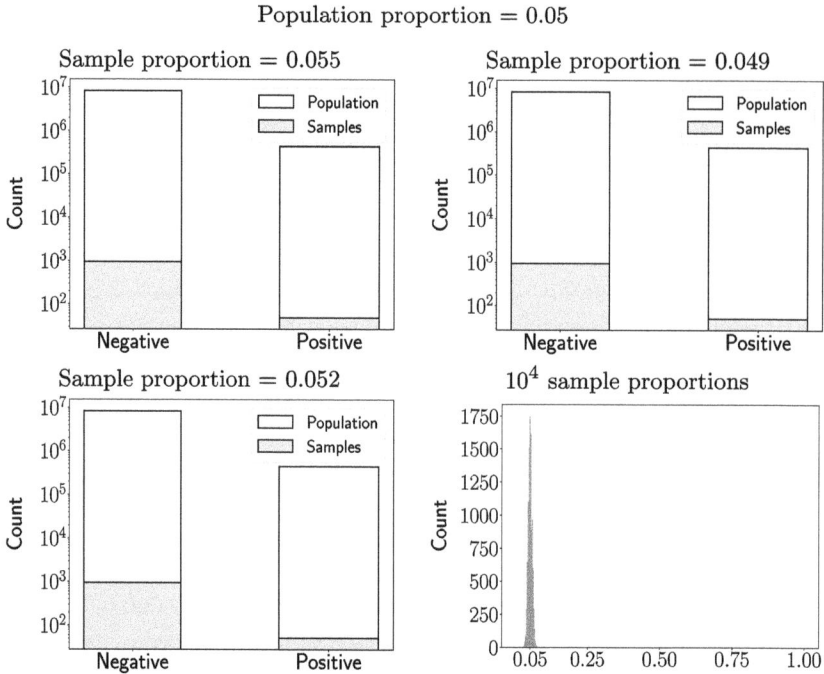

Figure 9.2 Estimating the prevalence of COVID-19 via random sampling. In the top row and the bottom left, the white bar plot indicates the number of people in New York with (positive) and without (negative) COVID-19 in a simulated scenario, where the COVID-19 prevalence is 5%. The gray bar plots represent the number of positive and negative tests obtained from 1,000 random samples measured uniformly at random with replacement out of the total population of 8.8 million people. The sample proportion of positive tests in the random samples is indicated at the top of each graph. The plot on the bottom right shows a histogram of the sample proportions of 10,000 sets of 1,000 samples obtained in the same way. The histogram concentrates tightly around the true prevalence.

Definition 9.3 to Example 9.1 by setting $a_i := h_i$, $1 \leq i \leq N$, then the sample-mean estimator used to approximate the population mean of the heights is the random variable

$$\widetilde{m} := \frac{1}{n} \sum_{j=1}^{n} \widetilde{x}_j, \tag{9.6}$$

where $\widetilde{x}_1, \widetilde{x}_2, \ldots, \widetilde{x}_n$ represent the n height measurements.

Example 9.4 (Prevalence of COVID-19: Probabilistic analysis). In Example 9.2, we encode the status of the ith individual, $1 \leq i \leq N$, in the notation of Definition 9.3 by setting $a_i := 1$ if they have COVID-19, and $a_i := 0$ if they don't. The data obtained via independent, uniform random sampling with replacement is represented by n random variables $\widetilde{x}_1$, $\widetilde{x}_2, \ldots, \widetilde{x}_n$, which are equal to one if the corresponding individual has COVID-19 and zero otherwise. Consequently, each of the random variables is Bernoulli with parameter equal to the ground-truth population prevalence θ_{pop} (see Definition 2.15) because that is the probability that we select a person with COVID-19 every time we sample. The fraction of positive tests among the n selected individuals is equal to the sample mean of these Bernoulli random variables,

$$\widetilde{m} := \frac{1}{n} \sum_{j=1}^{n} \tilde{x}_j. \tag{9.7}$$

. . . .

In the following sections, we study the probabilistic behavior of estimators of population parameters, with a particular focus on the sample mean. In our analysis, we interpret the parameters of interest, such as the population mean μ_{pop} in Example 9.1, as fixed deterministic quantities. In statistics, this is known as a *frequentist* viewpoint, in contrast to the Bayesian framework, described in Section 6.7, which models the parameters as random variables. We compare the two perspectives in Example 9.47.

9.2 The Bias

The *bias* of an estimator is its mean error. If the bias is zero, then the estimator is said to be *unbiased*. The mean of an unbiased estimator is the population parameter of interest.

Definition 9.5 (Bias). *Let $\tilde{x}_1$, $\tilde{x}_2$, ..., $\tilde{x}_n$ denote random variables representing measurements associated with a population parameter $\gamma \in \mathbb{R}$, and let $h \colon \mathbb{R}^n \to \mathbb{R}$ denote an estimator designed to approximate γ from these measurements. The bias of the estimator is the mean of the difference between γ and the random variable $h(\tilde{x}_1, \ldots, \tilde{x}_n)$,*

$$\text{Bias} := \mathbb{E}\left[\gamma - h(\tilde{x}_1, \ldots, \tilde{x}_n)\right] = \gamma - \mathbb{E}\left[h(\tilde{x}_1, \ldots, \tilde{x}_n)\right]. \tag{9.8}$$

If the mean is equal to γ,

$$\mathbb{E}\left[h(\tilde{x}_1, \ldots, \tilde{x}_n)\right] = \gamma, \tag{9.9}$$

so that the bias is zero, then the estimator is unbiased.

The following theorem establishes that the sample mean is an unbiased estimator of the population mean, when it is computed using random samples. This explains why the histogram of sample means is centered exactly at the population mean in the bottom right graph of Figure 9.1 (see also Figure 9.3).

Theorem 9.6 (The sample mean is unbiased). *Let a_1, a_2, ..., a_N denote a dataset of size N with population mean*

$$\mu_{\text{pop}} := \frac{1}{N} \sum_{i=1}^{N} a_i \tag{9.10}$$

and let $\tilde{x}_1$, $\tilde{x}_2$, ..., $\tilde{x}_n$ be independent, uniform random samples following Definition 9.3. The sample mean

$$\widetilde{m} := \frac{1}{n} \sum_{j=1}^{n} \widetilde{x}_j \tag{9.11}$$

is an unbiased estimator of μ_{pop},

$$\mathbb{E}\left[\widetilde{m}\right] = \mu_{\text{pop}}. \tag{9.12}$$

Proof   By Definition 9.3 the jth random sample $\widetilde{x}_j$ is a deterministic function of $\widetilde{k}_j$, which equals a_k if $\widetilde{k}_j = k$, so by Definition 7.4,

$$\mathbb{E}\left[\widetilde{x}_j\right] = \sum_{k=1}^{N} a_k p_{\widetilde{k}_j}(k) \tag{9.13}$$

$$= \frac{1}{N} \sum_{k=1}^{N} a_k \tag{9.14}$$

$$= \mu_{\text{pop}}. \tag{9.15}$$

The mean of each individual sample is equal to the population mean. By linearity of expectation (Theorem 7.18), this implies

$$\mathbb{E}\left[\widetilde{m}\right] = \mathbb{E}\left[\frac{1}{n} \sum_{j=1}^{n} \widetilde{x}_j\right] \tag{9.16}$$

$$= \frac{1}{n} \sum_{j=1}^{n} \mathbb{E}\left[\widetilde{x}_j\right] \tag{9.17}$$

$$= \mu_{\text{pop}}. \tag{9.18}$$

■

Example 9.7 (Prevalence of COVID-19: Bias). As explained in Example 9.4, we can model the tests in Example 9.2 as n Bernoulli random variables $\widetilde{x}_1, \dots, \widetilde{x}_n$ with parameter equal to the ground-truth population prevalence θ_{pop}. Our prevalence estimator is the sample proportion of positive tests, which equals the sample mean of these random variables. In this case, the population mean is equal to θ_{pop},

$$\frac{1}{N} \sum_{i=1}^{N} a_i = \frac{\text{Number of COVID-19 cases}}{N} := \theta_{\text{pop}}. \tag{9.19}$$

By Theorem 9.6, the mean of the sample proportion is therefore equal to θ_{pop}. This is why the histogram of sample proportions is centered at the ground-truth prevalence in the bottom right graph of Figure 9.2 (see also Figure 9.4).

....

In Definition 7.37, we define the sample variance of n data points, as an average of the squared deviation from the sample mean, where we divide by $n - 1$ instead of n. This ensures that the estimator is unbiased, under the assumption of independent, uniform random sampling with replacement.

Theorem 9.8 (The sample variance is unbiased). *Let $a_1, a_2, \ldots, a_N$ denote a dataset of size N with population mean μ_{pop}, defined in (9.10), and population variance*

$$\sigma_{\mathrm{pop}}^2 := \frac{1}{N} \sum_{i=1}^{N} (a_i - \mu_{\mathrm{pop}})^2, \tag{9.20}$$

and let $\tilde{x}_1, \tilde{x}_2, \ldots, \tilde{x}_n$ be independent, uniform random samples following Definition 9.3. The sample variance

$$\tilde{v} := \frac{1}{n-1} \sum_{j=1}^{n} (\tilde{x}_j - \tilde{m})^2, \tag{9.21}$$

where the sample mean $\tilde{m}$ is defined as in (9.11), is an unbiased estimator of σ_{pop}^2,

$$\mathbb{E}[\tilde{v}] = \sigma_{\mathrm{pop}}^2. \tag{9.22}$$

Proof We decompose the mean of the sample variance as follows, applying linearity of expectation (Theorem 7.18):

$$\mathbb{E}[\tilde{v}] = \mathbb{E}\left[\frac{1}{n-1} \sum_{j=1}^{n} (\tilde{x}_j - \tilde{m})^2\right] \tag{9.23}$$

$$= \frac{1}{n-1} \left(\sum_{j=1}^{n} \mathbb{E}[\tilde{x}_j^2] - 2\sum_{j=1}^{n} \mathbb{E}[\tilde{m}\tilde{x}_j] + \sum_{j=1}^{n} \mathbb{E}[\tilde{m}^2]\right). \tag{9.24}$$

Again by linearity of expectation,

$$\sum_{j=1}^{n} \mathbb{E}[\tilde{m}^2] = n\mathbb{E}\left[\tilde{m}\frac{1}{n}\sum_{j=1}^{n}\tilde{x}_j\right] = \sum_{j=1}^{n} \mathbb{E}[\tilde{m}\tilde{x}_j], \tag{9.25}$$

which in turn equals

$$\sum_{j=1}^{n} \mathbb{E}[\tilde{m}\tilde{x}_j] = \frac{1}{n}\sum_{j=1}^{n}\sum_{k=1}^{n} \mathbb{E}[\tilde{x}_j\tilde{x}_k] = \frac{1}{n}\sum_{j=1}^{n} \mathbb{E}[\tilde{x}_j^2] + \frac{1}{n}\sum_{j=1}^{n}\sum_{k\neq j} \mathbb{E}[\tilde{x}_j\tilde{x}_k] \tag{9.26}$$

$$= \frac{1}{n}\sum_{j=1}^{n} \left(\mathbb{E}[\tilde{x}_j^2] + (n-1)\mu_{\mathrm{pop}}^2\right) \tag{9.27}$$

because $\mathbb{E}[\tilde{x}_j\tilde{x}_k] = \mathbb{E}[\tilde{x}_j]\mathbb{E}[\tilde{x}_k] = \mu_{\mathrm{pop}}^2$ by the independence assumption and Theorem 7.20. Plugging (9.25) and (9.27) into (9.24) yields

$$\mathbb{E}[\tilde{v}] = \frac{1}{n-1}\left(\sum_{j=1}^{n} \mathbb{E}[\tilde{x}_j^2] - \sum_{j=1}^{n} \mathbb{E}[\tilde{m}\tilde{x}_j]\right) \tag{9.28}$$

$$= \frac{1}{n-1}\left(\sum_{j=1}^{n} \mathbb{E}[\tilde{x}_j^2] - \frac{1}{n}\sum_{j=1}^{n} \left(\mathbb{E}[\tilde{x}_j^2] + (n-1)\mu_{\mathrm{pop}}^2\right)\right) \tag{9.29}$$

$$= \frac{1}{n-1}\frac{n-1}{n}\sum_{j=1}^{n} \left(\mathbb{E}[\tilde{x}_j^2] - \mu_{\mathrm{pop}}^2\right) \tag{9.30}$$

$$= \frac{1}{n} \sum_{j=1}^{n} \left(\mathbb{E}\left[\tilde{x}_j^2\right] - \mu_{\text{pop}}^2 \right) \tag{9.31}$$

$$= \frac{1}{n} \sum_{j=1}^{n} \text{Var}\left[\tilde{x}_j\right], \tag{9.32}$$

by Lemma 7.33, since $\mathbb{E}\left[\tilde{x}_j\right] = \mu_{\text{pop}}$ by Theorem 9.6. This implies that $\mathbb{E}\left[\tilde{v}\right] = \sigma_{\text{pop}}^2$ because it follows from Definition 9.3 that the variance of each individual sample is equal to the population variance. In more detail, by Definition 9.3 $(\tilde{x}_j - \mathbb{E}\left[\tilde{x}_j\right])^2 = (\tilde{x}_j - \mu_{\text{pop}})^2$ is a deterministic function of $\tilde{k}_j$, which equals $(a_k - \mu_{\text{pop}})^2$ if $\tilde{k}_j = k$, so by Definition 7.4,

$$\text{Var}\left[\tilde{x}_j\right] := \mathbb{E}\left[(\tilde{x}_j - \mathbb{E}\left[\tilde{x}_j\right])^2\right] \tag{9.33}$$

$$= \mathbb{E}\left[(\tilde{x}_j - \mu_{\text{pop}})^2\right] \tag{9.34}$$

$$= \sum_{k=1}^{N} (a_k - \mu_{\text{pop}})^2 p_{\tilde{k}_j}(k) \tag{9.35}$$

$$= \frac{1}{N} \sum_{k=1}^{N} (a_k - \mu_{\text{pop}})^2 \tag{9.36}$$

$$= \sigma_{\text{pop}}^2. \tag{9.37}$$

∎

9.3 The Standard Error

As explained in Section 9.2, the mean of an unbiased estimator is equal to the population parameter of interest. However, this does not necessarily imply that the estimator is accurate: If its standard deviation is large, it will tend to be far from the population parameter, leading to considerable estimation errors. Because of this, the standard deviation is commonly referred to as the *standard error* of the estimator.

Definition 9.9 (Standard error). *Let $\tilde{x}_1$, $\tilde{x}_2$, ..., $\tilde{x}_n$ denote random variables representing measurements associated with a population parameter $\gamma \in \mathbb{R}$, and let $h\colon \mathbb{R}^n \to \mathbb{R}$ denote an unbiased estimator of γ. The standard error of h is the standard deviation of the random variable $h(\tilde{x}_1, \ldots, \tilde{x}_n)$,*

$$\text{se}\left[h(\tilde{x}_1, \ldots, \tilde{x}_n)\right] := \sqrt{\text{Var}\left[h(\tilde{x}_1, \ldots, \tilde{x}_n)\right]}. \tag{9.38}$$

The following simple lemma justifies calling the standard error an error: It is equal to the root mean square error between the estimator and the parameter of interest, as long as the estimator is unbiased.

Lemma 9.10. *Let $\tilde{x}_1$, $\tilde{x}_2$, ..., $\tilde{x}_n$ denote random variables representing measurements associated with a population parameter $\gamma \in \mathbb{R}$, and let $h\colon \mathbb{R}^n \to \mathbb{R}$ denote an unbiased estimator of γ. The standard error of the estimator is equal to the root mean square difference between the estimator and γ,*

$$\text{se}\left[h(\tilde{x}_1, \ldots, \tilde{x}_n)\right] = \sqrt{\mathbb{E}\left[(h(\tilde{x}_1, \ldots, \tilde{x}_n) - \gamma)^2\right]}. \tag{9.39}$$

Proof Since the estimator is unbiased, $\mathbb{E}\left[h(\tilde{x}_1, \ldots, \tilde{x}_n)\right] = \gamma$ by Definition 9.5. Consequently, by Definitions 9.9 and 7.32,

$$\mathrm{se}\left[h(\tilde{x}_1, \ldots, \tilde{x}_n)\right] := \sqrt{\mathrm{Var}\left[h(\tilde{x}_1, \ldots, \tilde{x}_n)\right]} \tag{9.40}$$

$$= \sqrt{\mathbb{E}\left[\left(h(\tilde{x}_1, \ldots, \tilde{x}_n) - \mathbb{E}\left[h(\tilde{x}_1, \ldots, \tilde{x}_n)\right]\right)^2\right]} \tag{9.41}$$

$$= \sqrt{\mathbb{E}\left[\left(h(\tilde{x}_1, \ldots, \tilde{x}_n) - \gamma\right)^2\right]}. \tag{9.42}$$

∎

In order to determine the standard error of estimators that involve averaging, we establish a key result: The variance of a sum of independent random variables is equal to the sum of their variances.

Theorem 9.11 (Variance of the sum of independent random variables). *Let $\tilde{a}_1, \tilde{a}_2, \ldots, \tilde{a}_n$ be independent random variables with finite variance belonging to the same probability space. The variance of their sum is equal to the sum of their variances,*

$$\mathrm{Var}\left[\sum_{k=1}^{n} \tilde{a}_k\right] = \sum_{k=1}^{n} \mathrm{Var}\left[\tilde{a}_k\right]. \tag{9.43}$$

Proof By the independence assumption, for any $1 \leq t \leq n$, the random variables $\tilde{a}_t$ and $\sum_{k=t+1}^{n} \tilde{a}_k$ are independent, and hence uncorrelated by Lemma 8.28. We can therefore apply Corollary 8.23 $n - 1$ times to prove the result:

$$\mathrm{Var}\left[\sum_{k=1}^{n} \tilde{a}_k\right] = \mathrm{Var}\left[\tilde{a}_1\right] + \mathrm{Var}\left[\sum_{k=2}^{n} \tilde{a}_k\right] \tag{9.44}$$

$$= \mathrm{Var}\left[\tilde{a}_1\right] + \mathrm{Var}\left[\tilde{a}_2\right] + \mathrm{Var}\left[\sum_{k=3}^{n} \tilde{a}_k\right] \tag{9.45}$$

$$= \cdots \tag{9.46}$$

$$= \sum_{k=1}^{n} \mathrm{Var}\left[\tilde{a}_k\right]. \tag{9.47}$$

∎

Theorem 9.11 can be leveraged to derive the variance of a binomial random variable.

Lemma 9.12 (Variance of a binomial random variable). *The variance of a binomial random variable $\tilde{a}$ with parameters n and θ equals $\mathrm{Var}[\tilde{a}] = n\theta(1 - \theta)$.*

Proof Recall that a binomial random variable with parameters n and θ can be represented as the sum of n independent Bernoulli random variables $\tilde{b}_1, \ldots, \tilde{b}_n$ with parameter θ, as explained in the proof of Lemma 7.19. By Theorem 9.11 and Lemma 7.34,

$$\mathrm{Var}\left[\tilde{a}\right] = \mathrm{Var}\left[\sum_{k=1}^{n} \tilde{b}_k\right] \tag{9.48}$$

$$= \sum_{k=1}^{n} \text{Var}[\tilde{b}_k] \tag{9.49}$$

$$= n\theta(1-\theta). \tag{9.50}$$

∎

The number of samples in a dataset is called the *sample size*. The following theorem establishes that the standard error of the sample mean decreases proportionally to the square root of the sample size, assuming independent and uniform random sampling with replacement. Interestingly, the standard error does *not* depend on the size of the population N.

Theorem 9.13 (Standard error of the sample mean). *Let* a_1, a_2, ..., a_N *denote a dataset of size* N *with population mean* μ_{pop}, *defined as in* (9.10), *and population variance* σ^2_{pop}, *defined as in* (9.20). *Let* $\tilde{x}_1$, $\tilde{x}_2$, ..., $\tilde{x}_n$ *be independent, uniform random samples following Definition 9.3. The standard error of the sample mean* $\widetilde{m} := \frac{1}{n}\sum_{j=1}^{n} \tilde{x}_j$ *equals*

$$\text{se}\,[\widetilde{m}] = \frac{\sigma_{\text{pop}}}{\sqrt{n}}. \tag{9.51}$$

Proof   By (9.37) the variance of each sample, $\text{Var}\,[\tilde{x}_j]$ for $1 \le j \le n$, equals σ^2_{pop}. The result then follows from Definition 9.9, Lemma 7.36 and Theorem 9.11,

$$\text{se}\,[\widetilde{m}]^2 = \text{Var}\,[\widetilde{m}] = \text{Var}\left[\frac{1}{n}\sum_{j=1}^{n} \tilde{x}_j\right] \tag{9.52}$$

$$= \frac{1}{n^2}\text{Var}\left[\sum_{j=1}^{n} \tilde{x}_j\right] \tag{9.53}$$

$$= \frac{1}{n^2}\sum_{j=1}^{n}\text{Var}\,[\tilde{x}_j] \tag{9.54}$$

$$= \frac{\sigma^2_{\text{pop}}}{n}. \tag{9.55}$$

∎

Figure 9.3 depicts the behavior of the sample mean estimator in Example 9.1 as a function of the sample size n. As predicted by Theorem 9.13, the sample means are increasingly concentrated around the population mean, and their average spread decreases proportionally to $\sqrt{n}$.

Example 9.14 (Prevalence of COVID-19: Standard error).   In this example, we analyze the standard error of the prevalence estimator in Example 9.2, assuming that the data follow the probabilistic model described in Example 9.4. In Example 9.7, we show that the population mean is equal to the population proportion θ_{pop}. The population variance therefore equals

$$\sigma^2_{\text{pop}} := \frac{1}{N}\sum_{k=1}^{N}(a_k - \theta_{\text{pop}})^2 \tag{9.56}$$

$$= \frac{1}{N}\sum_{k=1}^{N} a_k^2 - \frac{2\theta_{\text{pop}}}{N}\sum_{k=1}^{N} a_k + \frac{1}{N}\sum_{k=1}^{N}\theta^2_{\text{pop}} \tag{9.57}$$

Population mean $\mu_{\text{pop}} := 175.6$

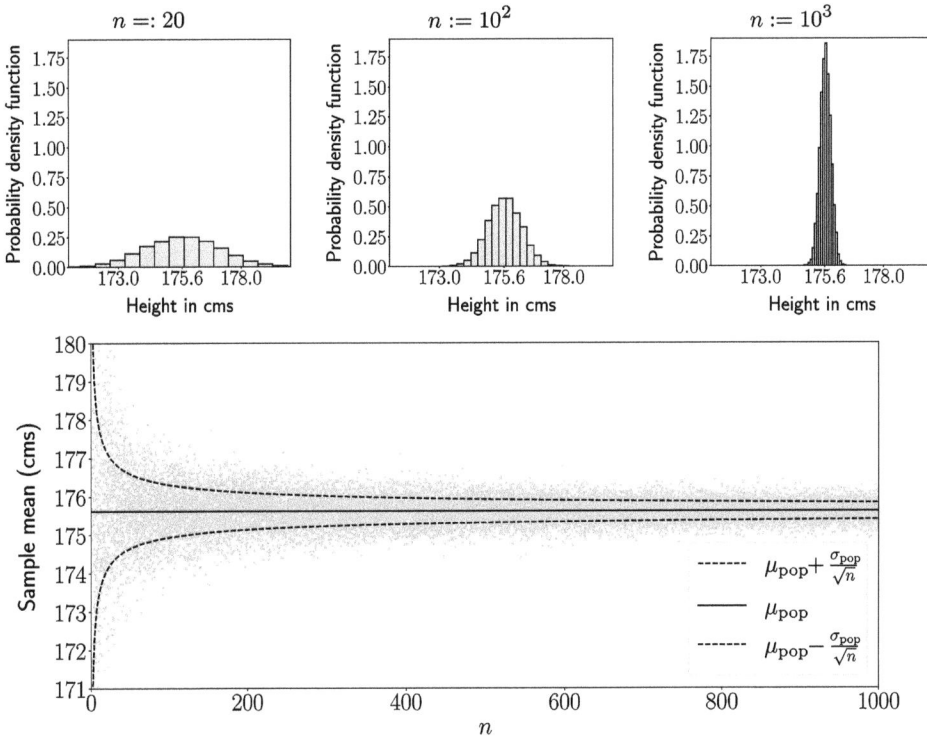

Figure 9.3 **Bias and standard error of the sample mean.** Analysis of the sample mean estimator applied to independent, uniform samples from the dataset in Example 9.1. The top row shows normalized histograms of the sample mean of n height measurements for different values of n. Each histogram is computed using 10^4 independent instances of the sample mean. The histograms are centered at the true population mean $\mu_{\text{pop}} = 175.6$ cm, indicating that the estimates are unbiased, in accordance with Theorem 9.6. The scatterplot shows that as n increases, the standard error decreases proportionally to $\sigma_{\text{pop}}/\sqrt{n}$, where $\sigma_{\text{pop}} = 6.85$ cm is the population standard deviation, as established in Theorem 9.13.

$$= \theta_{\text{pop}} - 2\theta_{\text{pop}}^2 + \theta_{\text{pop}}^2 \tag{9.58}$$

$$= \theta_{\text{pop}}(1 - \theta_{\text{pop}}). \tag{9.59}$$

Since the sample proportion is equal to the sample mean of the samples, by Theorem 9.13 its standard error equals

$$\frac{\sigma_{\text{pop}}}{\sqrt{n}} = \sqrt{\frac{\theta_{\text{pop}}(1 - \theta_{\text{pop}})}{n}}. \tag{9.60}$$

Figure 9.4 shows that this formula is consistent with the scaling observed in numerical simulations for different values of n.

. . . .

An important consequence of Theorem 9.13 is that the mean squared error of the sample mean (with respect to the population mean) tends to zero asymptotically, as the sample size tends to infinity. In probability theory, this is known as *convergence in mean square*.

Population proportion $\theta_{\text{pop}} := 0.05$

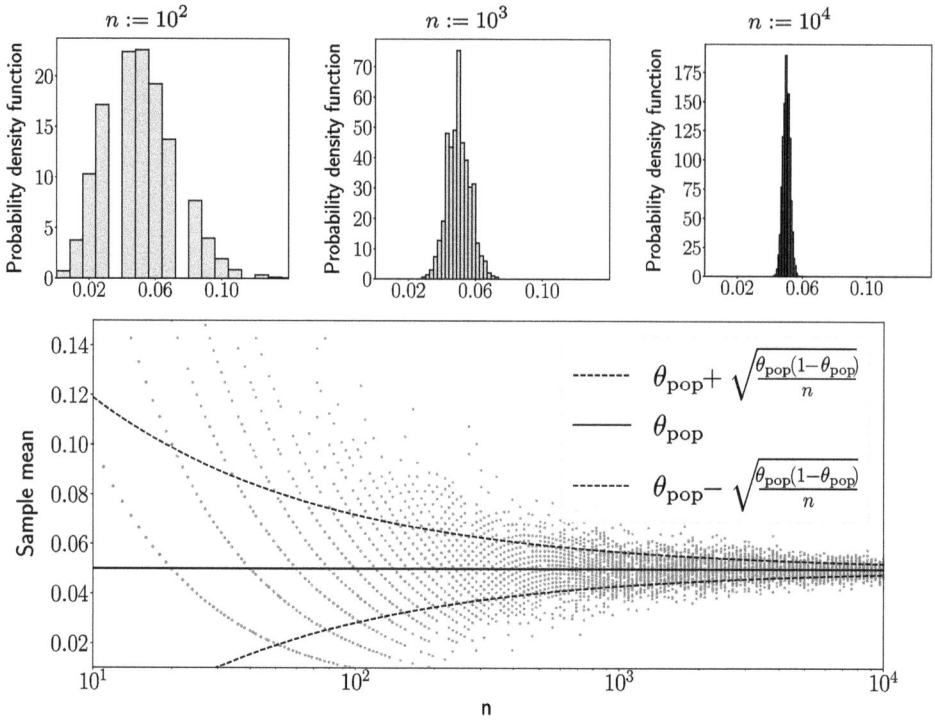

Figure 9.4 Bias and standard error of the sample proportion. Analysis of proportion estimates obtained from independent, uniform samples in the scenario described in Example 9.2. The top row shows normalized histograms of the sample proportion of positives out of n COVID-19 tests, for different values of n. Each histogram is computed using 10^4 independent instances of the sample proportion. The histograms are centered at the true population proportion $\theta_{\text{pop}} = 0.05$, indicating that the estimates are unbiased, in accordance with Theorem 9.6. The scatterplot shows that as n increases, the standard error decreases proportionally to $\sqrt{\theta_{\text{pop}}(1 - \theta_{\text{pop}})}/\sqrt{n}$, as derived in Example 9.14.

Corollary 9.15 (Convergence of the sample mean in mean square)**.** *Let* $a_1, a_2, \ldots, a_N$ *denote a dataset of size N with population mean μ_{pop}, defined in (9.10). Let $\tilde{x}_1, \tilde{x}_2, \ldots,$ be a sequence of independent, uniform random samples following Definition 9.3. The mean squared error between μ_{pop} and the sample mean $\tilde{m}_n := \frac{1}{n} \sum_{j=1}^{n} \tilde{x}_j$ converges to zero as the number of measurements tends to infinity,*

$$\lim_{n \to \infty} \mathbb{E}\left[(\tilde{m}_n - \mu_{\text{pop}})^2\right] = 0. \tag{9.61}$$

Proof By Theorem 9.6, the mean of $\tilde{m}_n$ is μ_{pop}, so by Definitions 7.32 and 9.9, and Theorem 9.13,

$$\mathbb{E}\left[(\tilde{m}_n - \mu_{\text{pop}})^2\right] = \text{Var}\left[\tilde{m}_n\right] \tag{9.62}$$

$$= \text{se}\left[\tilde{m}\right]^2 \tag{9.63}$$

$$= \frac{\sigma_{\text{pop}}^2}{n}, \tag{9.64}$$

which converges to zero as $n \to \infty$. ∎

9.4 Deviation Bounds: Markov's and Chebyshev's Inequalities

In this section, we take a small detour to derive deviation bounds, which allow us to make statements about the probabilistic behavior of a random variable just based on its mean and variance. These bounds are a fundamental tool in the theoretical analysis of statistical estimators. In Section 9.5, we use them to characterize the asymptotic behavior of the sample mean and prove the law of large numbers.

Markov's inequality states that, if a random variable is nonnegative and small, then it cannot take large values with high probability.

Theorem 9.16 (Markov's inequality). *Let $\tilde{a}$ be a nonnegative random variable with a pmf or pdf that is zero for negative values. For any positive constant $c > 0$,*

$$P\left(\tilde{a} \geq c\right) \leq \frac{\mathbb{E}\left[\tilde{a}\right]}{c}. \tag{9.65}$$

Proof We prove the result assuming $\tilde{a}$ is continuous and has a pdf. The proof for discrete variables is the same replacing integrals by sums and the pdf by the pmf. Since the pdf is zero for negative values of a, by Definition 7.8 and Theorem 3.14,

$$\mathbb{E}\left[\tilde{a}\right] = \int_{a \in \mathbb{R}} a f_{\tilde{a}}(a) \, \mathrm{d}a \tag{9.66}$$

$$= \int_{a=0}^{c} a f_{\tilde{a}}(a) \, \mathrm{d}a + \int_{a=c}^{\infty} a f_{\tilde{a}}(a) \, \mathrm{d}a \tag{9.67}$$

$$\geq \int_{a=0}^{c} a f_{\tilde{a}}(a) \, \mathrm{d}a + c \int_{a=c}^{\infty} f_{\tilde{a}}(a) \, \mathrm{d}a \tag{9.68}$$

$$\geq c \, P\left(\tilde{a} \geq c\right). \tag{9.69}$$

∎

Example 9.17 (Age of students). The mean age of students at a university is 20 years. We bound the fraction of students above 30 based on this information. Modeling age as a nonnegative random variable $\tilde{a}$, by Markov's inequality (Theorem 9.16),

$$P(\tilde{a} \geq 30) \leq \frac{\mathbb{E}\left[\tilde{a}\right]}{30} = \frac{2}{3}. \tag{9.70}$$

At most two thirds of the students are over 30.

· ·

The variance of a random variable is defined in Section 7.7 as the mean squared deviation from the mean. Consequently, when the variance is small, it seems plausible that the random variable cannot be far from its mean with high probability. This is indeed the case. The corresponding deviation bound is known as Chebyshev's inequality.

Theorem 9.18 (Chebyshev's inequality). *For any positive constant $c > 0$ and any random variable $\tilde{a}$ with mean μ and bounded variance,*

$$P\left(|\tilde{a} - \mu| \geq c\right) \leq \frac{\mathrm{Var}\left[\tilde{a}\right]}{c^2}. \tag{9.71}$$

Proof We apply Markov's inequality (Theorem 9.16) to the random variable $\tilde{b} := (\tilde{a} - \mu)^2$, which yields

$$P\left(|\tilde{a} - \mu| \geq c\right) = P\left(\tilde{b} \geq c^2\right) \leq \frac{\mathbb{E}\left[\tilde{b}\right]}{c^2} \tag{9.72}$$

$$= \frac{\mathbb{E}\left[(\tilde{a} - \mu)^2\right]}{c^2} \tag{9.73}$$

$$= \frac{\text{Var}\left[\tilde{a}\right]}{c^2} \tag{9.74}$$

by Definition 7.32. ∎

A corollary to Chebyshev's inequality is that if the variance of a random variable is zero, then the random variable is constant (the probability that it deviates from its mean is zero). This result is key in our geometric interpretation of the covariance as an inner product in Section 8.7 (see Theorem 8.35). To prove it, we first need to introduce the union bound, also known as Boole's inequality, which states that the probability of a union of events is always smaller than or equal to the sum of the individual probabilities.

Theorem 9.19 (Union bound). *Let $(\Omega, \mathcal{C}, P)$ be a probability space, and let A_1, A_2, ... A_k be k events in $\mathcal{C}$. Then,*

$$P\left(\cup_{i=1}^k A_i\right) \leq \sum_{i=1}^k P\left(A_i\right). \tag{9.75}$$

The result also holds for a countably infinite union of events, when $k \to \infty$.

Proof We define the events B_1, B_2, ... B_k, setting $B_1 := A_1$ and

$$B_i := A_i \cap \left(\cup_{j=1}^{i-1} A_j\right)^c \tag{9.76}$$

for $2 \leq i \leq k$. In words, B_i is the subset of A_i that does not belong to any A_j for $j < i$. Notice that these events are all disjoint, B_i is a subset of A_i and $\cup_{i=1}^k B_i = \cup_{i=1}^k A_i$. Consequently, by Axiom 3 in Definition 1.9 and Lemma 1.12,

$$P\left(\cup_{i=1}^k A_i\right) = P\left(\cup_{i=1}^k B_i\right) \tag{9.77}$$

$$= \sum_{i=1}^k P\left(B_i\right) \tag{9.78}$$

$$\leq \sum_{i=1}^k P\left(A_i\right). \tag{9.79}$$

The result still holds when $k \to \infty$ because Axiom 3 in Definition 1.9 holds for countably infinite sequences of disjoint events. ∎

Corollary 9.20. *If the variance of a random variable $\tilde{a}$ is zero, $\text{Var}\left[\tilde{a}\right] = 0$, then $\tilde{a}$ is equal to its mean μ with probability one, $P\left(\tilde{a} = \mu\right) = 1$. If the mean square is zero, then $\tilde{a}$ equals zero with probability one, $P\left(\tilde{a} = 0\right) = 1$.*

Proof We express the event $\tilde{a} \neq \mu$ as the union of the events $|\tilde{a} - \mu| > 1/i$ for $i = 1, 2, \ldots$ If the variance is zero, by Chebyshev's inequality (Theorem 9.18) the probability of each of these events is zero:

$$P\left(|\tilde{a} - \mu| \geq \frac{1}{i}\right) \leq i^2 \text{Var}\left[\tilde{a}\right] = 0. \tag{9.80}$$

By the union bound in Theorem 9.19, this implies that the probability of the union is also zero,

$$P\left(|\tilde{a} - \mu| \neq 0\right) = P\left(\cup_{i=1}^{\infty}\left\{|\tilde{a} - \mu| > \frac{1}{i}\right\}\right) \tag{9.81}$$

$$\leq \sum_{i=1}^{\infty} P\left(|\tilde{a} - \mu| \geq \frac{1}{i}\right) \tag{9.82}$$

$$= 0, \tag{9.83}$$

so the probability that $\tilde{a}$ equals μ is one.

By Lemma 7.33 the mean square is equal to the sum of the variance and the squared mean. Since these are nonnegative quantities, if their sum is zero, each of them must also equal zero. Consequently, if the mean square of a random variable is zero, its variance is zero. The random variable is therefore equal to its mean, which is zero, with probability one. ∎

Example 9.21 (Age of students: Improved bound). We are not very satisfied with our bound on the number of students older than 30 years in Example 9.17. Further research reveals that the standard deviation of student age is 3 years. This allows us to obtain a tighter bound using Chebyshev's inequality (Theorem 9.18):

$$P(\tilde{a} \geq 30) \leq P\left(|\tilde{a} - \mathbb{E}[\tilde{a}]| \geq 10\right) \tag{9.84}$$

$$\leq \frac{\operatorname{Var}[\tilde{a}]}{100} \tag{9.85}$$

$$= \frac{9}{100}. \tag{9.86}$$

At most 9% of the students are older than 30.

· ·

9.5 The Law of Large Numbers

In Section 7.1, we define the mean of a random variable as an averaging operation that can be applied to random variables. This motivates our definition of the sample mean in Section 7.1.4, which estimates the mean of a random variable by averaging its samples. In this section, we show that this estimator approximates the mean of a random variable with arbitrary accuracy, as long as we use enough samples, and these samples are independent. To make this statement precise, we introduce the concept of convergence in probability.

Let us consider a sequence of i.i.d. random variables $\tilde{x}_1, \tilde{x}_2, \ldots$ with mean μ, belonging to the same probability space. We denote the running average of the first n variables in the sequence by

$$\tilde{m}_n := \frac{1}{n}\sum_{j=1}^{n} \tilde{x}_j. \tag{9.87}$$

Consider the probability that $\tilde{m}_n$ is at a distance of more than ϵ from μ, where ϵ is a small positive constant:

$$p_n := P\left(|\tilde{m}_n - \mu| > \epsilon\right). \tag{9.88}$$

The sequence p_1, p_2, p_3, ... is a deterministic sequence of real numbers. If this sequence converges to zero for any $\epsilon > 0$, we say that $\widetilde{m}_n$ converges to μ *in probability*. This means that no matter how small ϵ is, we can increase n so that $\widetilde{m}_n$ is ϵ-close to μ with arbitrarily high probability. The celebrated law of large numbers states that this is guaranteed to occur, as long as the i.i.d. random variables have bounded variance.

Theorem 9.22 (The law of large numbers). *Let $\tilde{x}_1$, $\tilde{x}_2$, ... be a countably infinite sequence of i.i.d. random variables with mean μ and variance σ^2, belonging to the same probability space. The running average or sample mean $\widetilde{m}_n := \frac{1}{n} \sum_{j=1}^{n} \tilde{x}_j$ converges in probability to μ as $n \to \infty$, in the sense that for any $\epsilon > 0$,*

$$\lim_{n \to \infty} \mathrm{P}\left(|\widetilde{m}_n - \mu| > \epsilon\right) = 0. \tag{9.89}$$

Proof By linearity of expectation (Theorem 7.18), the mean of the sample mean is μ,

$$\mathbb{E}\left[\widetilde{m}_n\right] = \mathbb{E}\left[\frac{1}{n} \sum_{j=1}^{n} \tilde{x}_j\right] \tag{9.90}$$

$$= \frac{1}{n} \sum_{j=1}^{n} \mathbb{E}\left[\tilde{x}_j\right] \tag{9.91}$$

$$= \mu. \tag{9.92}$$

By Lemma 7.36 and Theorem 9.11, the variance equals

$$\mathrm{Var}\left[\widetilde{m}_n\right] = \mathrm{Var}\left[\frac{1}{n} \sum_{j=1}^{n} \tilde{x}_j\right] \tag{9.93}$$

$$= \frac{1}{n^2} \mathrm{Var}\left[\sum_{j=1}^{n} \tilde{x}_j\right] \tag{9.94}$$

$$= \frac{1}{n^2} \sum_{j=1}^{n} \mathrm{Var}\left[\tilde{x}_j\right] \tag{9.95}$$

$$= \frac{\sigma^2}{n}. \tag{9.96}$$

Consequently, for any $\epsilon > 0$, Chebyshev's inequality (Theorem 9.18) implies

$$\mathrm{P}\left(|\widetilde{m}_n - \mu| > \epsilon\right) \leq \frac{\mathrm{Var}\left[\widetilde{m}_n\right]}{\epsilon^2} \tag{9.97}$$

$$\leq \frac{\sigma^2}{n\epsilon^2}. \tag{9.98}$$

The limit when $n \to \infty$ is zero because ϵ is fixed (even if it is very small). ∎

An important consequence of the law of large numbers is that if random samples are drawn following Definition 9.3 from a fixed population, then the sample mean converges in probability to the population mean as the sample size increases. Since sample proportions can be interpreted as sample means (see Example 9.4), the sample proportion converges to the true population proportion under the same assumptions. In statistics, estimators that converge in probability to the parameter of interest are said to be *consistent*. Figure 9.5 illustrates the consistency of the sample mean for the data in Example 9.1. For any ϵ (in the

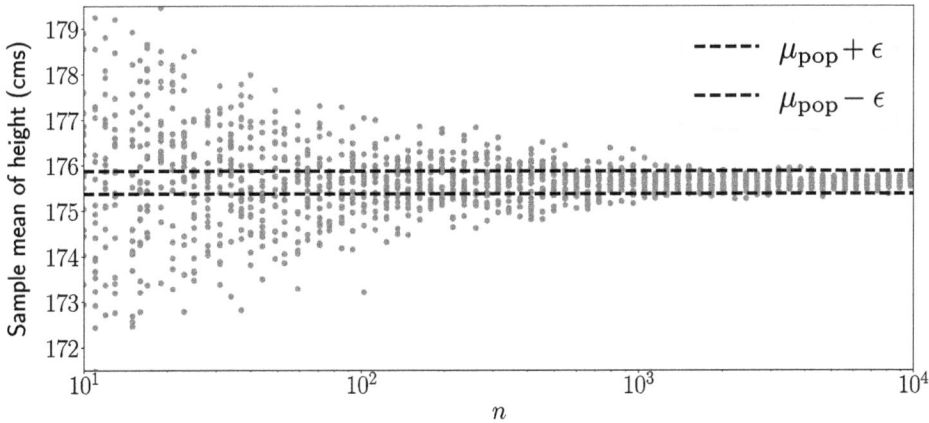

Figure 9.5 Consistency of the sample mean. Each point in the scatterplot corresponds to a sample mean computed using n independent, uniform samples from the dataset in Example 9.1. If the sample mean is in the region between the dashed lines, the difference with the population mean $\mu_{\text{pop}} = 175.6$ cm is less than $\epsilon := 0.25$ cm. When the sample size n grows sufficiently large, the fraction of sample means outside of the region vanishes, as predicted by Theorem 9.23, which establishes that the sample mean converges to the population mean in probability.

figure, $\epsilon := 0.25$ cms), the probability that the sample mean deviates from the population by more than ϵ eventually converges to zero when the sample size becomes sufficiently large.

Theorem 9.23 (Consistency of the sample mean). *Let a_1, a_2, ..., a_N denote a dataset of size N with population mean μ_{pop}, defined as in (9.10). Let $\tilde{x}_1$, $\tilde{x}_2$, ..., be a sequence of independent, uniform random samples following Definition 9.3. The sample mean $\tilde{m}_n := \frac{1}{n} \sum_{j=1}^{n} \tilde{x}_j$ converges to μ_{pop} in probability as the number of measurements tends to infinity. For any $\epsilon > 0$,*

$$\lim_{n \to \infty} \mathrm{P}\left(|\tilde{m}_n - \mu_{\text{pop}}| > \epsilon\right) = 0. \tag{9.99}$$

Proof The result follows directly from the law of large numbers (Theorem 9.22) because the mean of each random sample is equal to μ_{pop} by (9.15) and the variance of each random sample equals the population variance (which is a fixed constant, since N is finite) by (9.37). ∎

The law of large numbers enables us to establish the consistency of the empirical-probability estimator in Definition 1.22.

Theorem 9.24 (Consistency of the empirical-probability estimator). *Given a probability space $(\Omega, \mathcal{C}, \mathrm{P})$ (see Definition 1.15), let $\mathrm{P}(A)$ denote the probability of an event A belonging to the collection $\mathcal{C}$. Consider a dataset $X := \{x_1, x_2, \ldots, x_n\}$ of outcomes belonging to the sample space Ω. If the data are generated according to the probability measure P, then the probability that each data point is in A equals $\mathrm{P}(A)$. To capture this, we define the Bernoulli random variables*

$$\tilde{b}_i = \begin{cases} 1, & \text{if the ith data point is in } A, \\ 0, & \text{otherwise} \end{cases} \tag{9.100}$$

for $1 \leq i \leq n$, such that $\mathrm{P}\left(\tilde{b}_i = 1\right) = \mathrm{P}\left(A\right)$ and $\mathrm{P}\left(\tilde{b}_i = 0\right) = 1 - \mathrm{P}\left(A\right)$. If these random variables are independent, then the empirical-probability estimator in Definition 1.22

$$\widetilde{\mathrm{P}}_X(A) := \frac{1}{n} \sum_{i=1}^{n} \tilde{b}_i \tag{9.101}$$

converges to $\mathrm{P}\left(A\right)$ in probability as $n \to \infty$.

Proof The empirical-probability estimator is the sample mean of the Bernoulli random variables (see Definition 7.15). The mean of each Bernoulli variable is $\mathrm{P}\left(A\right)$ by Lemma 7.2 and its variance is $\mathrm{P}\left(A\right)\left(1 - \mathrm{P}\left(A\right)\right)$ by Lemma 7.34. The result follows from applying the law of large numbers (Theorem 9.22) to these variables. ∎

Our proof of the law of large numbers provides a general strategy to establish that an estimator is consistent:

1 Prove that the estimator is unbiased.
2 Prove that the variance of the estimator vanishes as the sample size tends to infinity.

Convergence in probability to the parameter of interest then follows from Chebyshev's inequality (Theorem 9.18). As an illustration, we apply this strategy to show that the sample variance is a consistent estimator of the variance of a sequence of i.i.d. random variables.

Theorem 9.25 (The sample variance converges to the variance). *Let $\tilde{x}_1, \tilde{x}_2, \ldots$ be a countably infinite sequence of i.i.d. random variables with finite variance σ^2, belonging to the same probability space. We assume that the fourth central moment is bounded, i.e.*

$$\kappa := \mathbb{E}\left[\left(\tilde{x}_i - \mathbb{E}\left[\tilde{x}_i\right]\right)^4\right] \leq c, \tag{9.102}$$

for a fixed positive constant c and all i. Then the sample variance

$$\tilde{v}_n := \frac{1}{n-1} \sum_{j=1}^{n} \left(\tilde{x}_j - \tilde{m}_n\right)^2, \qquad \tilde{m}_n := \frac{1}{n} \sum_{k=1}^{n} \tilde{x}_k, \tag{9.103}$$

converges in probability to σ^2 as $n \to \infty$.

In particular, if $\tilde{x}_1, \tilde{x}_2, \ldots$ are a sequence of random samples drawn from a dataset $a_1, a_2, \ldots, a_N$ of size N following Definition 9.3, then the sample variance is a consistent estimator of the population variance σ^2_{pop}, defined as in (9.20).

Proof We use the following expression for the variance of the sample variance, which is straightforward (but painful) to derive by repeatedly applying linearity of expectation (Theorem 7.18):

$$\mathrm{Var}\left[\tilde{v}_n\right] = \frac{1}{n}\left(\kappa - \frac{(n-3)\sigma^4}{n-1}\right), \qquad \kappa := \mathbb{E}\left[\left(\tilde{x}_i - \mu\right)^4\right], \tag{9.104}$$

where μ and κ are the mean and fourth central moment of the i.i.d. variables.[2] The sample variance is an unbiased estimator of the variance, $\mathbb{E}\left[\tilde{v}\right] = \sigma^2$ (we omit the derivations,

---

[2] In the literature, κ is sometimes also used to denote the kurtosis, which is the fourth central moment divided by the square of the variance.

which are identical to those in the proof of Theorem 9.8), so by Chebyshev's inequality (Theorem 9.18), for any $\epsilon > 0$,

$$P\left(\left|\tilde{v}_n - \sigma^2\right| > \epsilon\right) \leq \frac{\text{Var}\left[\tilde{v}_n\right]}{\epsilon^2} \tag{9.105}$$

$$= \frac{1}{n\epsilon^2}\left(\kappa - \frac{(n-3)\sigma^4}{n-1}\right) \tag{9.106}$$

$$\leq \frac{c}{n\epsilon^2}, \tag{9.107}$$

which converges to zero as $n \to \infty$.

We can directly apply this result to establish consistency of the sample variance computed from random samples $\tilde{x}_1, \tilde{x}_2, \ldots$ following Definition 9.3. The mean of each random sample equals the population mean μ_{pop} by (9.15), and its variance equals the population variance σ^2_{pop} by (9.37). By Definition 9.3 for all j $(\tilde{x}_j - \mathbb{E}\left[\tilde{x}_j\right])^4 = (\tilde{x}_j - \mu_{\text{pop}})^4$ is a deterministic function of $\tilde{k}_j$, which equals $(a_k - \mu_{\text{pop}})^4$ if $\tilde{k}_j = k$, so by Definition 7.4 the fourth central moment equals

$$\mathbb{E}\left[(\tilde{x}_j - \mathbb{E}\left[\tilde{x}_j\right])^4\right] = \mathbb{E}\left[(\tilde{x}_j - \mu_{\text{pop}})^4\right] \tag{9.108}$$

$$= \sum_{k=1}^{N}(a_k - \mu_{\text{pop}})^4 p_{\tilde{k}_j}(k) \tag{9.109}$$

$$= \frac{1}{N}\sum_{k=1}^{N}(a_k - \mu_{\text{pop}})^4, \tag{9.110}$$

which is a fixed constant. We conclude that the sample variance of the samples converges in probability to the population variance. ∎

To end this section, we show that the Chebyshev bound used in our proof of the law of the large numbers is usually very loose. Consequently, it does not provide an accurate quantitative description of the probabilistic behavior of the sample mean.

Example 9.26 (Estimating the mean height: Chebyshev bound)**.** The proof of the law of large numbers (Theorem 9.22) is based on the Chebyshev bound

$$P\left(\left|\tilde{m}_n - \mu\right| > \epsilon\right) \leq \frac{\sigma^2}{n\epsilon^2}. \tag{9.111}$$

Here we evaluate this bound for the height data in Example 9.1. We fix $\epsilon := 1$ cm and set μ and σ^2 equal to the true mean and variance of the random samples, which in this case are equal to the population mean and population variance by (9.15) and (9.37), respectively (recall that this is a controlled example, where we know the population parameters). For different values of the sample size n, the probability is estimated via the Monte Carlo method in Section 1.7. In more detail, we generate many datasets of n random samples, compute their sample mean and determine the fraction of datasets for which the absolute difference between the sample mean and the population mean is greater than 1 cm. This yields an approximation of the probability $P\left(\left|\tilde{m}_n - \mu\right| > \epsilon\right)$ that we can compare to the Chebyshev bound.

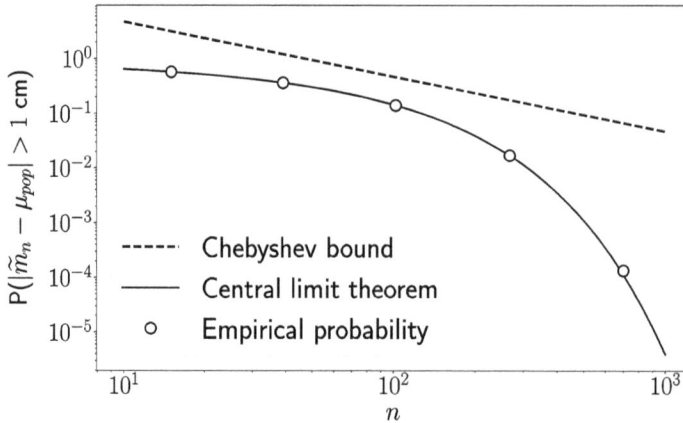

Figure 9.6 Chebyshev bound evaluated on real data. The dashed line depicts the Chebyshev bound $\sigma^2/n\epsilon^2$ in the proof of Theorem 9.22 for $\epsilon := 1$ cm and different values of the sample size n, using the height data in Example 9.1. The white markers indicate the probability that the sample mean is at a distance of more than 1 cm from the population mean, computed via 10^6 Monte Carlo simulations as described in Example 9.26. Both the bound and the empirical probability converge to zero as n increases, but the bound is extremely loose. The black curve represents a much more accurate approximation of the probability based on the central limit theorem (see Section 9.7).

Figure 9.6 shows the result of our analysis. Reassuringly, the Chebyshev bound is indeed greater than the approximated probability, and both converge to zero as n increases, in accordance with the law of large numbers. However, the bound does not provide an accurate approximation to the empirical probability: It is orders of magnitude larger. The central limit theorem, presented in Section 9.7, characterizes the probabilistic behavior of the sample mean much more precisely.

· ·

9.6 Some Averages Are Not to Be Trusted

The law of large numbers, described in Section 9.5, states that the average of a sequence of i.i.d. random variables converges to the mean. However, there are situations where the law of large numbers does not hold, as illustrated by the following two examples.

Example 9.27 (St Petersburg paradox). Zefi challenges her sister Anna to play a game. They will flip a fair coin until it lands on tails. Then, Zefi will pay Anna 2^k dollars, where k is the number of coin flips. Anna needs to decide how much money she is willing to pay to participate in the game. In order to make the decision, she simulates the game n times and computes the sample mean of her winnings. She gradually increases n, expecting the sample mean to converge to a fixed value, as predicted by the law of large numbers (Theorem 9.22). However, this never happens! To Anna's surprise, the sample mean just keeps growing, as shown in Figure 9.7.

To understand what is going on, Anna performs a probabilistic analysis of her simulations. The winnings can be modeled as a random variable $\tilde{w} := 2^{\tilde{k}}$, where $\tilde{k}$ represents the number of heads. Assuming the coin flips are independent, $\tilde{k}$ is a geometric random variable with parameter $1/2$ (see Exercise 2.1 and Definition 2.20), so its pmf equals $p_{\tilde{k}}(k) = 1/2^k$. By Definition 7.4 the mean of the winnings is

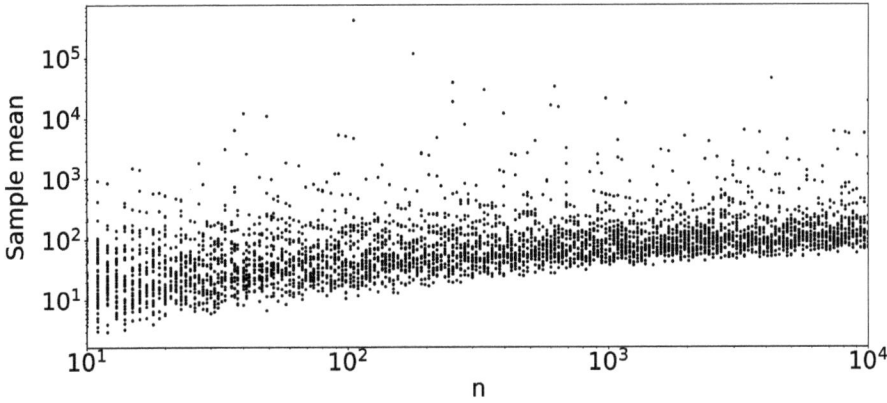

Figure 9.7 Average winnings in the St Petersburg paradox. The plot shows the sample mean of the winnings in n independent simulations of the game described in Example 9.27. Instead of converging to a finite value, as in Figures 9.3 and 9.4, the sample mean increases indefinitely as n grows, without converging to a specific value. The reason is that the mean of the winnings is infinite, so the law of large numbers does not hold.

$$\mathbb{E}\left[\tilde{w}\right] = \mathbb{E}\left[2^{\tilde{k}}\right] \tag{9.112}$$

$$= \sum_{k=1}^{\infty} 2^k p_{\tilde{k}}(k) \tag{9.113}$$

$$= \sum_{k=1}^{\infty} 2^k \cdot \frac{1}{2^k} \tag{9.114}$$

$$= \infty. \tag{9.115}$$

The sample mean of n i.i.d. random variables does indeed converge to their mean according to Theorem 9.22, *as long as their mean and variance are finite*. The catch is that the mean of the winnings is *infinite*! Consequently, the law of large numbers does not apply. Intuitively, the sample mean does not converge in this scenario because very long sequences yield astronomical winnings, which scale exponentially (2^k) with the length k. As the number of simulations n grows, there is an increasing chance of observing a very long sequence that single-handedly blows up the sample mean.

This phenomenon is known as the St Petersburg paradox. If Anna's criterion is to maximize the mean of her winnings, she should pay as much money as she can afford to play the game! However, that makes no sense because the long sequences that blow up the sample mean occur with very low probability. As suggested in Section 7.5, we can use the median to quantify the typical winnings without being overly influenced by extreme values. Half of the time the first coin flip lands on tails, so the median winnings are just two dollars (see Definition 3.10). Anna should not invest her life savings in the game.

· ·

Example 9.28 (Locating a radioactive source). Dimitra carries out an experiment, where a rod containing a single radioactive source is placed in front of a line of radiation detectors. The goal is to determine the location of the source on the rod. The source emits particles from time to time. When the particle hits a detector, she records the location of the detector. She reasons that since the radiation is isotropic – the particle is equally likely to be emitted in any direction – the average of the locations should provide a reasonable estimate of the position of the source. However, when she computes the sample mean of her data, she is surprised to find that it does not converge to a fixed value: As she gathers more measurements, the sample mean just fluctuates!

Let us model the location $\tilde{\ell}$ hit by the particle as a random variable. For simplicity, we measure the location in terms of its displacement from the source (i.e. the source is at location zero). If Dimitra's intuition is correct, then averaging $\tilde{\ell}$ should yield zero, and correctly locate the source. Figure 9.8 shows the result of averaging different numbers of i.i.d. samples from $\tilde{\ell}$. As we increase the sample size n, the sample mean does not converge

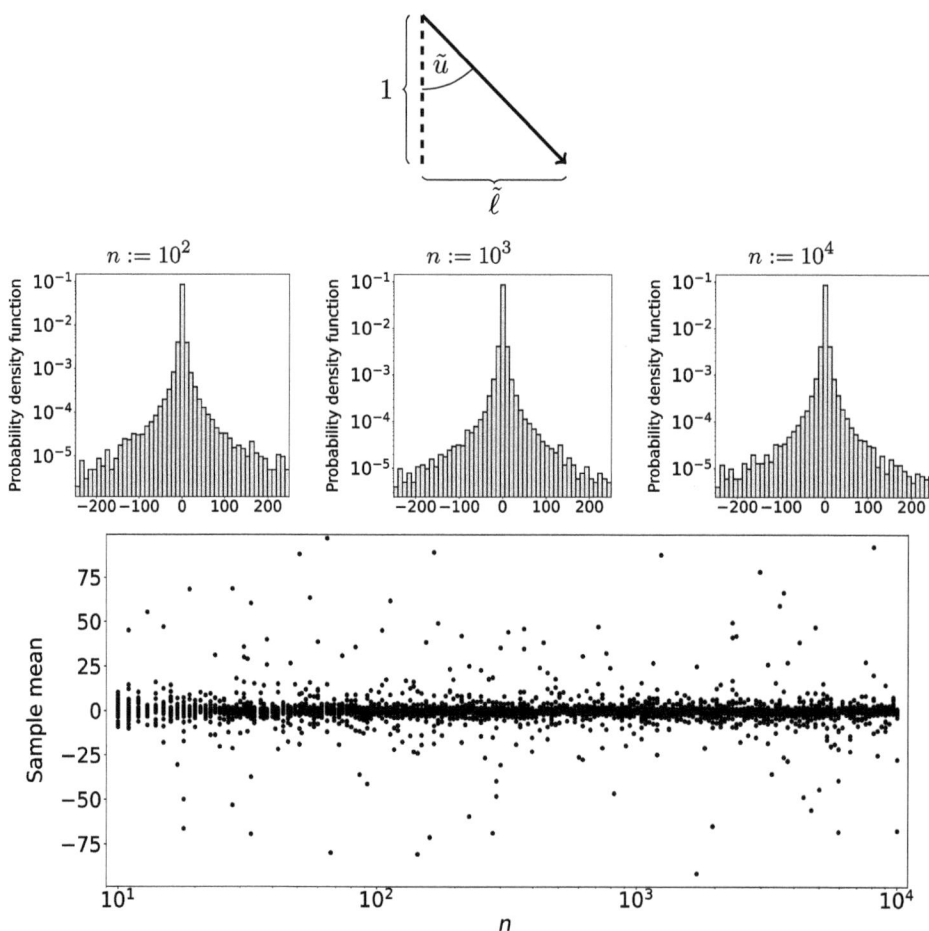

Figure 9.8 Locating a radioactive sample. The diagram at the top describes the probabilistic model for the trajectory of the radioactive particle emitted by the source in Example 9.28. The random variables $\tilde{u}$ and $\tilde{\ell}$ represent the angle of the trajectory and the location of the detector hit by the particle, respectively. The middle row shows the pdf of the sample mean of n i.i.d. samples of $\tilde{\ell}$ (approximated using histograms computed from 10^5 simulated datasets of size n, following Definition 3.23). The spread of the pdf does not decrease as the sample size n grows. This is confirmed by the bottom graph, which plots realizations of the sample mean as a function of n. In contrast to Figures 9.3 and 9.4, the sample mean does not converge to a fixed value because of the non-negligible probability of encountering values with extremely large magnitudes.

to zero: Some averages have very large negative or positive amplitudes, even when n is very large. This seems to contradict the law of large numbers (Theorem 9.22)!

In order to shed some light on this mystery, Dimitra decides to derive the mean of $\tilde{\ell}$. Under the assumption that the radiation is isotropic, the angle $\tilde{u}$ of the particle trajectory is uniformly distributed between $-\pi/2$ and $\pi/2$. The sensors are one meter away from the rod, as depicted in the diagram at the top of Figure 9.8, so $\tilde{\ell}$ is equal to the tangent of $\tilde{u}$. By Definition 3.3 and Theorem 3.14, the cdf of $\tilde{\ell}$ is

$$F_{\tilde{\ell}}(\ell) = P(\tilde{\ell} \leq \ell) \tag{9.116}$$

$$= P(\tan \tilde{u} \leq \ell) \tag{9.117}$$

$$= P(\tilde{u} \leq \arctan \ell) \tag{9.118}$$

$$= \frac{1}{\pi} \int_{-\pi/2}^{\arctan \ell} du \tag{9.119}$$

$$= \frac{1}{2} + \frac{\arctan \ell}{\pi}, \tag{9.120}$$

where (9.118) holds because the tangent is a strictly increasing function between $-\pi/2$ and $\pi/2$. Differentiating the cdf yields the pdf of $\tilde{\ell}$ by Definition 3.13,

$$f_{\tilde{\ell}}(\ell) = \frac{1}{\pi(1 + \ell^2)}. \tag{9.121}$$

By Definition 7.8, the mean of $\tilde{\ell}$ is

$$E[\tilde{\ell}] = \int_{-\infty}^{\infty} \frac{\ell}{\pi(1 + \ell^2)} \, d\ell \tag{9.122}$$

$$= \int_{0}^{\infty} \frac{\ell}{\pi(1 + \ell^2)} \, d\ell + \int_{-\infty}^{0} \frac{\ell}{\pi(1 + \ell^2)} \, d\ell \tag{9.123}$$

$$= \int_{0}^{\infty} \frac{\ell}{\pi(1 + \ell^2)} \, d\ell - \int_{0}^{\infty} \frac{a}{\pi(1 + a^2)} \, da, \tag{9.124}$$

where we apply the change of variables $a := -\ell$. The mean is equal to the difference between two identical terms. You may be tempted to cancel out these terms and declare that the mean is zero, but this is incorrect! The reason is that both terms diverge. By the change of variables $t := \ell^2$,

$$\int_{0}^{\infty} \frac{\ell}{\pi(1 + \ell^2)} \, d\ell = \int_{0}^{\infty} \frac{1}{2\pi(1 + t)} dt = \lim_{t \to \infty} \frac{\log(1 + t)}{2\pi} = \infty. \tag{9.125}$$

The tail of the pdf $f_{\tilde{\ell}}$ decays fast enough to ensure that it integrates to one, but not fast enough for the function $f_{\tilde{\ell}}(\ell)\ell$ to have a finite integral. From basic calculus, the difference of two terms that tend to infinity is not well defined because we can manipulate the expression to yield any value we want.

We conclude that the random variable $\tilde{\ell}$ does not have a mean, so the law of large numbers does not apply to it. When we sample from its distribution, there is a non-negligible probability of observing enormous positive or negative values, which dominate the sample mean. This prevents convergence to a fixed value, as we observe in Figure 9.8.

. . . .

The random variable $\tilde{\ell}$ in Example 9.28 follows a Cauchy distribution.

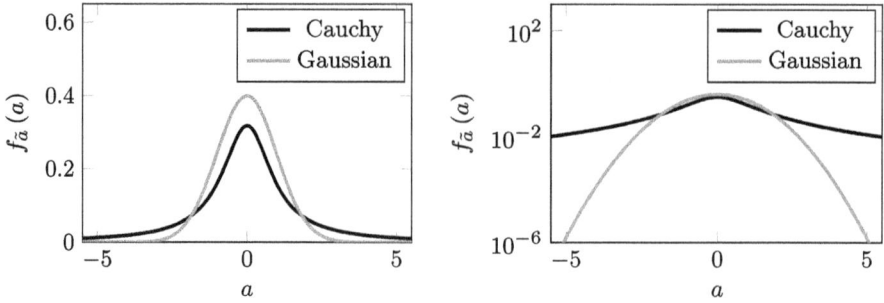

Figure 9.9 Comparison between the Cauchy and Gaussian pdfs. The plots show the pdfs of Cauchy and Gaussian distributions centered at the origin on a linear (left) and logarithmic (right) scale. The tail of the Cauchy distribution decays much more slowly than that of the Gaussian. As a result, extreme values have much higher probability.

Definition 9.29 (Cauchy distribution). *A Cauchy random variable $\tilde{a}$ has a pdf of the form*

$$f_{\tilde{a}}(a) = \frac{1}{\pi(1 + a^2)}. \tag{9.126}$$

Figure 9.9 compares the pdfs of a Gaussian (see Definition 3.30) and a Cauchy distribution. The tail of the Cauchy distribution decays much more slowly, so it takes extreme values with much higher probability than the Gaussian. As a result, its mean is not well defined, as we establish in Example 9.28. At first glance, this might seem a mathematical curiosity, but Figure 9.8 shows that it has a very practical implication: The sample mean of i.i.d. samples from this distribution does not converge to a fixed value, no matter how many samples we average.

The following example shows that the sample mean may be distorted by extreme values, even in situations where the law of large numbers does hold.

Example 9.30 (Local economic activity). The G-Econ research project at Yale developed a metric to measure local economic activity all over the world. The gross cell product (GCP) quantifies the economic output of small regions called cells, which partition the globe. In total, there are $N := 20,100$ cells. We consider the problem of estimating the population mean of the GCP, which equals 2 million dollars, from a small number of random samples. Figure 9.10 shows histograms of the GCP of 100 cells selected uniformly at random with replacement from Dataset 16 (top row and bottom left). The distribution of the GCP contains extreme outliers: The median GCP is 0.03 million, but multiple cells have a GCP of more than 200 million. When one of these outliers is selected, as in the bottom left graph of Figure 9.10, it blows up the sample mean to around 20 million, which is one order of magnitude larger than the population mean.

In general, the sample mean is not a reliable estimator of the population mean (see the bottom right graph in Figure 9.10), but this does *not* contradict the consistency of the sample mean (Theorem 9.23). We just require substantially more than 100 samples for the sample mean to concentrate near the population mean. Since the population standard deviation is 17.7 million, the standard error of the sample mean equals $17.7/\sqrt{n}$ million by Theorem 9.13, where n is the sample size. Consequently, if we only have 100 samples, the standard error is 1.77 million, which is almost as large as the population mean!

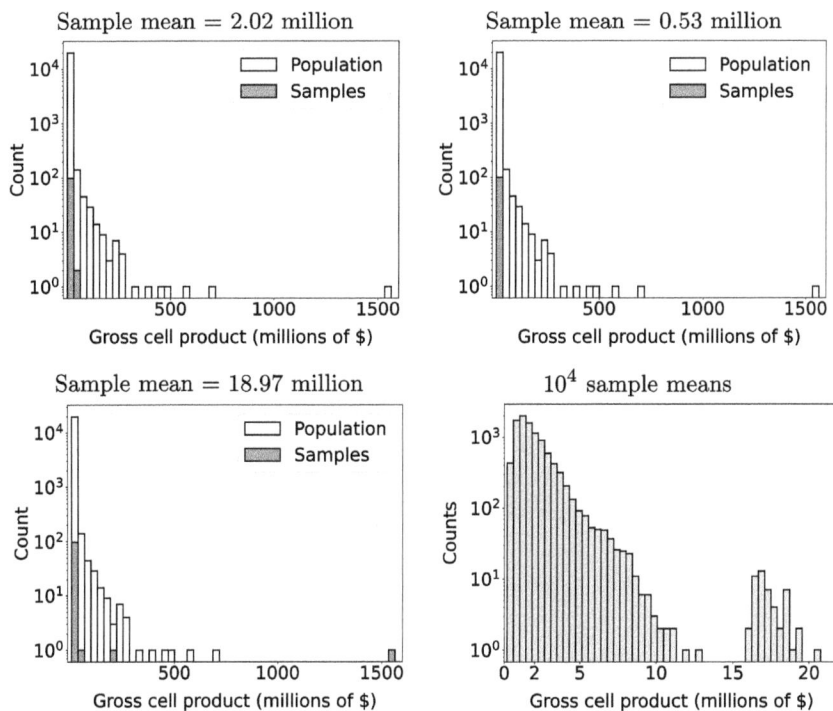

Figure 9.10 Sample mean of data with outliers. In the top row and the bottom left, the white histogram represents the local economic activity of 20,100 regions, quantified by the gross cell product (GCP) as described in Example 9.30. The gray histograms represent the GCP of 100 regions selected uniformly at random with replacement. The sample mean of the GCP of the selected regions is indicated at the top of each graph (for comparison, the population mean equals 2 million). The plot on the bottom right shows a histogram of 10^4 independent instances of the sample mean obtained in the same way. The histogram is skewed, and does not concentrate tightly around the population mean (compare to Figures 9.1 and 9.2) because the sample size is not large enough with respect to the population variance.

In summary, the sample mean is not to be trusted in the presence of extreme values. In such situations, it is usually advisable to not even attempt to estimate the mean in the first place. The median is a more reasonable description of a typical value, as discussed in Section 7.5.

9.7 The Central Limit Theorem

The central limit theorem is a fundamental result in probability, which has crucial implications in statistics. It states that sums of independent quantities tend to have a Gaussian distribution. In Section 9.7.1, we study the distribution of sums of independent random variables. In Section 9.7.2, we present the central limit theorem and use it to characterize the behavior of sample means computed from random samples. Section 9.7.3 provides a cautionary example inspired by the 2008 Financial Crisis, which illustrates situations where the central limit theorem does not hold.

9.7.1 Sums of Independent Random Variables

In this section, we study the sum of two independent random variables, showing that the resulting distribution can be obtained by convolving the individual pmfs or pdfs. We begin by analyzing sums of discrete variables.

Theorem 9.31 (Sum of two independent discrete random variables). *Let $\tilde{a}$ and $\tilde{b}$ be two independent discrete random variables with ranges A and B belonging to the same probability space. The pmf of their sum $\tilde{s} = \tilde{a} + \tilde{b}$ equals*

$$p_{\tilde{s}}(s) = \sum_{a \in A} p_{\tilde{a}}(a) \, p_{\tilde{b}}(s - a), \tag{9.127}$$

where $p_{\tilde{a}}$ and $p_{\tilde{b}}$ denote the pmfs of $\tilde{a}$ and $\tilde{b}$, respectively. Note that $p_{\tilde{s}}$ is nonzero only for values of s such that there exists some $a \in A$ for which $s - a \in B$.

If $\tilde{a}$ and $\tilde{b}$ are integer valued (A and B are subsets of the integers), then the pmf of their sum $\tilde{s} = \tilde{a} + \tilde{b}$ is equal to the convolution of their respective pmfs,

$$p_{\tilde{s}}(s) = p_{\tilde{a}} * p_{\tilde{b}}(s) := \sum_{a=-\infty}^{\infty} p_{\tilde{a}}(a) \, p_{\tilde{b}}(s - a). \tag{9.128}$$

Proof　We express the event $\tilde{a} + \tilde{b} = s$ as the union of the intersections of the events $\tilde{a} = a$ and $\tilde{b} = s - a$, for all $a \in A$. These intersections are disjoint events, so the probability of the union is equal to the sum of their individual probabilities by Axiom 3 in Definition 1.9. In addition, $\tilde{a}$ and $\tilde{b}$ are independent, so the events $\tilde{a} = a$ and $\tilde{b} = s - a$ are independent (see Section 4.4). Consequently, by Definitions 2.2 and 1.26,

$$p_{\tilde{s}}(s) = \mathrm{P}\left(\tilde{s} = s\right) \tag{9.129}$$

$$= \mathrm{P}\left(\cup_{a \in A} \{\tilde{a} = a\} \cap \{\tilde{b} = s - a\}\right) \tag{9.130}$$

$$= \sum_{a \in A} \mathrm{P}\left(\tilde{a} = a, \tilde{b} = s - a\right) \tag{9.131}$$

$$= \sum_{a \in A} \mathrm{P}\left(\tilde{a} = a\right) \mathrm{P}\left(\tilde{b} = s - a\right) \tag{9.132}$$

$$= \sum_{a \in A} p_{\tilde{a}}(a) \, p_{\tilde{b}}(s - a). \tag{9.133}$$

If A and B are subsets of the integers, then (9.133) can be rewritten as (9.128). ∎

For random variables with integer values, it follows immediately from Theorem 9.31 that the pmf of the sum of multiple discrete random variables is equal to the convolution of their individual pmfs.

Corollary 9.32 (Sum of multiple independent discrete random variables). *Let $\tilde{a}_1$, $\tilde{a}_2$, ..., $\tilde{a}_n$ be n independent discrete random variables with integer values belonging to the same probability space. The pmf of their sum $\tilde{s}_n = \sum_{i=1}^{n} \tilde{a}_i$ equals the convolution of their pmfs,*

$$p_{\tilde{s}_n}(s) = p_{\tilde{a}_1} * p_{\tilde{a}_2} * \cdots * p_{\tilde{a}_n}(s), \tag{9.134}$$

where the convolution operation $*$ is defined as in (9.128). Since convolution is associative, it does not matter in what order we compute the convolutions.

Example 9.33 (Soccer league). In Example 2.9, we derive the distribution of the points earned by Barcelona in a game against Atletico de Madrid. Here, we consider the problem of modeling the total points obtained by Barcelona over several similar games. We assume that in each individual game the distribution of the earned points is the same as in Example 2.9 and that the games are independent.

Let n be the number of games and let $\tilde{x}_1$, $\tilde{x}_2$, ..., $\tilde{x}_n$ denote i.i.d. random variables representing the points obtained in each game. We are interested in the pmf of the sum $\tilde{s}_n := \sum_{i=1}^n \tilde{x}_i$ for different values of n. From Example 2.9, the marginal pmf of $\tilde{x}_i$ is

$$p_{\tilde{x}_i}(0) = 0.3, \quad p_{\tilde{x}_i}(1) = 0.3, \quad p_{\tilde{x}_i}(3) = 0.4, \tag{9.135}$$

and $p_{\tilde{x}_i}(x) = 0$ for any other value of x. By Theorem 9.31, the pmf of $\tilde{s}_2$ is equal to the convolution of $p_{\tilde{x}_1}$ and $p_{\tilde{x}_2}$,

$$p_{\tilde{s}_2}(s) = p_{\tilde{x}_1} * p_{\tilde{x}_2}(s) \tag{9.136}$$

$$= \sum_{x=-\infty}^{\infty} p_{\tilde{x}_1}(x) \, p_{\tilde{x}_2}(s - x). \tag{9.137}$$

Only a few terms in the infinite sum are nonzero. For example,

$$p_{\tilde{s}_2}(1) = p_{\tilde{x}_1}(0) \, p_{\tilde{x}_2}(1) + p_{\tilde{x}_1}(1) \, p_{\tilde{x}_2}(0) \tag{9.138}$$

$$= 0.18. \tag{9.139}$$

Similar calculations yield

$$p_{\tilde{s}_2}(0) = 0.09, \quad p_{\tilde{s}_2}(1) = 0.18, \quad p_{\tilde{s}_2}(2) = 0.09, \tag{9.140}$$

$$p_{\tilde{s}_2}(3) = 0.24, \quad p_{\tilde{s}_2}(4) = 0.24, \quad p_{\tilde{s}_2}(6) = 0.16. \tag{9.141}$$

For all other values of s, $p_{\tilde{s}_2}(s) = 0$.

By Corollary 9.32, the pmf of $\tilde{s}_n$ is

$$p_{\tilde{s}_n}(s) = p_{\tilde{x}_1} * p_{\tilde{x}_2} * \cdots * p_{\tilde{x}_n}(s). \tag{9.142}$$

The random variables are identically distributed, so all the pmfs are the same. Consequently, $p_{\tilde{s}_n}$ is obtained by convolving the same pmf with itself $n - 1$ times. Figure 9.11 shows plots of $p_{\tilde{s}_n}$ for $1 \leq n \leq 9$. As n increases, the nonzero support of the pmf grows, and its shape becomes increasingly smooth and similar to a Gaussian pdf. To verify this, we compare the pmf of the sum to a Gaussian pdf with the same mean and variance, which equal

$$\mathbb{E}[\tilde{s}_n] = \sum_{i=1}^n \mathbb{E}[\tilde{x}_i] = 1.5n, \qquad \text{Var}[\tilde{s}_n] = \sum_{i=1}^n \text{Var}[\tilde{x}_i] = 1.65n \tag{9.143}$$

by linearity of expectation (Theorem 7.18) and Theorem 9.11, respectively. Figure 9.11 shows that the pmf of the sum and the pdf of the Gaussian are remarkably similar, even for small values of n.

....

The following theorem derives the distribution of the sum of two independent continuous random variables. In this case, the pdf of the sum is equal to the continuous convolution of the individual pdfs.

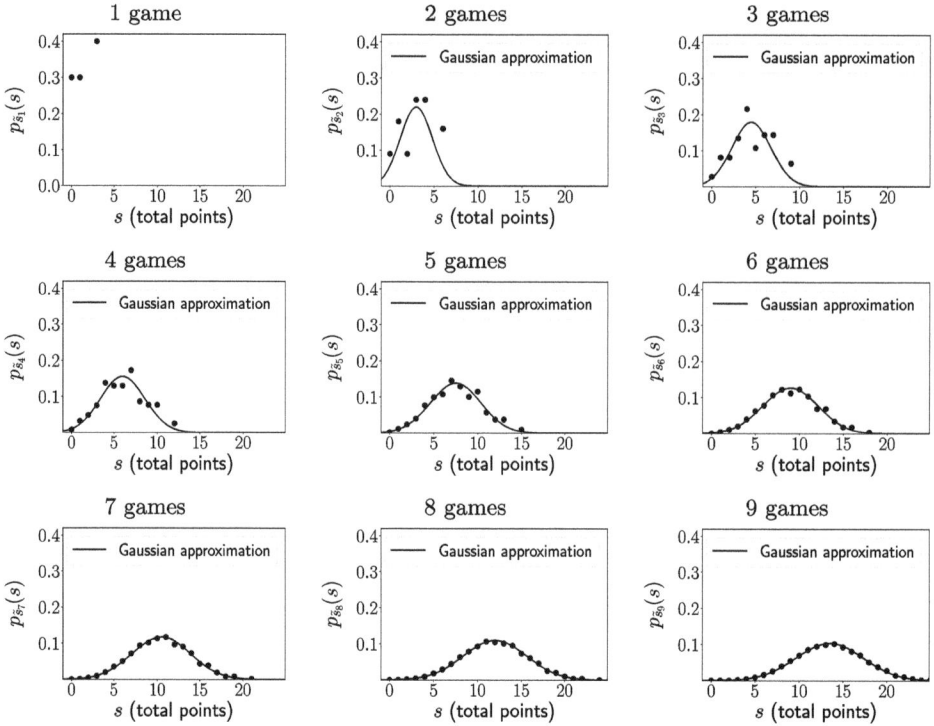

Figure 9.11 Points earned by soccer team. The circular markers represent the pmf of the sum of points earned by the soccer team in Example 9.33 for different numbers of games. The pmfs converge to a Gaussian pdf (black line), as predicted by the central limit theorem (Theorem 9.37).

Theorem 9.34 (Sum of independent continuous random variables). *Let $\tilde{a}$ and $\tilde{b}$ be two independent continuous random variables, belonging to the same probability space, with pdfs $f_{\tilde{a}}$ and $f_{\tilde{b}}$. The pdf of their sum $\tilde{s} = \tilde{a} + \tilde{b}$ equals the convolution of their pdfs,*

$$f_{\tilde{s}}(s) = f_{\tilde{a}} * f_{\tilde{b}}(s) := \int_{a=-\infty}^{\infty} f_{\tilde{a}}(a) \, f_{\tilde{b}}(s-a) \, \mathrm{d}a. \qquad (9.144)$$

Let $\tilde{a}_1, \tilde{a}_2, \ldots, \tilde{a}_n$ be n independent continuous random variables with pdfs denoted by $f_{\tilde{a}_1}, \ldots, f_{\tilde{a}_n}$ belonging to the same probability space. The pdf of their sum $\tilde{s}_n = \sum_{i=1}^{n} \tilde{a}_i$ equals the convolution of their pdfs,

$$f_{\tilde{s}_n}(s) = f_{\tilde{a}_1} * f_{\tilde{a}_2} * \cdots * f_{\tilde{a}_n}(s), \qquad (9.145)$$

where the convolution operation $*$ is defined as in (9.144). Since convolution is associative, it does not matter in what order we compute the convolutions.

Proof First we derive the cdf of $\tilde{s}$, integrating the joint pdf of $\tilde{a}$ and $\tilde{b}$ over the region corresponding to the event $\tilde{a} + \tilde{b} \leq s$. The independence assumption implies that the joint pdf is the product of the individual pdfs (see Definition 5.15), so by Definitions 3.3 and Theorem 3.14,

$$F_{\tilde{s}}(s) = P\left(\tilde{a} + \tilde{b} \leq s\right) \tag{9.146}$$

$$= \int_{a=-\infty}^{\infty} \int_{b=-\infty}^{s-a} f_{\tilde{a},\tilde{b}}(a,b) \, \mathrm{d}a \, \mathrm{d}b \tag{9.147}$$

$$= \int_{a=-\infty}^{\infty} \int_{b=-\infty}^{s-a} f_{\tilde{a}}(a) f_{\tilde{b}}(b) \, \mathrm{d}a \, \mathrm{d}b \tag{9.148}$$

$$= \int_{a=-\infty}^{\infty} f_{\tilde{a}}(a) F_{\tilde{b}}(s-a) \, \mathrm{d}a. \tag{9.149}$$

We now differentiate the cdf to obtain the pdf following Definition 3.13. This requires an interchange of a limit operator with a differentiation operator and another interchange of an integral operator with a differentiation operator, which are justified because the functions involved are bounded and integrable:

$$f_{\tilde{s}}(s) = \frac{\mathrm{d}}{\mathrm{d}s} \lim_{t\to\infty} \int_{a=-t}^{t} f_{\tilde{a}}(a) F_{\tilde{b}}(s-a) \, \mathrm{d}a \tag{9.150}$$

$$= \lim_{t\to\infty} \int_{a=-t}^{t} \frac{\mathrm{d}}{\mathrm{d}s} \left(f_{\tilde{a}}(a) F_{\tilde{b}}(s-a)\right) \, \mathrm{d}a \tag{9.151}$$

$$= \lim_{t\to\infty} \int_{a=-t}^{t} f_{\tilde{a}}(a) f_{\tilde{b}}(s-a) \, \mathrm{d}a. \tag{9.152}$$

The result for n random variables follows immediately, by applying (9.144) $n-1$ times. ■

Example 9.35 (Coffee supply). A coffee shop in Manhattan has access to many coffee suppliers around the world, but the supply from each of them is volatile because it depends on demand and weather conditions at that location. In order to protect themselves from volatility, the cafe makes a deal with n suppliers. They will buy a quantity of coffee from each supplier equal to their individual available supply divided by n.

We model the coffee available from the ith supplier as a random variable $\tilde{c}_i$, $1 \leq i \leq n$, which is uniformly distributed between 0 and 1 ton. The suppliers are at very different locations, so we assume that the variables are independent. The total coffee, aggregated across all the suppliers, is equal to the sum of n i.i.d. uniform random variables,

$$\tilde{s}_n := \sum_{i=1}^{n} \tilde{c}_i. \tag{9.153}$$

The coffee purchased by the cafe is $\tilde{m}_n := \tilde{s}_n / n$.

If there are only two suppliers, by Theorem 9.34, the pdf of the sum is equal to the convolution between two uniform pdfs,

$$f_{\tilde{s}_2}(s) = \int_{c=-\infty}^{\infty} f_{\tilde{c}_1}(c) f_{\tilde{c}_2}(s-c) \, \mathrm{d}c \tag{9.154}$$

$$= \int_{c=0}^{1} f_{\tilde{c}_2}(s-c) \, \mathrm{d}c. \tag{9.155}$$

The pdf $f_{\tilde{c}_2}(s-c)$ is equal to one, if $0 \leq s-c \leq 1$, i.e. if $s-1 \leq c \leq s$, and zero otherwise. Consequently,

$$f_{\tilde{s}_2}(s) = \int_{c=\max\{0,s-1\}}^{\min\{1,s\}} f_{\tilde{c}_2}(s-c) \, \mathrm{d}c, \tag{9.156}$$

which yields

$$f_{\tilde{s}_2}(s) = \begin{cases} s & \text{for } 0 \le s \le 1, \\ 2 - s & \text{for } 1 \le s \le 2, \\ 0 & \text{otherwise}. \end{cases} \tag{9.157}$$

The pdf of the sum is triangular between 0 and 2. By Theorem 3.19, the pdf of the average is also triangular. It equals

$$f_{\tilde{m}_2}(m) = 2f_{\tilde{s}_2}(2m) = \begin{cases} 4m & \text{for } 0 \le m \le 0.5, \\ 4(1 - m) & \text{for } 0.5 \le m \le 1, \\ 0 & \text{otherwise}. \end{cases} \tag{9.158}$$

By Theorem 9.34, the pdf of the total coffee from n suppliers is

$$f_{\tilde{s}_n}(s) = f_{\tilde{c}_1} * f_{\tilde{c}_2} * \cdots * f_{\tilde{c}_n}(s), \tag{9.159}$$

and, by Theorem 3.19, the pdf of the average is

$$f_{\tilde{m}_n}(m) = n f_{\tilde{s}_n}(nm) \tag{9.160}$$

$$= n f_{\tilde{c}_1} * f_{\tilde{c}_2} * \cdots * f_{\tilde{c}_n}(nm). \tag{9.161}$$

In words, the pdf is obtained by convolving the uniform pdf with itself $n - 1$ times, and then scaling it. Figure 9.12 shows the pdf for different values of n. The convolutions gradually smooth the shape of the pdf until it resembles a Gaussian. To verify this, we compare it to a Gaussian with the same mean and variance, which equal

$$\mathbb{E}[\tilde{m}_n] = \frac{1}{n} \sum_{i=1}^{n} \mathbb{E}[\tilde{c}_i] = 0.5, \qquad \text{Var}[\tilde{m}_n] = \frac{1}{n^2} \sum_{i=1}^{n} \text{Var}[\tilde{c}_i] = \frac{1}{12n} \tag{9.162}$$

by linearity of expectation (Theorem 7.18), Lemma 7.9, Theorem 9.11, and Lemma 7.35. The Gaussian approximation, depicted in Figure 9.12, is very accurate.

· ·

In Examples 9.33 and 9.35, we observe that repeatedly summing independent random variables has a smoothing effect on their pmf or pdf that results in a Gaussian-like distribution. The following theorem provides an additional connection between the convolution operation and the Gaussian distribution. The convolution of two Gaussians pdfs is Gaussian. This means that if we sum independent Gaussian random variables together, the result is Gaussian.

Theorem 9.36 (Sum of independent Gaussian random variables). *Let $\tilde{a}_1$ and $\tilde{a}_2$ be two independent continuous random variables with Gaussian distributions belonging to the same probability space. Their sum $\tilde{s} := \tilde{a}_1 + \tilde{a}_2$ is Gaussian with mean $\mu_{\tilde{s}} := \mu_1 + \mu_2$ and variance $\sigma_{\tilde{s}}^2 := \sigma_1^2 + \sigma_2^2$, where μ_1 and σ_1^2 denote the mean and variance of $\tilde{a}_1$ and μ_2 and σ_2^2 denote the mean and variance of $\tilde{a}_2$.*

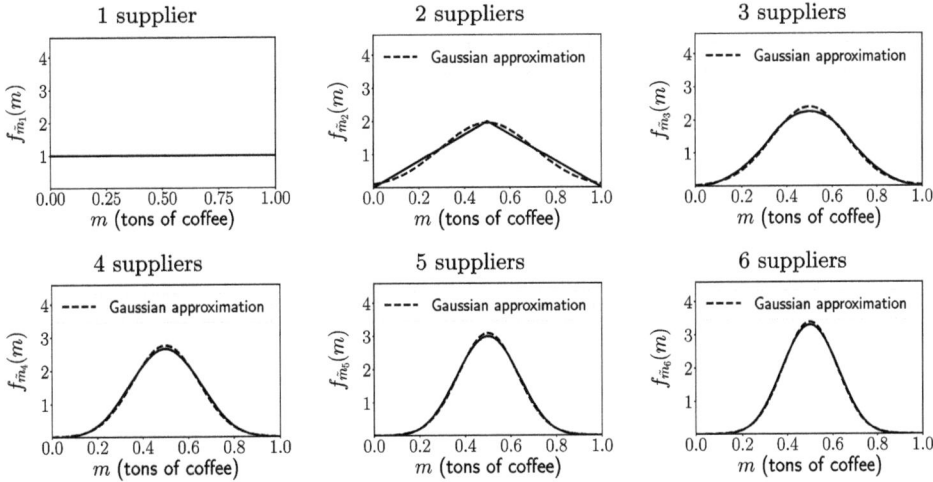

Figure 9.12 Coffee supply. The black curves in each graph represent the pdfs of the quantity of coffee purchased by the coffee shop in Example 9.35 for different numbers of suppliers. The dashed line depicts a Gaussian approximation, which converges to the pdf of the purchased coffee, as predicted by the central limit theorem (Theorem 9.37).

Proof By Theorem 9.34 and Definition 3.30, the pdf of $\tilde{s}$ equals

$$f_{\tilde{s}}(s) = \int_{a=-\infty}^{\infty} f_{\tilde{a}_1}(a) f_{\tilde{a}_2}(s-a) \, da \tag{9.163}$$

$$= \int_{a=-\infty}^{\infty} \frac{1}{\sqrt{2\pi}\sigma_1} \exp\left(-\frac{(a-\mu_1)^2}{2\sigma_1^2}\right) \frac{1}{\sqrt{2\pi}\sigma_2} \exp\left(-\frac{(s-a-\mu_2)^2}{2\sigma_2^2}\right) da$$

$$= \int_{a=-\infty}^{\infty} \frac{1}{2\pi\sigma_1\sigma_2} \exp\left(-\frac{1}{2}\left(\frac{(a-\mu_1)^2}{\sigma_1^2} + \frac{(s-a-\mu_2)^2}{\sigma_2^2}\right)\right) da. \tag{9.164}$$

Manipulating the expression in the exponent we obtain

$$\frac{(a-\mu_1)^2}{\sigma_1^2} + \frac{(s-a-\mu_2)^2}{\sigma_2^2} \tag{9.165}$$

$$= \frac{\sigma_2^2(a^2 + \mu_1^2 - 2\mu_1 a) + \sigma_1^2(s^2 + a^2 + \mu_2^2 - 2sa - 2s\mu_2 + 2a\mu_2)}{\sigma_1^2\sigma_2^2} \tag{9.166}$$

$$= \frac{\sigma_{\tilde{s}}^2 a^2 - 2(\sigma_2^2\mu_1 + \sigma_1^2(s-\mu_2))a + \sigma_2^2\mu_1^2 + \sigma_1^2(s^2 + \mu_2^2 - 2s\mu_2)}{\sigma_1^2\sigma_2^2} \tag{9.167}$$

$$= \frac{\sigma_{\tilde{s}}^2}{\sigma_1^2\sigma_2^2}\left(a^2 - 2ab + c\right) \tag{9.168}$$

$$= \frac{\sigma_{\tilde{s}}^2}{\sigma_1^2\sigma_2^2}\left((a-b)^2 - b^2 + c\right), \tag{9.169}$$

where

$$b := \frac{\sigma_2^2\mu_1 + \sigma_1^2(s-\mu_2)}{\sigma_{\tilde{s}}^2}, \tag{9.170}$$

$$c := \frac{\sigma_2^2\mu_1^2 + \sigma_1^2(s-\mu_2)^2}{\sigma_{\tilde{s}}^2}. \tag{9.171}$$

These operations are called *completing the square*. Some additional somewhat tedious algebraic manipulations yield

$$c - b^2 = \frac{\sigma_1^2 \sigma_2^2 (s - \mu_{\tilde{s}})^2}{\sigma_{\tilde{s}}^4}. \tag{9.172}$$

Combining (9.169), (9.164) and (9.172), we have

$$
\begin{aligned}
f_{\tilde{s}}(s) &= \int_{a=-\infty}^{\infty} \frac{1}{2\pi\sigma_1\sigma_2} \exp\left(-\frac{\sigma_{\tilde{s}}^2}{2\sigma_1^2\sigma_2^2} \left((a-b)^2 + \frac{\sigma_1^2\sigma_2^2(s-\mu_{\tilde{s}})^2}{\sigma_{\tilde{s}}^4} \right) \right) da \\
&= \frac{1}{\sqrt{2\pi}\sigma_{\tilde{s}}} \exp\left(-\frac{(s-\mu_{\tilde{s}})^2}{2\sigma_{\tilde{s}}^2} \right) \int_{a=-\infty}^{\infty} \frac{1}{\sqrt{2\pi}\frac{\sigma_1\sigma_2}{\sigma_{\tilde{s}}}} \exp\left(-\frac{(a-b)^2}{2\frac{\sigma_1^2\sigma_2^2}{\sigma_{\tilde{s}}^2}} \right) da \\
&= \frac{1}{\sqrt{2\pi}\sigma_{\tilde{s}}} \exp\left(-\frac{(s-\mu_{\tilde{s}})^2}{2\sigma_{\tilde{s}}^2} \right).
\end{aligned}
\tag{9.173}
$$

The integral is equal to one by Theorem 3.16 because the term inside is a Gaussian pdf with mean b and standard deviation $\sigma_1\sigma_2/\sigma_{\tilde{s}}$. ∎

9.7.2 Convergence to the Gaussian Distribution

In this section, we study the distribution of the sample mean of sequences of independent identically distributed (i.i.d.) random variables. Let $\tilde{x}_1, \tilde{x}_2, \ldots$ denote a sequence of i.i.d. random variables with mean μ and variance σ^2. The sample mean of the first n variables is

$$\tilde{m}_n := \frac{1}{n} \sum_{j=1}^{n} \tilde{x}_j. \tag{9.174}$$

As derived in the proof of Theorem 9.22, the mean and variance of the sample mean equal

$$\mathbb{E}\left[\tilde{m}_n\right] = \mu, \qquad \operatorname{Var}\left[\tilde{m}_n\right] = \frac{\sigma^2}{n}, \tag{9.175}$$

so the sample mean is an unbiased estimator of the true mean μ, and its variance decays linearly with n. Consequently, the sample mean converges to μ with high probability, as stated in the law of large numbers. However, this does not provide a precise characterization of the *distribution of the sample mean*. For example, we may be interested in the probability $P\left(|\tilde{m}_n - \mu| > \epsilon\right)$ that $\tilde{m}_n$ deviates from μ by some constant ϵ for a fixed value of n. As illustrated in Example 9.26, our bounds based on Chebyshev's inequality provide a terrible estimate of this probability.

Empirically, in Examples 9.33 and 9.35, we observe that the distribution of sums of independent random variables is close to being Gaussian. This suggests approximating $\tilde{m}_n$ using a Gaussian random variable with mean μ and variance σ^2/n. By Theorem 3.31, this is equivalent to approximating the standardized sample mean

$$s(\tilde{m}_n) := \frac{\tilde{m}_n - \mu}{\frac{\sigma}{\sqrt{n}}} \tag{9.176}$$

using a standard Gaussian with zero mean and unit variance. The central limit theorem establishes that the approximation becomes arbitrarily accurate as n tends to infinity.

Theorem 9.37 (Central limit theorem). *Let $\tilde{x}_1$, $\tilde{x}_2$, ... be a countably infinite sequence of independent identically distributed random variables with mean μ and variance σ^2 belonging to the same probability space. The standardized sample mean*

$$s(\widetilde{m}_n) := \frac{\widetilde{m}_n - \mu}{\frac{\sigma}{\sqrt{n}}}, \tag{9.177}$$

where $\widetilde{m}_n := \frac{1}{n}\sum_{j=1}^{n} \tilde{x}_j$, converges in distribution to a standard Gaussian random variable with zero mean and unit variance, in the sense that the cdf of $F_{s(\widetilde{m}_n)}$ converges to the cdf of a standard Gaussian as $n \to \infty$.

Proof The proof of the central limit theorem is beyond the scope of this book, as it requires introducing advanced concepts from probability theory and functional analysis. We refer to Chapter 3 in Durrett (2019) for the proof and the necessary mathematical background. ∎

It is important to emphasize that the central limit theorem states that the sample mean *converges in distribution*. In contrast to convergence in mean square or convergence in probability (see Corollary 9.15 and Theorem 9.22), this does not mean that the sample mean becomes arbitrarily close to a certain value. It means that its probabilistic behavior is increasingly well approximated as following a Gaussian distribution. Even though the theorem is asymptotic, the Gaussian approximation is often very accurate even for small n, as illustrated in Examples 9.33 and 9.35. Motivated by this, we define a Gaussian approximation to the sample mean inspired by the central limit theorem.

Definition 9.38 (Gaussian approximation to the sample mean). *Given n independent identically distributed random variables $\tilde{x}_1$, $\tilde{x}_2$, ..., $\tilde{x}_n$ with mean μ and variance σ^2, the Gaussian approximation of the sample mean*

$$\tilde{m} := \frac{1}{n}\sum_{i=1}^{n} \tilde{x}_i \tag{9.178}$$

is a Gaussian random variable with mean μ and variance $\frac{\sigma^2}{n}$.

A common application of the Gaussian approximation to the sample mean is to approximate the binomial distribution.

Definition 9.39 (Gaussian approximation to the binomial distribution). *Recall that a binomial random variable with parameters n and θ can be represented as the sum of n independent Bernoulli random variables $\tilde{b}_1$, ..., $\tilde{b}_n$ with parameter θ,*

$$\tilde{a} = \sum_{i=1}^{n} \tilde{b}_i, \tag{9.179}$$

as explained in the proof of Lemma 7.19. By Lemmas 7.2 and 7.34, the mean and variance of each Bernoulli are θ and $\theta(1 - \theta)$, respectively. Consequently, by the Gaussian approximation of the sample mean in Definition 9.38, we can approximate $\tilde{a}/n$ as a Gaussian random variable with mean θ and variance $\theta(1 - \theta)/n$. Equivalently, by Theorem 3.31, we can approximate $\tilde{a}$ as a Gaussian random variable with mean $n\theta$ and variance $n\theta(1 - \theta)$.

Example 9.40 (Basketball strategy). Logan is a data scientist working for a basketball team. He wants to convince the coach to increase the proportion of three-point shots, following the

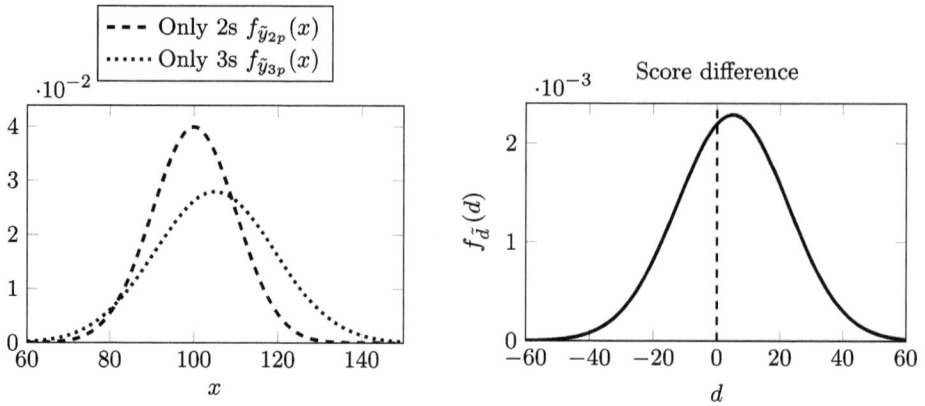

Figure 9.13 Basketball strategy. The graph on the left shows the Gaussian approximations derived in Example 9.40 for the distribution of points scored by a team that only attempts 2-point ($\tilde{y}_{2p}$) or 3-point ($\tilde{y}_{2p}$) shots. The graph on the right shows the Gaussian approximation to the score difference between the teams.

trend of NBA teams in recent years. To this end, he builds a probabilistic model of a game between two identical teams, where one only takes 2-point shots (*Strategy 2p*) and the other only takes 3-point shots (*Strategy 3p*).

From past data, Logan determines that the probability of making a 2-point and a 3-point shot are $\theta_2 := 0.5$ and $\theta_3 := 0.35$, respectively. In addition, he assumes that the different shots in a game are independent and that there are 100 possessions in a game. Under these assumptions, the number of shots made when following Strategy 2p can be modeled as a binomial random variable $\tilde{x}_{2p}$ with parameters $n := 100$ and $\theta_2 := 0.5$ (see Section 2.3.2). Similarly, the number of shots made following Strategy 3p can be modeled as a binomial random variable $\tilde{x}_{3p}$ with parameters $n := 100$ and $\theta_3 := 0.35$. Logan is interested in the distribution of the score difference $\tilde{d} := 3\tilde{x}_{3p} - 2\tilde{x}_{2p}$, and, in particular, in the probability that it is positive or negative, as this determines the outcome of the game. Deriving the exact pmf of $\tilde{d}$ is quite complicated. Instead, he approximates it based on the central limit theorem.

By Definition 9.39 $\tilde{x}_{2p}$ can be approximated as a Gaussian with mean $100\,\theta_2$ and variance $100\,\theta_2(1 - \theta_2)$. By Theorem 3.31, if $\tilde{x}_{2p}$ is Gaussian, then the random variable $\tilde{y}_{2p} := 2\tilde{x}_{2p}$, which represents the score for Strategy 2p, is Gaussian with mean $200\,\theta_2 = 100$ and variance $400\,\theta_2(1-\theta_2) = 100$. By the same reasoning, $\tilde{x}_{3p}$ is approximately Gaussian with mean $100\,\theta_3$ and variance $100\,\theta_3(1 - \theta_3)$. Consequently, the random variable $\tilde{y}_{3p} := 3\tilde{x}_{3p}$, representing the score for Strategy 3p, is approximately Gaussian with mean $300\,\theta_3 = 105$ and variance $900\,\theta_3(1 - \theta_3) = 204.75$.

The left graph in Figure 9.13 compares the approximate distributions of $\tilde{y}_{2p}$ and $\tilde{y}_{3p}$. Attempting 3-point shots results in a higher mean and a higher variance. By Theorem 9.36, if $\tilde{y}_{2p}$ and $\tilde{y}_{3p}$ followed their approximate Gaussian distributions exactly, then the score difference $\tilde{d} := \tilde{y}_{3p} - \tilde{y}_{2p}$ would be Gaussian with mean $105 - 100 = 5$ and variance $204.75 + 100 = 304.75$ (depicted on the right in Figure 9.13). By Theorem 3.31 we can express such a Gaussian as $\sqrt{304.75}\tilde{z} + 5$, where $\tilde{z}$ is a standard Gaussian with zero mean and unit variance. The probability that Strategy 3p beats Strategy 2p can therefore be approximated as

$$P(\text{Strategy 3p beats Strategy 2p}) \approx P\left(\sqrt{304.75}\tilde{z} + 5 > 0\right) \tag{9.180}$$

$$= P\left(\tilde{z} > -0.2864\right) = 0.613. \tag{9.181}$$

Just shooting 3 pointers wins approximately 60% of the time. To evaluate the approximation, Logan leverages the Monte Carlo method described in Section 1.7. He simulates one million games between a team following Strategy 2p and a team following Strategy 3p. Strategy 3p indeed wins close to 60% of the games (599,790), as predicted by his analysis. This example illustrates the usefulness of the central limit theorem for back-of-the-envelope calculations.

. . . .

The central limit theorem implies that the sample mean of random samples converges in distribution to a Gaussian random variable, as long as the samples are measured independently and uniformly at random. The mean of the Gaussian variable is the population mean and its standard deviation is equal to the standard error (see Definition 9.9).

Corollary 9.41 (Distribution of the sample mean). *Let a_1, a_2, ..., a_N denote a dataset of size N with population mean μ_{pop} and population variance σ^2_{pop}, defined as in (9.10) and (9.20), respectively. Let $\tilde{x}_1$, $\tilde{x}_2$, ..., be a sequence of independent, uniform random samples following Definition 9.3. We define the standardized sample mean as*

$$s(\tilde{m}_n) := \frac{\tilde{m}_n - \mu_{\text{pop}}}{\text{se}\left[\tilde{m}_n\right]}, \tag{9.182}$$

where $\tilde{m}_n := \frac{1}{n}\sum_{j=1}^{n}\tilde{x}_j$ and $\text{se}\left[\tilde{m}\right] = \sigma_{\text{pop}}/\sqrt{n}$ denotes the standard error (see Theorem 9.13). As $n \to \infty$, the standardized sample mean converges in distribution to a standard Gaussian random variable with zero mean and unit variance.

By Theorem 3.31, this suggests that the distribution of the sample mean $\tilde{m}_n$ can be approximated as being Gaussian with mean μ_{pop} and standard deviation $\text{se}\left[\tilde{m}\right] = \sigma_{\text{pop}}/\sqrt{n}$.

Proof The random variables $\tilde{x}_1$, $\tilde{x}_2$, ... are i.i.d. with mean μ_{pop} and variance σ^2_{pop} by (9.15) and (9.37), so the result follows directly from the central limit theorem (Theorem 9.37). ∎

Figure 9.14 shows histograms of the standardized sample mean $s(\tilde{m})$ of n independent, uniform samples from the dataset in Example 9.1 for different values of n. The Gaussian approximation provided by Corollary 9.41 is very accurate even for small values of n. The figure also shows histograms of the standardized sample proportion of positive tests among n individuals chosen independently and uniformly at random in Example 9.2. Corollary 9.41 holds for this setting because the sample proportion can be interpreted as a sample mean (see Example 9.4). We observe that the Gaussian approximation improves as we increase n and is already very accurate for $n := 10^4$.

9.7.3 The Financial Crisis and the Central Limit Theorem: How Not to Estimate Risk

The Gaussian approximation provided by the central limit theorem is widely used to model averages. However, we must not forget that the approximation assumes independence between the averaged quantities. In this section, we illustrate the dangers of blindly relying on Gaussian approximations, when this assumption does not hold.

Estimation of the mean height in a population

Estimation of the prevalence of COVID-19 in New York City

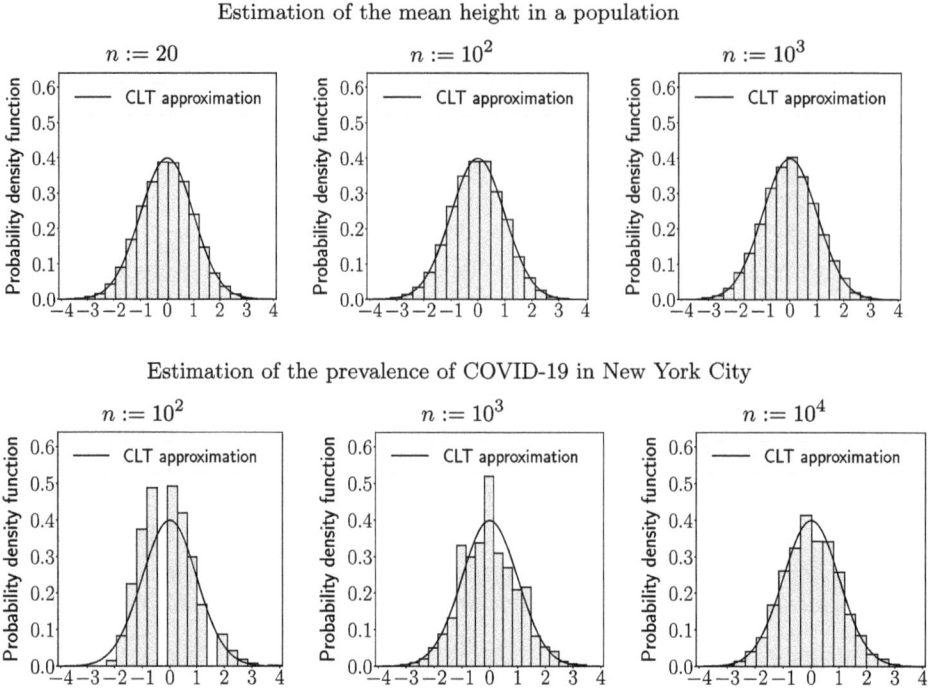

Figure 9.14 Distribution of the sample mean. The top row shows histograms of the standardized sample mean, defined in Corollary 9.41, of n independent, uniform samples from the dataset in Example 9.1 for different values of n. The bottom row shows histograms of the standardized sample proportion computed from n independent measurements in the scenario described in Example 9.2, also for different values of n. Each histogram is computed using 10^4 sample means or proportions. A Gaussian approximation based on the central limit theorem is shown superposed on each histogram. As n increases, the approximation becomes increasingly accurate, as predicted by Corollary 9.41.

We focus on a fictitious example inspired by the 2008 financial crisis. Our goal is to evaluate the risk of a collateralized debt obligation (CDO), a financial instrument that gained notorious fame due to the 2008 financial crisis. The CDO consists of a pool of n subprime mortgages from borrowers with a poor credit history (this is known as a collateralized mortgage obligation). The probability that each of the borrowers defaults and does not pay back their debt is 2/3.

In order to alleviate the risk, the CDO is divided into 10 different sections called *tranches*. When borrowers default, the tranches suffer losses sequentially. If 10% or less borrowers default, then only the first tranche is affected. If between 10% and 20% default, then the first and second tranches lose money. The last tranche, which is known as the *senior* tranche, is the most protected; more than 90% of the borrowers must default for it to suffer losses.

In the remainder of this section, we evaluate the risk of the senior tranche under two different modeling assumptions. Our analysis illustrates the dramatic consequences of inadequate independence assumptions in risk estimation.

Assumption 1: Borrowers Default Independently

Let us assume that the borrowers default independently. In that case, the number of defaults is distributed as a binomial random variable $\tilde{d}_1$ with parameters n and $\theta := 2/3$ (see

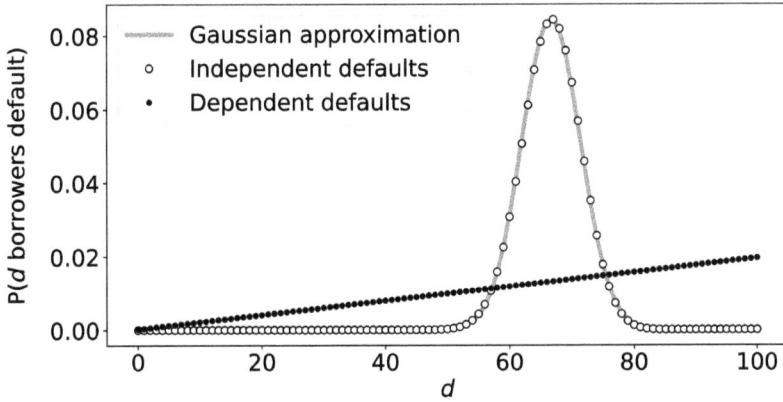

Figure 9.15 Effect of independence assumptions on risk estimates. The figure compares the estimated distributions of the number of defaults in a pool of 100 subprime mortgages for the two scenarios described in Section 9.7.3. The white markers depict the pmf of the number of borrowers that default under Assumption 1 (the borrowers default independently). The pmf is well approximated by a Gaussian (gray curve). Under this scenario, it is extremely unlikely that more than 80% of the borrowers default. The black markers depict the pmf of the number of borrowers that default under Assumption 2 (the default risk depends on a common latent variable representing economic context). Due to the dependence between borrowers, the probability that many of them default simultaneously is much higher than under Assumption 1.

Section 2.3.2). By Definition 9.39 and Theorem 3.31, the probability that more than 90% of the borrowers default can be approximated as

$$P\left(\tilde{d}_1 > 0.9n\right) = P\left(\frac{\tilde{d}_1 - \theta n}{\sqrt{\theta(1-\theta)n}} > \frac{0.9n - \theta n}{\sqrt{\theta(1-\theta)n}}\right) \tag{9.183}$$

$$\approx P\left(\tilde{z} > 0.49\sqrt{n}\right), \tag{9.184}$$

where $\tilde{z}$ is a Gaussian random variable with zero mean and unit variance. The probability that the senior tranche suffers losses decreases as we increase n. By selecting n large enough, we can make it into a low-risk investment. For instance, if there are $n := 100$ mortgages in the CDO, the risk of losing money from the senior tranche is essentially zero (less than 10^{-6})! Figure 9.15 shows the pmf of $\tilde{d}_1$ and the corresponding Gaussian approximation for $n := 100$.

Assumption 2: Borrowers Default Depending On Economic Context

We now assume that the probability of default depends on the state of the economy. We define a random variable $\tilde{r}$ that represents to what extent the economy is in recession: $\tilde{r} = 0$ indicates a strong economy, whereas $\tilde{r} = 1$ corresponds to economic disaster. The pdf of $\tilde{r}$ is equal to

$$f_{\tilde{r}}(r) := 2r, \qquad 0 \leq r \leq 1, \tag{9.185}$$

and zero otherwise. Conditioned on $\tilde{r} = r$, each borrower is modeled as a Bernoulli random variable $\tilde{b}_i$ with parameter r, where $\tilde{b}_i = 1$ indicates that the ith borrower defaults. Consequently, the conditional probability of default is equal to r, which makes sense since higher values of $\tilde{r}$ correspond to a worse economic context. By Theorem 6.5, the probability that each individual borrower defaults equals

$$p_{\tilde{b}_i}(1) = \int_{r=-\infty}^{\infty} p_{\tilde{b}_i \mid \tilde{r}}(1 \mid r) f_{\tilde{r}}(r) \, dr \tag{9.186}$$

$$= \int_{r=0}^{1} 2r^2 \, dr = \frac{2}{3}, \tag{9.187}$$

which is *exactly the same* as under Assumption 1.

We assume that there is no additional dependence between the borrowers, so that $\tilde{b}_1, \ldots, \tilde{b}_n$ are conditionally independent given $\tilde{r}$. As a result, conditioned on $\tilde{r} = r$, the number of defaults $\tilde{d}_2$ has a binomial distribution with parameters n and r. By Theorem 6.5, its pmf equals

$$p_{\tilde{d}_2}(d) = \int_{r=-\infty}^{\infty} f_{\tilde{r}}(r) p_{\tilde{d}_2 \mid \tilde{r}}(d \mid r) \, dr \tag{9.188}$$

$$= \int_{r=0}^{1} 2r \binom{n}{d} r^d (1-r)^{n-d} \, dr \tag{9.189}$$

$$= 2 \binom{n}{d} \int_{r=0}^{1} r^{d+1} (1-r)^{n-d} \, dr \tag{9.190}$$

$$= \frac{2n!}{d!(n-d)!} \frac{(d+1)!(n-d)!}{(n+2)!} \tag{9.191}$$

$$= \frac{2(d+1)}{(n+1)(n+2)}. \tag{9.192}$$

Figure 9.15 shows the pmf of $\tilde{d}_2$ for $n := 100$. It does not resemble a Gaussian at all! Note that this does *not* contradict the central limit theorem (Theorem 9.37) because the borrowers do not default independently.

By Theorem 2.5, the probability that more than 90% of the borrowers default under Assumption 2 is equal to

$$P\left(\tilde{d}_2 > 0.9n\right) = \sum_{d=0.9n+1}^{n} p_{\tilde{d}_2}(d) \tag{9.193}$$

$$= \frac{2}{(n+1)(n+2)} \left(0.1n + \sum_{d=0.9n+1}^{n} d\right) \tag{9.194}$$

$$= \frac{0.2n + 0.1n(1.9n+1)}{(n+1)(n+2)}, \tag{9.195}$$

which follows from the arithmetic series formula $\sum_{d=a}^{b} d = \frac{(a+b)(b-a+1)}{2}$. For $n := 100$, the probability is 0.187. This is dramatically higher than under Assumption 1. In this scenario, when the economy is in recession, many borrowers are likely to default *simultaneously*.

Under Assumption 1, we can reduce the risk of the senior tranche by including more mortgages in the CDO. Under Assumption 2, this is not possible: The limit of the probability in (9.195) does not converge to zero as n tends to infinity. The limit is 0.19. No matter how many mortgages we pool together, the senior tranche remains a risky investment. This is a cartoon example: The actual models used to evaluate the risk of CDOs prior to the

2008 financial crisis did incorporate dependence. However, they definitely underestimated the risk of CDOs based on subprime mortgages, which suffered enormous losses despite being considered low-risk investments.

9.8 Confidence Intervals

Estimates of population parameters computed from random samples are *uncertain*, as illustrated in Figures 9.1 and 9.2. In practice, it is important to quantify this uncertainty, which can be achieved by providing a *confidence interval* that contains the parameter with high probability, instead of a single point estimate. In Section 9.8.1, we explain how to build confidence intervals for the mean. Section 9.8.2 describes confidence intervals for proportions and probabilities. Finally, Section 9.8.3 discusses how to interpret confidence intervals, and cautions against applying them when the available data are not sampled independently.

9.8.1 Confidence Interval for the Mean

As explained in Section 9.1, estimates obtained from random samples are uncertain: If we repeat our measurements, we obtain a different estimate. Confidence intervals allow us to quantify this uncertainty by providing a range of values that contain the population parameter with high probability. In this section, we show how to construct confidence intervals for the population mean.

We consider the problem of estimating the population mean μ_{pop} of a population a_1, a_2, ..., a_N from random samples. The available measurements are modeled as n i.i.d. random variables $\tilde{x}_1, \tilde{x}_2, \ldots, \tilde{x}_n$ following Definition 9.3. Our estimator is the sample mean $\tilde{m} := \sum_{j=1}^{n} \tilde{x}_j$. In order to quantify the uncertainty associated with this estimator, we build an interval around $\tilde{m}$ that contains μ_{pop} with high probability.

By Corollary 9.41 $\tilde{m}$ is approximately Gaussian with mean μ_{pop}. Consequently, we can select a constant c such that $\tilde{m}$ belongs to the interval $[\mu_{\text{pop}} - c, \mu_{\text{pop}} + c]$ with high probability. This deterministic interval is depicted at the top of Figure 9.16. The key insight is that if $\tilde{m}$ is in $[\mu_{\text{pop}} - c, \mu_{\text{pop}} + c]$, then this implies that the population mean μ_{pop} is in $[\tilde{m} - c, \tilde{m} + c]$. Therefore, $[\tilde{m} - c, \tilde{m} + c]$ is a confidence interval for μ_{pop}! Since the sample mean is a random variable, the confidence interval is random. In practice, we compute a realization of the confidence interval from the available data, as illustrated in Figure 9.16. The following lemma derives the value of the constant c, as a function of the probability associated with the confidence interval, for any estimator that has a Gaussian distribution.

Lemma 9.42 (Confidence interval for unbiased Gaussian estimators). *Let $\tilde{g}$ be an unbiased estimator of a parameter γ with standard error* se. *If the distribution of $\tilde{g}$ is Gaussian, for any $\alpha \in (0, 1)$, the random interval*

$$\widetilde{\mathcal{I}}_{1-\alpha} := [\tilde{g} - c_\alpha \, \text{se}, \tilde{g} + c_\alpha \, \text{se}], \qquad c_\alpha := F_{\tilde{z}}^{-1}\left(1 - \frac{\alpha}{2}\right), \qquad (9.196)$$

is a 1-α confidence interval for γ, in the sense that

$$\text{P}\left(\gamma \in \widetilde{\mathcal{I}}_{1-\alpha}\right) = 1 - \alpha. \qquad (9.197)$$

In particular,

$$\widetilde{\mathcal{I}}_{0.95} := [\tilde{g} - 1.96 \, \mathrm{se}, \tilde{g} + 1.96 \, \mathrm{se}] \tag{9.198}$$

is a 0.95 confidence interval for γ. Here, $F_{\tilde{z}}$ denotes the cdf of a standard Gaussian random variable $\tilde{z}$ with zero mean and unit variance.

Proof The mean of $\tilde{g}$ is γ because it is unbiased (see Definition 9.5), and its standard deviation equals se (see Definition 9.9). Since $\tilde{g}$ is Gaussian, by Theorem 3.31 $\tilde{z} := (\tilde{g} - \gamma)/\,$se is Gaussian with zero mean and unit variance. The event $\gamma \in \widetilde{\mathcal{I}}_{1-\alpha}$ is the complement of the union of the events $\tilde{g} - c_\alpha \, \mathrm{se} > \gamma$ and $\tilde{g} + c_\alpha \, \mathrm{se} < \gamma$, which are disjoint, so by Lemma 1.11 and Axiom 3 in Definition 1.9,

$$P\left(\mu \in \widetilde{\mathcal{I}}_{1-\alpha}\right) = 1 - P\left(\tilde{g} - c_\alpha \, \mathrm{se} > \gamma\right) - P\left(\tilde{g} + c_\alpha \, \mathrm{se} < \gamma\right) \tag{9.199}$$

$$= 1 - P\left(\frac{\tilde{g} - \gamma}{\mathrm{se}} > c_\alpha\right) - P\left(\frac{\tilde{g} - \gamma}{\mathrm{se}} < -c_\alpha\right) \tag{9.200}$$

$$= 1 - P\left(\tilde{z} > c_\alpha\right) - P\left(\tilde{z} < -c_\alpha\right) \tag{9.201}$$

$$= 1 - 2P\left(\tilde{z} > c_\alpha\right), \tag{9.202}$$

where the last step follows from the symmetry of the Gaussian pdf. By the definition of c_α and Definition 3.3,

$$P\left(\tilde{z} > c_\alpha\right) = 1 - P\left(\tilde{z} \leq c_\alpha\right) \tag{9.203}$$

$$= 1 - F_{\tilde{z}}(c_\alpha) \tag{9.204}$$

$$= \frac{\alpha}{2}. \tag{9.205}$$

Plugging this into (9.202) yields $P(\mu \in \widetilde{\mathcal{I}}_{1-\alpha}) = 1 - \alpha$. The expression for $\widetilde{\mathcal{I}}_{0.95}$ follows from the fact that $c_{0.05} = 1.96$ because $F_{\tilde{z}}(1.96) = 1 - 0.05/2$. ∎

Armed with Lemma 9.42, we can build confidence intervals for the mean based on the Gaussian approximations of the sample mean in Definition 9.38 and Corollary 9.41.

Definition 9.43 (Approximate confidence interval for the mean). *Let $\tilde{x}_1$, $\tilde{x}_2$, ..., $\tilde{x}_n$ be n independent identically distributed random variables with mean μ and variance σ^2. By Definition 9.38 the sample mean $\tilde{m} := \frac{1}{n} \sum_{j=1}^{n} \tilde{x}_j$ can be approximated as an unbiased Gaussian estimator with mean μ and standard error $\sigma/\sqrt{n}$. For any $\alpha \in (0, 1)$, the random interval*

$$\widetilde{\mathcal{I}}_{1-\alpha} := \left[\tilde{m} - \frac{c_\alpha \sigma}{\sqrt{n}}, \tilde{m} + \frac{c_\alpha \sigma}{\sqrt{n}}\right], \qquad c_\alpha := F_{\tilde{z}}^{-1}\left(1 - \frac{\alpha}{2}\right), \tag{9.206}$$

is an approximate 1-α confidence interval for μ based on Lemma 9.42, where $F_{\tilde{z}}$ denotes the cdf of a standard Gaussian with zero mean and unit variance.

If $\tilde{x}_1$, $\tilde{x}_2$, ..., $\tilde{x}_n$ are independent, uniform random samples from a dataset $\{a_1, a_2, \ldots, a_N\}$ with population mean μ_{pop} and population variance σ_{pop}^2 (defined as in (9.10) and (9.20), respectively), by Corollary 9.41 the sample mean $\tilde{m}$ can be approximated as an unbiased Gaussian estimator of μ_{pop} with standard error $\sigma_{\mathrm{pop}}/\sqrt{n}$. The random interval

$$\widetilde{\mathcal{I}}_{1-\alpha} := \left[\tilde{m} - \frac{c_\alpha \sigma_{\text{pop}}}{\sqrt{n}}, \tilde{m} + \frac{c_\alpha \sigma_{\text{pop}}}{\sqrt{n}} \right] \tag{9.207}$$

is an approximate 1-α confidence interval for μ_{pop} inspired by Lemma 9.42.

The careful reader may have realized that the standard error required to compute the confidence intervals in Definition 9.43 is a function of the population variance, which is unknown! To build confidence intervals in practice, we can replace the population variance by an estimate or upper bound. A reasonable choice is the sample variance, which is a consistent estimator of the population variance by Theorem 9.25 and typically results in an accurate approximation, unless the sample size is very small. Figure 9.16 shows approximate 0.95 confidence intervals for the population mean in Example 9.1 computed using the sample variance for two different sample sizes. The intervals are narrower when the sample size is

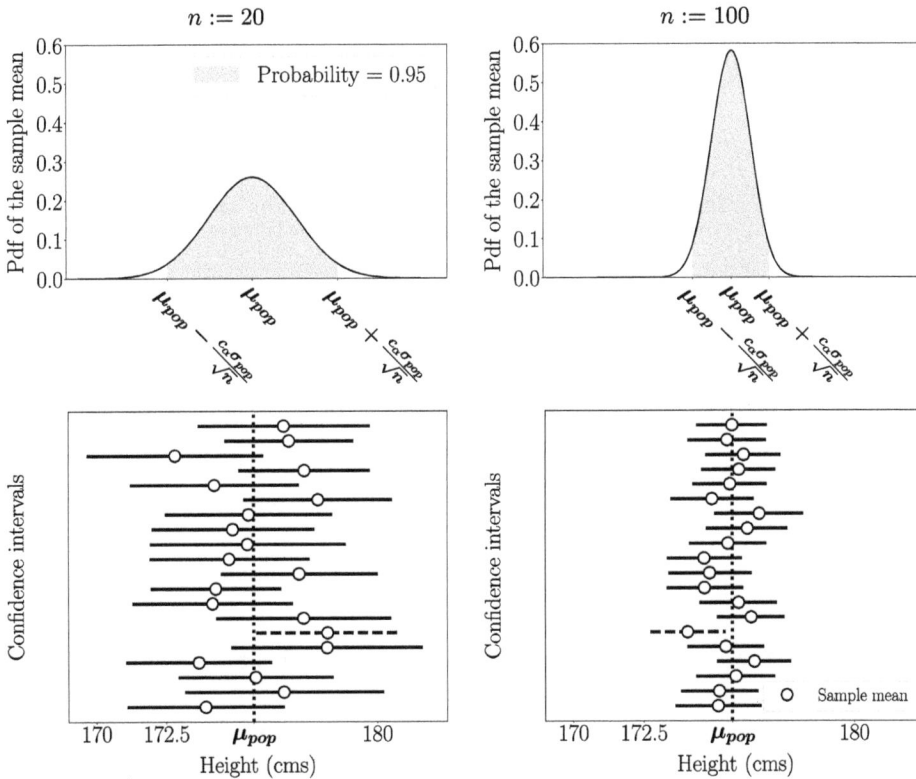

Figure 9.16 Confidence intervals based on the central limit theorem. The top row depicts the Gaussian approximation of the sample mean in Corollary 9.41 applied to the dataset in Example 9.1 for two different values of the sample size n. The sample mean belongs to the shaded deterministic interval with probability 0.95 for both values of n. The bottom row shows 0.95 confidence intervals computed from 20 different datasets of n random samples following Definition 9.43. Each interval is depicted as a black line centered at the corresponding sample mean, indicated by a white marker. The intervals have approximately the same width as the shaded region, but not exactly the same, because they are computed based on the sample variance of the samples, not the true population variance. Ninety-five percent (38/40) of the intervals contain the population mean (indicated by the dotted line). Those that do not are depicted by dashed lines.

larger, reflecting the decrease in uncertainty when we compute the sample mean averaging over more data.

To evaluate our confidence intervals, we can estimate their *coverage probability*, which is the probability that they contain the population mean. By the definition of confidence intervals, the coverage probability should be close to 0.95. In this case, the empirical coverage probability is exactly 0.95 because 95% of the intervals (38/40) contain the population mean (see Definition 1.22).

9.8.2 Confidence Intervals for Probabilities and Proportions

In this section we explain how to build confidence intervals for probabilities and proportions.

Definition 9.44 (Approximate confidence interval for a probability). *Let $\tilde{b}_1, \tilde{b}_2, \ldots, \tilde{b}_n$ be n independent Bernoulli random variables with parameter θ, and let $\tilde{m} := \frac{1}{n}\sum_{j=1}^{n} \tilde{b}_j$ be the proportion of these variables that are equal to one. For any $\alpha \in (0, 1)$, the random interval*

$$\widetilde{\mathcal{I}}_{1-\alpha} := \left[\tilde{m} - c_\alpha\sqrt{\frac{\theta(1-\theta)}{n}}, \tilde{m} + c_\alpha\sqrt{\frac{\theta(1-\theta)}{n}}\right] \tag{9.208}$$

is a 1-α confidence interval for θ, where $c_\alpha := F_{\tilde{z}}^{-1}\left(1 - \frac{\alpha}{2}\right)$ and $F_{\tilde{z}}$ denotes the cdf of a standard Gaussian with zero mean and unit variance. This interval is included in the following wider confidence interval, which does not depend on θ:

$$\widetilde{\mathcal{I}}_{1-\alpha} \subset \left[\tilde{m} - \frac{0.5c_\alpha}{\sqrt{n}}, \tilde{m} + \frac{0.5c_\alpha}{\sqrt{n}}\right]. \tag{9.209}$$

In particular,

$$\widetilde{\mathcal{I}}_{0.95} \subset \left[\tilde{m} - \frac{0.98}{\sqrt{n}}, \tilde{m} + \frac{0.98}{\sqrt{n}}\right]. \tag{9.210}$$

Derivation   The mean of each Bernoulli random variable $\tilde{b}_i$ is θ by Lemma 7.2 and the variance equals $\theta(1 - \theta)$ by Lemma 7.34. The confidence interval $\widetilde{\mathcal{I}}_{1-\alpha}$ is obtained by applying Definition 9.43 with $\tilde{x}_j := \tilde{b}_j, 1 \leq j \leq n$. The wider interval (9.209) is obtained from the bound $\theta(1 - \theta) \leq 0.25$, which holds because the maximum of $h(\theta) := \theta(1 - \theta)$ equals 0.25.[3]   ■

Example 9.45 (Prevalence of COVID-19: Sample size). In this example, we explain how to determine the number of tests needed to obtain an accurate estimate of the prevalence θ_{pop} in Example 9.2. We require that the error be no more than 1% with probability 0.95. Equivalently, there should exist a 0.95 confidence interval $\widetilde{\mathcal{I}}_{0.95}$ for θ_{pop}, such that the half-width of the interval is 0.01. If $\theta_{\text{pop}} \in \widetilde{\mathcal{I}}_{0.95}$ then the error is smaller than or equal to the half-width, so this indeed implies the desired accuracy with probability

$$\mathrm{P}\left(\theta_{\text{pop}} \in \widetilde{\mathcal{I}}_{0.95}\right) \approx 0.95. \tag{9.211}$$

Assuming that we test n individuals selected independently and uniformly at random from the population, the test results are independent Bernoulli random variables with parameter

---

[3]   The second derivative of $h(\theta)$ is -2, so the function is strictly concave. We can therefore set the first derivative $1 - 2\theta$ equal to zero to find the maximum.

θ_{pop}, as explained in Example 9.4. Therefore, by (9.210), the half-width of the approximate 0.95 confidence interval is smaller than 0.01 if

$$\frac{0.98}{\sqrt{n}} \leq 0.01, \tag{9.212}$$

which yields the following bound on n:

$$n \geq 9604 = \frac{0.98^2}{0.01^2}. \tag{9.213}$$

We need to test less than 10,000 people in a population of 8 million, which showcases the power of random sampling!

Our result is actually an overestimate, unless the prevalence is close to 50%. Otherwise, the bound $\theta_{\text{pop}}(1 - \theta_{\text{pop}}) \leq 0.25$ in Definition 9.44 is quite conservative. If $\theta_{\text{pop}} := 0.05$, plugging the true prevalence into (9.208) reduces the bound to 1,825 tests. Let us verify that this is precise. We make an error of less than 1%, if the sample proportion is between 4% and 6%. If we carry out 1,825 tests, this occurs when the number of positive tests is between 73 (4%) and 110 (5%). Under our assumptions, as explained in Example 9.4, each test can be modeled as a Bernoulli random variable with parameter equal to the population proportion $\theta_{\text{pop}} := 0.05$, so the total number of positive tests $\tilde{t}$ is binomial with parameters $n := 1825$ and $\theta_{\text{pop}} := 0.05$ (see Section 2.3.2). Consequently, by Definition 2.17 and Theorem 2.5, the probability that we make an error of less than 1% is

$$P(73 \leq \tilde{t} \leq 110) = \sum_{t=73}^{110} p_{\tilde{t}}(t) \tag{9.214}$$

$$= \sum_{t=73}^{110} \binom{1825}{t} 0.05^t \, 0.95^{1825-t} \tag{9.215}$$

$$= 0.959\%, \tag{9.216}$$

which is very close to 0.95.

. . . .

An important practical application of confidence intervals is to quantify the uncertainty in probability estimates obtained via the Monte Carlo method described in Section 1.7.

Example 9.46 (Confidence intervals for the Monte Carlo method). Confidence intervals allow us to perform uncertainty quantification when applying the Monte Carlo method in Definition 1.35. Let $\theta := P(A)$ be the true probability of an event of interest A, and let n be the number of Monte Carlo simulations. We assume that each simulation produces an outcome, which belongs to A with probability θ, independently from the other simulations. We define n independent Bernoulli random variables, which indicate whether the outcome of each simulation is in A. Applying Definition 9.44 to these Bernoulli random variables yields the approximate $1 - \alpha$ confidence interval

$$\left[P_{\text{MC}}(A) - \frac{0.5c_\alpha}{\sqrt{n}}, P_{\text{MC}}(A) + \frac{0.5c_\alpha}{\sqrt{n}} \right], \tag{9.217}$$

where the probability estimate $P_{\text{MC}}(A)$ is the proportion of simulated outcomes in A.

Figure 9.17 shows 0.95 confidence intervals for the probability that Serbia and Latvia win the gold medal in the 3x3 basketball tournament at the Tokyo Olympics, according to the model described in Example 1.36. For $n := 10^3$ the probability estimate for Latvia is higher

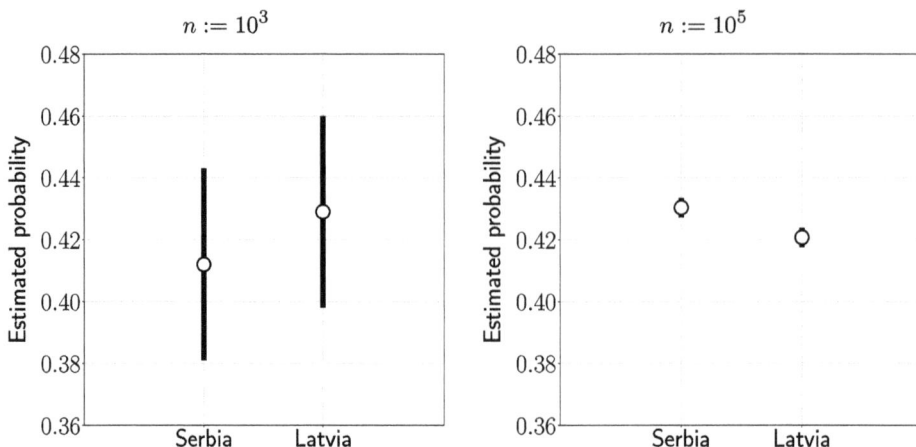

Figure 9.17 Confidence intervals for Monte Carlo simulations. Probability that Serbia and Latvia win the gold medal in the 3x3 basketball tournament at the Tokyo olympics, estimated via Monte Carlo simulation as described in Example 1.36. The graph shows the estimated probabilities and associated 0.95 confidence intervals for 10^3 (left) and 10^5 (right) simulations. On the left, the confidence intervals overlap, so even though the probability of Latvia winning is higher (compare the white markers), the result is inconclusive. On the right, Serbia's probability is higher and the confidence intervals do not overlap, indicating that we have performed enough simulations to distinguish the two probabilities.

than that of Serbia, but the confidence intervals overlap, so we cannot be sure that this is correct. In fact, it is wrong! Increasing the number of simulations to $n := 10^5$ results in much narrower confidence intervals, which no longer overlap and reveal that the probability that Serbia wins is actually higher. This illustrates how confidence intervals can help us decide whether the number of Monte Carlo simulations is enough to be confident of our conclusions.

··

9.8.3 Interpretation and Applicability of Confidence Intervals

Interpreting the meaning of confidence intervals is somewhat tricky. Imagine that we have computed a 0.95 confidence interval for the population mean of the height data in Figure 9.16 that equals $[174.6, 177.4]$. We might be tempted to state:

The probability that the mean height in the population is between 174.6 cms and 177.4 cms is 0.95.

However, the population mean is a deterministic quantity, so there are no random quantities in this statement! In our height example, we actually know the value of the population mean: It equals 175.6 cms. It clearly does not make any sense to state that 175.6 is in $[174.6, 177.4]$ with probability 0.95 (it always is!). The correct probabilistic statement describing the confidence interval should be:

If I were to gather many datasets of random samples and compute a 0.95 confidence interval corresponding to each dataset, then the population mean would belong to 95% of these confidence intervals.

Figure 9.16 illustrates this probabilistic statement. It shows 40 confidence intervals computed from different datasets of random samples. Ninety-five percent (38 out of 40) of the intervals contain the population mean.

Confidence intervals are challenging to interpret because they involve deterministic population parameters. In statistics, this is known as the frequentist framework, in contrast to the Bayesian framework presented in Section 6.7, where the parameters are modeled as random variables. The following example compares the two perspectives.

Example 9.47 (Poll in Pennsylvania: Confidence interval)**.** We consider the data reported in Example 6.19 from a poll in Pennsylvania for the 2020 US election, where 281 people intend to vote for Trump and 300 for Biden. The corresponding estimate for the fraction of people θ_{pop} that intend to vote for Trump is $281/581 = 48.4\%$. We assume that the people in the poll were selected independently and uniformly at random from the general population. Consequently, we can interpret the estimate as a sample proportion of i.i.d. Bernoulli random variables with parameter θ_{pop}. Applying Definition 9.44 to these variables yields the 0.95 confidence interval

$$\left[0.484 - \frac{0.98}{\sqrt{581}}, 0.484 + \frac{0.98}{\sqrt{581}}\right] = [0.443, 0.524]. \tag{9.218}$$

Unfortunately, within the frequentist framework, we cannot convert the confidence interval to a statement about the probability that either of the candidates wins because we are modeling θ_{pop} as a deterministic quantity. All we can say is that we cannot be very certain that Trump will not win. In particular, we *cannot* conclude that the confidence interval contains θ_{pop} with probability 0.95 because θ_{pop} is a constant, so such a statement doesn't make sense. From the frequentist viewpoint, we can only state that if we repeat the polling process over and over, the resulting confidence interval will contain θ_{pop} 95% of the time.

By contrast, the Bayesian analysis in Example 6.19 interprets the fraction of Trump voters as a random variable $\tilde{\theta}$ and is therefore able to generate probabilistic statements such as: *The probability that Biden wins in Pennsylvania is 0.75.* Remember, however, that there is a price to pay: These statements depend on the prior distribution chosen for $\tilde{\theta}$, and can change substantially for different priors, as demonstrated in Figure 6.13.
. . . .

When computing confidence intervals on real data, it is easy to forget that we are implicitly modeling the measurements as *independent*. In some situations, such as the Monte Carlo simulations in Example 9.46, it is straightforward to ensure that the independence assumption holds. However, in many others, it is not. In Example 9.47, we assume that the participants in the election poll are chosen independently at random from the whole population of Pennsylvania, but in practice this is very difficult to achieve. For instance, young urban voters are typically much easier to reach than old rural voters. The following example computes confidence intervals using dependent samples, with catastrophic results.

Example 9.48 (Confidence intervals require independent sampling)**.** We consider the problem of estimating the probability of precipitation in Coos Bay (Oregon) in 2015 using 500 hourly measurements from Dataset 9, each indicating the presence or absence of precipitation (this is the same data used in Example 4.34). We compare two sampling strategies: (1) using 500 successive measurements and (2) using 500 independent measurements sampled uniformly at random with replacement. We compute 0.95 confidence intervals for all

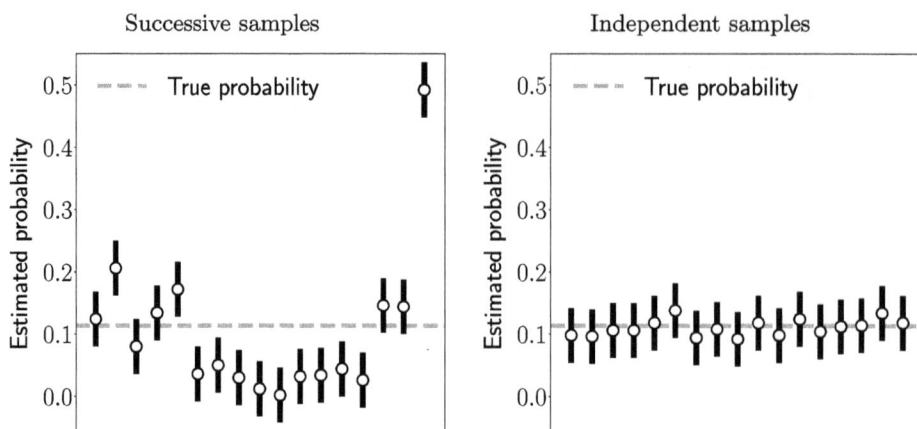

Figure 9.18 Confidence intervals require independent sampling. The white markers indicate estimates of the probability of precipitation in Coos Bay (Oregon) in 2015, obtained from 500 hourly measurements. The vertical black lines are 0.95 confidence intervals, computed based on Definition 9.44. On the left, the 500 measurements are sampled successively, and are therefore not independent. As a result, the assumptions of Definition 9.44 do not hold and very few of the confidence intervals contain the true probability of precipitation, represented by the horizontal gray dashed line. On the right, the 500 measurements are sampled independently, which results in valid confidence intervals that are likely to contain the true probability (in this case, all of them do).

estimates applying Definition 9.44. The *true* probability of precipitation is 0.113 because precipitation occurs in 11.3% of the total measurements.

Figure 9.18 shows estimates of the probability of precipitation, and their associated 0.95 confidence intervals, obtained from multiple datasets of samples acquired following our two sampling strategies. The estimates computed from successive samples are not very accurate. More worryingly, very few of the corresponding confidence intervals actually contain the true probability. They are not valid confidence intervals! The reason is that the measurements are not independent, which completely violates the assumptions of Definition 9.44. By contrast, the independent measurements sampled uniformly at random result in valid confidence intervals that contain the true probability of precipitation with high probability.

In this case, it is quite obvious why lack of independence is problematic: Precipitation tends to concentrate at certain times of the year. Over the summer it rains very little, so the probability estimated from successive samples is very low. Conversely, in the winter there are periods of intense precipitation, which results in a high estimated probability. As a result, the standard error of the empirical-probability estimator is much larger than the one assumed by Definition 9.44, when we compute it from successive samples, so the resulting confidence intervals are too narrow. Unfortunately, in practice, it is often difficult to implement truly independent sampling, and dependence can be much more challenging to spot than in this example.

. . . .

9.9 The Bootstrap

The bootstrap is a computational technique to perform uncertainty quantification of estimators of population parameters. Section 9.9.1 introduces the bootstrap and explains how to use it to estimate standard errors. Section 9.9.2 describes how to use the bootstrap to build confidence intervals and shows that the approach is effective for estimators that are approximately Gaussian after a monotonic transformation.

9.9.1 The Bootstrap Standard Error

As explained in Section 9.3, the standard error is the standard deviation of an estimator. Approximating the standard error from data makes it possible to build confidence intervals for unbiased estimators that are approximately Gaussian, leveraging Lemma 9.42. In this section, we present a computational approach to achieve this.

Let $\tilde{x}_1, \ldots, \tilde{x}_n$ be n random variables representing random samples from a population. By Definition 9.9, the standard error se $(\tilde{w})$ of an estimator $\tilde{w} := h(\tilde{x}_1, \ldots, \tilde{x}_n)$ is equal to its standard deviation. Here h is the deterministic function that implements the estimator, mapping n data points to the estimate of the population parameter of interest. In principle, the standard error can be estimated by computing the sample standard deviation of independent realizations $w_1, w_2, \ldots$ of the estimator (see Definition 7.37). By Theorem 9.25, the sample variance is a consistent estimator of the variance, so the sample standard deviation converges to se $(\tilde{w})$ in probability.[4] Unfortunately, this does not work because in practice we only have access to a single realization of the n data points $x_1, \ldots, x_n$, which yield a single realization of $\tilde{w}$, equal to $h(x_1, \ldots, x_n)$. Computing the sample variance would require access to independent realizations of the estimator, obtained by applying the estimating function h to additional datasets of n independent random samples from $\tilde{x}_1, \ldots, \tilde{x}_n$. The bootstrap method provides a pragmatic solution to this conundrum: *sampling from the available data*, in order to obtain additional resampled datasets with the same size as the original one. This is known as *bootstrapping*, from the expression *to pull yourself up by your own bootstraps*, because we are using the available data as a proxy for the population.

Definition 9.49 (Bootstrap samples). *To obtain bootstrap samples from a real-valued dataset $X := \{x_1, \ldots, x_n\}$, we produce bootstrap indices $\tilde{k}_1, \tilde{k}_2, \ldots, \tilde{k}_n$ by sampling independently and uniformly at random with replacement from the set of possible indices $\{1, \ldots, n\}$. These indices are mutually independent and satisfy*

$$\mathrm{P}\left(\tilde{k}_j = i\right) = \frac{1}{n}, \qquad 1 \le i, j \le n. \tag{9.219}$$

The bootstrap indices are used to select bootstrap samples $\tilde{b}_1, \ldots, \tilde{b}_n$ by setting

$$\tilde{b}_j = x_{\tilde{k}_j}, \qquad 1 \le j \le n. \tag{9.220}$$

To estimate the standard error via bootstrapping, we compute the standard deviation of the bootstrap estimator

$$\tilde{w}_{\mathrm{bs}} := h(\tilde{b}_1, \ldots, \tilde{b}_n), \tag{9.221}$$

where $\tilde{b}_1, \ldots, \tilde{b}_n$ are bootstrap samples obtained from the available data following Definition 9.49.

---

[4] Assuming that the fourth central moment of $\tilde{w}$ is finite, which is the case for most choices of the estimating function h, when the estimator is computed from random samples extracted from a fixed population.

Definition 9.50 (Bootstrap standard error). *Let $X := \{x_1, \ldots, x_n\}$ be a real-valued dataset and $h : \mathbb{R}^n \to \mathbb{R}$ an estimator of a parameter of interest. The bootstrap standard error of the estimator equals*

$$\mathrm{se}_{\mathrm{bs}} = \sqrt{\mathrm{Var}\left[h(\tilde{b}_1, \tilde{b}_2, \ldots, \tilde{b}_n)\right]}, \tag{9.222}$$

where $\tilde{b}_1, \ldots, \tilde{b}_n$ are bootstrap samples of X following Definition 9.49.

In practice, the bootstrap standard error is approximated computationally. We generate K batches of n bootstrap samples, $b_j^{[k]}$, $1 \le j \le n$, $1 \le k \le K$, and compute the sample standard deviation $\sqrt{v(W_{\mathrm{bs}})}$ of the parameter estimates,

$$W_{\mathrm{bs}} := \left\{w_1^{[\mathrm{bs}]}, w_2^{[\mathrm{bs}]}, \ldots, w_K^{[\mathrm{bs}]}\right\}, \qquad w_k^{[\mathrm{bs}]} := h(b_1^{[k]}, b_2^{[k]}, \ldots, b_n^{[k]}), \tag{9.223}$$

following Definition 7.37. If K is set sufficiently large, this yields an accurate approximation because $\sqrt{v(W)}$ converges to $\mathrm{se}_{\mathrm{bs}}$ when $K \to \infty$ by Theorem 9.25.

In order to illustrate the computation of the bootstrap standard error, we apply it to approximate the standard error of the sample-mean estimator in Example 9.1. Recall that the complete ground-truth population consists of $N := 4082$ height values. We consider a dataset of $n := 400$ samples $X := \{x_1, \ldots, x_{400}\}$, depicted in the top left plot of Figure 9.19, which are chosen independently and uniformly at random from the population, as in Example 9.1. From these samples, we estimate the population mean μ_{pop} via the sample-mean estimator,

$$m(X) := \frac{1}{400}\sum_{i=1}^{400} x_i = 175.8. \tag{9.224}$$

The top middle plot in Figure 9.19 shows the pdf of the sample mean, obtained by generating many datasets of 400 random samples and computing the histogram of their sample means (see Section 3.5.1), as in Figure 9.1. The true standard error $\mathrm{se}_{\mathrm{true}}$ is the standard deviation of this distribution. Our goal is to approximate this standard error from X, i.e. from a single fixed batch of 400 samples.

Following Definition 9.49, we sample from X to obtain K batches of 400 bootstrap samples $b_j^{[k]}$, $1 \le j \le 400$, $1 \le k \le K$. The bottom row of Figure 9.19 shows three of the batches. From each batch, we compute the sample mean,

$$m_{\mathrm{bs}}^{[k]} := \frac{1}{400}\sum_{i=1}^{400} b_i^{[k]}, \tag{9.225}$$

which yields K i.i.d. samples $M_{\mathrm{bs}} := \left\{m_{\mathrm{bs}}^{[1]}, m_{\mathrm{bs}}^{[2]}, \ldots, m_{\mathrm{bs}}^{[K]}\right\}$ from the distribution of the bootstrap sample mean

$$\tilde{m}_{\mathrm{bs}} := \frac{1}{400}\sum_{i=1}^{400} \tilde{b}_i, \tag{9.226}$$

where $\tilde{b}_1, \ldots, \tilde{b}_{400}$ are bootstrap samples following Definition 9.49. The top right plot in Figure 9.19 shows a histogram of M_{bs} (in white), which approximates the pdf of the bootstrap sample mean, alongside a histogram approximating the true pdf of the sample mean (in gray).

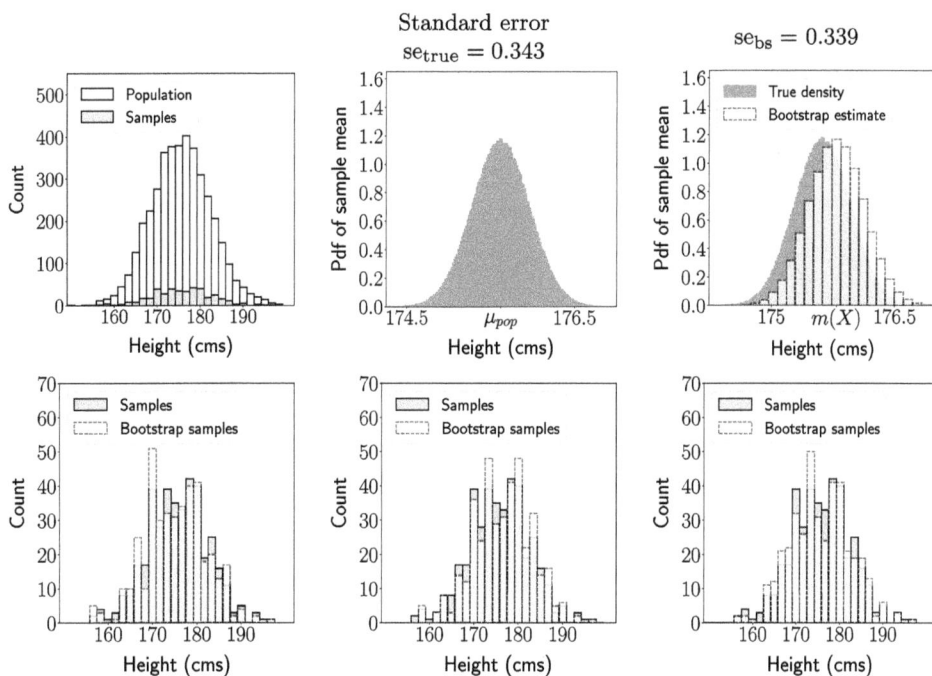

Figure 9.19 Bootstrap standard error of the sample mean. The white histogram in the top left plot represents the heights of a population of 4,082 individuals, extracted from Dataset 5. The gray histogram represents a dataset X of 400 heights sampled independently and uniformly at random from the population. The gray histogram in the top middle plot shows the pdf of the sample mean of 400 random samples, computed as in Figure 9.1. The true standard error se_{true} of the sample mean is the standard deviation of this distribution. The top right plot compares the pdf of the sample mean (gray histogram), with the pdf of the bootstrap sample mean (white histogram), obtained from one million batches of 400 bootstrap samples resampled from X following Definition 9.49. The bottom row shows the histograms of three such batches (gray), superposed onto the histogram of X (white). The bootstrap standard error se_{bs} is the standard deviation of the bootstrap distribution.

As established in Theorem 9.6, the distribution of the sample mean is centered at the population mean μ_{pop}, but this is *not* the case for the bootstrap distribution. To understand why, notice that the bootstrap samples $\tilde{b}_1, \ldots, \tilde{b}_{400}$ satisfy Definition 9.3, if we interpret X as a population. Consequently, by Theorem 9.6 the mean of the bootstrap sample mean $\widetilde{m}_{\text{bs}}$ is the population mean of X, which equals the sample mean $m(X)$, not μ_{pop} (compare (9.10) and Definition 7.15). The bootstrap distribution of the sample mean is therefore centered at $m(X)$, as we observe in the top right plot of Figure 9.19.

The bootstrap standard error is obtained by computing the sample standard deviation (see Definition 7.37) of the bootstrap sample means in M_{bs},

$$\text{se}_{\text{bs}} \approx \sqrt{v(M_{\text{bs}})} \tag{9.227}$$

$$= 0.339, \tag{9.228}$$

using a very large value of K (one million), to ensure convergence of the sample variance $v(M_{\text{bs}})$ to the variance of $\widetilde{m}_{\text{bs}}$ by Theorem 9.25. The following theorem provides an expression for the resulting value of se_{bs}.

Theorem 9.51 (Bootstrap standard error of the sample mean). *Let $X := \{x_1, \ldots, x_n\}$ be a real-valued dataset and let $\tilde{b}_1, \ldots, \tilde{b}_n$ be bootstrap samples of X following Definition 9.49. The bootstrap standard error of the sample mean computed from these bootstrap samples equals*

$$\mathrm{se}_{\mathrm{bs}} = \sqrt{\frac{n-1}{n}} \sqrt{\frac{v(X)}{n}}, \tag{9.229}$$

where $v(X)$ is the sample variance of X (see Definition 7.37).

Proof As mentioned earlier, the bootstrap samples $\tilde{b}_1, \ldots, \tilde{b}_n$ satisfy Definition 9.3, if we interpret X as a population, and the population mean $\mu_{\mathrm{pop}}(X)$ of X is the sample mean $m(X)$. Consequently, the population variance of X, defined as in (9.20), equals

$$\sigma_{\mathrm{pop}}(X)^2 := \frac{1}{n} \sum_{j=1}^{n} (x_j - \mu_{\mathrm{pop}}(X))^2 \tag{9.230}$$

$$= \frac{1}{n} \sum_{j=1}^{n} (x_j - m(X))^2 \tag{9.231}$$

$$= \frac{n-1}{n} v(X). \tag{9.232}$$

Applying Theorem 9.13 to the sample mean

$$\tilde{m}_{\mathrm{bs}} := \frac{1}{n} \sum_{i=1}^{n} \tilde{b}_i \tag{9.233}$$

yields

$$\mathrm{se}_{\mathrm{bs}}^2 := \mathrm{Var}\left[\tilde{m}_{\mathrm{bs}}\right] \tag{9.234}$$

$$= \frac{\sigma_{\mathrm{pop}}(X)^2}{n} \tag{9.235}$$

$$= \frac{n-1}{n^2} v(X). \tag{9.236}$$

■

By Theorem 9.13, the true standard error of the sample mean equals

$$\mathrm{se}_{\mathrm{true}} = \frac{\sigma_{\mathrm{pop}}}{\sqrt{n}}, \tag{9.237}$$

where σ_{pop} is the population standard deviation. Theorem 9.51 establishes that the bootstrap standard error is a very reasonable approximation to $\mathrm{se}_{\mathrm{true}}$, since the sample variance $v(X)$ is a consistent estimator of the population variance σ_{pop}^2 by Theorem 9.25. Consequently, for a sufficiently large sample size n,

$$\mathrm{se}_{\mathrm{bs}} = \sqrt{\frac{n-1}{n}} \sqrt{\frac{v(X)}{n}} \approx \frac{\sigma_{\mathrm{pop}}}{\sqrt{n}} = \mathrm{se}_{\mathrm{true}}. \tag{9.238}$$

In our example dataset, $\mathrm{se}_{\mathrm{true}} = 0.343$ and $\mathrm{se}_{\mathrm{bs}} = 0.339$.

In the case of the sample mean, we do not really need to leverage the bootstrap to approximate the standard error. It is easier to apply the formula in Theorem 9.13, replacing the population standard deviation by the sample standard deviation of the available data, as explained at the end of Section 9.8.1. This is how we estimate the standard error to generate the confidence intervals in Figure 9.16. The resulting estimate of the standard error $\mathrm{se}_{\mathrm{est}}$ is essentially the same as the bootstrap standard error, unless the sample size n is very small:

$$\mathrm{se}_{\mathrm{est}} = \sqrt{\frac{v(X)}{n}} \approx \sqrt{\frac{n-1}{n}} \sqrt{\frac{v(X)}{n}} = \mathrm{se}_{\mathrm{bs}}. \tag{9.239}$$

In our example dataset where $n := 400$, $\mathrm{se}_{\mathrm{est}} = 0.340$ and $\mathrm{se}_{\mathrm{bs}} = 0.339$.

The bootstrap standard error enables us to build confidence intervals for estimators that are approximately Gaussian and unbiased. By Lemma 9.42, if the distribution of an estimator $\tilde{g}$ with standard error se is well approximated by a Gaussian with mean equal to the population parameter γ, then $[\tilde{g} - c_\alpha \, \mathrm{se}, \tilde{g} + c_\alpha \, \mathrm{se}]$ is a 1-α confidence interval for γ, where c_α is an appropriately chosen constant. Plugging in the bootstrap standard error to replace the true standard error yields the bootstrap Gaussian confidence interval.

Definition 9.52 (Bootstrap Gaussian confidence interval). *Let* $X := \{x_1, \ldots, x_n\}$ *be a real-valued dataset, and let* $h(X)$ *denote an estimator of a parameter* γ *computed from the elements of* X. *For any* $\alpha \in (0,1)$, *the 1-α bootstrap Gaussian confidence interval for* γ *is*

$$\mathcal{I}_{1-\alpha}^{\mathrm{BSG}} := [h(X) - c_\alpha \, \mathrm{se}_{\mathrm{bs}}, h(X) + c_\alpha \, \mathrm{se}_{\mathrm{bs}}], \qquad c_\alpha := F_{\tilde{z}}^{-1}\left(1 - \frac{\alpha}{2}\right),$$

where $\mathrm{se}_{\mathrm{bs}}$ *is the bootstrap standard error of* h *(see Definition 9.50), and* $F_{\tilde{z}}$ *denotes the cdf of a standard Gaussian with zero mean and unit variance.*

It is important to bear in mind that Definition 9.52 assumes that the estimator is approximately Gaussian. If this is not the case, the intervals may not provide accurate uncertainty quantification, as illustrated by the following example.

Example 9.53 (Confidence interval for the correlation coefficient). We consider the problem of estimating the correlation coefficient between the height and the foot length of individuals in a population. We use the data in Figure 5.19, extracted from Dataset 5, as the complete ground-truth population. The population correlation coefficient equals

$$\rho_{\mathrm{pop}} := \frac{\mathrm{Cov}_{\mathrm{pop}}(\mathrm{height}, \mathrm{foot})}{\sigma_{\mathrm{pop}}(\mathrm{height})\sigma_{\mathrm{pop}}(\mathrm{foot})} = 0.718. \tag{9.240}$$

Here, $\sigma_{\mathrm{pop}}(\mathrm{height})^2$ and $\sigma_{\mathrm{pop}}(\mathrm{foot})^2$ denote the population variance of height and foot length, defined as in (9.20), and

$$\mathrm{Cov}_{\mathrm{pop}}(\mathrm{height}, \mathrm{foot}) := \frac{1}{N} \sum_{k=1}^{N} (h_i - \mu_{\mathrm{pop}}(\mathrm{height}))(l_i - \mu_{\mathrm{pop}}(\mathrm{foot})) \tag{9.241}$$

is the population covariance, where $(h_1, l_1), (h_2, l_2), \ldots, (h_N, l_N)$ denote the height and foot length of the $N := 4{,}082$ individuals in the population, and $\mu_{\mathrm{pop}}(\mathrm{height})$ and $\mu_{\mathrm{pop}}(\mathrm{foot})$ denote the population means of the height and the foot length, respectively.

We estimate the correlation coefficient by computing the sample correlation coefficient (see Definition 8.10) from the data $(x_1, y_1), (x_2, y_2), \ldots, (x_n, y_n)$ corresponding to n individuals selected independently and uniformly at random with replacement from the population. Figure 9.20 shows scatterplots of the samples, and plots of the distribution of the sample correlation coefficient, when we repeat the sampling process many times. For

Population correlation coefficient $\rho_{\text{pop}} = 0.718$

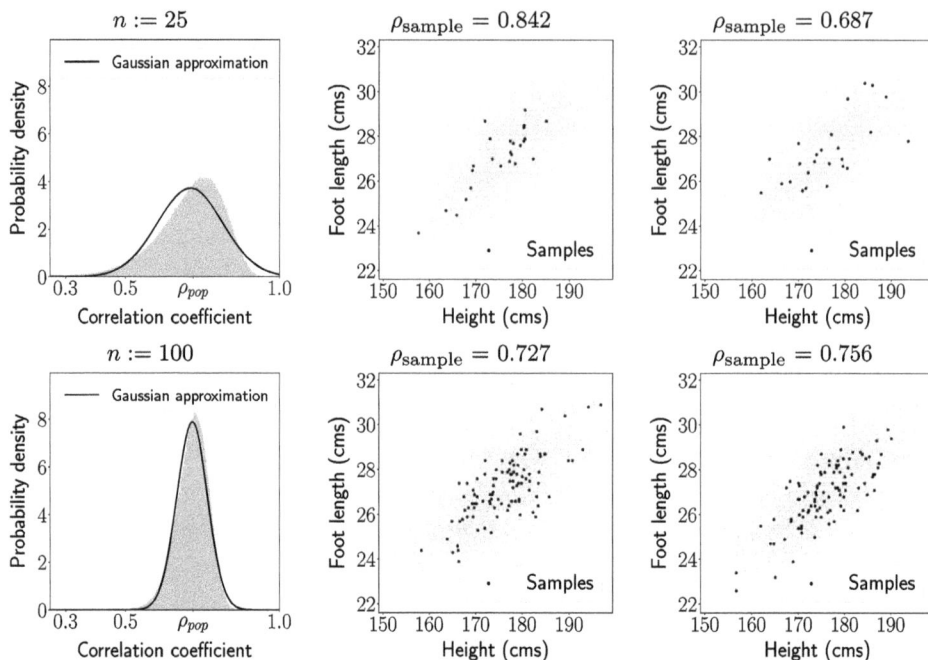

Figure 9.20 Distribution of the sample correlation coefficient. The left column shows the pdf of the sample correlation coefficient between height and foot length for two values of the sample size n, using the data in Example 9.53. The density is estimated using a normalized histogram of sample correlation coefficients computed from one million different datasets of independent, uniform samples. The black line shows a Gaussian fit to the histogram, which approximates the distribution well for $n := 100$, but not for $n := 25$. The middle and right columns show example scatterplots of some of the samples (black), as well as the underlying population (light gray). The corresponding sample correlation coefficient is reported at the top of each plot.

$n := 100$, the distribution is approximately Gaussian (although it is somewhat skewed), but for $n := 25$ the Gaussian approximation is not very accurate.

In order to quantify the uncertainty associated with the sample correlation coefficient using only n fixed samples, we build bootstrap Gaussian confidence intervals following Definition 9.52. Figure 9.21 shows examples of the pdf of the bootstrap sample correlation coefficient for two different values of the sample size n. By Definition 9.50, the bootstrap standard error is equal to the standard deviation of this distribution, approximated using a large number of bootstrap samples. Figure 9.22 shows 0.95 bootstrap Gaussian confidence intervals ($\alpha := 0.05$) based on the resulting estimate of the standard error.

We evaluate our Gaussian bootstrap confidence intervals by computing their coverage probability, which is the probability that they contain the population correlation coefficient. This is achieved via the Monte Carlo method (see Section 1.7). We repeatedly select datasets of n random samples from the population and compute the corresponding bootstrap Gaussian confidence intervals. The coverage probability is approximated by the fraction of such intervals that contain the population correlation coefficient. In theory, by the definition of confidence intervals (see Section 9.8), the coverage probability should equal 0.95. However, in our case it equals 93.7% for $n := 100$, and only 90.7% for $n := 25$. The problem is that

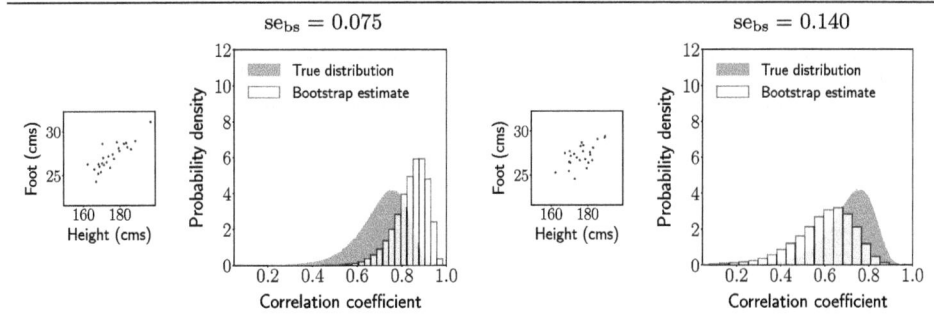

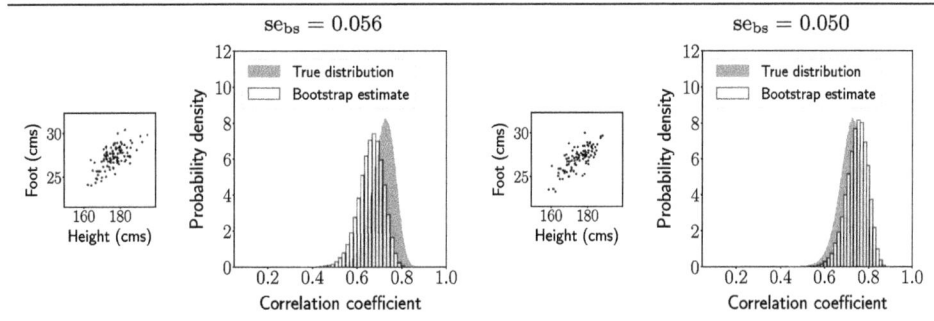

Figure 9.21 Bootstrap standard error of the sample correlation coefficient. The small scatterplots depict n samples of height and foot length from the dataset in Example 9.53 for $n := 25$ (top row) and $n := 100$ (bottom row). The graph to the right of each scatterplot compares the pdf of the sample correlation coefficient (gray histogram, computed as explained in the caption of Figure 9.20), with the pdf of the bootstrap sample correlation coefficient (white histogram, computed using 10^5 sample correlation coefficients obtained from bootstrap samples following Definition 9.49). The standard error is the standard deviation of the true distribution; the bootstrap standard error is the standard deviation of the bootstrap distribution.

the distribution of the sample correlation coefficient is not well approximated by a Gaussian, particularly for $n := 25$ (see Figure 9.20). In the following section, we discuss an alternative approach to build confidence intervals based on the bootstrap, which is better suited to non-Gaussian estimators.

9.9.2 The Bootstrap Percentile Confidence Interval

In this section, we show that the bootstrap can be used to build confidence intervals directly, without approximating the standard error of the estimator of interest. This alternative approach has the advantage that it provides valid confidence intervals for estimators, such as the sample correlation coefficient, which are non-Gaussian, but are rendered approximately Gaussian by a monotonic transformation. To explain the approach, we first focus on the sample mean.

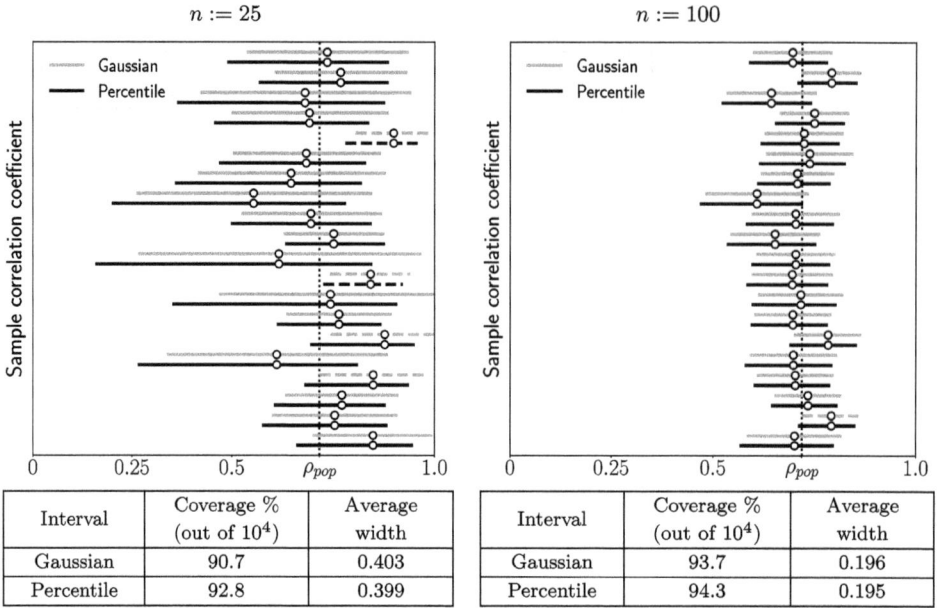

	$n := 25$				$n := 100$	
Interval	Coverage % (out of 10^4)	Average width		Interval	Coverage % (out of 10^4)	Average width
Gaussian	90.7	0.403		Gaussian	93.7	0.196
Percentile	92.8	0.399		Percentile	94.3	0.195

Figure 9.22 Bootstrap confidence intervals for the correlation coefficient. The top graph shows bootstrap Gaussian and percentile confidence intervals computed following Definitions 9.52 and 9.54, respectively, for the correlation coefficient of the data in Example 9.53 using two different sample sizes. The table reports the coverage probability (how often the intervals contain the population correlation coefficient) and the average width of 10^4 intervals. The percentile intervals have a higher coverage probability and smaller average width because they are able to automatically adapt to the skewness of the distribution of the sample correlation coefficient.

Consider the sample mean

$$\tilde{m} := \frac{1}{n} \sum_{i=1}^{n} \tilde{x}_i \tag{9.242}$$

of n independent, uniform random samples $\tilde{x}_1, \tilde{x}_2, \ldots, \tilde{x}_n$ following Definition 9.3. The top left plot in Figure 9.23 shows the pdf of $\tilde{m}$ for sample size $n := 25$, when the underlying population is the height data in Example 9.1. The pdf is centered at the population mean μ_{pop} by Theorem 9.6, and is approximately Gaussian (see Definition 9.38). As a result, for any $\alpha \in (0, 1)$ by Lemma 9.42, the random interval

$$\tilde{\mathcal{I}}_{1-\alpha} := [\tilde{m} - c_\alpha \, \text{se}_{\text{true}}, \tilde{m} + c_\alpha \, \text{se}_{\text{true}}] \tag{9.243}$$

is an approximate confidence interval for the population mean, in the sense that

$$P\left(\mu_{\text{pop}} \in \tilde{\mathcal{I}}_{1-\alpha}\right) \approx 1 - \alpha. \tag{9.244}$$

Here, se_{true} is the standard error or standard deviation of $\tilde{m}$, and $c_\alpha := F_{\tilde{z}}^{-1}\left(1 - \frac{\alpha}{2}\right)$, where $F_{\tilde{z}}$ denotes the cdf of a Gaussian random variable $\tilde{z}$ with zero mean and unit variance.

In practice, we need to build confidence intervals from datasets with only n fixed samples. Given a dataset $X := \{x_1, \ldots, x_n\}$, this requires approximating the realization of the confidence interval (9.243),

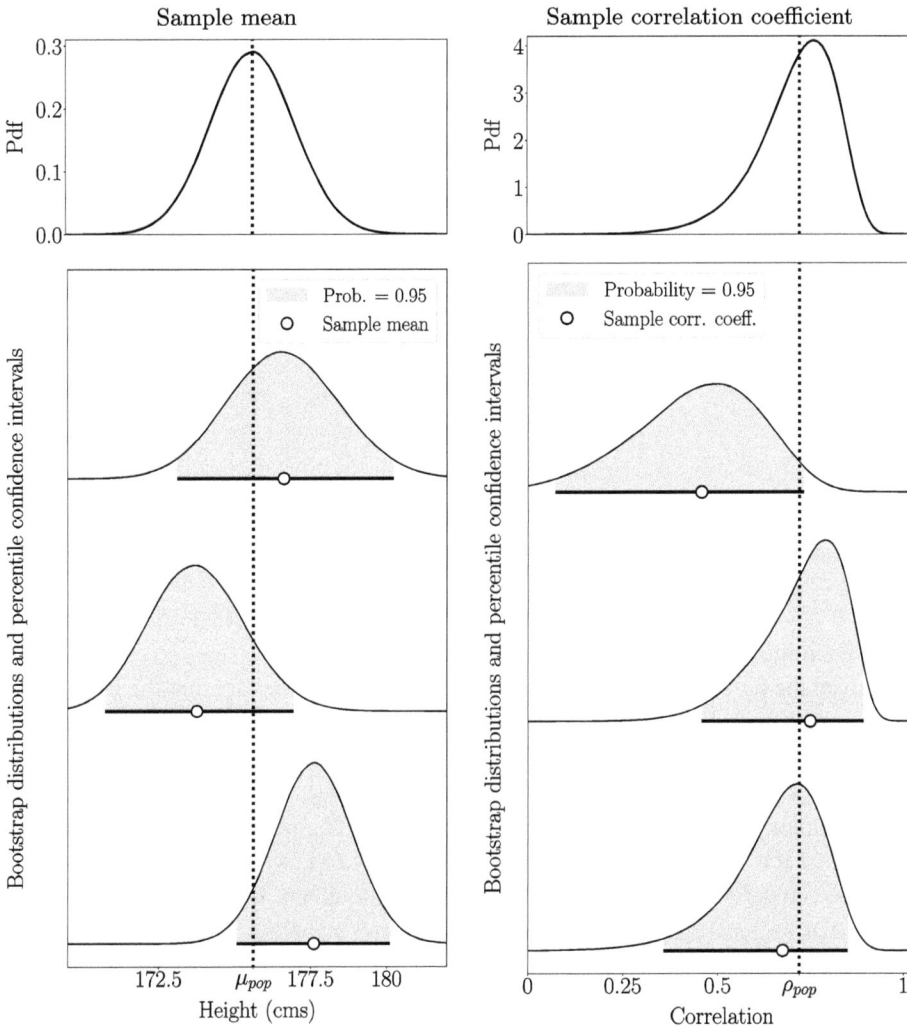

Figure 9.23 Bootstrap percentile confidence intervals. The top row shows the pdfs of the sample mean (left) and the sample correlation coefficient (right) of 25 random samples from the height data in Example 9.1 and the height and foot-length data in Example 9.53, respectively. The remaining plots depict the pdfs of the sample means (left) and sample correlation coefficients (right) of bootstrap samples from three datasets with 25 fixed samples. The sample means and sample correlation coefficients of the datasets are indicated by white markers, and the corresponding 0.95 percentile confidence intervals by horizontal black lines. By design, the probability that the bootstrap sample mean and bootstrap sample correlation coefficient belong to the intervals is 0.95, as indicated by the shaded regions under the pdfs. By Theorems 9.55 and 9.56, the intervals contain the population mean and the population correlation coefficient (indicated by vertical dotted lines) with probability approximately equal to 0.95.

$$\mathcal{I}_{1-\alpha} := [m(X) - c_\alpha \, \mathrm{se}_{\mathrm{true}}, m(X) + c_\alpha \, \mathrm{se}_{\mathrm{true}}], \tag{9.245}$$

corresponding to the realization $m(X)$ of the sample mean $\tilde{m}$ computed from the data. In Sections 9.8.1 and 9.9.1, we approximate $\mathcal{I}_{1-\alpha}$ by plugging an estimate of the standard error $\mathrm{se}_{\mathrm{true}}$ into (9.245), obtained either by estimating the population variance (see the end of

Section 9.8.1) or via bootstrapping (see Section 9.9.1). Here, we present a different approach, which does not require an explicit estimate of the standard error.

Our approach relies on the bootstrap sample mean

$$\widetilde{m}_{\mathrm{bs}} := \frac{1}{n} \sum_{i=1}^{n} \widetilde{b}_i, \tag{9.246}$$

where $\widetilde{b}_1, \ldots, \widetilde{b}_n$ are bootstrap samples from the dataset X following Definition 9.49. For any α, we define the $\alpha/2$ and $1-\alpha/2$ quantiles $q_{\alpha/2}(X)$ and $q_{1-\alpha/2}(X)$ of $\widetilde{m}_{\mathrm{bs}}$ as constants such that

$$\mathrm{P}\left(\widetilde{m}_{\mathrm{bs}} \leq q_{\alpha/2}(X)\right) = \frac{\alpha}{2}, \tag{9.247}$$
$$\mathrm{P}\left(\widetilde{m}_{\mathrm{bs}} \leq q_{1-\alpha/2}(X)\right) = 1 - \frac{\alpha}{2}.$$

The values are chosen so that the probability that the bootstrap sample mean $\widetilde{m}_{\mathrm{bs}}$ is in the interval $[q_{\alpha/2}(X), q_{1-\alpha/2}(X)]$ is $1 - \alpha$. This interval is our proposed approximation to the confidence interval $\mathcal{I}_{1-\alpha}$. It is called a *percentile* interval because if we express $q_{\alpha/2}(X)$ and $q_{1-\alpha/2}(X)$ as percentages, then they are the $100\,\alpha/2$ and $100(1 - \alpha/2)$ percentiles of $\widetilde{m}_{\mathrm{bs}}$ (see Definition 3.9).

To analyze the percentile confidence interval, we leverage the fact that the bootstrap samples $\widetilde{b}_1, \ldots, \widetilde{b}_n$ satisfy Definition 9.3, if we interpret X as the population. As a result, the pdf of $\widetilde{m}_{\mathrm{bs}}$ is approximately Gaussian (by Definition 9.38) with mean $m(X)$ (by Theorem 9.6). The bottom left plot of Figure 9.23 shows the pdf of $\widetilde{m}_{\mathrm{bs}}$ for three different datasets of $n := 25$ samples from the height data in Example 9.1. The pdfs indeed look Gaussian and are centered at the sample mean $m(X)$ of each dataset. As explained in Section 9.9.1 (see (9.238)) the standard deviation of the bootstrap sample mean, which we call the bootstrap standard error $\mathrm{se}_{\mathrm{bs}}$, is approximately equal to the true standard error of the sample-mean estimator $\mathrm{se}_{\mathrm{true}}$.

The properties of the bootstrap sample mean, combined with Theorem 3.31, imply that $(\widetilde{m}_{\mathrm{bs}} - m(X))/\mathrm{se}_{\mathrm{true}} \approx (\widetilde{m}_{\mathrm{bs}} - m(X))/\mathrm{se}_{\mathrm{bs}}$ is approximately Gaussian with mean zero and unit variance, which allows us to establish that the quantile $q_{1-\frac{\alpha}{2}}$ is approximately equal to the upper limit $m(X) + c_\alpha\,\mathrm{se}_{\mathrm{true}}$ of $\mathcal{I}_{1-\alpha}$:

$$\mathrm{P}\left(\widetilde{m}_{\mathrm{bs}} \leq m(X) + c_\alpha\,\mathrm{se}_{\mathrm{true}}\right) = \mathrm{P}\left(\frac{\widetilde{m}_{\mathrm{bs}} - m(X)}{\mathrm{se}_{\mathrm{true}}} \leq c_\alpha\right) \tag{9.248}$$
$$\approx \mathrm{P}\left(\frac{\widetilde{m}_{\mathrm{bs}} + m(X)}{\mathrm{se}_{\mathrm{bs}}} \leq c_\alpha\right) \tag{9.249}$$
$$\approx \mathrm{P}\left(\widetilde{z} \leq c_\alpha\right) \tag{9.250}$$
$$= F_{\widetilde{z}}\left(c_\alpha\right) \tag{9.251}$$
$$= 1 - \frac{\alpha}{2}. \tag{9.252}$$

The same argument can be used to establish that $q_{\frac{\alpha}{2}}$ is approximately equal to $m(X) - c_\alpha\,\mathrm{se}_{\mathrm{true}}$ because

$$\mathrm{P}\left(\widetilde{m}_{\mathrm{bs}} \leq m(X) - c_\alpha\,\mathrm{se}_{\mathrm{true}}\right) \approx \frac{\alpha}{2}. \tag{9.253}$$

We conclude that the percentile confidence interval $[q_{\alpha/2}(X), q_{1-\alpha/2}(X)]$ is an approximation of $\mathcal{I}_{1-\alpha}$ and should therefore contain the population mean μ_{pop} with probability $1 - \alpha$. The bottom left plot of Figure 9.23 shows examples of 0.95 bootstrap percentile confidence intervals for the population mean of the height data in Example 9.1, represented by the shaded regions under the corresponding pdfs of the bootstrap sample mean.

The following definition generalizes our approach for building bootstrap percentile confidence intervals to other estimators beyond the sample mean, and explains how to compute the intervals in practice.

Definition 9.54 (Bootstrap percentile confidence interval). *Let* $X := \{x_1, \ldots, x_n\}$ *be a real-valued dataset, and let* $h(X)$ *denote an estimator of a parameter* γ *computed from the elements of* X. *For any* $\alpha \in (0, 1)$, *we define the* $\alpha/2$ *and* $1 - \alpha/2$ *quantiles* $q_{\alpha/2}(X)$ *and* $q_{1-\alpha/2}(X)$ *as the values satisfying*

$$P\left(h(\tilde{b}_1, \tilde{b}_2, \ldots, \tilde{b}_n) \leq q_{\alpha/2}(X)\right) = \frac{\alpha}{2}, \tag{9.254}$$

$$P\left(h(\tilde{b}_1, \tilde{b}_2, \ldots, \tilde{b}_n) \leq q_{1-\alpha/2}(X)\right) = 1 - \frac{\alpha}{2}, \tag{9.255}$$

where $\tilde{b}_1, \ldots, \tilde{b}_n$ *are bootstrap samples of* X *following Definition 9.49. The 1-α bootstrap percentile confidence interval for* γ *is*

$$\mathcal{I}_{1-\alpha}^{\text{BSP}} := [q_{\alpha/2}(X), q_{1-\alpha/2}(X)]. \tag{9.256}$$

In practice, the bootstrap percentiles are approximated computationally, generating K *batches of* n *bootstrap samples,* $b_i^{[k]}$, $1 \leq k \leq K$, *and setting* $q_{\alpha/2}$ *and* $q_{1-\alpha/2}$ *to be the* $\alpha/2$ *and* $1 - \alpha/2$ *quantiles of the parameter estimates*

$$W_{\text{bs}} := \left\{w_1^{[\text{bs}]}, w_2^{[\text{bs}]}, \ldots, w_K^{[\text{bs}]}\right\}, \qquad w_k^{[\text{bs}]} := h(b_1^{[k]}, b_2^{[k]}, \ldots, b_n^{[k]}). \tag{9.257}$$

For large enough K, *this yields an accurate approximation to the quantiles because the empirical pmf of* W_{bs} *is a consistent estimator of the pmf of the estimator by Theorem 9.24.*

The following theorem shows that bootstrap percentile confidence intervals are indeed valid confidence intervals for estimators that are unbiased and have Gaussian distributions.

Theorem 9.55 (Percentile confidence interval under Gaussian assumptions). *Let* $\tilde{x}_1$, $\tilde{x}_2$, $\ldots$, $\tilde{x}_n$ *be* n *random samples from a population following Definition 9.3, and let* $\tilde{g} := h(\tilde{x}_1, \ldots, \tilde{x}_n)$ *be an estimator of a population parameter* $\gamma \in \mathbb{R}$. *We denote by* $q_{\alpha/2}(X)$ *and* $q_{1-\alpha/2}(X)$, *the* $\alpha/2$ *and* $1 - \alpha/2$ *quantiles of the bootstrap estimator computed from a fixed dataset of* n *samples* $X := \{x_1, \ldots, x_n\}$, *as in Definition 9.54. For any* $\alpha \in (0, 1)$, *the bootstrap percentile confidence interval*

$$\widetilde{\mathcal{I}}_{1-\alpha}^{\text{BSP}} := [q_{\alpha/2}(\{\tilde{x}_1, \ldots, \tilde{x}_n\}), q_{1-\alpha/2}(\{\tilde{x}_1, \ldots, \tilde{x}_n\})] \tag{9.258}$$

is a valid $1 - \alpha$ *confidence interval for the population parameter* γ, *in the sense that*

$$P\left(\gamma \in \widetilde{\mathcal{I}}_{1-\alpha}^{\text{BSP}}\right) = 1 - \alpha, \tag{9.259}$$

under the following assumptions:

- *The estimator $\tilde{g}$ has a Gaussian distribution with mean γ.*
- *Conditioned on $\{\tilde{x}_1, \ldots, \tilde{x}_n\} = X$, the bootstrap estimator $\tilde{g}_{bs} := h(\tilde{b}_1, \tilde{b}_2, \ldots, \tilde{b}_n)$ is Gaussian with mean $h(X)$. Here $\tilde{b}_1, \ldots, \tilde{b}_n$ are bootstrap samples of X following Definition 9.49.*
- *The standard deviations or standard errors of $\tilde{g}$ and $\tilde{g}_{bs}$ are the same.*

Proof   The argument is the same as the one used earlier to justify why the bootstrap percentile confidence interval works for the sample mean. Let $c_\alpha := F_{\tilde{z}}^{-1}\left(1 - \frac{\alpha}{2}\right)$, where $\tilde{z}$ is a standard Gaussian random variable with zero mean and unit variance, and let se denote the standard error of $\tilde{g}$ and $\tilde{g}_{bs}$. Since the conditional distribution of $\tilde{g}_{bs}$ given $\{\tilde{x}_1, \ldots, \tilde{x}_n\} = X$ is assumed to be Gaussian, by Theorem 3.31 $(\tilde{g}_{bs} - h(X))/$ se is conditionally Gaussian with mean zero and unit standard deviation, so

$$\mathrm{P}\left(\tilde{g}_{bs} \leq h(X) + c_\alpha \,\mathrm{se} \mid \{\tilde{x}_1, \ldots, \tilde{x}_n\} = X\right)$$

$$= \mathrm{P}\left(\frac{\tilde{g}_{bs} - h(X)}{\mathrm{se}} \leq c_\alpha \,\Bigg|\, \{\tilde{x}_1, \ldots, \tilde{x}_n\} = X\right) \tag{9.260}$$

$$= F_{\tilde{z}}(c_\alpha) = 1 - \frac{\alpha}{2}. \tag{9.261}$$

By symmetry of the Gaussian pdf,

$$\mathrm{P}\left(\tilde{g}_{bs} \leq h(X) - c_\alpha \,\mathrm{se} \mid \{\tilde{x}_1, \ldots, \tilde{x}_n\} = X\right) = \alpha/2. \tag{9.262}$$

Consequently, by definition of the quantiles (see Definition 9.54),

$$q_{\alpha/2}(X) = h(X) - c_\alpha \,\mathrm{se}, \tag{9.263}$$
$$q_{1-\alpha/2}(X) = h(X) + c_\alpha \,\mathrm{se}, \tag{9.264}$$

so

$$\tilde{\mathcal{I}}_{1-\alpha}^{\mathrm{BSP}} := [q_{\alpha/2}(\{\tilde{x}_1, \ldots, \tilde{x}_n\}), q_{1-\alpha/2}(\{\tilde{x}_1, \ldots, \tilde{x}_n\})] \tag{9.265}$$
$$= [\tilde{g} - c_\alpha \,\mathrm{se}, \tilde{g} + c_\alpha \,\mathrm{se}], \tag{9.266}$$

which is a 1-α confidence interval for γ by Lemma 9.42, as long as $\tilde{g}$ is Gaussian with mean γ and standard error se.   ∎

An important property of the bootstrap percentile confidence interval is that it can be applied effectively to certain estimators that are non-Gaussian. The right column of Figure 9.23 depicts the computation of bootstrap percentile confidence intervals for the population correlation coefficient of the height and foot-length data in Example 9.53. The top graph shows the pdf of the sample correlation coefficient, which is clearly skewed and hence non-Gaussian. The remaining graphs show three percentile confidence intervals corresponding to three fixed datasets with $n := 25$ random samples. In contrast with the bootstrap Gaussian confidence intervals (see Definition 9.52), these intervals are not symmetric around the observed sample correlation coefficient, which enables them to adapt to the skewness of the pdf of the estimator. Figure 9.22 compares the two types of bootstrap confidence intervals. The coverage probability of the percentile intervals is closer to $1 - \alpha := 0.95$, even though their average length is smaller. The improvement is more noticeable for $n := 25$, where the Gaussian approximation for the sample correlation coefficient is less accurate (see Figure 9.20).

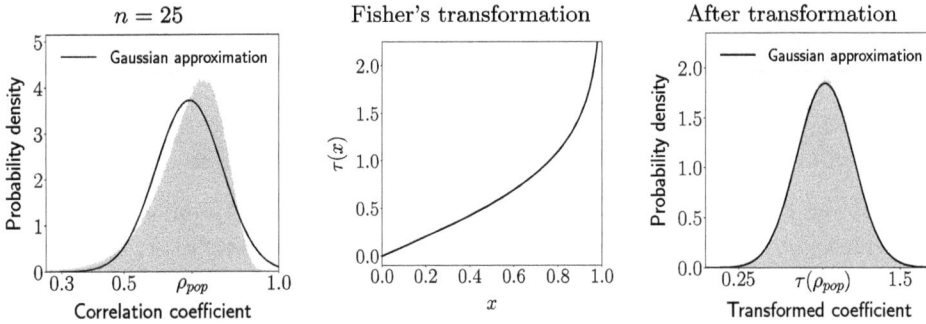

Figure 9.24 Fisher's transformation of the sample correlation coefficient. The plot on the left shows the normalized histogram of the sample correlation coefficient between height and foot length, computed using 25 random samples from the height and foot-length data in Example 9.53. The distribution is skewed, and not well approximated as Gaussian. The plot in the middle shows Fisher's transformation τ given by (9.267). The plot on the right shows the normalized histogram of the sample correlation coefficient after applying Fisher's transformation: The resulting distribution is well approximated as a Gaussian centered at $\tau(\rho_{\text{pop}})$, where ρ_{pop} is the population correlation coefficient.

The reason why bootstrap percentile confidence intervals work for the sample correlation coefficient is that there exists a monotonic transformation that renders the distribution of the sample correlation coefficient approximately Gaussian:

$$\tau(\rho) := \frac{1}{2} \log\left(\frac{1+\rho}{1-\rho}\right). \tag{9.267}$$

The transformation, which is known as Fisher's transformation (Fisher, 1915), is depicted in Figure 9.24. The following theorem establishes that bootstrap percentile confidence intervals are valid confidence intervals for non-Gaussian estimators, as long as they are rendered Gaussian by a monotonic transformation. Notice that the transformation must exist, but we *don't actually need to know it* in order to build the interval! The proof of the theorem relies on the fact that the quantiles of a random variable transformed by a deterministic monotonic function are equal to its transformed quantiles because monotonic functions preserve the order of their inputs.

Theorem 9.56 (Percentile confidence intervals and monotonic transformations). *Let $\tilde{x}_1$, $\tilde{x}_2$, ..., $\tilde{x}_n$ be n random samples from a population following Definition 9.3, and let $\tilde{w} := h(\tilde{x}_1, \ldots, \tilde{x}_n)$ be an estimator of a population parameter $\gamma \in \mathbb{R}$. We denote by $q_{\alpha/2}(X)$ and $q_{1-\alpha/2}(X)$, the $\alpha/2$ and $1 - \alpha/2$ quantiles of the bootstrap estimator computed from a fixed dataset of n samples $X := \{x_1, \ldots, x_n\}$, as in Definition 9.54. For any $\alpha \in (0, 1)$, the bootstrap percentile confidence interval*

$$\widetilde{\mathcal{I}}_{1-\alpha}^{\text{BSP}} := [q_{\alpha/2}(\{\tilde{x}_1, \ldots, \tilde{x}_n\}), q_{1-\alpha/2}(\{\tilde{x}_1, \ldots, \tilde{x}_n\})] \tag{9.268}$$

is a valid $1 - \alpha$ confidence interval for the population parameter γ, in the sense that

$$\mathrm{P}\left(\gamma \in \widetilde{\mathcal{I}}_{1-\alpha}^{\text{BSP}}\right) = 1 - \alpha \tag{9.269}$$

if there exists a strictly monotonic transformation τ satisfying the following conditions:

- *The transformed estimator $\tilde{g} := \tau(\tilde{w})$ has a Gaussian distribution with mean $\tau(\gamma)$.*
- *Conditioned on $\{\tilde{x}_1, \ldots, \tilde{x}_n\} = X$, the transformed bootstrap estimator $\tilde{g}_{\text{bs}} := \tau(h(\tilde{b}_1, \tilde{b}_2, \ldots, \tilde{b}_n))$ is Gaussian with mean $\tau(h(X))$. Here $\tilde{b}_1, \ldots, \tilde{b}_n$ are bootstrap samples of X following Definition 9.49.*
- *The standard deviations or standard errors of $\tilde{g}$ and $\tilde{g}_{\text{bs}}$ are the same.*

Proof   To simplify the exposition, we assume that τ is increasing. A similar argument holds if it is decreasing. First, we realize that, for any fixed dataset of n samples X, if we apply the transformation τ to $q_{\alpha/2}(X)$ and $q_{1-\alpha/2}(X)$, we obtain $\alpha/2$ and $1 - \alpha/2$ quantiles for the transformed bootstrap estimator $\tau(h(\tilde{b}_1, \tilde{b}_2, \ldots, \tilde{b}_n))$. The reason is that if τ is increasing, then the events

$$h(\tilde{b}_1, \tilde{b}_2, \ldots, \tilde{b}_n) \leq q_{\alpha/2}(X) \tag{9.270}$$

and

$$\tau\left(h(\tilde{b}_1, \tilde{b}_2, \ldots, \tilde{b}_n)\right) \leq \tau\left(q_{\alpha/2}(X)\right) \tag{9.271}$$

are equivalent. Consequently,

$$\mathrm{P}\left(\tau\left(h(\tilde{b}_1, \tilde{b}_2, \ldots, \tilde{b}_n)\right) \leq \tau\left(q_{\alpha/2}(X)\right)\right) = \mathrm{P}\left(h(\tilde{b}_1, \tilde{b}_2, \ldots, \tilde{b}_n) \leq q_{\alpha/2}(X)\right)$$
$$= \frac{\alpha}{2}, \tag{9.272}$$

and by the same argument,

$$\mathrm{P}\left(\tau\left(h(\tilde{b}_1, \tilde{b}_2, \ldots, \tilde{b}_n)\right) \leq \tau\left(q_{1-\alpha/2}(X)\right)\right) = \mathrm{P}\left(h(\tilde{b}_1, \tilde{b}_2, \ldots, \tilde{b}_n) \leq q_{1-\alpha/2}(X)\right)$$
$$= 1 - \frac{\alpha}{2}. \tag{9.273}$$

If we interpret $\tilde{g}$ as an estimator of the transformed parameter $\tau(\gamma)$, we can apply Theorem 9.55 to the transformed estimator because $\tilde{g}$ and the transformed bootstrap estimator $\tilde{g}_{\text{bs}}$ satisfy the necessary conditions. As a result, the probability of the event

$$\tau\left(q_{\alpha/2}(\{\tilde{x}_1, \ldots, \tilde{x}_n\})\right) \leq \tau(\gamma) \leq \tau\left(q_{1-\alpha/2}(\{\tilde{x}_1, \ldots, \tilde{x}_n\})\right) \tag{9.274}$$

is $1 - \alpha$ because the transformed quantiles are equal to the quantiles of the transformed bootstrap estimator, as explained earlier. By the assumption that τ is increasing, the event (9.274) is equivalent to

$$q_{\alpha/2}(\{\tilde{x}_1, \ldots, \tilde{x}_n\}) \leq \gamma \leq q_{1-\alpha/2}(\{\tilde{x}_1, \ldots, \tilde{x}_n\}), \tag{9.275}$$

i.e. to $\gamma \in \widetilde{\mathcal{I}}_{1-\alpha}^{\text{BSP}}$, so

$$\mathrm{P}\left(\gamma \in \widetilde{\mathcal{I}}_{1-\alpha}^{\text{BSP}}\right) = 1 - \alpha. \tag{9.276}$$

∎

Exercises

9.1   (Markov's and Chebyshev's inequalities are tight) Markov's and Chebyshev's inequalities cannot be improved without further assumptions because there exist random variables for which they are tight.

a For any $c > 0$ and any $0 < \theta < 1$, build a nonnegative random variable $\tilde{a}$ such that

$$P(\tilde{a} \geq c) = \theta = \frac{\mathbb{E}[\tilde{a}]}{c}. \qquad (9.277)$$

b For any $c > 0$, any $0 < \theta < 1$ and any $\mu \in \mathbb{R}$, build a random variable $\tilde{b}$ with mean μ and finite variance such that

$$P(|\tilde{b} - \mu| \geq c) = \theta = \frac{\text{Var}[\tilde{b}]}{c^2}. \qquad (9.278)$$

9.2 (Online poll) In online polls, young people are often overrepresented. In this problem, we show how to correct for this. Our goal is to analyze an online poll for a fictitious election where there are two candidates, a Democrat and a Republican. The poll consists of n_1 young people and n_2 old people. We denote by y and o the number of young and old Democrat voters in the poll, respectively. We assume that the fraction of young voters in the population, denoted by α, is known.

a Derive an estimator of the proportion of Democrat voters in the election, as a function of y, o, n_1 and n_2. As a hint, it may be useful to consider the population parameters θ_{young} and θ_{old}, representing the proportion of Democrats among young and old voters in the population, respectively.

b Apply your estimator to a poll where 60 out of 100 participants intend to vote for the Democratic candidate. 70 out of the 100 participants are young, and 50 of these 70 are Democratic voters. The fraction of young voters in the overall population is 25%.

c Under what assumptions is your estimator unbiased? Justify your answer mathematically.

d Show that your estimator is consistent as $n_1 \to \infty$ and $n_2 \to \infty$.

9.3 (Participation) The participants in a poll are asked about their voting preferences, and whether they actually intend to vote in the election. Here are the results:

	Democrat	Republican
Certain to vote	20	30
Likely to vote	40	10
Not likely to vote	30	20

A study shows that 80% of the people that are *certain to vote* end up voting, 10% of the people who are *likely to vote* end up voting, and none of the people who are *not likely to vote* actually vote.

a Let α_R denote the fraction of people in the population, who are certain to vote, and will vote Republican. Assuming that the people in the poll were selected independently and with replacement from the population, propose an unbiased estimator for α_R, derive its standard error for the available sample size, and compute the estimator using the available data.

b Provide an estimate for the fraction of voters, who will vote for the Republican candidate in the election. State your assumptions.

9.4 (The sample covariance is unbiased) Let (a_1, b_1), (a_2, b_2), ..., (a_N, b_N) be two quantities measured in a population of size N with population covariance

$$c_{\text{pop}} := \frac{1}{N} \sum_{i=1}^{N} (a_i - \mu_a)(b_i - \mu_b), \qquad (9.279)$$

where μ_a and μ_b denote the population means of the first and second quantity, respectively. We consider n independent, uniform random samples

$$\tilde{x}_j = a_{\tilde{k}_j}, \qquad \tilde{y}_i = b_{\tilde{k}_j}, \qquad 1 \leq j \leq n, \qquad (9.280)$$

where $\tilde{k}_j$ is defined as in Definition 9.3. In this exercise, we prove that the sample covariance

$$\tilde{c} := \frac{1}{n-1} \sum_{i=1}^{n} (\tilde{x}_i - \tilde{m}_1)(\tilde{y}_i - \tilde{m}_2), \qquad (9.281)$$

where

$$\tilde{m}_1 := \frac{1}{n} \sum_{i=1}^{n} \tilde{x}_i, \qquad \tilde{m}_2 := \frac{1}{n} \sum_{i=1}^{n} \tilde{y}_i \qquad (9.282)$$

is an unbiased estimator of c_{pop}.

a Show that

$$\sum_{i=1}^{n} (\tilde{x}_i - \tilde{m}_1)(\tilde{y}_i - \tilde{m}_2) = \sum_{i=1}^{n} \tilde{x}_i \tilde{y}_i - n\tilde{m}_1 \tilde{m}_2. \qquad (9.283)$$

b Show that for any $1 \leq i \leq n$,

$$\mathbb{E}\left[\tilde{x}_i \tilde{y}_i\right] = c_{\text{pop}} + \mu_a \mu_b. \qquad (9.284)$$

c Use linearity of the covariance operator (see the proof of Theorem 8.35) to show that

$$\mathbb{E}\left[\tilde{m}_1 \tilde{m}_2\right] = \frac{c_{\text{pop}}}{n} + \mu_a \mu_b. \qquad (9.285)$$

d Use the previous results to prove that the sample covariance is unbiased,

$$\mathbb{E}\left[\tilde{c}\right] = c_{\text{pop}}. \qquad (9.286)$$

9.5 (Herding) *Herding* is the act of manipulating poll results, so that they are more similar to other polls. Let $\tilde{m}_1$, $\tilde{m}_2$, and $\tilde{m}_3$ be random variables representing the sample proportion of voters supporting a certain candidate in three polls. The polls are independent and have the same standard error s. Poll 1 is released first. After seeing it, the second pollster uses its result to perform herding via averaging, obtaining the *herded* estimator,

$$\tilde{h}_2 := \frac{\tilde{m}_1 + \tilde{m}_2}{2}. \qquad (9.287)$$

After Poll 2 is released, the third pollster also performs herding using the average of the two first polls,

$$\tilde{h}_3 := \frac{1}{2}\left(\frac{\tilde{m}_1 + \tilde{h}_2}{2} + \tilde{m}_3\right). \qquad (9.288)$$

a Assuming $\tilde{m}_1$, $\tilde{m}_2$ and $\tilde{m}_3$ are unbiased, are $\tilde{h}_2$ and $\tilde{h}_3$ unbiased? Does herding produce bias in this situation?

b Derive the standard error of the herded estimators $\tilde{h}_2$ and $\tilde{h}_3$ as a function of the standard error s. Does herding improve or degrade the estimators?

c Derive the standard error of the estimator obtained by averaging the three non-herded estimators,

$$\tilde{m}_{\text{all}} := \frac{\tilde{m}_1 + \tilde{m}_2 + \tilde{m}_3}{3}, \qquad (9.289)$$

and the estimator obtained by averaging the three herded estimators,

$$\tilde{h}_{\text{all}} := \frac{\tilde{m}_1 + \tilde{h}_2 + \tilde{h}_3}{3}, \qquad (9.290)$$

as a function of the standard error s. Why is herding problematic?

9.6   (Multiplicative noise) Consider the random variables

$$\tilde{x}_i = \gamma \tilde{a}_i, \qquad 1 \leq i \leq n, \tag{9.291}$$

which represent a dataset. Our goal is to estimate the population parameter γ from these data. The random variables $\tilde{a}_1, \ldots, \tilde{a}_n$ represent multiplicative noise and are i.i.d. with mean μ and variance σ^2. Assume that we are able to estimate μ using additional data.

a Derive an unbiased estimator of γ given $\tilde{x}_1, \ldots, \tilde{x}_n$. Can μ equal any value?
b Derive the standard error of the estimator.
c Is the estimator consistent?

9.7   (Consistency of the sample median) Let $\tilde{x}_1, \tilde{x}_2, \ldots$ be a sequence of i.i.d. random variables from a distribution with median γ. We assume that γ is the only point that satisfies $F_{\tilde{x}_i}(\gamma) = 1/2$. Let $\widetilde{\mathrm{md}}_n$ denote the median of the n first elements of the sequence. In this problem we establish that for any $\epsilon > 0$,

$$\lim_{n \to \infty} \mathrm{P}\left(\left| \widetilde{\mathrm{md}}_n - \gamma \right| \geq \epsilon \right) = 0, \tag{9.292}$$

so the sample median is a consistent estimator of the median. Specifically, the goal is to prove

$$\lim_{n \to \infty} \mathrm{P}\left(\widetilde{\mathrm{md}}_n \geq \gamma + \epsilon \right) = 0 \tag{9.293}$$

because the same argument can be used to prove

$$\lim_{n \to \infty} \mathrm{P}\left(\widetilde{\mathrm{md}}_n \leq \gamma - \epsilon \right) = 0, \tag{9.294}$$

and combining (9.293) and (9.294) yields (9.292).

a Let $\tilde{b}$ be the number of elements in $\{\tilde{x}_1, \ldots, \tilde{x}_n\}$ greater than $\gamma + \epsilon$. Explain why

$$\mathrm{P}\left(\widetilde{\mathrm{md}}_n \geq \gamma + \epsilon \right) \leq \mathrm{P}\left(\tilde{b} \geq \frac{n}{2} \right). \tag{9.295}$$

b Use Chebyshev's inequality and (9.295) to prove (9.293). (Hint: By the assumption that γ is the only point that satisfies $F_{\tilde{x}_i}(\gamma) = 1/2$, there exists a constant $\epsilon' > 0$ such that for any i, the probability that $\tilde{x}_i > \gamma + \epsilon$ is $\mathrm{P}(\tilde{x}_i > \gamma + \epsilon) = 1/2 - \epsilon'$.)

9.8   (Watering a plant) Two absentminded roommates share a plant. One of them gives the plant an amount of water that is uniformly distributed between 0 and 1 liter. The other one gives the plant an amount of water that is uniformly distributed between 0 and 2 liters, independently from the first. Derive the pdf of the water received by the plant and plot it.

9.9   (Confidence intervals) The following are $1 - \alpha$ confidence intervals for a certain quantity μ, estimated using the sample mean of the observed data.

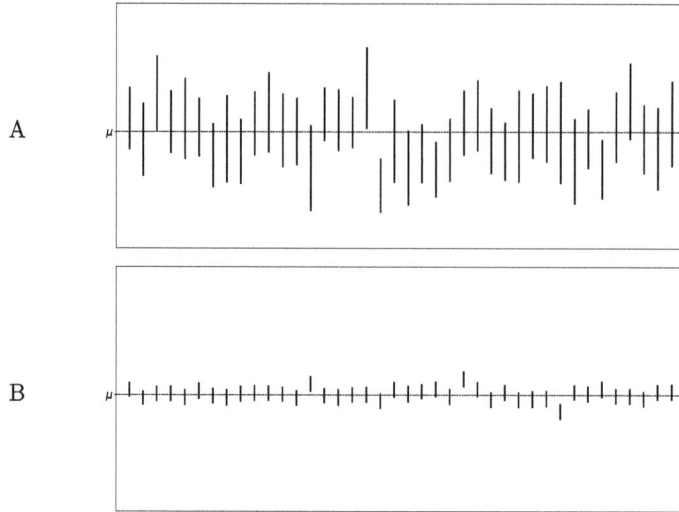

a Estimate α.

b In each scenario, choose whether you think $X = A$ and $Y = B$, or $X = B$ and $Y = A$.
   - In plot X, the confidence intervals are based on the central limit theorem. In plot Y, they are built using Chebyshev's inequality (everything else is the same).
   - In plot X, the confidence intervals are built using more data than in plot Y (everything else is the same).
   - In plot X, the data are sampled from a distribution that has higher variance than plot Y (everything else is the same).

9.10 (Cholesterol) We want to determine whether men have higher cholesterol than women in a certain population. We pick 200 men and 100 women independently at random with replacement from the population, and compute their average cholesterol levels.

   a Let σ_{men} and σ_{women} denote the standard deviations of the cholesterol level for the men and women in the population, respectively. What is the standard error of the difference between the average cholesterol levels computed from our data?

   b The average cholesterol levels we observe are 200 for the men and 180 for the women. Assuming $\sigma_{\text{men}} = \sigma_{\text{women}} = 10$, compute a 95% confidence interval for the difference between the mean cholesterol level of the men and the women in the population.

9.11 (Difference between proportions) In this exercise we derive a confidence interval for the difference between two proportions. We consider two binary datasets of sizes n_A and n_B. The number of instances equal to one in each dataset is k_A and k_B, respectively. We assume that the datasets are sampled i.i.d. from Bernoulli distributions with parameters θ_A and θ_B. Our goal is to construct a confidence interval for $\theta_A - \theta_B$.

   a What is the standard error of the difference between the observed proportions?

   b Explain how to construct a $1 - \alpha$ confidence interval for the difference in proportions $\theta_A - \theta_B$, without using bootstrapping.

   c Construct 0.95 confidence intervals for the difference between cure rates in the two trials from Example 10.36.

   d Imagine that we don't know how to derive the standard error of the difference between observed proportions. Explain how to estimate the standard error via bootstrapping instead.

    e Explain two different ways to build confidence intervals for the difference between proportions via bootstrapping. Do you expect to obtain very different intervals from the ones that don't use bootstrapping?

9.12   (Exact confidence interval) The confidence interval for the mean in Definition 9.43 is approximate because it relies on the central limit theorem, which is an asymptotic result. In this exercise we derive an exact confidence interval based on Chebyshev's inequality.

    a Let $\tilde{x}_1, \tilde{x}_2, \ldots, \tilde{x}_n$ be i.i.d. random variables with mean μ and variance $\sigma^2 \leq b^2$ for some $b > 0$. Prove that for any $0 < \alpha < 1$,

$$\left[\tilde{m} - \frac{b}{\sqrt{\alpha\, n}}, \tilde{m} + \frac{b}{\sqrt{\alpha\, n}} \right], \qquad \tilde{m} := \frac{1}{n} \sum_{i=1}^{n} \tilde{x}_i, \qquad (9.296)$$

    is a 1-α confidence interval for μ.

    b If we use the exact confidence interval instead of the approximate confidence interval in Example 9.45, how many tests are required so that the half-width of a 0.95 confidence interval for the fraction of positive tests is 0.01 or less?

9.13   (Modified Gaussian bootstrap confidence interval) Theorem 9.56 establishes that bootstrap percentile confidence intervals work for estimators that are rendered approximately Gaussian by a strictly monotonic transformation, even if the transformation is not known to us. Propose an alternative approach, based on the bootstrap standard error, to build confidence intervals when the transformation is known (as in the case of Fisher's transformation).

10

Hypothesis Testing

Overview

Hypothesis testing is a fundamental tool in data science. It enables us to determine whether the available data provide sufficient evidence to support a certain hypothesis. Section 10.1 explains why hypotheses should be selected before looking at the data and introduces two running examples, used throughout the chapter. Section 10.2 describes the main idea in hypothesis testing: Interpreting the data under the *null hypothesis* that the hypothesis of interest does not hold. In Sections 10.3 and 10.4, we explain how to use parametric modeling to perform hypothesis testing, and introduce the p-value. A small p-value indicates that the data is not consistent with the null hypothesis, which is evidence in favor of the original hypothesis of interest. Section 10.5 shows that this procedure is guaranteed to control the probability of endorsing a false finding. Section 10.6 defines the power of a test, which is the probability of correctly rejecting the null hypothesis, when it does not hold. In Section 10.7, we explain how to perform hypothesis testing without parametric models, focusing on the permutation test. Section 10.8 describes multiple testing, a challenging setting where many tests are performed simultaneously. In Section 10.9, we discuss the interplay between hypothesis testing and causal inference. Finally, Section 10.10 warns against relying solely on p-values to evaluate the significance of a finding.

10.1 Selecting a Hypothesis

The goal of hypothesis testing is to evaluate whether a hypothesis is supported by data. This is achieved by interpreting the data under the assumption that the hypothesis does not hold. If this interpretation is implausible, then we conclude that the data supports our hypothesis. For this logic to be sound, it is crucial to choose the hypothesis *beforehand*. It may be tempting to use the same data to select the hypothesis and also back it up, but this is a serious mistake! To see why, imagine that you are strolling down the street in New York with a friend, and they tell you:

Look, a car with license plate number EMC6055! About one million cars drive through Manhattan every day. Isn't it amazing that we saw this particular one? It's a one-in-a-million chance!

No, it is not amazing. The probability of seeing that particular license plate is indeed one in a million. However, your friend did not specify the number beforehand, so they could have said the same about any other number! Consequently, their hypothesis was not really *seeing the license plate number EMC6055*, but rather *seeing a car with any license plate number*, which is much less interesting. If they had told you the number *before seeing the car*, then the data would indeed support their claimed hypothesis, and you would be very impressed.

There are always many ways to interpret a dataset. If we select a hypothesis because it is consistent with the data, then of course we will find that our hypothesis is supported by the data! In order to avoid this circular logic, it is crucial to *first* define a hypothesis, and *then* look at the data.

Example 10.1 (Unfair die: Hypothesis). In Example 1.23, we study a toy die. Rolling it 60 times yielded substantially more threes (18) than any other number. This suggests the following hypothesis:

The probability of rolling a three is greater than 1/6.

We might be tempted to use the data in Example 1.23 as evidence to support this hypothesis, but we shouldn't! We used the data to come up with the hypothesis, so we need additional data to test it. To this end, I gathered new measurements by rolling the toy die 100 more times. We analyze these data in the following sections.

....

Example 10.2 (Free throws under pressure). Giannis Antetokounmpo is an NBA superstar who (at the time of writing) plays for the Milwaukee Bucks. One of the few weaknesses in his game is free-throw shooting. Before each free throw, Antetokounmpo performs a long routine, which often exceeds the official 10-second limit. In the 2020/2021 season, fans from opposing teams started counting loudly during the routine in order to request a 10-second violation from the referees. While watching the 2021 NBA playoffs, I wondered whether this affected Antetokounmpo's free-throw percentage. It seemed to me that he was shooting worse at away games than at home games, when the fans were mostly silent during his free throws. My hypothesis was:

Antetokounmpo's free-throw percentage is better at home than away.

I could not test this hypothesis based on the games I had already watched, since they had influenced my choice of hypothesis. Instead, I decided to use the games from the NBA finals, which had not yet occurred. We analyze these data in the following sections.

....

10.2 The Null Hypothesis and the Test Statistic

When we perform hypothesis testing, we take a skeptical point of view and try to explain the data under the assumption that our hypothesis of interest does not hold. This assumption is known as the *null hypothesis*. By contrast, the hypothesis of interest is called the *alternative hypothesis*. If the data are inconsistent with the null hypothesis, then this is interpreted as evidence supporting the alternative hypothesis.

Example 10.3 (Unfair die: Null and alternative hypotheses). In Example 10.1, we conjecture that the die is not fair. Therefore, a natural choice for the null hypothesis is:

The probability of rolling a three is equal to 1/6.

The corresponding alternative hypothesis is that the probability of rolling a three is different from 1/6.

....

Example 10.4 (Free throws under pressure: Null and alternative hypotheses). In Example 10.2, we suspect that Antetokounmpo does not shoot free throws with equal accuracy at home and away games. Therefore, a reasonable null hypothesis is:

Antetokounmpo's free-throw percentage is the same at home and away.

The corresponding alternative hypothesis is that the free-throw percentages at home and away are different.
....

Hypothesis tests are based on a predefined quantity, called the *test statistic*, which summarizes the available data. The test statistic is designed so that it is likely to be small, under the assumption that the null hypothesis holds. If the test statistic computed from the observed data is large, this is interpreted as evidence against the null hypothesis.

Example 10.5 (Unfair die: Test statistic). To test the hypothesis in Example 10.1, we need a test statistic that tends to be small, if the null hypothesis holds, i.e. if the probability of rolling a three is 1/6. A natural choice is the number of rolled threes. In fact, we were originally motivated to investigate the die because this quantity was abnormally large in our preliminary data (see Example 1.23). In our new data, 21 out 100 rolls are threes, so the test statistic t_{data} equals 21. In the following sections, we explain how to quantify whether this is consistent with the null hypothesis.
....

Example 10.6 (Free throws under pressure: Test statistic). To test the hypothesis in Example 10.2, the test statistic should have a high chance of being small, if Antetokounmpo's free-throw percentage is the same at home and away. We choose the difference between the fraction of made free throws at home and the fraction of made free throws away:

$$t_{\text{data}} := \frac{\text{Made free throws at home}}{\text{Attempted free throws at home}} - \frac{\text{Made free throws away}}{\text{Attempted free throws away}}.$$

During the finals, Antetokounmpo made 34 out of 44 free throws in home games, and 22 out of 41 free throws in away games, so $t_{\text{data}} = 0.236$.
....

10.3 Parametric Testing and the P-Value

As explained in Section 10.2, our goal in hypothesis testing is to determine whether the data are consistent with the null hypothesis. To this end, we compute a predefined test statistic from the data, which is likely to be small under the null hypothesis. If the test statistic is abnormally large, then this is evidence against the null hypothesis (and in favor of the alternative hypothesis). Determining what values qualify as *abnormally large* requires characterizing the probabilistic behavior of the test statistic under the null hypothesis. In this section, we leverage parametric models (introduced in Sections 2.3 and 3.6) for this purpose. Section 10.7 describes an alternative nonparametric approach.

In parametric hypothesis testing, we interpret the test statistic as a random variable with a distribution that depends on a small number of deterministic parameters. This allows us to define the null and alternative hypotheses in terms of these parameters. A *simple* null hypothesis states that each model parameter has a single fixed value. A *composite* null hypothesis states that each parameter is in a predefined set.

Example 10.7 (Unfair die: Parametric model). In Example 10.5, we choose the number of rolled threes as the test statistic for our die example. To derive a parametric model for the test statistic, we assume that the rolls are independent and that the probability of rolling a three is equal to a fixed parameter θ. In that case, the distribution of the test statistic is binomial with parameters $n := 100$, equal to the total number of rolls, and θ (see Section 2.3.2). We denote by $\tilde{t}_\theta$ the random variable representing the test statistic in order to emphasize its dependence on θ. By Definition 2.17 the pmf of $\tilde{t}_\theta$ is

$$p_{\tilde{t}_\theta}(t) = \binom{100}{t} \theta^t (1 - \theta)^{(100-t)}, \quad t = 0, 1, \dots, 100. \tag{10.1}$$

The null hypothesis defined in Example 10.3 is a simple null hypothesis, which states that $\theta = 1/6$. Since our original hypothesis is that threes are more likely, we could also have chosen the composite null hypothesis $\theta \in [0, 1/6]$, stating that the probability of rolling a three is less than or equal to 1/6.

. . . .

To evaluate the consistency between the data and the null hypothesis, we define a random variable that follows the assumed parametric distribution of the test statistic, setting the parameters equal to those associated with the null hypothesis. Then, we determine the probability that the random variable is greater than or equal to the observed test statistic. This probability is called the *p*-value. If the *p*-value is small, then the observed test statistic is abnormally large because the probability of observing such a large value under the null hypothesis is low. Consequently, we *reject* the null hypothesis, declaring it to be inconsistent with the data. If the *p*-value is not small, we do not reject the null hypothesis. In such cases, the hypothesis test is inconclusive, since the observed value would not be unlikely under the null hypothesis. However, this does not mean that the null hypothesis necessarily holds, so we should *not* interpret a large *p*-value as evidence in favor of the null hypothesis.

Definition 10.8 (P-value function and *p*-value). *The p-value is the probability that the test statistic under the null hypothesis is greater than or equal to the test statistic computed from the available data. We define the p-value in terms of the p-value function, which is a function mapping every possible value of the test statistic to the corresponding p-value.*

Let $\tilde{t}_\theta$ be a random variable representing a test statistic with a parametric distribution that depends on a parameter (or vector of parameters) θ. Consider the simple null hypothesis $\theta = \theta_{\text{null}}$, for some constant $\theta_{\text{null}} \in \mathbb{R}$. The p-value function pv associated with this null hypothesis maps t to the probability that $\tilde{t}_{\theta_{\text{null}}}$ is greater than or equal to t,

$$\text{pv}(t) := \text{P}\left(\tilde{t}_{\theta_{\text{null}}} \geq t\right). \tag{10.2}$$

Consider a composite null hypothesis of the form $\theta \in \Theta_{\text{null}}$ for a certain predefined set Θ_{null}. The corresponding p-value function pv maps each t to the supremum over Θ_{null} of the probability that $\tilde{t}_{\theta_{\text{null}}}$ is greater than or equal to t,

$$\text{pv}(t) := \sup_{\theta \in \Theta_{\text{null}}} \text{P}\left(\tilde{t}_\theta \geq t\right). \tag{10.3}$$

The p-value of the available data is equal to $\text{pv}(t_{\text{data}})$, where t_{data} denotes the observed test statistic.

Example 10.9 (Unfair die: P-value). Following Definition 10.8, we compute the p-value function corresponding to the simple null hypothesis $\theta = \theta_{\mathrm{null}} := 1/6$, using the pmf of $\tilde{t}_{\theta_{\mathrm{null}}}$ derived in Example 10.7. By Theorem 2.5,

$$\mathrm{pv}(t) := \mathrm{P}\left(\tilde{t}_{\theta_{\mathrm{null}}} \geq t\right) \tag{10.4}$$

$$= \sum_{i=t}^{100} p_{\tilde{t}_{\theta_{\mathrm{null}}}}(i) \tag{10.5}$$

$$= \sum_{i=t}^{100} \binom{100}{i} \left(\frac{1}{6}\right)^i \left(\frac{5}{6}\right)^{100-i}, \quad t = 0, 1, \ldots, 100. \tag{10.6}$$

The p-value function under the composite null hypothesis $\theta \in \Theta_{\mathrm{null}} := [0, 1/6]$ is the same as under the simple null hypothesis because the supremum is attained exactly at $\theta = 1/6$:

$$\mathrm{pv}(t) := \sup_{\theta \in \Theta_{\mathrm{null}}} \mathrm{P}\left(\tilde{t}_{\theta_{\mathrm{null}}} \geq t\right) \tag{10.7}$$

$$= \sup_{\theta \in [0, 1/6]} \sum_{i=t}^{100} \binom{100}{i} \theta^i (1 - \theta)^{100-i} \tag{10.8}$$

$$= \sum_{i=t}^{100} \binom{100}{i} \left(\frac{1}{6}\right)^i \left(\frac{5}{6}\right)^{100-i}, \quad t = 0, 1, \ldots, 100. \tag{10.9}$$

The left graph in Figure 10.1 shows the p-value function. It is close to one for small values of the test statistic because we are almost certain to observe a larger test statistic under the null hypothesis. Then it decreases until it reaches the p-value for the observed test statistic $t_{\mathrm{data}} := 21$, which equals

$$\mathrm{pv}(t_{\mathrm{data}}) = \sum_{i=21}^{100} \binom{100}{i} \left(\frac{1}{6}\right)^i \left(\frac{5}{6}\right)^{100-i} = 0.15, \tag{10.10}$$

as depicted in the right graph of Figure 10.1.

· ·

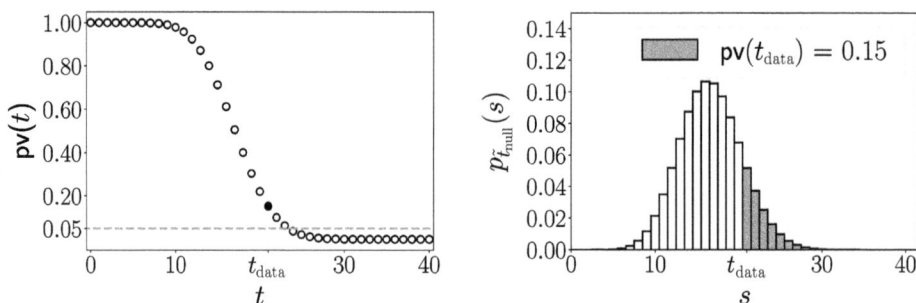

Figure 10.1 P-**value for a (possibly) unfair die.** The left plot shows the p-value function pv derived in Example 10.9. The p-value corresponding to the observed test statistic $t_{\mathrm{data}} := 21$ is equal to 0.15. This is above 0.05, a popular threshold for statistical significance (see Section 10.5) represented by the dashed gray line. The right plot depicts the pmf of the test statistic (the number of rolled threes) under the null hypothesis ($\theta = \theta_{\mathrm{null}} = 1/6$). The p-value at t_{data} is the sum of the pmf over all values greater than or equal to t_{data}.

A common misconception is that the *p*-value represents the probability that the null hypothesis holds. In Example 10.9, this would translate to the statement:

The probability that the die is fair (or equivalently that θ = 1/6) is 0.15.

However, this is incorrect because we are assuming that the model parameters are deterministic. Under this frequentist perspective, the null hypothesis is not random: It either holds or not. In the case of the die, θ is either equal to 1/6 or it is not. In order to make probabilistic statements about the null hypothesis, we can take a Bayesian perspective (see Section 6.7). However, as discussed in Example 9.47, these statements would then depend on the prior distribution assigned to the parameters.

From our frequentist viewpoint, the correct statement associated with the *p*-value computed in Example 10.9 is:

If the die is fair (θ = 1/6), the probability of observing 21 threes or more is 0.15.

If this probability is small, then we interpret it as evidence against the null hypothesis. To determine whether the *p*-value is small enough to be convincing, we compare it to a predetermined threshold, which is often set equal to 0.05 (but can vary depending on the domain). Section 10.5 shows that this allows us to control the probability of endorsing a false finding. The *p*-value in Example 10.9 is larger than 0.05. The observed data are not very unlikely given the null hypothesis. If the die is indeed fair and we repeat the experiment over and over, then we will observe 21 threes or more about 15% of the time. In conclusion, we cannot reject the null hypothesis based on the data. This does not mean that we believe that the null hypothesis is correct; it just means that we do not have enough evidence to rule it out. In fact, I personally still think that the die is not fair.

10.4 Two-Sample Tests

In the hypothesis test described in Example 10.9, all the available data are interpreted as samples from the same distribution. Such tests are called *one-sample* tests. By contrast, in Example 10.4, the data are divided into two groups, corresponding to home and away games. The alternative hypothesis is that the groups have different distributions, whereas the null hypothesis is that they are sampled from the same distribution. Such tests are known as *two-sample* tests. The following definition describes a parametric two-sample test called a z test, which leverages the Gaussian approximation to the sample mean based on the central limit theorem (Definition 9.38).

Definition 10.10 (Two-sample z test). *Given two binary datasets of sizes n_A and n_B, the null hypothesis of the two-sample z test is that all the data are generated as i.i.d. samples from the same Bernoulli distribution. If the test is chosen to be one-tailed, the test statistic is the difference between the sample proportions,*

$$t_{\text{data}} := \frac{k_A}{n_A} - \frac{k_B}{n_B}, \tag{10.11}$$

where k_A and k_B are the number of instances equal to one in the first and second datasets, respectively. The corresponding p-value is approximated as follows:

$$\text{pv}(t_{\text{data}}) \approx 1 - F_{\tilde{z}}\left(\frac{t_{\text{data}}}{\sigma_{\text{null}}}\right). \tag{10.12}$$

$F_{\tilde{z}}$ is the cdf of a standard Gaussian random variable with mean zero and unit variance, and

$$\sigma_{\text{null}}^2 := \frac{k(n-k)}{n^2}\left(\frac{1}{n_A} + \frac{1}{n_B}\right), \tag{10.13}$$

where $n := n_A + n_B$ and $k := k_A + k_B$.

If the test is chosen to be two-tailed, then the test statistic is the absolute value of the difference between the sample proportions,

$$t_{\text{data}} := \left|\frac{k_A}{n_A} - \frac{k_B}{n_B}\right|, \tag{10.14}$$

and the p-value is approximated as follows:

$$\text{pv}(t_{\text{data}}) \approx 2\left(1 - F_{\tilde{z}}\left(\frac{t_{\text{data}}}{\sigma_{\text{null}}}\right)\right). \tag{10.15}$$

Derivation   Let the random variables $\tilde{x}_1, \ldots, \tilde{x}_n$ represent the data. We denote by $\mathcal{A}$ and $\mathcal{B}$ the indices of the data in the first and second groups, respectively. For the one-tailed test, the test statistic is represented by the random variable,

$$\tilde{t}_{\text{1-tail}} := \frac{1}{n_A}\sum_{i \in \mathcal{A}} \tilde{x}_i - \frac{1}{n_B}\sum_{i \in \mathcal{B}} \tilde{x}_i. \tag{10.16}$$

Under the null hypothesis, the random variables $\tilde{x}_i$, $1 \leq i \leq n$, are i.i.d. Bernoulli with the same parameter, which we denote by θ_{null}. The sum of Bernoulli random variables is a binomial random variable, as explained in the proof of Lemma 7.19. Therefore, $\sum_{i \in \mathcal{A}} \tilde{x}_i$ is binomial with parameters θ_{null} and n_A, and $\sum_{i \in \mathcal{B}} \tilde{x}_i$ is binomial with parameters θ_{null} and n_B. By the Gaussian approximation to the binomial distribution (Definition 9.39), $\sum_{i \in \mathcal{A}} \tilde{x}_i$ is approximately Gaussian with mean $n_A \theta_{\text{null}}$ and variance $n_A \theta_{\text{null}}(1 - \theta_{\text{null}})$. Equivalently, by Theorem 3.31, $\frac{1}{n_A}\sum_{i \in \mathcal{A}} \tilde{x}_i$ is approximately Gaussian with mean θ_{null} and variance $\theta_{\text{null}}(1 - \theta_{\text{null}})/n_A$. Similarly, $-\frac{1}{n_B}\sum_{i \in \mathcal{B}} \tilde{x}_i$ is approximately Gaussian with mean $-\theta_{\text{null}}$ and variance $\theta_{\text{null}}(1 - \theta_{\text{null}})/n_B$.

We can therefore approximate $\tilde{t}_{\text{1-tail}}$ as the sum of two independent Gaussian random variables with means that cancel out. By Theorem 9.36, the difference is Gaussian with zero mean and variance equal to the sum of the individual variances,

$$\theta_{\text{null}}(1 - \theta_{\text{null}})\left(\frac{1}{n_A} + \frac{1}{n_B}\right). \tag{10.17}$$

The null hypothesis does not specify the value of θ_{null} (just that it's the same for the two datasets), so we need to approximate it from the data. Under the null hypothesis that the data are i.i.d. Bernoulli with parameter θ_{null}, the maximum-likelihood estimator of θ_{null} is the total number of positive instances $k := k_A + k_B$ divided by the total number of data

$n := n_A + n_B$ (see Example 2.26). Replacing θ_{null} by k/n in (10.17) provides an estimate of the variance of $\tilde{t}_{\text{1-tail}}$ that can be computed from the data:

$$\sigma^2_{\text{null}} := \frac{k(n-k)}{n^2} \left(\frac{1}{n_A} + \frac{1}{n_B} \right). \tag{10.18}$$

If $\tilde{t}_{\text{1-tail}}$ is Gaussian with mean zero and variance σ^2_{null}, then $\tilde{t}_{\text{1-tail}}/\sigma_{\text{null}}$ is a standard Gaussian random variable $\tilde{z}$ with mean zero and unit variance (see Theorem 3.31). This yields the following approximation of the p-value function by Definition 3.3:

$$\text{pv}(t) := \text{P}\left(\tilde{t}_{\text{1-tail}} \geq t \right) \tag{10.19}$$

$$\approx \text{P}\left(\tilde{z} \geq \frac{t}{\sigma_{\text{null}}} \right) \tag{10.20}$$

$$= 1 - F_{\tilde{z}}\left(\frac{t}{\sigma_{\text{null}}} \right). \tag{10.21}$$

The test statistic for the two-tailed test is equal to the absolute value of the test statistic for the one-tailed test,

$$\tilde{t}_{\text{2-tails}} := \left| \tilde{t}_{\text{1-tail}} \right|, \tag{10.22}$$

so the corresponding approximation of the p-value function equals

$$\text{pv}(t) := \text{P}\left(\tilde{t}_{\text{2-tails}} \geq t \right) \tag{10.23}$$

$$= \text{P}\left(\tilde{t}_{\text{1-tail}} \geq t \right) + \text{P}\left(\tilde{t}_{\text{1-tail}} \leq -t \right) \tag{10.24}$$

$$\approx \text{P}\left(\tilde{z} \geq \frac{t}{\sigma_{\text{null}}} \right) + \text{P}\left(\tilde{z} \leq -\frac{t}{\sigma_{\text{null}}} \right) \tag{10.25}$$

$$= 2\text{P}\left(\tilde{z} \geq \frac{t}{\sigma_{\text{null}}} \right) \tag{10.26}$$

$$= 2\left(1 - F_{\tilde{z}}\left(\frac{t}{\sigma_{\text{null}}} \right) \right) \tag{10.27}$$

by symmetry of the zero-mean Gaussian pdf. ∎

The p-value for the *one-tailed* test in Definition 10.10 is the probability that the difference between the sample proportions is large *and positive* under the null hypothesis. Therefore, the test does not reject the null hypothesis, when the difference is *negative*, even if it is very extreme. By contrast, the *two-tailed* test takes into account both possibilities. The bottom two plots in Figure 10.2 illustrate the difference between the two tests.

Example 10.11 (Free throws under pressure: P-value). We apply the two-sample z test in Definition 10.10 to obtain a p-value for the test statistic in Example 10.6, assuming that the free throws are independent. We begin by applying a one-tailed test, which is only sensitive to positive differences between the home and away free-throw percentages, and is oblivious to situations where the away percentage is higher. Setting $n_A := 44$, $k_A := 34$, $n_B := 41$, $k_B := 22$, we obtain $\sigma_{\text{null}} = 0.103$, which yields the p-value function

$$\text{pv}(t) = 1 - F_{\tilde{z}}\left(\frac{t}{0.103} \right), \tag{10.28}$$

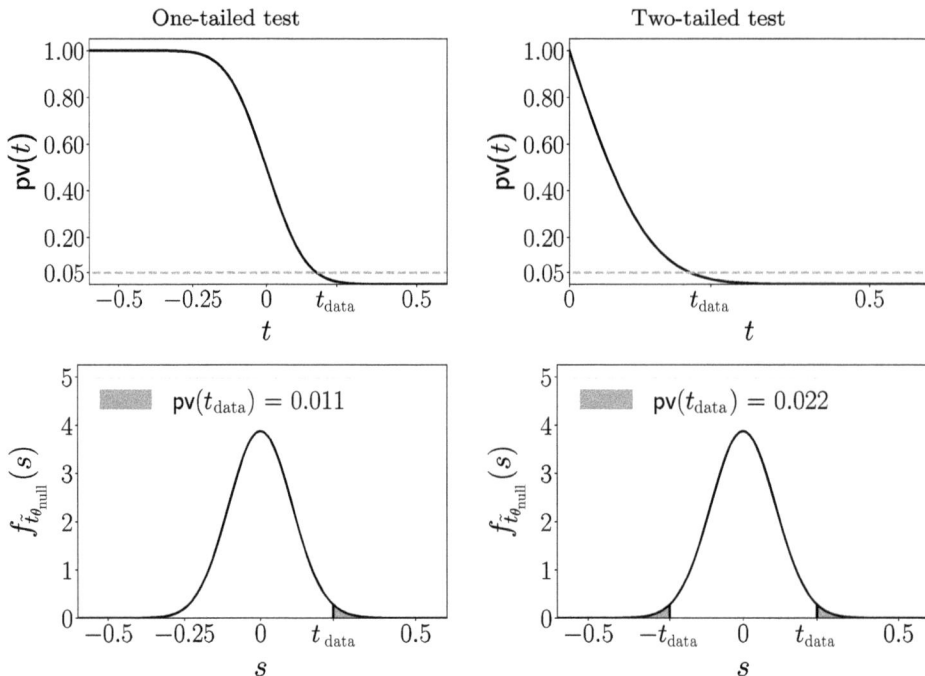

Figure 10.2 P-value for free throws under pressure. The top row shows the p-value functions for the one-tailed (left column) and two-tailed (right column) tests in Example 10.11. The p-value corresponding to the observed test statistic is equal to 0.011 and 0.022, respectively. This is below 0.05, a popular threshold for statistical significance (see Section 10.5) represented by the dashed gray line. The bottom row depicts the Gaussian approximation of the test-statistic distribution under the null hypothesis, derived in Definition 10.10. The p-value for the one-tailed test is obtained by integrating the right tail of the pdf (left). The p-value for the two-tailed test is obtained by integrating both tails of the pdf (right).

depicted in the top left graph of Figure 10.2. The bottom left graph shows the Gaussian approximation of the test-statistic distribution under the null hypothesis. The p-value corresponding to the observed test statistic $t_{\mathrm{data}} = 0.236$ is equal to 0.011. The data are not consistent with the null hypothesis, in the sense that if it holds (and our independence assumptions are correct), then we would observe results like this only around 1% of the time.

The one-tailed test derived in Example 10.11 ignores the possibility that Antetokounmpo might actually shoot better at away games (perhaps motivated by the fans taunting him). If we do want to take that possibility into account, we can instead apply the two-tailed test in Definition 10.10. The corresponding p-value function equals

$$\mathrm{pv}(t) = 2\left(1 - F_{\tilde{z}}\left(\frac{t}{0.103}\right)\right), \tag{10.29}$$

as depicted in the top right graph of Figure 10.2. The p-value function is obtained by integrating both tails of the same Gaussian pdf used for the one-tailed test, as shown in the bottom right graph of Figure 10.2. The resulting p-value is 0.022, double that of the one-tailed test, but still very small. Example 10.17 compares the properties of the one-tailed and two-tailed tests in more detail.

10.5 Statistical Significance

As described in Section 10.3, the p-value quantifies the consistency between the null hypothesis and the available data. In the scientific literature, p-values are typically compared to a predetermined threshold called the *significance level*. If the p-value is below the threshold, then we reject the null hypothesis, meaning that it is inconsistent with the data, and declare the result to be *statistically significant*. If the p-value is above the threshold, then we do not reject the null hypothesis, concluding that the evidence against it is insufficient. The following definition summarizes the different steps in a hypothesis test.

Definition 10.12 (Hypothesis test). *To perform a hypothesis test we:*

1 *Choose a conjecture.*
2 *Determine the corresponding null hypothesis.*
3 *Design a test statistic and model its distribution under the null hypothesis.*
4 *Decide on a significance level α.*
5 *Gather the data and compute the test statistic.*
6 *Compute the p-value.*
7 *Reject the null hypothesis, if the p-value is less than or equal to α.*

A decision based on a hypothesis test can be wrong in two different ways. A *false positive* or *type 1 error* occurs, if the null hypothesis holds, but we reject it. We call this a false positive because rejecting the null hypothesis supports our original conjecture. Conversely, a *false negative* or *type 2 error* occurs, if we fail to reject the null hypothesis, when it does not hold. In hypothesis testing, the main priority is to avoid the first type of error. In fact, the procedure described in Definition 10.12 is designed specifically to control the probability of a false positive. To understand why, we need to study the behavior of the test statistic under the null hypothesis.

Let us assume that the null hypothesis is simple: It states that $\theta = \theta_{null}$ for a certain constant θ_{null}. The random variable $\tilde{t}_{\theta_{null}}$ represents the test statistic under this null hypothesis. A false positive happens if (1) the null hypothesis holds and (2) the p-value is below the significance level α. The probability of this event is

$$\mathrm{P}\left(\text{False positive}\right) = \mathrm{P}\left(\mathrm{pv}(\tilde{t}_{\theta_{null}}) \leq \alpha\right). \tag{10.30}$$

To make sense of this expression, recall that the p-value function is a deterministic function, which maps any possible value of the test statistic to the corresponding p-value. Therefore, plugging the random variable $\tilde{t}_{\theta_{null}}$ into it yields another random variable, which we call

$$\tilde{u} := \mathrm{pv}(\tilde{t}_{\theta_{null}}). \tag{10.31}$$

A false positive occurs when $\tilde{u} \leq \alpha$, so the probability of a false positive equals the cdf of $\tilde{u}$ at α,

$$\mathrm{P}\left(\text{False positive}\right) = \mathrm{P}\left(\tilde{u} \leq \alpha\right) = F_{\tilde{u}}(\alpha). \tag{10.32}$$

Figure 10.3 shows the cdf $F_{\tilde{u}}$ of $\tilde{u}$ for the p-value functions derived in Examples 10.9 and 10.11. Notice that in both cases $F_{\tilde{u}}(\alpha) \leq \alpha$ for all α. This is no coincidence. Although it is not obvious at first sight, the p-value function is defined so that

$$F_{\tilde{u}}(\alpha) \leq \alpha, \tag{10.33}$$

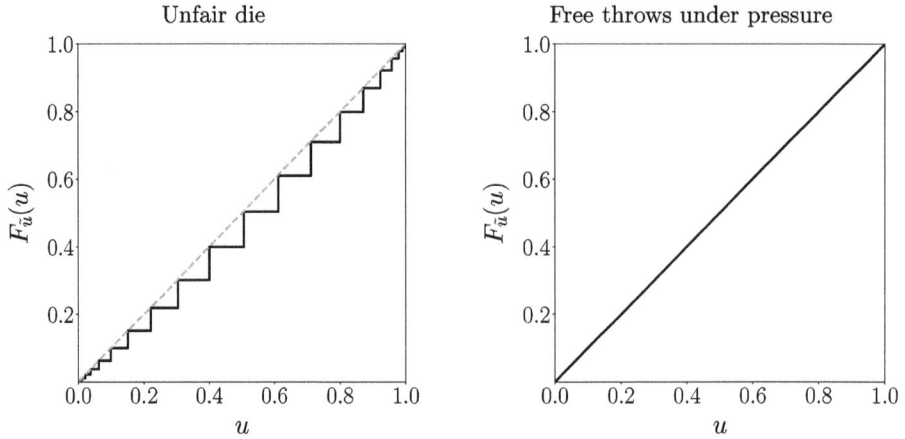

Figure 10.3 P-value under the null hypothesis. The plots show the cdf of the random variable $\tilde{u} := \mathrm{pv}(\tilde{t}_{\theta_{\mathrm{null}}})$ for the p-value functions in Examples 10.9 (left) and 10.11 (right). The cdf on the left lies below the diagonal line where $F_{\tilde{u}}(u) = u$. The cdf on the right coincides with the diagonal line, which implies that the p-value is uniformly distributed under the null hypothesis, as established by Theorem 10.13.

precisely in order to ensure that *the probability of a false positive is bounded by the significance level.* We dedicate the rest of the section to proving this rigorously.

The following theorem characterizes the distribution of the p-value under the null hypothesis.

Theorem 10.13 (Distribution of the p-value under the null hypothesis). *Let $\tilde{t}_\theta$ be a random variable representing a test statistic with a parametric distribution that depends on a parameter (or vector of parameters) θ. For any θ_{null}, let pv be the p-value function associated with the simple null hypothesis $\theta = \theta_{\mathrm{null}}$. We define the random variable $\tilde{u} := \mathrm{pv}(\tilde{t}_{\theta_{\mathrm{null}}})$ to represent the p-value under the null hypothesis. If $\tilde{t}_{\theta_{\mathrm{null}}}$ is continuous, $\tilde{u}$ is uniformly distributed in $[0, 1]$. If $\tilde{t}_{\theta_{\mathrm{null}}}$ is discrete, for any $u \in [0, 1]$,*

$$F_{\tilde{u}}(u) \leq u, \tag{10.34}$$

with equality if there is a t such that $\mathrm{pv}(t) = u$ and $\mathrm{P}\left(\tilde{t}_{\theta_{\mathrm{null}}} = t\right) \neq 0$.

Proof If $\tilde{t}_{\theta_{\mathrm{null}}}$ is continuous, the distribution of $F_{\tilde{t}_{\theta_{\mathrm{null}}}}(\tilde{t}_{\theta_{\mathrm{null}}})$ is uniformly distributed in $[0, 1]$ by Theorem 3.22. By definition, the p-value function equals

$$\mathrm{pv}(t) := \mathrm{P}\left(\tilde{t}_{\theta_{\mathrm{null}}} \geq t\right) = 1 - F_{\tilde{t}_{\theta_{\mathrm{null}}}}(t), \tag{10.35}$$

so $1 - \tilde{u} = 1 - \mathrm{pv}(\tilde{t}_{\theta_{\mathrm{null}}}) = F_{\tilde{t}_{\theta_{\mathrm{null}}}}(\tilde{t}_{\theta_{\mathrm{null}}})$ is uniformly distributed in $[0, 1]$, and therefore so is $\tilde{u}$ (see Exercise 3.10). Consequently, the cdf of $\tilde{u}$ is linear (see Example 3.7 and Figure 3.3), as illustrated in the right graph of Figure 10.3.

If $\tilde{t}_{\theta_{\mathrm{null}}}$ is discrete, then so is $\tilde{u}$. Let $t_1 < t_2 < \ldots$ denote the (countable) points for which $\mathrm{P}\left(\tilde{t}_{\theta_{\mathrm{null}}} = t_i\right) \neq 0$, $i = 1, 2, \ldots$ The corresponding values $u_i := \mathrm{pv}(t_i)$ that $\tilde{u}$ takes with nonzero probability are decreasing: $u_1 > u_2 > \ldots$ because for $t_i < t_j$,

$$u_i := \mathrm{pv}(t_i) = \mathrm{P}\left(\tilde{t}_{\theta_{\mathrm{null}}} \geq t_i\right) \tag{10.36}$$

$$= \mathrm{P}\left(\tilde{t}_{\theta_{\mathrm{null}}} = t_i\right) + \mathrm{P}\left(t_i < \tilde{t}_{\theta_{\mathrm{null}}} < t_j\right) + \mathrm{P}\left(\tilde{t}_{\theta_{\mathrm{null}}} \geq t_j\right) \tag{10.37}$$

$$> \mathrm{pv}(t_j) := u_j. \tag{10.38}$$

Consequently, for any $i = 1, 2, \ldots$ the events $\tilde{u} \leq u_i$ and $\tilde{t}_{\theta_{\mathrm{null}}} \geq t_i$ are equivalent, which implies

$$F_{\tilde{u}}(u_i) = \mathrm{P}\left(\tilde{u} \leq u_i\right) = \mathrm{P}\left(\tilde{t}_{\theta_{\mathrm{null}}} \geq t_i\right) \tag{10.39}$$

$$= \mathrm{pv}(t_i) = u_i. \tag{10.40}$$

Finally, for $u_i < u < u_{i+1}$, $\mathrm{P}\left(u_i < \tilde{u} \leq u\right) = 0$, so

$$F_{\tilde{u}}(u) = \mathrm{P}\left(\tilde{u} \leq u\right) \tag{10.41}$$

$$= \mathrm{P}\left(\tilde{u} \leq u_i\right) \tag{10.42}$$

$$= F_{\tilde{u}}(u_i) \tag{10.43}$$

$$= u_i < u. \tag{10.44}$$

The cdf therefore looks like the one depicted on the left of Figure 10.3. ∎

For simple null hypotheses, Theorem 10.13 immediately implies that the significance level bounds the probability of a false positive. The following theorem provides a proof and establishes that the same is true for composite null hypotheses.

Theorem 10.14 (The significance level bounds the probability of a false positive). *Let $\tilde{t}_\theta$ be a random variable representing a test statistic with a parametric distribution that depends on a parameter (or vector of parameters) θ. For any $0 < \alpha < 1$, if the null hypothesis is rejected when the p-value is smaller than or equal to the significance level α, then the probability of a false positive is bounded by α.*

Proof If the null hypothesis is simple, and can be expressed as $\theta = \theta_{\mathrm{null}}$ for some θ_{null}, then by Theorem 10.13,

$$F_{\tilde{u}}(u) \leq u \tag{10.45}$$

for $\tilde{u} := \mathrm{pv}(\tilde{t}_{\theta_{\mathrm{null}}})$ and any $u \in [0, 1]$. Consequently,

$$\mathrm{P}\left(\text{False positive}\right) = \mathrm{P}\left(\mathrm{pv}(\tilde{t}_{\theta_{\mathrm{null}}}) \leq \alpha\right) \tag{10.46}$$

$$= \mathrm{P}\left(\tilde{u} \leq \alpha\right) \tag{10.47}$$

$$= F_{\tilde{u}}(\alpha) \tag{10.48}$$

$$\leq \alpha. \tag{10.49}$$

Now, let us assume that the null hypothesis is composite and can be expressed as $\theta \in \Theta_{\mathrm{null}}$ for some set Θ_{null}. By definition of the p-value function for composite hypotheses, $\mathrm{pv}(t) = \sup_{\theta \in \Theta_{\mathrm{null}}} \mathrm{pv}_\theta(t)$. For a false positive to occur, the null hypothesis must hold, so $\theta = \theta_0$ for some $\theta_0 \in \Theta_{\mathrm{null}}$. To ease notation, we define $\mathrm{pv}_{\theta_0}(t) := \mathrm{P}\left(\tilde{t}_{\theta_0} \geq t\right)$, which is the p-value function under the null hypothesis $\theta = \theta_0$. For any $\theta_0 \in \Theta_{\mathrm{null}}$,

$$P\left(\text{False positive}\right) = P\left(\text{pv}(\tilde{t}_{\theta_0}) \leq \alpha\right) \tag{10.50}$$

$$= P\left(\sup_{\theta \in \Theta_{\text{null}}} \text{pv}_\theta(\tilde{t}_{\theta_0}) \leq \alpha\right) \tag{10.51}$$

$$\leq P\left(\text{pv}_{\theta_0}(\tilde{t}_{\theta_0}) \leq \alpha\right) \tag{10.52}$$

$$= F_{\tilde{u}}(\alpha) \tag{10.53}$$

$$\leq \alpha \tag{10.54}$$

by Theorem 10.13, setting $\tilde{u} := \text{pv}_{\theta_0}(\tilde{t}_{\theta_0})$. The inequality (10.52) holds because θ_0 belongs to Θ_{null} by assumption. This implies $\sup_{\theta \in \Theta_{\text{null}}} \text{pv}_\theta(\tilde{t}_{\theta_0}) \geq \text{pv}_{\theta_0}(\tilde{t}_{\theta_0})$, so the event $\sup_{\theta \in \Theta_{\text{null}}} \text{pv}_\theta(\tilde{t}_{\theta_0}) \leq \alpha$ is a subset of the event $\text{pv}_{\theta_0}(\tilde{t}_{\theta_0}) \leq \alpha$ and hence has smaller or equal probability by Lemma 1.12. ∎

10.6 The Power

As explained in Section 10.5, the hypothesis-testing framework is designed to control the probability of false positives. However, in order to be useful, hypothesis tests must also find true positives! The *power* of a test is the probability of a true positive, i.e. of rejecting the null hypothesis when it does *not* hold.

To formally define the power for a parametric test, we introduce the power function, which maps each possible value of the parameter to the corresponding probability of rejecting the null hypothesis.

Definition 10.15 (Power function). *Let $\tilde{t}_\theta$ be a random variable representing a test statistic with a parametric distribution that depends on a parameter (or vector of parameters) θ, and let pv be the p-value function associated with a certain null hypothesis. As prescribed in Definition 10.12 we reject the null hypothesis, if the p-value is smaller than or equal to the significance level $\alpha \in [0, 1]$. The power function maps each possible value of θ to the probability of rejecting the null hypothesis, if the test statistic is generated by the parametric distribution with parameter equal to θ,*

$$\text{pow}(\theta) := P\left(\text{pv}\left(\tilde{t}_\theta\right) \leq \alpha\right). \tag{10.55}$$

The term *power function* is somewhat misleading because the function is also defined for values of the parameters associated with the null hypothesis. Let Θ_{null} and Θ_{alt} be the set of values of θ associated with the null and alternative hypotheses, respectively. For $\theta \in \Theta_{\text{alt}}$, $\text{pow}(\theta)$ is equal to the power of the test, defined as the probability of a true positive. By contrast, for $\theta \in \Theta_{\text{null}}$, $\text{pow}(\theta)$ is the probability of a false positive. Therefore, we would like $\text{pow}(\theta)$ to be close to one for $\theta \in \Theta_{\text{alt}}$, and close to zero for $\theta \in \Theta_{\text{null}}$.

Example 10.16 (Unfair die: Power function). In this example, we derive the power function associated with the hypothesis test in Example 10.9 for an arbitrary number of die rolls n. We begin by defining the set of values of the test statistic for which we reject the null hypothesis. This set is known as the *rejection region*. As prescribed in Definition 10.12, we reject the null hypothesis, if the p-value is smaller than or equal to the significance level $\alpha \in [0, 1]$, so the rejection region equals

$$\mathcal{R} := \{\tau : \text{pv}(\tau) \leq \alpha\}. \tag{10.56}$$

By (10.6) (replacing 100 with n), the minimum value τ_{thresh} in the rejection region equals

$$\tau_{\text{thresh}} := \min_{0 \leq \tau \leq n} \{\tau : \text{pv}(\tau) \leq \alpha\} \tag{10.57}$$

$$= \min_{0 \leq \tau \leq n} \left\{\tau : \sum_{i=\tau}^{n} \binom{n}{i} \theta_{\text{null}}^{i} (1 - \theta_{\text{null}})^{n-i} \leq \alpha\right\}, \tag{10.58}$$

where $\theta_{\text{null}} := 1/6$. If $\tau \geq \tau_{\text{thresh}}$, then $\tau \in \mathcal{R}$ since

$$\text{pv}(\tau) := \text{P}\left(\tilde{t}_{\theta_{\text{null}}} \geq \tau\right) \tag{10.59}$$

$$\leq \text{P}\left(\tilde{t}_{\theta_{\text{null}}} \geq \tau_{\text{thresh}}\right) \tag{10.60}$$

$$= \text{pv}(\tau_{\text{thresh}}) \tag{10.61}$$

$$\leq \alpha \tag{10.62}$$

because the event $\tilde{t}_{\theta_{\text{null}}} \geq \tau$ is a subset of the event $\tilde{t}_{\theta_{\text{null}}} \geq \tau_{\text{thresh}}$ and hence has smaller or equal probability by Lemma 1.12. Conversely, if $\tau < \tau_{\text{thresh}}$, then $\tau \notin \mathcal{R}$ because τ_{thresh} is the minimum of $\mathcal{R}$. Therefore, we reject the null hypothesis, if and only if the test statistic is greater than or equal to the threshold τ_{thresh}. A lower significance level corresponds to a higher threshold, and vice versa. If $n := 100$, by (10.58) the threshold τ_{thresh} equals 27 for $\alpha := 0.01$, 24 for $\alpha := 0.05$, and 20 for $\alpha := 0.25$. The test statistic computed from the observed data is 21, so we would reject the null hypothesis for $\alpha := 0.25$, but not for $\alpha := 0.05$.

Since the events $\text{pv}\left(\tilde{t}_{\theta}\right) \leq \alpha$ and $\tilde{t}_{\theta} \geq \tau_{\text{thresh}}$ are equivalent, we can express the power function in terms of τ_{thresh}. The test statistic is binomial with parameters n and θ, as explained in Example 10.7. Consequently, by Definition 2.17 and Theorem 2.5, the power function equals

$$\text{pow}(\theta) := \text{P}\left(\text{pv}\left(\tilde{t}_{\theta}\right) \leq \alpha\right) \tag{10.63}$$

$$= \text{P}\left(\tilde{t}_{\theta} \geq \tau_{\text{thresh}}\right) \tag{10.64}$$

$$= \sum_{i=\tau_{\text{thresh}}}^{n} \binom{n}{i} \theta^{i} (1 - \theta)^{n-i}. \tag{10.65}$$

The graph on the left of Figure 10.4 shows the power function for different values of α when $n := 100$. In all cases, the power function is bounded by α for $\theta \leq \theta_{\text{null}}$. This should come as no surprise. The p-value function is associated with the composite null hypothesis $\theta \in \Theta_{\text{null}} := [0, 1/6]$ (see Example 10.9), so by Theorem 10.14 the probability of rejecting the null hypothesis must be bounded by α for any $\theta \in \Theta_{\text{null}}$.

Increasing the significance level α, increases the power function for any fixed value of the parameter θ because for any $\alpha_1 \leq \alpha_2$, the event $\text{pv}\left(\tilde{t}_{\theta}\right) \leq \alpha_1$ is a subset of the event $\text{pv}\left(\tilde{t}_{\theta}\right) \leq \alpha_2$, so by Lemma 1.12.

$$\text{P}\left(\text{pv}\left(\tilde{t}_{\theta}\right) \leq \alpha_1\right) \leq \text{P}\left(\text{pv}\left(\tilde{t}_{\theta}\right) \leq \alpha_2\right). \tag{10.66}$$

Therefore higher α results in more power, as we can see in the left graph of Figure 10.4. For instance, at $\theta := 1/4$, the power equals 0.358 for $\alpha := 0.01$, 0.629 for $\alpha := 0.05$ and 0.9 for $\alpha := 0.25$. In other words, if the probability of rolling a three is $1/4$, the test correctly rejects the null hypothesis one third of the time if $\alpha := 0.01$, a bit less than two thirds of the time if $\alpha := 0.05$, and 90% of the time if $\alpha := 0.25$. However, increasing α is not a

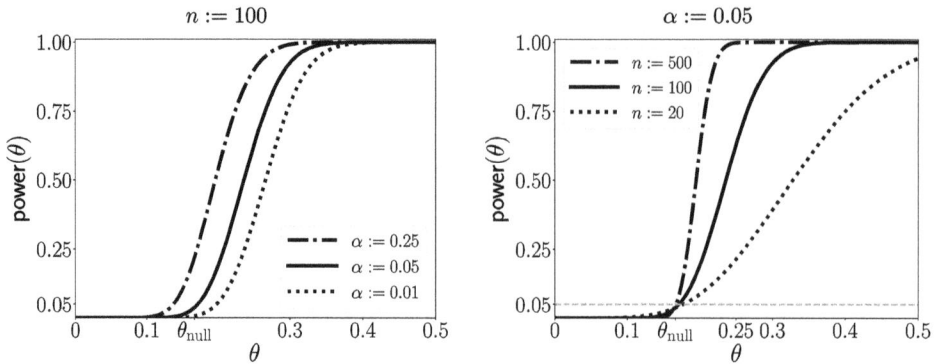

Figure 10.4 Power function for a (possibly) unfair die. The graphs show the power function derived in Example 10.16 for different values of the significance level α and number of data n. The power function stays below α for $\theta \leq \theta_{\text{null}}$ in all cases. Increasing α shifts the whole curve upwards (left plot): The probability of true positives is higher (because the power function is larger for $\theta > \theta_{\text{null}}$), but so is the probability of false positives (because the power function is larger for $\theta \leq \theta_{\text{null}}$). By contrast, increasing the number of data while keeping α constant (right plot) increases the power function for the alternative hypothesis $\theta > \theta_{\text{null}}$, but not for the null hypothesis $\theta \leq \theta_{\text{null}}$, so it increases the probability of true positives, without increasing the probability of false positives.

viable strategy to increase the power because it comes at the cost of raising the probability of a false positive. If the die is fair ($\theta := 1/6$) and we set $\alpha := 0.25$, the probability of a false positive is around 1/4!

Often, the only way to increase the power without increasing the probability of false positives is to use more data. The graph on the right of Figure 10.4 shows the power function (10.65) for $\alpha := 0.05$ and $n \in \{20, 100, 500\}$. The power function remains bounded by 0.05 in the null-hypothesis region $\theta \leq \theta_{\text{null}}$ for all values of n, which means that the probability of a false positive is less than or equal to 0.05, as guaranteed by Theorem 10.14. However, the power increases dramatically as a function of n over the alternative-hypothesis region $\theta \geq \theta_{\text{null}}$. At $\theta := 1/4$, the power equals 0.214 for $n := 20$, 0.629 for $n := 100$, and 0.998 for $n := 500$. This means that for $n := 20$ we fail to reject the null hypothesis 78.6% of the time, and for $n := 100$, we fail to reject more than one third of the time. Increasing the number of data solves the problem. For $n := 500$, if the probability of rolling a three is 1/4, we are almost guaranteed to correctly reject the null hypothesis; the probability of a false negative is just 0.2%.

A hypothesis test with low power is said to be *underpowered*. Example 10.16 shows that this is the case for our die example. In order to ensure sufficient power to detect that the die is unfair, 100 die rolls are not enough. We need to gather more data.

The power function is useful to compare different hypothesis tests designed for the same scenario. In the following example, we leverage it to compare the one-tailed and two-tailed tests in Example 10.11.

Example 10.17 (Free throws under pressure: Power function). In this example, we compare the one-tailed and two-tailed tests in Example 10.11 using their respective power functions. For simplicity, we fix the home free-throw percentage to $\theta_{\text{home}} := 0.685$, which was Antetokounmpo's free-throw percentage during the 2020/2021 regular season, and compare the power functions as a function of the away free-throw percentage θ_{away}. To derive each power function, we follow a similar strategy to Example 10.16. First, we compute the smallest value of the test statistic for which we reject the null hypothesis. Then, we compute the probability that the test statistic exceeds that value as a function of the parameters.

By Definition 10.10 the p-value function for the one-tailed test in Example 10.11 is

$$\text{pv}(t) = 1 - F_{\tilde{z}}\left(\frac{t}{\sigma_{\text{null}}}\right), \tag{10.67}$$

where $\tilde{z}$ is a standard Gaussian random variable with zero mean and unit variance, and σ^2_{null} is the variance of the test statistic $\tilde{t}_{\text{1-tail}}$ under the null hypothesis that $\theta_{\text{away}} = \theta_{\text{home}}$.

The cdf $F_{\tilde{z}}$ of the Gaussian random variable $\tilde{z}$ is invertible, so from (10.67) we can find the threshold $\tau_{1-\text{tail}}$, at which we reject the null hypothesis for a significance level α, by solving the equation

$$\text{pv}(\tau_{1-\text{tail}}) = 1 - F_{\tilde{z}}\left(\frac{\tau_{1-\text{tail}}}{\sigma_{\text{null}}}\right) = \alpha, \tag{10.68}$$

which yields

$$\tau_{1-\text{tail}} = \sigma_{\text{null}} F_{\tilde{z}}^{-1}(1 - \alpha). \tag{10.69}$$

In our observed data $\sigma_{\text{null}} = 0.103$ (see Example 10.11). For $\alpha = 0.05$, we obtain $\tau_{1-\text{tail}} = 0.166$. Recall that the observed test statistic is $0.236 > \tau_{1-\text{tail}}$, so we reject the null hypothesis, in accordance with Example 10.11.

By the same reasoning, the threshold for the test statistic in the two-tailed test from Example 10.11 is obtained by solving the equation

$$\text{pv}(\tau_{\text{2-tails}}) = 2\left(1 - F_{\tilde{z}}\left(\frac{\tau_{2-\text{tails}}}{\sigma_{\text{null}}}\right)\right) = \alpha, \tag{10.70}$$

which yields

$$\tau_{2-\text{tails}} = \sigma_{\text{null}} F_{\tilde{z}}^{-1}\left(1 - \frac{\alpha}{2}\right). \tag{10.71}$$

In our observed data, for $\alpha = 0.05$, $\tau_{2-\text{tails}} = 0.198$. We also reject the null hypothesis for this test, since the observed test statistic is $0.236 > \tau_{2-\text{tails}}$.

In order to derive the power function, we model the test statistic for the one-tailed test as the random variable

$$\tilde{t}_\theta := \frac{1}{n_{\text{home}}} \sum_{i=1}^{n_{\text{home}}} \tilde{h}_i - \frac{1}{n_{\text{away}}} \sum_{i=1}^{n_{\text{away}}} \tilde{a}_i, \tag{10.72}$$

where $\tilde{h}_i$ and $\tilde{a}_i$ are independent Bernoulli random variables representing the ith free throw attempted at home and away, respectively. We assume that all home games share the same Bernoulli parameter θ_{home} and all away games share the same Bernoulli parameter θ_{away}, so the distribution only depends on the parameter vector

$$\theta := \begin{bmatrix} \theta_{\text{home}} \\ \theta_{\text{away}} \end{bmatrix}. \tag{10.73}$$

By similar arguments to the ones used to derive Definition 10.10, $\tilde{t}_\theta$ is approximately Gaussian with mean $\theta_{home} - \theta_{away}$ and variance

$$\sigma^2_{pow} := \frac{\theta_{home}(1 - \theta_{home})}{n_{home}} + \frac{\theta_{away}(1 - \theta_{away})}{n_{away}}. \tag{10.74}$$

See Exercise 10.5 for more details. Equivalently, by Theorem 3.31, $\tilde{t}_\theta$ can be approximated by $\sigma_{pow}\tilde{z} + \theta_{home} - \theta_{away}$, where $\tilde{z}$ is a standard Gaussian random variable with mean zero and unit variance, with cdf denoted by $F_{\tilde{z}}$. This yields the following approximation for the power function of the one-tailed test:

$$\text{pow}(\theta) := P\left(\text{pv}\left(\tilde{t}_\theta\right) \leq \alpha\right) \tag{10.75}$$

$$= P\left(\tilde{t}_\theta \geq \tau_{1-tail}\right) \tag{10.76}$$

$$\approx P\left(\sigma_{pow}\tilde{z} + \theta_{home} - \theta_{away} \geq \tau_{1-tail}\right) \tag{10.77}$$

$$= P\left(\tilde{z} \geq \frac{\tau_{1-tail} - (\theta_{home} - \theta_{away})}{\sigma_{pow}}\right) \tag{10.78}$$

$$= 1 - F_{\tilde{z}}\left(\frac{\tau_{1-tail} - (\theta_{home} - \theta_{away})}{\sigma_{pow}}\right). \tag{10.79}$$

By the same argument, the approximation for the two-tailed test is

$$\text{pow}(\theta) = P\left(\left|\tilde{t}_\theta\right| \geq \tau_{2-tails}\right) \tag{10.80}$$

$$\approx P\left(\sigma_{pow}\tilde{z} + \theta_{home} - \theta_{away} \geq \tau_{2-tails}\right) + P\left(\sigma_{pow}\tilde{z} + \theta_{home} - \theta_{away} \leq -\tau_{2-tails}\right)$$

$$= P\left(\tilde{z} \geq \frac{\tau_{2-tails} - (\theta_{home} - \theta_{away})}{\sigma_{pow}}\right) + P\left(\tilde{z} \leq \frac{-\tau_{2-tails} - (\theta_{home} - \theta_{away})}{\sigma_{pow}}\right)$$

$$= 2 - F_{\tilde{z}}\left(\frac{\tau_{2-tails} - (\theta_{home} - \theta_{away})}{\sigma_{pow}}\right) - F_{\tilde{z}}\left(\frac{\tau_{2-tails} + (\theta_{home} - \theta_{away})}{\sigma_{pow}}\right) \tag{10.81}$$

by symmetry of the Gaussian pdf. A subtle detail is that computing τ_{1-tail} and $\tau_{2-tails}$ requires knowing the variance of the test statistic under the null hypothesis σ^2_{null}, which is typically estimated from the observed data using the formula (10.13). Here, we instead set

$$\sigma^2_{null} := \theta_{home}(1 - \theta_{home})\left(\frac{1}{n_{home}} + \frac{1}{n_{away}}\right) \tag{10.82}$$

since this is the true variance of the test statistic under the null hypothesis $\theta_{home} = \theta_{away}$; see (10.17) in the derivation of Definition 10.10.

Figure 10.5 shows the two power functions for $\alpha := 0.05$ as a function of θ_{away}, fixing $\theta_{home} := 0.685$. Reassuringly, both power functions are bounded by α when the null hypothesis holds ($\theta_{away} = \theta_{home}$), as guaranteed by Theorem 10.14. For $\theta_{away} < \theta_{home}$, the power function of both tests increases as a function of θ_{away}, but it is clearly higher for the one-tailed test. Consequently, the one-tailed test has higher power when Antetokounmpo shoots worse in away games. By contrast, if he shoots *better* in away games, then the one-tailed test is completely underpowered: For $\theta_{away} > \theta_{home}$, its power function rapidly decays toward zero. By contrast, within that region the power function of the two-tailed test increases as θ_{away} decreases, indicating that the test is able to detect situations where the free-throw percentage at away games is higher than at home, as it was designed to do.

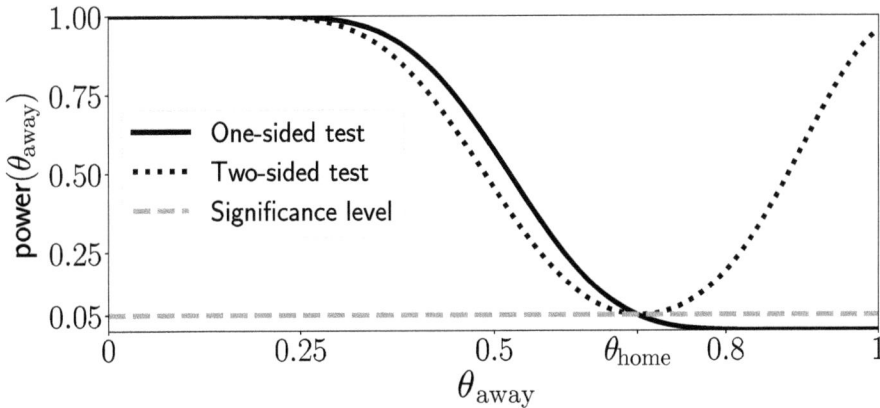

Figure 10.5 Power function for free throws under pressure. The graph shows the power functions of the one-tailed and two-tailed tests in Example 10.11 as a function of the free-throw percentage in away games θ_{away}, when the percentage at home games is fixed to equal $\theta_{\text{home}} := 0.685$. For $\theta_{\text{away}} < \theta_{\text{home}}$, the power function of both tests increases away from θ_{home}, but it is higher for the one-tailed test. By contrast, for $\theta_{\text{away}} > \theta_{\text{home}}$, the power function of the two-tailed test again increases away from θ_{home}, but the power function of the one-tailed test does not.

Computing the power function of a hypothesis test analytically can be complicated or even intractable. An alternative is to leverage the Monte Carlo method from Section 1.7. The key insight is that the power is a probability of a certain event (rejection of the null hypothesis), so it can be approximated via the Monte Carlo method, as long as we can simulate the event.

Definition 10.18 (Monte Carlo power estimation). *Let $\tilde{t}_\theta$ denote a test statistic following a known parametric distribution with parameters θ. To estimate the power function at θ, we first simulate k independent samples of $\tilde{t}_\theta$: $t_1, \ldots, t_k$. Then we apply the p-value function pv of the hypothesis test to each of the samples. By Definition 10.15, the power function at θ is the probability of rejecting the null hypothesis, or equivalently the probability that the p-value is smaller than or equal to the significance level α. We approximate this probability by the corresponding empirical probability, which equals the fraction of p-values smaller than or equal to α,*

$$\text{pow}(\theta) := P\left(\text{pv}\left(\tilde{t}_\theta\right) \leq \alpha\right) \approx \frac{1}{k} \sum_{i=1}^{k} 1(\text{pv}\left(t_i\right) \leq \alpha), \tag{10.83}$$

where $1(\text{pv}\left(t_i\right) \leq \alpha)$ is an indicator function that is equal to one if $\text{pv}\left(t_i\right) \leq \alpha$ and to zero otherwise.

In Figure 10.6, we show the results of applying the Monte Carlo method to approximate the power function of the one-tailed test in Example 10.17 for $\theta_{\text{home}} := 0.685$ and several values of θ_{away}. Following Definition 10.18, we repeatedly sample the test statistic of the one-tailed test, defined in (10.72), by simulating the home free throws using n_{home} independent Bernoulli random variables with parameter θ_{home} and the away free throws using n_{away} Bernoulli random variables with parameter θ_{away}. Then, we compute the corresponding p-values using the p-value function derived in Example 10.11. The estimated power is the fraction of simulations for which the p-value is below the significance level $\alpha := 0.05$.

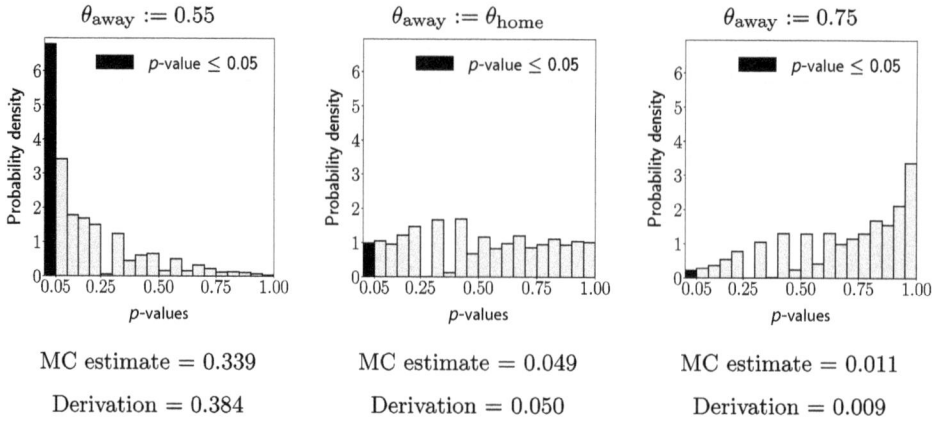

Figure 10.6 Monte Carlo power-function estimation. The plots depict a Monte Carlo estimate of the power function of the one-tailed test in Example 10.17 for $\theta_{\text{home}} := 0.685$ and different values of the parameter θ_{away}. Each plot shows the histogram of one million p-values computed from independent samples of the test statistic. The fraction of p-values below the significance threshold (represented by the black bar) approximate the probability of rejecting the null hypothesis, and therefore the power function of the test for that value of θ. At the bottom of each graph, we compare the Monte Carlo (MC) estimates to the values computed from the power function derived analytically in Example 10.17.

Interestingly, this is *more accurate* than our theoretical derivation in Example 10.17 (but also much more computationally expensive). The reason is that in the theoretical derivation, we plug the true parameter θ_{home} into (10.82) in order to compute the variance of the test statistic under the null hypothesis σ^2_{null}. However, in practice σ^2_{null} is estimated from the data via the formula (10.13). When performing the Monte Carlo simulations, we take this into account, computing σ^2_{null} by applying (10.13) to the simulated free throws.

10.7 Nonparametric Testing: The Permutation Test

Sections 10.3 and 10.4 present parametric hypothesis tests, which rely on a parametric model to describe the distribution of the test statistic. In certain situations, it can be challenging to identify a parametric model that is appropriate for the available data. Fortunately, it is possible to perform hypothesis testing without a parametric model, as explained in this section. We focus on the permutation test, which is one of the most popular nonparametric tests. Section 10.7.1 provides an intuitive introduction to permutation testing based on a toy example. In Section 10.7.2, we define the p-value of the permutation test formally and explain how to compute it in practice. Finally, Section 10.7.3 shows that permutation tests allow us to control the probability of a false positive, just like parametric tests.

10.7.1 Intuition

In this section, we use a toy example to explain the logic underlying permutation tests. My friends from Spain often complain about prices in New York, claiming that it is more expensive than Madrid. Our goal is to verify this via hypothesis testing, focusing on hamburger prices.

Following the steps in Definition 10.12, we begin by defining the null hypothesis. Our original hypothesis is that prices are different in both cities, so we choose the null hypothesis that *the distribution of burger prices is the same in both cities*. Then, we select a test statistic that should be small if the null hypothesis holds. A natural choice is the difference between the mean burger price in New York and in Madrid. Our data are the price of a cheeseburger at two restaurants in New York (NY) and at two restaurants in Madrid:

$$\text{NY}: \$16, \quad \text{NY}: \$18, \quad \text{Madrid}: \$13, \quad \text{Madrid}: \$12. \tag{10.84}$$

The observed value of the test statistic equals

$$t_{\text{data}} := m(\text{NY}) - m(\text{Madrid}) = \frac{16 + 18}{2} - \frac{13 + 12}{2} = 4.5, \tag{10.85}$$

where $m(\text{NY})$ and $m(\text{Madrid})$ denote the sample mean of burger prices in New York and Madrid respectively (see Definition 7.15). This value seems pretty high, but we don't have a lot of data, so it could just be large by chance. To settle the question, we need to model the test statistic under the null hypothesis and compute the p-value. Our goal is to do this without a parametric model of the test statistic.

In order to bypass the use of a parametric model, we exploit a key implication of the null hypothesis. If the distribution of burger prices is the same in New York and Madrid, then the label indicating the city associated with each data point is *meaningless*. Therefore, permuting the data while fixing the labels *should not change the behavior of the test statistic*. If the null hypothesis holds, we would have been equally likely to observe

$$\text{NY}: \$13, \quad \text{NY}: \$18, \quad \text{Madrid}: \$12, \quad \text{Madrid}: \$16, \tag{10.86}$$

or any of the $4! = 24$ possible permutations listed in Figure 10.7. Permuting the data changes the value of the test statistic. For example, the test statistic for (10.86) equals

$$t := m(\text{NY}) - m(\text{Madrid}) = \frac{18 + 13}{2} - \frac{16 + 12}{2} = 1.5. \tag{10.87}$$

If the labels are meaningless, the test statistic associated with each permutation should occur with the same probability. In that case, it is unlikely for t_{data} to be larger than most of them. Consequently, an abnormally large t_{data} is evidence against the null hypothesis. To determine whether t_{data} is abnormally large, we compute the fraction of permutations that result in test statistics greater than or equal to t_{data}. In our example, there are four such permutations (see Figure 10.7). This means that, under the null hypothesis, the test statistic is greater than or equal to the observed value $4/24 = 16.7\%$ of the time. The probability is not very small, so we cannot conclude that the data contradict the null hypothesis.

10.7.2 The P-Value of the Permutation Test

The approach described in Section 10.7.1 yields a probability quantifying the consistency between the null hypothesis and the observed test statistic. In other words, it allows us to compute a p-value without a parametric model! The formal definition of this p-value is similar to Definition 10.8, except that in this case the p-value is a conditional probability.

Definition 10.19 (*P*-value function and p-value of a permutation test)**.** *We define Π_x to be the set of vectors that can be obtained by permuting the entries of the vector x. For example, if $a \neq b \neq c$,*

NY	NY	M	M	t
12	13	16	18	-4.5
12	13	18	16	-4.5
12	16	13	18	-1.5
12	16	18	13	-1.5
12	18	13	16	0.5
12	18	16	13	0.5
13	12	16	18	-4.5
13	12	18	16	-4.5
13	16	12	18	-0.5
13	16	18	12	-0.5
13	18	12	16	1.5
13	18	16	12	1.5

NY	NY	M	M	t
16	12	13	18	-1.5
16	12	18	13	-1.5
16	13	12	18	-0.5
16	13	18	12	-0.5
16	**18**	**12**	**13**	**4.5**
16	**18**	**13**	**12**	**4.5**
18	12	16	13	0.5
18	12	13	16	0.5
18	**16**	**12**	**13**	**4.5**
18	**16**	**13**	**12**	**4.5**
18	13	12	16	1.5
18	13	16	12	1.5

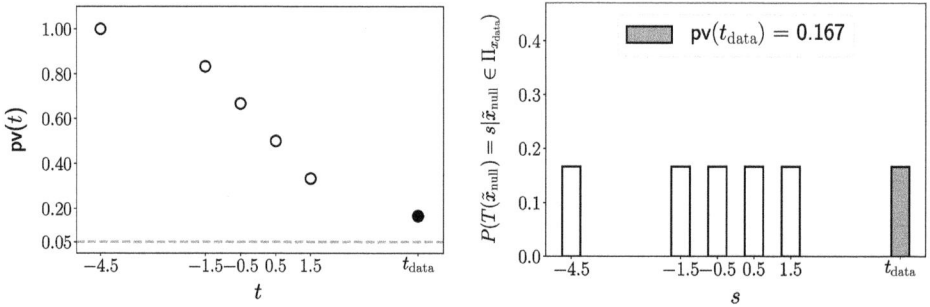

Figure 10.7 *P*-value for difference in burger prices. The table shows all the possible permutations of the data x_{data} in Section 10.7.1 and the corresponding values of the test statistic t. The plots show the p-value function (left) and the conditional pmf of the test statistic given that the data are a permutation of x_{data} (right). The p-value is computed by summing the pmf over all values greater than or equal to the observed test statistic t_{data}. This corresponds to the four permutations highlighted in bold in the table. The resulting p-value equals 4/24, which is above the significance level of 0.05 indicated by the dashed line.

$$x := \begin{bmatrix} a \\ b \\ c \end{bmatrix} \qquad then \qquad \Pi_x = \left\{ \begin{bmatrix} a \\ b \\ c \end{bmatrix}, \begin{bmatrix} a \\ c \\ b \end{bmatrix}, \begin{bmatrix} b \\ a \\ c \end{bmatrix}, \begin{bmatrix} b \\ c \\ a \end{bmatrix}, \begin{bmatrix} c \\ a \\ b \end{bmatrix}, \begin{bmatrix} c \\ b \\ a \end{bmatrix} \right\}, \qquad (10.88)$$

assuming $a \neq b \neq c$, so that all the elements in Π_x are distinct.

Let $x_{\text{data}} \in \mathbb{R}^n$ be a vector of data, and let $T \colon \mathbb{R}^n \to \mathbb{R}$ be a function that maps any n-dimensional data vector x to a test statistic $T(x)$. We model the data under the null hypothesis as an n-dimensional random vector $\tilde{x}_{\text{null}}$. The p-value function pv of the permutation test maps every possible value t of the test statistic to the conditional probability that $\tilde{t}_{\text{null}} := T(\tilde{x}_{\text{null}})$ is greater than or equal to t, given that $\tilde{x}_{\text{null}}$ is in $\Pi_{x_{\text{data}}}$:

$$\text{pv}(t) := \text{P}\left(\tilde{t}_{\text{null}} \geq t \mid \tilde{x}_{\text{null}} \in \Pi_{x_{\text{data}}} \right). \qquad (10.89)$$

The p-value of the observed data x_{data} is equal to $\text{pv}(t_{\text{data}})$, where $t_{\text{data}} := T(x_{\text{data}})$ denotes the observed test statistic.

Our reasoning in Section 10.7.1 hinges on the assumption that the data are *exchangeable* under the null hypothesis, which means that if we permute their order (with respect to the labels indicating the city), their joint distribution remains unchanged.

Definition 10.20 (Exchangeability). *The entries of a random vector $\tilde{x}$ are exchangeable, if permuting them does not change the joint distribution of $\tilde{x}$. Let Π_x be the set of vectors obtained by permuting the entries of the vector x. The entries of a discrete random vector $\tilde{x}$ are exchangeable, if for all x in the range of $\tilde{x}$,*

$$p_{\tilde{x}}(x) = p_{\tilde{x}}(v) \qquad for\ all\ v \in \Pi_x, \tag{10.90}$$

where $p_{\tilde{x}}$ denotes the joint pmf of $\tilde{x}$. Similarly, the entries of a d-dimensional continuous random vector $\tilde{x}$ are exchangeable, if for all $x \in \mathbb{R}^d$,

$$f_{\tilde{x}}(x) = f_{\tilde{x}}(v) \qquad for\ all\ v \in \Pi_x, \tag{10.91}$$

where $f_{\tilde{x}}$ denotes the joint pdf of $\tilde{x}$.

The following lemma shows that i.i.d. random variables are exchangeable.

Lemma 10.21 (Exchangeability of i.i.d. random variables). *The elements of any sequence of d i.i.d. random variables $\tilde{x}_1, \tilde{x}_2, \ldots, \tilde{x}_d$ are exchangeable.*

Proof Let us assume that the random vector is continuous. The same argument can be applied to discrete random vectors replacing the joint pdf with the joint pmf. By the i.i.d. assumption,

$$f_{\tilde{x}}(x) = \prod_{i=1}^{d} f_{\tilde{x}_i}(x_i) = \prod_{i=1}^{d} f_{\mathrm{marg}}(x_i), \tag{10.92}$$

where f_{marg} denotes the marginal pdf of the random variables. Therefore, for any $v \in \Pi_x$,

$$f_{\tilde{x}}(v) = \prod_{i=1}^{d} f_{\mathrm{marg}}(v_i) = \prod_{i=1}^{d} f_{\mathrm{marg}}(x_i) = f_{\tilde{x}}(x) \tag{10.93}$$

because the values of the entries of v are the same as those of x (although their order can be different). $\blacksquare$

If we assume exchangeability under the null hypothesis in a permutation test, then any permutation of the observed data is equally likely. This allows us to compute the p-value of the test, without access to a model of the marginal distribution of the test statistic.

Theorem 10.22 (*P*-value function of a permutation test). *Let $x_{\mathrm{data}} \in \mathbb{R}^d$ be a vector of observed data, and $T: \mathbb{R}^n \to \mathbb{R}$ a predefined test statistic. Under the null hypothesis that the data can be represented as an n-dimensional random vector $\tilde{x}_{\mathrm{null}}$ with exchangeable entries, the p-value function in Definition 10.19 is*

$$\mathrm{pv}(t) = \frac{1}{|\Pi_{x_{\mathrm{data}}}|} \sum_{v \in \Pi_{x_{\mathrm{data}}}} 1\left(T(v) \geq t\right), \tag{10.94}$$

where $\Pi_{x_{\mathrm{data}}}$ is the set of vectors obtained by permuting the entries of x_{data}, and $|\Pi_{x_{\mathrm{data}}}|$ is its cardinality (i.e. the number of different permutations, which is $n!$ if the entries of x_{data}

are distinct). $1\left(T(v) \geq t\right)$ *is an indicator function that is equal to one, if* $T(v) \geq t$, *and to zero otherwise.*

Proof If $\tilde{x}_{\text{null}}$ is discrete, then exchangeability implies that for any $v_1, v_2 \in \Pi_{x_{\text{data}}}$ $p_{\tilde{x}_{\text{null}}}(v_1) = p_{\tilde{x}_{\text{null}}}(v_2)$, so by Definition 1.16,

$$P\left(\tilde{x}_{\text{null}} = v_1 \,\middle|\, \tilde{x}_{\text{null}} \in \Pi_{x_{\text{data}}}\right) = \frac{P\left(\tilde{x}_{\text{null}} = v_1, \tilde{x}_{\text{null}} \in \Pi_{x_{\text{data}}}\right)}{P\left(\tilde{x}_{\text{null}} \in \Pi_{x_{\text{data}}}\right)} \tag{10.95}$$

$$= \frac{p_{\tilde{x}_{\text{null}}}(v_1)}{P\left(\tilde{x}_{\text{null}} \in \Pi_{x_{\text{data}}}\right)} \tag{10.96}$$

$$= \frac{p_{\tilde{x}_{\text{null}}}(v_2)}{P\left(\tilde{x}_{\text{null}} \in \Pi_{x_{\text{data}}}\right)} \tag{10.97}$$

$$= P\left(\tilde{x}_{\text{null}} = v_2 \,\middle|\, \tilde{x}_{\text{null}} \in \Pi_{x_{\text{data}}}\right). \tag{10.98}$$

Consequently, $P\left(\tilde{x}_{\text{null}} = v \,\middle|\, \tilde{x}_{\text{null}} \in \Pi_{x_{\text{data}}}\right)$ is the same for all $v \in \Pi_{x_{\text{data}}}$. These $|\Pi_{x_{\text{data}}}|$ probabilities need to add up to one because the corresponding events are a partition of $\tilde{x}_{\text{null}} \in \Pi_{x_{\text{data}}}$. Indeed, by Axiom 3 in Definition 1.9 applied to the conditional probability measure $P\left(\cdot \,\middle|\, \tilde{x}_{\text{null}} \in \Pi_{x_{\text{data}}}\right)$,

$$\sum_{v \in \Pi_{x_{\text{data}}}} P\left(\tilde{x}_{\text{null}} = v \,\middle|\, \tilde{x}_{\text{null}} \in \Pi_{x_{\text{data}}}\right) = P\left(\cup_{v \in \Pi_{x_{\text{data}}}} \tilde{x}_{\text{null}} = v \,\middle|\, \tilde{x}_{\text{null}} \in \Pi_{x_{\text{data}}}\right)$$

$$= P\left(\tilde{x}_{\text{null}} \in \Pi_{x_{\text{data}}} \,\middle|\, \tilde{x}_{\text{null}} \in \Pi_{x_{\text{data}}}\right) \tag{10.99}$$

$$= 1. \tag{10.100}$$

Since the probabilities are all equal,

$$P\left(\tilde{x}_{\text{null}} = v \,\middle|\, \tilde{x}_{\text{null}} \in \Pi_{x_{\text{data}}}\right) = \frac{1}{|\Pi_{x_{\text{data}}}|}. \tag{10.101}$$

The same holds if $\tilde{x}_{\text{null}}$ is continuous. The proof is similar, but requires taking limits, as in the proof of Theorem 6.6, since the probability that $\tilde{x}_{\text{null}} \in \Pi_{x_{\text{data}}}$ is zero because $\Pi_{x_{\text{data}}}$ is a discrete set (see Section 3.1).

The p-value function can be derived directly from (10.101), as it is the conditional probability of the union of the disjoint events $\tilde{x}_{\text{null}} = v$ for every v such that $T(v) \geq t$. Consequently, by Axiom 3 in Definition 1.9,

$$\text{pv}(t) := P\left(T\left(\tilde{x}_{\text{null}}\right) \geq t \,\middle|\, \tilde{x}_{\text{null}} \in \Pi_{x_{\text{data}}}\right) \tag{10.102}$$

$$= P\left(\cup_{\{v \in \Pi_{x_{\text{data}}} \,:\, T(v) \geq t\}} \{\tilde{x}_{\text{null}} = v\} \,\middle|\, \tilde{x}_{\text{null}} \in \Pi_{x_{\text{data}}}\right) \tag{10.103}$$

$$= \sum_{\{v \in \Pi_{x_{\text{data}}} \,:\, T(v) \geq t\}} P\left(\tilde{x}_{\text{null}} = v \,\middle|\, \tilde{x}_{\text{null}} \in \Pi_{x_{\text{data}}}\right) \tag{10.104}$$

$$= \frac{1}{|\Pi_{x_{\text{data}}}|} \sum_{v \in \Pi_{x_{\text{data}}}} 1\left(T(v) \geq t\right). \tag{10.105}$$

∎

Theorem 10.22 establishes that the p-value of the permutation test equals the fraction of permutations with a test statistic greater than or equal to the observed test statistic t_{data}, which coincides with our intuitive calculation in Section 10.7.1 and Figure 10.7. A key question is whether thresholding this p-value enables us to control the probability of a false positive, as in parametric testing (see Section 10.5). Section 10.7.3 proves that this is indeed the case: Rejecting the null hypothesis when the p-value of a permutation test is below the significance level α guarantees that the probability of a false positive is bounded by α.

It is usually impossible to compute the exact p-value associated with a permutation test because the number of possible permutations is too large. For instance, if there are $n := 40$ distinct data points, the number of permutations is $40! > 10^{47}$. In practice, the p-value must be approximated computationally. This can be achieved by estimating the conditional probability of the event $T(\tilde{x}_{\text{null}}) \geq t_{\text{data}}$ given $\tilde{x}_{\text{null}} \in \Pi_{x_{\text{data}}}$ via the Monte Carlo method in Section 1.7.

Definition 10.23 (Permutation test via the Monte Carlo method). *Let $x_{\text{data}} \in \mathbb{R}^n$ be a vector of data and $T \colon \mathbb{R}^n \to \mathbb{R}$ a test statistic. To perform a permutation test associated with the null hypothesis that the data points are exchangeable, we:*

1. *Generate k independent permutations $v_1, \ldots, v_k \in \Pi_{x_{\text{data}}}$ of the entries of x_{data}.*
2. *Compute the test statistic corresponding to each permutation, $t_i = T(v_i)$, $1 \leq i \leq k$.*
3. *Compute the p-value estimate*

$$\text{pv}(t_{\text{data}}) = \frac{1}{k} \sum_{i=1}^{k} 1\left(t_i \geq t_{\text{data}}\right), \qquad (10.106)$$

where $t_{\text{data}} := T(x_{\text{data}})$ is the observed test statistic and $1\left(t_i \geq t_{\text{data}}\right)$ is an indicator function that is equal to one, if $t_i \geq t_{\text{data}}$, and to zero otherwise.
4. *Reject the null hypothesis if the p-value is less than or equal to a predefined significance level α.*

Permutation tests based on the Monte Carlo method are close in spirit to the bootstrap technique described in Section 9.9, as they both involve resampling the available data. In the bootstrap, we sample the data with replacement. In the permutation test, we sample the data without replacement.

Example 10.24 (Free throws under pressure: Permutation test). In this example, we apply a permutation test to the data in Example 10.2. Under the null hypothesis that the free-throw percentage is the same at home and away (see Example 10.4), the data can be modeled as exchangeable by Lemma 10.21, as long as we assume that the free throws are independent. The number of data is 85, so we cannot consider all possible permutations. We instead resort to the Monte Carlo approach in Definition 10.23. We set $k := 10^6$ and use the same test statistic as in the one-tailed test described in Example 10.6: The difference between the free-throw percentage at home and away.

The left graph in Figure 10.8 shows the estimated conditional pmf of the test statistic under the null hypothesis that the data are exchangeable. The conditional pmf is quite different from the continuous pdf of the test statistic in the one-tailed parametric test of Example 10.11 (bottom left graph of Figure 10.2). However, the resulting p-values are quite close: 0.019 for the permutation test and 0.011 for the parametric test. The p-value functions of both tests, depicted in Figure 10.9, are very similar.

. . . .

Example 10.25 (Tom Brady and Category 5 hurricanes: Permutation test). The data in Example 1.30 seem to indicate that when Tom Brady wins the Super Bowl, Category 5 hurricanes in the North Atlantic Ocean are more likely. No matter how you feel about Tom

Brady, this doesn't make a lot of sense. In order to check whether this pattern is due to random chance, we apply a permutation test. Our null hypothesis is that the distribution of hurricanes is the same, whether Tom Brady wins the Super Bowl or not, which implies that the data are exchangeable.

A reasonable test statistic is the difference between the fraction of years with hurricanes when Tom Brady wins the Super Bowl (denoted by $m(x_W)$) and when he does not (denoted by $m(x_L)$):

$$t_{data} := m(x_W) - m(x_L) = 0.264. \tag{10.107}$$

The graph on the right of Figure 10.8 depicts the conditional pmf of the test statistic under our null hypothesis, computed following Definition 10.23 with $k := 10^6$. The p-value equals 0.251: One fourth of the random permutations result in a test statistic that is greater than or equal to the observed test statistic. This shows that even though the test statistic looks quite large, it is not inconsistent with the null hypothesis. Consequently, the data do not support a connection between Brady's success and Category 5 hurricanes.

It is important to point out that in this case a small p-value would actually *not* be convincing evidence against the null hypothesis. The reason is that we are using the same data that prompted us to define our conjecture of interest, instead of gathering additional data. This invalidates the test, as explained in Section 10.1.

. . . .

Example 10.26 (School grades). Permutation tests are particularly useful when it is challenging to derive a parametric model for the test statistic. The graph on the left of Figure 10.10 shows histograms of the grades for a course on Portuguese language in two Portuguese schools, extracted from Dataset 15. There is a large jump at 10, presumably because this is the grade needed to pass the course. Our goal is to determine whether the grades in the two schools are systematically different. Our null hypothesis is that the grade

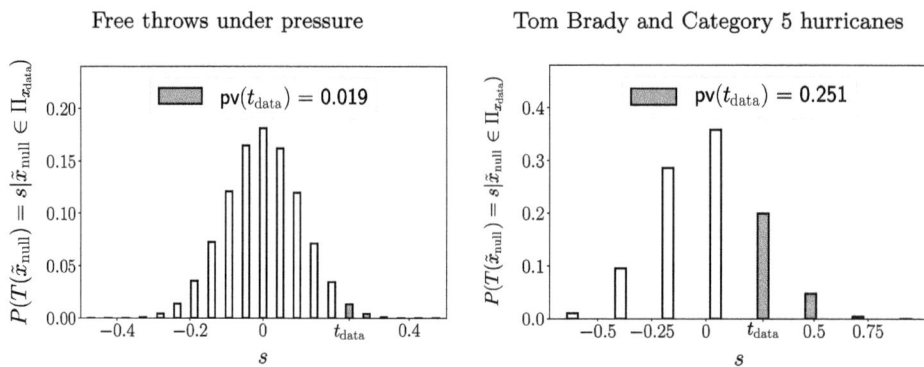

Figure 10.8 Permutation tests. This figure illustrates the p-value calculation for the permutation test in Definition 10.23, applied to the data in Examples 10.2 and 1.30 (see Example 10.25). The graphs depict the conditional pmf of the test statistic under the null hypothesis, given that we observe a permutation of the available data. The probability is estimated via the Monte Carlo method in Definition 10.23 using $k := 10^6$ independent permutations. The p-value is the conditional probability of observing a test statistic that is greater than or equal to the observed test statistic t_{data}, so it equals the sum of the corresponding entries in the conditional pmf (gray bars). On the left, the p-value is small enough to reject the null hypothesis based on a significance level of 0.05. On the right, the p-value is quite large, so we do not reject the null hypothesis.

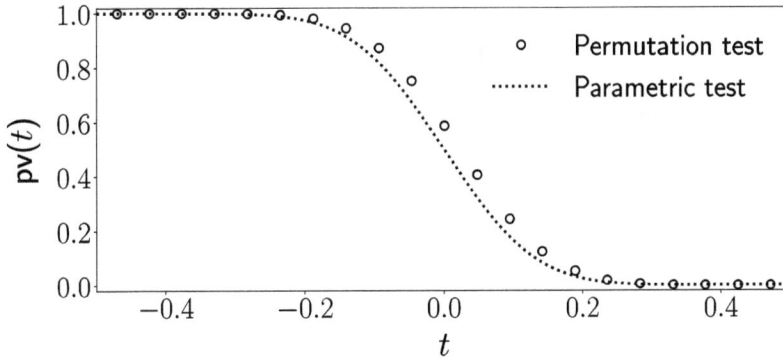

Figure 10.9 Comparison of *p*-value function for permutation and parametric tests. The graph compares the *p*-value functions corresponding to the permutation test in Example 10.24 (circular markers) and the one-tailed parametric test in Example 10.11 (dotted line). The permutation test produces a similar *p*-value function, without requiring a parametric model.

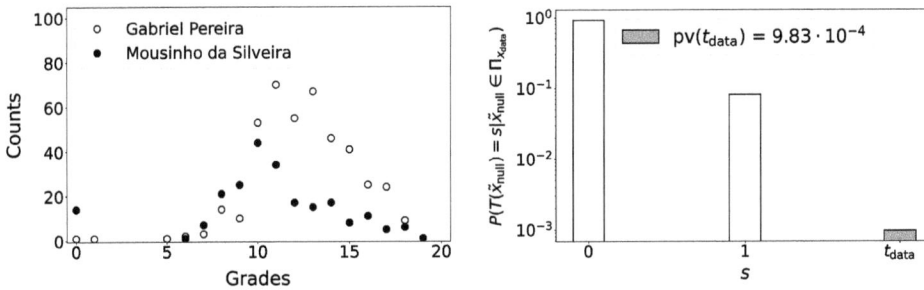

Figure 10.10 Permutation test for school grades. The graph on the left shows histograms of the grades for a course on Portuguese language in two different Portuguese schools, extracted from Dataset 15. The graph on the right depicts the conditional pmf of the test statistic under the null hypothesis that the data are exchangeable, given that we observe a permutation of the available data. The conditional pmf is estimated via the Monte Carlo method in Definition 10.23 using $k := 10^6$ independent permutations. The *p*-value corresponding to the observed test statistic t_{data} (gray bar) is smaller than 10^{-3}, indicating that the data are inconsistent with the null hypothesis.

distributions in both schools are the same. We choose the absolute difference between medians as our test statistic. It is not obvious how to design a parametric model for this test statistic, but thanks to the permutation test we don't need to!

The observed test statistic equals

$$t_{data} := |\text{median}(x_{\text{GP}}) - \text{median}(x_{\text{MS}})| = 2, \tag{10.108}$$

where x_{GP} are the grades from the school Gabriel Pereira and x_{MS} are the grades from the school Mousinho da Silveira. The graph on the right of Figure 10.10 depicts the distribution of the test statistic, under the null hypothesis that the data are exchangeable, computed following Definition 10.23 with $k := 10^6$. The *p*-value is smaller than 10^{-3}. Consequently, the observed data are extremely unlikely given the null hypothesis. We conclude that the difference between the grade distributions of both schools is statistically significant.

10.7.3 Statistical Significance

In Definition 10.23, we compare the p-value of a permutation test with a predefined significance level α, in order to decide whether to reject the null hypothesis. In this section, we prove that this allows us to control the probability of a false positive, as in the case of parametric testing (see Section 10.5). Recall that the null hypothesis in permutation testing is that the entries of the data are exchangeable. We begin by showing that, if this null hypothesis holds, the conditional probability of a false positive is bounded by α. The probability is conditioned on the event that we observe a permutation of the available data. Exercise 10.9 illustrates the result using the toy example from Section 10.7.1.

Theorem 10.27 (The significance level bounds the conditional probability of a false positive). *Consider a permutation test, where we reject the null hypothesis, if the p-value in Definition 10.19 is smaller than or equal to the significance level $\alpha \in [0, 1]$. Under the null hypothesis that the data are samples from a random vector with exchangeable entries, the conditional probability of a false positive, given that we observe a permutation of the available data, is bounded by α.*

Proof   Let x_{data} represent the available data, and let $\tilde{x}_{\mathrm{null}}$ be a random vector used to model the data under the null hypothesis. We define the random variable $\tilde{t}_{\mathrm{null}} := T(\tilde{x}_{\mathrm{null}})$. The p-value function is the deterministic function

$$\mathrm{pv}(t) := \mathrm{P}\left(\tilde{t}_{\mathrm{null}} \geq t \,\middle|\, \tilde{x}_{\mathrm{null}} \in \Pi_{x_{\mathrm{data}}}\right), \tag{10.109}$$

so the random variable

$$\tilde{u} := \mathrm{pv}(\tilde{t}_{\mathrm{null}}) \tag{10.110}$$

$$= 1 - F_{\tilde{t}_{\mathrm{null}}}\left(\tilde{t}_{\mathrm{null}} \,\middle|\, \tilde{x}_{\mathrm{null}} \in \Pi_{x_{\mathrm{data}}}\right) \tag{10.111}$$

captures the conditional behavior of the p-value under the null hypothesis, where

$$F_{\tilde{t}_{\mathrm{null}}}\left(t \,\middle|\, \tilde{x}_{\mathrm{null}} \in \Pi_{x_{\mathrm{data}}}\right) := \mathrm{P}\left(\tilde{t}_{\mathrm{null}} \leq t \,\middle|\, \tilde{x}_{\mathrm{null}} \in \Pi_{x_{\mathrm{data}}}\right) \tag{10.112}$$

is the conditional cdf of $\tilde{t}_{\mathrm{null}}$ given the event $\tilde{x}_{\mathrm{null}} \in \Pi_{x_{\mathrm{data}}}$. By the same argument as in the proof of Theorem 10.13 (replacing the cdfs by conditional cdfs), for any u in $[0, 1]$,

$$F_{\tilde{u}}(u \,|\, \tilde{x}_{\mathrm{null}} \in \Pi_{x_{\mathrm{data}}}) \leq u. \tag{10.113}$$

Consequently,

$$\mathrm{P}\left(\text{False positive} \,\middle|\, \tilde{x}_{\mathrm{null}} \in \Pi_{x_{\mathrm{data}}}\right) = \mathrm{P}\left(\mathrm{pv}(\tilde{t}_{\mathrm{null}}) \leq \alpha \,\middle|\, \tilde{x}_{\mathrm{null}} \in \Pi_{x_{\mathrm{data}}}\right) \tag{10.114}$$

$$= F_{\tilde{u}}(\alpha \,|\, \tilde{x}_{\mathrm{null}} \in \Pi_{x_{\mathrm{data}}}) \leq \alpha. \tag{10.115}$$

∎

Our bound on the conditional probability in Theorem 10.27 allows us to also bound the corresponding *marginal* probability, which establishes that we can indeed control the probability of a false positive by thresholding the p-value of the nonparametric test.

Corollary 10.28 (The significance level bounds the probability of a false positive). *Consider a permutation test, where we reject the null hypothesis, if the p-value in Definition 10.19 is smaller than or equal to the significance level $\alpha \in [0, 1]$. Under the null hypothesis that the* *data are samples from a random vector with exchangeable entries, then the probability of a false positive is bounded by* α.

Proof Let $\tilde{x}_{\text{null}}$ be an n-dimensional random vector representing the data under the null hypothesis. We define a random vector $\tilde{w} := \text{sort}(\tilde{x}_{\text{null}})$, obtained by sorting the entries of $\tilde{x}_{\text{null}}$. For example,

$$\text{sort}\left(\begin{bmatrix} 1 \\ 0 \\ 3 \end{bmatrix}\right) = \begin{bmatrix} 0 \\ 1 \\ 3 \end{bmatrix}. \tag{10.116}$$

The key idea is that $\tilde{w}$ encodes the entries of $\tilde{x}_{\text{null}}$, but not their order. Consequently the event $\tilde{x}_{\text{null}} \in \Pi_w$ (where Π_w contains all the unique permutations of the vector w) is equivalent to $\tilde{w} = w$. Let $\widehat{\text{fp}}$ be a Bernoulli random variable that is equal to one, if a false positive occurs, and to zero otherwise. If $\tilde{x}_{\text{null}}$ is discrete, by Theorems 4.10 and 4.14 and Lemma 2.6,

$$\text{P (False positive)} = \sum_{w \in \mathcal{W}} p_{\widehat{\text{fp}} \mid \tilde{w}}(1 \mid w) p_{\tilde{w}}(w) \tag{10.117}$$

$$= \sum_{w \in \mathcal{W}} \text{P (False positive} \mid \tilde{w} = w) \, p_{\tilde{w}}(w) \tag{10.118}$$

$$= \sum_{w \in \mathcal{W}} \text{P (False positive} \mid \tilde{x}_{\text{null}} \in \Pi_w) \, p_{\tilde{w}}(w) \tag{10.119}$$

$$\leq \alpha \sum_{w \in \mathcal{W}} p_{\tilde{w}}(w) = \alpha, \tag{10.120}$$

where $\mathcal{W}$ denotes the range of $\tilde{w}$ (the set of all possible sorted entries of $\tilde{x}_{\text{null}}$) and the inequality follows from Theorem 10.27. If $\tilde{x}_{\text{null}}$ is continuous, the proof is very similar. By Theorems 6.5 and 3.16,

$$\text{P (False positive)} = \int_{w \in \mathbb{R}^n} p_{\widehat{\text{fp}} \mid \tilde{w}}(1 \mid w) f_{\tilde{w}}(w) \, dw \tag{10.121}$$

$$\leq \alpha \int_{w \in \mathbb{R}^n} f_{\tilde{w}}(w) \, dw \tag{10.122}$$

$$= \alpha. \tag{10.123}$$

■

10.8 Multiple Testing

In some domains, it is commonplace to perform many hypothesis tests at the same time. This is known as *multiple testing*. An important example is computational genomics, where thousands of genetic markers may be evaluated to determine whether they are associated with a disease. Sections 10.5 and 10.7.3 establish that the probability of a false positive in a hypothesis test is controlled by the significance level α. However, this only holds *for a single hypothesis test*. If we perform many such tests simultaneously, then the probability of a false positive can be dramatically higher.

Imagine that we carry out k independent hypothesis tests with significance level α, and that the probability of a false positive in each test equals α. Then, by Definition 1.28, the probability of incurring at least one false positive is

$$1 - \text{P (No false positives)} = 1 - (1 - \alpha)^k. \tag{10.124}$$

For $\alpha := 0.05$ and $k := 100$, the probability is 0.99, so false positives are essentially guaranteed to occur!

The following example illustrates how easy it is to find apparent evidence for false findings in multiple testing scenarios.

Example 10.29 (3-point shooting in the clutch). Basketball analysts often praise or criticize players based on their performance *in the clutch*, which refers to the final moments that decide a game. Players that are able to play better at that crucial time are said to be *clutch*. Here we study clutch 3-point shooting during the 2014/2015 NBA season, using data extracted from Dataset 10. We define the clutch as the fourth quarter of close games, with a final point differential below 10.

To determine whether a player shoots better in the clutch, we apply a parametric hypothesis test. The parameter of interest is the player's clutch 3-point percentage. Our null hypothesis is that the parameter is equal to the player's season 3-point percentage θ_{season}, i.e. that the player's accuracy is the same in the clutch as at any other time. We choose the number of shots made in the clutch as the test statistic t_{data}.

If we assume that the shots are independent, the distribution of the made shots under the null hypothesis, represented by the random variable $\tilde{t}_{\theta_{\text{season}}}$, is binomial with parameters n and θ_{season}, where n denotes the number of clutch shots taken by the player (see Section 2.3.2). Therefore, by Definition 2.17, the p-value equals

$$\text{pv}(t_{\text{data}}) := \text{P}\left(\tilde{t}_{\theta_{\text{season}}} \geq t_{\text{data}}\right) \tag{10.125}$$

$$= \sum_{i=t_{\text{data}}}^{n} \binom{n}{i} \theta_{\text{season}}^{i} \left(1 - \theta_{\text{season}}\right)^{n-i}. \tag{10.126}$$

Notice that p-value function is the same as in Example 10.9.

We apply the hypothesis test to the 146 players who shot at least 100 3-pointers during the season, computing the p-value of their clutch 3-point shots during the first half of the season. For example, Robert Covington's season percentage was 38.2% and he made 11 out of 15 3-point shots in the clutch, so we set $\theta_{\text{season}} := 0.382$, $n := 15$ and $t_{\text{data}} := 11$ in (10.126) to obtain a p-value of 0.006. A histogram of all the p-values is shown in the top half of Figure 10.11. Only five players (including Covington), have p-values below 0.05. They are listed in the bottom half of Figure 10.11.

If we were sports journalists, we would go ahead and write an article about how Covington is clutch because of his tireless work ethic or his killer instinct. Instead, let us use some held-out data to double check whether what we are seeing is real. The table in the bottom half of Figure 10.11 also shows the clutch 3-point percentages for our five clutch players during the second half of the season, and the corresponding p-values. In the second half, Covington and Butler shot *worse* in the clutch than their season percentages! The remaining players shot better, but all the p-values are much higher than 0.05. From the second-half data, we cannot conclude that any of the players are actually clutch.
··

Example 10.29 illustrates the key challenge of multiple testing: The probability that a single player overperforms in the clutch by sheer chance is low, but the probability that *one out of the 146* players gets lucky is very high. Because of this, a p-value below the significance level is not necessarily compelling evidence against the null hypothesis. This is reflected in the distribution of the p-values, shown in the histogram at the top of Figure 10.11.

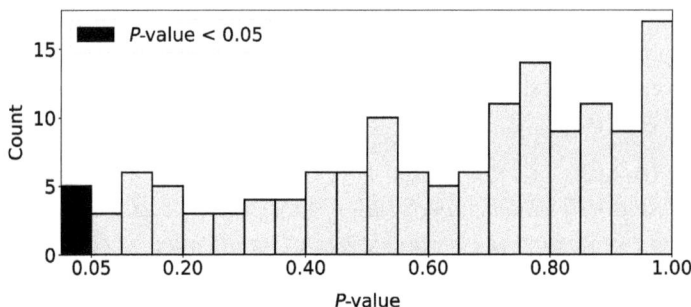

		First half		Second half	
	Season %	Clutch %	*P*-value	Clutch %	*P*-value
Rob. Covington	38.2	73.3 (11/15)	0.006	31.8 (7/22)	0.796
Nikola Mirotic	34.1	62.5 (10/16)	0.019	37.5 (6/16)	0.478
Caron Butler	32.1	61.5 (8/13)	0.027	25.0 (2/8)	0.783
Mike Conley	39.2	60.9 (14/23)	0.029	50.0 (8/16)	0.262
Kirk Hinrich	31.7	52.4 (11/21)	0.039	37.5 (3/8)	0.491

Figure 10.11 3-point shooting in the clutch. The graph at the top shows a histogram of the *p*-values corresponding to the 3-point clutch shooting of 146 NBA players in the first half of the 2014/2015 season, computed as described in Example 10.29. The five players with a *p*-value below 0.05 are listed in the table. The table shows their season 3-point percentage, their clutch 3-point shooting in the first and second half of the season, and the associated *p*-values.

Every interval of length 0.05 between 0 and 1 contains at least three *p*-values. It is therefore not surprising that five of them end up in $[0, 0.05]$ by pure accident.

In Theorem 10.13 we show that under a simple null hypothesis, the *p*-value of a continuous test statistic has a uniform distribution. If we perform a single test, that is good news: the *p*-value falls in the interval $[0, 0.05]$ (assuming a significance level of 0.05), resulting in a false positive, only 5% of the time. However, if we perform a large number of tests, then approximately 5% end up in $[0, 0.05]$ and result in false positives.

In multiple testing, a small *p*-value is not convincing evidence, but a *very small* *p*-value can be. It just needs to be so small that it is unlikely to be observed *even if we account for the number of simultaneous tests.* Bonferroni's correction provides a threshold to decide when this is the case.

Definition 10.30 (Bonferroni's correction). *When performing k simultaneous hypothesis tests, Bonferroni's correction consists of rejecting the null hypothesis for each test, if the p-value is smaller than or equal to $\alpha_B := \alpha/k$, where α is the desired significance level.*

Imagine again that we perform k independent hypothesis tests, but compare each of the *p*-values to $\alpha_B := \alpha/k$, instead of the desired significance level α. By Theorems 10.14 and Corollary 10.28, the probability of a false positive in each individual hypothesis test is bounded by α_B. Consequently, the probability of incurring at least one false positive is

$$1 - P\left(\text{No false positives}\right) = 1 - (1 - \alpha_B)^k \tag{10.127}$$

$$= 1 - \left(1 - \frac{\alpha}{k}\right)^k. \tag{10.128}$$

For $\alpha := 0.05$ and $k := 100$, the probability equals $0.049 \leq \alpha$, so the correction indeed allows us to keep the false-positive rate below the target significance level. In fact, the correction works even if the tests are not independent, as a consequence of the union bound (Theorem 9.19).

Theorem 10.31 (Bonferroni's correction works). *If we perform Bonferroni's correction, described in Definition 10.30, when performing k hypothesis tests, then the probability of at least one false positive occurring is bounded by the significance level α.*

Proof By Theorems 10.14 and Corollary 10.28, the probability of a false positive in each individual test is bounded by $\alpha_{\mathrm{B}} := \alpha/k$. Consequently, by Theorem 9.19,

$$P\left(\text{At least one false positive}\right) = P\left(\cup_{i=1}^{k} \text{False positive in test i}\right) \tag{10.129}$$

$$\leq \sum_{i=1}^{k} P\left(\text{False positive in test i}\right) \tag{10.130}$$

$$\leq k \cdot \frac{\alpha}{k} = \alpha. \tag{10.131}$$

∎

If we apply Bonferroni's correction to the $k := 146$ hypothesis tests in Example 10.29, then the Bonferroni threshold α_{B} corresponding to a significance level of $\alpha := 0.05$ equals $3.42 \cdot 10^{-4}$. All observed p-values are larger than the threshold, so we don't reject the null hypothesis for any of them. Thanks to the correction, we avoid the false positives that occur, if we interpret the tests as being applied in isolation.

A downside of Bonferroni's correction is that it can dramatically reduce our power to detect true positives. This is the main challenge of multiple testing: Adapting the testing strategy to control the overall probability of false positives comes at the expense of reducing power and increasing false negatives.

Example 10.32 (Evaluating NBA players). Measuring the impact of individual players is challenging in any team sport. In this example, we consider the problem of quantifying the *added value* that a specific player provides to a team. Our strategy is to compare the performance of the team with and without the player. To this end, we use Dataset 17, which consists of box scores of NBA regular-season games between 2012 and 2018.

For each player, we compute the point differential of their team in the games with and without the player. We then apply hypothesis testing to evaluate whether the difference is statistically significant. Our null hypothesis is that the point differentials from all the games belong to the same distribution (whether the player plays or not). Our alternative hypothesis is that the point differential with the player is higher than the point differential without the player (on average). Therefore, we choose the following test statistic,

$$t_{\mathrm{data}} := m_{\mathrm{with}} - m_{\mathrm{without}}, \tag{10.132}$$

where m_{with} and m_{without} denote the sample mean of point differentials with and without the player, respectively. We compute the test statistic separately for each team that the player belonged to. For example, during Lebron James's second stint with the Cleveland Cavaliers, the Cavs played 301 games with James and 27 games without him. The difference between the mean point differentials in those games is $t_{\mathrm{data}} = 16.6$, which suggests that he was very valuable to the Cavs. As a comparison, the test statistic for his time with the Miami Heat is just $t_{\mathrm{data}} = 3.7$.

It is not obvious how to derive a parametric model for the distribution of the point differential under the null hypothesis, so we resort to a nonparametric test. We apply the permutation test based on the Monte Carlo method from Definition 10.23 to compute the p-value. For James's time with the Cavs, the p-value is less than 10^{-7}. By contrast, the p-value for his time with the Heat is 0.180. This does not mean that he was not valuable to the Heat (he surely was!), just that we do not have enough data to contradict the null hypothesis (he just missed 9 games).

In total, we consider 1,397 player/team pairs. The list on the left of Table 10.1 shows the 20 players with the largest values of the test statistic. Surprisingly, they are mostly role players who played very few minutes. It seems quite plausible that their associated test statistics are large by random chance. If we perform Bonferroni's correction, the threshold α_B to reject the null hypothesis in order to ensure a significance level $\alpha := 0.05$ is $3.58 \cdot 10^{-5}$. Most of the players in the top 20 have p-values above the Bonferroni threshold α_B, but below the uncorrected threshold α. This illustrates the effectiveness of Bonferroni's correction in avoiding false positives.

The list on the right of Table 10.1 shows the 20 player/team pairs with the smallest p-values. The list looks much more meaningful than the top 20 obtained just based on the test statistic; it includes many well known stars. However, there are only eight that are below the Bonferroni threshold. This illustrates the main drawback of Bonferroni's correction. It ensures that we avoid false positives with high probability, but it also hinders the detection of true positives. Here we incur seven clear false negatives: Kevin Durant, Chris Paul, Stephen Curry, Anthony Davis, Marc Gasol, Kawhi Leonard, and Klay Thompson. In case you don't follow the NBA, there is no doubt that these players were crucial to their respective teams, but their p-values are just above the Bonferroni threshold.

The histogram at the top of Figure 10.12 shows the p-values of all 1,397 player/team pairs. We can interpret the distribution as a mixture of *alternative-hypothesis players*, who are truly important to their team, and *null-hypothesis players*, who are not. We expect the p-values of the null-hypothesis players to be evenly spaced out in the unit interval by Theorem 10.13. Indeed, in the histogram we observe a *floor* of around 10 players in every interval of length 0.01, which probably corresponds to null-hypothesis players. By contrast, we expect the alternative-hypothesis players to be concentrated close to zero, since their test statistic should be abnormally large provided there are enough data points associated with them. Otherwise, as in the case of LeBron James in the Miami Heat, they will be lost in the sea of null-hypothesis players. Bonferroni's hypothesis tells us how to threshold the p-value, in order to filter out the null-hypothesis players with high probability. The histogram at the bottom of Figure 10.12 shows p-values in the interval $[0, 0.01]$. Most of the p-values that are closer to zero probably correspond to alternative-hypothesis players, but only the smallest ones (indicated by the black bar) are below the Bonferroni threshold.

....

Example 10.32 showcases the difficulty of multiple testing: A small threshold misses some alternative-hypothesis players, resulting in false negatives, whereas a high threshold increases the probability of false positives. In many practical applications of multiple testing, it is often advisable to allow for some false positives, in order to detect more true positives. This can be achieved by thresholding the p-values using rules that are less strict than Bonferroni's correction (see (Benjamini and Hochberg, 1995) for a procedure with theoretical guarantees).

Table 10.1 **Evaluating NBA players.** *The list on the left contains the top 20 out of 1,397 NBA players between 2012 and 2018 according to the mean point differential between games where their team played with them, and games played without them. Abbreviations indicating the players' teams are included in parenthesis. Most of the players played very few minutes per game, which suggests that they are not actually important to their teams. The list on the right shows a ranking based on the p-value associated with the mean point differential, computed via a permutation test. Most of the players on the right are well-known stars, who play a lot of minutes. Players with p-values below the Bonferroni threshold are highlighted in bold.*

Top 20 mean point differential

	Mean point diff.	P-value	Mins per game
Marcus Paige (CHA)	28.5	$2 \cdot 10^{-4}$	5.4
N. Mohammed (OKC)	18.5	$3 \cdot 10^{-3}$	4.0
Georges Niang (UTA)	17.1	$2 \cdot 10^{-4}$	3.7
L. James (CLE)	16.7	$< 10^{-7}$	36.6
A. Goudelock (HOU)	16.5	$3 \cdot 10^{-2}$	6.4
B. Caboclo (TOR)	16.4	$< 10^{-7}$	4.6
Roy Hibbert (DEN)	16.1	$3 \cdot 10^{-3}$	2.0
Brandon Knight (DET)	16.1	$2 \cdot 10^{-3}$	31.5
Michael Gbinije (DET)	15.8	$5 \cdot 10^{-3}$	3.4
DeMarre Carroll (BKN)	15.7	$2 \cdot 10^{-3}$	29.9
Enes Kanter (NY)	15.5	$4 \cdot 10^{-3}$	25.8
MarShon Brooks (GS)	15.4	$5 \cdot 10^{-2}$	2.4
Victor Oladipo (IND)	15.0	$2 \cdot 10^{-3}$	34.1
Ronnie Brewer (OKC)	13.7	$2 \cdot 10^{-3}$	10.1
J. Cunningham (ATL)	12.5	$3 \cdot 10^{-2}$	4.6
Steve Novak (MIL)	11.9	$4 \cdot 10^{-3}$	3.9
Ronnie Brewer (NY)	10.7	$5 \cdot 10^{-2}$	15.4
Jonas Jerebko (UTA)	10.5	$9 \cdot 10^{-2}$	15.3
Eric Maynor (OKC)	10.5	$2 \cdot 10^{-3}$	10.6
A. McKinnie (TOR)	10.3	$2 \cdot 10^{-3}$	3.9

Bottom 20 p-values

	Mean point diff.	P-value	Mins per game
L. James (CLE)	16.7	$< 10^{-7}$	36.6
B. Caboclo (TOR)	16.4	$< 10^{-7}$	4.6
N. Mirotic (CHI)	10.3	$3 \cdot 10^{-7}$	23.1
C. Anthony (NY)	8.1	$5 \cdot 10^{-7}$	36.3
Ricky Rubio (MIN)	7.6	$7 \cdot 10^{-7}$	31.4
James Jones (MIA)	8.2	$6 \cdot 10^{-6}$	7.8
Brandon Rush (GS)	6.7	$6 \cdot 10^{-6}$	12.6
Joel Embiid (PHI)	8.7	$2 \cdot 10^{-5}$	28.7
Kevin Durant (OKC)	6.9	$1 \cdot 10^{-4}$	37.3
Kevin Garnett (MIN)	9.2	$2 \cdot 10^{-4}$	15.3
Marcus Paige (CHA)	28.5	$2 \cdot 10^{-4}$	5.4
Georges Niang (UTA)	17.1	$2 \cdot 10^{-4}$	3.7
Chris Paul (LAC)	6.8	$2 \cdot 10^{-4}$	33.6
Stephen Curry (GS)	8.2	$3 \cdot 10^{-4}$	34.6
Anthony Davis (NO)	5.1	$4 \cdot 10^{-4}$	34.8
Marc Gasol (MEM)	5.5	$5 \cdot 10^{-4}$	33.9
DeMarre Carroll (ATL)	10.1	$5 \cdot 10^{-4}$	31.5
Kawhi Leonard (SA)	4.7	$6 \cdot 10^{-4}$	31.6
Nikola Pekovic (MIN)	5.0	$8 \cdot 10^{-4}$	28.7
Klay Thompson (GS)	10.0	$9 \cdot 10^{-4}$	34.1

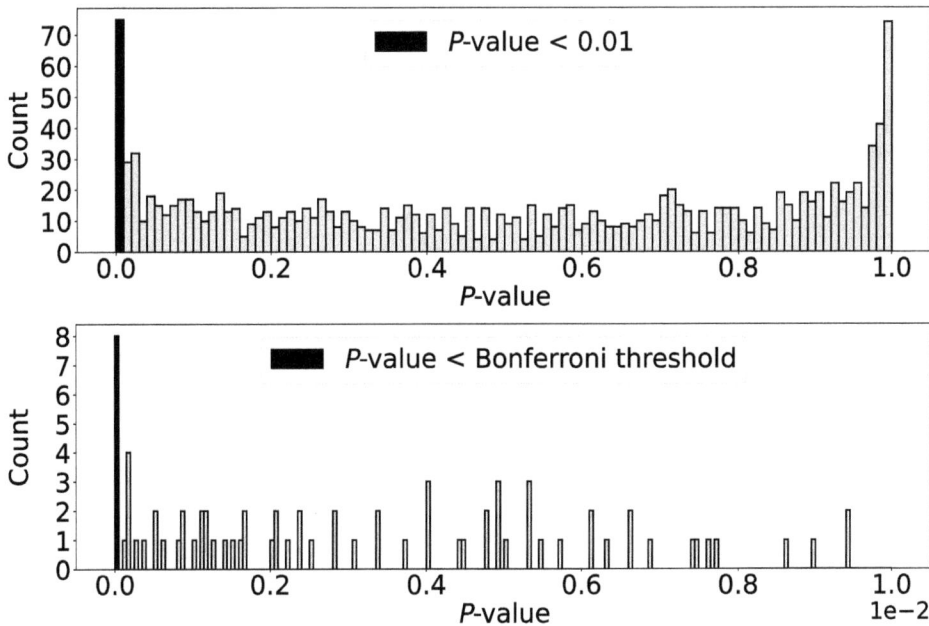

Figure 10.12 *P*-value distribution in multiple testing. The histogram at the top shows the *p*-values of all 1,397 player/team pairs in Example 10.32. The histogram at the bottom shows the *p*-values smaller than 0.01, corresponding to the black bar in the top histogram. At this higher resolution, we see some separation between a group of players with very small *p*-values and the rest, but the separation is not completely clear cut. The subset of *p*-values below the Bonferroni threshold is indicated by a black bar.

10.9 Hypothesis Testing and Causal Inference

An important consideration when performing hypothesis testing is whether the results can be interpreted in terms of causal effects. In Example 10.2, our motivation to analyze Antetokounmpo's free-throw shooting is that fans were loudly chanting while he shot free throws in away games. The hypothesis test in Example 10.11 establishes that there is indeed a statistically significant difference in free-throw percentage between home and away games; the data strongly support that

$$p_{\tilde{y}\,|\,\tilde{t}}(1\,|\,0) > p_{\tilde{y}\,|\,\tilde{t}}(1\,|\,1). \tag{10.133}$$

The random variable $\tilde{y}$ represents the outcome of interest: Whether a free throw is made ($\tilde{y} = 1$) or not ($\tilde{y} = 0$). The random variable $\tilde{t}$ represents the treatment, indicating whether fans were taunting Antetokounmpo ($\tilde{t} = 1$) or not ($\tilde{t} = 0$). Does this imply that the taunts caused the difference in free-throw percentage? No! Statistical significance does *not* imply the existence of a causal effect. As explained in Section 4.6, the difference could be due to confounding factors. For instance, Antetokounmpo may have been more tired at away games due to traveling.

To further emphasize the distinction between statistical significance and causal effects, let us delve deeper into the results in Example 10.32, where we evaluate NBA players by comparing the games played by their teams with and without them. Bruno Caboclo from the Toronto Raptors has an impressive difference between mean point differentials (16.4, sixth overall). However, he played less than five minutes per game, which makes it unlikely that his presence could have been that impactful. Surprisingly, the difference is highly statistically significant, even if we account for multiple testing: The corresponding p-value is the second smallest overall! There is overwhelming evidence that the conditional mean $\mu_{\tilde{y}\mid\tilde{t}}(1)$ of the point differential (see Definition 7.44) given that Caboclo plays, is greater than the conditional mean $\mu_{\tilde{y}\mid\tilde{t}}(0)$ if he does not play:

$$\mu_{\tilde{y}\mid\tilde{t}}(1) > \mu_{\tilde{y}\mid\tilde{t}}(0). \tag{10.134}$$

The random variables $\tilde{y}$ and $\tilde{t}$ again represent the outcome of interest (the point differential in a basketball game) and the treatment (whether Caboclo plays the game or not), respectively. However, as discussed in Section 7.9, the difference in conditional means does *not* imply that Caboclo's presence *causes* an increase in point differential.

In fact, Caboclo played only 24 games for the Raptors over 4 years (missing more than 200). Since the Raptors won these games by a very large margin on average and Caboclo played very few minutes, it seems plausible that he was included in games only when they were clearly decided in favor of the Raptors. In other words, the Raptors did not win those games by a wide margin because Caboclo played, but rather *Caboclo played those games because the Raptors were winning by a wide margin.*

As explained in Sections 4.6.3 and 7.9, randomizing the treatment guarantees that differences in conditional probabilities or conditional means reflect causal effects. In such situations, the results of hypothesis tests *can* be interpreted causally. In Example 4.25, we describe the randomized controlled trial used to evaluate Pfizer's COVID-19 vaccine. Randomization and hypothesis testing play complementary roles in the analysis of the data. Hypothesis testing allows us to determine that there is a statistically-significant difference between the control and vaccine groups. Randomization allows us to conclude that the difference reflects a causal effect.

Example 10.33 (COVID-19 vaccine: P value). We apply the two-sample z test in Definition 10.10 to the data in Example 4.25, choosing the one-tailed version of the test, because we are not interested in detecting a situation where the vaccine results in more COVID-19 cases. The null hypothesis is that the probability of contracting COVID-19, represented by the parameter θ_{null}, is the same in the control and the vaccine group. The test statistic is the difference between the fraction of positives in the vaccine group and in the control group,

$$t_{\text{data}} = \frac{162}{21728} - \frac{8}{21720} = 7.09 \cdot 10^{-3}. \tag{10.135}$$

Setting $n_A := 21728$, $k_A := 162$, $n_B := 21720$, $k_B := 8$ in Definition 10.10, we obtain $\sigma_{\text{null}} = 5.99 \cdot 10^{-4}$, which yields a p-value equal to

$$\text{pv}(t_{\text{data}}) = 1 - F_{\tilde{z}}\left(\frac{7.09 \cdot 10^{-3}}{5.99 \cdot 10^{-4}}\right) < 10^{-23}. \tag{10.136}$$

The p-value is extremely small, so the data are completely inconsistent with the null hypothesis. Moreover, the treatment is randomized, so the statistically significant difference is due to a causal effect. We conclude that the vaccine works.
....

In nonmedical applications, the combination of randomized experiments and hypothesis testing is typically known as A/B testing.

Definition 10.34 (A/B testing). *In A/B testing, users are randomly assigned to a control group (A) or a treatment group (B). Users in group A are exposed to a product without a specific modification. Users in group B are exposed to the product with the modification. Hypothesis testing is then applied to determine whether the difference in user response between the two groups, measured by an appropriate metric, is statistically significant.*

In the Internet era, A/B testing has become a fundamental tool in the tech industry and beyond. Famously, it played a key role in Barack Obama's presidential campaigns.

Example 10.35 (Obama's presidential campaign). Data scientists working for the Obama campaign used A/B testing to design the campaign website. Specifically, they ran A/B testing to determine what images or videos would be more effective in getting viewers to sign up for the campaign mailing list (Siroker, 2010). For simplicity, we focus on a comparison between images and videos. The sign-up rate for the images was 14,016 out of 155,280 users (9.03%). The sign-up rate for the videos was 10,337 out of 155,102 (6.66%). To determine whether this difference is statistically significant, we apply the two-tailed version of the two-sample z test in Definition 10.10.

The null hypothesis is that the sign-up rate θ_{null} is the same for all users. The test statistic is the absolute value of the difference between sign-up rates,

$$t_{data} = \left| \frac{14016}{155280} - \frac{10337}{155102} \right| = 2.36 \cdot 10^{-2}. \tag{10.137}$$

Setting $n_A := 155280$, $k_A := 14016$, $n_B := 155102$, $k_B := 10337$ in Definition 10.10, we obtain $\sigma_{null} = 9.65 \cdot 10^{-4}$, which yields a p-value equal to

$$\mathrm{pv}(t_{data}) = 2 \left(1 - F_{\tilde{z}} \left(\frac{2.36 \cdot 10^{-2}}{9.65 \cdot 10^{-4}} \right) \right) < 10^{-80}. \tag{10.138}$$

The p-value is minuscule, so we conclude that the difference is indeed statistically significant. Given that the users were randomly assigned to the two groups, we can safely assume that the difference is due to the choice between images and videos, so we conclude that images are more effective than videos.

Extremely small p-values, like the one obtained in this example, are typical in online applications of A/B testing, where the number of subjects tends to be orders of magnitude larger than in other domains. As we discuss in Section 10.10.1, this can be a mixed blessing because tiny differences without any practical impact may be statistically significant.

10.10 *P*-Value Abuse

In many scientific domains, reporting a p-value below a predefined significance level (usually 0.05) is a requisite for publication. As a result, statistical significance is often interpreted as a *stamp of approval* for scientific discoveries. This is problematic for several reasons. First, as described in Section 10.9, statistical significance does not provide insight into causal effects in observational studies where data acquisition is not randomized. In such cases, any claim implying causality must be justified separately. Second, statistical significance does not necessarily imply practical significance, as we explain in Section 10.10.1. Third, there is a strong incentive to achieve artificially small p-values by cherry-picking results, as described in Section 10.10.2. Consequently, statistical significance should not be the end goal of a scientific study, but rather a sanity check complementing a more comprehensive analysis.

10.10.1 Practical Significance

Statistical significance does not automatically imply *practical significance*. A result is practically significant if it is meaningful or useful for the application of interest. A vaccine that works 1% of the time is not very useful. A change in a website that increases the click-through rate by 0.001% is probably not worth it. Practical significance cannot be directly assessed using p-values because they only quantify to what extent the observed data is inconsistent with the null hypothesis.

Example 10.36 (Statistical vs. practical significance). We consider a fictional scenario where two drugs are being evaluated via randomized controlled trials. Both drugs produce side effects and are expensive, so medical experts decide that a drug should produce an increase of at least 5% in the cure rate (the probability that a patient recovers) in order to be approved.

In the first trial, 52 out of 100 patients in the treatment group and 30 out of 100 patients in the control group are cured. We apply the one-tailed two-sample z test in Definition 10.10 to determine whether the result is statistically significant. The null hypothesis is that the control and treatment groups have the same distribution. The test statistic is the difference in the cure rate in the two groups, which is the fraction of cured subjects. Setting $n_A := 100$, $k_A := 52, n_B := 100, k_B := 30$, we obtain $t_{\text{data}} = 0.22$ and $\sigma_{\text{null}} = 6.96 \cdot 10^{-2}$, so the p-value equals

$$\text{pv}(t_{\text{data}}) = 1 - F_{\tilde{z}}\left(\frac{0.22}{6.96 \cdot 10^{-2}}\right) = 7.8 \cdot 10^{-4}. \qquad (10.139)$$

In the second trial, 30,650 out of 100,000 patients in the treatment group and 30,000 out of 100,000 patients in the control group are cured. In this case, the difference in the cure rate is minuscule: $t_{\text{data}} = 6.5 \cdot 10^{-3}$. However, the trial has so many subjects that the standard deviation of the test statistic under the null hypothesis is also very small: $\sigma_{\text{null}} = 2.06 \cdot 10^{-3}$. As a result, the p-value is the same as in the first study,

$$\text{pv}(t_{\text{data}}) = 1 - F_{\tilde{z}}\left(\frac{6.5 \cdot 10^{-3}}{2.06 \cdot 10^{-3}}\right) = 7.8 \cdot 10^{-4}. \qquad (10.140)$$

The p-value is very small for both studies, so we can be very certain that both drugs increase the cure rate with respect to the control group. However, the p-value does not tell us *by how much*, which is critical to determine the practical significance of the results.

To quantify the practical significance of each trial, we compute 0.95 confidence intervals for the cure rates, following Definition 9.44, and for the difference in cure rate between the control and treatment groups, as explained in Exercise 9.11. The confidence intervals for the difference, shown in Figure 10.13, equal $[8.71\%, 35.3\%]$ for the first drug and $[0.25\%, 1.05\%]$ for the second drug. The difference is positive in both trials, but it is only practically significant (above 5%) for the first drug. Despite sharing the same p-value, the practical significance of the two results is very different.

· ·

Example 10.36 shows that tiny differences between two groups can be statistically significant when data are very plentiful. This can easily occur in large randomized controlled

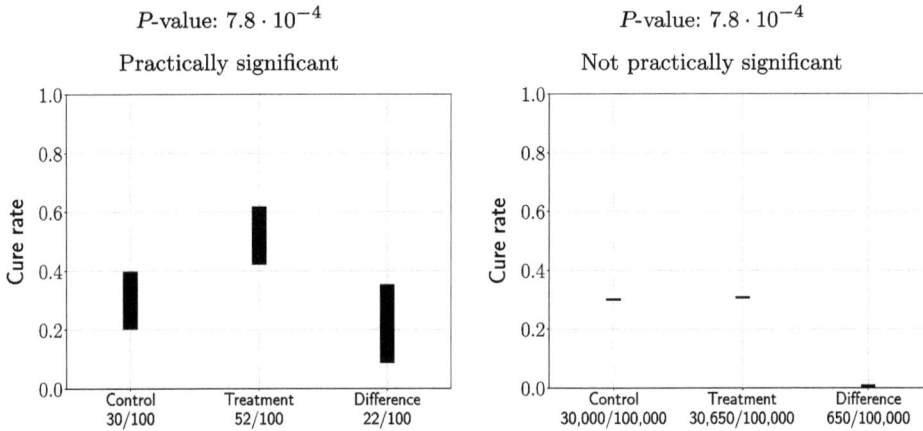

P-value: $7.8 \cdot 10^{-4}$ P-value: $7.8 \cdot 10^{-4}$

Practically significant   Not practically significant

Figure 10.13 Statistical vs. practical significance. The two fictional randomized controlled trials in Example 10.36 have the same p-value. In both, the difference between the control and treatment groups is equally statistically significant, but the difference is only practically significant for the trial on the left. The difference is obvious in the confidence intervals shown in the graphs. The improvement in the cure rate is much larger for the trial on the left. The trial on the right has the same p-value because of the large number of subjects, which renders the tiny improvement statistically significant. The confidence intervals are built following Definition 9.44 and Exercise 9.11.

trials and in online A/B tests. In Example 10.33, imagine there were 120 positive cases in the vaccine group (instead of 8). The corresponding p-value would be 0.006, so the difference would still be statistically significant at a significance level of 0.05 (or even 0.01), but the ratio between the positive cases in the vaccine and in the control group would only be 3/4 (in the real data it is 1/20). Such a modest reduction in the infection rate might not be worth the cost of producing and administering the vaccine.

Similarly, if in Example 10.35 the users that sign up after viewing the videos were 13,650 (instead of 10,337), then the p-value for the hypothesis test would be 0.027. This difference is statistically significant for the typical significance level of 0.05. However, the observed difference in sign-up rate would just be 0.2%, which is probably negligible in practice.

These examples hopefully make it clear that small p-values are not sufficient evidence to establish practical significance. Figure 10.14 uses confidence intervals to show that the results in Examples 10.33 and 10.35 are indeed practically significant.

10.10.2 Publication Bias and P-Hacking

P-values are often interpreted as a guarantee that a published result is meaningful. Unfortunately, this is not necessarily warranted. Imagine that 100 studies are performed by different research groups around the world to test whether eating pizza cures COVID-19. This is a multiple testing scenario, where many hypothesis tests are carried out simultaneously. As explained in Section 10.8, if the null hypothesis holds for the 100 tests, because pizza does not cure COVID-19, then around 5 of them will have p-values below 0.05 by random chance. If all of these results were made public, this would not be a big deal. Unfortunately, positive

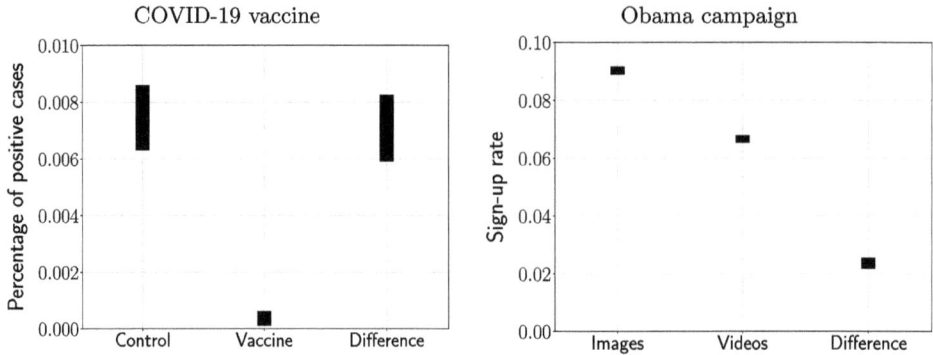

Figure 10.14 Confidence intervals establish practical significance. The graphs show 0.95 confidence intervals computed from the data in Examples 10.33 and 10.35. In both cases, the confidence intervals establish that there is a substantial difference between the two groups, so the results are practically significant, as well as statistically significant. The confidence intervals are built following Definition 9.44 and Exercise 9.11.

findings are much more likely to be published than negative results. This is known as *publication bias* and it is somewhat understandable: In the null hypothesis framework, a negative result just indicates that there is not sufficient evidence against the null hypothesis, *not that it holds*. Therefore, it may be difficult to get reviewers and editors excited about it. The headline *Pizza cures COVID-19!* is much more attractive than *There is no conclusive evidence that pizza cures COVID-19...*

Beyond publication bias, the acceptance of statistical significance as a sufficient condition for publication is problematic because it may motivate researchers to cherry-pick their results. For instance, consider a researcher trying to determine whether different food additives are harmful. They test thousands of additives by feeding them to mice and recording any adverse effects. One of the additives results in a small p-value, but not small enough to be statistically significant after applying Bonferroni's correction. The researcher has two options: (1) Gather additional data and perform another hypothesis test focused on the detected additive. (2) Publish the result without mentioning that thousands of additives were tested. The second option is known as *p-hacking* because it is a fraudulent manipulation of the hypothesis-testing framework. Unfortunately, researchers may have strong incentives to recur to p-hacking, such as graduating on time, or publishing a high-impact paper.

Figure 10.15 shows a histogram of the p-values reported in the abstracts of open-access articles in PubMed, a popular biomedical database, extracted from Dataset 18. We round the p-values to two decimal places, in order to account for the different precisions at which they are reported. We observe a clear decreasing trend: 12,074 abstracts report p-values in $[0, 0.005)$, 7,966 in $[0.005, 0.015)$, 6,224 in $[0.015, 0.025)$, and 4,281 in $[0.025, 0.035)$. This is expected: For truly significant results, the p-values should be heavily skewed toward very small values (see the p-value distribution for $\theta_{\mathrm{away}} := 0.55$ in Figure 10.6 or the histograms in Figure 10.12). However, just below the standard threshold for statistical significance the trend is broken: The p-values reported in $[0.035, 0.045)$ are 4,173, almost the same as in $[0.025, 0.035)$. This is very suspicious: 0.05 is an arbitrary value, except for the fact that it is the standard significance threshold. P-hacking is a plausible explanation for the inflated number of p-values just below the threshold.

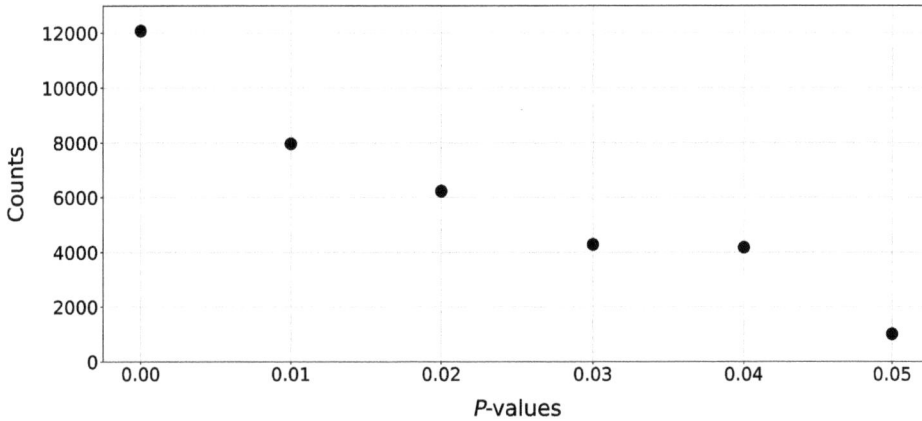

Figure 10.15 **P-value distribution in published papers.** The graph shows a histogram of the p-values reported in the abstracts of open-access articles in PubMed, a popular biomedical database, collected by the authors of (Head et al., 2015). The p-values are rounded to two decimal places, in order to account for the different precisions at which they are reported. There is a smooth decreasing trend in the p-values, which is to be expected if most results are truly significant. However, at 0.04 the trend is broken: The number of p-values is almost the same as for 0.03. P-hacking may be artificially inflating the number of p-values just below the standard significance threshold of 0.05.

Exercises

10.1   (Test choice) Test 1 has very high power and a very high significance level. Test 2 has very low power and a very low significance level. Choose one of these two tests for each of the two applications below and justify your answer.

   a   Null hypothesis: Water from a fountain in a school is not safe to drink.
   b   Null hypothesis: A piece of rock in a mine does not contain gold.

10.2   (Scout) A basketball scout invites a player for a workout. She doesn't have much time, so in order to evaluate the player's 3-point shooting, she asks the player to shoot 3 pointers until they miss. She wants to be pretty sure that the player's 3-point shooting percentage is 40% or less, in order to recruit them.

   a   How can the scout use parametric hypothesis testing to determine whether to recruit the player? What is the parameter of interest, the null hypothesis and the test statistic?
   b   Assuming that the scout tolerates a false-positive rate of 5%, how many 3 points does the player have to make to be recruited? State any assumptions you make.
   c   If the player's 3-point shooting percentage is 50%, what is the probability that they are not recruited?

10.3   (Road renovation) A small town decides to renovate a 10-mile road. According to certain reports, there is a specific 2.5-mile section which is very dangerous. Completely renovating that section would use up all of the budget, so the engineer in charge of the renovation wants to make sure that this is a priority. She decides to perform a hypothesis test using the next 4 accidents reported on the road. Her null hypothesis is that the accidents are independent and uniformly distributed on the 10-mile road, which would imply that the *dangerous* section is not that dangerous. The test statistic is the number of accidents that occur in the dangerous 2.5-mile section. The significance level is set to $\alpha := 0.05$.

a Derive the p-value function of the test for all possible values of the test statistic.

b What is the probability of a false positive?

c Let θ denote the probability that an accident occurs in the *dangerous* section. Derive and plot the power function of the test as a function of θ, under the assumption that accidents occur independently.

d What is the minimum value of θ for which the probability of a true positive is at least 50%?

10.4 (Computer component) A computer manufacturer wants to make sure that a certain component will last on average more than a year. They decide to apply a hypothesis test, where the null hypothesis is that the mean duration is less than a year. The data correspond to n units of the component, which can be assumed to be independent. The test statistic is the minimum duration of the n units. If the time until the component fails is modeled using an exponential distribution, what is the power function of the test as a function of the exponential parameter λ and the significance level α? What is the power at $\lambda = 1$? What is the limit of the power as $\lambda \to 0$?

10.5 (Test statistic in two-sample test) Explain why the test statistic

$$\tilde{t}_\theta := \frac{1}{n_{\text{home}}} \sum_{i=1}^{n_{\text{home}}} \tilde{h}_i - \frac{1}{n_{\text{away}}} \sum_{i=1}^{n_{\text{away}}} \tilde{a}_i \tag{10.141}$$

in Example 10.17 is approximately Gaussian with mean $\theta_{\text{home}} - \theta_{\text{away}}$ and variance

$$\sigma_{\text{pow}}^2 := \frac{\theta_{\text{home}}(1 - \theta_{\text{home}})}{n_{\text{home}}} + \frac{\theta_{\text{away}}(1 - \theta_{\text{away}})}{n_{\text{away}}}. \tag{10.142}$$

10.6 (Tom Brady and hurricanes) In Example 10.25, we analyze the data in Example 1.30 via permutation testing. Here we use a z test instead.

a Compute the p-value of a one-tailed two-sample z test, where the null hypothesis is that hurricanes have the same distribution when Brady wins and when he doesn't.

b If the p-value had been extremely small, would this be convincing evidence that hurricanes occur more often when Brady wins? Justify your answer.

10.7 (Unemployment in Spain) Figure 8.12 shows that unemployment and temperature in Spain seem to be negatively correlated. The correlation coefficient computed from 92 data points is -0.21.

a Explain how to build a hypothesis test based on Fisher's transformation, defined in equation (9.267) (see also Figure 9.24), to evaluate whether a correlation coefficient is different from zero. Use the fact that the standard error of the transformed correlation coefficient can be approximated by $1/\sqrt{n-1}$, where n is the number of samples. Derive an expression for the p-value function as a function of the test statistic, the cdf of a standard Gaussian with zero mean and unit variance, and the number of data.

b Is the correlation between unemployment and temperature in Spain statistically significant at a significance level of 0.05 according to your test?

c Do the data satisfy the assumptions of the test exactly? Are you convinced that the correlation is not just due to small sample size? Assuming you are convinced, can we conclude that higher temperatures cause a decrease in unemployment?

10.8 (Median cholesterol) The following table shows the cholesterol levels of three people that are healthy (label H) and three people that are ill (label I)

H	H	H	I	I	I
150	250	260	400	240	300

Apply a permutation test to the data to test whether the median of the cholesterol level is higher for the ill people using the following random permutations of the labels: HIHIIH, HIIIHH, IIHHHI, IHHIHI, HHIIIH. What is the p-value?

10.9 (False positives of a permutation test) In this exercise, we illustrate the implications of Theorem 10.27 using the toy example from Section 10.7.1. Let x_{data} represent the four data points. We model the data under the null hypothesis as a random vector $\tilde{x}_{\text{null}}$ with exchangeable entries. The random variable $\tilde{t}_{\text{null}} := T(\tilde{x}_{\text{null}})$ is the corresponding test statistic, which equals the difference between the sample mean of the prices in New York, and the sample mean of the prices in Madrid. The random variable $\tilde{u}$ represents the p-value under the null hypothesis (conditioned on the event $\tilde{x}_{\text{null}} \in \Pi_{x_{\text{data}}}$), defined as in (10.111):

$$\tilde{u} := \text{pv}(\tilde{t}_{\text{null}}) \tag{10.143}$$

$$= 1 - F_{\tilde{t}_{\text{null}}}\left(\tilde{t}_{\text{null}} \mid \tilde{x}_{\text{null}} \in \Pi_{x_{\text{data}}}\right), \tag{10.144}$$

where $\Pi_{x_{\text{data}}}$ is the set of vectors obtained by permuting the entries of x_{data}.

a What values of $\tilde{u}$ occur with nonzero probability, given $\tilde{x}_{\text{null}} \in \Pi_{x_{\text{data}}}$?

b Under the null hypothesis that the entries of $\tilde{x}_{\text{null}}$ are exchangeable, what is the conditional cdf of $\tilde{u}$ given $\tilde{x}_{\text{null}} \in \Pi_{x_{\text{data}}}$?

c We decide to reject the null hypothesis, if the p-value is below a significance level equal to $\alpha := 0.2$. What is the conditional probability of a false positive given $\tilde{x}_{\text{null}} \in \Pi_{x_{\text{data}}}$?

d Is the *marginal* probability of a false positive bounded by the significance level?

10.10 (Sign test) The sign test is a nonparametric two-sample test. Given n pairs of data (x_1, y_1), (x_2, y_2), ..., (x_n, y_n), the null hypothesis is that the two entries of each pair belong to the same distribution. The test statistic is

$$t = \sum_{i=1}^{n} 1\left(y_i > x_i\right), \tag{10.145}$$

where $1\left(y_i > x_i\right)$ is an indicator function that is equal to one, if $y_i > x_i$, and to zero otherwise.

a If $\tilde{a}$ and $\tilde{b}$ are independent continuous random variables with the same distribution, what is the probability that $\tilde{a} > \tilde{b}$?

b Derive the distribution of the test statistic of the sign test under the null hypothesis that the data are independent samples from the same continuous distribution.

c Your friend is convinced that, in general, the left ear of most people is longer than the right ear. You measure the ears of some of your other friends and obtain the following data (in inches).

Left	2.4	2.7	3.2	2.3	2.0	2.6	3.2	2.3	2.9	2.3
Right	2.2	2.6	3.3	2.2	2.1	2.5	3.1	2.5	2.7	2.2

Apply the sign test to compute a p-value associated with the null hypothesis that the left and right ear have the same length.

d Is the sign test still valid for the null hypothesis that x_i and y_i are sampled from the same continuous distribution, but these distributions are different for different values of i?

10.11 (Drug design) A company is trying to design a drug for a certain disease. From past data, the survival time of the patients with the disease is known to be exponential with parameter equal to one. The company has a large number of promising drug candidates. When a drug candidate is effective, we assume that it modifies the survival time to be exponential with parameter equal to a fixed constant θ. If it is not effective, it does not affect the survival time at all. In order to evaluate the candidates, each one is given to a different patient. A hypothesis test is then applied, where the test statistic is the survival time of the patient. The null hypothesis is that the drug does not work and therefore the survival time is exponential with parameter equal to one.

a Derive the pdf of the p-value, when the drug is effective, as a function of the parameter θ of the exponential survival distribution. Plot it for θ equal to 0.1. What happens if θ is equal to 1?

b Derive the pdf of the p-value, if the probability that a drug candidate is effective is 1/20 (i.e. only 5% of the candidates are effective), as a function of θ. Plot the pdf for $\theta := 0.1$.

c If we reject the null hypothesis, what is the conditional probability that there is a false positive and the drug candidate doesn't actually work? Derive the conditional probability as a function of the significance level α and of θ and report it for $\alpha = 0.05$ and $\theta := 0.1$.

d Derive the conditional probability that the drug candidate doesn't work given that the p-value equals c, as a function of c and θ. Plot the function for $0 \leq x \leq 0.05$ and $\theta := 0.1$. Based on your result, suggest a strategy to select candidates sequentially for follow-up testing.

10.12   (P-hacking)

a If $\tilde{u}$ is a uniform random variable between 0 and 1, derive the conditional pdf of $\tilde{u}$ conditioned on the event $\tilde{u} \leq \alpha$ for $0 < \alpha < 1$.

b The histogram of the p-values in the publications of a research group looks like this:

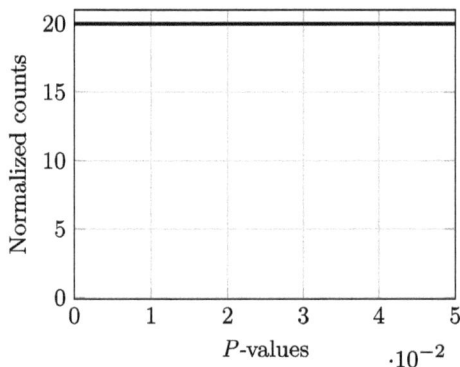

What does this suggest?

c If the total number of reported p-values is 100, estimate the number of p-values that the research group has not reported.

11

Principal Component Analysis
and Low-Rank Models

Overview

In this chapter, we describe techniques to analyze high-dimensional datasets, where each data point consists of multiple features. Such data are naturally modeled as vectors or matrices. Section 11.1 defines the mean of random vectors and random matrices. Section 11.2 introduces the covariance matrix, which encodes the variance of any linear combination of the entries in a random vector. In Section 11.3, we explain how to estimate the covariance matrix from data. Section 11.4 describes principal component analysis (PCA), a popular method to extract the directions of maximum variance in a dataset. Section 11.5 discusses how to use PCA to find optimal low-dimensional representations of high-dimensional data. In Section 11.6, we introduce low-rank models for matrix-valued data and explain how to fit them using the singular-value decomposition. Finally, in Section 11.7, we explain how to use low-rank models to estimate missing entries in a matrix, motivated by an application to movie-rating prediction.

11.1 The Mean in Multiple Dimensions

Consider a d-dimensional random vector

$$\tilde{x} := \begin{bmatrix} \tilde{x}[1] \\ \tilde{x}[2] \\ \dots \\ \tilde{x}[d] \end{bmatrix}, \tag{11.1}$$

which models a dataset with d features. We define the mean of $\tilde{x}$ as the vector formed by the means of its entries. This is a multidimensional generalization of the mean of a random variable, defined in Section 7.1.

Definition 11.1 (Mean of a random vector). *The mean of a d-dimensional random vector $\tilde{x}$ is the vector of entry-wise means:*

$$\mathbb{E}[\tilde{x}] := \begin{bmatrix} \mathbb{E}[\tilde{x}[1]] \\ \mathbb{E}[\tilde{x}[2]] \\ \dots \\ \mathbb{E}[\tilde{x}[d]] \end{bmatrix}. \tag{11.2}$$

As established in the following theorem, the mean is the point in d-dimensional space that minimizes the average squared Euclidean distance to the random vector. It can therefore be interpreted as the center of the distribution of the random vector (with the caveat that it may be distorted by extreme values, as discussed in Section 7.5). We often refer to the operation of subtracting the mean as *centering*.

Theorem 11.2. *For any d-dimensional random vector $\tilde{x}$ with finite mean,*

$$\mathbb{E}\left[\tilde{x}\right] = \arg\min_{a \in \mathbb{R}^d} \mathbb{E}\left[||\tilde{x} - a||_2^2\right]. \tag{11.3}$$

Proof   By linearity of expectation (Theorem 7.18), the mean squared ℓ_2 distance between $\tilde{x}$ and any d-dimensional vector a can be decomposed into d terms equal to the mean squared difference between $\tilde{x}[i]$ and $a[i]$ for $1 \leq i \leq d$,

$$\mathbb{E}\left[||\tilde{x} - a||_2^2\right] = \mathbb{E}\left[\sum_{i=1}^{d}(\tilde{x}[i] - a[i])^2\right] \tag{11.4}$$

$$= \sum_{i=1}^{d}\mathbb{E}\left[(\tilde{x}[i] - a[i])^2\right]. \tag{11.5}$$

Consequently, we can minimize each term separately. The minimum is achieved by the mean of each entry of the random vector by Theorem 7.30, so the result follows from Definition 11.1.   ∎

The mean of a Gaussian random vector equals its mean parameter. For these random vectors, the mean is at the center of the contour surfaces of the joint pdf, as illustrated in Figure 11.1.

Lemma 11.3 (Mean of a Gaussian random vector). *The mean of a d-dimensional Gaussian random vector $\tilde{x}$ is equal to its mean parameter.*

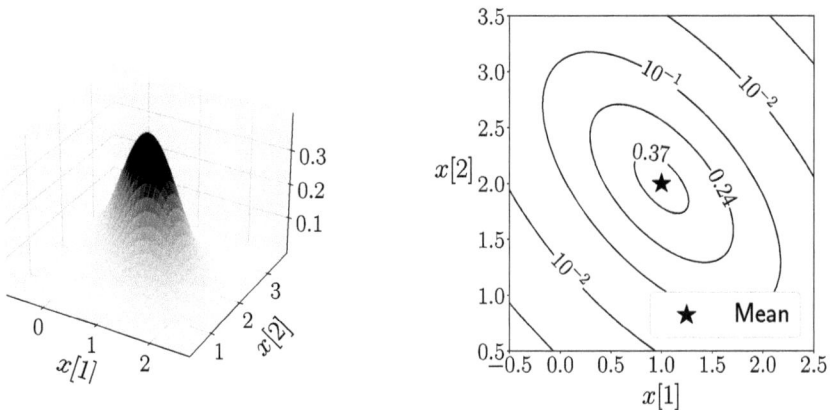

Figure 11.1 Mean of a Gaussian random vector. The graphs depict the joint pdf of a two-dimensional Gaussian random vector (left) and two-dimensional plot of its contour lines (right). The mean of the random vector (indicated by the star marker) is at the center of the contour lines.

Proof Let $\mu \in \mathbb{R}^d$ denote the mean parameter of $\tilde{x}$. For any $1 \leq i \leq d$, by Theorem 5.23 the ith entry $\tilde{x}[i]$ is a Gaussian random variable with mean parameter $\mu[i]$. Consequently, the mean of $\tilde{x}[i]$ is $\mu[i]$ by Lemma 7.25. ∎

To estimate the mean of a random vector, we compute its sample mean, which consists of the sample mean of each entry.

Definition 11.4 (Sample mean of multidimensional data). *Let* $X := \{x_1, x_2, \ldots, x_n\}$ *denote a dataset of d-dimensional real-valued vectors. The sample mean is the arithmetic average*

$$m(X) := \frac{1}{n} \sum_{i=1}^{n} x_i. \tag{11.6}$$

Equivalently, each entry of the sample mean is equal to the sample mean of the corresponding entry of the vectors in X (see Definition 7.15).

Example 11.5 (Canadian cities: Sample mean). Dataset 19 contains the locations (latitude and longitude) of cities in Canada with more than 1,000 inhabitants. The left plot in Figure 11.2 shows a scatterplot of the data. The sample mean is

$$m(X) = \begin{bmatrix} -91.9 \\ 52.9 \end{bmatrix}, \tag{11.7}$$

where the first entry is the sample mean of the longitude and the second entry is the sample mean of the latitude.
....

Example 11.6 (Faces: Sample mean). The Olivetti Faces dataset (Dataset 20) contains 400 64×64 images of 40 different subjects (10 per subject). We interpret each image as a 4,096-dimensional vector, where each entry corresponds to a pixel. Figure 11.2 shows the sample mean of the dataset (on the right). This is the *average face* in the dataset.
....

As we show in Section 11.2, manipulating random vectors often produces *random matrices* with entries that are random variables. For example, if $\tilde{x}$ is a d-dimensional random vector, then $\tilde{x}\tilde{x}^T$ is a $d \times d$ random matrix. We define the mean of a random matrix as the matrix of entry-wise means.

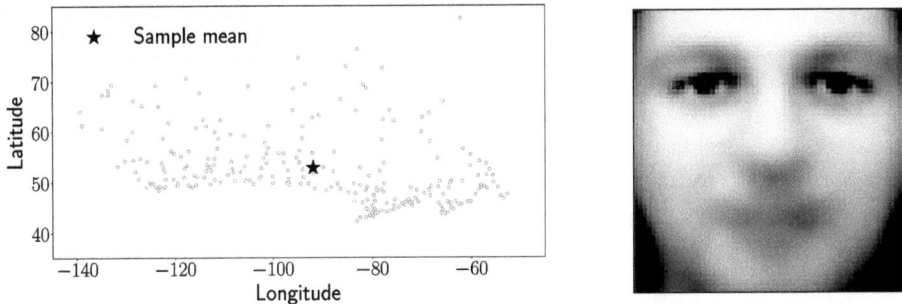

Figure 11.2 Sample mean of a dataset. The left plot shows a scatterplot of the latitude and longitude of 248 cities in Canada, represented by two-dimensional vectors as explained in Example 11.5. The star indicates the sample mean of the dataset. The right image is the sample mean of the dataset of face images in Example 11.6.

Definition 11.7 (Mean of a random matrix). *Let* $\widetilde{M}$ *be a* $d_1 \times d_2$ *random matrix, where each entry* $\widetilde{M}[i, j]$ *is a random variable for* $1 \leq i \leq d_1$ *and* $1 \leq j \leq d_2$. *The mean of* $\widetilde{M}$ *is the deterministic matrix*

$$
\mathbb{E}[\widetilde{M}] := \begin{bmatrix} \mathbb{E}\left[\widetilde{M}[1, 1]\right] & \mathbb{E}\left[\widetilde{M}[1, 2]\right] & \cdots & \mathbb{E}\left[\widetilde{M}[1, d_2]\right] \\ \mathbb{E}\left[\widetilde{M}[2, 1]\right] & \mathbb{E}\left[\widetilde{M}[2, 2]\right] & \cdots & \mathbb{E}\left[\widetilde{M}[2, d_2]\right] \\ & & \cdots & \\ \mathbb{E}\left[\widetilde{M}[d_1, 1]\right] & \mathbb{E}\left[\widetilde{M}[d_1, 2]\right] & \cdots & \mathbb{E}\left[\widetilde{M}[d_1, d_2]\right] \end{bmatrix}. \tag{11.8}
$$

Linearity of expectation holds for random vectors and random matrices. This follows directly from linearity of expectation for scalars.

Theorem 11.8 (Linearity of expectation for random vectors and matrices). *Let* $\tilde{x}$ *be a* d-*dimensional random vector. For any deterministic vector* $b \in \mathbb{R}^k$ *and any deterministic matrix* $A \in \mathbb{R}^{k \times d}$, *where* k *is a positive integer,*

$$
\mathbb{E}\left[A\tilde{x} + b\right] = A\mathbb{E}\left[\tilde{x}\right] + b. \tag{11.9}
$$

Similarly, let $\widetilde{M}$ *be a* $d_1 \times d_2$ *random matrix. For any deterministic matrices* $B \in \mathbb{R}^{k \times d_2}$ *and* $A \in \mathbb{R}^{k \times d_1}$,

$$
\mathbb{E}[A\widetilde{M} + B] = A\mathbb{E}[\widetilde{M}] + B. \tag{11.10}
$$

Proof   We prove the result for vectors, the argument for matrices is the same. By linearity of expectation for scalars (Theorem 7.18) and Definition 11.1, the ith entry of $\mathbb{E}\left[A\tilde{x} + b\right]$ equals

$$
\mathbb{E}\left[A\tilde{x} + b\right][i] = \mathbb{E}\left[(A\tilde{x} + b)[i]\right] \tag{11.11}
$$

$$
= \mathbb{E}\left[\sum_{j=1}^{d} A[i, j]\tilde{x}[j] + b[i]\right] \tag{11.12}
$$

$$
= \sum_{j=1}^{d} A[i, j]\mathbb{E}\left[\tilde{x}[j]\right] + b[i] \tag{11.13}
$$

$$
= (A\mathbb{E}\left[\tilde{x}\right] + b)[i]. \tag{11.14}
$$

$\blacksquare$

11.2 The Covariance Matrix

As explained in Section 7.7, the variance of a random variable quantifies how much it varies, on average, around its mean. In order to characterize the behavior of a random vector $\tilde{x}$, it is often useful to compute the variance of linear combinations of its entries. Any such linear combination can be expressed as a random variable

$$
\sum_{i=1}^{d} a[i]\tilde{x}[i] = a^T \tilde{x} \tag{11.15}
$$

for some deterministic vector $a \in \mathbb{R}^d$, where d is the number of entries in $\tilde{x}$. By Definition 7.32 and linearity of expectation (Theorem 11.8), the variance of $a^T \tilde{x}$ equals

$$\mathrm{Var}\left[a^T \tilde{x}\right] = \mathbb{E}\left[\left(a^T \tilde{x} - \mathbb{E}\left[a^T \tilde{x}\right]\right)^2\right] \tag{11.16}$$

$$= \mathbb{E}\left[\left(a^T \tilde{x} - a^T \mathbb{E}\left[\tilde{x}\right]\right)^2\right] \tag{11.17}$$

$$= \mathbb{E}\left[\left(a^T \,\mathrm{ct}\,(\tilde{x})\right)^2\right] \tag{11.18}$$

$$= \mathbb{E}\left[a^T \,\mathrm{ct}\,(\tilde{x})\,\mathrm{ct}\,(\tilde{x})^T\, a\right] \tag{11.19}$$

$$= a^T \mathbb{E}\left[\mathrm{ct}\,(\tilde{x})\,\mathrm{ct}\,(\tilde{x})^T\right] a, \tag{11.20}$$

where $\mathrm{ct}\,(\tilde{x}) := \tilde{x} - \mathbb{E}\left[\tilde{x}\right]$ is the result of centering $\tilde{x}$ by subtracting its mean. The matrix $\mathbb{E}[\mathrm{ct}\,(\tilde{x})\,\mathrm{ct}\,(\tilde{x})^T]$ in (11.20) does not depend on the vector a; it is always the same. We call this matrix the *covariance matrix* of the vector. It can be used to compute the variance of *any* linear combination of the entries of the random vector. By Definition 7.32, each diagonal entry of the covariance matrix is equal to the variance of the corresponding entry of $\tilde{x}$,

$$\mathbb{E}\left[\left(\mathrm{ct}\,(\tilde{x})\,\mathrm{ct}\,(\tilde{x})^T\right)[i,i]\right] = \mathbb{E}\left[\mathrm{ct}\,(\tilde{x}[i])^2\right] \tag{11.21}$$

$$= \mathrm{Var}\left[\tilde{x}[i]\right], \qquad 1 \leq i \leq d. \tag{11.22}$$

By Definition 8.4, the off-diagonal entries are equal to the covariance between the corresponding entries of $\tilde{x}$,

$$\mathbb{E}\left[\left(\mathrm{ct}\,(\tilde{x})\,\mathrm{ct}\,(\tilde{x})^T\right)[i,j]\right] = \mathbb{E}\left[\mathrm{ct}\,(\tilde{x}[i])\,\mathrm{ct}\,(\tilde{x}[j])\right] \tag{11.23}$$

$$= \mathrm{Cov}\left[\tilde{x}[i], \tilde{x}[j]\right], \quad 1 \leq i, j \leq d.$$

Definition 11.9 (Covariance matrix). *The covariance matrix of a d-dimensional random vector $\tilde{x}$ is the $d \times d$ matrix*

$$\Sigma_{\tilde{x}} := \mathbb{E}\left[\mathrm{ct}\,(\tilde{x})\,\mathrm{ct}\,(\tilde{x})^T\right] \tag{11.24}$$

$$= \begin{bmatrix} \mathrm{Var}\left[\tilde{x}[1]\right] & \mathrm{Cov}\left[\tilde{x}[1], \tilde{x}[2]\right] & \cdots & \mathrm{Cov}\left[\tilde{x}[1], \tilde{x}[d]\right] \\ \mathrm{Cov}\left[\tilde{x}[1], \tilde{x}[2]\right] & \mathrm{Var}\left[\tilde{x}[2]\right] & \cdots & \mathrm{Cov}\left[\tilde{x}[2], \tilde{x}[d]\right] \\ \vdots & \vdots & \ddots & \vdots \\ \mathrm{Cov}\left[\tilde{x}[1], \tilde{x}[d]\right] & \mathrm{Cov}\left[\tilde{x}[2], \tilde{x}[d]\right] & \cdots & \mathrm{Var}\left[\tilde{x}[d]\right] \end{bmatrix}, \tag{11.25}$$

where $\mathrm{ct}\,(\tilde{x}) := \tilde{x} - \mathbb{E}\left[\tilde{x}\right]$.

Reassuringly, the covariance-matrix parameter of a Gaussian random vector is equal to its covariance matrix.

Theorem 11.10 (Covariance matrix of a Gaussian random vector). *The covariance matrix of a Gaussian random vector is equal to its covariance-matrix parameter.*

Proof Let $\Sigma \in \mathbb{R}^{d \times d}$ denote the covariance-matrix parameter of $\tilde{x}$, and $\Sigma_{\tilde{x}} \in \mathbb{R}^{d \times d}$ its covariance matrix, as defined in Definition 11.9. For any $1 \leq i \leq d$, by Theorem 5.23 the ith entry $\tilde{x}[i]$ is a Gaussian random variable with variance parameter $\Sigma[i, i]$. By Lemma 7.43, this implies that each diagonal entry $\Sigma[i, i]$ is equal to the variance of $\tilde{x}[i]$, and therefore to the corresponding diagonal entry $\Sigma_{\tilde{x}}[i, i]$ of the covariance matrix.

For any $1 \leq i, j \leq d$, by Theorem 5.23 the joint pdf of the subvector

$$\begin{bmatrix} \tilde{x}[i] \\ \tilde{x}[j] \end{bmatrix} \tag{11.26}$$

is Gaussian with covariance-matrix parameter

$$\begin{bmatrix} \Sigma[i,i] & \Sigma[i,j] \\ \Sigma[j,i] & \Sigma[j,j] \end{bmatrix}. \tag{11.27}$$

In the proof of Theorem 8.16 (see also Example 8.8), we show that the correlation coefficient of $\tilde{x}[i]$ and $\tilde{x}[j]$ is equal to $\Sigma[i,j]$ divided by the standard deviations of $\tilde{x}[i]$ and $\tilde{x}[j]$. Therefore, by Definition 8.6 $\Sigma[i,j]$ is equal to the covariance of $\tilde{x}[i]$ and $\tilde{x}[j]$, and consequently to $\Sigma_{\tilde{x}}[i,j]$ by Definition 11.9. ∎

As explained at the beginning of this section, the covariance matrix enables us to compute the variance of any linear combination of the entries of a random vector.

Theorem 11.11. *For any random vector $\tilde{x}$ with covariance matrix $\Sigma_{\tilde{x}}$, and any deterministic vector $a \in \mathbb{R}^d$,*

$$\mathrm{Var}\left[a^T \tilde{x}\right] = a^T \Sigma_{\tilde{x}} a. \tag{11.28}$$

Proof The result follows directly from (11.20). ∎

Example 11.12 (Cheese sandwich). A deli in New York is worried about the fluctuations in the cost of their signature cheese sandwich. The ingredients of the sandwich are bread, a local cheese, and an imported cheese. They model the price of each ingredient as an entry in a three-dimensional random vector $\tilde{x}$. The entries $\tilde{x}[1]$, $\tilde{x}[2]$, and $\tilde{x}[3]$ represent the price (in cents per gram) of the bread, the local cheese and the imported cheese, respectively. From past data, the covariance matrix of $\tilde{x}$ is estimated to equal

$$\Sigma_{\tilde{x}} = \begin{bmatrix} 1 & 0.8 & 0 \\ 0.8 & 1 & 0 \\ 0 & 0 & 1.2 \end{bmatrix}. \tag{11.29}$$

The deli considers two recipes. Recipe 1 uses 100g of bread, 50g of local cheese, and 50g of imported cheese. Recipe 2 uses 100g of bread, 100g of local cheese, and no imported cheese. By Theorem 11.11, the standard deviation in the price of Recipe 1 is

$$\sigma_{100\tilde{x}[1]+50\tilde{x}[2]+50\tilde{x}[3]} = \sqrt{\begin{bmatrix} 100 & 50 & 50 \end{bmatrix} \Sigma_{\tilde{x}} \begin{bmatrix} 100 \\ 50 \\ 50 \end{bmatrix}} = 153 \text{ cents.} \tag{11.30}$$

The standard deviation in the price of Recipe 2 is

$$\sigma_{100\tilde{x}[1]+100\tilde{x}[2]} = \sqrt{\begin{bmatrix} 100 & 100 & 0 \end{bmatrix} \Sigma_{\tilde{x}} \begin{bmatrix} 100 \\ 100 \\ 0 \end{bmatrix}} = 190 \text{ cents.} \tag{11.31}$$

Even though the price of the imported cheese has a higher variance (1.2) than that of the local cheese (1), adding it to the recipe lowers the variance of the overall cost because the

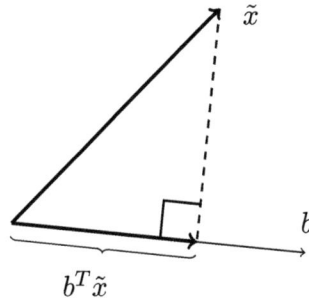

Figure 11.3 Variance in a certain direction. The diagram depicts the decomposition in (11.32). To quantify the variance of a zero-mean random vector $\tilde{x}$ in the direction of a deterministic vector b with unit ℓ_2 norm, we compute the variance of the amplitude $b^T \tilde{x}$ of the orthogonal projection of $\tilde{x}$ onto the line spanned by b.

imported cheese is uncorrelated with the other ingredients (the corresponding off-diagonal entry in the covariance matrix is zero).

Theorem 11.11 enables us to determine the variance of a random vector in a specific direction from its covariance matrix. Let $b \in \mathbb{R}^d$ be a deterministic vector with unit ℓ_2 norm[1] and let $\tilde{x}$ be a zero-mean d-dimensional random vector $\tilde{x}$. We decompose $\tilde{x}$ into a component collinear with b and a component orthogonal to b:

$$\tilde{x} = \underbrace{(b^T \tilde{x})b}_{\text{collinear with b}} + \underbrace{\tilde{x} - (b^T \tilde{x})b}_{\text{orthogonal to b}} . \tag{11.32}$$

The decomposition is depicted in Figure 11.3. Geometrically, $(b^T \tilde{x})b$ is the orthogonal projection of $\tilde{x}$ onto the line spanned by b. The variance of $\tilde{x}$ in the direction of b is the variance of the amplitude $b^T \tilde{x}$ of this projection. By Theorem 11.11, it equals

$$\mathrm{Var}[b^T \tilde{x}] = b^T \Sigma_{\tilde{x}} b. \tag{11.33}$$

If the random vector $\tilde{x}$ has nonzero mean, then you might be tempted to center the vector by removing its mean, before considering its variance in a specific direction. However, this does not affect the computation, since the covariance matrices of $\tilde{x}$ and of the centered vector $\mathrm{ct}\,(\tilde{x}) := \tilde{x} - \mathbb{E}\,[\tilde{x}]$ are the same (see Definition 11.9), so by Theorem 11.11,

$$\mathrm{Var}[b^T \mathrm{ct}\,(\tilde{x})] = b^T \Sigma_{\mathrm{ct}(\tilde{x})} b \tag{11.34}$$
$$= b^T \Sigma_{\tilde{x}} b. \tag{11.35}$$

Example 11.13 (Variance of a random vector in a specific direction). We consider the two-dimensional Gaussian random vector $\tilde{x}$ depicted in Figure 11.1. Its mean and covariance-matrix parameters equal

---

[1] The ℓ_2 or Euclidean norm of a d-dimensional vector b is the square root of the sum of its squared entries, $||b||_2 := \sqrt{\sum_{j=1}^d b[j]^2}$.

$$\mu := \begin{bmatrix} 1 \\ 2 \end{bmatrix}, \qquad \Sigma := \begin{bmatrix} 0.5 & -0.3 \\ -0.3 & 0.5 \end{bmatrix}. \tag{11.36}$$

Our goal is to quantify the variation of $\tilde{x}$ in the direction of the vector

$$b := \begin{bmatrix} 1 \\ 0 \end{bmatrix}. \tag{11.37}$$

We begin by removing the mean, to obtain the centered random vector $\mathrm{ct}\,(\tilde{x}) := \tilde{x} - \mu$. The top left graph in Figure 11.4 shows the line spanned by b, superposed onto the contour lines of the pdf of $\mathrm{ct}\,(\tilde{x})$. The amplitude $b^T\,\mathrm{ct}\,(\tilde{x})$ of the orthogonal projection of $\mathrm{ct}\,(\tilde{x})$ onto the line is a random variable, whose pdf is shown in the top right graph of Figure 11.4. By (11.35), the variance of this random variable equals

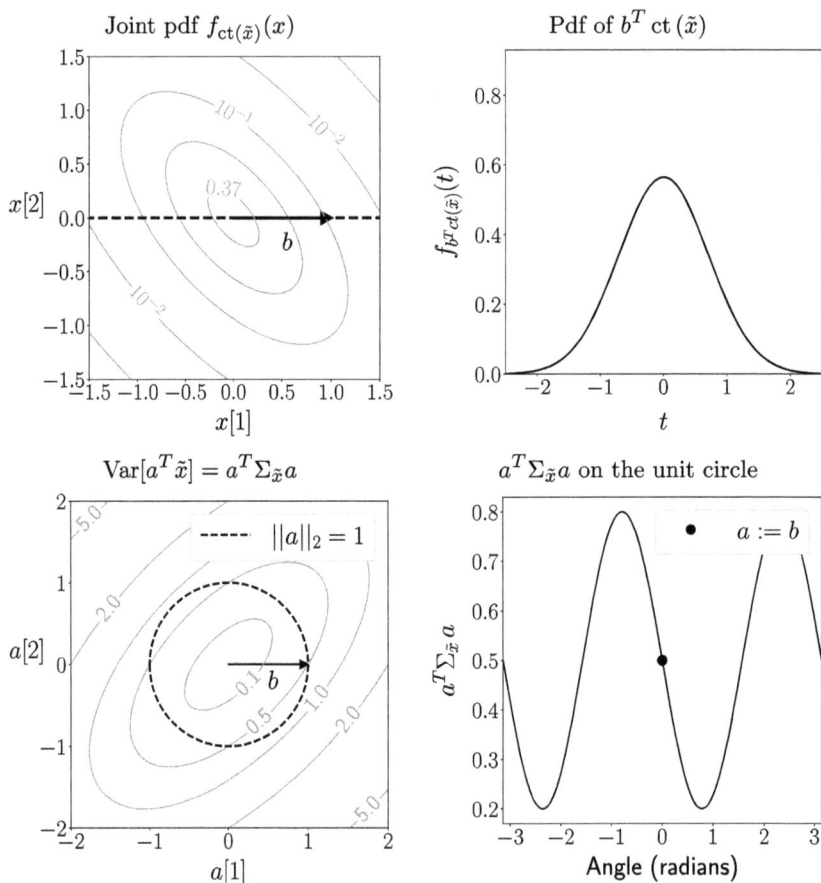

Figure 11.4 Variance in a specific direction. The top left graph depicts the unit-norm vector b in Example 11.13, superposed onto the contour lines of the joint pdf of the centered vector $\mathrm{ct}\,(\tilde{x}) := \tilde{x} - \mu$. The top right graph shows the pdf of the amplitude $b^T\,\mathrm{ct}\,(\tilde{x})$ of the orthogonal projection of $\mathrm{ct}\,(\tilde{x})$ onto the line spanned by b. The bottom left graph shows the contour lines of the function $a^T\Sigma_{\tilde{x}}a$, as well as the vector b. The values of this function on the dashed unit circle encode the variance of $\tilde{x}$ in every possible direction. These values are plotted as a function of the angle with the horizontal axis in the bottom right graph. The circular marker corresponds to the direction of the vector b.

$$\text{Var}\left[b^T \text{ct}\left(\tilde{x}\right)\right] = b^T \Sigma_{\tilde{x}}\, b \tag{11.38}$$

$$= b^T \Sigma\, b \tag{11.39}$$

$$= 0.5. \tag{11.40}$$

The bottom left graph in Figure 11.4 depicts the contour lines of the function,

$$q(a) := a^T \Sigma_{\tilde{x}}\, a. \tag{11.41}$$

The values of this function on the unit circle $\left\{a \in \mathbb{R}^d : ||a||_2 = 1\right\}$ encode the variance of $\tilde{x}$ in every possible direction. They are plotted in the bottom right graph of Figure 11.4, as a function of the angle with respect to the horizontal axis.

· ·

11.3 The Sample Covariance Matrix

In order to analyze a dataset with d features, we can model the data as samples from a d-dimensional random vector and approximate its mean and covariance matrix. A reasonable estimator for the covariance matrix is the sample covariance matrix, which contains the sample variances and sample covariances of the features.

Definition 11.14 (Sample covariance matrix). *Let $X := \{x_1, x_2, \ldots, x_n\}$ denote n vectors containing d features, and let $X[j] := \{x_1[j], \ldots, x_n[j]\}$ denote the entries corresponding to the jth feature for $1 \leq j \leq d$. The sample covariance matrix of the data equals*

$$\Sigma_X := \frac{1}{n-1} \sum_{i=1}^{n} \text{ct}\left(x_i\right) \text{ct}\left(x_i\right)^T \tag{11.42}$$

$$= \begin{bmatrix} v(X[1]) & c(X[1], X[2]) & \cdots & c(X[1], X[d]) \\ c(X[1], X[2]) & v(X[2]) & \cdots & c(X[2], X[d]) \\ \vdots & \vdots & \ddots & \vdots \\ c(X[1], X[d]) & c(X[2], X[d]) & \cdots & v(X[d]) \end{bmatrix}, \tag{11.43}$$

where $\text{ct}\left(x_i\right) := x_i - m(X)$ is the result of centering the ith vector using the sample mean $m(X)$ for $1 \leq i \leq d$ (see Definition 11.4), $v(X[j])$ is the sample variance of $X[j]$ for $1 \leq j \leq d$ (see Definition 7.37), and $c(X[j], X[k])$ is the sample covariance of $X[j]$ and $X[k]$ for $1 \leq j, k \leq d$ (see Definition 8.9).

A similar estimator for the covariance matrix can be obtained by fitting a Gaussian distribution to the data using maximum likelihood. By Theorem 5.24, the maximum-likelihood estimator of the covariance-matrix parameter of a Gaussian random vector equals

$$\Sigma_{\text{ML}} = \frac{1}{n} \sum_{i=1}^{n} \text{ct}\left(x_i\right) \text{ct}\left(x_i\right)^T, \tag{11.44}$$

which is essentially equal to the sample covariance matrix (unless n is very small).

Example 11.15 (Canadian cities: Sample covariance matrix)**.** The sample covariance matrix of the Canadian-city data in Example 11.5 is

$$\Sigma_X = \begin{bmatrix} 524.9 & -59.8 \\ -59.8 & 53.7 \end{bmatrix}. \tag{11.45}$$

The longitude has much higher variance than the latitude. Latitude and longitude are negatively correlated because people at higher longitudes (in the east) tend to live at lower latitudes (in the south).

....

Example 11.16 (Temperatures in the United States). In this example, we use the sample covariance matrix to describe a dataset of hourly temperatures at four weather stations in the United States, extracted from Dataset 9. The sample covariance matrix of the temperatures is shown in Table 11.1. The diagonal entries of the covariance matrix reveal that the temperature in Hawaii has much lower variance than the rest, whereas the temperature at Ithaca in upstate New York has very high variance. The off-diagonal entries show that all the temperatures are positively correlated.

Table 11.1 also shows the sample correlation matrix of the data, which is defined as the matrix containing the sample correlation coefficients between every pair of features in the data (see Definition 8.10). The closer the stations are geographically, the higher the correlation. For instance, the temperatures in Ithaca and Durham are very correlated, whereas the temperature in Hawaii has relatively low correlation with either of them.

....

Just like the covariance matrix encodes the variance of any linear combination of the entries of a random vector, the sample covariance matrix of a dataset encodes the sample variance of any linear combination of the features in a dataset.

Table 11.1 *Sample covariance matrix and correlation of temperature data. The sample covariance matrix of the data in Example 11.16 contains the sample variance of the temperature at each location, as well as the sample covariances between the temperatures. The correlation matrix contains the corresponding sample correlation coefficients.*

	Covariance matrix			
	Tucson, AZ	Hilo, HI	Durham, NC	Ithaca, NY
Tucson, AZ	78.6	14.7	54.8	65.0
Hilo, HI	14.7	8.4	9.5	11.8
Durham, NC	54.8	9.5	89.4	97.4
Ithaca, NY	65.0	11.8	97.4	137.3

	Correlation matrix			
	Tucson, AZ	Hilo, HI	Durham, NC	Ithaca, NY
Tucson, AZ	1	0.57	0.65	0.63
Hilo, HI	0.57	1	0.35	0.35
Durham, NC	0.65	0.35	1	0.88
Ithaca, NY	0.63	0.35	0.88	1

Theorem 11.17. *For any dataset $X = \{x_1, \ldots, x_n\}$ of d-dimensional data and any vector $a \in \mathbb{R}^d$, let*

$$X_a := \{a^T x_1, \ldots, a^T x_n\} \tag{11.46}$$

be the inner products between a and the data points in X. The sample variance of X_a equals

$$v(X_a) = a^T \Sigma_X a. \tag{11.47}$$

Proof The proof mimics the proof of Theorem 11.11. By Definitions 7.37 and 11.14,

$$v(X_a) = \frac{1}{n-1} \sum_{i=1}^{n} \left(a^T x_i - \frac{1}{n} \sum_{j=1}^{n} a^T x_j \right)^2 \tag{11.48}$$

$$= \frac{1}{n-1} \sum_{i=1}^{n} \left(a^T \left(x_i - \frac{1}{n} \sum_{j=1}^{n} x_j \right) \right)^2 \tag{11.49}$$

$$= \frac{1}{n-1} \sum_{i=1}^{n} (a^T \operatorname{ct}(x_i))^2 \tag{11.50}$$

$$= \frac{1}{n-1} \sum_{i=1}^{n} a^T \operatorname{ct}(x_i) \operatorname{ct}(x_i)^T a \tag{11.51}$$

$$= a^T \left(\frac{1}{n-1} \sum_{i=1}^{n} \operatorname{ct}(x_i) \operatorname{ct}(x_i)^T \right) a \tag{11.52}$$

$$= a^T \Sigma_X a, \tag{11.53}$$

where $\operatorname{ct}(x_i) := x_i - m(X)$ and $m(X)$ is the sample mean of the dataset. ∎

In order to describe the geometry of a dataset, we can estimate its variance in different directions. Analogously to the approach described at the end of Section 11.2, we first project the data points onto a specific direction of interest by computing the inner product between them and a unit-norm vector in that direction. Then, we estimate the variance by computing the sample variance of the inner products.

Example 11.18 (Canadian cities: Variance in a specific direction). Our goal in this example is to quantify how the distribution of Canadian cities varies in the southwest-northeast direction corresponding to the unit ℓ_2 norm vector

$$b := \frac{1}{\sqrt{2}} \begin{bmatrix} 1 \\ 1 \end{bmatrix}. \tag{11.54}$$

The top left graph in Figure 11.5 depicts the line spanned by b superposed onto a scatterplot of the data. The data are centered, so their component in the direction of interest is captured by the inner product of each data point with b. Geometrically, this is the amplitude of the orthogonal projection of the data point onto the line spanned by b. We denote the one-dimensional dataset of inner products by

$$X_b := \{b^T x_1, \ldots, b^T x_n\}. \tag{11.55}$$

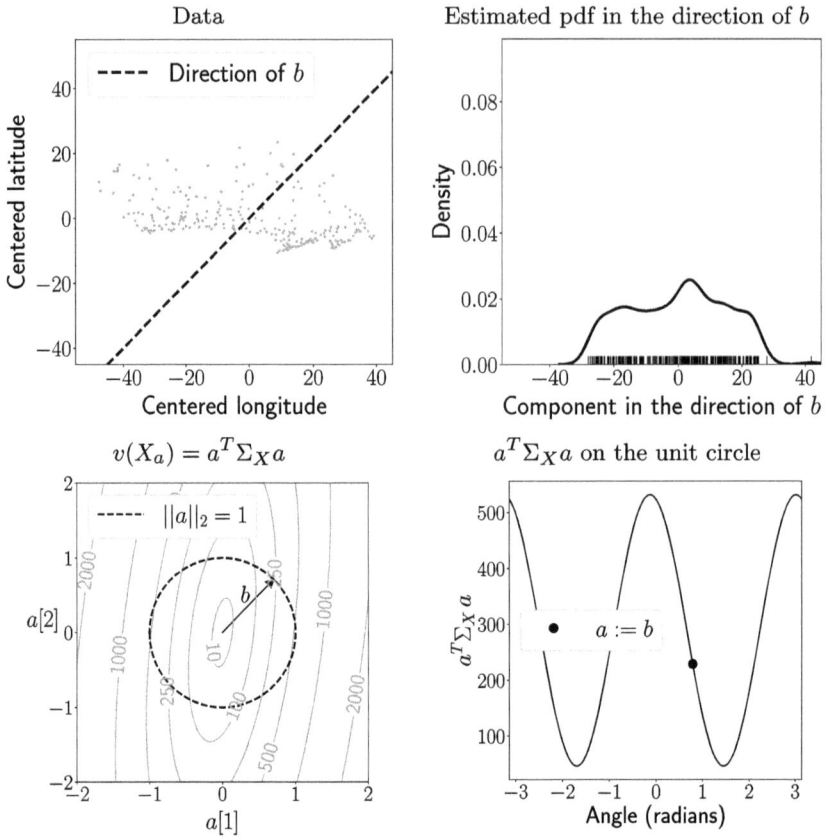

Figure 11.5 Sample variance in a specific direction. The top left plot shows the dashed line spanned by the vector b in Example 11.18 superposed onto a scatterplot of the data. The top right plot shows a rug plot of the set X_b containing the amplitudes of the orthogonal projections of the data onto that line, and an estimate of its probability density (obtained via kernel density estimation, as described in Section 3.5.2). The bottom left plot shows the contours of the function $a^T \Sigma_X a$, as well as the vector b (represented by an arrow). The values on the dashed unit circle are the sample variances of the dataset in every possible direction. These values are plotted as a function of the angle in the bottom right graph. The circular marker corresponds to the direction of the vector b.

The top left graph in Figure 11.5 shows an estimate of the probability density of this dataset. By Theorem 11.17, we can compute its sample variance directly from the sample covariance matrix Σ_X in Example 11.15:

$$v(X_b) = \frac{1}{\sqrt{2}} \begin{bmatrix} 1 & 1 \end{bmatrix} \Sigma_X \frac{1}{\sqrt{2}} \begin{bmatrix} 1 \\ 1 \end{bmatrix} \tag{11.56}$$

$$= 229. \tag{11.57}$$

The bottom left graph in Figure 11.5 shows the contour lines of the function

$$q(a) := a^T \Sigma_X a. \tag{11.58}$$

The values of this function on the circle $\{a \in \mathbb{R}^d : ||a||_2 = 1\}$ encode the sample variance of the dataset in every possible direction. They are depicted in the bottom right graph of Figure 11.5.

··

11.4 Principal Component Analysis

Principal component analysis (PCA) is a very popular technique to analyze data with multiple features. If we interpret such data as points in a multidimensional space, we expect directions in which the data exhibit more variation to be informative. PCA identifies the directions of maximum variance and decomposes the data into components corresponding to each of these directions. Section 11.4.1 defines PCA for random vectors. Section 11.4.2 describes how to perform PCA on a dataset. Section 11.4.3 provides an intuitive explanation of the mathematics behind PCA.

11.4.1 Principal Component Analysis of a Random Vector

As explained in Section 11.2, the covariance matrix $\Sigma_{\tilde{x}}$ of a random vector $\tilde{x}$ encodes the variance of the vector in every direction. This can be leveraged to find the direction of highest variance. By Theorem 11.11, the variance in the direction of a unit-norm vector a is equal to $a^T \Sigma_{\tilde{x}} a$. Therefore, in order to maximize the variance, we need to maximize the function

$$q(a) := a^T \Sigma_{\tilde{x}} a, \tag{11.59}$$

over the set of unit ℓ_2 norm vectors $\{a : ||a||_2 = 1\}$. We can solve this constrained optimization problem by applying the spectral theorem, which is a fundamental result in linear algebra.

Theorem 11.19 (Spectral theorem for symmetric matrices). *If $M \in \mathbb{R}^{d \times d}$ is symmetric, then it has an eigendecomposition of the form*

$$M = \begin{bmatrix} u_1 & u_2 & \cdots & u_d \end{bmatrix} \begin{bmatrix} \lambda_1 & 0 & \cdots & 0 \\ 0 & \lambda_2 & \cdots & 0 \\ & & \cdots & \\ 0 & 0 & \cdots & \lambda_d \end{bmatrix} \begin{bmatrix} u_1 & u_2 & \cdots & u_d \end{bmatrix}^T, \tag{11.60}$$

where the eigenvalues $\lambda_1 \geq \lambda_2 \geq \cdots \geq \lambda_d$ are real and the eigenvectors u_1, u_2, ..., $u_d \in \mathbb{R}^d$ are orthogonal to each other, and normalized to have unit ℓ_2 norm.

The eigenvalues of M are extrema of the function $a^T M a$ constrained to the unit sphere $\{a : ||a||_2 = 1\}$, which are attained at the corresponding eigenvectors:

$$\lambda_1 = \max_{||a||_2=1} a^T M a, \tag{11.61}$$

$$u_1 = \arg \max_{||a||_2=1} a^T M a, \tag{11.62}$$

$$\lambda_k = \max_{||a||_2=1, a \perp u_1, \ldots, u_{k-1}} a^T M a, \quad 2 \leq k \leq d-1, \tag{11.63}$$

$$u_k = \arg \max_{||a||_2=1, a \perp u_1, \ldots, u_{k-1}} a^T M a, \quad 2 \le k \le d-1, \tag{11.64}$$

$$\lambda_d = \min_{||a||_2=1} a^T M a, \tag{11.65}$$

$$u_d = \arg \min_{||a||_2=1} a^T M a. \tag{11.66}$$

In Section 11.4.3, we provide a detailed explanation of why the extrema of the function $a^T M a$ on the unit sphere are reached at eigenvectors of M. For a formal proof of the spectral theorem, we refer to any advanced textbook on linear algebra.

By the spectral theorem and Theorem 11.11, the eigendecomposition of the covariance matrix of a random vector yields a *ranking* of orthogonal directions. The largest eigenvalue λ_1 is the highest variance of the random vector in any direction, attained in the direction of the first eigenvector u_1. In directions orthogonal to u_1, the highest variance is the second largest eigenvalue λ_2, attained in the direction of the second eigenvector u_2. In general, when restricted to the orthogonal complement of the span of $u_1, \ldots, u_k$ for $1 \le k \le d-1$, the variance is highest in the direction of the $k+1$th eigenvector u_{k+1}, and equals the corresponding eigenvalue. Finally, the random vector has lowest variance, equal to the smallest eigenvalue λ_d, in the direction of the last eigenvector u_d.

Theorem 11.20 (Directions of maximum and minimum variance). *Let $\tilde{x}$ be a d-dimensional random vector with covariance matrix $\Sigma_{\tilde{x}}$. The eigenvectors $u_1, u_2, \ldots, u_d$ (normalized to have unit ℓ_2 norm) and the corresponding eigenvalues $\lambda_1 \ge \lambda_2 \ge \cdots \ge \lambda_d$ of $\Sigma_{\tilde{x}}$ satisfy*

$$\lambda_1 = \max_{||a||_2=1} \mathrm{Var}[a^T \tilde{x}], \tag{11.67}$$

$$u_1 = \arg \max_{||a||_2=1} \mathrm{Var}[a^T \tilde{x}], \tag{11.68}$$

$$\lambda_k = \max_{||a||_2=1, a \perp u_1, \ldots, u_{k-1}} \mathrm{Var}[a^T \tilde{x}], \quad 2 \le k \le d-1, \tag{11.69}$$

$$u_k = \arg \max_{||a||_2=1, a \perp u_1, \ldots, u_{k-1}} \mathrm{Var}[a^T \tilde{x}], \quad 2 \le k \le d-1, \tag{11.70}$$

$$\lambda_d = \min_{||a||_2=1} \mathrm{Var}[a^T \tilde{x}], \tag{11.71}$$

$$u_d = \arg \min_{||a||_2=1} \mathrm{Var}[a^T \tilde{x}]. \tag{11.72}$$

Proof By Definition 11.7, the transpose of the mean of a matrix equals the mean of its transpose. Consequently, by Definition 11.9,

$$\Sigma_{\tilde{x}}^T = \left(\mathbb{E} \left[\mathrm{ct}\,(\tilde{x})\, \mathrm{ct}\,(\tilde{x})^T \right] \right)^T \tag{11.73}$$

$$= \mathbb{E} \left[\left(\mathrm{ct}\,(\tilde{x})\, \mathrm{ct}\,(\tilde{x})^T \right)^T \right] \tag{11.74}$$

$$= \mathbb{E} \left[\mathrm{ct}\,(\tilde{x})\, \mathrm{ct}\,(\tilde{x})^T \right] = \Sigma_{\tilde{x}}, \tag{11.75}$$

where $\mathrm{ct}\,(\tilde{x}) := \tilde{x} - \mathbb{E}\,[\tilde{x}]$. Therefore, covariance matrices are symmetric. By Theorem 11.11 $\mathrm{Var}[a^T \tilde{x}] = a^T \Sigma_{\tilde{x}} a$ for any $a \in \mathbb{R}^d$, so the result follows from Theorem 11.19, setting $M := \Sigma_{\tilde{x}}$. ∎

The *principal directions* of a random vector are the eigenvectors of its covariance matrix, or, more precisely, the directions aligned with them. The principal directions are orthogonal and span the entire d-dimensional space, so they are an orthonormal basis of $\mathbb{R}^d$ (assuming they are normalized to have unit ℓ_2 norm). The components of the centered random vector $\text{ct}\,(\tilde{x}) := \tilde{x} - \mathbb{E}[\tilde{x}]$ in this basis are called *principal components*,

$$\tilde{w}_j := u_j^T \,\text{ct}\,(\tilde{x})\,, \quad 1 \leq j \leq d. \tag{11.76}$$

Computing the principal components can be interpreted as a *change of basis*, which rotates the centered random vector. The rotation aligns the axes with the principal directions, so that the first entry (equal to the first principal component) captures the most variance, the second entry captures the second most variance, and so on. The variance captured by each principal component is equal to the eigenvalue of the covariance matrix associated with the corresponding principal direction.

Lemma 11.21 (Variance of the principal components). *Let $\tilde{x}$ be a d-dimensional random vector with covariance matrix $\Sigma_{\tilde{x}}$. The variance of the jth principal component $\tilde{w}_j$ equals the corresponding eigenvalue λ_j of $\Sigma_{\tilde{x}}$.*

Proof The principal direction u_j is a unit ℓ_2-norm eigenvector of the covariance matrix associated with the eigenvalue λ_j, so by Theorem 11.11,

$$\text{Var}\,[\tilde{w}_j] = \text{Var}\,[u_j^T \,\text{ct}\,(\tilde{x})] \tag{11.77}$$
$$= u_j^T \Sigma_{\tilde{x}} u_j \tag{11.78}$$
$$= \lambda_j u_j^T u_j \tag{11.79}$$
$$= \lambda_j. \tag{11.80}$$

∎

The principal components of a random vector are uncorrelated.

Lemma 11.22 (Principal components are uncorrelated). *Let $\tilde{x}$ be a d-dimensional random vector with covariance matrix $\Sigma_{\tilde{x}}$. The principal components $\tilde{w}_1$, $\tilde{w}_2$, ..., $\tilde{w}_d$ of $\tilde{x}$ are uncorrelated.*

Proof Let u_i be the eigenvector of the covariance matrix corresponding to the ith principal component. By Theorem 11.20, the eigenvectors are orthogonal. The mean of the principal components is zero by linearity of expectation (Theorem 11.8),

$$\mathbb{E}\,[\tilde{w}_i] = \mathbb{E}\,[u_i^T \,\text{ct}\,(\tilde{x})] \tag{11.81}$$
$$= u_i^T \mathbb{E}\,[\text{ct}\,(\tilde{x})] = 0, \tag{11.82}$$

because $\mathbb{E}\,[\text{ct}\,(\tilde{x})] = \mathbb{E}\,[\tilde{x}] - \mathbb{E}\,[\tilde{x}] = 0$. Consequently, if $i \neq j$, by Definition 11.9,

$$\text{Cov}[\tilde{w}_i \tilde{w}_j] = \mathbb{E}[\tilde{w}_i \tilde{w}_j] = \mathbb{E}\left[u_i^T \,\text{ct}\,(\tilde{x})\,\text{ct}\,(\tilde{x})^T u_j\right] \tag{11.83}$$
$$= u_i^T \mathbb{E}[\text{ct}\,(\tilde{x})\,\text{ct}\,(\tilde{x})^T] u_j \tag{11.84}$$
$$= u_i^T \Sigma_{\tilde{x}} u_j \tag{11.85}$$
$$= \lambda_j u_i^T u_j \tag{11.86}$$
$$= 0. \tag{11.87}$$

∎

Example 11.23 (Principal component analysis of a Gaussian random vector). We apply PCA to the Gaussian random vector $\tilde{x}$ in Example 11.13, by computing the eigendecomposition of its covariance matrix,

$$\Sigma_{\tilde{x}} = \begin{bmatrix} u_1 & u_2 \end{bmatrix} \begin{bmatrix} \lambda_1 & 0 \\ 0 & \lambda_2 \end{bmatrix} \begin{bmatrix} u_1 & u_2 \end{bmatrix}^T. \tag{11.88}$$

The top row of Figure 11.6 shows the two principal directions superposed onto the contour lines of the function $a^T \Sigma_{\tilde{x}} a$. The second row shows the values of the function restricted to the unit circle, where $\|a\|_2 = 1$. The maximum of the restricted function is reached at the first principal direction

$$u_1 = \begin{bmatrix} \frac{1}{\sqrt{2}} \\ -\frac{1}{\sqrt{2}} \end{bmatrix}, \tag{11.89}$$

which is therefore the direction of maximum variance, as established in Theorem 11.20. By Lemma 11.21 the corresponding variance is equal to the first eigenvalue $\lambda_1 = 0.8$. Conversely, the minimum of the restricted function is reached at the second (and last) principal direction

$$u_2 = \begin{bmatrix} \frac{1}{\sqrt{2}} \\ \frac{1}{\sqrt{2}} \end{bmatrix}, \tag{11.90}$$

which is the direction of minimum variance. The corresponding variance is equal to the second eigenvalue $\lambda_2 = 0.2$.

The bottom row of the figure shows the two principal directions superposed onto the contour lines of the joint pdf of the centered random vector $\mathrm{ct}(\tilde{x})$. As described in Section 5.10.1, the contour lines of Gaussian random vectors are ellipsoidal. PCA enables us to find the axes of these ellipsoids (after centering to remove the mean). The major axis is equal to the first principal direction because it is the direction in which the density is the most *stretched* and hence has the highest variance. The second major axis is equal to the second principal direction, and so on. The analysis of the ellipsoidal contour lines of the joint pdf of $\tilde{x}$ in Example 5.20 confirms that the axes of the ellipses indeed coincide with the principal directions.

The top row of Figure 11.7 shows the joint pdf of the principal components $\tilde{w}_1 := u_1^T \mathrm{ct}(\tilde{x})$ and $\tilde{w}_2 := u_2^T \mathrm{ct}(\tilde{x})$, and compares it to the joint pdf of $\tilde{x}$. To obtain the joint pdf of the principal components, we first center the joint pdf of $\tilde{x}$ by subtracting the mean, and then we rotate it so that the axes of the ellipsoidal contour lines align with the coordinate axes. This maximizes the variance of the marginal distribution along the horizontal axis and minimizes the variance of the marginal distribution along the vertical axis. The corresponding marginal pdfs are shown on the bottom row of Figure 11.7. After the rotation, the horizontal and vertical components of the rotated vector are uncorrelated, as established in Lemma 11.22. By Theorem 8.29, they are also independent because the random vector is Gaussian.

....

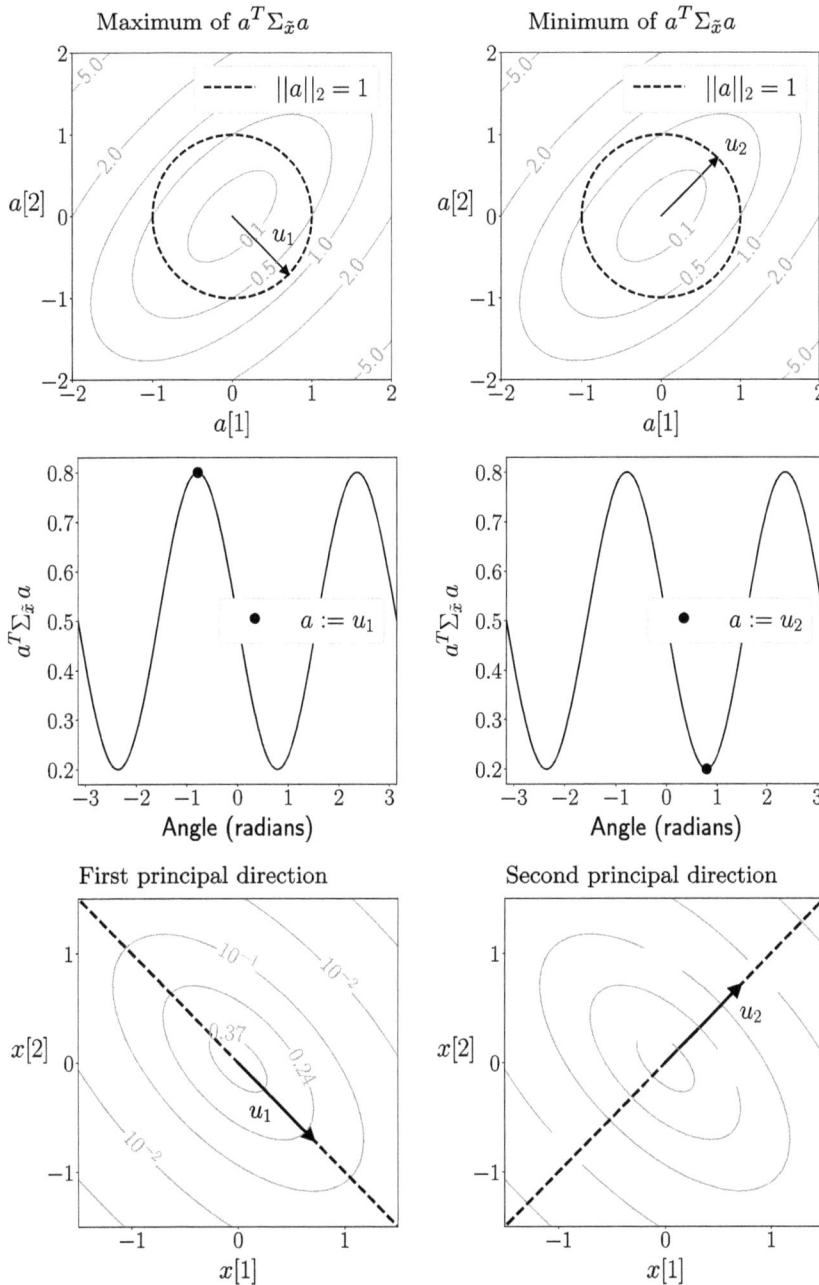

Figure 11.6 Principal directions of a Gaussian random vector. The top row shows the first (left) and second (right) principal directions u_1 and u_2 of the Gaussian random vector $\tilde{x}$ in Example 11.23. The principal directions are superposed onto the contour lines of the function $a^T \Sigma_{\tilde{x}} a$. The values of this function on the dashed unit circle are plotted as a function of the angle in the middle row. The restricted function reaches its maximum and minimum (depicted by circular markers) at the first and second principal directions, respectively, which are therefore the directions of maximum and minimum variance, as established in Theorem 11.20. The bottom row shows that the principal directions are aligned with the axes of the ellipsoidal contour lines of the joint pdf of the centered vector ct $(\tilde{x})$.

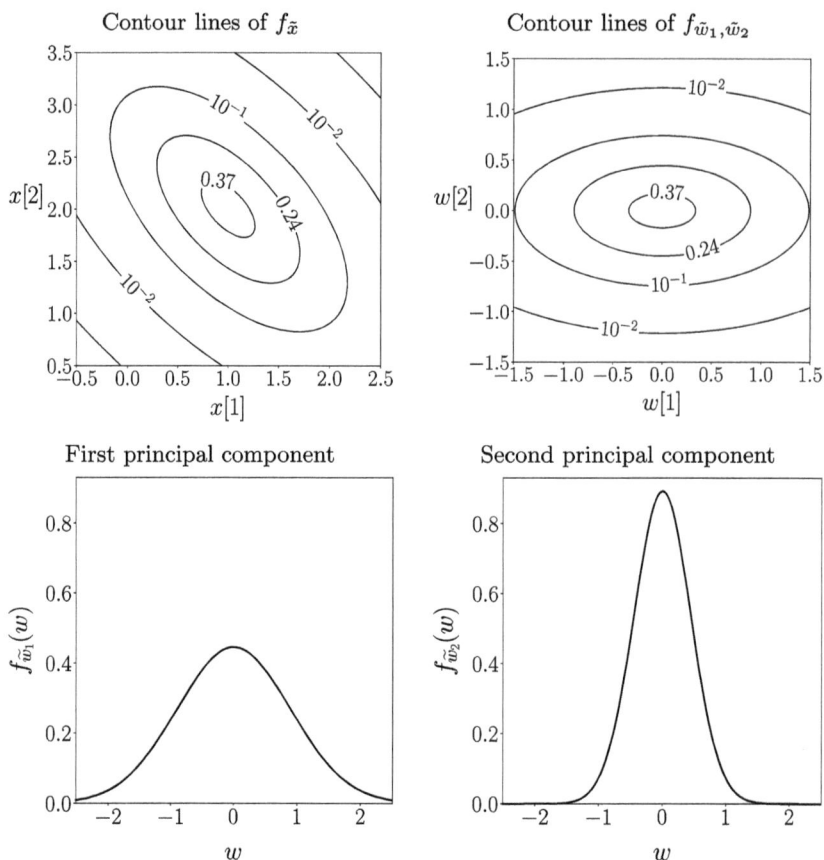

Figure 11.7 Principal components of a Gaussian random vector. The top left graph shows the contour lines of the joint pdf $f_{\tilde{x}}$ of the Gaussian random vector in Example 11.13. The top right graph shows the contour lines of the joint pdf of its first and second principal components $\tilde{w}_1$ and $\tilde{w}_2$, which are obtained by centering the ellipsoidal contour lines of $f_{\tilde{x}}$ and then rotating them so that their axes align with the coordinate axes. This yields a horizontal component with maximum variance ($\tilde{w}_1$) and a vertical component with minimum variance ($\tilde{w}_2$). The bottom row shows the marginal pdfs of $\tilde{w}_1$ and $\tilde{w}_2$.

11.4.2 Principal Component Analysis of a Dataset

As explained in Section 11.4.1, principal component analysis (PCA) extracts the directions of maximum (and minimum variance) of a random vector from the eigendecomposition of its covariance matrix. The same procedure can be applied to the sample covariance matrix of a dataset.

Definition 11.24 (Principal component analysis). *To perform principal component analysis of a dataset X containing n vectors $x_1, x_2, \ldots, x_n \in \mathbb{R}^d$ with d features each, we apply the following steps:*

1 We compute the sample covariance matrix of the data Σ_X following Definition 11.14.
2 We compute the eigendecomposition of Σ_X. The principal directions of the data are the eigenvectors $u_1, u_2, \ldots, u_d$ of Σ_X, ordered according to the corresponding eigenvalues $\lambda_1 \geq \lambda_2 \geq \cdots \geq \lambda_d$ and normalized to have unit ℓ_2 norm.

3 We center the data and compute the principal components

$$w_j[i] := u_j^T \, \text{ct}\,(x_i), \quad 1 \le i \le n, \, 1 \le j \le d, \tag{11.91}$$

where $\text{ct}\,(x_i) := x_i - m(X)$, *and* $m(X)$ *denotes the sample mean (see Definition 11.4).*

When we perform PCA on a dataset, the resulting principal directions are the direction of maximum (and minimum) sample variance.

Theorem 11.25 (Directions of maximum and minimum sample variance). *Let X be a dataset containing n vectors $x_1, x_2, \ldots, x_n \in \mathbb{R}^d$ with sample covariance matrix Σ_X. The eigenvectors $u_1, u_2, \ldots, u_d$ (normalized to have unit ℓ_2 norm) and the corresponding eigenvalues $\lambda_1 \ge \lambda_2 \ge \cdots \ge \lambda_d$ of Σ_X satisfy*

$$\lambda_1 = \max_{\|a\|_2=1} v(X_a), \tag{11.92}$$

$$u_1 = \arg\max_{\|a\|_2=1} v(X_a), \tag{11.93}$$

$$\lambda_k = \max_{\|a\|_2=1, a \perp u_1, \ldots, u_{k-1}} v(X_a), \quad 2 \le k \le d-1, \tag{11.94}$$

$$u_k = \arg\max_{\|a\|_2=1, a \perp u_1, \ldots, u_{k-1}} v(X_a), \quad 2 \le k \le d-1, \tag{11.95}$$

$$\lambda_d = \min_{\|a\|_2=1} v(X_a), \tag{11.96}$$

$$u_d = \arg\min_{\|a\|_2=1} v(X_a), \tag{11.97}$$

where, for any vector $a \in \mathbb{R}^d$, $v(X_a)$ is the sample variance (see Definition 7.37) of X_a, which contains the inner products between a and the data,

$$X_a := \left\{ a^T x_1, \ldots, a^T x_n \right\}. \tag{11.98}$$

Proof Sample covariance matrices are symmetric:

$$\Sigma_X^T = \left(\frac{1}{n-1} \sum_{i=1}^{n} \text{ct}\,(x_i)\,\text{ct}\,(x_i)^T \right)^T \tag{11.99}$$

$$= \frac{1}{n-1} \sum_{i=1}^{n} \text{ct}\,(x_i)\,\text{ct}\,(x_i)^T = \Sigma_X. \tag{11.100}$$

By Theorem 11.17, $v(X_a) = a^T \Sigma_X a$ for any $a \in \mathbb{R}^d$, so the result follows from Theorem 11.19, setting $M := \Sigma_X$. $\blacksquare$

In words, the first principal direction u_1 of a dataset is the direction of maximum sample variance, the second principal direction u_2 is the direction of maximum sample variance orthogonal to u_1, and the kth principal direction u_k is the direction of maximum sample variance that is orthogonal to $u_1, u_2, \ldots, u_{k-1}$. Finally, the last principal direction u_d is the direction of minimum sample variance. The sample variances in each of these directions are equal to the corresponding eigenvalues of the sample covariance matrix.

Figure 11.8 shows the two principal directions of the two-dimensional data in Example 11.18, computed following Definition 11.24. As established by Theorem 11.25, the function $a^T \Sigma_X \, a$ reaches its maximum and minimum on the unit circle in these directions, which are therefore the directions in which the data have maximum and minimum sample

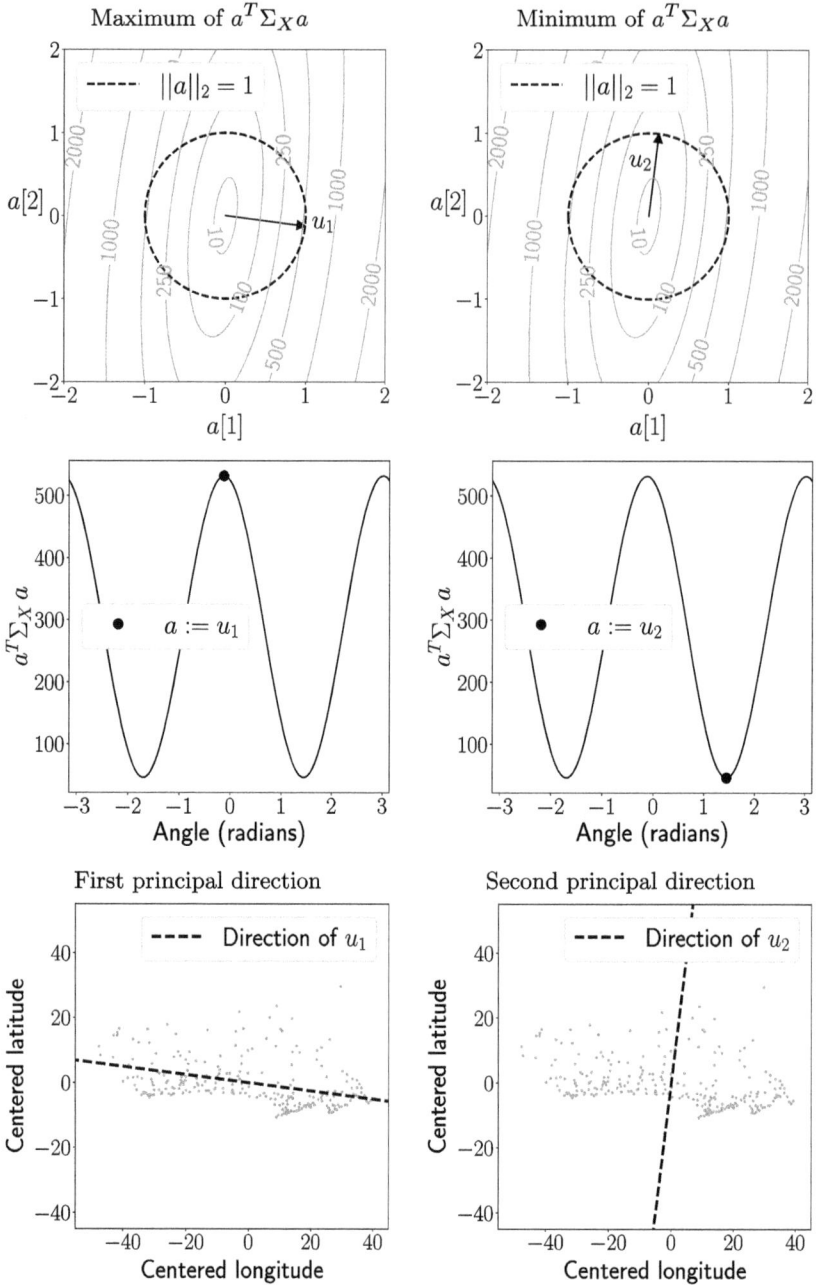

Figure 11.8 Principal directions of a dataset. The top row shows the first (left) and second (right) principal directions u_1 and u_2 of the data in Example 11.18, computed following Definition 11.24. The principal directions are superposed onto the contour lines of the function $a^T \Sigma_X a$. The values on the dashed unit circle are plotted as a function of the angle in the middle row. The restricted function reaches its maximum and minimum (depicted by circular markers) at the first and second principal directions, respectively. The bottom row shows that the principal directions are indeed aligned with the directions of maximum and minimum variation of the data.

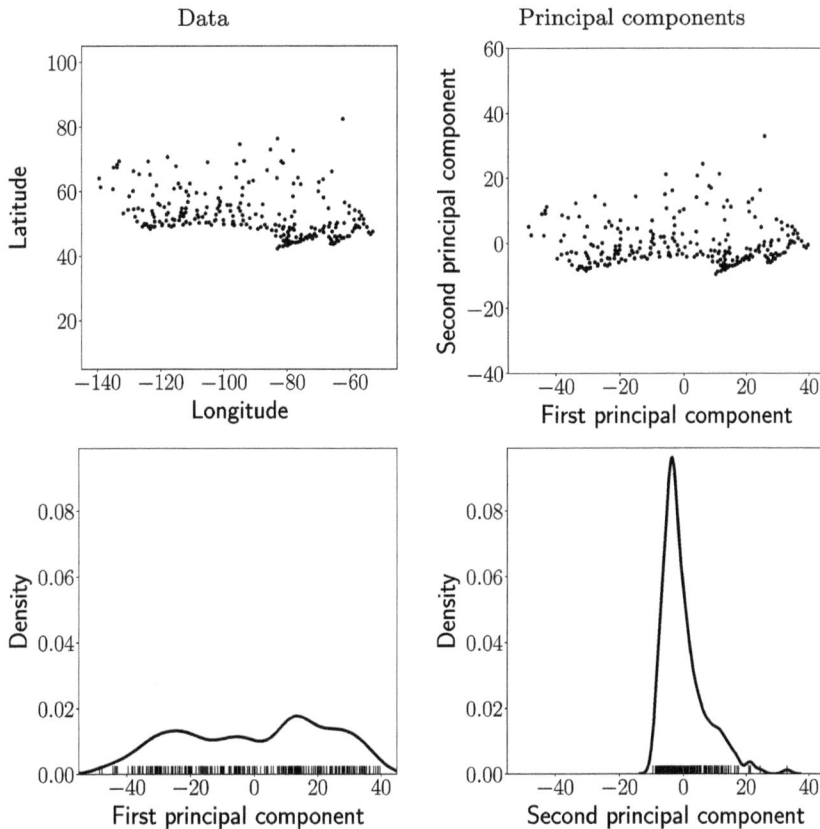

Figure 11.9 Principal components of a dataset. The top left graph shows the data in Example 11.18. The top right graph shows the scatterplot of the first and second principal components, obtained by centering the data and then rotating them, so that their axes align with the principal directions. This yields a horizontal component with maximum sample variance and a vertical component with minimum sample variance. The bottom row shows the marginal pdfs of the principal components, estimated via kernel density estimation, as described in Section 3.5.2.

variance. The principal components are obtained by centering the data and rotating it, so that the coordinate axes align with the principal directions, as illustrated in Figure 11.9. After the rotation, the sample variance is maximized along the horizontal axis and minimized along the vertical axis. Analogously to Lemma 11.22, the sample correlation between the principal components is zero (see Exercise 11.4).

Up to now we have focused on examples with two features because they are easy to visualize. However, in practice, we don't really need PCA to analyze these data: We can just plot them and take a look! By contrast, analyzing data with more features is much more challenging. The following example considers a dataset with 4,096 features. We can plot data in two or three dimensions, but 4,096 dimensions is out of the question. PCA allows us to identify and visualize the directions of maximum variance of such high-dimensional data.

Example 11.26 (Faces: Principal component analysis)**.** We use PCA to analyze Dataset 20, which contains 400 64×64 images of 40 different subjects. As explained in Example 11.6,

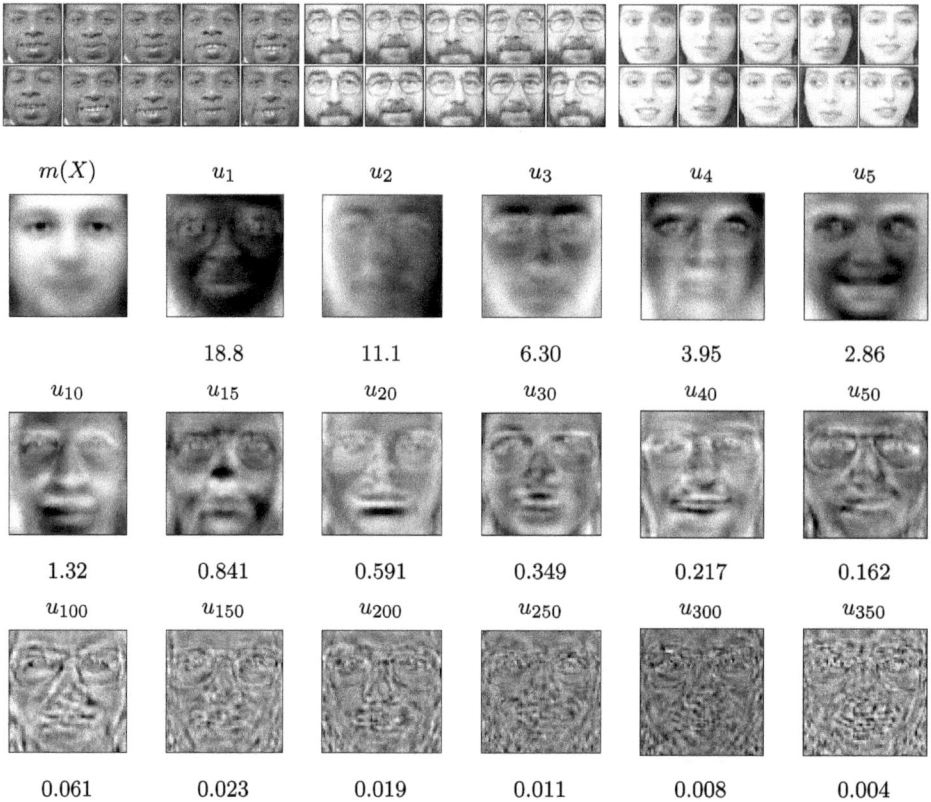

Figure 11.10 Principal directions of face data. The top row shows three out of the 40 individuals from the dataset in Example 11.26. The sample mean $m(X)$ and some of the principal directions u_i, $1 \leq i \leq d$, obtained by applying PCA to the whole dataset are depicted, reshaped as images for ease of visualization. The sample variance of each principal component is listed under the corresponding principal direction.

we can represent each image as a 4,096-dimensional vector. To perform PCA, we apply Definition 11.24 to these vectors. Figure 11.10 depicts the resulting principal directions, and reports the sample variance of the corresponding principal components. The first principal directions capture coarse-level face structure (eyebrows, nose, mouth, face contour) and also the illumination of the photographs. These are the characteristics that account for most of the variance in the dataset.

···

11.4.3 Why Eigenvectors? The Mathematics behind PCA

In this section, we explain how to prove the spectral theorem (Theorem 11.19), which is the theoretical underpinning of principal component analysis. Our goal is to establish that, for any $d \times d$ symmetric matrix M, the maximum of the function

$$q(a) := a^T M a \tag{11.101}$$

on the unit sphere (where the ℓ_2 norm of a equals one) is an eigenvector of M. If the matrix is a covariance matrix of a random vector or the sample covariance matrix of a dataset, this means that we can identify the direction of maximum variance via eigendecomposition, as explained in Sections 11.4.1 and 11.4.2.

The function q is guaranteed to reach a maximum value on the unit sphere because of the extreme value theorem. The unit sphere is a closed and bounded set. Since q is continuous, the set of values taken by q, when its argument is on the unit sphere, must also be closed and bounded. Consequently, this set cannot grow toward a limit value that it does not contain, so it must include its maximum.

In order to characterize the location of the maximum of q on the unit sphere, we study the gradient of q, which is differentiable because it is a second-order polynomial. The gradient at a is

$$\nabla q(a) = 2Ma. \tag{11.102}$$

Figure 11.11 shows the direction of the gradient on the unit circle when M is the sample covariance matrix in Example 11.15. The gradient encodes the rate of change of q in different directions. More specifically, at a point b, the rate of change in the direction of a unit vector h is equal to the inner product between $\nabla q(b)$ and h. This rate is the directional derivative of q at b

$$q_h'(b) := \lim_{\epsilon \to 0} \frac{q(b + \epsilon h) - q(b)}{\epsilon} \tag{11.103}$$

$$= \nabla q(b)^T h. \tag{11.104}$$

If the derivative is positive, $q_h'(b) > 0$, then the function increases in that direction, i.e. for small enough $\epsilon > 0$, $q(b + \epsilon h) > q(b)$. Let u_1 be the point at which q attains its maximum on the unit sphere. At u_1, the directional derivative *cannot be positive* for any direction that stays in the unit sphere because this would imply that the unit sphere contains another point where q is larger than $q(u_1)$. In order to leverage this fact to identify u_1, we have to resolve a minor difficulty: Since u_1 is on the sphere, $u_1 + \epsilon h$ can never be on the sphere because the

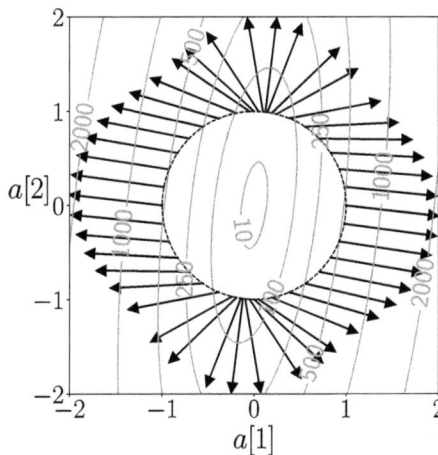

Figure 11.11 Gradients of a quadratic form. The arrows indicate the direction of the gradient of the function $q(a) := a^T \Sigma_X a$, where Σ_X is the sample covariance matrix in Example 11.15, at different points of the unit circle. The contour lines of q are shown in the background (gray lines).

sphere is a curved surface! Instead, we need to consider directions that *almost* stay on the sphere, in the sense that they belong to its *tangent plane*.

The unit sphere is a level surface of the function $s(a) := ||a||_2^2 = a^T a$ because it contains every point a such that $s(a) = 1$. The tangent plane of the level surface of a differentiable function is the set of vectors orthogonal to its gradient. In the case of the unit sphere, the tangent plane $\mathcal{T}$ at a point c is the set of vectors orthogonal to the gradient of s at c. A point y belongs to $\mathcal{T}$ if

$$\nabla s(c)^T (y - c) = 0. \tag{11.105}$$

For such points, if $y - c$ is small, then by Taylor expansion $s(y) \approx s(c) + \nabla s(c)^T (y-c) = s(c)$, so y is *almost* on the level surface.

The key insight is that if $q(u_1)$ is the maximum value of q on the unit sphere, then there cannot be any point b on the tangent plane of the sphere at u_1 such that

$$q_h'(u_1) = \nabla q(u_1)^T h > 0, \qquad h := b - u_1. \tag{11.106}$$

Otherwise, for small enough ϵ, there is a point $x := u_1 + \epsilon h$ such that $q(x) > q(u_1)$ because the directional derivative of q in that direction is positive. If b is on the tangent plane at u_1, then so is x:

$$\nabla s(u_1)^T (x - u_1) = \nabla s(u_1)^T (u_1 + \epsilon(b - u_1) - u_1) \tag{11.107}$$
$$= \epsilon \nabla s(u_1)^T (b - u_1) = 0. \tag{11.108}$$

Since x is on the tangent plane, by making ϵ small enough, we can find a point y on the sphere close enough to x, so that

$$q(y) \approx q(x) > q(u_1), \tag{11.109}$$

and therefore u_1 cannot be the maximum of q on the unit sphere!

In order to ensure that there is no point b in the tangent plane of the sphere at u_1, such that $\nabla q(u_1)^T (b - u_1) > 0$, the inner product $\nabla q(u_1)^T (b - u_1)$ should be zero for any b in the tangent plane. In other words, $\nabla q(u_1)$ must be *orthogonal* to the tangent plane at u_1. By definition of the tangent plane, this implies that $\nabla q(u_1)$ is collinear with the gradient of s. Figure 11.12 illustrates this in two dimensions. On the left, the gradients of q and s do not align at a certain point b, so we can find a direction in which q increases locally on the unit circle. On the right, the gradients are shown to be collinear at the two maxima and the two minima of q on the unit circle (our argument also applies to the minimum of q on the unit sphere).

Collinearity of the gradients of q and s implies that there exists a constant $\lambda_1 \in \mathbb{R}$ such that

$$\nabla q(u_1) = \lambda_1 \nabla s(u_1). \tag{11.110}$$

Since $\nabla q(u_1) = 2 M u_1$ and $\nabla s(u_1) = 2 u_1$, we conclude that

$$M u_1 = \lambda_1 u_1. \tag{11.111}$$

The maximum is attained at an eigenvector of the matrix M! The same argument can be used to establish that the minimum is also attained at an eigenvector. To complete the proof

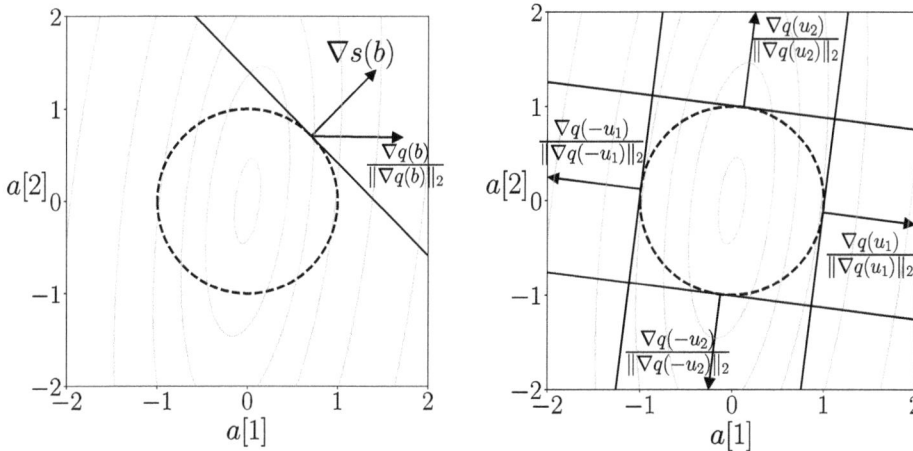

Figure 11.12 Proof of the spectral theorem. The diagrams illustrate the proof of the spectral theorem in Section 11.4.3, using the sample covariance matrix Σ_X in Example 11.15. On the left, the gradient $\nabla q(b)$ of the function $q(a) := a^T \Sigma_X a$ at b is not orthogonal to the tangent plane (black solid line) of the unit circle (dashed circle). Consequently, we can find a point on the circle where the value of q is higher than at b. The right plot shows four points at which the gradient is orthogonal to the tangent plane. These are the principal directions $\pm u_1$ and $\pm u_2$ of the data, where q (and therefore the sample variance) attains its maximum and minimum values. In both diagrams, the contour lines of q are shown in the background (light gray lines).

of the spectral theorem, we can apply the same argument to the orthogonal complement of u_1 (and then to the orthogonal complement of the span of u_1 and u_2, and so on).

11.5 Dimensionality Reduction

Data with a large number of features can be difficult to analyze and process. The goal of dimensionality-reduction techniques is to embed such data in a low-dimensional space, where they can be described in terms of a small number of variables. This is a crucial preprocessing step in many applications.

A popular choice is to perform *linear* dimensionality reduction, where the lower-dimensional representation is obtained by computing the inner products of each data point with a small number of basis vectors. Let us interpret the data as samples from a d-dimensional random vector $\tilde{x}$. Dimensionality reduction is typically applied after centering the data by subtracting their mean, so we assume that $\tilde{x}$ has zero mean. Consider an orthonormal basis of $\mathbb{R}^d$ formed by the vectors $b_1, b_2, \ldots, b_d$. If we express $\tilde{x}$ as a linear combination of the basis vectors,

$$\tilde{x} = \sum_{i=1}^{d} \tilde{a}[i] b_i, \qquad \tilde{a}[i] := b_i^T \tilde{x}, \tag{11.112}$$

then the coefficients $\tilde{a}[1], \ldots, \tilde{a}[d]$ are a d-dimensional representation of $\tilde{x}$. In order to compute a representation of lower dimensionality, we can truncate the coefficients and only use the first k. A key challenge is how to select the k vectors $b_1, \ldots, b_k$, so that the representation preserves as much information about $\tilde{x}$ as possible.

The best dimensionality-reduction scheme may vary depending on the specific downstream task of interest. However, a reasonable general-purpose criterion is to minimize the approximation error between the original vector and the k-dimensional approximation

$$\operatorname*{approx}_{b_1,\dots,b_k}(\tilde{x}) := \sum_{i=1}^{k} \tilde{a}[i]b_i \tag{11.113}$$

provided by the truncated coefficients. Consider the decomposition of $\tilde{x}$ into the approximation and the corresponding error:

$$\underbrace{\sum_{i=1}^{d} \tilde{a}[i]b_i}_{\tilde{x}} = \underbrace{\sum_{i=1}^{k} \tilde{a}[i]b_i}_{\substack{\operatorname{approx}(\tilde{x}) \\ b_1,\dots,b_k}} + \underbrace{\sum_{i=k+1}^{d} \tilde{a}[i]b_i}_{\widetilde{\operatorname{error}}}. \tag{11.114}$$

A popular metric to evaluate the approximation error is its squared ℓ_2-norm, which is the sum of squared entries. The approximation and the error are orthogonal, as they are linear combinations of orthogonal vectors, so by the Pythagorean theorem,

$$||\tilde{x}||_2^2 = \left|\left| \sum_{i=1}^{k} \tilde{a}[i]b_i \right|\right|_2^2 + ||\widetilde{\operatorname{error}}||_2^2, \tag{11.115}$$

which implies

$$||\widetilde{\operatorname{error}}||_2^2 = ||\tilde{x}||_2^2 - \left|\left| \sum_{i=1}^{k} \tilde{a}[i]b_i \right|\right|_2^2 \tag{11.116}$$

$$= ||\tilde{x}||_2^2 - \sum_{i=1}^{k} \tilde{a}[i]^2 \tag{11.117}$$

$$= ||\tilde{x}||_2^2 - \sum_{i=1}^{k} \left(b_i^T \tilde{x}\right)^2, \tag{11.118}$$

where (11.117) holds because $b_1, \dots, b_k$ are orthonormal. By linearity of expectation (Theorem 7.18), Definition 7.32 and the assumption that the mean of $\tilde{x}$ is zero, the mean of the squared ℓ_2 norm of the error equals

$$\mathbb{E}\left[||\widetilde{\operatorname{error}}||_2^2\right] = \mathbb{E}\left[||\tilde{x}||_2^2\right] - \sum_{i=1}^{k} \mathbb{E}\left[\left(b_i^T \tilde{x}\right)^2\right] \tag{11.119}$$

$$= \mathbb{E}\left[||\tilde{x}||_2^2\right] - \sum_{i=1}^{k} \operatorname{Var}\left[b_i^T \tilde{x}\right]. \tag{11.120}$$

The first term $\mathbb{E}\left[||\tilde{x}||_2^2\right]$ does not depend on the chosen basis. Consequently, in order to minimize the error, we should minimize the second term, choosing the k basis vectors that make the sum of the variances in the corresponding directions as large as possible. This suggests selecting the first k principal directions of the random vector $\tilde{x}$, since they are the directions of maximum variance by Theorem 11.20. The resulting low-dimensional representation of $\tilde{x}$ consists of its first k principal components. The following theorem confirms that this choice is optimal in terms of mean squared ℓ_2-norm error, for any k.

Theorem 11.27 (Linear dimensionality reduction via PCA is optimal). *Let $\tilde{x}$ be a d-dimensional random vector with zero mean and covariance matrix $\Sigma_{\tilde{x}}$ and let k be a positive integer smaller than d. We denote the k-dimensional linear approximation of $\tilde{x}$ in a basis of k orthonormal vectors $b_1, \ldots, b_k$ by*

$$\operatorname*{approx}_{b_1,\ldots,b_k}(\tilde{x}) := \sum_{i=1}^{k} b_i^T \tilde{x}\, b_i. \tag{11.121}$$

The first k principal directions associated with the k largest eigenvalues of $\Sigma_{\tilde{x}}$ yield the optimal k-dimensional linear approximation in terms of mean squared ℓ_2-norm error,

$$\{u_1, \ldots, u_k\} = \arg \min_{\substack{\{b_1,\ldots,b_k\} \\ ||b_i||_2=1, 1\le i \le k \\ b_i \perp b_j, i \ne j}} \mathbb{E}\left[\left|\left|\tilde{x} - \operatorname*{approx}_{b_1,\ldots,b_k}(\tilde{x})\right|\right|_2^2\right]. \tag{11.122}$$

The optimal linear k-dimensional approximation is therefore

$$\operatorname*{approx}_{u_1,\ldots,u_k}(\tilde{x}) := \sum_{i=1}^{k} \tilde{w}_i u_i, \tag{11.123}$$

where $\tilde{w}_i := u_i^T \tilde{x}$ is the ith principal component of $\tilde{x}$.

Proof By (11.120),

$$\arg \min_{\substack{\{b_1,\ldots,b_k\} \\ ||b_i||_2=1\ 1\le i \le k \\ b_i \perp b_j, i \ne j}} \mathbb{E}\left[\left|\left|\tilde{x} - \operatorname*{approx}_{b_1,\ldots,b_k}(\tilde{x})\right|\right|_2^2\right] = \arg \max_{\substack{\{b_1,\ldots,b_k\} \\ ||b_i||_2=1\ 1\le i \le k \\ b_i \perp b_j, i \ne j}} \sum_{i=1}^{k} \operatorname{Var}\left[b_i^T \tilde{x}\right],$$

so to prove the result, we need to show that the first k principal directions maximize the sum of the variances. We prove this by induction on k. The base case $k := 1$ follows immediately from (11.68) in Theorem 11.20, which implies

$$u_1 = \arg \max_{||b||_2=1} \operatorname{Var}[b^T \tilde{x}]. \tag{11.124}$$

To complete the proof, we establish that if the induction hypothesis

$$\{u_1, \ldots, u_{k-1}\} = \arg \max_{\substack{\{b_1,\ldots,b_{k-1}\} \\ ||b_i||_2=1, 1\le i \le k-1 \\ b_i \perp b_j, i \ne j}} \sum_{i=1}^{k-1} \operatorname{Var}\left[b_i^T \tilde{x}\right] \tag{11.125}$$

holds, then

$$\{u_1, \ldots, u_k\} = \arg \max_{\substack{\{b_1,\ldots,b_k\} \\ ||b_i||_2=1, 1\le i \le k \\ b_i \perp b_j, i \ne j}} \sum_{i=1}^{k} \operatorname{Var}\left[b_i^T \tilde{x}\right]. \tag{11.126}$$

To prove this, we set $b_1, \ldots, b_k$ to be an arbitrary fixed set of k orthonormal vectors, and show that they cannot capture more variance than the principal directions $u_1, \ldots, u_k$, if the induction hypothesis holds. Consider the subspace $\mathcal{S} := \operatorname{span}(b_1, \ldots, b_k)$ spanned by $b_1, \ldots, b_k$. We are interested in the projection $\mathcal{P}_{\mathcal{S}}\, \tilde{x} := \sum_{i=1}^{k} b_i^T \tilde{x} b_i$ of $\tilde{x}$ onto this subspace, because by linearity of expectation (Theorem 7.18),

$$\sum_{i=1}^{k} \text{Var}\left[b_i^T \tilde{x} \right] = \sum_{i=1}^{k} \mathbb{E}\left[\left(b_i^T \tilde{x} \right)^2 \right] \tag{11.127}$$

$$= \mathbb{E}\left[\sum_{i=1}^{k} \left(b_i^T \tilde{x} \right)^2 \right] \tag{11.128}$$

$$= \mathbb{E}\left[\left\| \sum_{i=1}^{k} b_i^T \tilde{x} b_i \right\|_2^2 \right] \tag{11.129}$$

$$= \mathbb{E}\left[\| \mathcal{P}_S \tilde{x} \|_2^2 \right]. \tag{11.130}$$

We can represent this projection using any arbitrary orthonormal basis of S. Indeed, any set of k orthonormal vectors $a_1, \ldots, a_k$ spanning S satisfy

$$\mathcal{P}_S \tilde{x} := \sum_{i=1}^{k} b_i^T \tilde{x} b_i = \sum_{i=1}^{k} a_i^T \tilde{x} a_i. \tag{11.131}$$

The key is to choose the basis wisely. S has dimension k, so it must contain at least one vector $a_\perp$ that is orthogonal to the first $k-1$ principal directions $u_1, u_2, \ldots, u_{k-1}$. By (11.70) in Theorem 11.20, the variance in that direction cannot be higher than in the kth principal direction,

$$\text{Var}[u_k^T \tilde{x}] \geq \text{Var}[a_\perp^T \tilde{x}]. \tag{11.132}$$

In order to exploit this, we select the orthonormal basis $a_1, a_2, \ldots, a_k$ for S so that $a_k := a_\perp$ (such a basis can be constructed applying the Gram-Schmidt process[2], starting with $a_\perp$). By the induction hypothesis,

$$\sum_{i=1}^{k-1} \text{Var}\left[u_i^T \tilde{x} \right] \geq \sum_{i=1}^{k-1} \text{Var}\left[a_i^T \tilde{x} \right]. \tag{11.133}$$

Combining (11.133), (11.132), and (11.130) yields

$$\sum_{i=1}^{k} \text{Var}\left[u_i^T \tilde{x} \right] \geq \sum_{i=1}^{k} \text{Var}\left[a_i^T \tilde{x} \right] \tag{11.134}$$

$$= \mathbb{E}\left[\| \mathcal{P}_S \tilde{x} \|_2^2 \right] \tag{11.135}$$

$$= \sum_{i=1}^{k} \text{Var}\left[b_i^T \tilde{x} \right]. \tag{11.136}$$

Since this holds for any choice of $b_1, \ldots, b_k$, the proof is complete. ∎

Example 11.28 (Dimensionality reduction for visualization). In order to visualize a dataset, we need to embed the data in two or three dimensions, while preserving their structure as much as possible. In this example, we consider Dataset 21, which describes the geometry of wheat seeds. Each data point consists of seven features: Area, perimeter, compactness, length of kernel, width of kernel, asymmetry coefficient, and length of kernel groove. The features have different units, so we normalize each of them, dividing by the corresponding

---

[2] We refer the reader to any text on linear algebra for a description of this method.

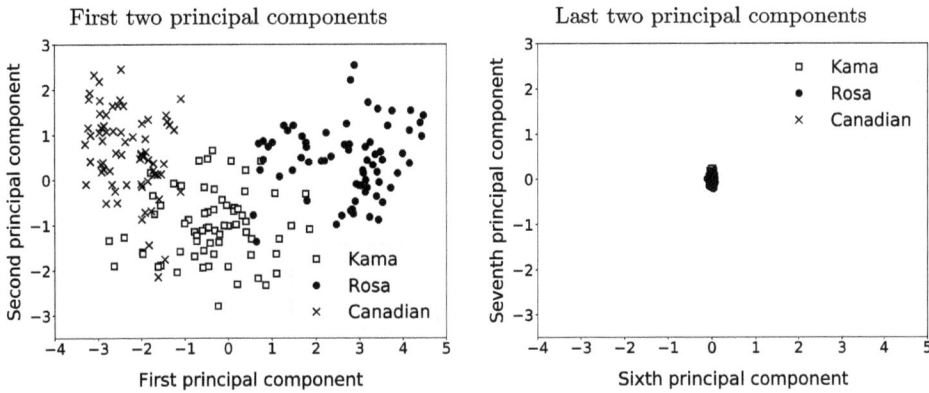

Figure 11.13 Dimensionality reduction for visualization. The left graph shows a scatterplot of the two first principal components of the data in Example 11.28. Each marker indicates a different variety of wheat. The low-dimensional representation makes it possible to visualize meaningful structure; the data points corresponding to each variety cluster together. The right graph shows a scatterplot of the two last principal components, which capture very little variance.

sample standard deviation after centering using their sample means. The seeds belong to three different varieties of wheat: Kama, Rosa, and Canadian.

Motivated by Theorem 11.27, in order to visualize the data, we reduce their dimensionality via PCA down to two dimensions. The left plot in Figure 11.13 shows a scatterplot of the two first principal components, obtained as described in Definition 11.24. The three varieties of wheat form three distinct clusters. This suggests that the two-dimensional representation preserves meaningful structure, which could be useful for downstream tasks such as clustering or classification. The right plot in Figure 11.13 shows a scatterplot of the two last principal components. These are the components with the lowest variance, so they are very close to the origin.

··

Example 11.29 (Faces: Dimensionality reduction via PCA). Example 11.26 applies PCA to Dataset 20, which consists of face images. In order to visualize the k-dimensional representation of a face provided by its first k principal components we compute the corresponding approximation in the original high-dimensional space of 64×64 images. The approximation is obtained by summing the first k principal directions $u_1, \ldots, u_k$ weighted by the respective principal components $w_1, \ldots, w_k$, and adding the sample mean $m(X)$:

$$\operatorname*{approx}_{u_1,\ldots,u_k}(x) := m(X) + \sum_{j=1}^{k} w_j u_j, \quad w_j := u_j^T \operatorname{ct}(x), \qquad (11.137)$$

where x is the chosen face, $\operatorname{ct}(x) := x - m(X)$ is the centered face, and X is the whole dataset. Figure 11.14 illustrates the decomposition of the resulting approximation into its different components. Figure 11.15 shows k-dimensional approximations of a face for different values of k. As suggested by the visualization of the principal directions in Figure 11.10, the first principal directions capture coarse-level characteristics, so when k is small, the approximations are quite blurry. This is a shortcoming of dimensionality reduction via PCA for image processing: it is optimal in terms of preserving the variance in the dataset, but it suppresses fine-scale details that may be crucial for downstream tasks such as classification.

··

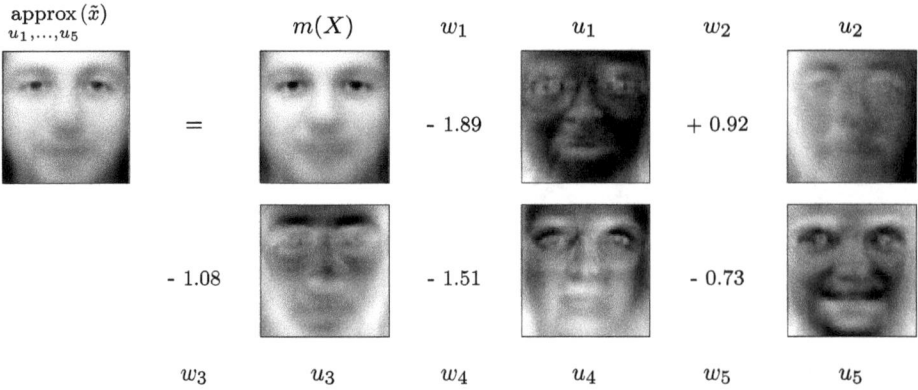

Figure 11.14 Dimensionality reduction of faces based on PCA. The figure provides a visualization of the five-dimensional representation of a face from Dataset 20 based on PCA. To visualize the representation, we project it onto the image space by summing the sample mean and the first five principal directions weighted by the corresponding principal components, as described by (11.137). This reveals what characteristics of the original image are preserved by the first five principal components and the mean.

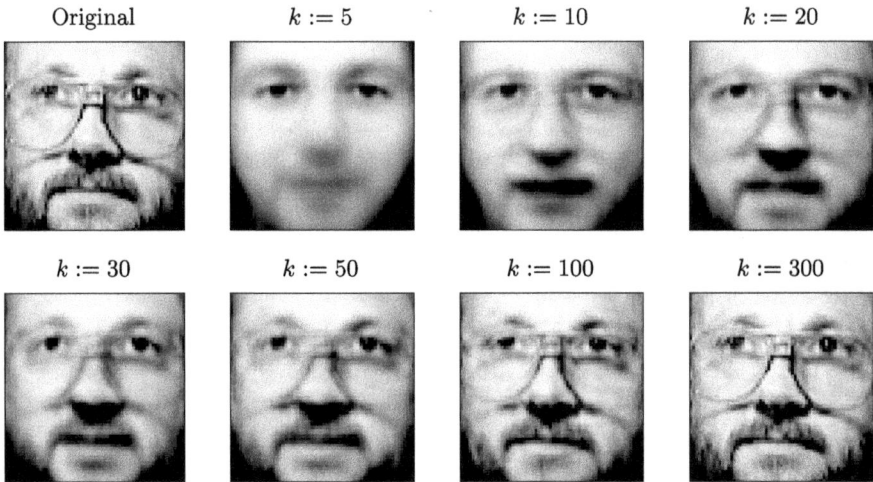

Figure 11.15 Low-dimensional approximations of face data. The images show k-dimensional approximations of a face from Dataset 20 based on PCA (computed following (11.137), as illustrated in Figure 11.14), for different values of k. The approximation is very blurry for small k because the principal directions mainly capture coarse-level structure. As k increases, the approximation improves.

Example 11.30 (Nearest neighbors in principal-component space). PCA-based dimensionality reduction is often used as a preprocessing step when analyzing data with many features. In this example, we combine it with the nearest-neighbor method to perform face classification.

Assume that we have access to a training set of n pairs of data encoded as vectors in $\mathbb{R}^d$ along with their corresponding labels: $\{x_1, y_1\}, \ldots, \{x_n, y_n\}$. To classify a test data point x_{test} using the nearest-neighbor method, we find the element in the training set that is closest to x_{test},

$$i^* := \arg \min_{1 \le i \le n} \left\| x_{\text{test}} - x_i \right\|_2 , \tag{11.138}$$

and assign the corresponding label y_{i^*} to x_{test}.

Every time we classify a new point, we compute n distances in a d-dimensional space. To obtain each distance, we need to consider the d entries of the vectors, so the overall computational cost is proportional to d, and can be high if d is very large. This cost can be alleviated via dimensionality reduction, representing each point in the training data by its first k principal components. The distances are then computed in the space of reduced dimensionality. To classify a test data point, we:

1 Center the test data point by subtracting the sample mean of the training data.
2 Compute the inner product of the centered test data point with the first k principal directions of the training data in order to obtain a k-dimensional representation w_{test}.
3 Find the nearest neighbor in the reduced space:

$$i^*_{[k]} := \arg \min_{1 \le i \le n} \left\| w_{\text{test}} - w_{[1:\,k]}[i] \right\|_2 , \tag{11.139}$$

where $w_{[1:\,k]}[i]$ denotes the k first principal components of the ith training data point.

Since the distances are computed between k-dimensional points, the cost of searching for the closest data point is now proportional to k, instead of d. Computing the eigendecomposition of the sample covariance matrix is costly, but it only needs to be done once.

We apply this approach to identify what subjects appear in the face images from Dataset 20. We divide the dataset into a training set of 360 images from 40 different subjects (nine per subject), and a test set with 40 images (one for each subject). Each image has 4,096 pixels, so the full dimensionality of each data point is $d := 4096$. Without dimensionality reduction, the nearest-neighbor method classifies 36 of the 40 test images correctly. Figure 11.16 shows the results with dimensionality reduction. For $k \ge 9$, the performance is almost the same as without dimensionality reduction, and for $k := 42$ it is exactly the same,

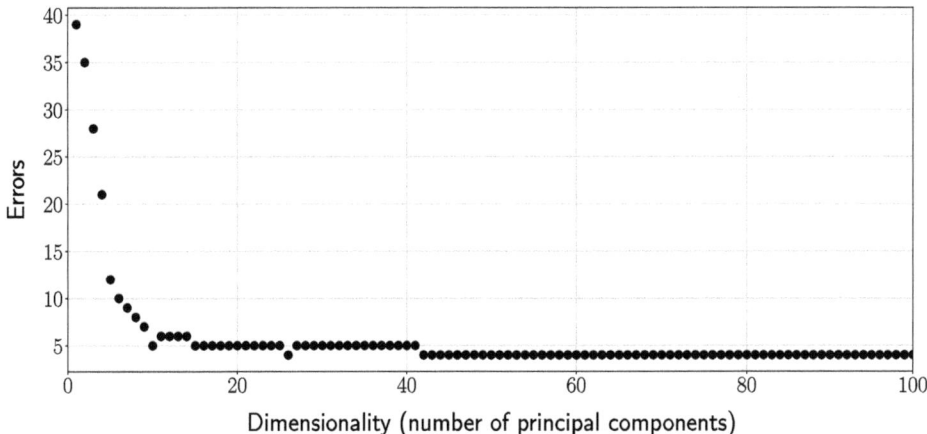

Figure 11.16 Classification in a space of reduced dimensionality. The plot shows the test error of the reduced-dimensionality nearest-neighbor classification approach in Example 11.30 applied to the face data in Dataset 20, for different values of the reduced dimensionality k. The performance for $k \ge 9$ is very close to the performance achieved without dimensionality reduction, and the performance for $k \ge 42$ is the same.

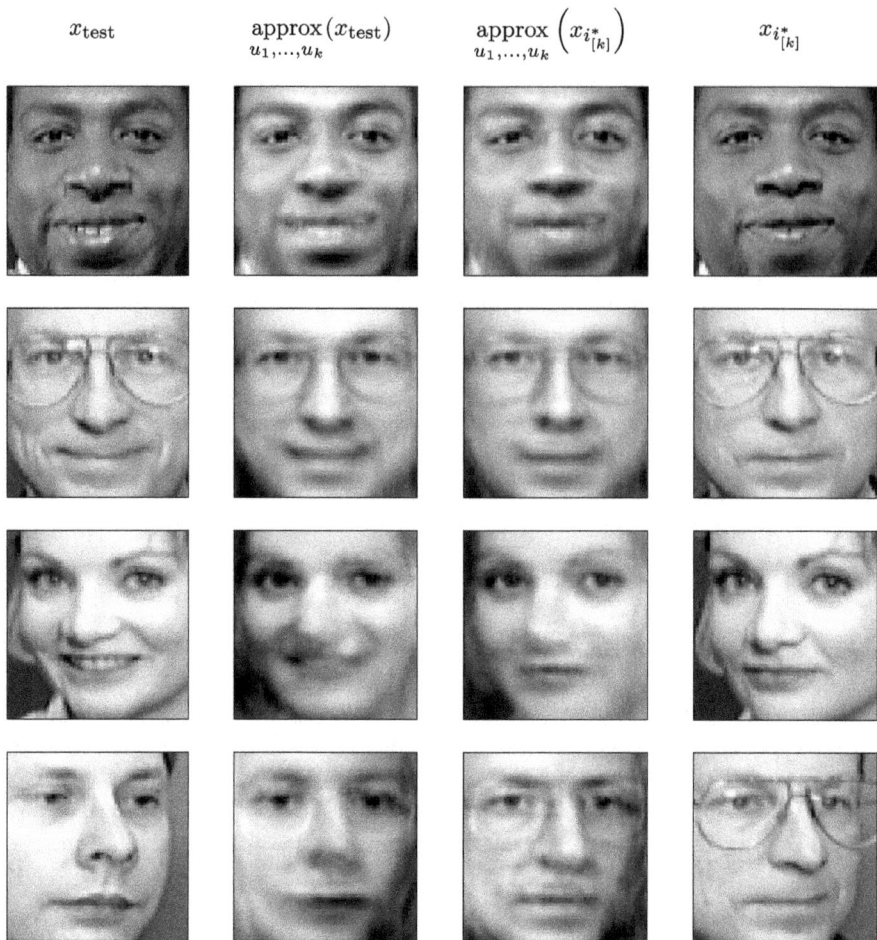

x_{test}　　　approx(x_{test})　　approx$\left(x_{i^*_{[k]}}\right)$　　　$x_{i^*_{[k]}}$
　　　　　　$u_1,...,u_k$　　　　　$u_1,...,u_k$

Figure 11.17 Face classification in a space of reduced dimensionality. The figure shows the results of nearest-neighbor classification performed after PCA-based dimensionality reduction with $k := 42$ for four test images from Example 11.30. The first column shows the test image x_{test}. The second column shows the corresponding k-dimensional approximation $\text{approx}_{u_1,...,u_k}(x_{\text{test}})$, computed as described in (11.137) and Figure 11.14. The third column shows the k-dimensional approximation of the image in the training set that is closest to $\text{approx}_{u_1,...,u_k}(x_{\text{test}})$ denoted by $\text{approx}_{u_1,...,u_k}(x_{i^*_{[k]}})$, where $i^*_{[k]}$ is defined in (11.139). The fourth column shows the corresponding training image $x_{i^*_{[k]}}$. The assignments of the first three examples are correct, but the fourth is wrong.

even though k is two orders of magnitude smaller than the full dimensionality of the data. Figure 11.17 shows some test examples along with a visualization of their nearest neighbors in the k-dimensional space.

· ·

11.6 Low-Rank Models

In Sections 11.1–11.5, we consider datasets where each data point is associated with a single entity. For instance, in Example 11.26 each data point is the image of a person, and in Example 11.28 each data point represents the measurements corresponding to a seed. In this

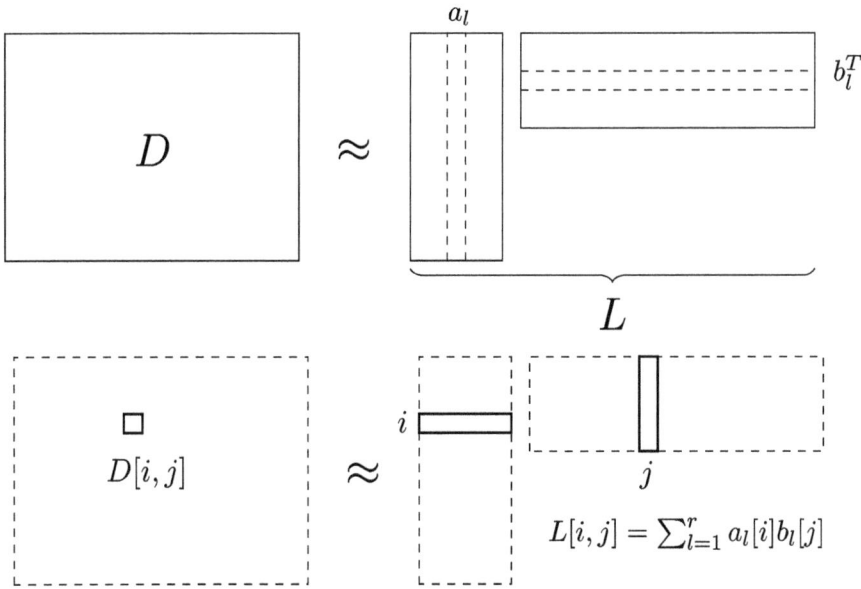

Figure 11.18 Low-rank model. The diagram illustrates a rank-r approximation L to a data matrix $D \in \mathbb{R}^{n_1 \times n_2}$. The parameters of the rank-r model are r n_1-dimensional vectors $a_1, \ldots,$ a_r spanning the columns of L, and r n_2-dimensional vectors $b_1, \ldots, b_r$ spanning the rows of L. To fit the low-rank model, we approximate each entry $D[i, j]$ of the data matrix by $L[i, j]$, which is the sum of the products of the ith entry of a_l and the jth entry of b_l for $1 \leq l \leq r$.

section, we explain how to model datasets where the data are associated with *two* different entities. We represent such data as entries in a matrix D. The column i and the row j of each entry $D[i, j]$ indicate the entities associated with the data point. For example, $D[i, j]$ could be the rating given to a product i by a user j, the expression level of a gene i in a cell j, or the temperature measured at a location i and a time j.

In order to analyze and manipulate a data matrix, it is often useful to compute a low-rank approximation consisting of a small number of components. Let us consider a matrix of movie ratings $D \in \mathbb{R}^{n_1 \times n_2}$, where the entry $D[i, j]$ is the rating given to movie i by user j. To model D, we approximate $D[i, j]$ as a sum of r components, where r is much smaller than n_1 or n_2. The lth component is the product of a coefficient $a_l[i]$ associated with movie i and a coefficient $b_l[j]$ associated with user j. The resulting approximation can be represented as a matrix $L \in \mathbb{R}^{n_1 \times n_2}$, where

$$L[i, j] := \sum_{l=1}^{r} a_l[i]b_l[j], \qquad 1 \leq i \leq n_1, 1 \leq j \leq n_2, \tag{11.140}$$

as illustrated by the diagram in Figure 11.18.

We can interpret each term in (11.140) as a factor that influences the rating. The coefficient $a_l[i]$ determines whether the association of movie i to factor l is positive (> 0), negative (< 0) or negligible (≈ 0). Similarly, the coefficient $b_l[j]$ determines whether the association of user j to factor l is positive (> 0), negative (< 0) or negligible (≈ 0). The model is *bilinear* because the approximation is a bilinear function of the coefficients: if the user coefficients are fixed, the model is linear in the movie coefficients; if the movie coefficients are fixed, the model is linear in the user coefficients.

The columns of the matrix L are spanned by the r n_1-dimensional vectors $a_1, a_2, \ldots, a_r$ and its rows are spanned by the r n_2-dimensional vectors $b_1, b_2, \ldots, b_r$ (see Figure 11.18). Consequently, as long as these vectors are linearly independent, the rank of L is equal to r. Since r is typically chosen to be much smaller than the rank of the data matrix D, the resulting model is known as a *low-rank model*. In Section 11.6.1, we explore the connection between low-rank models and PCA. Section 11.6.2 introduces the singular-value decomposition (SVD), a fundamental tool from linear algebra that decomposes a matrix into rank-one components. In Section 11.6.3, we explain how to fit low-rank models using the SVD. Finally, Section 11.6.4 proves that SVD-based low-rank approximations are optimal.

11.6.1 Low-Rank Models and Principal Component Analysis

In the low-rank model (11.140), each column of L is a linear combination of the same r vectors $a_1, \ldots, a_r$. In order to fit the model from data, we need to estimate these vectors, and also the corresponding coefficients of the linear combination, which are the entries of $b_1, \ldots, b_r$. In this section, we show that this can be achieved via PCA.

Our strategy is to interpret the columns of the data matrix $D \in \mathbb{R}^{n_1 \times n_2}$ as a set of n_1-dimensional data points, and reduce their dimensionality from n_1 to r via PCA, as explained in Section 11.5. The first step is to center the columns by subtracting the column mean, obtained by averaging each row. This yields the centered data matrix $D_{\text{ct-cols}}$.

Let us denote each of the n_2 columns in the centered matrix by $D_{\text{ct-cols}}[:, j] \in \mathbb{R}^{n_1}$, $1 \le j \le n_2$, using Python notation. In order to apply PCA following Definition 11.24, we compute the eigendecomposition of the sample covariance matrix of the columns,

$$\Sigma_{\text{cols}} := \frac{1}{n_2 - 1} \sum_{j=1}^{n_2} D_{\text{ct-cols}}[:, j] D_{\text{ct-cols}}[:, j]^T \tag{11.141}$$

$$= \frac{1}{n_2 - 1} D_{\text{ct-cols}} D_{\text{ct-cols}}^T. \tag{11.142}$$

Let $u_1, u_2, \ldots, u_r$ be the eigenvectors of Σ_{cols} corresponding to the r largest eigenvalues, normalized to have unit ℓ_2 norm. The lth principal component of the jth column $D_{\text{ct-cols}}[:, j]$ equals

$$w_l[j] := u_l^T D_{\text{ct-cols}}[:, j], \qquad 1 \le l \le r, 1 \le j \le n_2. \tag{11.143}$$

The r principal components of the centered column $D_{\text{ct-cols}}[:, j]$ weighted by the corresponding principal directions yield the approximation

$$D_{\text{ct-cols}}[:, j] \approx \sum_{l=1}^{r} w_l[j] u_l. \tag{11.144}$$

If we group the approximations of all the columns, as the columns of an $n_1 \times n_2$ matrix $L_{\text{PCA-cols}}$, we obtain a rank-r model of $D_{\text{ct-cols}}$,

$$L_{\text{PCA-cols}}[i, j] := \sum_{l=1}^{r} u_l[i] w_l[j], \qquad 1 \le i \le n_1, 1 \le j \le n_2, \tag{11.145}$$

as desired.

You might be wondering why we interpret the *columns* as data points, and not the *rows*. Indeed, we can apply the same reasoning to the rows. First, we center the rows by subtracting the row mean, obtained by averaging each column, which yields the centered data matrix $D_{\text{ct-rows}}$. Let us denote each of the n_1 rows by $D_{\text{ct-rows}}[i,:] \in \mathbb{R}^{1 \times n_2}$, $1 \le i \le n_1$, again using Python notation.[3] We apply PCA to the rows by computing the eigendecomposition of their sample covariance matrix,

$$\Sigma_{\text{rows}} := \frac{1}{n_1 - 1} \sum_{i=1}^{n_1} D_{\text{ct-rows}}[i,:]^T D_{\text{ct-rows}}[i,:] \tag{11.146}$$

$$= \frac{1}{n_1 - 1} D_{\text{ct-rows}}^T D_{\text{ct-rows}}. \tag{11.147}$$

Let $v_1, v_2, \ldots, v_r$ be the eigenvectors of Σ_{rows} associated with the r largest eigenvalues, normalized to have unit ℓ_2 norm. The lth principal component of the ith row $D_{\text{ct-rows}}[i,:]$ equals

$$z_l[i] := D_{\text{ct-rows}}[i,:]v_l, \qquad 1 \le l \le r, 1 \le i \le n_1. \tag{11.148}$$

We can approximate $D_{\text{ct-rows}}[i,:]$ as a linear combination of the first r principal directions, weighted by the corresponding principal components,

$$D_{\text{ct-rows}}[i,:] \approx \sum_{l=1}^{r} z_l[i]v_l^T. \tag{11.149}$$

If we stack the approximations of all rows as a $n_1 \times n_2$ matrix, we obtain a rank-r model of $D_{\text{ct-rows}}$:

$$L_{\text{PCA-rows}}[i,j] := \sum_{l=1}^{r} z_l[i]v_l[j], \qquad 1 \le i \le n_1, 1 \le j \le n_2. \tag{11.150}$$

In summary, PCA yields two alternative low-rank approximations to the data, depending on whether we apply it to the columns or to the rows. It turns out that these low-rank representations are essentially equivalent! The only difference is the initial centering operation. Theorem 11.35 shows that this is the case using the singular value decomposition, which is introduced in the following section.

11.6.2 The Singular-Value Decomposition

The singular-value decomposition (SVD) of a matrix is a fundamental tool in linear algebra and applied mathematics. It decomposes a matrix into the product of a matrix with orthonormal columns, a diagonal matrix, and a matrix with orthonormal rows. The following theorem shows that every matrix has an SVD, as a consequence of the spectral theorem.

---

[3] To be clear, we interpret $D_{\text{ct-rows}}[i,:]$ as a row vector, in contrast to all the other vectors in this book, which are interpreted as column vectors.

Theorem 11.31 (Singular-value decomposition). *Any matrix $A \in R^{n_1 \times n_2}$, $n_1 \leq n_2$, with rank t has a singular-value decomposition (SVD) of the form*

$$
A = \underbrace{\begin{bmatrix} u_1 & u_2 & \cdots & u_{n_1} \end{bmatrix}}_{U} \underbrace{\begin{bmatrix} s_1 & 0 & \cdots & 0 & 0 & \cdots & 0 \\ 0 & s_2 & \cdots & 0 & 0 & \cdots & 0 \\ \cdots & \cdots & \ddots & \cdots & \cdots & \cdots & \cdots \\ 0 & 0 & \cdots & s_t & 0 & \cdots & 0 \\ 0 & 0 & \cdots & 0 & 0 & \cdots & 0 \\ \cdots & \cdots & \cdots & \cdots & \cdots & \ddots & \cdots \\ 0 & 0 & \cdots & 0 & 0 & \cdots & 0 \end{bmatrix}}_{S} \underbrace{\begin{bmatrix} v_1 & v_2 & \cdots & v_{n_1} \end{bmatrix}}_{V^T}^{T},
$$

where the singular values $s_1 \geq s_2 \geq \cdots \geq s_t$ are positive real numbers, the left singular vectors u_1, u_2, $\ldots u_{n_1} \in \mathbb{R}^{n_1}$ are orthonormal, and the right singular vectors v_1, v_2, $\ldots v_{n_1} \in \mathbb{R}^{n_2}$ are also orthonormal.

Proof By the spectral theorem (Theorem 11.19), the symmetric matrix $M := AA^T \in \mathbb{R}^{n_1 \times n_1}$ has n_1 orthonormal eigenvectors $u_1, \ldots, u_{n_1}$. The corresponding eigenvalues $\lambda_1, \ldots, \lambda_{n_1}$ are nonnegative, since for $1 \leq j \leq n_1$, $AA^T u_j = \lambda_j u_j$, so

$$\lambda_j = \lambda_j u_j^T u_j \tag{11.151}$$

$$= u_j^T (\lambda_j u_j) \tag{11.152}$$

$$= u_j^T AA^T u_j \tag{11.153}$$

$$= \left\| A^T u_j \right\|_2^2 \geq 0. \tag{11.154}$$

The number of nonzero eigenvalues is equal to the rank t of A because A and AA^T have the same rank. For $1 \leq l \leq t$, we define $s_l := \sqrt{\lambda_l}$ and

$$v_l := \frac{1}{s_l} A^T u_l. \tag{11.155}$$

These vectors have unit ℓ_2 norm,

$$\|v_l\|_2^2 = \frac{1}{s_l^2} u_l^T AA^T u_l \tag{11.156}$$

$$= \frac{\lambda_l}{\lambda_l} u_l^T u_l = 1, \tag{11.157}$$

and are orthogonal to each other if $l \neq k$ because $u_l^T u_k = 0$, which implies

$$v_l^T v_k = \frac{u_l^T AA^T u_k}{s_l s_k} \tag{11.158}$$

$$= \frac{\lambda_k u_l^T u_k}{s_l s_k} = 0. \tag{11.159}$$

We now define the matrices

$$U := \begin{bmatrix} u_1 & u_2 & \cdots & u_{n_1} \end{bmatrix}, \tag{11.160}$$

$$V := \begin{bmatrix} v_1 & v_2 & \cdots & v_{n_1} \end{bmatrix}, \tag{11.161}$$

where we choose v_{t+1}, v_{t+2}, ..., v_{n_1} to be an orthonormal set of vectors, which are also orthogonal to v_1, ..., v_t. By (11.155),

$$U^T A = S V^T. \tag{11.162}$$

Notice that U is an orthogonal or unitary matrix because it is square and has orthonormal columns. As a result, UU^T is equal to the identity, so

$$A = UU^T A \tag{11.163}$$
$$= USV^T. \tag{11.164}$$

∎

The rank of a matrix is equal to the number of nonzero singular values. Indeed, if a matrix has t nonzero singular values, then its columns are all linear combinations of the corresponding t left singular vectors and its rows are all linear combinations of the corresponding t right singular vectors.

We can decompose any matrix $D \in \mathbb{R}^{n_1 \times n_2}$ into a sum of rank-one components using its SVD:

$$D = USV^T \tag{11.165}$$

$$= \sum_{l=1}^{n_1} s_l u_l v_l^T \tag{11.166}$$

$$= \sum_{l=1}^{n_1} s_l K_l, \qquad K_l := u_l v_l^T, \tag{11.167}$$

where we assume $n_1 \leq n_2$ (otherwise, we can apply the same reasoning to the transpose of D). The rank of each matrix component K_l is one because it is the outer product of the left singular vector u_l and the right singular vector v_l (its columns are scaled copies of u_l and its rows are scaled copies of v_l). In order to gain intuition about this decomposition, we consider the vector space of matrices, where each matrix is interpreted as a vector. To be clear, here we mean the abstract notion of *vector*, as an object that satisfies the properties in Definition 8.32. It turns out that the rank-one matrices K_1, ..., K_{n_1} in (11.167) are orthogonal under the Frobenius inner product for matrices and have unit norm.

Definition 11.32 (Frobenius inner product and norm). *The Frobenius inner product between two matrices $A \in \mathbb{R}^{n_1 \times n_2}$ and $B \in \mathbb{R}^{n_1 \times n_2}$ equals*

$$\langle A, B \rangle := \sum_{i=1}^{n_1} \sum_{j=1}^{n_2} A[i,j] B[i,j] \tag{11.168}$$

$$= \mathrm{Trace}\left(A^T B\right), \tag{11.169}$$

where the trace of a matrix is the sum of its diagonal entries. If we vectorize two matrices (e.g. by concatenating the columns into a vector of length $n_1 n_2$), the standard Euclidean inner product between the resulting vectors is equal to the Frobenius inner product between the matrices.

The norm associated with the Frobenius inner product is known as the Frobenius norm,

$$||A||_{\mathrm{F}} := \sqrt{\langle A, A \rangle} \tag{11.170}$$

$$= \sqrt{\sum_{i=1}^{n_1} \sum_{j=1}^{n_2} A[i,j]^2} \tag{11.171}$$

$$= \sqrt{\mathrm{Trace}\,(A^T A)}. \tag{11.172}$$

If we vectorize a matrix, the ℓ_2 norm of the resulting vector is equal to the Frobenius norm of the matrix.

Theorem 11.33 (Orthonormality of rank-one matrices). *For $n_1 \leq n_2$, let $u_1, \ldots, u_{n_1}$ be an orthonormal set of vectors in $\mathbb{R}^{n_1}$ and let $v_1, \ldots, v_{n_1}$ be an orthonormal set of vectors in $\mathbb{R}^{n_2}$. The rank-one matrices $K_1, \ldots, K_{n_1}$, where*

$$K_l := u_l v_l^T, \qquad 1 \leq l \leq n_1, \tag{11.173}$$

form an orthonormal set with respect to the Frobenius inner product and norm.

Proof For any matrices $A \in \mathbb{R}^{n_1 \times n_2}$ and $B \in \mathbb{R}^{n_1 \times n_2}$, we have $\mathrm{Trace}\,(A^T B) = \mathrm{Trace}\,(BA^T)$ (you can check by writing out both expressions). As a result, for $1 \leq l \leq n_1$,

$$||K_l||_{\mathrm{F}}^2 = ||u_l v_l^T||_{\mathrm{F}}^2 = \mathrm{Trace}\,(v_l u_l^T u_l v_l^T) \tag{11.174}$$

$$= \mathrm{Trace}\,(v_l^T v_l u_l^T u_l) \tag{11.175}$$

$$= \mathrm{Trace}\,(1) = 1, \tag{11.176}$$

and for any $k \neq l, 1 \leq k \leq n_1$,

$$\langle K_l, K_k \rangle = \langle u_l v_l^T, u_k v_k^T \rangle = \mathrm{Trace}\,(v_l u_l^T u_k v_k^T) \tag{11.177}$$

$$= \mathrm{Trace}\,(v_k^T v_l u_l^T u_k) \tag{11.178}$$

$$= \mathrm{Trace}\,(0) = 0 \tag{11.179}$$

by orthonormality of $u_1, \ldots, u_{n_1}$ (or of $v_1, \ldots, v_{n_1}$). ∎

In the SVD decomposition (11.167), each rank-one matrix K_l with unit Frobenius norm is weighted by the corresponding singular value s_l. The following lemma shows that the sum of squared singular values is equal to the squared Frobenius norm of the whole matrix.

Theorem 11.34 (Frobenius norm and singular values). *For any matrix $A \in \mathbb{R}^{n_1 \times n_2}$, with singular values $s_1, \ldots, s_{\min\{n_1, n_2\}}$,*

$$||A||_{\mathrm{F}}^2 = \sum_{l=1}^{\min\{n_1, n_2\}} s_l^2. \tag{11.180}$$

Proof Assume $n_1 \leq n_2$ (for $n_1 > n_2$, the same argument applies to the transpose of A). We denote the SVD of A by $\sum_{i=1}^{n_1} s_l K_l$, where $K_l := u_l v_l^T$, as in (11.167). Theorem 11.33 establishes that the rank-one matrices K_l are orthogonal and have unit Frobenius norm, so by the Pythagorean theorem,

$$\|A\|_F^2 = \left\| \sum_{i=1}^{n_1} s_l K_l \right\|_F^2 \tag{11.181}$$

$$= \sum_{i=1}^{n_1} \left\| s_l K_l^T \right\|_F^2 \tag{11.182}$$

$$= \sum_{i=1}^{n_1} s_l^2 \left\| K_l \right\|_F^2 \tag{11.183}$$

$$= \sum_{i=1}^{n_1} s_l^2. \tag{11.184}$$

∎

In summary, the SVD provides a decomposition in terms of orthogonal rank-one matrices, where the squared norm of each component is equal to the corresponding squared singular value. This suggests a way of defining when a matrix is *approximately* rank r: r of its singular values are much larger than the rest (see Exercise 11.9 for more details). For such matrices, most of the energy (measured in Frobenius norm) is concentrated in the r rank-one components corresponding to the largest singular values.

11.6.3 Truncating the Singular-Value Decomposition

In this section, we show how to obtain a low-rank model for a matrix by leveraging its SVD. According to our analysis in Section 11.6.2, we can decompose any matrix into a sum of rank-one components, weighted by the corresponding singular values. The singular values allow us to quantify the contribution of each component because the squared Frobenius norm of the matrix is equal to the sum of the squared singular values according to Theorem 11.34. Motivated by this, we obtain a rank-r approximation of a matrix $D \in \mathbb{R}^{n_1 \times n_2}$ by computing its SVD, and keeping only the rank-one components corresponding to the r largest singular values. The resulting truncated-SVD model, depicted in Figure 11.19, is

$$L_{\text{SVD}} := \sum_{l=1}^{r} s_l K_l \tag{11.185}$$

$$= \sum_{l=1}^{r} s_l u_l v_l^T. \tag{11.186}$$

By Theorem 11.34, the squared approximation error of the model is

$$\|D - L_{\text{SVD}}\|_F^2 = \left\| \sum_{l=r+1}^{n_1} s_l u_l v_l^T \right\|_F^2 \tag{11.187}$$

$$= \sum_{l=r+1}^{n_1} s_l^2 \tag{11.188}$$

because $s_{r+1}, \ldots, s_{r+n_1}$ are the singular values of $D - L_{\text{SVD}}$.[4] Section 11.6.4 establishes that this is optimal. No other rank-r approximation can have a smaller Frobenius-norm error.

---

[4] We assume $n_1 \leq n_2$. Otherwise the same argument applies to the transpose of D.

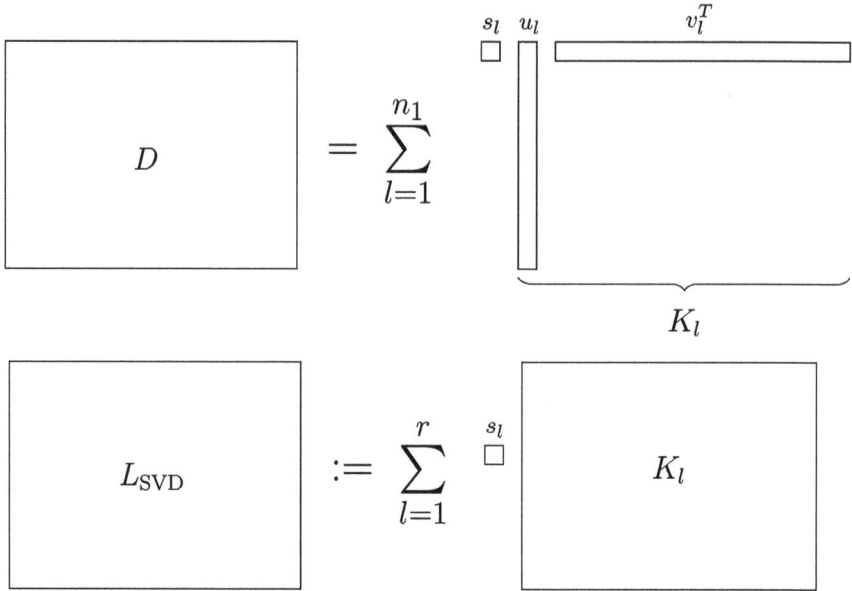

Figure 11.19 Low-rank approximation via SVD truncation. As explained in Section 11.6.3, the SVD of a matrix can be interpreted as a sum of the rank-one matrices K_l depicted at the top, weighted by the corresponding singular values. In order to obtain a rank-r approximation, we truncate the sum, retaining the components corresponding to the top r singular values.

Although it may not be obvious at first glance, the truncated-SVD model is closely related to the PCA-based low-rank approximations defined in Section 11.6.1. The column-based and row-based approximations differ in how the data are centered: We either subtract the column mean from the columns or the row mean from the rows. However, let us assume that we center the matrix in the same way. For example, we can subtract the sample mean of all entries (as in Example 11.36) or apply double centering to ensure that the sample mean of the rows and of the columns are both zero (see Exercise 11.10). Then, the resulting column-wise and row-wise PCA-based low-rank approximation are *exactly the same*, and they are also equal to the low-rank approximation obtained via SVD truncation. The diagram in Figure 11.20 depicts the three low-rank approximations.

Theorem 11.35 (PCA and SVD yield the same low-rank approximation, up to centering). *Let D be a real-valued $n_1 \times n_2$ matrix, and let r be a positive integer smaller than n_1 and n_2. Note that D is not necessarily centered in any specific way: it can be column centered, row centered, double centered, centered using the average of all entries, or not centered at all. We define the column-wise PCA rank-r approximation of D by*

$$L'_{\text{PCA-cols}} := \sum_{l=1}^{r} u_l w_l^T, \tag{11.189}$$

$$w_l[j] := u_l^T D[:, j], \quad 1 \le l \le r, 1 \le j \le n_2, \tag{11.190}$$

where the principal directions u_l, $1 \le l \le r$, are the eigenvectors of the pseudo sample covariance matrix

$$\Sigma'_{\text{cols}} := \frac{1}{n_2 - 1} \sum_{j=1}^{n_2} D[:, j] D[:, j]^T \tag{11.191}$$

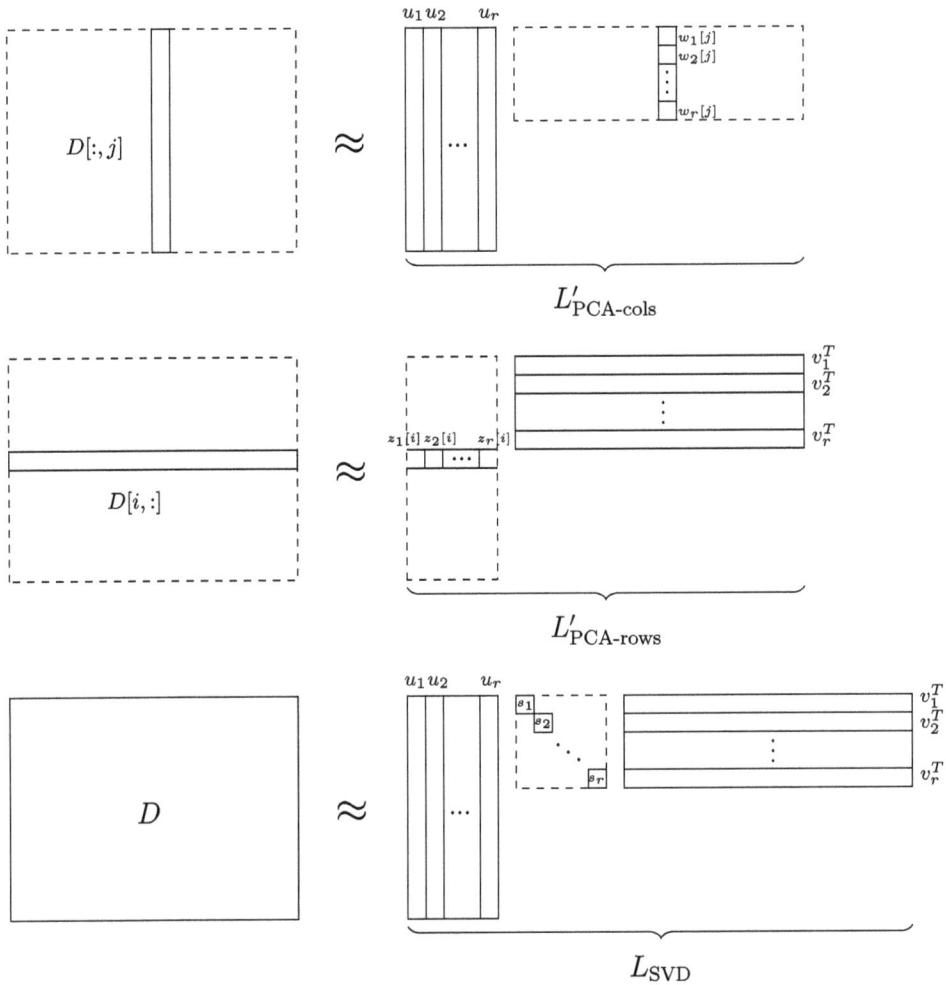

Figure 11.20 Low-rank models computed via PCA and truncated SVD are equivalent. The top diagram shows the column-wise PCA low-rank approximation $L'_{\text{PCA-cols}}$, obtained by performing dimensionality reduction of each column $D[:,j]$ ($1 \leq j \leq n_2$) of the data matrix D via PCA. The middle diagram shows the row-wise PCA low-rank approximation $L'_{\text{PCA-rows}}$, obtained by performing dimensionality reduction of the each row $D[i,:]$ ($1 \leq i \leq n_1$) of D via PCA. Theorem 11.35 establishes that both approximations are exactly the same, and also equal to the low-rank approximation L_{SVD} in the bottom diagram, obtained by truncating the singular-value-decomposition of D, provided the initial centering operation applied to the matrix is the same for the three approximations.

associated with the r largest eigenvalues. We call Σ'_{cols} a pseudo sample covariance matrix because the columns of D are not necessarily centered to have zero sample mean. The notation $L'_{\text{PCA-cols}}$ is meant to distinguish the approximation from $L_{\text{PCA-cols}}$ in (11.145), which does assume that the columns are centered.

Similarly, we denote the row-wise PCA rank-r approximation of D by

$$L'_{\text{PCA-rows}} := \sum_{l=1}^{r} z_l v_l^T, \tag{11.192}$$

$$z_l[i] := D[i,:]v_l, \quad 1 \leq l \leq r, 1 \leq i \leq n_1, \tag{11.193}$$

where v_l, $1 \leq l \leq r$, are the eigenvectors of the pseudo sample covariance matrix

$$\Sigma'_{\text{rows}} := \frac{1}{n_1 - 1} \sum_{i=1}^{n_1} D[i,:]^T D[i,:] \tag{11.194}$$

associated with the r largest eigenvalues. Once again, Σ'_{rows} may not be a true sample covariance matrix because the rows of D are not necessarily centered to have zero sample mean, and $L'_{\text{PCA-rows}}$ may differ from $L_{\text{PCA-rows}}$ in (11.150) for the same reason.

The rank-r SVD-based approximation L_{SVD}, obtained by computing the SVD of D and truncating it to keep only the r rank-one components associated with the r largest singular values, as in (11.186), is equal to both $L'_{\text{PCA-cols}}$ and $L'_{\text{PCA-rows}}$.

Proof We assume $n_1 \leq n_2$, otherwise the same argument applies to the transpose of D. Let $s_1, \ldots, s_{n_1}$ be the singular values of D, $u_1, \ldots, u_{n_1}$ the corresponding left singular vectors, and $v_1, \ldots, v_{n_1}$ the corresponding right singular vectors. The pseudo sample covariance matrix of the columns equals

$$\Sigma'_{\text{cols}} := \frac{1}{n_2 - 1} D D^T. \tag{11.195}$$

In the proof of Theorem 11.31, we define the left singular vectors of D to equal the eigenvectors of DD^T, normalized to have unit ℓ_2 norm, and the singular values to equal the square roots of the corresponding eigenvalues. Consequently, the r eigenvectors of Σ'_{cols} with the largest eigenvalues, normalized to have unit ℓ_2 norm, are equal to the first r left singular vectors $u_1, \ldots, u_r$ of D. If we represent the jth column $D[:,j]$ in terms of the singular value decomposition of D, the lth principal component of the jth column $D[:,j]$ equals

$$w_l[j] := u_l^T D[:,j] \tag{11.196}$$

$$= u_l^T \sum_{k=1}^{n_1} s_k u_k v_k[j] \tag{11.197}$$

$$= s_l v_l[j]. \tag{11.198}$$

As a result, the column-based PCA approximation is the same as the truncated-SVD approximation,

$$L'_{\text{PCA-cols}}[i,j] := \sum_{l=1}^{r} u_l[i] w_l[j] \tag{11.199}$$

$$= \sum_{l=1}^{r} s_l u_l[i] v_l[j] = L_{\text{SVD}}[i,j]. \tag{11.200}$$

The pseudo sample covariance matrix of the rows equals

$$\Sigma'_{\text{rows}} := \frac{1}{n_1 - 1} D^T D. \tag{11.201}$$

Again by our definition of the left singular vectors in the proof of Theorem 11.31, the right singular vectors of D are the eigenvectors of $D^T D$, and the singular values are the square roots of the singular values.[5] Consequently, the r eigenvectors of Σ'_{rows} with the largest

---

[5] In more detail, the right singular vectors of D are the left singular vectors of D^T, which are defined to equal the eigenvectors of $D^T (D^T)^T = D^T D$. In addition, the eigenvalues of $D^T D$ are the same as those of DD^T, which can be easily shown using the SVD. Note that all eigenvectors are normalized to have unit ℓ_2 norm.

eigenvalues are equal to the first r right singular vectors $v_1, \ldots, v_r$ of D. The lth principal component of the ith row $D[i, :]$ equals

$$z_l[i] := D[i, :]v_l \tag{11.202}$$

$$= \left(\sum_{k=1}^{r} s_k u_k[i] v_k^T \right) v_l \tag{11.203}$$

$$= s_l u_l[i]. \tag{11.204}$$

We conclude that the row-based PCA approximation is also the same as the truncated-SVD approximation,

$$L'_{\text{PCA-rows}}[i, j] := \sum_{l=1}^{r} z_l[i] v_l[j] \tag{11.205}$$

$$= \sum_{l=1}^{r} s_l u_l[i] v_l[j] = L_{\text{SVD}}[i, j]. \tag{11.206}$$

∎

Example 11.36 (Rank-one model for movie ratings). Bob, Molly, Mary, and Larry rate the following six movies from 1 to 5,

$$D := \begin{array}{cccc} \text{Bob} & \text{Molly} & \text{Mary} & \text{Larry} \\ \begin{pmatrix} 1 & 1 & 5 & 4 \\ 2 & 1 & 4 & 5 \\ 4 & 5 & 2 & 1 \\ 5 & 4 & 2 & 1 \\ 4 & 5 & 1 & 2 \\ 1 & 2 & 5 & 5 \end{pmatrix} & \begin{array}{l} \text{The Dark Knight} \\ \text{Spiderman 3} \\ \text{Love Actually} \\ \text{Bridget Jones's Diary} \\ \text{Pretty Woman} \\ \text{Superman 2} \end{array} \end{array} \tag{11.207}$$

Our goal is to fit a low-rank model to these data using the SVD. We begin by subtracting the sample mean of all the ratings,

$$m(D) := \frac{1}{24} \sum_{i=1}^{6} \sum_{j=1}^{4} D[i, j] \tag{11.208}$$

$$= 3, \tag{11.209}$$

from each entry to obtain the centered matrix $D_{\text{ct}} := D - m(D)$. The SVD of D_{ct} equals

$$D_{\text{ct}} = USV^T = U \begin{bmatrix} 7.79 & 0 & 0 & 0 \\ 0 & 1.62 & 0 & 0 \\ 0 & 0 & 1.55 & 0 \\ 0 & 0 & 0 & 0.62 \end{bmatrix} V^T. \tag{11.210}$$

The first singular value is much larger than the rest, which suggests that the data can be well approximated by a rank-one matrix. To compute the approximation, we truncate the SVD of D_{ct}, retaining only the rank-one component corresponding to the largest singular value. We then add back the mean rating to approximate the original non-centered matrix D,

$$L[i, j] := m(D) + s_1 u_1[i] v_1[j], \quad 1 \le i \le 6, 1 \le j \le 4. \tag{11.211}$$

This yields the low-rank approximation,

$$
L = \begin{pmatrix}
1.34\,(1) & 1.19\,(1) & 4.66\,(5) & 4.81\,(4) \\
1.55\,(2) & 1.42\,(1) & 4.45\,(4) & 4.58\,(5) \\
4.45\,(4) & 4.58\,(5) & 1.55\,(2) & 1.42\,(1) \\
4.43\,(5) & 4.56\,(4) & 1.57\,(2) & 1.44\,(1) \\
4.43\,(4) & 4.56\,(5) & 1.57\,(1) & 1.44\,(2) \\
1.34\,(1) & 1.19\,(2) & 4.66\,(5) & 4.81\,(5)
\end{pmatrix}
\begin{array}{l}
\text{The Dark Knight} \\
\text{Spiderman 3} \\
\text{Love Actually} \\
\text{Bridget Jones's Diary} \\
\text{Pretty Woman} \\
\text{Superman 2}
\end{array}
$$

with columns labeled Bob, Molly, Mary, Larry.

For ease of comparison, the values of D are shown in parenthesis. The model is able to approximate the ratings quite well.

We now take a closer look at the components of our low-rank model. The first left singular vector is equal to

	D. Knight	Spiderman 3	Love Act.	B.J.'s Diary	P. Woman	Superman 2
$u_1 = ($	-0.45	-0.39	0.39	0.38	0.38	-0.45 $).$

The entries of u_1 are negative for action movies (The Dark Knight, Spiderman 3, Superman 2) and positive for romantic movies (Love Actually, Bridget Jones's Diary, Pretty Woman). The estimate of the rating given by user j to movie i according to the low rank model equals

$$L[i,j] := m(D) + s_1 u_1[i] v_1[j] \tag{11.212}$$
$$= 3 + 7.79\, u_1[i] v_1[j]. \tag{11.213}$$

Assume $v_1[j]$ is positive. Then action movies receive lower ratings because $u_1[i]$ is negative, so $u_1[i]v_1[j]$ is negative. Conversely, romantic movies receive higher ratings because $u_1[i]$ is positive, so $u_1[i]v_1[j]$ is positive. By contrast, if $v_1[j]$ is negative, action movies receive higher ratings ($u_1[i]v_1[j]$ is positive) and romantic movies receive lower ratings ($u_1[i]v_1[j]$ is negative). This means that we can use the first right singular value to cluster the users according to their preferences:

$$
\begin{array}{ccccc}
 & \text{Bob} & \text{Molly} & \text{Mary} & \text{Larry} \\
v_1 = (& 0.48 & 0.52 & -0.48 & -0.52).
\end{array}
\tag{11.214}
$$

Negative entries indicate users that rate action movies higher than romantic movies (Mary and Larry), whereas positive entries indicate the contrary (Bob and Molly).

··

Example 11.37 (Low-rank model of temperature). In this example, we analyze the temporal patterns of temperature in the United States using a low-rank model. The data are extracted from Dataset 9. They are hourly temperatures from 2015, measured at 134 weather stations in the United States. We represent the data as a matrix with 134 rows, each corresponding to a weather station, and $24 \cdot 365 = 8{,}760$ columns, one for each hour in the year. In Example 7.16, we compute the sample means of the temperatures at each location, which reveal that temperatures are colder at higher latitudes and warmer at lower latitudes. Here, we are interested in the temporal structure of the data, so we subtract the sample mean of the temperatures at each location, which is equivalent to centering the rows of the data matrix.

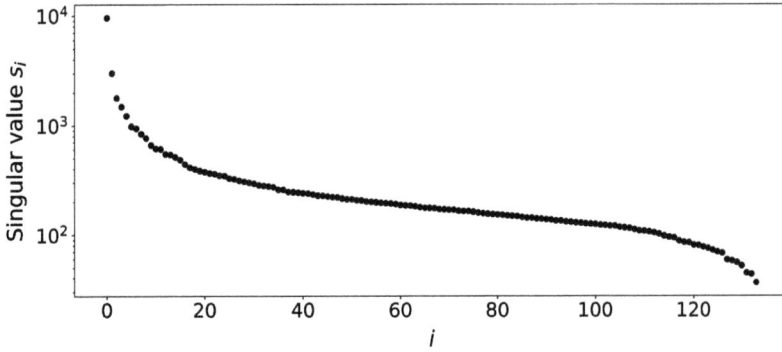

Figure 11.21 Singular values of temperature data. Singular values of the row-centered data matrix in Example 11.37. The two largest singular values are larger than the rest, suggesting that the matrix can be approximated by a rank-two matrix via SVD truncation.

The singular values of the row-centered matrix are shown in Figure 11.21. The two largest singular values are much larger than the rest. Motivated by this, we compute a rank-two approximation to the matrix via SVD truncation. The approximation yields the following low-rank model $L[\ell, t]$ for the temperature at location ℓ and time t:

$$L[\ell, t] = m_\ell + s_1 u_1[\ell] v_1[t] + s_2 u_2[\ell] v_2[t], \tag{11.215}$$

where m_ℓ is the sample mean temperature at the ℓth location, s_1 and s_2 are the two largest singular values of the row-centered matrix, u_1 and u_2 are the corresponding 134-dimensional left singular vectors, and v_1 and v_2 the corresponding 8760-dimensional right singular vectors. Since we have centered the rows of the matrix, by Theorem 11.35, this model is equivalent to the row-wise PCA model $L_{\text{PCA-rows}}$ defined in (11.150). The principal directions are the two right singular vectors, and the principal components are the left singular vectors scaled by the corresponding left singular values.

Figure 11.22 shows a visualization of the first left and right singular vectors u_1 and v_1, which reveals that the first component $s_1 u_1[\ell] v_1[t]$ of the low-rank model captures seasonality and daily periodicity. The first right singular vector v_1 follows a yearly pattern where the summer is warmer than the winter (see the graph at the center of Figure 11.22), and a daily pattern where the day is warmer than the night (see the zoomed-in graphs at the bottom of Figure 11.22). Each entry of the first left singular vector u_1 weights the contribution of v_1 to the low-rank temperature estimate for the corresponding location. These entries are depicted on the graph at the top of Figure 11.22. The radius of the circle at each location is proportional to the magnitude of the entry. The magnitude is largest for locations in the interior of the United States, where temperatures experience substantial seasonal variation, and smallest in Hawaii, the West Coast, or Florida, where the temperature is relatively constant throughout the year.

Figure 11.23 shows a visualization of the second left and right singular vectors u_2 and v_2. Just by looking at v_2, it is not immediately obvious what temporal structure it captures. However, the entries of u_2 form a striking pattern: They are negative in the west, very small at the center of the United States, and positive in the east. To understand what is going on, we take a closer look at our rank-two model for the temperature at Corvallis (Oregon) and Kingston (Rhode Island):

Left singular vector u_1

Right singular vector v_1

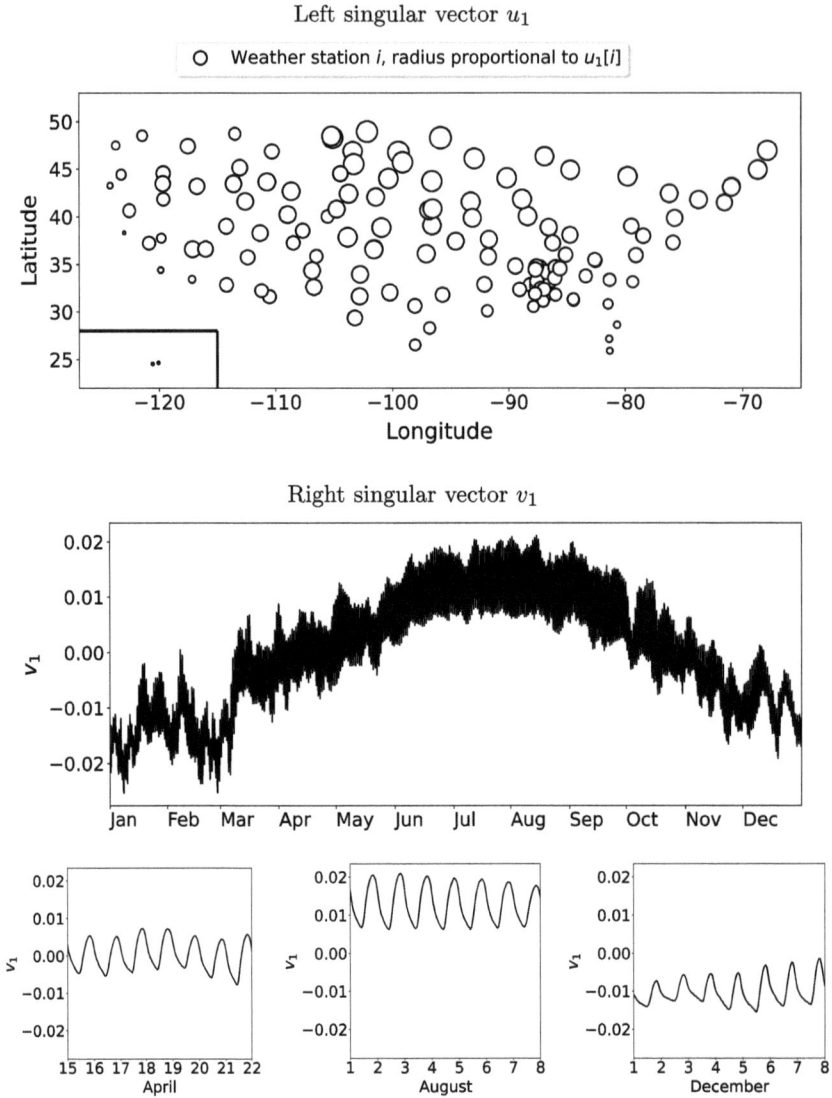

Figure 11.22 SVD analysis automatically reveals seasonality and daily periodicity. The graph at the top shows a visualization of the first left singular vector u_1 of the temperature matrix in Example 11.37. The radius of each circle is proportional to the magnitude of the corresponding entry of u_1 (all entries are positive). The right singular vector v_1 reveals that the first rank-one component of the SVD captures annual seasonality (see the graph at the center) and daily periodicity (see the zoomed-in graphs underneath).

$$L_2[\text{Corvallis}, \text{t}] = m_{\text{Corvallis}} + s_1 u_1[\text{Corvallis}]v_1[t] + s_2 u_2[\text{Corvallis}]v_2[t] \quad (11.216)$$

$$= 12.3 + 554\, v_1[t] - \mathbf{218}\, v_2[t], \quad (11.217)$$

$$L_2[\text{Kingston}, \text{t}] = m_{\text{Kingston}} + s_1 u_1[\text{Kingston}]v_1[t] + s_2 u_2[\text{Kingston}]v_2[t] \quad (11.218)$$

$$= 9.7 + 853\, v_1[t] + \mathbf{311}\, v_2[t]. \quad (11.219)$$

Figure 11.24 compares the data at both stations (solid lines) with the rank-two models (dashed lines). In addition, we plot rank-one models (dotted lines), which only depend on

Left singular vector u_2

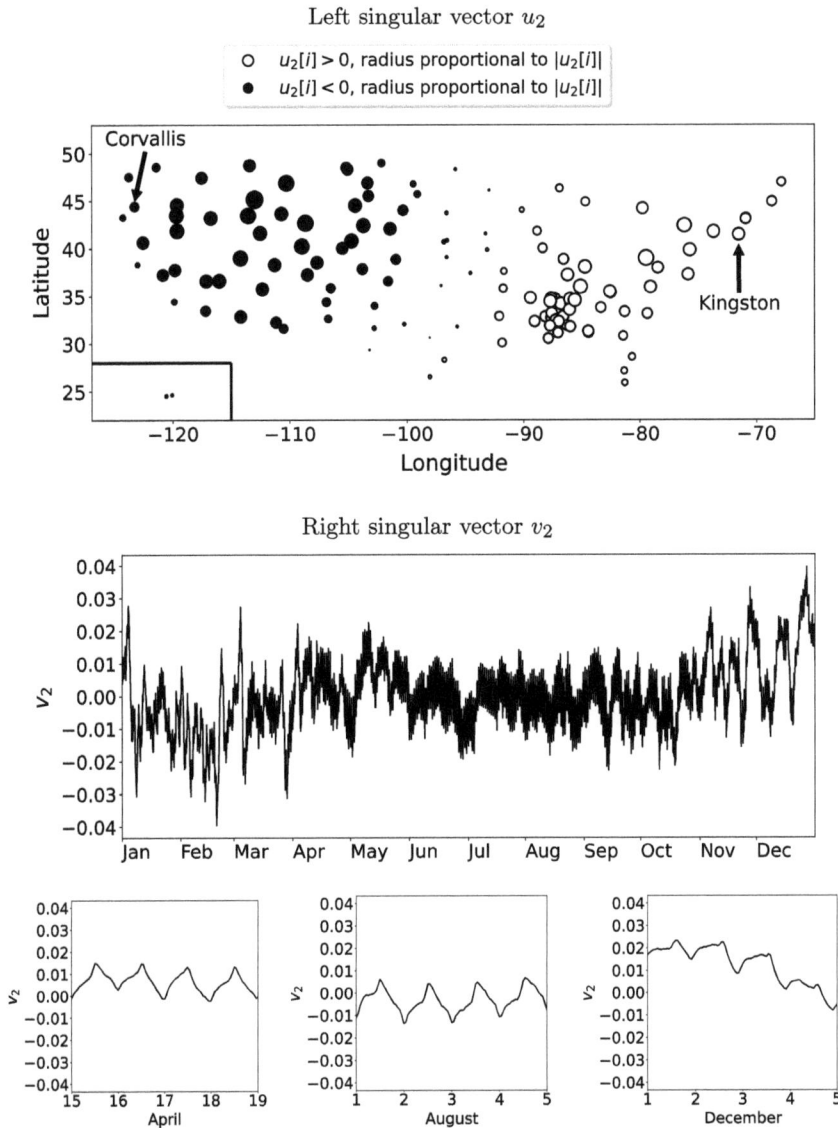

Figure 11.23 **SVD analysis reveals the effect of longitude on temperature.** The graph at the top shows a visualization of the second left singular vector u_2 of the temperature matrix in Example 11.37. Black circles represent negative entries and white circles represent positive entries at the corresponding location. The radius of each circle is proportional to the magnitude. The right singular vector v_2 is depicted by the center and bottom graphs. The corresponding component in the low-rank model shifts the daily pattern associated with the first component (shown in the bottom row of Figure 11.22) to the left or right, to account for the sun rising earlier in the east than in the west. This is why the coefficients in the top graph change from negative to positive, as we move from west to east.

the first SVD component,

$$L_1[\text{Corvallis}, t] = m_{\text{Corvallis}} + s_1 u_1[\text{Corvallis}]v_1[t] \tag{11.220}$$
$$= 12.3 + 554\, v_1[t], \tag{11.221}$$
$$L_1[\text{Kingston}, t] = m_{\text{Kingston}} + s_1 u_1[\text{Kingston}]v_1[t] \tag{11.222}$$
$$= 9.7 + 853\, v_1[t]. \tag{11.223}$$

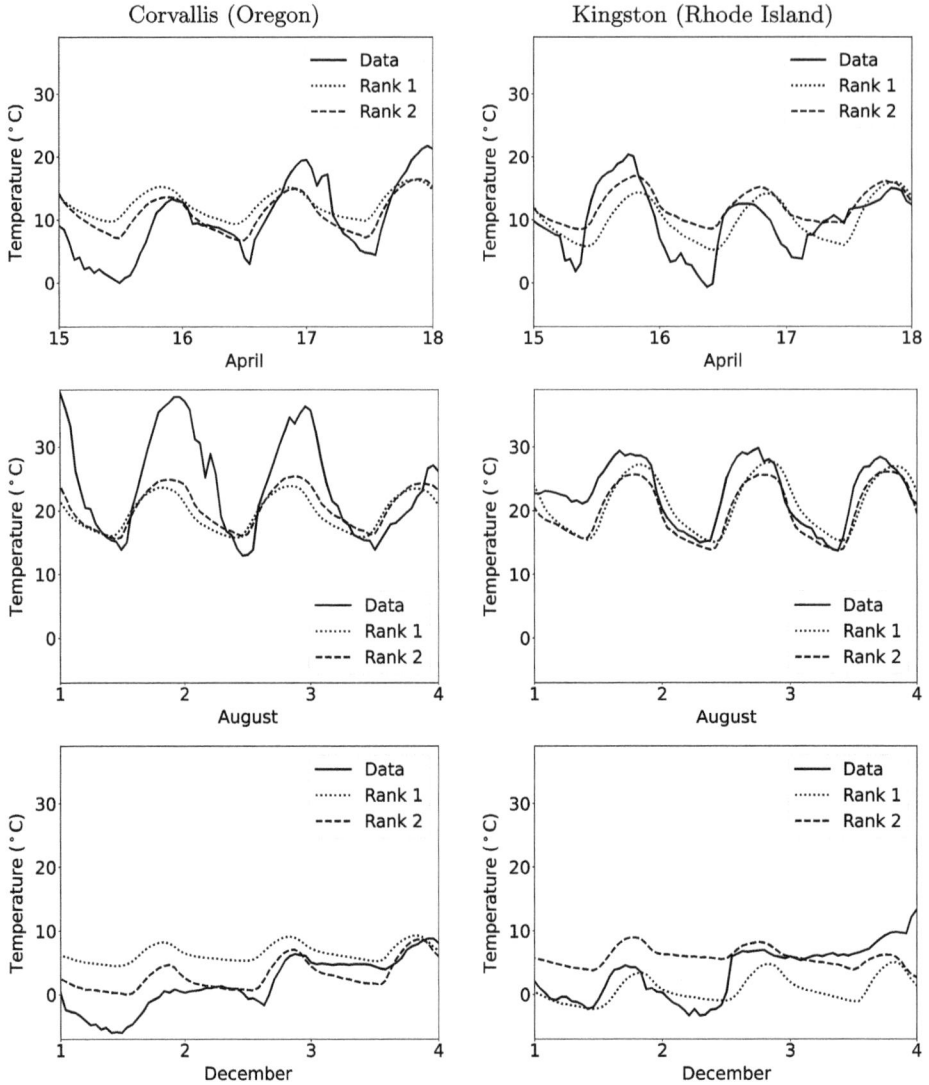

Figure 11.24 Low-rank models for temperature. The graphs compare the temperature data (solid lines) observed in Corvallis (left column) and Kingston (right column) with the rank-one (dotted lines) and rank-two (dashed lines) models described in Example 11.37 at different times of the year (different rows). The rank-two model for Corvallis is shifted to the right with respect to the rank-one model. Conversely, the rank-two model for Kingston is shifted to the left. These shifts achieve better alignment with the data because the sun rises three hours earlier in Kingston than in Corvallis, due to their geographic locations (see the graph at the top of Figure 11.23).

Compared to the corresponding rank-one model, the rank-two model is shifted to the left in the case of Corvallis, and to the right in the case of Kingston. This improves the fit to the data because the sun rises approximately three hours earlier in Kingston than in Corvallis, due to their geographic locations (see the graph at the top of Figure 11.23), which *causes a delay between their daily temperature cycles*. The delay is not accounted for in the rank-one model. The coefficients that multiply v_1 are similar for both locations because v_1 captures seasonal and daily variability that is similar at both locations. As a result, the daily periodicity patterns of the rank-one models coincide, but are misaligned with the observed temperatures. The rank-two model improves the alignment by introducing an additional coefficient (highlighted in bold in (11.217) and (11.219)) multiplied by the second right singular vector v_2. The coefficient is proportional to the corresponding entry of the second left singular vector u_2. Negative entries in u_2 shift the pattern to the right, accounting for delayed diurnal cycles in the west. Conversely, positive entries shift it to the left, accounting for earlier diurnal cycles in the east. This explains why the entries change from negative to positive, as we move from west to east, in the graph at the top of Figure 11.23.

. . . .

11.6.4 Optimal Low-Rank Matrix Estimation

In this section, we prove that the SVD-truncation approach in Section 11.6.3 yields an optimal rank-r estimator, which minimizes the approximation error in Frobenius norm (see Definition 11.32). The following lemma is key to the proof.

Lemma 11.38. *If the column spaces of any pair of matrices $A, B \in \mathbb{R}^{n_1 \times n_2}$ are orthogonal, then*

$$||A + B||_{\mathrm{F}}^2 = ||A||_{\mathrm{F}}^2 + ||B||_{\mathrm{F}}^2 . \tag{11.224}$$

Proof The matrices are orthogonal with respect to the Frobenius product introduced in Definition 11.32, so the result follows directly from the Pythagorean theorem. To establish orthogonality, we express the product as a sum of products between the columns of A and B. If the column spaces are orthogonal, then all of these products equal zero:

$$\langle A, B \rangle := \mathrm{Trace}\left(A^T B\right) \tag{11.225}$$

$$= \sum_{i=1}^{n} \langle A[:, i], B[:, i] \rangle = 0, \tag{11.226}$$

where $A[:, i]$ and $B[:, i]$ denote the ith column of A and B, respectively. ∎

Theorem 11.39 (Optimal rank-r approximation). *For any rank $r \leq \min\{n_1, n_2\}$, truncating the SVD achieves the best rank-r approximation of a matrix $D \in \mathbb{R}^{n_1 \times n_2}$ in terms of Frobenius-norm error:*

$$L_{\mathrm{SVD}} := \sum_{l=1}^{r} s_l u_l v_l^T = \arg \min_{\mathrm{rank}(L)=r} ||D - L||_{\mathrm{F}} , \tag{11.227}$$

where $s_1, \ldots, s_r$ are the r largest singular values of D, $u_1, \ldots, u_r$ the corresponding left singular vectors, and $v_1, \ldots, v_r$ the corresponding right singular vectors.

Proof Let L be an arbitrary matrix in $\mathbb{R}^{n_1 \times n_2}$ with rank r. Let U_L be a matrix with r orthonormal columns that span the column space of L, such that $U_L U_L^T L = L$. The low-rank approximation $U_L U_L^T D$ obtained by projecting the columns of D onto the column space $\mathrm{col}(L)$ of L can only improve the approximation error with respect to L. This follows from Lemma 11.38. The column space of $D_\perp := D - U_L U_L^T D$ is orthogonal to $\mathrm{col}(L)$

because each column of $D_\perp$ is the projection of the corresponding column of D onto the orthogonal complement of $\mathrm{col}(L)$, so

$$||D - L||_F^2 = ||D - U_L U_L^T D||_F^2 + ||L - U_L U_L^T D||_F^2 \qquad (11.228)$$

$$\geq ||D - U_L U_L^T D||_F^2. \qquad (11.229)$$

We conclude that it suffices to show that the approximation error of L_{SVD} is not larger than that of $U_L U_L^T D$.

Let U_* be a matrix with r columns equal to the first r left singular vectors of D. The SVD-based low-rank approximation can be obtained by applying $U_* U_*^T$ to each column of D. To see why, let S_* be a diagonal matrix, containing the first r singular values of D, and V_* a matrix with rows equal to the corresponding r right singular vectors. Then, $L_{\mathrm{SVD}} = U_* S_* V_*^T$. We denote by $U_\perp, S_\perp$, and $V_\perp$ the matrices containing the rest of singular vectors and singular values,

$$D = \begin{bmatrix} U_* & U_\perp \end{bmatrix} \begin{bmatrix} S_* & 0 \\ 0 & S_\perp \end{bmatrix} \begin{bmatrix} V_* & V_\perp \end{bmatrix}^T, \qquad (11.230)$$

where 0 denotes a matrix of zeros with the appropriate dimensions. By orthogonality of the left singular vectors, the columns of U_* are orthogonal to those of $U_\perp$, so

$$U_* U_*^T D = U_* U_*^T \begin{bmatrix} U_* & U_\perp \end{bmatrix} \begin{bmatrix} S_* & 0 \\ 0 & S_\perp \end{bmatrix} \begin{bmatrix} V_* & V_\perp \end{bmatrix}^T \qquad (11.231)$$

$$= \begin{bmatrix} U_* & 0 \end{bmatrix} \begin{bmatrix} S_* & 0 \\ 0 & S_\perp \end{bmatrix} \begin{bmatrix} V_* & V_\perp \end{bmatrix}^T \qquad (11.232)$$

$$= U_* S_* V_*^T \qquad (11.233)$$

$$= L_{\mathrm{SVD}}. \qquad (11.234)$$

Consequently, $D = U_* U_*^T D + U_\perp S_\perp V_\perp^T$, and the two components have orthogonal column spaces. By Lemma 11.38,

$$||D||_F^2 = ||U_* U_*^T D||_F^2 + ||U_\perp S_\perp V_\perp^T||_F^2, \qquad (11.235)$$

so the approximation error incurred by L_{SVD} equals

$$||D - L_{\mathrm{SVD}}||_F^2 = ||U_\perp S_\perp V_\perp^T||_F^2 \qquad (11.236)$$

$$= ||D||_F^2 - ||U_* U_*^T D||_F^2. \qquad (11.237)$$

Also, by Lemma 11.38 and the Pythagorean theorem, the error incurred by $U_L U_L^T D$ equals

$$||D - U_L U_L^T D||_F^2 = ||D||_F^2 - ||U_L U_L^T D||_F^2. \qquad (11.238)$$

To complete the proof we need to show

$$||U_* U_*^T D||_F^2 \geq ||U_L U_L^T D||_F^2 \qquad (11.239)$$

for any possible choice of U_L. We can express these quantities in terms of the pseudo sample covariance matrix

$$\Sigma'_{\mathrm{cols}} := \frac{1}{n_2 - 1} \sum_{j=1}^{n_2} D[:, j] D[:, j]^T \qquad (11.240)$$

of the columns of D. This is not necessarily a true sample covariance matrix because we don't assume that the columns of D are centered. Since the Frobenius norm is the sum of the squared entries of a matrix by Definition 11.32,

$$\left\|U_* U_*^T D\right\|_F^2 = \sum_{j=1}^{n_2} \left\|U_* U_*^T D[:, j]\right\|_2^2 \tag{11.241}$$

$$= \sum_{j=1}^{n_2} D[:, j]^T U_* U_*^T U_* U_*^T D[:, j] \tag{11.242}$$

$$= \sum_{j=1}^{n_2} D[:, j]^T U_* U_*^T D[:, j] \tag{11.243}$$

$$= \sum_{l=1}^{r} \sum_{j=1}^{n_2} u_l^T D[:, j] D[:, j]^T u_l \tag{11.244}$$

$$= (n_2 - 1) \sum_{l=1}^{r} u_l^T \left(\frac{1}{n_2 - 1} \sum_{j=1}^{n_2} D[:, j] D[:, j]^T\right) u_l \tag{11.245}$$

$$= (n_2 - 1) \sum_{l=1}^{r} u_l^T \Sigma'_{\text{cols}} u_l, \tag{11.246}$$

where $u_1, \ldots, u_r$ are the first left singular vectors of D and $D[:, j]$ denotes the jth column of D. By the same argument,

$$\left\|U_L U_L^T D\right\|_F^2 = (n_2 - 1) \sum_{l=1}^{r} (u_l^{[L]})^T \Sigma'_{\text{cols}} u_l^{[L]}. \tag{11.247}$$

If we assume that D has centered columns, then by Theorem 11.17 (11.246) is equal to the sum of the sample variances of the data in the direction of the orthonormal vectors $u_1, \ldots, u_r$, and (11.247) equals the sum of the sample variances of the data in the direction of $u_1^{[L]}, \ldots, u_r^{[L]}$. In Theorem 11.27, we establish that the first r eigenvectors of a covariance matrix are the set of r orthonormal vectors that capture the most variance. The same holds for sample covariance matrices (see Exercise 11.7): The eigenvectors corresponding to the r largest eigenvalues capture the most sample variance. These eigenvectors are exactly equal to the first r left singular vectors $u_1, \ldots, u_r$ (see the proof of Theorem 11.35). Therefore (11.246) is larger than (11.247) for any value of $r \leq \min\{n_1, n_2\}$, which completes the proof. To prove the general result, we can use the same arguments outlined in Exercise 11.7 to establish that (11.246) is larger than (11.247), even when the columns of D are not centered. ∎

11.7 Matrix Completion for Collaborative Filtering

The goal of *collaborative filtering* in recommender systems is to estimate user preferences from past data. In this section, we cast collaborative filtering as a *matrix completion* problem. We use prediction of movie ratings as a motivating example, inspired by the Netflix Prize, a competition held by Netflix between 2006 and 2009, which rewarded the best collaborative-filtering method with one million dollars.

Let us consider a set of observed ratings, stored as entries of a matrix, where each user is assigned a column, and each movie is assigned a row. The missing entries correspond to movies that have not been rated by the users associated with those columns (see Example 11.43). Predicting the missing ratings is equivalent to completing the matrix.

At first glance, completing a matrix such as this one

$$\begin{bmatrix} 1 & ? & 1 \\ ? & 6 & 3 \\ ? & 4 & 2 \end{bmatrix} \tag{11.248}$$

looks like an ill-posed problem. The missing entries can be filled in arbitrarily! For the problem to be well posed, we need to make assumptions about the relationship between the missing and observed entries. A natural assumption is that the entries depend on a small number of factors, as in Example 11.36. Under this assumption, the ratings matrix $D \in \mathbb{R}^{n_1 \times n_2}$ is low rank, with a rank r equal to the number of factors. Let us express each observed entry $D[i, j]$ in terms of the coefficients of the low-rank decomposition,

$$L[i, j] := \sum_{l=1}^{r} a_l[i] b_l[j], \qquad 1 \le i \le n_1, 1 \le j \le n_2. \tag{11.249}$$

If the number of observed entries is sufficiently large, with respect to the assumed rank r, we can estimate the coefficients $a_1, \ldots, a_r \in \mathbb{R}^{n_1}$ and $b_1, \ldots, b_r \in \mathbb{R}^{n_2}$ from the data. Then, we use the coefficients to estimate the missing entries. For instance, if the entry with indices (k, m) is missing, we approximate it by

$$L[k, m] := \sum_{l=1}^{r} a_l[k] b_l[m]. \tag{11.250}$$

Figure 11.25 illustrates this matrix-completion approach.

Example 11.40 (Completing a rank-one matrix). Let us fit a rank-one model to the matrix in (11.248):

$$\begin{bmatrix} 1 & ? & 1 \\ ? & 6 & 3 \\ ? & 4 & 2 \end{bmatrix} = a_1 b_1^T = \begin{bmatrix} a_1[1] b_1[1] & a_1[1] b_1[2] & a_1[1] b_1[3] \\ a_1[2] b_1[1] & a_1[2] b_1[2] & a_1[2] b_1[3] \\ a_1[3] b_1[1] & a_1[3] b_1[2] & a_1[3] b_1[3] \end{bmatrix}. \tag{11.251}$$

The rank-one model has six coefficients, corresponding to the entries of a_1 and b_1. Our goal is to estimate these coefficients from the six observed entries. The assumption that the rank

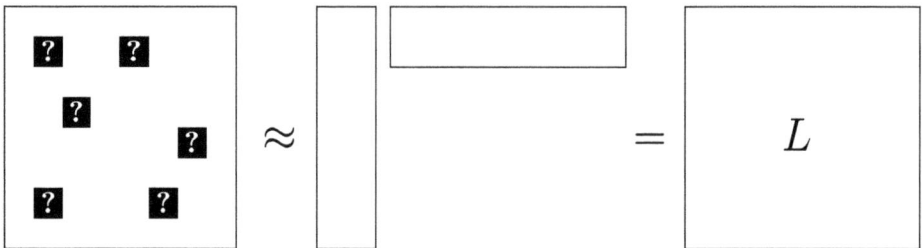

Figure 11.25 Low-rank matrix completion. The diagram shows how to leverage low-rank models to estimate the missing entries in a matrix. The parameters of the model are estimated using the observed data, and then used to form the low-rank estimate L, which provides an estimate of the unobserved entries.

of the matrix is one yields six equations:

$$a_1[1]\, b_1[1] = 1, \qquad a_1[1]\, b_1[3] = 1, \qquad a_1[2]\, b_1[2] = 6, \qquad (11.252)$$
$$a_1[2]\, b_1[3] = 3, \qquad a_1[3]\, b_1[2] = 4, \qquad a_1[3]\, b_1[3] = 2. \qquad (11.253)$$

For any choice of a_1 and b_1, we can obtain an equivalent model by multiplying all entries of a_1 by a constant α, and all entries of b_1 by $1/\alpha$. To avoid this ambiguity, we set $a_1[1] := 1$ (another option could have been to impose $\|a_1\|_2 = 1$). From the equations, we deduce

$$b_1[1] = \frac{1}{a_1[1]} = 1, \qquad b_1[3] = \frac{1}{a_1[1]} = 1, \qquad a_1[2] = \frac{3}{b_1[3]} = 3, \qquad (11.254)$$
$$b_1[2] = \frac{6}{a_1[2]} = 2, \qquad a_1[3] = \frac{4}{b_1[2]} = 2. \qquad (11.255)$$

The resulting completed matrix is

$$L = a_1 b_1^T = \begin{bmatrix} 1 \\ 3 \\ 2 \end{bmatrix} \begin{bmatrix} 1 & 2 & 1 \end{bmatrix} = \begin{bmatrix} 1 & \mathbf{2} & 1 \\ 3 & \mathbf{6} & 3 \\ \mathbf{2} & \mathbf{4} & \mathbf{2} \end{bmatrix}, \qquad (11.256)$$

where the estimated missing entries are highlighted in bold.

· ·

Even under a low-rank assumption, matrix completion can be an ill-posed problem, as illustrated by the following example.

Example 11.41 (Matrix completion can be ill posed)**.** If a whole row or a whole column is missing from a data matrix, then the matrix-completion problem is ill posed. The corresponding entry of the low-rank parameters cannot possibly be determined from the observed entries. For instance, consider the problem of fitting a rank-one model to the following data

$$\begin{bmatrix} 1 & 1 & 1 \\ ? & ? & ? \\ 1 & 1 & 1 \end{bmatrix} = a_1 b_1^T = \begin{bmatrix} 1 \\ ? \\ 1 \end{bmatrix} \begin{bmatrix} 1 & 1 & 1 \end{bmatrix}. \qquad (11.257)$$

The second entry of a_1 cannot be recovered because it is only encoded in the missing entries. In order to avoid such situations, we need to observe some entries in every column and every row. In the context of movie-rating prediction this makes sense: How can we produce a rating prediction tailored to the preferences of a specific user, if we don't observe any of their ratings?

Even if we observe every row and column, some low-rank matrices cannot be recovered from a subset of their entries. Consider the rank-one matrix

$$\begin{bmatrix} 0 & 0 & 0 \\ 0 & 23 & 0 \\ 0 & 0 & 0 \end{bmatrix}. \qquad (11.258)$$

If we do not observe the nonzero entry 23, there isn't a unique rank-one model that fits the data:

$$\begin{bmatrix} 0 & 0 & 0 \\ 0 & ? & 0 \\ 0 & 0 & 0 \end{bmatrix} = a_1 b_1^T = \begin{bmatrix} 0 \\ ? \\ 0 \end{bmatrix} \begin{bmatrix} 0 & ? & 0 \end{bmatrix}. \qquad (11.259)$$

The second entries of a_1 and b_1 are only encoded in the missing entry, so we cannot recover them from the observed entries. In movie-rating prediction, we cannot expect to predict highly idiosyncratic preferences that are unique to a specific movie and user. For example, imagine that a user loves Star Wars movies, except for the first one, because their girlfriend broke up with them after watching it. Then, we won't be able to predict the unusual rating if it is missing because it is completely unrelated to the observed ratings.

. . . .

In Example 11.40, we complete the matrix by assuming that its rank is equal to one. In practice, we cannot expect the data to exactly follow a low-rank model. Instead, we can fit a low-rank model that provides a good approximation of the data. A reasonable criterion to evaluate the approximation is the sum of squared errors between the observed entries and the low-rank estimator,

$$\sum_{(i,j)\in\text{observed}} \left(D[i,j] - \sum_{l=1}^{r} a_l[i]b_l[j] \right)^2. \tag{11.260}$$

Ideally, we would fit the model by selecting the coefficients that minimize this cost function. However, this is easier said than done: The cost function is non-convex, so it is very challenging to find a global minimum. Fortunately, there exist several approaches that identify local minima, which is often sufficient to obtain a useful model. These approaches include variants of gradient descent (Keshavan et al., 2009) and alternating least squares (Jain et al., 2013) (see also (Candès and Recht, 2009) for a method that uses a convex regularization term to promote low-rank structure).

Here, we present a simple heuristic procedure to minimize (11.260), which is often effective in practice. Section 11.6.4 shows that truncating the SVD of a matrix yields an optimal rank-r approximation. However, this assumes that all the entries are available, which is not the case in matrix completion. Without the missing entries, we cannot compute the SVD. A pragmatic solution is to impute those entries, for example using the average of the observed data (or the average of the corresponding row or column). Then we can compute a truncated SVD to obtain a low-rank approximation of the imputed matrix. Unfortunately, the resulting low-rank model *also approximates the imputed entries*. Nonetheless, these estimated entries tend to be better estimates of the missing entries than our naive initial imputation. Consequently, we can use them to re-impute the missing entries and re-compute the SVD in order to obtain an even better low-rank estimate. These two steps (imputation and SVD truncation) can be repeated over and over until the estimate converges. This method is known as *hard-impute* in the literature (Mazumder et al., 2010) because we alternately impute the missing entries and hard-threshold the singular values. It is usually applied after centering the data, as described in the following definition.

Definition 11.42 (SVD-based low-rank matrix completion). *Let $D \in \mathbb{R}^{n_1 \times n_2}$ denote a data matrix with incomplete entries. We begin by imputing the missing entries, using the mean of all the observed entries, or the mean of the corresponding columns or rows. This yields the imputed matrix $D_{\text{imputed}} \in \mathbb{R}^{n_1 \times n_2}$. Then, we center D_{imputed}, subtracting the sample mean of all its entries, denoted by $m(D_{\text{imputed}})$,*

$$D_{\text{ct}}[i,j] := D_{\text{imputed}}[i,j] - m(D_{\text{imputed}}), \quad 1 \le i \le n_1, 1 \le j \le n_2. \tag{11.261}$$

In order to obtain a rank-r approximation to D_{ct}, we initialize the auxiliary matrix M, setting it equal to D_{ct}, and repeat the following steps until convergence:

1 *Compute the SVD of M and truncate it to obtain the rank-r matrix*

$$L_{ct} := \sum_{l=1}^{r} s_l u_l v_l^T, \tag{11.262}$$

where $s_1, \ldots, s_r$ are the first r singular values of M, $u_1, \ldots, u_r$ the corresponding left singular vectors, and $v_1, \ldots, v_r$ the corresponding right singular vectors.

2 *For $1 \leq i \leq n_1$ and $1 \leq j \leq n_2$, re-impute the missing entries using the corresponding entries of L_{ct}:*

$$M[i,j] := \begin{cases} D_{ct}[i,j] & \text{if(i, j) is observed,} \\ L_{ct}[i,j] & \text{if(i, j) is missing.} \end{cases} \tag{11.263}$$

After convergence, the final estimator is obtained by adding back the subtracted mean to the rank-r matrix L_{ct},

$$L[i,j] := m(D_{\text{imputed}}) + L_{ct}[i,j], \quad 1 \leq i \leq n_1, 1 \leq j \leq n_2. \tag{11.264}$$

As illustrated in Example 11.44, the rank r of a low-rank model for matrix completion can be chosen by evaluating the model on held-out validation data.

Example 11.43 (Rank-one model for matrix completion). The following data correspond to a subset of the movie ratings in Example 11.36:

$$D := \begin{array}{c} \begin{array}{cccc} \text{Bob} & \text{Molly} & \text{Mary} & \text{Larry} \end{array} \\ \left(\begin{array}{cccc} ? & ? & 5 & 4 \\ ? & 1 & 4 & ? \\ 4 & 5 & 2 & ? \\ ? & 4 & 2 & 1 \\ 4 & ? & 1 & 2 \\ 1 & 2 & ? & 5 \end{array} \right) \end{array} \begin{array}{l} \\ \text{The Dark Knight} \\ \text{Spiderman 3} \\ \text{Love Actually} \\ \text{Bridget Jones's Diary} \\ \text{Pretty Woman} \\ \text{Superman 2} \end{array} \tag{11.265}$$

Our goal is to estimate the missing ratings, performing collaborative filtering via the SVD-based matrix-completion method in Definition 11.42. We first impute the missing entries using the sample mean of the observed ratings, $m(D_{\text{imputed}}) = 2.94$, and subtract the overall mean, to obtain the imputed centered matrix

$$D_{ct} := \begin{bmatrix} 0 & 0 & 2.06 & 1.06 \\ 0 & -1.94 & 1.06 & 0 \\ 1.06 & 2.06 & -0.94 & 0 \\ 0 & 1.06 & -0.94 & -1.94 \\ 1.06 & 0 & -1.94 & -0.94 \\ -1.94 & -0.94 & 0 & 2.06 \end{bmatrix}. \tag{11.266}$$

We then compute the SVD of D_{ct}. The singular values equal

$$s_1 = 4.8, \; s_2 = 2.48, \; s_3 = 2.36, \; s_4 = 1.46. \tag{11.267}$$

The first singular value is larger than the rest, suggesting that a rank-one model might fit the data well, although this is much less obvious now that we are missing entries (compare to (11.210)). To compute the rank-one approximation, we combine the first singular value s_1 and the singular vectors u_1 and v_1. The resulting estimator of the ratings, obtained by adding the rank-one approximation and the subtracted mean, is

$$L_{\text{initial}}[i, j] := m(D_{\text{imputed}}) + s_1 u_1[i] v_1[j], \quad 1 \leq i \leq 6, 1 \leq j \leq 4. \qquad (11.268)$$

This yields the low-rank model

$$L_{\text{initial}} = \begin{pmatrix} \mathbf{2.28}\ (1) & \mathbf{2.08}\ (1) & 3.91\ (5) & 3.88\ (4) \\ \mathbf{2.35}\ (2) & \mathbf{2.16}\ (1) & 3.81\ (4) & \mathbf{3.79}\ (5) \\ 3.67\ (4) & 3.91\ (5) & 1.85\ (2) & \mathbf{1.87}\ (1) \\ \mathbf{3.73}\ (5) & 3.99\ (4) & 1.76\ (2) & 1.79\ (1) \\ 3.69\ (4) & \mathbf{3.93}\ (5) & 1.82\ (1) & 1.85\ (2) \\ 2.06\ (1) & 1.78\ (2) & \mathbf{4.24}\ (5) & 4.21\ (5) \end{pmatrix} \begin{matrix} \text{The Dark Knight} \\ \text{Spiderman 3} \\ \text{Love Actually} \\ \text{Bridget Jones's Diary} \\ \text{Pretty Woman} \\ \text{Superman 2} \end{matrix}$$

$$\begin{matrix} \text{Bob} & \text{Molly} & \text{Mary} & \text{Larry} \end{matrix}$$

The original ratings are in parentheses and the missing ratings are highlighted in bold.

This initial low-rank model is clearly influenced by the imputed values for the missing entries: The corresponding estimates are biased toward the mean rating $m(D_{\text{imputed}}) = 2.94$. We address this by applying the iterative algorithm from Definition 11.42. Figure 11.26 shows the values of the missing entries as the iterations proceed. They move away from the mean and eventually converge to fixed values. Let L_{cnt} denote the centered rank-one estimate produced by the algorithm. We add the subtracted mean rating, to obtain our final estimator

$$L_{\text{final}}[i, j] := m(D_{\text{imputed}}) + L_{\text{cnt}}[i, j], \quad 1 \leq i \leq 6, 1 \leq j \leq 4, \qquad (11.269)$$

which yields the low-rank model

$$L_{\text{final}} = \begin{pmatrix} \mathbf{1.48}\ (1) & \mathbf{1.38}\ (1) & 4.45\ (5) & 4.52\ (4) \\ \mathbf{1.50}\ (2) & \mathbf{1.41}\ (1) & 4.42\ (4) & \mathbf{4.50}\ (5) \\ 4.26\ (4) & 4.34\ (5) & 1.57\ (2) & \mathbf{1.51}\ (1) \\ \mathbf{4.18}\ (5) & 4.26\ (4) & 1.65\ (2) & 1.59\ (1) \\ 4.2\ (4) & \mathbf{4.28}\ (5) & 1.64\ (1) & 1.57\ (2) \\ 1.37\ (1) & 1.27\ (2) & \mathbf{4.55}\ (5) & 4.63\ (5) \end{pmatrix} \begin{matrix} \text{The Dark Knight} \\ \text{Spiderman 3} \\ \text{Love Actually} \\ \text{Bridget Jones's Diary} \\ \text{Pretty Woman} \\ \text{Superman 2} \end{matrix}$$

$$\begin{matrix} \text{Bob} & \text{Molly} & \text{Mary} & \text{Larry} \end{matrix}$$

The original ratings are again shown in parentheses and the missing ratings in bold. The low-rank model clearly improves after the iterative procedure, yielding a very reasonable estimate of the missing ratings.

· ·

Example 11.44 (Real movie ratings). The Movielens dataset (Dataset 8) contains ratings given by a group of users to popular movies. The ratings are integers between 1 and 5. For this example, we select the 1,000 users and 100 movies with the most ratings. The number of observed ratings is 30,055. We divide these data at random into disjoint training, validation, and test sets. The validation and test sets contain 1,000 ratings each. Our goal is to estimate

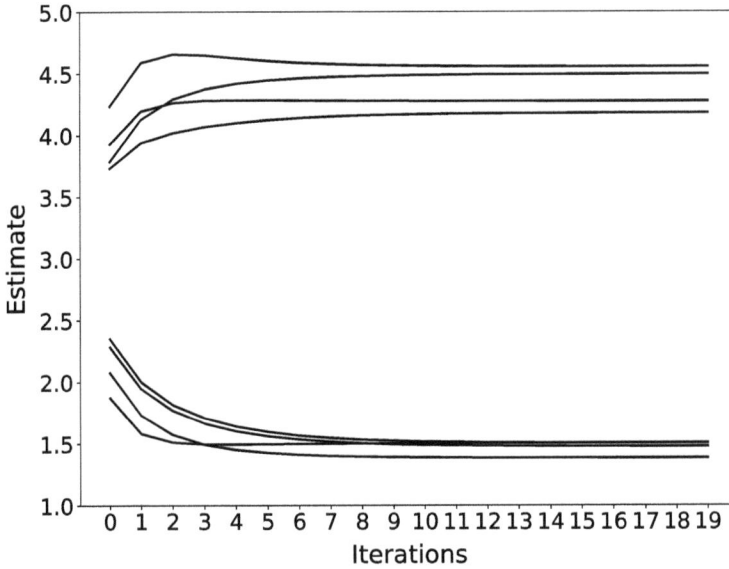

Figure 11.26 SVD-based matrix completion. Evolution of the estimates for the missing entries in Example 11.43 during the iterations of the SVD-based matrix completion method from Definition 11.42. The estimates are initially close to the mean rating (2.94) because this is the value used to impute the missing entries. As the iterations proceed, the estimates become more accurate, moving away from the mean rating.

the ratings in the test set from a model trained on the training set. We evaluate the result using the root mean square error (RMSE), which is the square root of the average squared difference between the estimated and the ground-truth ratings.

In order to estimate the ratings, we fit a low-rank model via the SVD-based matrix-completion method in Definition 11.42 to the training data. We impute the missing ratings (including the ones corresponding to the validation and test set) with the mean rating of each movie computed from the training data (this yields better results than using the overall mean rating, or the mean rating of each user). We select the rank r by evaluating the RMSE on the validation set. Figure 11.27 shows the training and validation error for different values of r. The parameters of the rank-r model are the r singular values, the 100-dimensional r left singular vectors, and the 1,000-dimensional r right singular vectors. Consequently, the number of parameters is proportional to r. If r is too large, the model is able to *overfit* the training data, which hurts its generalization to the validation data. If r is too small, the model is not complex enough and *underfits* the training data. The best trade-off is achieved by $r := 3$, which yields a model with 3,303 parameters (roughly 9 times less than the training data).

Our rank-3 model achieves an RMSE of 0.89 on the test set. For comparison, a baseline that estimates each rating using the mean rating for the corresponding movie (computed from the training data) yields a test RMSE of 0.97. In order to interpret our low-rank model, we examine the singular vectors. The estimated rating for movie i and user j is equal to

$$L[i, j] := m(D_{\text{imputed}}) + \sum_{l=1}^{3} s_l u_l[i] v_l[j], \tag{11.270}$$

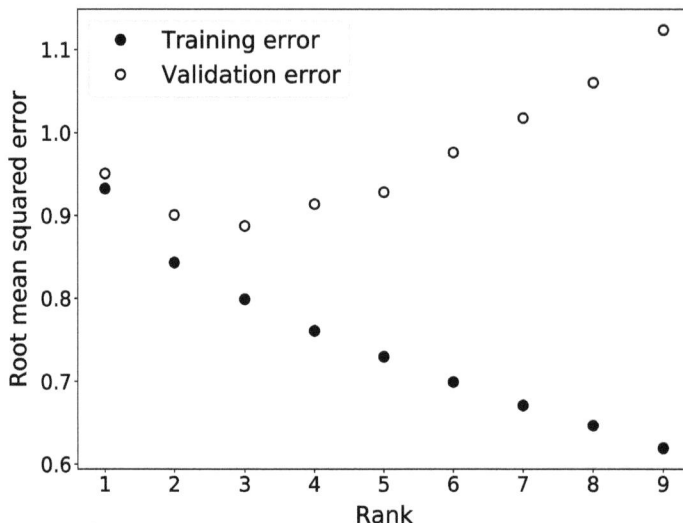

Figure 11.27 Training and validation error of movie-rating estimates for different model ranks. The graph shows the root mean square error of the rank-r model in Example 11.44 for different values of r on the training set and on the validation set. On the training set, the error decreases with r because higher-rank models are able to fit the training data better. On the validation set, the error initially decreases with r, and then increases. For small r (1 and 2) the model underfits the training data. For large r (above 3) it overfits, since its performance no longer generalizes to the validation set.

where $m(D_{\mathrm{imputed}})$ is the sample mean rating in the initial imputed matrix, and s_l, u_l and v_l are the lth singular value and singular vectors of the low-rank approximation L_{ct} in Definition 11.42 (after convergence). As explained in Section 11.6 and Example 11.36, we can interpret each of the three components of the low-rank approximation as representing a *factor*. The value $u_l[i]$ determines how movie i is associated with the lth factor. Table 11.2 shows the movies corresponding to the most positive and most negative values of u_1, u_2, u_3. Even though they are learned in a completely data-driven fashion, these singular vectors capture intuitive properties:

1 The most positive entries of u_1 correspond to three terrible movies: Dante's Peak (FilmAffinity score: 4.6/10), Volcano (FilmAffinity score: 4.2/10), The Saint (FilmAffinity score: 5.5/10). The most negative entries correspond to classics that received critical acclaim: One Flew Over The Cuckoo's Nest (5 Oscars), The Godfather (3 Oscars), Casablanca (3 Oscars).

2 The most positive entries of u_2 are again terrible movies, which were box-office flops: Dante's Peak, Evita (FilmAffinity score: 5.3/10), Volcano. The most negative entries are some of the highest-grossing movies of all time: Titanic, Star Wars, Raiders of the Lost Ark.

3 The most positive entries of u_3 are popular adventure movies: Return of the Jedi, Star Wars, Raiders of the Lost Ark. The most negative entries are also very popular movies, but edgier and less commercial: Leaving Las Vegas, Trainspotting, A Clockwork Orange.

Table 11.2 *Interpreting a collaborative-filtering model for movie ratings. The entries of u_1, u_2 and u_3 in (11.270) reveal the factors learned by the collaborative-filtering model in Example 11.44. The first factor seems to capture critical acclaim. The second seems to capture popularity and box-office success. The third differentiates between adventure blockbusters and darker, less commercial films.*

l	1	2	3
Positive entries of u_l	Dante's Peak (0.29) Volcano (0.28) The Saint (0.24)	Volcano (0.07) Evita (0.05) Dante's Peak (0.04)	Return of the Jedi (0.14) Star Wars (0.13) Raiders of the Lost Ark (0.13)
Negative entries of u_l	One Flew Over The Cuckoo's Nest (−0.10) The Godfather (−0.12) Casablanca (−0.13)	Titanic (−0.16) Star Wars (−0.16) Raiders of the Lost Ark (−0.17)	Leaving Las Vegas (−0.28) Trainspotting (−0.29) A Clockwork Orange (−0.30)

We can analyze the tastes of different users by examining the entries of the right singular vectors corresponding to these factors. Since v_1 is multiplied with u_1 in the estimated rating, large negative entries of v_1 indicate that the user prefers classical movies such as The Godfather or Casablanca. Large negative entries of v_2 indicate that the user likes box-office superhits like Titanic or Star Wars. Conversely, large positive entries of v_1 and v_2 indicate that the user has questionable taste, and enjoys terrible movies. Finally, large positive entries of v_3 suggest a preference for adventure blockbusters such as Star Wars, over less commercial, edgier movies like Trainspotting.

Exercises

11.1 (Random vector) A random vector $\tilde{x}$ with zero mean has a covariance matrix $\Sigma_{\tilde{x}}$ with the following eigendecomposition:

$$\Sigma_{\tilde{x}} = \begin{bmatrix} 1 & 0 & 0 \\ 0 & \frac{1}{\sqrt{2}} & \frac{1}{\sqrt{2}} \\ 0 & \frac{1}{\sqrt{2}} & -\frac{1}{\sqrt{2}} \end{bmatrix} \begin{bmatrix} 1 & 0 & 0 \\ 0 & 0.5 & 0 \\ 0 & 0 & 0 \end{bmatrix} \begin{bmatrix} 1 & 0 & 0 \\ 0 & \frac{1}{\sqrt{2}} & \frac{1}{\sqrt{2}} \\ 0 & \frac{1}{\sqrt{2}} & -\frac{1}{\sqrt{2}} \end{bmatrix}. \quad (11.271)$$

a What is the variance of each of the entries of the random vector $\tilde{x}[1]$, $\tilde{x}[2]$, and $\tilde{x}[3]$?

b Is it possible to find a unit-norm vector u such that the inner product between $\tilde{x}$ and u (i.e. the amplitude of the projection of $\tilde{x}$ onto that direction) has variance greater than one?

c Find three constants a_1, a_2 and a_3, such that at least one of them is nonzero and $P(a_1\tilde{x}[1] + a_2\tilde{x}[2] + a_3\tilde{x}[3] = 0) = 1$. Justify your answer mathematically, and interpret it geometrically.

11.2 (Basketball team) The coach of a basketball team needs to select two out of three players to complete the line-up for the last quarter of a game. The covariance matrix of the points scored by the players per quarter is the following:

	Player A	Player B	Player C
Player A	100	−80	10
Player B	−80	81	50
Player C	10	50	100

Compute the variance of the total number of points, if the coach plays the three possible combinations of two players. Assuming all three players score the same number of points on average, which combination would you recommend to the coach, if they are winning by a wide margin? Which would you recommend, if they are losing by a wide margin?

11.3   (Correlation and PCA) The entries of a two-dimensional random vector have a correlation coefficient equal to one. What is the variance of the second principal component? Explain why this makes intuitive sense.

11.4   (Principal components are uncorrelated) Show that the principal components of a dataset are uncorrelated.

11.5   (Estimating a direction) We consider a dataset of d-dimensional vectors modeled as samples from a random vector

$$\tilde{y} := \tilde{x}v + \tilde{z}, \tag{11.272}$$

where $\tilde{x} \in R$ is a random variable with zero mean and variance σ_{signal}^2, $v \in R^d$ is a fixed deterministic unit-norm vector, and $\tilde{z} \in R^d$ is a Gaussian random vector with independent entries, each of which has mean zero and variance σ_{noise}^2. Assume that $\tilde{x}$ and $\tilde{z}$ are independent.

a   Sketch some samples of $\tilde{y}$ for $d = 2$ when σ_{signal} is much larger than σ_{noise}. You can assume any v for the diagram.

b   For the v you picked in part (a), sketch some samples of $\tilde{y}$ for $d = 2$ when σ_{signal} is much smaller than σ_{noise}.

c   Is averaging the dataset a good algorithm for estimating v?

d   Compute the covariance matrix of $\tilde{y}$.

e   Express the eigendecomposition of the covariance matrix in terms of σ_{signal}, σ_{noise}, v, and a set of vectors $u_2, \ldots, u_d$, which have unit ℓ_2-norm and are orthogonal to v and each other.

f   Suggest an algorithm to estimate the direction of v from the data.

11.6   (Normalization) In this problem we study the effect of a common preprocessing procedure on PCA. The procedure is to divide each feature by its standard deviation, in order to normalize it. Let $\tilde{x}$ be a zero-mean three-dimensional random vector with covariance matrix

$$\Sigma_{\tilde{x}} := \begin{bmatrix} 100 & 25 & 0 \\ 25 & 400 & 0 \\ 0 & 0 & 0.16 \end{bmatrix}. \tag{11.273}$$

We define the normalized vector $\tilde{y}$ as

$$\tilde{y}[i] := \frac{\tilde{x}[i]}{\sqrt{\text{Var}[\tilde{x}[i]]}} \quad 1 \leq i \leq 3. \tag{11.274}$$

a   Compute the covariance matrix of $\tilde{y}$.

b   Is the directional variance of $\tilde{y}$ equal to one in every direction?

c   We decide to reduce the dimensionality of $\tilde{x}$ and $\tilde{y}$ to two dimensions using PCA. Report what directions are selected for each of the random vectors. (Feel free to use a computer for your calculations, but explain what you are doing.)

d Explain whether normalizing the data as above before PCA-based dimensionality reduction makes sense in the following two situations: (1) The entries of $\tilde{x}$ represent the weight (in kilograms), heart rate (in beats per minute), and height (in meters) of a hospital patient. (2) The entries of $\tilde{x}$ represent the length, width and height (all in centimeters) of a car.

11.7 (PCA and sample variance) Prove that the eigenvectors corresponding to the r largest eigenvalues of the sample covariance matrix of a dataset are the r orthogonal directions that capture the most sample variance. For simplicity, assume that the data are centered, so that the columns have zero sample mean.

11.8 (Whitening) Whitening or sphering is a useful preprocessing step in many applications. The goal is to render the entries of a random vector uncorrelated, which can be useful to emphasize nonlinear structure. Let $\tilde{x}$ be a d-dimensional random vector with full-rank covariance matrix $\Sigma_{\tilde{x}} \in \mathbb{R}^{m \times m}$. Derive a matrix $A \in \mathbb{R}^{m \times m}$ that whitens $\tilde{x}$, so that the entries of $\tilde{y} := A\tilde{x}$ are all uncorrelated and have unit variance.

11.9 (Numerical rank) The rank of a matrix is not numerically stable, in the sense that even the tiniest of perturbations will render the matrix full rank. A more stable definition of rank is the numerical rank

$$\text{rank}_{\text{num}}(M) := \min_{||L-M||_{\text{F}} \leq \epsilon} \text{rank}(L), \tag{11.275}$$

where ϵ is a fixed constant. Explain how to compute the numerical rank and justify why your method works.

11.10 (Double centering) Given a matrix $D \in \mathbb{R}^{n_1 \times n_2}$, the column sample mean, row sample mean, and grand sample mean are defined as follows:

$$\mu_{\text{row}} := \frac{1}{n_1} \sum_{i=1}^{n_1} D[i,:], \tag{11.276}$$

$$\mu_{\text{col}} := \frac{1}{n_2} \sum_{j=1}^{n_2} D[:,j], \tag{11.277}$$

$$\mu_{\text{all}} := \frac{1}{n_1 n_2} \sum_{i=1}^{n_1} \sum_{j=1}^{n_2} D[i,j], \tag{11.278}$$

where $D[i,:] \in \mathbb{R}^{n_2}$ and $D[:,j] \in \mathbb{R}^{n_1}$ denote the ith row and the jth column of D, respectively.

a What is the average of the entries of the column sample mean, and of the row sample mean?

b If we subtract the row sample mean from the columns of a matrix and the row column mean from the rows of the same matrix, do we obtain a matrix with centered columns and rows (i.e. with zero column sample mean, and also zero row sample mean)?

c Explain how to obtain a matrix with centered columns and centered rows.

11.11 (Mutant mosquito) The group of friends in Example 11.36 watch a new movie called Mutant Mosquito, but Molly misses it because she is out of town. Bob, Mary, and Larry give the movie a rating of 2, 4, and 4, respectively.

a Consider the rank-one model for movie ratings of the form

$$\text{rating} := c + a[\text{movie}]b[\text{user}], \tag{11.279}$$

where c is a constant, $a[\text{movie}]$ is a coefficient that only depends on the movie, and $b[\text{user}]$ is a coefficient that only depends on the user. Based on the SVD analysis in Example 11.36, what are reasonable values for c and $b[\text{user}]$?

b Estimate the coefficient $a[\text{MM}]$ corresponding to Mutant Mosquito, minimizing the residual sum of squared errors between the rating estimates provided by the rank-one model and the observed ratings for Mutant Mosquito.

c Estimate Molly's rating for Mutant Mosquito.

11.12 (Exact matrix completion) Find the matrix with the lowest possible rank that is compatible with the following observed entries:

$$D := \begin{bmatrix} ? & ? & 2 & 0 \\ 2 & 2 & 3 & ? \\ 2 & 3 & 4 & -2 \\ 2 & 4 & ? & -3 \end{bmatrix}. \tag{11.280}$$

11.13 (Topic modeling) Topic modeling aims to learn the thematic structure of a text corpus automatically. In this example, we take six newspaper articles and compute the frequency of a list of key words. The following matrix, which we denote by D, contains the counts of each word in every article. The entry $D[i, j]$ is the number of times that the jth word is mentioned in the ith article.

singer	GDP	senate	election	vote	stock	concert	market	band	*Articles*
6	1	1	0	0	1	9	0	8	a
1	0	9	5	8	1	0	1	0	b
8	1	0	1	0	0	9	1	7	c
0	7	1	0	0	9	1	7	0	d
0	5	6	7	5	6	0	7	2	e
1	0	8	5	9	2	0	0	1	f

We consider a low-rank topic model, where the counts $D[i, j]$ for article i and word j are approximated by

$$D[i, j] \approx \sum_{l=1}^{r} a_l[i]b_l[j], \qquad 1 \le i \le 6, 1 \le j \le 9. \tag{11.281}$$

Intuitively, each term $a_l b_l$ is associated with a different *topic*.

a Compute the SVD of the matrix. Based on the results, what do you think could be the number of topics r in the model?

b Ideally, we would like the coefficients $a_l[i]$ and $b_l[j]$ to be nonnegative, since a word contributes to the count if it is present, and does not contribute if it isn't, but negative contributions don't make much sense. Why is it difficult for several left or right singular vectors of a matrix to all have nonnegative entries?

c Fit the low-rank model using nonnegative matrix factorization, which constrains the coefficients $a_l[i]$ and $b_l[j]$ to be nonnegative (we suggest that you use the *NMF* function in the Scikit-learn library). Set r following your answer to part (a). Report the coefficients and use them to assign each word and article to one or more topics.

d Compare your model to a model with the same rank obtained by truncating the SVD of the matrix. Which low-rank model achieves better approximation error to the original matrix in Frobenius norm? Was this to be expected? Does this mean that it is a better model?

12

Regression and Classification

Overview

In this chapter, we study the problems of regression and classification, where the goal is to characterize the relationship between several observed features and a quantity of interest, called the response, as illustrated by the examples in Figure 12.1. In regression, the response is modeled as a numerical variable. In classification, the response belongs to a finite set of predetermined classes. In Chapters 4, 6, and 7, we derive optimal estimators for both problems: Theorem 7.59 establishes that the conditional mean is optimal for regression, and Theorems 4.30 and 6.11 prove that maximum a posteriori estimation is optimal for classification. However, it is usually intractable to compute these estimators from data in practical scenarios due to the curse of dimensionality (see Sections 4.7 and 4.8, and the end of Section 7.8.3). The chapter covers practical methods for regression and classification, which make different assumptions to bypass the curse of dimensionality.

Section 12.1 provides a comprehensive description of linear regression models. Section 12.2 discusses generalization to held-out data, explaining when linear models can be expected to generalize or to overfit. Section 12.3 describes ridge regression and introduces the concept of regularization. Section 12.4 considers the problem of sparse regression, where the goal is to fit a linear model that uses a small subset of the available features. Sections 12.5 and 12.6 present linear models for binary and multiclass classification, respectively. Section 12.7 describes nonlinear models for regression and classification based on trees. Section 12.8 introduces neural networks and the deep-learning framework. Finally, Section 12.9 discusses how to evaluate classification performance. Throughout the chapter, we illustrate the different methods via computational examples, which leverage the Scikit-learn (Pedregosa et al., 2011) and PyTorch (Paszke et al., 2019) software libraries.

12.1 Linear Regression

Linear regression models are ubiquitous in data-science applications because they are simple, interpretable, and often surprisingly effective. In Section 12.1.1, we derive the optimal linear regression estimator, under the assumption that we have access to the true joint statistics of the features and the response. In Section 12.1.2, we explain how to fit linear regression models to data via least-squares estimation. Section 12.1.3 describes a decomposition of variance in linear regression, which is useful for evaluation. Finally, Section 12.1.4 explains how to perform causal inference leveraging linear models.

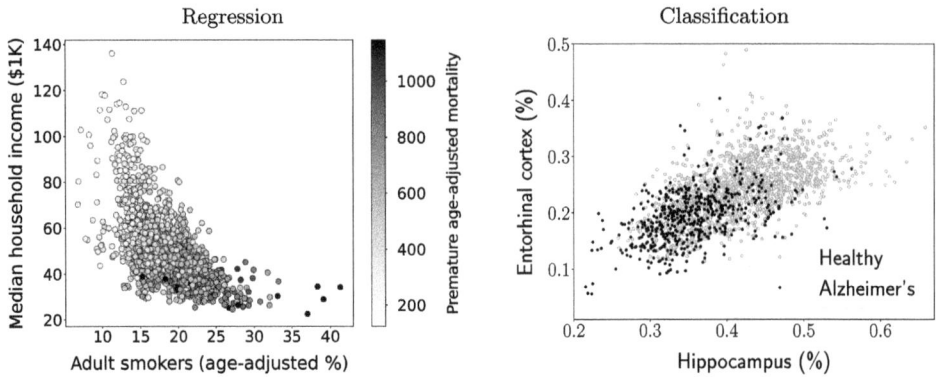

Figure 12.1 Regression and classification problems. The left scatterplot shows the median household income and the fraction of adult smokers (adjusted by age) for each county in the United States in 2019, extracted from Dataset 22. The response of interest is the premature mortality in each county (also adjusted by age), represented by the color of each marker. Since the response is a numerical quantity, estimating it from the features is a regression problem. The right scatterplot shows two features, extracted from magnetic-resonance scans in Dataset 12, which equal the normalized volume of their hippocampus and of their entorhinal cortex. The data are divided into two classes, depending on whether the corresponding subject suffers from Alzheimer's disease (black) or not (light gray). Estimating the class associated with a subject, based on the features, is a classification problem.

12.1.1 Linear Minimum Mean-Squared-Error Estimation

In linear regression, we assume that the dependence between the response and the features is linear. This section presents a derivation of the optimal linear estimator when the joint statistics of the response and the features are known. Let us represent the response by a random variable $\tilde{y}$ and the features as entries of a random vector $\tilde{x}$. A linear estimator approximates the response as a linear combination of the features,

$$\tilde{y} \approx \ell(\tilde{x}) := \sum_{j=1}^{d} \beta[j]\tilde{x}[j] + \alpha \tag{12.1}$$

$$= \beta^T \tilde{x} + \alpha, \tag{12.2}$$

where d is the number of features, and the vector of linear coefficients $\beta \in \mathbb{R}^d$ determines the contribution of each feature to the estimate. The additive constant or intercept $\alpha \in \mathbb{R}$ is needed when the mean of the response or the features is nonzero (see Theorem 12.2). Strictly speaking, this makes the estimator affine instead of linear, but such models are usually called *linear* nevertheless.

In order to fit a linear estimator, we need to find the linear coefficients β and the intercept α that best approximate the response. A popular fitting criterion is to minimize the mean squared error (MSE), introduced in Definition 7.29, between the response and the estimator. The corresponding estimator is known as the linear minimum MSE (MMSE) estimator.

To gain some geometric intuition about the linear MMSE estimator, we consider the case where the features and response are centered to have zero mean. In that case, the intercept is zero (see (12.16) in Theorem 12.2), so the linear MMSE estimator does not include an additive constant. As explained in Section 8.7, we can interpret the zero-mean random variables

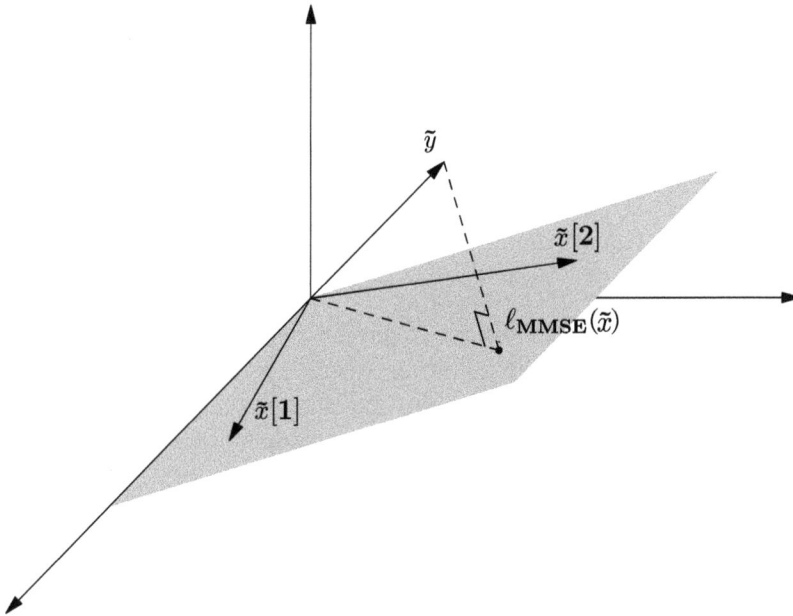

Figure 12.2 The linear MMSE estimator as an orthogonal projection. Following the geometric interpretation of zero-mean random variables as vectors from Section 8.7, we represent the response $\tilde{y}$ and the two features $\tilde{x}[1]$ and $\tilde{x}[2]$ in a regression problem as three-dimensional vectors. Any linear combination of the features must lie in the plane spanned by the corresponding vectors. In this vector space, the squared distance between vectors is the MSE between the corresponding random variables, so the linear MMSE estimator $\ell_{\mathrm{MMSE}}(\tilde{x})$ is the vector in the plane that is closest to $\tilde{y}$, i.e. the orthogonal projection of $\tilde{y}$ onto the plane.

representing the response and the features as vectors in a vector space, where the covariance is a valid inner product and the norm of a random variable is its standard deviation.

Within the vector space of zero-mean random variables, all possible linear estimators

$$\beta^T \tilde{x} = \sum_{j=1}^d \beta[j]\tilde{x}[j], \qquad \beta \in \mathbb{R}^d, \tag{12.3}$$

form a d-dimensional subspace or hyperplane of the vector space spanned by the random variables $\tilde{x}[1], \ldots, \tilde{x}[d]$ representing the features (see Figure 12.2). The squared distance between any element of this subspace and the response is the variance of the difference between the corresponding random variables, which is equal to the MSE of the linear estimator because all variables have zero mean,

$$\left\| \tilde{y} - \beta^T \tilde{x} \right\|^2 = \mathrm{Var}\left[\tilde{y} - \beta^T \tilde{x} \right] = \mathbb{E}\left[\left(\tilde{y} - \beta^T \tilde{x} \right)^2 \right]. \tag{12.4}$$

Therefore, the linear MMSE estimator $\ell_{\mathrm{MMSE}}(\tilde{x})$ of $\tilde{y}$ given $\tilde{x}$ is the vector in the subspace of linear estimators that is *closest* to the response, according to this distance. Equivalently, the linear MMSE estimator is the *orthogonal projection* of $\tilde{y}$ onto the subspace, so that its residual $\tilde{y} - \ell_{\mathrm{MMSE}}(\tilde{x})$ is orthogonal to the subspace of linear estimators, as depicted in Figure 12.2.

To see why the linear MMSE is indeed optimal, notice that the difference $\ell_{\mathrm{MMSE}}(\tilde{x}) - \beta^T \tilde{x}$ between $\ell_{\mathrm{MMSE}}(\tilde{x})$ and any other arbitrary linear estimator $\beta^T \tilde{x}$ is a vector that also belongs to the subspace of linear estimators (it is a linear combination of vectors in the subspace). This means that the residual of the linear MMSE estimator $\tilde{y} - \ell_{\mathrm{MMSE}}(\tilde{x})$ is orthogonal to that vector. As a result, the MSE of $\beta^T \tilde{x}$ cannot be smaller than that of $\ell_{\mathrm{MMSE}}(\tilde{x})$ for any $\beta \in \mathbb{R}^d$ because by the Pythagorean theorem,

$$\left|\left|\tilde{y} - \beta^T \tilde{x}\right|\right|^2 = \left|\left|\tilde{y} - \ell_{\mathrm{MMSE}}(\tilde{x}) + \ell_{\mathrm{MMSE}}(\tilde{x}) - \beta^T \tilde{x}\right|\right|^2 \tag{12.5}$$

$$= \left|\left|\tilde{y} - \ell_{\mathrm{MMSE}}(\tilde{x})\right|\right|^2 + \left|\left|\ell_{\mathrm{MMSE}}(\tilde{x}) - \beta^T \tilde{x}\right|\right|^2 \tag{12.6}$$

$$\geq \left|\left|\tilde{y} - \ell_{\mathrm{MMSE}}(\tilde{x})\right|\right|^2. \tag{12.7}$$

The orthogonality between the subspace of linear estimators and the residual of the linear MMSE estimator is captured by the following equation, which can be solved to derive the linear MMSE coefficients β_{MMSE}. Recall that the inner product is equal to the covariance, which is just the mean of the product for zero-mean random variables. For any $\beta \in \mathbb{R}^d$, by linearity of expectation (Theorem 11.8),

$$0 = \left\langle \beta^T \tilde{x}, \tilde{y} - \ell_{\mathrm{MMSE}}(\tilde{x}) \right\rangle \tag{12.8}$$

$$= \mathbb{E}\left[\beta^T \tilde{x} \left(\tilde{y} - \tilde{x}^T \beta_{\mathrm{MMSE}}\right)\right] \tag{12.9}$$

$$= \beta^T \left(\mathbb{E}\left[\tilde{x}\tilde{y}\right] - \mathbb{E}\left[\tilde{x}\tilde{x}^T\right] \beta_{\mathrm{MMSE}}\right). \tag{12.10}$$

Crucially, we do not need to know the whole joint distribution of the features and the response to compute the coefficients. The equation only involves the covariance matrix of the features $\Sigma_{\tilde{x}} := \mathbb{E}\left[\tilde{x}\tilde{x}^T\right]$ (see Definition 11.9) and the *cross-covariance* between features and the response.

Definition 12.1 (Cross-covariance). *The cross-covariance vector $\Sigma_{\tilde{x}\tilde{y}}$ between a d-dimensional random vector $\tilde{x}$ and a random variable $\tilde{y}$ is the d-dimensional vector with entries equal to the covariances (see Definition 8.4) between $\tilde{y}$ and the entries of $\tilde{x}$:*

$$\Sigma_{\tilde{x}\tilde{y}} := \mathbb{E}\left[\mathrm{ct}\left(\tilde{x}\right) \mathrm{ct}\left(\tilde{y}\right)\right] = \begin{bmatrix} \mathrm{Cov}\left[\tilde{x}[1], \tilde{y}\right] \\ \mathrm{Cov}\left[\tilde{x}[2], \tilde{y}\right] \\ \cdots \\ \mathrm{Cov}\left[\tilde{x}[d], \tilde{y}\right] \end{bmatrix}, \tag{12.11}$$

where $\mathrm{ct}\left(\tilde{x}\right) := \tilde{x} - \mathbb{E}\left[\tilde{x}\right]$ and $\mathrm{ct}\left(\tilde{y}\right) := \tilde{y} - \mathbb{E}\left[\tilde{y}\right]$ are obtained by centering $\tilde{x}$ and $\tilde{y}$ using their respective means.

Rewriting equation (12.10) in terms of the covariance matrix $\Sigma_{\tilde{x}}$ of the features and the cross-covariance $\Sigma_{\tilde{x}\tilde{y}}$ between the features and the response yields

$$\beta^T \left(\Sigma_{\tilde{x}\tilde{y}} - \Sigma_{\tilde{x}}\beta_{\mathrm{MMSE}}\right) = 0 \tag{12.12}$$

for any $\beta \in \mathbb{R}^d$. This can only hold if $\Sigma_{\tilde{x}\tilde{y}} = \Sigma_{\tilde{x}}\beta_{\mathrm{MMSE}}$, or equivalently $\beta_{\mathrm{MMSE}} = \Sigma_{\tilde{x}}^{-1}\Sigma_{\tilde{x}\tilde{y}}$, assuming the covariance matrix of the features is invertible. The following theorem establishes that these linear coefficients are indeed optimal and generalizes the result to features and responses with nonzero means. As promised, when $\tilde{x}$ and $\tilde{y}$ have zero mean, the intercept α_{MMSE} is zero.

Theorem 12.2 (Linear MMSE estimator). *Let $\tilde{x}$ and $\tilde{y}$ be a d-dimensional random vector and a random variable, representing the features and response in a regression problem. We* *denote the means of $\tilde{x}$ and $\tilde{y}$ by $\mu_{\tilde{x}}$ and $\mu_{\tilde{y}}$, respectively. If the covariance matrix $\Sigma_{\tilde{x}}$ of $\tilde{x}$ is invertible, the linear MMSE estimator of $\tilde{y}$ given $\tilde{x}$ is*

$$\ell_{\mathrm{MMSE}}(\tilde{x}) := \beta_{\mathrm{MMSE}}^T \tilde{x} + \alpha_{\mathrm{MMSE}} \tag{12.13}$$
$$= \Sigma_{\tilde{x}\tilde{y}}^T \Sigma_{\tilde{x}}^{-1} (\tilde{x} - \mu_{\tilde{x}}) + \mu_{\tilde{y}}, \tag{12.14}$$

where $\Sigma_{\tilde{x}\tilde{y}}$ is the cross-covariance between $\tilde{x}$ and $\tilde{y}$ (see Definition 12.1). The model parameters

$$\beta_{\mathrm{MMSE}} := \Sigma_{\tilde{x}}^{-1} \Sigma_{\tilde{x}\tilde{y}}, \tag{12.15}$$
$$\alpha_{\mathrm{MMSE}} := \mu_{\tilde{y}} - \beta_{\mathrm{MMSE}}^T \mu_{\tilde{x}} \tag{12.16}$$

are optimal, in the sense that they minimize the MSE cost function

$$(\beta_{\mathrm{MMSE}}, \alpha_{\mathrm{MMSE}}) = \arg \min_{\alpha \in \mathbb{R}, \beta \in \mathbb{R}^d} \mathbb{E}\left[\left(\tilde{y} - \beta^T \tilde{x} - \alpha\right)^2\right]. \tag{12.17}$$

Proof To prove the result, we generalize the argument in the proof of Theorem 8.14 to multiple features. We denote the MSE of an affine estimator with coefficients $\beta \in \mathbb{R}^d$ and $\alpha \in \mathbb{R}$ by

$$\mathrm{MSE}(\beta, \alpha) := \mathbb{E}\left[(\tilde{y} - \beta^T \tilde{x} - \alpha)^2\right]. \tag{12.18}$$

Let $\alpha^*(\beta)$ be the value of α that minimizes the MSE for a fixed $\beta \in \mathbb{R}^d$. Equivalently, $\alpha^*(\beta)$ is the best constant estimate of the random variable $\tilde{y} - \beta^T \tilde{x}$. By Theorem 7.30 and linearity of expectation (Theorem 11.8),

$$\alpha^*(\beta) = \arg \min_{\alpha \in \mathbb{R}} \mathrm{MSE}(\beta, \alpha) \tag{12.19}$$
$$= \mathbb{E}\left[\tilde{y} - \beta^T \tilde{x}\right] \tag{12.20}$$
$$= \mu_{\tilde{y}} - \beta^T \mu_{\tilde{x}}. \tag{12.21}$$

For any $\beta \in \mathbb{R}^d$ and any $\alpha \in \mathbb{R}$, $\mathrm{MSE}(\beta, \alpha) \geq \mathrm{MSE}(\beta, \alpha^*(\beta))$. To obtain the optimal coefficient vector β_{MMSE}, we can therefore set α equal to $\alpha^*(\beta)$ and minimize the resulting MSE,

$$\beta_{\mathrm{MMSE}} = \arg \min_{\beta \in \mathbb{R}^d} \mathrm{MSE}(\beta, \alpha^*(\beta)). \tag{12.22}$$

To alleviate notation, we denote the centered response and features by $\mathrm{ct}\,(\tilde{y}) := \tilde{y} - \mu_{\tilde{y}}$ and $\mathrm{ct}\,(\tilde{x}) := \tilde{x} - \mu_{\tilde{x}}$, respectively. By linearity of expectation, (12.21) and Definitions 11.9 and 12.1,

$$\mathrm{MSE}(\beta, \alpha^*(\beta)) = \mathbb{E}\left[(\tilde{y} - \beta^T \tilde{x} - \alpha^*(\beta))^2\right] \tag{12.23}$$
$$= \mathbb{E}\left[(\tilde{y} - \mu_{\tilde{y}} - \beta^T (\tilde{x} - \mu_{\tilde{x}}))^2\right] \tag{12.24}$$
$$= \mathbb{E}\left[(\mathrm{ct}\,(\tilde{y}) - \beta^T \mathrm{ct}\,(\tilde{x}))^2\right] \tag{12.25}$$
$$= \mathbb{E}\left[\mathrm{ct}\,(\tilde{y})^2\right] + \beta^T \mathbb{E}\left[\mathrm{ct}\,(\tilde{x})\,\mathrm{ct}\,(\tilde{x})^T\right]\beta - 2\beta^T \mathbb{E}\left[\mathrm{ct}\,(\tilde{x})\,\mathrm{ct}\,(\tilde{y})\right]$$
$$= \sigma_{\tilde{y}}^2 + \beta^T \Sigma_{\tilde{x}}\beta - 2\beta^T \Sigma_{\tilde{x}\tilde{y}}. \tag{12.26}$$

As a function of β, the MSE is a quadratic function. Its gradient and Hessian with respect to β equal

$$\nabla_\beta \mathrm{MSE}(\beta, \alpha^*(\beta)) = 2\Sigma_{\tilde{x}}\beta - 2\Sigma_{\tilde{x}\tilde{y}}, \tag{12.27}$$

$$\nabla_\beta^2 \mathrm{MSE}(\beta, \alpha^*(\beta)) = 2\Sigma_{\tilde{x}}. \tag{12.28}$$

Covariance matrices are positive semidefinite because by Theorem 11.11, for any vector $a \in \mathbb{R}^d$,

$$a^T \Sigma_{\tilde{x}} a = \mathrm{Var}\left[a^T \tilde{x}\right] \geq 0. \tag{12.29}$$

Since $\Sigma_{\tilde{x}}$ is invertible, there cannot be a nonzero vector such that $\Sigma_{\tilde{x}}a$ equals the zero vector, so the inequality is strict as long as a is not the zero vector. This means that the quadratic function is strictly convex. Consequently, we can set its gradient to zero to find the value of β that achieves the unique minimum:

$$\beta_{\mathrm{MMSE}} = \Sigma_{\tilde{x}}^{-1} \Sigma_{\tilde{x}\tilde{y}}. \tag{12.30}$$

By (12.21), the corresponding optimal intercept is

$$\alpha_{\mathrm{MMSE}} = \alpha^*(\beta_{\mathrm{MMSE}}) \tag{12.31}$$

$$= \mu_{\tilde{y}} - \beta_{\mathrm{MMSE}}^T \mu_{\tilde{x}}. \tag{12.32}$$

We conclude that the optimal linear estimator is

$$\ell_{\mathrm{MMSE}}(\tilde{x}) := \beta_{\mathrm{MMSE}}^T \tilde{x} + \alpha_{\mathrm{MMSE}} \tag{12.33}$$

$$= \Sigma_{\tilde{x}\tilde{y}}^T \Sigma_{\tilde{x}}^{-1} (\tilde{x} - \mu_{\tilde{x}}) + \mu_{\tilde{y}}. \tag{12.34}$$

∎

When there is a single feature ($d := 1$), the cross-covariance $\Sigma_{\tilde{x}\tilde{y}}$ is the covariance between the feature and the response, and the covariance matrix $\Sigma_{\tilde{x}}$ is just equal to the variance of the feature, so

$$\ell_{\mathrm{MMSE}}(\tilde{x}) = \frac{\mathrm{Cov}[\tilde{x}, \tilde{y}]}{\mathrm{Var}[\tilde{x}]} (\tilde{x} - \mu_{\tilde{x}}) + \mu_{\tilde{y}} \tag{12.35}$$

$$= \sigma_{\tilde{y}} \rho_{\tilde{x},\tilde{y}} \left(\frac{\tilde{x} - \mu_{\tilde{x}}}{\sigma_{\tilde{x}}}\right) + \mu_{\tilde{y}}, \tag{12.36}$$

where $\sigma_{\tilde{x}}$ and $\sigma_{\tilde{y}}$ are the standard deviations of $\tilde{x}$ and $\tilde{y}$, respectively, and $\rho_{\tilde{x},\tilde{y}}$ is the correlation coefficient between $\tilde{x}$ and $\tilde{y}$. Reassuringly, we recover the expression for the simple-linear-regression MMSE estimator derived in Theorem 8.14 (set $\tilde{a} := \tilde{x}$ and $\tilde{b} := \tilde{y}$).

Example 12.3 (Uncorrelated features). Consider a regression problem where the d features are all uncorrelated. For simplicity, we assume that the features and the response $\tilde{y}$ are centered, so their mean is zero. We denote the variance of each feature $\tilde{x}[i]$ by $\sigma_{\tilde{x}[i]}^2$ for $1 \leq i \leq d$, and the covariance between $\tilde{x}[i]$ and $\tilde{y}$ by $\sigma_{\tilde{x}[i],\tilde{y}}$. By Theorem 12.2, the linear MMSE estimator of the response $\tilde{y}$ given the d-dimensional feature vector $\tilde{x}$ equals

$$\ell_{\text{MMSE}}(\tilde{x}) = \Sigma_{\tilde{x}\tilde{y}}^T \Sigma_{\tilde{x}}^{-1} \tilde{x} \tag{12.37}$$

$$= \begin{bmatrix} \sigma_{\tilde{x}[1],\tilde{y}} \\ \sigma_{\tilde{x}[2],\tilde{y}} \\ \cdots \\ \sigma_{\tilde{x}[d],\tilde{y}} \end{bmatrix}^T \begin{bmatrix} \sigma_{\tilde{x}[1]}^2 & 0 & \cdots & 0 \\ 0 & \sigma_{\tilde{x}[2]}^2 & \cdots & 0 \\ \cdots & \cdots & \ddots & \cdots \\ 0 & 0 & \cdots & \sigma_{\tilde{x}[d]}^2 \end{bmatrix}^{-1} \tilde{x} \tag{12.38}$$

$$= \sum_{i=1}^{d} \frac{\sigma_{\tilde{x}[i],\tilde{y}}}{\sigma_{\tilde{x}[i]}^2} \tilde{x}[i] \tag{12.39}$$

$$= \sum_{i=1}^{d} \ell_{\text{MMSE}}(\tilde{x}[i]). \tag{12.40}$$

The estimator is simply the sum of the linear MMSE estimators of the response given each individual feature. Intuitively, since the features are uncorrelated, there is no need to correct for linear dependence between them. In terms of the geometric viewpoint in Section 8.7, the features are orthogonal in the vector space of zero-mean random variables endowed with the covariance inner product. Consequently, the projection of the response onto their span can be obtained by summing the individual projections onto the subspaces spanned by each feature.

When the features are correlated, the linear MMSE estimator is *not* just a sum of the individual linear MMSE estimators. The inverse covariance matrix $\Sigma_{\tilde{x}}^{-1}$ in (12.15) adjusts the linear coefficients to account for the inter-feature correlations.

....

Example 12.4 (Noise cancellation). We are interested in recording the voice of a pilot in a helicopter. To this end, we place a microphone inside her helmet and another microphone outside. We model the measurements as

$$\tilde{x}[1] = \tilde{y} + h\tilde{z}, \tag{12.41}$$
$$\tilde{x}[2] = h\tilde{y} + \tilde{z}, \tag{12.42}$$

where $\tilde{y}$ and $\tilde{z}$ are random variables representing the voice of the pilot and the noise in the helicopter, respectively. The constant $h \in [0, 1]$ models the effect of the helmet. We assume that $\tilde{y}$ and $\tilde{z}$ have zero mean and are uncorrelated with each other. The variances of $\tilde{y}$ and $\tilde{z}$ are equal to 1 and 100, respectively (the helicopter is much louder than the pilot).

Our goal is to estimate $\tilde{y}$ from $\tilde{x}$. By Theorem 12.2, the linear MMSE estimator of $\tilde{y}$ given $\tilde{x}$ equals

$$\ell_{\text{MMSE}}(\tilde{x}) = \Sigma_{\tilde{x}\tilde{y}}^T \Sigma_{\tilde{x}}^{-1} \tilde{x}. \tag{12.43}$$

Since $\tilde{y}$ and $\tilde{z}$ are uncorrelated, by Corollary 8.23 and Lemma 7.36,

$$\text{Var}[\tilde{x}[1]] = \text{Var}[\tilde{y} + h\tilde{z}] \tag{12.44}$$
$$= \text{Var}[\tilde{y}] + h^2\text{Var}[\tilde{z}] \tag{12.45}$$
$$= 1 + 100h^2, \tag{12.46}$$
$$\text{Var}[\tilde{x}[2]] = \text{Var}[h\tilde{y} + \tilde{z}] \tag{12.47}$$
$$= h^2\text{Var}[\tilde{y}] + \text{Var}[\tilde{z}] \tag{12.48}$$
$$= h^2 + 100. \tag{12.49}$$

In addition, by Definition 8.4 and linearity of expectation (Theorem 7.18),

$$\text{Cov}[\tilde{x}[1], \tilde{x}[2]] = \mathbb{E}\left[\tilde{x}[1]\tilde{x}[2]\right] \tag{12.50}$$

$$= \mathbb{E}\left[h\tilde{y}^2 + h\tilde{z}^2 + (1 + h^2)\tilde{y}\tilde{z}\right] \tag{12.51}$$

$$= h\mathbb{E}\left[\tilde{y}^2\right] + h\mathbb{E}\left[\tilde{z}^2\right] + (1 + h^2)\mathbb{E}\left[\tilde{y}\right]\mathbb{E}\left[\tilde{z}\right] \tag{12.52}$$

$$= 101h. \tag{12.53}$$

By Definition 11.9, the covariance matrix of the features therefore equals

$$\Sigma_{\tilde{x}} = \begin{bmatrix} 1 + 100h^2 & 101h \\ 101h & h^2 + 100 \end{bmatrix}. \tag{12.54}$$

Similarly,

$$\text{Cov}[\tilde{x}[1], \tilde{y}] = \mathbb{E}\left[\tilde{x}[1]\tilde{y}\right] \tag{12.55}$$

$$= \mathbb{E}\left[\tilde{y}^2 + h\tilde{y}\tilde{z}\right] \tag{12.56}$$

$$= \mathbb{E}\left[\tilde{y}^2\right] + h\mathbb{E}\left[\tilde{y}\right]\mathbb{E}\left[\tilde{z}\right] \tag{12.57}$$

$$= 1, \tag{12.58}$$

$$\text{Cov}[\tilde{x}[2], \tilde{y}] = \mathbb{E}\left[\tilde{x}[2]\tilde{y}\right] \tag{12.59}$$

$$= \mathbb{E}\left[h\tilde{y}^2 + \tilde{y}\tilde{z}\right] \tag{12.60}$$

$$= h\mathbb{E}\left[\tilde{y}^2\right] + \mathbb{E}\left[\tilde{y}\right]\mathbb{E}\left[\tilde{z}\right] \tag{12.61}$$

$$= h, \tag{12.62}$$

so by Definition 12.1, the cross-covariance equals

$$\Sigma_{\tilde{x}\tilde{y}} = \begin{bmatrix} 1 \\ h \end{bmatrix}. \tag{12.63}$$

Plugging everything into (12.43) yields

$$\ell_{\text{MMSE}}(\tilde{x}) = \begin{bmatrix} 1 & h \end{bmatrix} \begin{bmatrix} 1 + 100h^2 & 101h \\ 101h & h^2 + 100 \end{bmatrix}^{-1} \tilde{x} \tag{12.64}$$

$$= \begin{bmatrix} 1 & h \end{bmatrix} \frac{1}{100(1 - h^2)^2} \begin{bmatrix} h^2 + 100 & -101h \\ -101h & 1 + 100h^2 \end{bmatrix} \tilde{x} \tag{12.65}$$

$$= \frac{1}{100(1 - h^2)^2} \begin{bmatrix} 100(1 - h^2) & -100h(1 - h^2) \end{bmatrix} \tilde{x} \tag{12.66}$$

$$= \frac{\tilde{x}[1] - h\tilde{x}[2]}{1 - h^2}. \tag{12.67}$$

In order to evaluate the estimate, we express it in terms of $\tilde{y}$ and $\tilde{z}$:

$$\ell_{\text{MMSE}}(\tilde{x}) = \frac{\tilde{y} + h\tilde{z} - h(h\tilde{y} + \tilde{z})}{1 - h^2} \tag{12.68}$$

$$= \tilde{y}. \tag{12.69}$$

The estimator is exactly equal to the response, so its MSE is zero! The estimator is able to cancel out the noise completely by scaling the second measurement and subtracting it from the first one.

Linear estimation is optimal when the response and the features are jointly Gaussian, as established in the following theorem, which is a generalization of Theorem 8.16 to multiple features.

Theorem 12.5 (Linear estimation is optimal for Gaussian random vectors). *Let $\tilde{x}$ and $\tilde{y}$ be a d-dimensional random vector and a random variable, representing the features and response in a regression problem. We assume that $\tilde{x}$ and $\tilde{y}$ are jointly Gaussian, meaning that*

$$\begin{bmatrix} \tilde{x} \\ \tilde{y} \end{bmatrix} \tag{12.70}$$

is a Gaussian random vector with mean and covariance-matrix parameters

$$\mu := \begin{bmatrix} \mu_{\tilde{x}} \\ \mu_{\tilde{y}} \end{bmatrix}, \qquad \Sigma := \begin{bmatrix} \Sigma_{\tilde{x}} & \Sigma_{\tilde{x}\tilde{y}} \\ \Sigma_{\tilde{x}\tilde{y}}^T & \sigma_{\tilde{y}}^2 \end{bmatrix}, \tag{12.71}$$

where $\mu_{\tilde{x}}$ and $\mu_{\tilde{y}}$ are the means of $\tilde{x}$ and $\tilde{y}$, respectively, $\Sigma_{\tilde{x}}$ is the covariance matrix of $\tilde{x}$, $\sigma_{\tilde{y}}^2$ is the variance of $\tilde{y}$ and $\Sigma_{\tilde{x}\tilde{y}}$ is the cross-covariance between $\tilde{y}$ and $\tilde{x}$. Under these assumptions, the minimum MSE (MMSE) estimator of $\tilde{y}$ given $\tilde{x}$ is the linear MMSE estimator

$$\ell_{\mathrm{MMSE}}(\tilde{x}) := \beta_{\mathrm{MMSE}}^T \tilde{x} + \alpha_{\mathrm{MMSE}} \tag{12.72}$$

$$= \Sigma_{\tilde{x}\tilde{y}}^T \Sigma_{\tilde{x}}^{-1} (\tilde{x} - \mu_{\tilde{x}}) + \mu_{\tilde{y}}. \tag{12.73}$$

Proof By Theorem 5.23, the conditional distribution of $\tilde{y}$ given $\tilde{x} = x$ is Gaussian with mean parameter equal to

$$\mu_{\mathrm{cond}} = \Sigma_{\tilde{x}\tilde{y}}^T \Sigma_{\tilde{x}}^{-1} (x - \mu_{\tilde{x}}) + \mu_{\tilde{y}}. \tag{12.74}$$

By Definition 7.44 and Lemma 11.3, this is the conditional mean function $\mu_{\tilde{y}\,|\,\tilde{x}}$ of $\tilde{y}$ given $\tilde{x} = x$. Consequently, the linear MMSE estimator is equal to the conditional mean, obtained by plugging $\tilde{x}$ into $\mu_{\tilde{y}\,|\,\tilde{x}}$ (see Definition 7.51), which is the MMSE estimator of $\tilde{y}$ given $\tilde{x}$ by Theorem 7.59. ∎

In general, the linear MMSE estimator is not necessarily optimal because it cannot capture nonlinear dependence between the features and the response. This is illustrated by Example 8.18 and the temperature example in Section 12.7.1 (see Figure 12.26).

12.1.2 Ordinary Least Squares

In this section, we explain how to perform linear regression using data. Our strategy is to approximate the linear minimum MSE estimator derived in Theorem 12.2. The estimator only depends on the covariance matrix of the features, the cross-covariance between the features and the response, and the means of the features and the response. Estimating the cross-covariance from data is straightforward. Since each entry is equal to the covariance between a feature and the response, we approximate it using the corresponding sample covariance, introduced in Definition 8.9.

Definition 12.6 (Sample cross-covariance). *Consider a dataset containing n pairs (x_1, y_1), (x_2, y_2), ..., (x_n, y_n), each consisting of a response y_i and a corresponding d-dimensional* *feature vector x_i for $1 \leq i \leq n$. We denote the responses by $Y := \{y_1, y_2, \ldots, y_n\}$, the features by $X := \{x_1, x_2, \ldots, x_n\}$, and the entries corresponding to the jth feature by $X[j] := \{x_1[j], \ldots, x_n[j]\}$, for $1 \leq j \leq d$. The sample cross-covariance between the response and the features is*

$$\Sigma_{XY} := \frac{1}{n-1} \sum_{i=1}^{n} \text{ct}\,(x_i)\,\text{ct}\,(y_i) = \begin{bmatrix} c(X[1], Y) \\ c(X[2], Y) \\ \cdots \\ c(X[d], Y) \end{bmatrix}, \quad (12.75)$$

where $\text{ct}\,(x_i) := x_i - m(X)$ and $\text{ct}\,(y_i) - m(Y)$ are the result of centering the features and the response using the corresponding sample means $m(X)$ and $m(Y)$ for $1 \leq i \leq n$, and $c(X[j], Y)$ is the sample covariance of $X[j]$ and Y for $1 \leq j \leq d$.

To estimate the linear minimum MSE estimator defined in (12.14), we replace the means of the response and features with their respective sample means $m(Y)$ and $m(X)$ (see Definition 7.15), the covariance matrix of the features with the sample covariance matrix Σ_X (see Definition 11.14), and the cross-covariance of the features and the response with the sample cross-covariance Σ_{XY}:

$$\ell_{\text{MMSE}}(x) \approx \ell_{\text{OLS}}(x) := \Sigma_{XY}^T \Sigma_X^{-1} (x - m(X)) + m(Y), \quad (12.76)$$

where x is a feature vector. This is known as the *ordinary-least-squares* (OLS) estimator because it minimizes the sum of squared errors with respect to the observed response, as established in the following theorem.

Theorem 12.7 (Linear regression via ordinary least squares). *Given a training dataset formed by n pairs of a response y_i and a corresponding vector x_i containing d features ($1 \leq i \leq n$), we denote the responses by $Y := \{y_1, y_2, \ldots, y_n\}$ and the features by $X := \{x_1, x_2, \ldots, x_n\}$. If the sample covariance matrix Σ_X of the features is invertible, the ordinary least-squares estimator (OLS) of the response given the features is*

$$\ell_{\text{OLS}}(x) := \beta_{\text{OLS}}^T x + \alpha_{\text{OLS}}, \quad (12.77)$$

$$\beta_{\text{OLS}} := \Sigma_X^{-1} \Sigma_{XY}, \quad (12.78)$$

$$\alpha_{\text{OLS}} := m(Y) - \beta_{\text{OLS}}^T m(X) \quad (12.79)$$

for any feature vector $x \in \mathbb{R}^d$, where Σ_{XY} is the sample cross-covariance of X and Y, and $m(X)$ and $m(Y)$ are the sample means of X and Y.

The OLS estimator produces an optimal affine estimate, in the sense that it minimizes the residual sum of squares over the dataset,

$$(\beta_{\text{OLS}}, \alpha_{\text{OLS}}) = \arg \min_{\alpha \in \mathbb{R}, \beta \in \mathbb{R}^d} \sum_{i=1}^{n} \left(y_i - \beta^T x_i - \alpha\right)^2. \quad (12.80)$$

Proof The result can be established by generalizing the proof of Theorem 8.17 to multiple features, following a similar argument to the proof of Theorem 12.2 (see Exercise 12.4). ∎

For large-scale regression problems with many features, it may be computationally expensive to obtain the OLS estimator using the formula in Theorem 12.7, due to the cost of inverting the sample covariance matrix of the features. In such cases, iterative optimization methods, such as conjugate gradients (Shewchuk *et al.*, 1994), can be applied to directly minimize the residual sum of squares.

It is often convenient to express the OLS estimator in terms of a *design matrix*, containing the features, and a vector representing the response. We denote the design matrix by X_{train} and the response vector by y_{train}, to emphasize that they contain the data used to *train* the linear model.

Lemma 12.8 (OLS estimator in matrix-vector form). *Let (x_1, y_1), (x_2, y_2), ..., (x_n, y_n) be a dataset formed by n pairs of a response $y_i \in \mathbb{R}$ and a corresponding vector $x_i \in \mathbb{R}^d$ containing d features (where $1 \le i \le n$). We arrange the feature vectors to form the $n \times d$ design matrix*

$$X_{\text{train}} := \begin{bmatrix} x_1^T \\ x_2^T \\ \dots \\ x_n^T \end{bmatrix} \tag{12.81}$$

and the response values to form the n-dimensional response vector

$$y_{\text{train}} = \begin{bmatrix} y_1 \\ y_2 \\ \dots \\ y_n \end{bmatrix}. \tag{12.82}$$

If the features and responses are centered to have zero sample mean, then the OLS estimator of the response given the features equals

$$\beta_{\text{OLS}} = \arg \min_{\beta \in \mathbb{R}^d} ||X_{\text{train}}\beta - y_{\text{train}}||_2^2 \tag{12.83}$$

$$= \left(X_{\text{train}}^T X_{\text{train}} \right)^{-1} X_{\text{train}}^T y_{\text{train}}. \tag{12.84}$$

Proof By definition of the design matrix,

$$X_{\text{train}}[i, j] := x_i[j]. \tag{12.85}$$

Consequently, if the features have zero sample mean, by Definition 11.14, each entry of the sample covariance matrix of the features equals

$$\Sigma_X[l, k] = \frac{1}{n-1} \sum_{i=1}^{n} x_i[l]x_i[k]^T \tag{12.86}$$

$$= \frac{1}{n-1} \sum_{i=1}^{n} X_{\text{train}}^T[l, i]X_{\text{train}}[i, k] \tag{12.87}$$

$$= \frac{1}{n-1} \left(X_{\text{train}}^T X_{\text{train}} \right)[l, k], \qquad 1 \le l, k \le d, \tag{12.88}$$

so

$$\Sigma_X = \frac{1}{n-1} X_{\text{train}}^T X_{\text{train}}. \tag{12.89}$$

The inverse of the sample covariance matrix is therefore $(n-1)\left(X_{\text{train}}^T X_{\text{train}}\right)^{-1}$. If the sample mean of the response values is also zero, by Definition 12.6, the sample cross-covariance between the features and the response equals

$$\Sigma_{XY} = \frac{1}{n-1} \sum_{i=1}^{n} x_i y_i \tag{12.90}$$

$$= \frac{1}{n-1} X_{\text{train}}^T y_{\text{train}}. \tag{12.91}$$

The result then follows from Theorem 12.7.   ■

Example 12.9 (Premature mortality: OLS model)**.** We consider the problem of estimating premature mortality in 2,091 United States counties based on tobacco use and income, using data extracted from Dataset 22. Our response of interest is the number of deaths among residents under age 75 per 100,000 people, which is a common measure of premature mortality. The features are the percentage of adult smokers and the median annual household income in each county. Both premature mortality and tobacco use depend highly on the age distribution of the population (older people are more likely to die and also more likely to smoke), so the corresponding data are adjusted to correct for age.

The sample mean of the premature-mortality measure is 408 and the sample standard deviation is 116. The sample mean and sample covariance matrix of tobacco use (first entry) and income (second entry) are

$$m(X) = \begin{bmatrix} 18.0 \\ 50.9 \end{bmatrix}, \qquad \Sigma_X = \begin{bmatrix} 13.6 & -30.6 \\ -30.6 & 190 \end{bmatrix}. \tag{12.92}$$

The sample cross-covariance between the response and the two features is

$$\Sigma_{XY} = \begin{bmatrix} 306 \\ -1057 \end{bmatrix}. \tag{12.93}$$

By Theorem 12.7, the coefficients and additive constant of the OLS estimator of premature mortality given tobacco use and income are

$$\beta_{\text{OLS}} = \Sigma_X^{-1} \Sigma_{XY} = \begin{bmatrix} 15.7 \\ -3.04 \end{bmatrix}, \tag{12.94}$$

$$\alpha_{\text{OLS}} = m(Y) - \Sigma_{XY}^T \Sigma_X^{-1} m(X) = 281, \tag{12.95}$$

which yields the OLS model

$$\ell_{\text{OLS}}\left(x_{\text{tobacco}}, x_{\text{income}}\right) = 15.7\, x_{\text{tobacco}} - 3.04\, x_{\text{income}} + 281, \tag{12.96}$$

where x_{tobacco} represents tobacco use and x_{income} represents household income. According to the model, counties with higher tobacco use and lower household income have higher premature mortality. Figure 12.3 provides two different visualizations of the model.
· ·

The linear coefficient associated with each feature in a linear regression model represents the rate of change of the response as a function of the feature, *when the rest of the features are fixed.* For instance, in Example 12.9, if the annual household income in a county is unchanged, the model estimates 15.7 more premature deaths (per 100,000 people) for every additional 1% of adult smokers. Conversely, if the percentage of smokers is fixed, the model estimates 3.04 less premature deaths for every additional $1,000 of median annual household income.

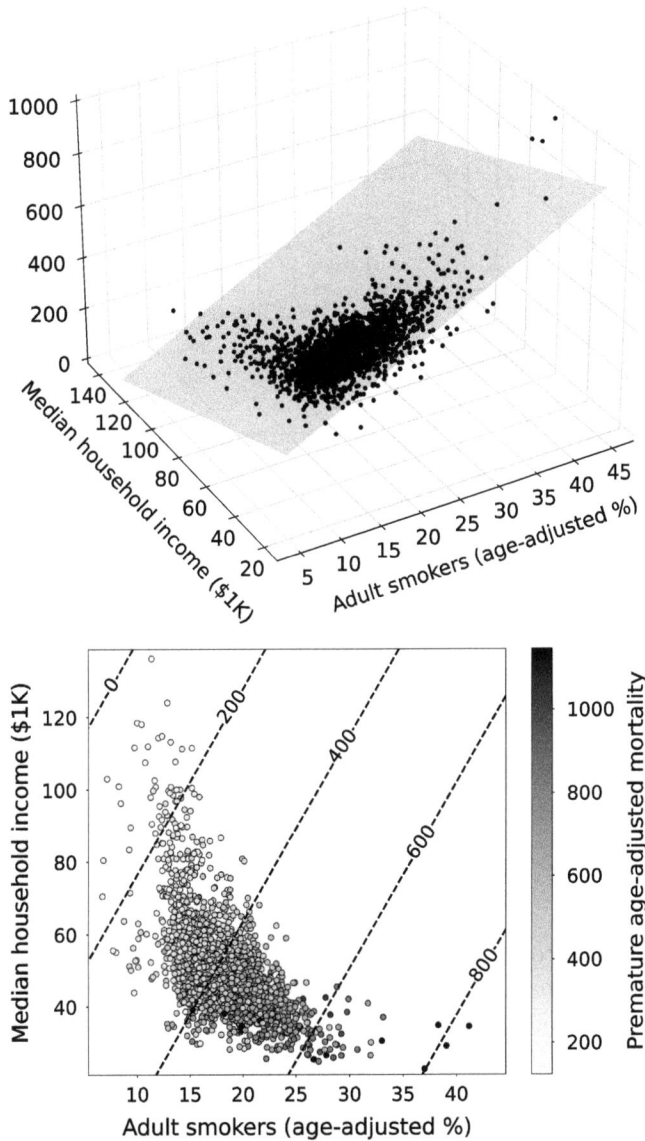

Figure 12.3 Linear model for premature mortality. The figure provides two visualizations of the linear-regression OLS model described in Example 12.9, which estimates premature mortality from tobacco use and income in United States counties. On the top, the data are represented in 3D, with the features as the horizontal coordinates and the response as the vertical coordinate. The OLS model is the plane that minimizes the sum of squared vertical distances to the data. On the bottom, a scatterplot of the data, color coded to show the corresponding response, is superposed onto the contour lines of the OLS model.

12.1.3 Explained Variance

In this section, we present a decomposition of variance for linear regression and explain how to use it to evaluate linear models. The key insight behind the decomposition is that, as in simple linear regression (see Section 8.5.3), the linear minimum MSE (MMSE) estimator is uncorrelated with its residual. This enables us to decompose the variance of the response as the sum of the variances of the estimator and the residual.

Theorem 12.10 (Linear MMSE estimator and residual are uncorrelated). *Let $\tilde{x}$ and $\tilde{y}$ be a d-dimensional random vector and a random variable, representing the features and response in a regression problem. The linear MMSE estimator $\ell_{\mathrm{MMSE}}(\tilde{x})$ of $\tilde{y}$ given $\tilde{x}$ (see Theorem 12.2) is uncorrelated with the residual $\tilde{y} - \ell_{\mathrm{MMSE}}(\tilde{x})$.*

Proof By linearity of expectation (Theorem 11.8), centering the linear MMSE estimator yields

$$\mathrm{ct}\left(\ell_{\mathrm{MMSE}}(\tilde{x})\right) = \beta_{\mathrm{MMSE}}^{T}\tilde{x} + \alpha_{\mathrm{MMSE}} - \mathbb{E}\left[\beta_{\mathrm{MMSE}}^{T}\tilde{x} + \alpha_{\mathrm{MMSE}}\right] \tag{12.97}$$

$$= \beta_{\mathrm{MMSE}}^{T}\tilde{x} + \alpha_{\mathrm{MMSE}} - \beta_{\mathrm{MMSE}}^{T}\mathbb{E}\left[\tilde{x}\right] - \alpha_{\mathrm{MMSE}} \tag{12.98}$$

$$= \beta_{\mathrm{MMSE}}^{T}\,\mathrm{ct}\left(\tilde{x}\right), \tag{12.99}$$

where $\mathrm{ct}\,(\tilde{x}) := \tilde{x} - \mathbb{E}\,[\tilde{x}]$ is the centered feature vector. Similarly, we denote the centered response by $\mathrm{ct}\,(\tilde{y}) := \tilde{y} - \mathbb{E}\,[\tilde{y}]$. By Definitions 8.4, 12.1, and 11.9, and linearity of expectation,

$$\mathrm{Cov}\left[\ell_{\mathrm{MMSE}}\left(\tilde{x}\right), \tilde{y} - \ell_{\mathrm{MMSE}}\left(\tilde{x}\right)\right] \tag{12.100}$$

$$= \mathbb{E}\left[\mathrm{ct}\left(\ell_{\mathrm{MMSE}}\left(\tilde{x}\right)\right)\mathrm{ct}\left(\tilde{y} - \ell_{\mathrm{MMSE}}\left(\tilde{x}\right)\right)\right] \tag{12.101}$$

$$= \mathbb{E}\left[\beta_{\mathrm{MMSE}}^{T}\,\mathrm{ct}\left(\tilde{x}\right)\left(\mathrm{ct}\left(\tilde{y}\right) - \mathrm{ct}\left(\tilde{x}\right)^{T}\beta_{\mathrm{MMSE}}\right)\right] \tag{12.102}$$

$$= \beta_{\mathrm{MMSE}}^{T}\mathbb{E}\left[\mathrm{ct}\left(\tilde{x}\right)\mathrm{ct}\left(\tilde{y}\right)\right] - \beta_{\mathrm{MMSE}}^{T}\mathbb{E}\left[\mathrm{ct}\left(\tilde{x}\right)\mathrm{ct}\left(\tilde{x}\right)^{T}\right]\beta_{\mathrm{MMSE}} \tag{12.103}$$

$$= \Sigma_{\tilde{x}\tilde{y}}^{T}\Sigma_{\tilde{x}}^{-1}\Sigma_{\tilde{x}\tilde{y}} - \Sigma_{\tilde{x}\tilde{y}}^{T}\Sigma_{\tilde{x}}^{-1}\Sigma_{\tilde{x}}\Sigma_{\tilde{x}}^{-1}\Sigma_{\tilde{x}\tilde{y}} \tag{12.104}$$

$$= \Sigma_{\tilde{x}\tilde{y}}^{T}\Sigma_{\tilde{x}}^{-1}\Sigma_{\tilde{x}\tilde{y}} - \Sigma_{\tilde{x}\tilde{y}}^{T}\Sigma_{\tilde{x}}^{-1}\Sigma_{\tilde{x}\tilde{y}} \tag{12.105}$$

$$= 0, \tag{12.106}$$

where we have plugged in the coefficients of the linear MMSE estimator derived in Theorem 12.2. ∎

Theorem 12.11 (Decomposition of variance). *Let $\tilde{x}$ and $\tilde{y}$ be a d-dimensional random vector and a random variable, representing the features and response in a regression problem. The variance of $\tilde{y}$ can be decomposed into the sum of the variance of the linear MMSE estimator $\ell_{\mathrm{MMSE}}(\tilde{x})$ of $\tilde{y}$ given $\tilde{x}$ and the MSE incurred by the estimator,*

$$\mathrm{Var}\left[\tilde{y}\right] = \mathrm{Var}\left[\ell_{\mathrm{MMSE}}(\tilde{x})\right] + \mathrm{MSE}, \qquad \mathrm{MSE} := \mathbb{E}\left[\left(\tilde{y} - \ell_{\mathrm{MMSE}}(\tilde{x})\right)^{2}\right]. \tag{12.107}$$

Proof By Theorem 12.10, the linear MMSE estimator $\ell_{\mathrm{MMSE}}(\tilde{x})$ and the residual $\tilde{y} - \ell_{\mathrm{MMSE}}(\tilde{x})$ are uncorrelated. Consequently, by Corollary 8.23,

$$\mathrm{Var}\left[\tilde{y}\right] = \mathrm{Var}\left[\ell_{\mathrm{MMSE}}(\tilde{x}) + \tilde{y} - \ell_{\mathrm{MMSE}}(\tilde{x})\right] \tag{12.108}$$

$$= \mathrm{Var}\left[\ell_{\mathrm{MMSE}}(\tilde{x})\right] + \mathrm{Var}\left[\tilde{y} - \ell_{\mathrm{MMSE}}(\tilde{x})\right]. \tag{12.109}$$

The variance of the residual is equal to the MSE by Lemma 7.33 because the mean of the residual is zero. This follows from the expression for the intercept $\alpha_{\mathrm{MMSE}} = \mathbb{E}\,[\tilde{y}] - \beta_{\mathrm{MMSE}}^{T}\mathbb{E}\,[\tilde{x}]$, derived in Theorem 12.2, and linearity of expectation (Theorem 11.8):

$$\mathbb{E}\left[\tilde{y} - \ell_{\mathrm{MMSE}}\left(\tilde{x}\right)\right] = \mathbb{E}\left[\tilde{y} - \beta_{\mathrm{MMSE}}^{T}\tilde{x} - \alpha_{\mathrm{MMSE}}\right] \tag{12.110}$$

$$= \mathbb{E}\left[\tilde{y}\right] - \beta_{\mathrm{MMSE}}^{T}\mathbb{E}\left[\tilde{x}\right] - \alpha_{\mathrm{MMSE}} \tag{12.111}$$

$$= 0. \tag{12.112}$$

∎

The decomposition of variance in Theorem 12.11 has an intuitive geometric explanation, if we interpret the features and response as vectors in the vector space of zero-mean random variables defined in Section 8.7.2 (assuming they are centered). As discussed in Section 12.1.1 and depicted in Figure 12.2, the residual $\tilde{y} - \ell_{\mathrm{MMSE}}(\tilde{x})$ of the linear MMSE estimator $\ell_{\mathrm{MMSE}}(\tilde{x})$ is orthogonal to the subspace spanned by the features, and consequently also to $\ell_{\mathrm{MMSE}}(\tilde{x})$ itself. The sum of $\ell_{\mathrm{MMSE}}(\tilde{x})$ and $\tilde{y} - \ell_{\mathrm{MMSE}}(\tilde{x})$ is equal to the response $\tilde{y}$, so by the Pythagorean theorem, the squared length of $\tilde{y}$ is the sum of the squared lengths of the estimator and the residual. The squared length in this space is equal to the variance because it is the square of the norm induced by the covariance inner product, as explained in Section 8.7.2. Consequently, the variance of $\tilde{y}$ is the sum of the variance of the estimator and the MSE, which equals the variance of the residual.

A useful metric to evaluate the linear MMSE estimator is the fraction of variance *explained* by the estimator. This metric is called the coefficient of determination R^2, as in simple linear regression (see Definition 8.26).

Definition 12.12 (Coefficient of determination). *Let $\tilde{x}$ and $\tilde{y}$ be a d-dimensional random vector and a random variable, representing the features and response in a regression problem. The coefficient of determination is the ratio between the variance of the linear MMSE estimator $\ell_{\mathrm{MMSE}}(\tilde{x})$ of $\tilde{y}$ given $\tilde{x}$ and the variance of $\tilde{y}$,*

$$R^2 := \frac{\mathrm{Var}\left[\ell_{\mathrm{MMSE}}(\tilde{x})\right]}{\mathrm{Var}\left[\tilde{y}\right]}. \tag{12.113}$$

As in the case of simple linear regression (see Theorem 8.27), the coefficient of determination can be expressed in terms of the MSE and is bounded between zero and one. When it equals one, the MSE is zero, so the linear MMSE estimator is perfect. When it equals zero, the MSE is equal to the variance of the response, so the linear MMSE estimator does not explain any variance in the response.

Theorem 12.13 (Properties of the coefficient of determination). *Let $\tilde{x}$ and $\tilde{y}$ be a d-dimensional random vector and a random variable, representing the features and response in a regression problem. The coefficient of determination equals*

$$R^2 := 1 - \frac{\mathrm{MSE}}{\mathrm{Var}\left[\tilde{y}\right]}, \qquad \mathrm{MSE} := \mathbb{E}\left[(\tilde{y} - \ell_{\mathrm{MMSE}}(\tilde{x}))^2\right]. \tag{12.114}$$

R^2 is always bounded between zero and one:

$$0 \leq R^2 \leq 1. \tag{12.115}$$

Proof By Theorem 12.11,

$$\mathrm{Var}\left[\tilde{y}\right] = \mathrm{Var}\left[\ell_{\mathrm{MMSE}}(\tilde{x})\right] + \mathrm{MSE}, \tag{12.116}$$

which directly implies

$$R^2 := \frac{\mathrm{Var}\left[\ell_{\mathrm{MMSE}}(\tilde{x})\right]}{\mathrm{Var}\left[\tilde{y}\right]} \tag{12.117}$$

$$= \frac{\mathrm{Var}\left[\tilde{y}\right] - \mathrm{MSE}}{\mathrm{Var}\left[\tilde{y}\right]} \tag{12.118}$$

$$= 1 - \frac{\mathrm{MSE}}{\mathrm{Var}\left[\tilde{y}\right]}. \tag{12.119}$$

By Definition 12.12, R^2 is a ratio of two nonnegative quantities, so it is nonnegative. By (12.116), $\text{Var}\left[\ell_{\text{MMSE}}(\tilde{x})\right] \leq \text{Var}\left[\tilde{y}\right]$ (because the MSE cannot be negative), so $R^2 \leq 1$. ∎

The following example uses the coefficient of determination to compare different linear regression models.

Example 12.14 (Premature mortality: Model evaluation). In Example 12.9, we build a linear model to estimate premature mortality using tobacco use and income. The sample variance of the response is $1.35 \cdot 10^4$. The sample variance of the model estimate is $8.00 \cdot 10^3$. The coefficient of determination R^2 is the ratio between these sample variances, which equals 0.59. The linear model explains 59% of the variance in the data.

We can use the coefficient of determination to compare our model to OLS models based on a single feature,

$$\ell_{\text{OLS}}\left(x_{\text{tobacco}}\right) = 22.5 \, x_{\text{tobacco}} + 2, \tag{12.120}$$

$$\ell_{\text{OLS}}\left(x_{\text{income}}\right) = -5.57 \, x_{\text{income}} + 692. \tag{12.121}$$

As depicted in Figure 12.4, the coefficient of determination for the tobacco model ($R^2 = 0.51$) is larger than for the income model ($R^2 = 0.44$). The linear association of premature mortality with tobacco is therefore stronger than with income. Notice that the fraction of variance explained by the two-feature model is only 8% larger than that of the tobacco model. The reason is that the two features are very correlated (their sample correlation coefficient is -0.6).

....

12.1.4 Estimation of Linear Causal Effects

As explained in Section 8.8, confounders make it challenging to estimate linear causal effects from data. Here, we show that under certain assumptions, fitting a linear regression model

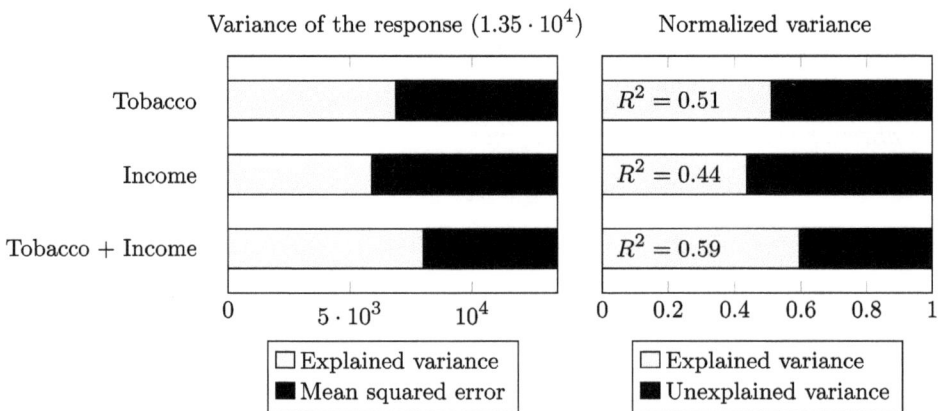

Figure 12.4 Explained variance and coefficient of determination. The left graph depicts the variance decomposition in Theorem 12.11 for the three linear models described in Example 12.14. For each model, the variance of the response is equal to the sum of the variance of the OLS estimator and the MSE. The right graph shows the result of normalizing the decomposition to obtain the corresponding coefficients of determination following Definition 12.12.

automatically adjusts for confounders, *as long as we include them in the model*. We consider the same scenario as in Theorem 8.37, where the potential outcome $\widetilde{\text{po}}_t$ is a linear combination of the treatment t, a confounder $\tilde{c}$, and an additional nonconfounding term $\tilde{z}$:

$$\widetilde{\text{po}}_t := \beta_{\text{treat}}t + \beta_{\text{conf}}\tilde{c} + \tilde{z}. \tag{12.122}$$

Theorem 8.37 establishes that the covariance between the treatment $\tilde{t}$ and the observed outcome

$$\tilde{y} = \beta_{\text{treat}}\tilde{t} + \beta_{\text{conf}}\tilde{c} + \tilde{z} \tag{12.123}$$

is not a reliable estimator of the causal-effect coefficient β_{treat} because of the presence of the confounder. In order to adjust for the confounder, we model its influence on the observed outcome explicitly, by fitting a linear regression model, where the response is the observed outcome and the features are the treatment and the confounder. The following theorem shows that the coefficient corresponding to the treatment in the resulting model recovers the true linear causal-effect coefficient β_{treat}.

Theorem 12.15 (Estimation of linear causal effects via linear regression). *Let $\widetilde{\text{po}}_t$ denote a potential outcome associated with a certain treatment $\tilde{t}$, given by*

$$\widetilde{\text{po}}_t := \beta_{\text{treat}}t + \beta_{\text{conf}}\tilde{c} + \tilde{z}, \tag{12.124}$$

where β_{treat} and β_{conf} are real-valued constants, and $\tilde{c}$ and $\tilde{z}$ are random variables. The observed outcome $\tilde{y}$ equals $\widetilde{\text{po}}_t$, if the treatment $\tilde{t}$ equals t, so

$$\tilde{y} = \beta_{\text{treat}}\tilde{t} + \beta_{\text{conf}}\tilde{c} + \tilde{z}. \tag{12.125}$$

We assume that $\tilde{t}$ and $\tilde{c}$ are centered, so that their mean is zero. If $\tilde{z}$ is independent from $\tilde{t}$ and $\tilde{c}$, then the coefficients of the linear MMSE estimator of the observed outcome $\tilde{y}$ given $\tilde{t}$ and $\tilde{c}$ equal

$$\beta_{\text{MMSE}} = \begin{bmatrix} \beta_{\text{treat}} \\ \beta_{\text{conf}} \end{bmatrix}. \tag{12.126}$$

Proof The features and the response satisfy the assumptions of Theorem 12.18, which derives the linear MMSE coefficients when the response is a linear function of the features. Consequently, setting

$$\tilde{x} := \begin{bmatrix} \tilde{t} \\ \tilde{c} \end{bmatrix}, \qquad \beta_{\text{true}} := \begin{bmatrix} \beta_{\text{treat}} \\ \beta_{\text{conf}} \end{bmatrix} \tag{12.127}$$

in the theorem proves the desired result. ∎

By Theorem 12.15, we can automatically adjust for a confounder by fitting a *long* regression model, which includes the confounder as a feature (as long as the assumptions in the theorem hold). By contrast, a *short* regression model, without the confounder, does *not* reveal the correct average causal effect, as established in Theorem 8.37.

Example 12.16 (Guinea-pig rescue: Short vs long regression). In Example 8.38, the potential outcome $\widetilde{\text{po}}_t$ associated with the weight gain of the guinea pigs does not depend on the treatment t, which represents an ineffective nutritional supplement. Instead, $\widetilde{\text{po}}_t$ is a function of a confounder (the food intake) and a non-confounding variable $\tilde{z}$, uncorrelated with the treatment:

$$\widetilde{\text{po}}_t := \tilde{c} + \tilde{z}. \tag{12.128}$$

If we fit a simple-linear-regression model of the observed weight gain $\tilde{y}$ given the treatment $\tilde{t}$, the linear coefficient β_{short} is equal to the covariance between $\tilde{y}$ and $\tilde{t}$, as established in Theorem 8.14 (setting $\tilde{b} := \tilde{y}$ and $\tilde{a} := \tilde{t}$). By Theorem 8.37, the covariance, and hence the linear-regression coefficient, are equal to the covariance $\sigma_{\tilde{t},\tilde{c}}$ between $\tilde{c}$ and $\tilde{t}$,

$$\beta_{\text{short}} = \text{Cov}\left[\tilde{y}, \tilde{t}\right] = \sigma_{\tilde{t},\tilde{c}}, \tag{12.129}$$

because in this case $\beta_{\text{treat}} := 0$ and $\beta_{\text{conf}} := 1$. Consequently, the coefficient in the short regression model is completely distorted by the correlation between the confounder and the treatment. This is apparent in the left column of Figure 12.5: The slope of the linear model changes dramatically for different values of $\sigma_{\tilde{t},\tilde{c}}$.

By contrast, a long regression model, which includes the confounder $\tilde{c}$ as an additional feature, correctly identifies that there is no true causal effect associated with the nutritional supplement. By Theorem 12.15 the linear coefficients of the long regression model equal

$$\beta_{\text{long}} = \begin{bmatrix} \beta_{\text{treat}} \\ \beta_{\text{conf}} \end{bmatrix} = \begin{bmatrix} 0 \\ 1 \end{bmatrix}, \tag{12.130}$$

where the first entry corresponds to the treatment, and the second to the confounder. Consequently the coefficient associated with the treatment is zero *regardless of the covariance between the confounder and the treatment*. The right column of Figure 12.5 confirms this numerically.
....

Example 12.17 (Unemployment in Spain). We consider the data in Figure 8.12, which shows that unemployment in Spain is negatively correlated with temperature. As explained in Section 8.8, we suspect that this correlation is due to a confounder: The number of tourists visiting Spain. Inspired by Theorem 12.15, we fit a long regression model where the response is unemployment and the features are the temperature and the number of tourists. Figure 12.6 compares this linear model to a short linear regression model where the only feature is the temperature. The coefficient corresponding to temperature in the long regression model is very small (and positive, instead of negative). This suggests that, if we adjust for the number of tourists, there is no linear causal effect of temperature on unemployment.

To further analyze the linear relationship between temperature and unemployment, we compute the coefficient of determination (see Definition 12.12) of the short regression model ($R^2 = 0.042$), the long regression model ($R^2 = 0.12026$), and a simple linear regression model, where the only feature is the number of tourists ($R^2 = 0.12024$). The coefficients of determination reveal that the number of tourists explains much more variance than the temperature. More importantly, if we incorporate temperature to the model that only relies on the number of tourists, the increase in explained variance is minuscule ($2 \cdot 10^{-5}$). This is strong evidence that the observed linear dependence between unemployment and temperature is due to the confounder.
....

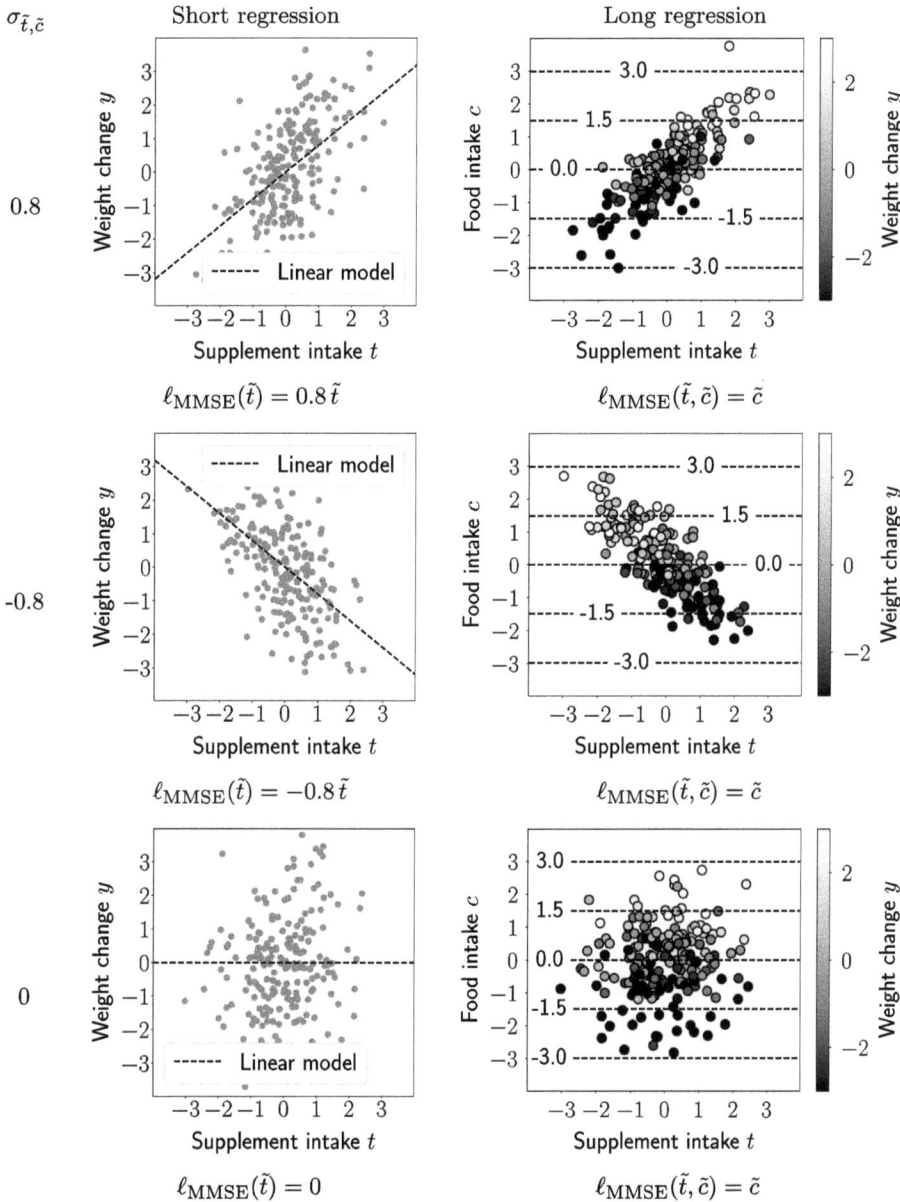

Figure 12.5 Adjusting for a confounder via long regression. The left column shows scatterplots of 200 simulated samples, corresponding to the supplement intake $\tilde{t}$ and the weight gain $\tilde{y}$ of 200 guinea pigs in Example 8.38, assuming that $\tilde{t}$ and the confounder $\tilde{c}$ are jointly Gaussian, and that the variance of $\tilde{z}$ equals one. The dashed line is the *short* linear-regression estimator of $\tilde{y}$ given $\tilde{t}$, which depends on the covariance $\sigma_{\tilde{t},\tilde{c}}$ between $\tilde{t}$ and $\tilde{c}$, as established in Theorem 8.37. The right column shows scatterplots of $\tilde{t}$ and $\tilde{c}$. The marker representing each sample is color coded to indicate the corresponding value of $\tilde{y}$. The dashed lines show the contour lines of the *long* linear-regression estimator of $\tilde{y}$ given $\tilde{t}$ and $\tilde{c}$. In accordance with Theorem 12.15, the coefficient corresponding to the treatment is zero, regardless of $\sigma_{\tilde{t},\tilde{c}}$, which correctly reveals that the supplement is useless.

Short regression   Long regression

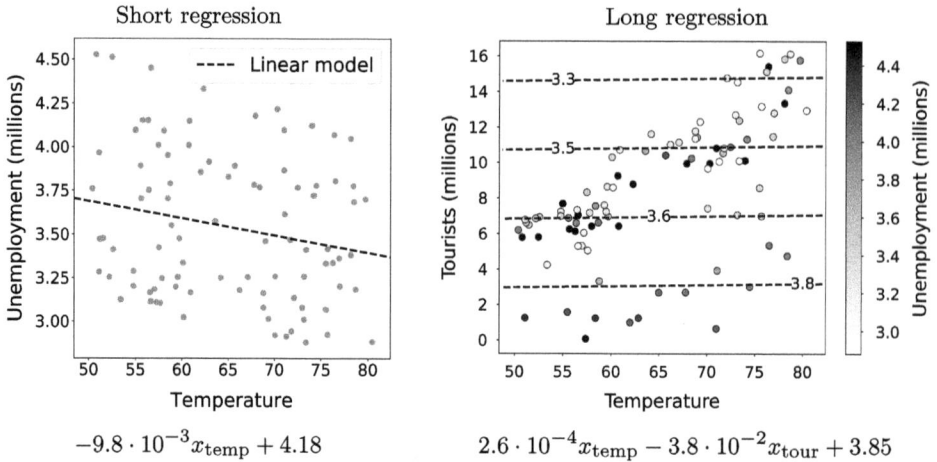

$$-9.8 \cdot 10^{-3} x_{\text{temp}} + 4.18 \qquad\qquad 2.6 \cdot 10^{-4} x_{\text{temp}} - 3.8 \cdot 10^{-2} x_{\text{tour}} + 3.85$$

Figure 12.6 Unemployment, temperature, and tourists. The left graph shows a simple linear regression model where the response is the unemployment in Spain between 2015 and 2022, and the feature is the corresponding average temperature (in °F). This short regression model indicates a negative association between unemployment and temperature. The right graph shows a long linear regression model, which also includes the number of tourists visiting Spain. In the long regression model, the coefficient corresponding to temperature is much smaller and positive, which suggests that the negative association observed in the short regression is an artifact of the correlation between the temperature and the number of tourists.

12.2 Generalization and Overfitting in Linear Regression

One of the main goals in regression and classification is to develop models that can predict the response for new feature values. It is therefore important to perform evaluation on held-out test data, distinct from the training data used to fit the models. A model that fits the training data well, but fails to *generalize* to the test data, is said to *overfit*. Figure 12.7 provides a cartoon illustration of generalization and overfitting. In this section, we study under what conditions linear regression models generalize effectively, or overfit the training data.

12.2.1 Linear Response with Additive Noise

In order to study the generalization behavior of linear regression models, we assume that the data satisfy the fundamental premise of these models: The response is approximately equal to a linear function of the features. More precisely, we consider a response that can be decomposed into a *signal* component, which is a linear function of the features, and a *noise* component, which is independent from the features. If we represent the features as a d-dimensional random vector $\tilde{x}$, the response equals

$$\tilde{y} := \tilde{x}^T \beta_{\text{true}} + \tilde{z}, \tag{12.131}$$

where $\beta_{\text{true}} \in \mathbb{R}^d$ is a vector of *true* linear coefficients, and $\tilde{z}$ is a random variable representing the noise, which is independent from $\tilde{x}$. The following theorem shows that the linear minimum MSE (MMSE) estimator of $\tilde{y}$ given $\tilde{x}$ recovers the true coefficients perfectly. The corresponding mean squared error of the estimator is equal to the variance of the noise.

Theorem 12.18 (Linear MMSE estimator for linear response with additive noise). *Let $\tilde{x}$ be a d-dimensional random vector of features with an invertible covariance matrix $\Sigma_{\tilde{x}}$, and let*

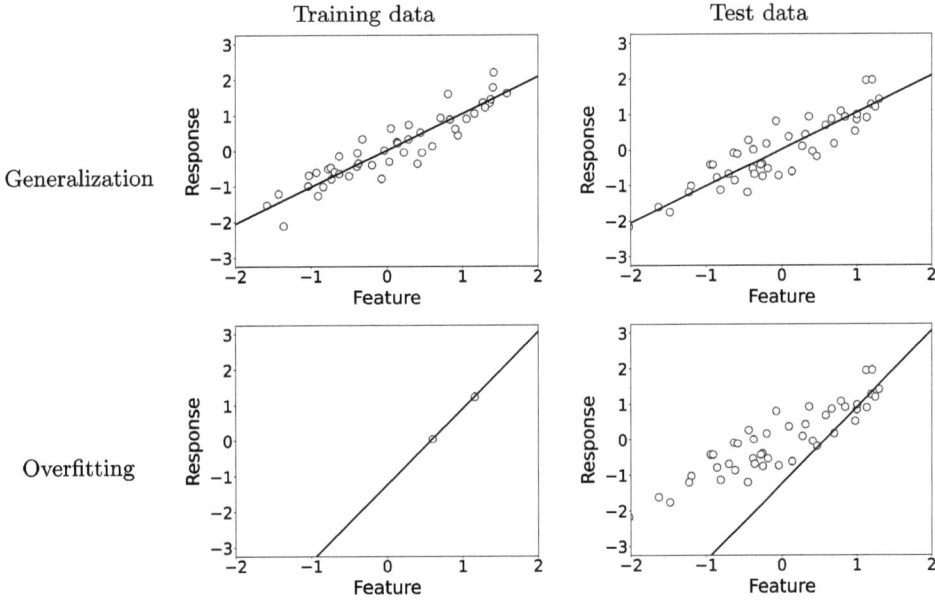

Figure 12.7 Generalization and overfitting. The left column depicts two training sets, consisting of a single feature and a response. The straight lines represent linear regression models, obtained by fitting the training data. The right column shows held-out test data, not used to train the models. The model in the top row generalizes well: It approximates the test data as well as the training data. By contrast, the model in the bottom row overfits: It yields a perfect estimate of the training data, but has terrible performance on the test data.

$\tilde{y} := \tilde{x}^T \beta_{\text{true}} + \tilde{z}$ *be the corresponding response, where $\tilde{z}$ has variance equal to σ^2 and is independent from $\tilde{x}$. We assume that the mean of the features and the noise is zero. Under these assumptions, the coefficients of the linear MMSE estimator of $\tilde{y}$ given $\tilde{x}$ are equal to the true linear coefficients,*

$$\beta_{\text{MMSE}} = \beta_{\text{true}}, \tag{12.132}$$

and the MSE is equal to the variance of $\tilde{z}$,

$$\text{MSE} = \sigma^2. \tag{12.133}$$

Proof Since the mean of the features is zero, the mean of the response is also zero by linearity of expectation (Theorem 11.8). The cross-covariance between the features and the response therefore equals

$$\Sigma_{\tilde{x}\tilde{y}} = \mathbb{E}\left[\tilde{x}\tilde{y}\right] \tag{12.134}$$

$$= \mathbb{E}\left[\tilde{x}\left(\tilde{x}^T \beta_{\text{true}} + \tilde{z}\right)\right] \tag{12.135}$$

$$= \mathbb{E}\left[\tilde{x}\tilde{x}^T\right] \beta_{\text{true}} + \mathbb{E}\left[\tilde{x}\tilde{z}\right] \tag{12.136}$$

$$= \Sigma_{\tilde{x}}\beta_{\text{true}} \tag{12.137}$$

by Definition 12.1, linearity of expectation, and the assumption that $\tilde{z}$ and the entries of $\tilde{x}$ are independent and have zero mean, which implies

$$\mathbb{E}\left[\tilde{x}\tilde{z}\right] = \mathbb{E}\left[\tilde{x}\right]\mathbb{E}\left[\tilde{z}\right] = 0. \tag{12.138}$$

Consequently, by Theorem 12.2,

$$\beta_{\text{MMSE}} = \Sigma_{\tilde{x}}^{-1} \Sigma_{\tilde{x}\tilde{y}} \tag{12.139}$$

$$= \Sigma_{\tilde{x}}^{-1} \Sigma_{\tilde{x}} \beta_{\text{true}} \tag{12.140}$$

$$= \beta_{\text{true}}. \tag{12.141}$$

The linear MMSE estimator $\ell_{\text{MMSE}}(\tilde{x}) := \beta_{\text{MMSE}}^T \tilde{x}$ equals $\beta_{\text{true}}^T \tilde{x}$, so the MSE is

$$\text{MSE} = \mathbb{E}\left[\tilde{y} - \ell_{\text{MMSE}}(\tilde{x})\right] \tag{12.142}$$

$$= \mathbb{E}\left[\left(\beta_{\text{true}}^T \tilde{x} + \tilde{z} - \beta_{\text{true}}^T \tilde{x}\right)^2\right] \tag{12.143}$$

$$= \mathbb{E}\left[\tilde{z}^2\right] \tag{12.144}$$

$$= \sigma^2. \tag{12.145}$$

∎

Theorem 12.18 establishes that the linear MMSE estimator recovers the true linear coefficients, as long as we have access to the true covariance matrix of the features and the true cross-covariance between the features and the response. However, in practice, we perform linear regression from finite data, as explained in Section 12.1.2, so this is not the case. In order to study the generalization properties of linear regression models in a more realistic scenario, we consider a training set with finite data, where the features are n deterministic vectors, and the response is equal to a linear combination of the features corrupted by additive noise.

Definition 12.19 (Finite-data linear response with additive noise). *Given n deterministic d-dimensional feature vectors $x_1, \ldots, x_n \in \mathbb{R}^d$, and a d-dimensional true coefficient vector $\beta_{\text{true}} \in \mathbb{R}^d$, we define the training design matrix as*

$$X_{\text{train}} := \begin{bmatrix} x_1^T \\ x_2^T \\ \ldots \\ x_n^T \end{bmatrix} \tag{12.146}$$

and the corresponding response as a linear function of the features perturbed by additive noise. If the noise is modeled as a deterministic n-dimensional vector $z_{\text{train}} \in \mathbb{R}^n$, then the training response is the deterministic vector

$$y_{\text{train}} := X_{\text{train}} \beta_{\text{true}} + z_{\text{train}}. \tag{12.147}$$

If the noise is modeled as an n-dimensional random vector $\tilde{z}_{\text{train}}$, then the response is the n-dimensional random vector

$$\tilde{y}_{\text{train}} := X_{\text{train}} \beta_{\text{true}} + \tilde{z}_{\text{train}}. \tag{12.148}$$

Definition 12.19 yields an alternative interpretation for the ordinary-least-squares (OLS) estimator introduced in Section 12.1.2. If we assume that the entries of the noise vector $\tilde{z}_{\text{train}}$ in (12.148) are i.i.d. Gaussian random variables, then the OLS estimator is the maximum-likelihood estimator of the true linear coefficients (see Exercise 12.6).

The following sections provide an analysis of the OLS estimator under the assumptions in Definition 12.19, which reveals when the estimator generalizes well to test data, and when it overfits the training data. Section 12.2.2 analyzes the OLS coefficient estimate, and Sections 12.2.3 and 12.2.4 characterize the OLS training and test error, respectively.

12.2.2 Coefficient Estimate

In this section, we study the linear coefficient estimate produced by the OLS estimator. The following example illustrates the behavior of OLS coefficients computed from real data, as we vary the number of training data.

Example 12.20 (Temperature prediction: OLS coefficients). We consider the problem of estimating the temperature in Versailles (Kentucky) from the temperatures at 133 other weather stations in the United States, using hourly temperature data extracted from Dataset 9. To solve the regression problem, we fit a linear regression model, where the response is the temperature in Versailles and the $d := 133$ features are the rest of the temperatures.

Figure 12.8 shows the OLS coefficients of the linear model (see Section 12.1.2) for different values of the number of training data n. When n is large with respect to the number of features d, three of the coefficients are clearly larger than the rest. They correspond to stations that are in geographical proximity to Versailles: Bowling Green (Kentucky), Bedford (Indiana), and Elkins (West Virginia). These coefficients are positive, so the OLS estimator for large n is approximately equal to a weighted average of the temperatures at these locations. This is a very reasonable way to estimate the temperature at Versailles. By contrast, for small n, many of the remaining coefficients have wildly fluctuating amplitudes, and the geographically meaningful coefficients are no longer the most prominent. As reported in Figure 12.10, the resulting models overfit the training data and do not generalize well to held-out data.

. . . .

Figure 12.8 illustrates an important phenomenon in linear regression: When the number of training data is small, OLS coefficients can be very noisy. In order to understand why, we derive a closed-form expression for the OLS coefficients, when the data are generated according to our finite-data linear-response assumption.

Theorem 12.21 (OLS coefficients for finite-data linear response with additive noise). *Let* $X := \{x_1, ..., x_n\}$ *be a dataset of n d-dimensional feature vectors with an invertible sample covariance matrix Σ_X, and let $\beta_{\text{true}} \in \mathbb{R}^d$ be a fixed vector of linear coefficients. If the response $y_{\text{train}} \in \mathbb{R}^n$ is defined as in (12.147) for a deterministic noise vector $z_{\text{train}} \in \mathbb{R}^n$, the OLS coefficients equal*

$$\beta_{\text{OLS}} = \beta_{\text{true}} + \Sigma_X^{-1} \Sigma_{XZ}, \tag{12.149}$$

where Σ_{XZ} denotes the sample cross-covariance (see Definition 12.6) between the features and the noise:

$$\Sigma_{XZ} := \frac{1}{n-1} \sum_{i=1}^{n} x_i z_{\text{train}}[i]. \tag{12.150}$$

OLS coefficients for large n

OLS coefficients

Smallest eigenvalue of feature sample covariance matrix

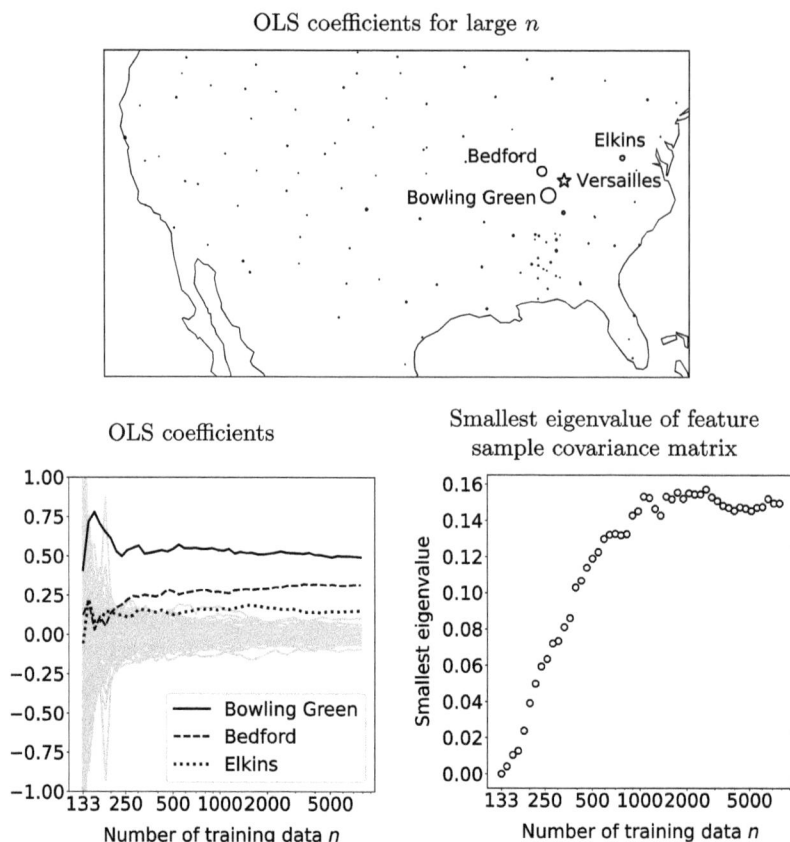

Figure 12.8 OLS coefficients for temperature estimation. The map at the top depicts the OLS coefficients in Example 12.20 when the number of training data n is large. The radius of the circular marker at each location is proportional to the corresponding OLS coefficient. The star indicates the location associated with the response. The plot on the bottom left shows the OLS coefficients for different values of n. For large n, the largest coefficients correspond to locations close to Versailles. For small n, the coefficients fluctuate wildly. The plot on the bottom right shows the smallest eigenvalue of the sample covariance matrix of the features as a function of n. The wild fluctuations of the OLS coefficients occur when the smallest eigenvalue is small.

Proof By Definition 12.6, the sample cross-covariance between the features and the response equals

$$\Sigma_{XY} = \frac{1}{n-1} \sum_{i=1}^{n} x_i y_{\text{train}}[i] \tag{12.151}$$

$$= \frac{1}{n-1} \sum_{i=1}^{n} x_i (x_i^T \beta_{\text{true}} + z_{\text{train}}[i]) \tag{12.152}$$

$$= \left(\frac{1}{n-1} \sum_{i=1}^{n} x_i x_i^T \right) \beta_{\text{true}} + \frac{1}{n-1} \sum_{i=1}^{n} x_i z_{\text{train}}[i] \tag{12.153}$$

$$= \Sigma_X \beta_{\text{true}} + \Sigma_{XZ}. \tag{12.154}$$

The result then follows from Theorem 12.7,

$$\beta_{\text{OLS}} = \Sigma_X^{-1} \Sigma_{XY} \tag{12.155}$$
$$= \beta_{\text{true}} + \Sigma_X^{-1} \Sigma_{XZ}. \tag{12.156}$$

∎

The key difference between the coefficient estimates in Theorems 12.18 and 12.21 is that the cross-covariance between the features and the noise is not necessarily zero for the finite-data model. Indeed, for finite data, we cannot expect the cross-covariance to be exactly zero even if the features are samples from a random vector $\tilde{x}$ that is independent from $\tilde{z}$ (this is the same phenomenon as in Example 1.30). The following example shows that the spurious correlation between the features and the noise can be amplified dramatically by *multicollinearity* in the design matrix. Multicollinearity occurs if a subset of the features is close to being linearly dependent.

Example 12.22 (Ordinary least squares and feature multicollinearity). Consider the following data, satisfying the assumptions in Theorem 12.21,

$$\underbrace{\begin{bmatrix} 0.33 \\ 0.91 \\ -1.51 \\ -0.10 \end{bmatrix}}_{y_{\text{train}}} := \underbrace{\begin{bmatrix} 0.46 & 0.44 \\ 0.97 & 1.03 \\ -1.52 & -1.51 \\ 0.09 & 0.04 \end{bmatrix}}_{X_{\text{train}}} \underbrace{\begin{bmatrix} 0.75 \\ 0.25 \end{bmatrix}}_{\beta_{\text{true}}} + \underbrace{\begin{bmatrix} -0.13 \\ -0.08 \\ 0.01 \\ -0.18 \end{bmatrix}}_{z_{\text{train}}}. \tag{12.157}$$

The features and noise are sampled from random vectors that are independent. Consequently, the sample cross-covariance between them is very small,

$$\Sigma_{XZ} = \begin{bmatrix} -0.055 \\ -0.053 \end{bmatrix}. \tag{12.158}$$

However, it is not exactly zero because the number of data is finite. The sample covariance matrix of the features is

$$\Sigma_X = \begin{bmatrix} 1.15 & 1.16 \\ 1.16 & 1.17 \end{bmatrix} = \underbrace{\begin{bmatrix} 0.70 & -0.71 \\ 0.71 & 0.70 \end{bmatrix}}_{U} \underbrace{\begin{bmatrix} 2.33 & 0 \\ 0 & 9.68 \cdot 10^{-4} \end{bmatrix}}_{\Lambda} \underbrace{\begin{bmatrix} 0.70 & 0.71 \\ -0.71 & 0.70 \end{bmatrix}}_{U^T},$$

where $U \Lambda U^T$ is the eigendecomposition of Σ_X, which encodes the principal directions of the features and their corresponding sample variance in those directions (see Theorem 11.25). The two features are very correlated. As a result, the first principal direction captures most of the variance in the data, while the variance in the second principal direction (equal to the second eigenvalue, highlighted in bold) is very small.

By Theorem 12.21, the OLS coefficients are

$$\beta_{\text{OLS}} = \beta_{\text{true}} + \Sigma_X^{-1} \Sigma_{XZ} \tag{12.159}$$
$$= \beta_{\text{true}} + U \Lambda^{-1} U^T \Sigma_{XZ} \tag{12.160}$$
$$= \begin{bmatrix} 0.75 \\ 0.25 \end{bmatrix} + \begin{bmatrix} 0.70 & -0.71 \\ 0.71 & 0.70 \end{bmatrix} \begin{bmatrix} 0.43 & 0 \\ 0 & \mathbf{1033} \end{bmatrix} \begin{bmatrix} 0.70 & 0.71 \\ -0.71 & 0.70 \end{bmatrix} \begin{bmatrix} -0.055 \\ -0.053 \end{bmatrix}$$
$$= \begin{bmatrix} -0.71 \\ 1.65 \end{bmatrix}. \tag{12.161}$$

The contribution of the noise to the coefficients is amplified by the inverse of the second eigenvalue of the sample covariance matrix (in bold), resulting in an estimate that is completely different from the true linear coefficients. This noise amplification allows the OLS estimator to overfit the component of the noise in the direction of least variance of the features. As a result, the OLS response estimate $y_{\text{OLS}} := X_{\text{train}}\beta_{\text{OLS}}$ is a closer approximation to the observed response y_{train} than the ideal response estimate $y_{\text{ideal}} := X_{\text{train}}\beta_{\text{true}}$ based on the true coefficients,

$$||y_{\text{OLS}} - y_{\text{train}}||_2^2 = 0.036 < 0.055 = ||y_{\text{ideal}} - y_{\text{train}}||_2^2 = ||z_{\text{train}}||_2^2. \qquad (12.162)$$

In other words, the linear model overfits the training data.
. . . .

Theorem 12.21 and Example 12.22 shed light onto the behavior of the OLS coefficients in Figure 12.8. As the number of training data n decreases, so does the smallest eigenvalue in the sample covariance matrix of the features (bottom right), indicating that there is multicollinearity among the features. This results in large noisy fluctuations of the OLS coefficients (bottom left), which enable the OLS model to overfit the training data.

In order to gain further insight into the typical behavior of OLS coefficients, we study their distribution under our finite-data linear-response assumption, when the additive noise is random. Figure 12.9 shows that the distribution of the resulting OLS coefficients is centered at the true coefficients. This is no coincidence: If the noise is independent and zero mean, then the OLS coefficient estimate is guaranteed to be unbiased.

Theorem 12.23 (The OLS coefficient estimate is unbiased). *Let x_1, ..., x_n be d-dimensional feature vectors with an invertible sample covariance matrix, and let $\beta_{\text{true}} \in \mathbb{R}^d$ be a fixed vector of linear coefficients. If the response $\tilde{y}_{\text{train}} \in \mathbb{R}^n$ is defined as in (12.148) for a noise vector $\tilde{z}_{\text{train}}$ with n independent entries that have zero mean, then OLS yields an unbiased estimate of the true linear coefficients,*

$$\mathbb{E}\left[\tilde{\beta}_{\text{OLS}}\right] = \beta_{\text{true}}. \qquad (12.163)$$

Proof Under the assumptions in Definition 12.19, the sample cross-covariance between the features and the noise is a random vector that equals

$$\tilde{\Sigma}_{XZ} := \frac{1}{n-1}\sum_{i=1}^{n} x_i \tilde{z}_{\text{train}}[i]. \qquad (12.164)$$

By Theorem 12.21,

$$\tilde{\beta}_{\text{OLS}} = \beta_{\text{true}} + \Sigma_X^{-1}\tilde{\Sigma}_{XZ}, \qquad (12.165)$$

so by linearity of expectation (Theorem 11.8) and the assumption that each noise entry has zero mean,

$$\mathbb{E}\left[\tilde{\beta}_{\text{OLS}}\right] = \beta_{\text{true}} + \Sigma_X^{-1}\mathbb{E}\left[\tilde{\Sigma}_{XZ}\right] \qquad (12.166)$$

$$= \beta_{\text{true}} + \Sigma_X^{-1}\frac{1}{n-1}\sum_{i=1}^{n} x_i \mathbb{E}\left[\tilde{z}_{\text{train}}[i]\right] \qquad (12.167)$$

$$= \beta_{\text{true}}. \qquad (12.168)$$

∎

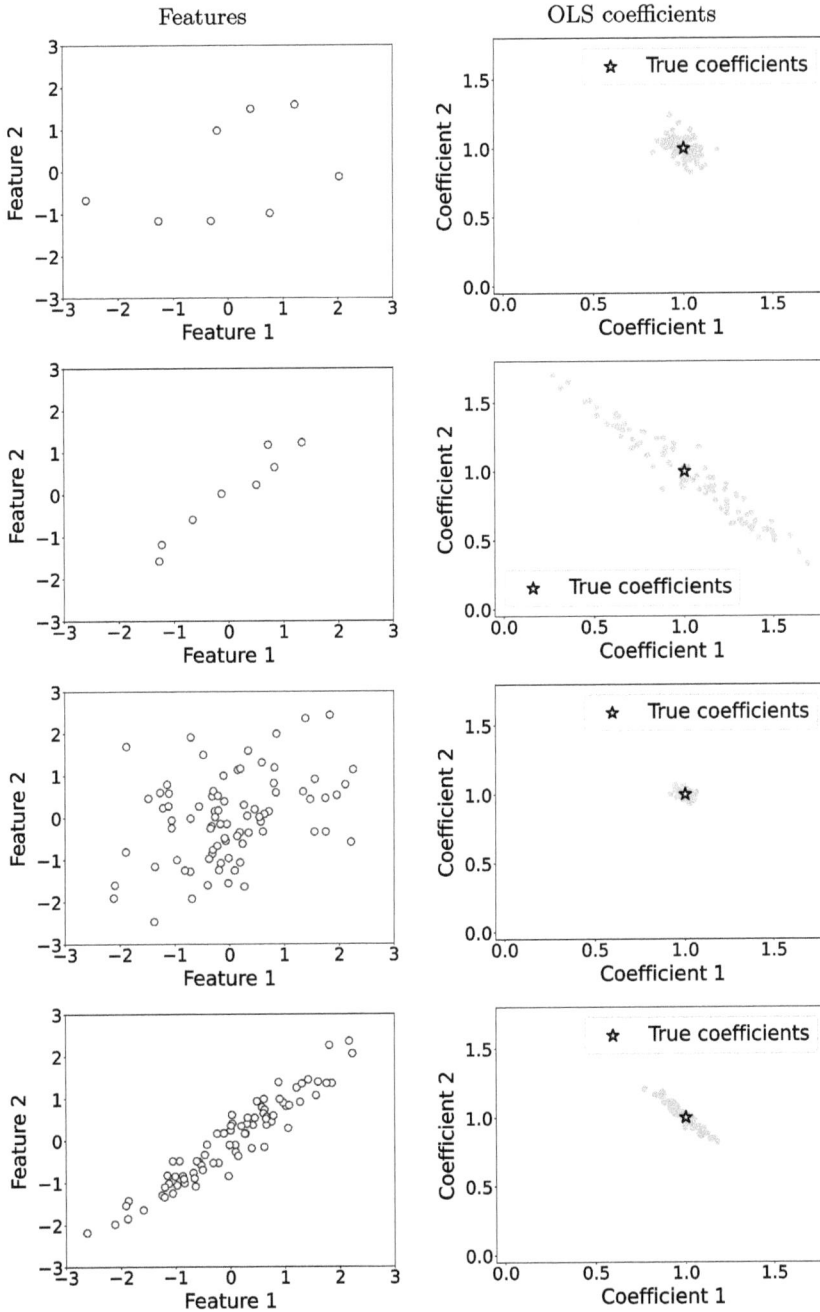

Figure 12.9 OLS coefficients for a linear response with additive noise. The left column shows two sets of $n := 8$ (top two rows) and $n := 80$ (bottom two rows) feature vectors. The right column shows 100 OLS coefficient estimates computed from 100 different realizations of the linear response (12.148) for each set of feature vectors. The true linear coefficients (represented by a star) are fixed, and the noise component in the response is sampled from a fixed-variance i.i.d. Gaussian distribution. As established in Theorem 12.23, the estimates are unbiased: Their distribution is centered at the true coefficients. As established in Theorem 12.24, the variance of the OLS coefficients is inversely proportional to the number of training data n and is highest in the direction of lowest variance of the features.

By Theorem 12.23, the OLS coefficients are centered at the true coefficients, but this does not necessarily imply that they are *close* to them. The following theorem derives the covariance matrix of the OLS coefficients, which quantifies to what extent they fluctuate around the true coefficients.

Theorem 12.24 (Covariance matrix of OLS coefficients)**.** *Let $x_1, \ldots, x_n$ be d-dimensional feature vectors, and let $\beta_{\text{true}} \in \mathbb{R}^d$ be a fixed vector of linear coefficients. If the response $\tilde{y}_{\text{train}} \in \mathbb{R}^n$ is defined as in (12.148) for a noise vector $\tilde{z}_{\text{train}}$ with n independent entries that have zero mean and variance σ^2, then the covariance matrix of the OLS coefficients equals*

$$\Sigma_{\tilde{\beta}_{\text{OLS}}} = \frac{\sigma^2}{n-1}\Sigma_X^{-1}, \tag{12.169}$$

where Σ_X is the sample covariance matrix of the features, which we assume to be invertible.

Proof By Theorem 12.21, the OLS coefficients equal

$$\tilde{\beta}_{\text{OLS}} = \beta_{\text{true}} + \Sigma_X^{-1}\widetilde{\Sigma}_{XZ}, \tag{12.170}$$

where $\widetilde{\Sigma}_{XZ}$ is the sample cross-covariance between the features and the noise, defined as in (12.164). Consequently, by Theorem 12.23, the centered OLS coefficients equal

$$\text{ct}\left(\tilde{\beta}_{\text{OLS}}\right) := \tilde{\beta}_{\text{OLS}} - \mathbb{E}\left[\tilde{\beta}_{\text{OLS}}\right] \tag{12.171}$$

$$= \beta_{\text{true}} + \Sigma_X^{-1}\widetilde{\Sigma}_{XZ} - \beta_{\text{true}} \tag{12.172}$$

$$= \Sigma_X^{-1}\widetilde{\Sigma}_{XZ}. \tag{12.173}$$

By Definitions 11.9 and 11.14, and linearity of expectation (Theorem 11.8), the covariance matrix of the sample cross-covariance equals

$$\mathbb{E}\left[\widetilde{\Sigma}_{XZ}\widetilde{\Sigma}_{XZ}^T\right] = \mathbb{E}\left[\frac{1}{n-1}\sum_{i=1}^n x_i \tilde{z}_{\text{train}}[i]\frac{1}{n-1}\sum_{j=1}^n \tilde{z}_{\text{train}}[j]x_j^T\right] \tag{12.174}$$

$$= \frac{1}{(n-1)^2}\sum_{i=1}^n\sum_{j=1}^n x_i \mathbb{E}\left[\tilde{z}_{\text{train}}[i]\tilde{z}_{\text{train}}[j]\right]x_j^T \tag{12.175}$$

$$= \frac{\sigma^2}{(n-1)^2}\sum_{i=1}^n x_i x_i^T \tag{12.176}$$

$$= \frac{\sigma^2}{n-1}\Sigma_X \tag{12.177}$$

because the noise entries are uncorrelated and have zero mean, so $\mathbb{E}\left[\tilde{z}_{\text{train}}[i]\tilde{z}_{\text{train}}[j]\right]$ equals $\mathbb{E}\left[\tilde{z}_{\text{train}}[i]^2\right] = \sigma^2$ if $i = j$, and zero otherwise.

We conclude that, by Definition 11.9 and linearity of expectation, the covariance matrix of the OLS coefficients is

$$\Sigma_{\tilde{\beta}_{\mathrm{OLS}}} = \mathbb{E}\left[\mathrm{ct}\left(\tilde{\beta}_{\mathrm{OLS}}\right)\mathrm{ct}\left(\tilde{\beta}_{\mathrm{OLS}}\right)^T\right] \tag{12.178}$$

$$= \mathbb{E}\left[\Sigma_X^{-1}\widetilde{\Sigma}_{XZ}\widetilde{\Sigma}_{XZ}^T\Sigma_X^{-1}\right] \tag{12.179}$$

$$= \Sigma_X^{-1}\mathbb{E}\left[\widetilde{\Sigma}_{XZ}\widetilde{\Sigma}_{XZ}^T\right]\Sigma_X^{-1} \tag{12.180}$$

$$= \frac{\sigma^2}{n-1}\Sigma_X^{-1}\Sigma_X\Sigma_X^{-1} \tag{12.181}$$

$$= \frac{\sigma^2}{n-1}\Sigma_X^{-1}. \tag{12.182}$$

∎

By Theorem 11.11, the variance of the coefficients in the direction of a normalized vector a is equal to $a^T\Sigma_{\tilde{\beta}_{\mathrm{OLS}}}a$. Consequently, Theorem 12.24 implies that the variance of the coefficients in every direction is proportional to the noise variance σ^2 and inversely proportional to the number of data n, as we observe in Figure 12.9. As $n \to \infty$, the variance tends to zero, which means that the OLS coefficients converge to the true coefficients. This establishes that OLS is a *consistent* estimator under our assumptions, in the sense that it recovers the true coefficients as the number of data tend to infinity.

Theorem 12.24 reveals the effect of the covariance structure of the features on the OLS coefficients. By Theorem 11.25, the eigendecomposition of the sample covariance matrix of the features $\Sigma_X = U\Lambda U^T$ encodes the principal directions of the features (columns of U) and their corresponding sample variance in those directions (diagonal entries of Λ). The eigendecomposition of the covariance matrix of the OLS coefficients equals

$$\Sigma_{\tilde{\beta}_{\mathrm{OLS}}} = \frac{\sigma^2}{n-1}\Sigma_X^{-1} \tag{12.183}$$

$$= \frac{\sigma^2}{n-1}U\Lambda^{-1}U^T. \tag{12.184}$$

By Theorem 11.20, the principal directions of the coefficients are equal to the columns of U, and hence are the same as the principal directions of the features. The sample variance in each direction is equal to the corresponding eigenvalue, stored in the respective diagonal entry of Λ^{-1}, and is *inversely* proportional to the sample variance of the features in that direction. When there is multicollinearity in the feature design matrix, some of the eigenvalues in Λ are small, so in the corresponding directions, the features have low variance and the OLS coefficients have high variance. This is why in the second and fourth rows of Figure 12.9 the variance of the OLS coefficients is highest (right plot) precisely in the direction of lowest variance of the features (left plot).

12.2.3 Training Error

The training error of a model quantifies how closely it approximates the data used to fit it. Given a dataset formed by n feature vectors $x_1, \ldots, x_n$ with d features and a corresponding response vector $y_{\mathrm{train}} \in \mathbb{R}^n$, we define the training error of the OLS estimator as the root average residual sum of squares,

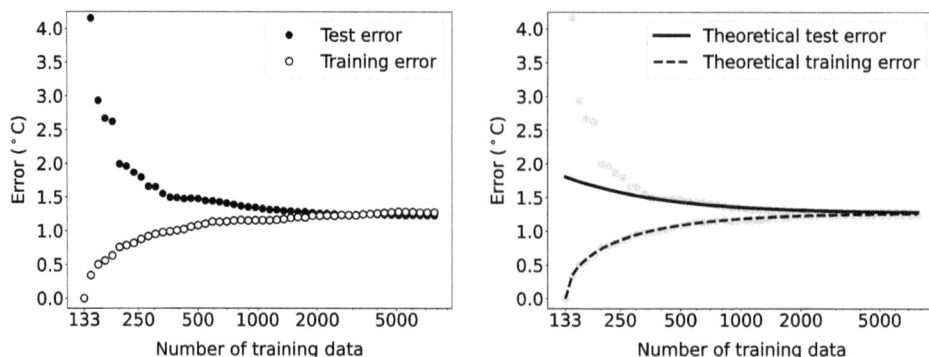

Figure 12.10 OLS training and test error for weather data. The left plot shows the performance of OLS models for the temperature-estimation problem described in Example 12.20 as
a function of the number of training data. The test data is fixed and consists of 1,000 held-
out data points. The error metric is the square root of the average squared error, defined as
in (12.185). The right plot shows the theoretical expressions of the training and test error derived
in Theorems 12.27 and 12.28, superposed onto the observed errors (gray markers). The standard
deviation of the noise σ in the theoretical expressions is set equal to the training error when the
number of training data is large.

$$\text{Training error} := \sqrt{\frac{1}{n} \sum_{i=1}^{n} (y_{\text{train}}[i] - \ell_{\text{OLS}}(x_i))^2}, \qquad (12.185)$$

where $\ell_{\text{OLS}}(x_i) := \beta_{\text{OLS}}^T x_i + \alpha_{\text{OLS}}$ is the OLS estimate of the ith response, β_{OLS} denotes
the OLS coefficients, and α_{OLS} is the OLS intercept. We apply the square root, so that we
can express the error in the same units as the response. Figure 12.10 shows the OLS training
error for the weather data in Example 12.20 as a function of the number of training data n.
When n is close to the number of features d, the training error is very small, which indicates
overfitting. As n increases, so does the training error, converging to a fixed value. The goal
of this section is to provide a theoretical explanation of this behavior.

In order to analyze the training error, we consider a regression problem with d features
and n training data, where the features and responses are centered to have zero sample
mean, so that we can restrict our attention to linear estimators without an intercept. Any
linear estimate of the response given the features can be expressed in matrix-vector form:

$$X_{\text{train}}\beta = \begin{bmatrix} x_1[1] & x_1[2] & \cdots & x_1[d] \\ x_2[1] & x_2[2] & \cdots & x_2[d] \\ \cdots & \cdots & \cdots & \cdots \\ x_n[1] & x_n[2] & \cdots & x_n[d] \end{bmatrix} \beta \qquad (12.186)$$

$$= \begin{bmatrix} X_1 & X_2 & \cdots & X_d \end{bmatrix} \beta \qquad (12.187)$$

$$= \sum_{j=1}^{d} X_j\beta[j], \qquad (12.188)$$

where $\beta \in \mathbb{R}^d$ is a vector of coefficients, and $x_i[j]$ denotes the jth feature in the ith example. The vector X_j corresponds to the jth column of the training design matrix X_{train}; it
contains the n examples of the jth feature. Since all linear estimates are linear combinations of the vectors $X_1, \ldots, X_d$, the set of possible linear estimates is equal to the column

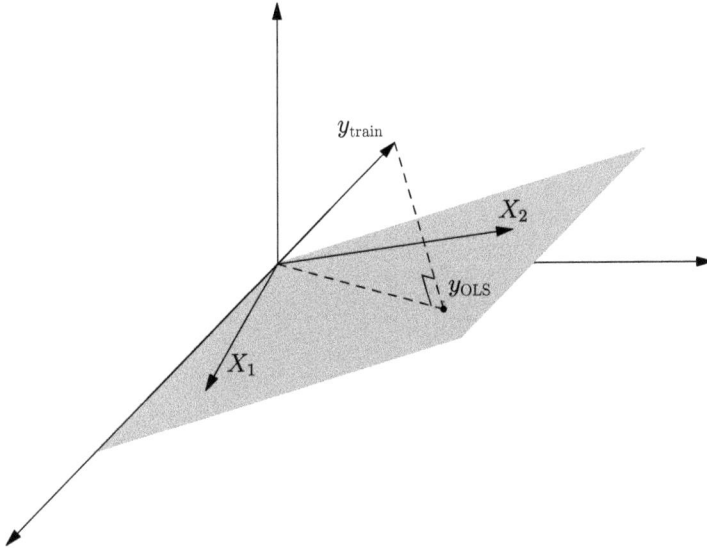

Figure 12.11 OLS response estimate as a projection. Illustration of Theorem 12.25 for a regression problem with $n := 3$ training examples and $d := 2$ features. The OLS response estimate is the orthogonal projection of the three-dimensional training response y_{train} onto the two-dimensional subspace spanned by the columns X_1 and X_2 of the design matrix X_{train}.

space $\text{col}(X_{\text{train}})$ of X_{train}, which is the subspace or hyperplane spanned by the columns of X_{train}.

The training error of a linear estimate $X_{\text{train}}\beta$ is proportional to the Euclidean distance between $X_{\text{train}}\beta$ and the response vector y_{train},

$$\text{Training error} := \sqrt{\frac{1}{n}\sum_{i=1}^{n}\left(y_{\text{train}}[i] - \sum_{j=1}^{d}x_i[j]\beta[j]\right)^2} = \frac{1}{\sqrt{n}}\left\|y_{\text{train}} - X_{\text{train}}\beta\right\|_2.$$

By Theorem 12.7, the OLS estimator minimizes the training error. Therefore, the OLS response estimate is the vector in $\text{col}(X_{\text{train}})$ that is *closest* to y_{train} in ℓ_2 norm. Geometrically, it is the orthogonal projection of y_{train} onto $\text{col}(X_{\text{train}})$, as illustrated by Figure 12.11. The following theorem formalizes this reasoning, which is analogous to our derivation of the linear MMSE estimator in Section 12.1.1. The only difference is that here the features and the response are represented by n-dimensional vectors, instead of random variables (compare Figures 12.2 and 12.11).

Theorem 12.25 (The OLS estimator is a projection). *Let $X_{\text{train}} \in \mathbb{R}^{n \times d}$ be a design matrix containing n d-dimensional vectors corresponding to the features in a regression problem, and let $y_{\text{train}} \in \mathbb{R}^n$ be the corresponding response vector. Assuming that the features and the response are centered to have zero sample mean and the sample covariance matrix of the features is invertible, the OLS response estimate is the projection of y_{train} onto the column space $\text{col}(X_{\text{train}})$ of X_{train},*

$$y_{\text{OLS}} := X_{\text{train}}\beta_{\text{OLS}} \tag{12.189}$$

$$= \mathcal{P}_{\text{col}(X_{\text{train}})}\, y_{\text{train}}. \tag{12.190}$$

Proof   Let $X_{\text{train}}^T = USV^T$ be the singular-value decomposition of the transposed design matrix (see Section 11.6.2), so that VSU^T is the singular-value decomposition of X_{train}. By Definition 12.6 the sample cross-covariance between the features and the response is

$$\Sigma_{XY} = \frac{1}{n-1}X_{\text{train}}^T y_{\text{train}} \tag{12.191}$$

$$= \frac{1}{n-1}USV^T y_{\text{train}}. \tag{12.192}$$

By Definition 11.14 and (12.89), the sample covariance matrix of the features equals

$$\Sigma_X = \frac{1}{n-1}X_{\text{train}}^T X_{\text{train}} \tag{12.193}$$

$$= \frac{1}{n-1}USV^T VSU^T \tag{12.194}$$

$$= \frac{1}{n-1}USSU^T \tag{12.195}$$

because $V^T V = I$ by Theorem 11.31, which also establishes that U has orthonormal columns. Consequently, the inverse of the sample covariance matrix equals

$$\Sigma_X^{-1} = (n-1)US^{-1}S^{-1}U^T. \tag{12.196}$$

As a result, by Theorem 12.7 the OLS response estimate is

$$y_{\text{OLS}} := X_{\text{train}}\beta_{\text{OLS}} \tag{12.197}$$

$$= X_{\text{train}}\Sigma_X^{-1}\Sigma_{XY} \tag{12.198}$$

$$= VSU^T US^{-1}S^{-1}U^T USV^T y_{\text{train}} \tag{12.199}$$

$$= VV^T y_{\text{train}} \tag{12.200}$$

because $U^T U = I$ (since U is a square orthogonal matrix). By Theorem 11.31, V is a matrix with d orthonormal columns that span the row space of X_{train}^T, which is the column space of X_{train}. Therefore, VV^T projects the response vector y_{train} onto the column space of X_{train}.   ∎

Leveraging Theorem 12.25, we derive the training residual of the OLS response estimate when the response is a linear function of the features contaminated by additive noise. The residual is the projection of the noise component onto the orthogonal complement of the column space of the feature design matrix.

Theorem 12.26 (Training residual for linear response with additive noise).   *Let the response* $y_{\text{train}} \in \mathbb{R}^n$ *in a regression problem be defined as in* (12.147), *for a deterministic noise vector* $z_{\text{train}} \in \mathbb{R}^n$. *Assuming that the features and the response are centered to have zero sample mean, and that the sample covariance matrix of the features is invertible, the training residual of the OLS coefficient estimate* $\beta_{\text{OLS}} \in \mathbb{R}^d$ *is equal to the projection of the noise onto the orthogonal complement* $\text{col}(X_{\text{train}})^\perp$ *of the column space of the feature design matrix* X_{train},

$$y_{\text{train}} - X_{\text{train}}\beta_{\text{OLS}} = \mathcal{P}_{\text{col}(X_{\text{train}})^\perp}\, z_{\text{train}}. \tag{12.201}$$

Proof By Theorem 12.25,

$$y_{\text{train}} - X_{\text{train}}\beta_{\text{OLS}} = y_{\text{train}} - \mathcal{P}_{\text{col}(X_{\text{train}})}\, y_{\text{train}} \tag{12.202}$$

$$= \mathcal{P}_{\text{col}(X_{\text{train}})^\perp}\, y_{\text{train}} \tag{12.203}$$

$$= \mathcal{P}_{\text{col}(X_{\text{train}})^\perp}\, (X_{\text{train}}\beta_{\text{true}} + z_{\text{train}}) \tag{12.204}$$

$$= \mathcal{P}_{\text{col}(X_{\text{train}})^\perp}\, z_{\text{train}} \tag{12.205}$$

because $X_{\text{train}}\beta_{\text{true}}$ belongs to the column space of X_{train} and, hence, is orthogonal to its orthogonal complement. ∎

Theorem 12.26 enables us to characterize the training error of the OLS estimator, under the linear-response assumption in Definition 12.19, as a function of the number of features d and the number of training data n. The feature design matrix has d columns of dimension n. Assuming these columns are linearly independent, the column space is a subspace of dimension d, so its orthogonal complement has dimension $n - d$. If the noise component is *isotropic*, meaning that it has the same variance in every direction, then projecting it onto the orthogonal complement should (on average) preserve a fraction of variance proportional to the dimension $n - d$ of the orthogonal complement. The following theorem shows that this intuitive back-of-the-envelope calculation is correct.

Theorem 12.27 (Training error of the OLS estimator for linear response with additive noise). *Let $x_1, \ldots, x_n$ be d-dimensional feature vectors with an invertible sample covariance matrix, and let $\beta_{\text{true}} \in \mathbb{R}^d$ be a fixed vector of linear coefficients. We assume that the response $\tilde{y}_{\text{train}}$ is defined as in (12.148) for a noise vector $\tilde{z}_{\text{train}}$ with independent entries that have zero mean and variance σ^2, and define the residual sum of squares as the random variable,*

$$\widetilde{\text{RSS}}_{\text{train}} := \sum_{i=1}^{n} \left(\tilde{y}_{\text{train}}[i] - x_i^T \tilde{\beta}_{\text{OLS}} \right)^2. \tag{12.206}$$

The mean of the residual sum of squares per example equals

$$\mathbb{E}\left[\frac{1}{n} \widetilde{\text{RSS}}_{\text{train}} \right] = \left(1 - \frac{d}{n} \right) \sigma^2. \tag{12.207}$$

Proof Let $V_\perp$ be a matrix with $n - d$ orthonormal columns $v_{\perp i}$, $1 \leq i \leq n - d$, spanning the orthogonal complement $\text{col}(X_{\text{train}})^\perp$ of the column space of X_{train}, so that $V_\perp^T V_\perp = I$ and $V_\perp V_\perp^T$ implements a projection onto $\text{col}(X_{\text{train}})^\perp$. By Theorem 12.26,

$$\widetilde{\text{RSS}}_{\text{train}} = \left\| \mathcal{P}_{\text{col}(X_{\text{train}})^\perp}\, \tilde{z}_{\text{train}} \right\|_2^2 \tag{12.208}$$

$$= \left\| V_\perp V_\perp^T \tilde{z}_{\text{train}} \right\|_2^2 \tag{12.209}$$

$$= \tilde{z}_{\text{train}}^T V_\perp V_\perp^T V_\perp V_\perp^T \tilde{z}_{\text{train}} \tag{12.210}$$

$$= \tilde{z}_{\text{train}}^T V_\perp V_\perp^T \tilde{z}_{\text{train}} \tag{12.211}$$

$$= \sum_{i=1}^{n-d} \left(v_{\perp i}^T \tilde{z}_{\text{train}} \right)^2 \tag{12.212}$$

$$= \sum_{i=1}^{n-d} v_{\perp i}^T \tilde{z}_{\text{train}} \tilde{z}_{\text{train}}^T v_{\perp i}. \tag{12.213}$$

Since the entries of $\tilde{z}_{\text{train}}$ are uncorrelated and have variance σ^2, its covariance matrix equals $\Sigma_{\tilde{z}_{\text{train}}} = \sigma^2 I$ by Definition 11.9. Consequently, by linearity of expectation (Theorem 11.8),

$$\mathbb{E}\left[\widetilde{\text{RSS}}_{\text{train}}\right] = \sum_{i=1}^{n-d} v_{\perp i}^T \mathbb{E}\left[\tilde{z}_{\text{train}}\tilde{z}_{\text{train}}^T\right] v_{\perp i} \tag{12.214}$$

$$= \sum_{i=1}^{n-d} v_{\perp i}^T \Sigma_{\tilde{z}_{\text{train}}} v_{\perp i} \tag{12.215}$$

$$= \sigma^2 \sum_{i=1}^{n-d} v_{\perp i}^T v_{\perp i} \tag{12.216}$$

$$= (n-d)\sigma^2. \tag{12.217}$$

∎

Figure 12.10 shows that Theorem 12.27 provides a very precise characterization of the dependence between the training error and the number of training data for the real-world dataset in Example 12.20. The training error is proportional to $\sqrt{1 - d/n}$, as predicted by the theorem.

When $n \gg d$, the residual sum of squares per example in Theorem 12.27 converges to the variance of the noise σ^2. Consequently, when the number of training data is large with respect to the number of features, the OLS estimator behaves like the linear MMSE estimator in Theorem 12.18, which has an MSE equal to the variance of the noise. However, when $n \approx d$, the training error of the OLS estimator approaches zero, so the estimator is able to perfectly approximate the response, even though it is corrupted with noise that is *independent from the features*. In other words, when the number of data is close to the number of features, the OLS model overfits the noise in the training data.

12.2.4 Test Error

The test error of an estimator quantifies its performance on held-out data, which have not been used to fit the model. In this section, we study the test error of the OLS estimator under the assumption that the response is a linear function of the features corrupted by additive noise, as in Definition 12.19. The training response vector $\tilde{y}_{\text{train}}$ is the n-dimensional random vector

$$\tilde{y}_{\text{train}} := X_{\text{train}}\beta_{\text{true}} + \tilde{z}_{\text{train}}, \tag{12.218}$$

where n is the number of training data, $X_{\text{train}} \in \mathbb{R}^{n \times d}$ is the training design matrix, and $\beta_{\text{true}} \in \mathbb{R}^d$ contains the ground-truth linear coefficients. We model the noise as an n-dimensional random vector $\tilde{z}_{\text{train}}$, which is i.i.d. with zero mean and variance σ^2.

Our goal is to analyze the performance of the OLS estimator computed from these training data, when we apply it to a test example. We represent the test features by a d-dimensional random vector $\tilde{x}_{\text{test}}$, which is independent from $\tilde{z}_{\text{train}}$. The unobserved test response follows a linear model with the same coefficients as the training data,

$$\tilde{y}_{\text{test}} := \tilde{x}_{\text{test}}^T \beta_{\text{true}} + \tilde{z}_{\text{test}}, \tag{12.219}$$

where $\tilde{z}_{\text{test}}$ is a zero-mean random variable representing the test noise. The linear coefficients β_{true} are shared by the training set and the test example, but the features and noise are not.

The OLS response estimate for the test data $\tilde{x}_{\text{test}}^T \tilde{\beta}_{\text{OLS}}$ is computed using the OLS coefficient estimator $\tilde{\beta}_{\text{OLS}}$, which *only depends on the training data*.

Under our assumptions, $\mathbb{E}[\tilde{\beta}_{\text{OLS}}] = \beta_{\text{true}}$ by Theorem 12.23, so we can express the difference between the true test response and the OLS estimate as follows:

$$\tilde{y}_{\text{test}} - \tilde{x}_{\text{test}}^T \tilde{\beta}_{\text{OLS}} = \tilde{y}_{\text{test}} - \tilde{x}_{\text{test}}^T \left(\beta_{\text{true}} + \text{ct}(\tilde{\beta}_{\text{OLS}}) \right) \tag{12.220}$$

$$= \tilde{x}_{\text{test}}^T \beta_{\text{true}} + \tilde{z}_{\text{test}} - \tilde{x}_{\text{test}}^T \beta_{\text{true}} - \tilde{x}_{\text{test}}^T \text{ct}(\tilde{\beta}_{\text{OLS}}) \tag{12.221}$$

$$= \tilde{z}_{\text{test}} - \tilde{x}_{\text{test}}^T \text{ct}(\tilde{\beta}_{\text{OLS}}), \tag{12.222}$$

where $\text{ct}(\tilde{\beta}_{\text{OLS}}) := \tilde{\beta}_{\text{OLS}} - \mathbb{E}[\tilde{\beta}_{\text{OLS}}]$ is the centered OLS coefficient estimate.

Since $\text{ct}(\tilde{\beta}_{\text{OLS}})$ has zero mean and is independent from $\tilde{x}_{\text{test}}$ (as it only depends on $\tilde{z}_{\text{train}}$), the mean of $\tilde{x}_{\text{test}}^T \text{ct}(\tilde{\beta}_{\text{OLS}})$ is also zero by Theorem 7.20. Consequently, the mean of the random variable $\tilde{z}_{\text{test}} - \tilde{x}_{\text{test}}^T \text{ct}(\tilde{\beta}_{\text{OLS}})$ is zero by linearity of expectation (Theorem 7.18), so its mean square is equal to its variance (see Lemma 7.33). If the test noise $\tilde{z}_{\text{test}}$ is independent from the other variables, together with Theorem 9.11, this implies that the test mean square error (MSE) equals

$$\text{MSE}_{\text{test}} := \mathbb{E}\left[\left(\tilde{y}_{\text{test}} - \tilde{x}_{\text{test}}^T \tilde{\beta}_{\text{OLS}} \right)^2 \right] \tag{12.223}$$

$$= \text{Var}\left[\tilde{y}_{\text{test}} - \tilde{x}_{\text{test}}^T \tilde{\beta}_{\text{OLS}} \right] \tag{12.224}$$

$$= \text{Var}\left[\tilde{z}_{\text{test}} - \tilde{x}_{\text{test}}^T \text{ct}(\tilde{\beta}_{\text{OLS}}) \right] \tag{12.225}$$

$$= \text{Var}\left[\tilde{z}_{\text{test}} \right] + \text{Var}\left[\tilde{x}_{\text{test}}^T \text{ct}(\tilde{\beta}_{\text{OLS}}) \right]. \tag{12.226}$$

The test MSE depends on the variance of the inner product between the OLS coefficients and the test features. By Theorem 12.24, the OLS coefficients may have high variance, if there is multicollinearity between the features (see Figure 12.9). However, this does *not* necessarily translate to a high test error, as long as the test features have *the same covariance matrix* as the training features. If this holds, whenever the training features have low variance in a certain direction, then so do the test features. These are precisely the directions in which the OLS coefficients have high variance (again by Theorem 12.24). Consequently, the high variance of the OLS coefficients is neutralized by the low variance of the test features, when we take the inner product in (12.226). The following theorem formalizes this intuition, providing a precise characterization of the test MSE in this scenario.

Theorem 12.28 (Test error of the OLS estimator for linear response with additive noise). *Let $x_1, \ldots, x_n$ be d-dimensional training feature vectors with an invertible sample covariance matrix Σ_X, and let $\beta_{\text{true}} \in \mathbb{R}^d$ be a fixed vector of linear coefficients. We assume that the training response vector $\tilde{y}_{\text{train}}$ is defined as in (12.148) for a noise vector $\tilde{z}_{\text{train}}$ with independent entries that have zero mean and variance σ^2.*

We consider a d-dimensional test feature vector $\tilde{x}_{\text{test}}$ and a corresponding test response $\tilde{y}_{\text{test}}$, defined as in (12.219), where the test noise $\tilde{z}_{\text{test}}$ has zero mean and variance σ^2. The test features $\tilde{x}_{\text{test}}$ and the training and test noise $\tilde{z}_{\text{train}}$ and $\tilde{z}_{\text{test}}$ are all mutually independent. If the covariance matrix of the test features is the same as the sample covariance matrix of the training feature vectors, $\Sigma_{\tilde{x}_{\text{test}}} = \Sigma_X$, then the mean squared test error of the OLS estimator equals

$$\text{MSE}_{\text{test}} := \mathbb{E}\left[\left(\tilde{y}_{\text{test}} - \tilde{x}_{\text{test}}^T \tilde{\beta}_{\text{OLS}}\right)^2\right] \tag{12.227}$$

$$= \sigma^2\left(1 + \frac{d}{n_{\text{train}} - 1}\right), \tag{12.228}$$

where $\tilde{\beta}_{\text{OLS}}$ denotes the OLS coefficient estimate obtained using the training data, and $n_{\text{train}} := n$ is the number of training examples.

Proof Note that $\tilde{x}_{\text{test}}$ and $\text{ct}(\tilde{\beta}_{\text{OLS}})$ are independent because $\text{ct}(\tilde{\beta}_{\text{OLS}})$ is a deterministic function of $\tilde{z}_{\text{train}}$, which is independent from $\tilde{x}_{\text{test}}$. Under the assumptions, (12.226) holds, as explained earlier, so

$$\text{MSE}_{\text{test}} = \text{Var}\left[\tilde{z}_{\text{test}}\right] + \text{Var}\left[\tilde{x}_{\text{test}}^T \text{ct}(\tilde{\beta}_{\text{OLS}})\right] \tag{12.229}$$

$$= \sigma^2 + \mathbb{E}\left[\left(\tilde{x}_{\text{test}}^T \text{ct}(\tilde{\beta}_{\text{OLS}})\right)^2\right]. \tag{12.230}$$

The variance of the second term equals the mean square by Lemma 7.33 because

$$\mathbb{E}\left[\tilde{x}_{\text{test}}^T \text{ct}(\tilde{\beta}_{\text{OLS}})\right] = \mathbb{E}\left[\tilde{x}_{\text{test}}\right]^T \mathbb{E}\left[\text{ct}(\tilde{\beta}_{\text{OLS}})\right] \tag{12.231}$$

$$= 0 \tag{12.232}$$

by Theorem 7.20 and independence of $\tilde{x}_{\text{test}}$ and $\text{ct}(\tilde{\beta}_{\text{OLS}})$.

In order to analyze the mean square of $\tilde{x}_{\text{test}}^T \text{ct}(\tilde{\beta}_{\text{OLS}})$, we use the fact that for any matrices $A \in \mathbb{R}^{n_1 \times n_2}$ and $B \in \mathbb{R}^{n_1 \times n_2}$, $\text{Trace}\left(A^T B\right) = \text{Trace}\left(B A^T\right)$, and also that for any random matrix $\widetilde{M}$,

$$\mathbb{E}[\text{Trace}(\widetilde{M})] = \text{Trace}(\mathbb{E}[\widetilde{M}]) \tag{12.233}$$

by linearity of expectation (Theorem 11.8). This allows us to decouple $\tilde{x}_{\text{test}}$ and $\text{ct}(\tilde{\beta}_{\text{OLS}})$. By Theorem 7.20 and independence of $\tilde{x}_{\text{test}}\tilde{x}_{\text{test}}^T$ and $\text{ct}(\tilde{\beta}_{\text{OLS}}) \text{ct}(\tilde{\beta}_{\text{OLS}})^T$,

$$\mathbb{E}\left[\left(\tilde{x}_{\text{test}}^T \text{ct}(\tilde{\beta}_{\text{OLS}})\right)^2\right] = \mathbb{E}\left[\text{Trace}\left(\tilde{x}_{\text{test}}^T \text{ct}(\tilde{\beta}_{\text{OLS}}) \text{ct}(\tilde{\beta}_{\text{OLS}})^T \tilde{x}_{\text{test}}\right)\right] \tag{12.234}$$

$$= \mathbb{E}\left[\text{Trace}\left(\tilde{x}_{\text{test}}\tilde{x}_{\text{test}}^T \text{ct}(\tilde{\beta}_{\text{OLS}}) \text{ct}(\tilde{\beta}_{\text{OLS}})^T\right)\right] \tag{12.235}$$

$$= \text{Trace}\left(\mathbb{E}\left[\tilde{x}_{\text{test}}\tilde{x}_{\text{test}}^T \text{ct}(\tilde{\beta}_{\text{OLS}}) \text{ct}(\tilde{\beta}_{\text{OLS}})^T\right]\right) \tag{12.236}$$

$$= \text{Trace}\left(\mathbb{E}\left[\tilde{x}_{\text{test}}\tilde{x}_{\text{test}}^T\right] \mathbb{E}\left[\text{ct}(\tilde{\beta}_{\text{OLS}}) \text{ct}(\tilde{\beta}_{\text{OLS}})^T\right]\right) \tag{12.237}$$

$$= \text{Trace}\left(\Sigma_{\tilde{x}_{\text{test}}} \Sigma_{\tilde{\beta}_{\text{OLS}}}\right). \tag{12.238}$$

Plugging in $\Sigma_{\tilde{x}_{\text{test}}} = \Sigma_X$ and the expression for the covariance matrix of the OLS coefficients derived in Theorem 12.24 yields

$$\mathbb{E}\left[\left(\tilde{x}_{\text{test}}^T \text{ct}(\tilde{\beta}_{\text{OLS}})\right)^2\right] = \text{Trace}\left(\Sigma_X \frac{\sigma^2}{n-1} \Sigma_X^{-1}\right) \tag{12.239}$$

$$= \frac{\sigma^2}{n-1} \text{Trace}\left(I\right) \tag{12.240}$$

$$= \frac{d\sigma^2}{n-1} \tag{12.241}$$

since $\Sigma_X \Sigma_X^{-1} = I$ is a $d \times d$ identity matrix. ∎

Given a test dataset formed by n_{test} feature vectors $x_1, \ldots, x_{n_{\text{test}}}$ and a response vector $y_{\text{test}} \in \mathbb{R}^{n_{\text{test}}}$, we can approximate the test MSE using the root average residual sum of squares:

$$\text{Test error} := \sqrt{\frac{1}{n_{\text{test}}} \sum_{i=1}^{n} \left(y_{\text{test}}[i] - \ell_{\text{OLS}}(x_i)\right)^2}. \tag{12.242}$$

From now on, we denote the number of training data by n_{train}, instead of n, to avoid confusion. If the data are sampled from a distribution satisfying the assumptions of Theorem 12.28, the error should scale proportionally to $\sqrt{1 + \frac{d}{n_{\text{train}}-1}}$. This is the case for our real-data temperature example in Figure 12.10, but only when the number of training data n_{train} is large with respect to the number of features d.

When n_{train} is close to d, the test error in Figure 12.10 is much larger than the theoretical prediction in Theorem 12.28. This is not surprising: We cannot expect Theorem 12.28 to provide an accurate approximation for small n_{train}, even if the response is actually linear as in (12.148). The reason is that the theorem assumes that the sample covariance matrix of the training data is equal to the true covariance matrix of the test data. For large n_{train}, this is approximately true if the training and test features are sampled from the same distribution because the sample covariance matrix is a consistent estimator of the covariance matrix (this follows from Theorem 9.25 and an analogous result for the sample covariance). However, for small n_{train} the test sample covariance matrix is likely to be different from the true covariance matrix. As a result, the directions of high variance of the OLS coefficients no longer coincide with directions of low variance in the test data, which translates into a large test error. In this regime, the OLS estimator overfits the training data, generalizing poorly to held-out test data.

For large n_{train}, Theorem 12.28 establishes that the test MSE converges to the noise variance σ^2. In that regime, the test error essentially matches the training error (see Theorem 12.27 and also Figure 12.10) and the error of the linear MMSE estimator in Theorem 12.18. Consequently, the OLS estimator generalizes robustly to test data. This justifies our analysis in Examples 12.9 and 12.17, where we evaluate the OLS models on the same data used to train them. In both cases, the number of training data is large and there are only two features, so there is no danger of overfitting.

12.3 Ridge Regression

As described in Section 12.2.2, linear regression coefficient estimates obtained via ordinary least squares (OLS) suffer from noise amplification, when there is multicollinearity among the features. In such cases, the OLS estimator overfits the training set and generalizes poorly to held-out test data, as explained in Sections 12.2.3 and 12.2.4. Ridge regression addresses this issue by adding a *regularization* term to the OLS cost function, equal to the sum of squared amplitudes of the coefficients. The goal is to penalize the large coefficients that occur when the OLS estimator overfits.

Definition 12.29 (Ridge regression). *Let* (x_1, y_1), (x_2, y_2), $\ldots$, (x_n, y_n) *be a dataset formed by n pairs of a response y_i and a corresponding vector x_i containing d features*

$(1 \leq i \leq n)$. *We assume that the features and the response are centered, so that their sample mean is zero. The ridge-regression estimator of the response given the features is*

$$\ell_{\mathrm{RR}}(x_i) := \beta_{\mathrm{RR}}^T x_i, \quad 1 \leq i \leq n, \tag{12.243}$$

where the linear coefficients minimize the regularized least-squares cost function,

$$\beta_{\mathrm{RR}} = \arg\min_{\beta \in \mathbb{R}^d} \sum_{i=1}^{n} \left(y_i - \beta^T x_i\right)^2 + \lambda \sum_{j=1}^{d} \beta[j]^2, \tag{12.244}$$

and λ is a nonnegative regularization parameter.

The regularization parameter λ governs the trade-off between the OLS data-fidelity term $\sum_{i=1}^{n} \left(y_i - \beta^T x_i\right)^2$ and the regularization term $\sum_{j=1}^{d} \beta[j]^2$ in the ridge-regression cost function. When λ is small, the data-fidelity term dominates, so the ridge-regression estimator is close to the OLS estimator. When λ is large, the regularization term dominates, resulting in very small coefficients. This is illustrated by Figure 12.12, which shows an application of ridge regression to the temperature-estimation problem from Example 12.20, in a regime where the OLS estimator overfits, due to the small number of training data. For small λ, the ridge-regression estimator overfits, just like the OLS estimator: The training error is very small, the test error is substantially larger, and many of the coefficients have large amplitudes, which drown out the geographically meaningful coefficients associated with Bedford and Elkins (see Figure 12.8). As we increase λ, the magnitudes of the noisy coefficients decrease, the meaningful coefficients rise above the rest, and the test error improves. However, if we increase λ too much, the coefficients eventually become very small, causing the model to underfit the training data, and the test error to deteriorate.

A key challenge is how to select the parameter λ in order to strike the right balance between data fidelity and regularization. Our ultimate goal is to minimize the test error, but we cannot use the test data to choose λ. The test set should be used *only* for evaluation.

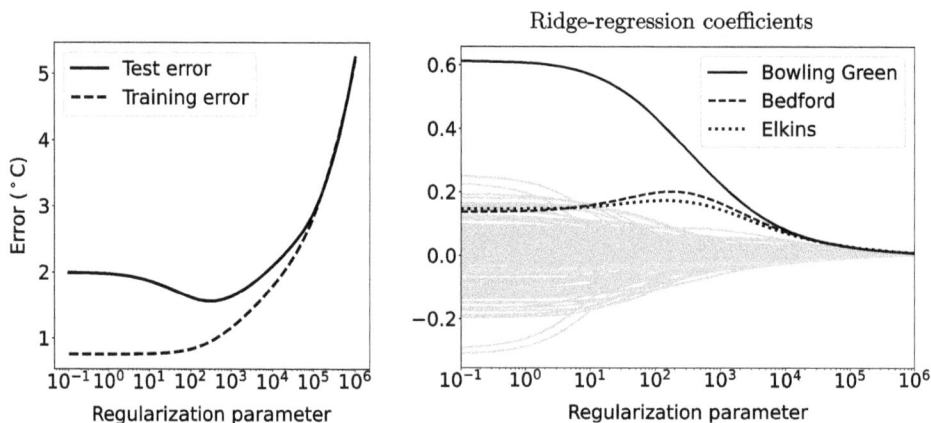

Figure 12.12 Effect of ridge-regression regularization. The left graph shows the training and test errors of the ridge-regression estimator applied to the temperature-estimation problem in Example 12.20 for different values of the regularization parameter λ. The number of training data is fixed to $n := 200$. The right graph shows the ridge-regression coefficients as a function of λ. Three coefficients corresponding to locations close to Versailles (see Figure 12.8), are highlighted in black, the rest are represented by gray lines.

Instead, we separate the training data into a training set and a held-out validation set. The training set is used to fit ridge-regression models with multiple values of λ (typically on a logarithmic grid). The validation set is used to evaluate these models and find the best λ value. This approach can be repeated multiple times using different training-validation splits, a technique known as *cross-validation.*

Figure 12.13 compares the performance of OLS and ridge regression on the temperature-estimation task from Example 12.20 for different values of the number of training data n. The regularization parameter is set via cross-validation for each n. When n is small, ridge-regression regularization prevents overfitting by neutralizing noise amplification in the coefficients. This results in a worse fit to the training data than OLS (higher training error) and much better generalization to the test data (lower test error, close to the training error). For large n, OLS does not overfit, and both methods yield similar results.

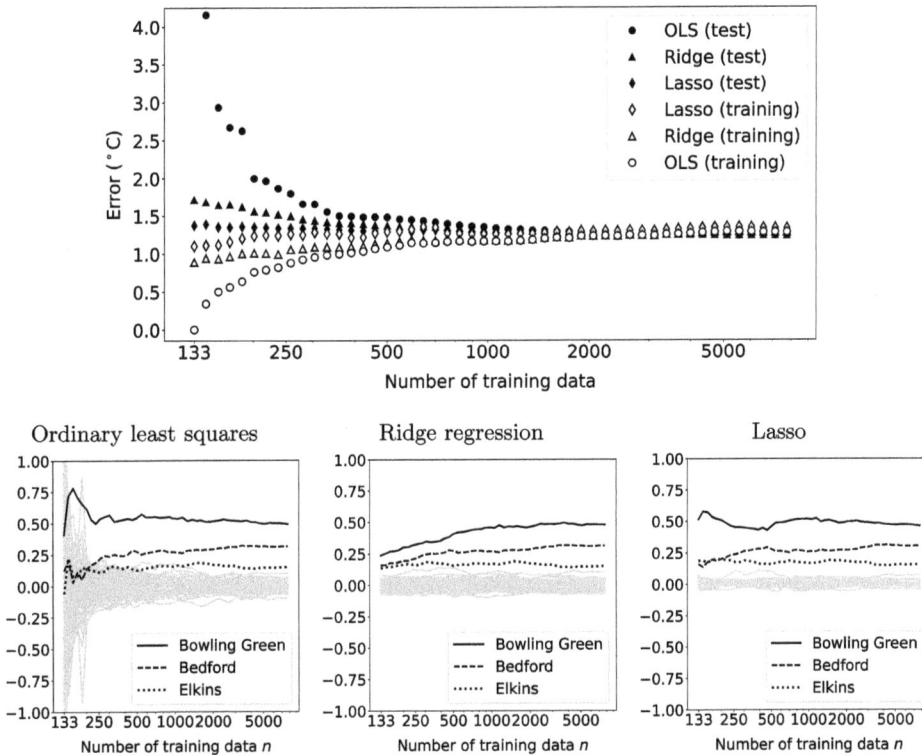

Figure 12.13 Comparison between ordinary least squares, ridge regression and the lasso. The top graph shows the training and test errors of OLS, ridge regression and the lasso applied to the temperature-estimation problem in Example 12.20, as a function of the number of training data n. The linear coefficients of the three models are depicted in the graphs at the bottom. For small n, OLS overfits the training data and has many large noisy coefficients (gray lines), which drown out the three geographically meaningful coefficients corresponding to Bowling Green, Bedford and Elkins (black lines). Ridge regression controls noise amplification in the coefficients, so that the three meaningful coefficients have the largest magnitude, resulting in better generalization to the test data and less overfitting (see the top graph). The lasso yields sparse coefficients, making the meaningful coefficients even more prominent, which further improves the test error. For large n, the three models have similar performance.

As in the case of OLS (see Theorem 12.7), we can derive an explicit formula for the ridge-regression coefficients that only depends on the sample covariance matrix of the features, and on the sample cross-covariance between the features and the response.

Theorem 12.30 (Ridge-regression estimator). *Let* (x_1, y_1), (x_2, y_2), ..., (x_n, y_n) *be a dataset formed by* n *pairs of a response* y_i *and a corresponding vector* x_i *containing* d *features* $(1 \leq i \leq n)$. *If the features and the response are centered to have zero sample mean, the linear coefficients of the ridge-regression estimator equal*

$$\beta_{\mathrm{RR}} = \left(\Sigma_X + \frac{\lambda}{n-1}I \right)^{-1} \Sigma_{XY}, \qquad (12.245)$$

where Σ_X *is the sample covariance matrix of the features,* Σ_{XY} *the sample cross-covariance of the features and the response, and* λ *the ridge-regression regularization parameter.*

Proof   Let us denote the feature design matrix by $X_{\mathrm{train}} \in \mathbb{R}^{n \times d}$ and the response vector by $y_{\mathrm{train}} \in \mathbb{R}^n$, defined as in (12.81) and (12.82). The ridge-regression cost function is equivalent to the cost function of a modified OLS estimator,

$$\sum_{i=1}^{n} \left(y_i - \beta^T x_i \right)^2 + \lambda \sum_{j=1}^{d} \beta[j]^2 = ||y_{\mathrm{train}} - X_{\mathrm{train}}\beta||_2^2 + \left\| 0 - \sqrt{\lambda}I\beta \right\|_2^2 \qquad (12.246)$$

$$= \left\| \begin{bmatrix} y_{\mathrm{train}} \\ 0 \end{bmatrix} - \begin{bmatrix} X_{\mathrm{train}} \\ \sqrt{\lambda}I \end{bmatrix} \beta \right\|_2^2 \qquad (12.247)$$

$$= ||y_{\mathrm{RR}} - X_{\mathrm{RR}}\beta||_2^2, \qquad (12.248)$$

with design matrix

$$X_{\mathrm{RR}} := \begin{bmatrix} X_{\mathrm{train}} \\ \sqrt{\lambda}I \end{bmatrix} \qquad (12.249)$$

and response vector

$$y_{\mathrm{RR}} := \begin{bmatrix} y_{\mathrm{train}} \\ 0 \end{bmatrix}. \qquad (12.250)$$

Here $0 \in \mathbb{R}^d$ denotes a vector containing d zeros and I is a $d \times d$ identity matrix. By Lemma 12.8, the coefficients of the modified OLS estimator equal

$$\beta_{\mathrm{RR}} := \arg \min_{\beta \in \mathbb{R}^d} ||y_{\mathrm{RR}} - X_{\mathrm{RR}}\beta||_2^2 \qquad (12.251)$$

$$= \left(X_{\mathrm{RR}}^T X_{\mathrm{RR}} \right)^{-1} X_{\mathrm{RR}}^T y_{\mathrm{RR}} \qquad (12.252)$$

$$= \left(\begin{bmatrix} X_{\mathrm{train}}^T & \sqrt{\lambda}I \end{bmatrix} \begin{bmatrix} X_{\mathrm{train}} \\ \sqrt{\lambda}I \end{bmatrix} \right)^{-1} \begin{bmatrix} X_{\mathrm{train}}^T & \sqrt{\lambda}I \end{bmatrix} \begin{bmatrix} y_{\mathrm{train}} \\ 0 \end{bmatrix} \qquad (12.253)$$

$$= \left(X_{\mathrm{train}}^T X_{\mathrm{train}} + \lambda I \right)^{-1} X_{\mathrm{train}}^T y_{\mathrm{train}} \qquad (12.254)$$

$$= \left(\Sigma_X + \frac{\lambda}{n-1}I \right)^{-1} \Sigma_{XY} \qquad (12.255)$$

because $\Sigma_X = \frac{1}{n-1}X_{\mathrm{train}}^T X_{\mathrm{train}}$ by (12.89) and $\Sigma_{XY} = \frac{1}{n-1}X_{\mathrm{train}}^T y_{\mathrm{train}}$ by (12.91). ∎

Comparing Theorems 12.30 and 12.7, the difference between the ridge-regression and the OLS coefficients is that the matrix multiplying the sample cross-covariance Σ_{XY} is the inverse of $\Sigma_X + \frac{\lambda}{n-1}I$, instead of the inverse of the sample covariance matrix Σ_X. This results in a simple relationship between the OLS and ridge-regression coefficients, which becomes apparent when we express them in the basis of principal directions of the features. The component of the ridge-regression coefficients in each principal direction is equal to the corresponding OLS component, shrunk by a factor that depends on the regularization parameter λ and the sample variance of the features in that direction.

Theorem 12.31 (Relationship between ridge-regression and OLS coefficients). *Let* (x_1, y_1), $(x_2, y_2), \ldots, (x_n, y_n)$ *be a dataset formed by* n *pairs of a response* y_i *and a d-dimensional feature vector* x_i $(1 \le i \le n)$, *with an invertible sample covariance matrix* Σ_X. *We assume that the features and the response are centered to have zero sample mean. Let*

$$\beta_{\text{OLS}} = \sum_{j=1}^{d} c_{\text{OLS}}[j] u_j, \tag{12.256}$$

$$c_{\text{OLS}}[j] := u_j^T \beta_{\text{OLS}}, \qquad 1 \le j \le d, \tag{12.257}$$

be the representation of the OLS coefficients in the basis of principal directions of the features $u_1, u_2, \ldots, u_d$, *which are the eigenvectors of* Σ_X *normalized to have unit* ℓ_2 *norm. The ridge-regression coefficients equal*

$$\beta_{\text{RR}} = \sum_{j=1}^{d} c_{\text{RR}}[j] u_j, \tag{12.258}$$

$$c_{\text{RR}}[j] := \frac{c_{\text{OLS}}[j]}{1 + \lambda_n/\xi_j}, \tag{12.259}$$

where $\lambda_n := \frac{\lambda}{n-1}$, λ *is the ridge-regression regularization parameter, and* $\xi_1 \ge \xi_2 \ge \cdots \ge \xi_d$ *are the sample variances of the features in each of their principal directions (the eigenvalues of* Σ_X).

Proof Let $U := \begin{bmatrix} u_1 & u_2 & \cdots & u_d \end{bmatrix}$ be the matrix of eigenvectors of Σ_X, and Λ the diagonal matrix of eigenvalues $\xi_1 \ge \xi_2 \ge \cdots \ge \xi_d$, so that $\Sigma_X = U \Lambda U^T$. The eigenvectors are orthonormal by Theorem 11.19, so the inverse of U is U^T and vice versa. Consequently, the inverse of Σ_X equals

$$\Sigma_X^{-1} = U \Lambda^{-1} U^T = \sum_{j=1}^{d} \frac{1}{\xi_j} u_j u_j^T. \tag{12.260}$$

By Theorem 12.7,

$$\beta_{\text{OLS}} = \Sigma_X^{-1} \Sigma_{XY} \tag{12.261}$$

$$= \sum_{j=1}^{d} \frac{1}{\xi_j} u_j u_j^T \Sigma_{XY}, \tag{12.262}$$

which implies

$$c_{\text{OLS}}[j] = \frac{u_j^T \Sigma_{XY}}{\xi_j}. \tag{12.263}$$

Since $UU^T = I$,

$$\Sigma_X + \lambda_n I = U \Lambda U^T + \lambda_n U U^T \tag{12.264}$$
$$= U \left(\Lambda + \lambda_n I \right) U^T, \tag{12.265}$$

so

$$\left(\Sigma_X + \lambda_n I \right)^{-1} = U \left(\Lambda + \lambda_n I \right)^{-1} U^T = \sum_{j=1}^{d} \frac{1}{\xi_j + \lambda_n} u_j u_j^T. \tag{12.266}$$

By Theorem 12.30,

$$\beta_{\text{RR}} = \left(\Sigma_X + \lambda_n I \right)^{-1} \Sigma_{XY} \tag{12.267}$$

$$= \sum_{j=1}^{d} \frac{1}{\xi_j + \lambda_n} u_j u_j^T \Sigma_{XY}, \tag{12.268}$$

which implies

$$c_{\text{RR}}[j] = \frac{u_j^T \Sigma_{XY}}{\xi_j + \lambda_n} \tag{12.269}$$

$$= \frac{\xi_j}{\xi_j + \lambda_n} \frac{u_j^T \Sigma_{XY}}{\xi_j} \tag{12.270}$$

$$= \frac{c_{\text{OLS}}[j]}{1 + \lambda_n/\xi_j}. \tag{12.271}$$

∎

Theorem 12.31 reveals why ridge regularization is able to alleviate noise amplification in OLS coefficients. As explained in Section 12.2.2, noise amplification occurs when the features have low variance in certain directions because there is multicollinearity in the design matrix. Consider a regression problem where the features have high variance ξ_1 in a certain principal direction u_1 and very low variance ξ_2 in another principal direction u_2. As a result, the component of the OLS coefficient in the high-variance direction $c_{\text{OLS}}[1] := u_1^T \beta_{\text{OLS}}$ is mostly unaffected by the noise, whereas the component in the low-variance direction $c_{\text{OLS}}[2] := u_2^T \beta_{\text{OLS}}$ is noisy and unreliable (see Example 12.22). To reduce noise amplification, we would like to suppress $c_{\text{OLS}}[2]$, but not $c_{\text{OLS}}[1]$. Ridge regression achieves this, as long as λ_n is selected so that $\lambda_n/\xi_2 \gg 1$ and $\lambda_n/\xi_1 \ll 1$, which is possible since ξ_1 is much larger than ξ_2. By Theorem 12.31, for such λ_n,

$$c_{\text{RR}}[1] = \frac{c_{\text{OLS}}[1]}{1 + \lambda_n/\xi_1} \approx c_{\text{OLS}}[1], \tag{12.272}$$

$$c_{\text{RR}}[2] = \frac{c_{\text{OLS}}[2]}{1 + \lambda_n/\xi_2} \approx 0, \tag{12.273}$$

so ridge regression selectively suppresses the noisier component, while preserving the cleaner component, as desired. The following example shows that this can substantially improve the coefficient error when the response is a linear function of the features contaminated with additive noise.

Example 12.32 (Ridge regression and feature multicollinearity). Example 12.22 illustrates the effect of multicollinearity in the OLS coefficients for a regression problem with two features, where the response follows Definition 12.19. Inspired by Theorem 12.31, we represent the true coefficients β_{true} in the orthonormal basis of principal directions u_1 and u_2 of the features, which are the eigenvectors of the sample covariance matrix Σ_X:

$$\beta_{\text{true}} := \begin{bmatrix} 0.75 \\ 0.25 \end{bmatrix} = c_{\text{true}}[1]u_1 + c_{\text{true}}[2]u_2, \tag{12.274}$$

$$c_{\text{true}} = \begin{bmatrix} u_1 & u_2 \end{bmatrix}^T \beta_{\text{true}} = \begin{bmatrix} 0.71 \\ -0.36 \end{bmatrix}. \tag{12.275}$$

The components of the cross-covariance between the features and the noise in the basis of principal directions are

$$u_1^T \Sigma_{XZ} := -0.076, \qquad u_2^T \Sigma_{XZ} := 0.002. \tag{12.276}$$

The variance of the features in the first principal direction is much larger than in the second principal direction, due to the multicollinearity of the training data:

$$\xi_1 := 2.33, \qquad \xi_2 := 9.68 \cdot 10^{-4}. \tag{12.277}$$

By Theorem 12.21 and (12.260), the OLS coefficients equal

$$\beta_{\text{OLS}} = \beta_{\text{true}} + \Sigma_X^{-1}\Sigma_{XZ} \tag{12.278}$$

$$= \sum_{j=1}^{2} c_{\text{true}}[j]u_j + \sum_{j=1}^{2} \frac{1}{\xi_j} u_j u_j^T \Sigma_{XZ} \tag{12.279}$$

$$= \sum_{j=1}^{2} \left(c_{\text{true}}[j] + \frac{u_j^T \Sigma_{XZ}}{\xi_j} \right) u_j. \tag{12.280}$$

Consequently, the OLS coefficients in the basis of principal directions are

$$c_{\text{OLS}} = c_{\text{true}} + \begin{bmatrix} \frac{u_1^T \Sigma_{XZ}}{\xi_1} \\ \frac{u_2^T \Sigma_{XZ}}{\xi_2} \end{bmatrix} \tag{12.281}$$

$$= c_{\text{true}} + \begin{bmatrix} -0.03 \\ 2.03 \end{bmatrix}. \tag{12.282}$$

The first component $c_{\text{OLS}}[1]$ is very close to $c_{\text{true}}[1]$ because the variance in the first principal direction is large, but the second component $c_{\text{OLS}}[2]$ is completely distorted by the noise because the variance in the second principal direction is small. By Theorem 12.31 the corresponding components of the ridge-regression coefficients equal

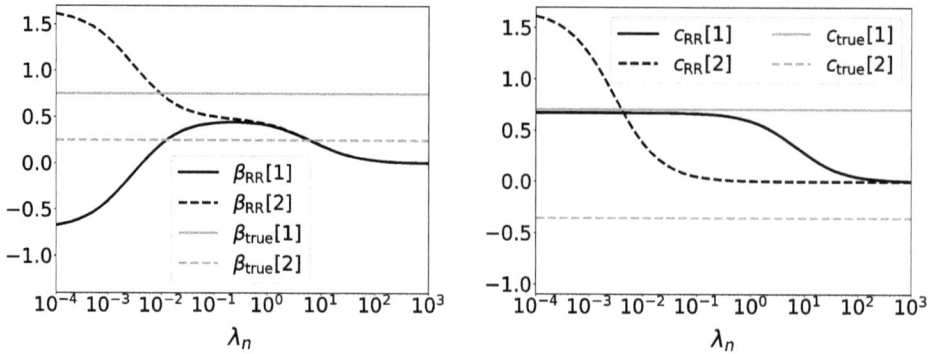

Figure 12.14 Effect of regularization in ridge regression. The left plot shows the ridge-regression coefficients (in black) as a function of the scaled regularization parameter λ_n for the regression problem in Example 12.32, compared to the true coefficients (in gray). The right plot shows the two components of the ridge-regression coefficients (in black) and the true coefficients (in gray) in the basis of principal directions of the features. Without regularization, the first OLS component is close to the truth (solid lines) because the features have high variance in that direction, whereas the second is completely distorted by the noise (dashed lines) because the features have low variance in that direction. The magnitude of both ridge-regression components decreases monotonically as λ_n increases, but the decrease is much faster for the noisy component. Consequently, for intermediate values of λ_n (e.g. $\lambda_n := 0.1$), this component is suppressed, while the first one is preserved, resulting in a better approximation to the true coefficients.

$$c_{\mathrm{RR}}[1] = \frac{c_{\mathrm{OLS}}[1]}{1 + \lambda_n/\xi_1} \tag{12.283}$$

$$= \frac{c_{\mathrm{true}}[1] - 0.03}{1 + 0.43\lambda_n}, \tag{12.284}$$

$$c_{\mathrm{RR}}[2] = \frac{c_{\mathrm{OLS}}[2]}{1 + \lambda_n/\xi_2} \tag{12.285}$$

$$= \frac{c_{\mathrm{true}}[2] + 2.03}{1 + 1033\lambda_n}, \tag{12.286}$$

so their magnitudes decrease monotonically as a function of λ_n, as depicted in the right plot of Figure 12.14. By contrast, the relationship between the coefficients β_{RR} and λ_n is much more complicated, as evinced by the trajectory of $\beta_{\mathrm{RR}}[1]$ in the left plot of Figure 12.14. This highlights the importance of analyzing the ridge-regression coefficients in the basis of principal directions.

Crucially, $c_{\mathrm{RR}}[2]$ decreases much faster than $c_{\mathrm{RR}}[1]$ as a function of λ_n. This makes it possible to choose a value of λ_n that neutralizes noise amplification. Setting $\lambda_n := 0.1$, so that

$$\frac{\lambda_n}{\xi_2} = 103.3 \gg 1 \gg 0.043 = \frac{\lambda_n}{\xi_1}, \tag{12.287}$$

suppresses the noisier coefficient, while preserving the cleaner one:

$$c_{\mathrm{RR}}[1] = \frac{c_{\mathrm{true}}[1] - 0.03}{1.043} = 0.65 \approx c_{\mathrm{true}}[1], \tag{12.288}$$

$$c_{\mathrm{RR}}[2] = \frac{c_{\mathrm{true}}[2] + 2.03}{104.3} = 0.02 \approx 0. \tag{12.289}$$

The resulting ridge-regression coefficients equal

$$\beta_{\text{RR}} = c_{\text{RR}}[1]u_1 + c_{\text{RR}}[2]u_2 \tag{12.290}$$

$$= \begin{bmatrix} 0.44 \\ 0.47 \end{bmatrix}, \tag{12.291}$$

which are much closer to $\beta_{\text{true}} := \begin{bmatrix} 0.75 \\ 0.25 \end{bmatrix}$ than the OLS coefficients $\beta_{\text{OLS}} = \begin{bmatrix} -0.71 \\ 1.65 \end{bmatrix}$. Figure 12.15 shows the contour lines of the OLS cost function (top left), the ridge-regression regularization term (top right) and the ridge-regression cost function (bottom left) for $\lambda_n := 0.1$. Regularization shifts the ridge-regression minimum toward the origin, closer to the true coefficients. Consistent with our analysis, the shift occurs mostly in the direction of u_2 (bottom right), canceling out the corresponding component.

. . . .

To gain further insight into the effect of regularization in ridge regression, we consider the linear-response assumption in Definition 12.19, when the additive noise is random. In that case, the component $\tilde{c}_{\text{OLS}}[j]$ of the OLS coefficients in the jth principal direction u_j of the features is unbiased. By Theorem 12.23 and linearity of expectation (Theorem 11.8),

$$\mathbb{E}\left[\tilde{c}_{\text{OLS}}[j]\right] = \mathbb{E}\left[u_j^T \tilde{\beta}_{\text{OLS}}\right] \tag{12.292}$$

$$= u_j^T \mathbb{E}\left[\tilde{\beta}_{\text{OLS}}\right] \tag{12.293}$$

$$= u_j^T \beta_{\text{true}} := c_{\text{true}}[j]. \tag{12.294}$$

By Theorems 11.11 and 12.24 and (12.260), the variance of the component equals

$$\text{Var}\left[\tilde{c}_{\text{OLS}}[j]\right] = \text{Var}\left[u_j^T \tilde{\beta}_{\text{OLS}}\right] \tag{12.295}$$

$$= u_j^T \Sigma_{\tilde{\beta}_{\text{OLS}}} u_j \tag{12.296}$$

$$= \frac{\sigma^2}{n-1} u_j^T \Sigma_X^{-1} u_j \tag{12.297}$$

$$= \frac{\sigma^2}{n-1} \sum_{l=1}^{d} \frac{1}{\xi_l} u_j^T u_l u_l^T u_j \tag{12.298}$$

$$= \frac{\sigma^2}{(n-1)\xi_j} \tag{12.299}$$

because $u_j^T u_l$ equals one when $j = l$, and zero otherwise. Here, $\xi_1 \geq \xi_2 \geq \cdots \geq \xi_d$ are the sample variances of the features in each of their principal directions. The jth component suffers from noise amplification, if the feature variance ξ_j in that direction is small, as depicted in Figure 12.9.

Figure 12.16 shows ridge-regression coefficients corresponding to multiple responses sampled from (12.148) for fixed features, that are the same as in the second row of Figure 12.9. As we increase the regularization parameter λ, the coefficients move toward the origin, resulting in a biased estimate that is not centered at the true coefficients (unlike the OLS coefficients, see Theorem 12.23 and Figure 12.9). However, this is accompanied by a decrease in the coefficient variance, when compared to the OLS coefficients, shown in the second row of Figure 12.9. As a result, the ridge-regression estimator tends to provide a better approximation of the true coefficients than the OLS estimator for intermediate values of λ. Theorem 12.33 provides an explanation for this. Combined with (12.299), it establishes

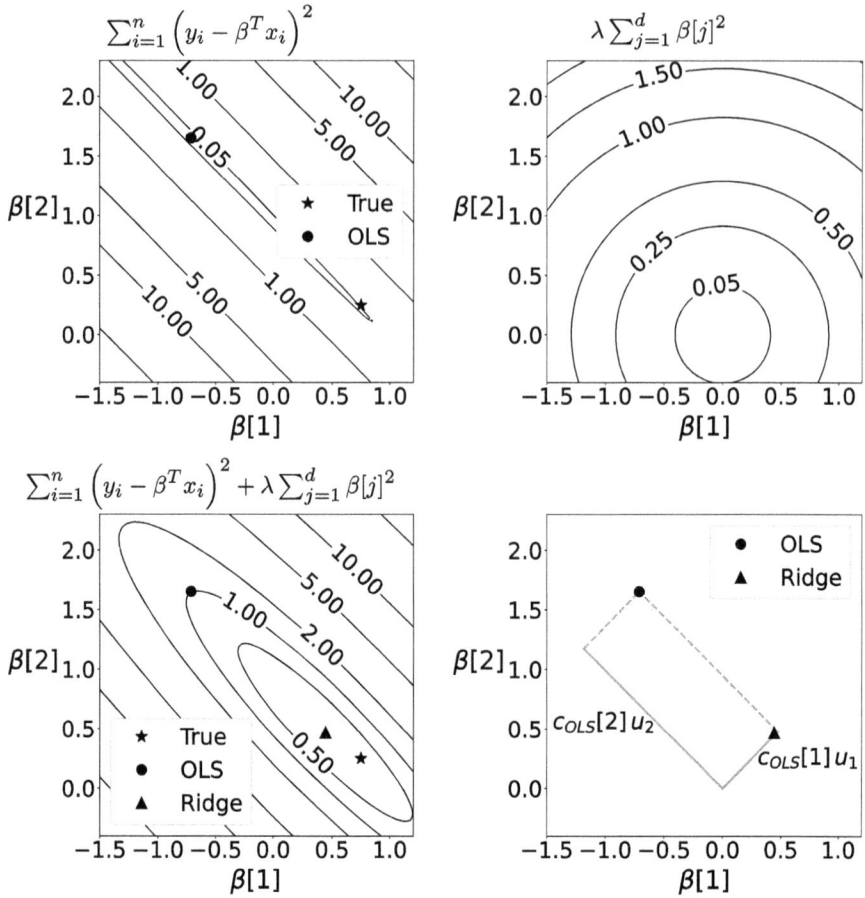

Figure 12.15 Optimization landscape of ridge regression. The top left plot shows the contour lines of the OLS cost function for the data in Examples 12.22 and 12.32. Due to multicollinearity in the feature design matrix, the minimum of the OLS cost function (circle marker) is far from the true coefficients (star marker). The top right plot shows the contour lines of the regularization term in ridge regression for $\lambda := \frac{0.1}{n-1}$. The bottom left plot shows the contour lines of the ridge-regression cost function, resulting from summing the two terms in the top row. Regularization shifts the minimum toward the origin, closer to the true coefficients (triangle marker). The bottom right plot depicts the components of the OLS coefficients in the basis of principal directions of the features, revealing that this shift takes place almost entirely in the second principal direction, as described in Example 12.32.

that the mean and variance of the component of the ridge-regression coefficients in the jth principal direction equal

$$\mathbb{E}\left[\tilde{c}_{\text{RR}}[j]\right] = \frac{c_{\text{true}}[j]}{1 + \lambda_n/\xi_j}, \tag{12.300}$$

$$\text{Var}\left[\tilde{c}_{\text{RR}}[j]\right] = \frac{\text{Var}\left[\tilde{c}_{\text{OLS}}[j]\right]}{\left(1 + \lambda_n/\xi_j\right)^2}, \tag{12.301}$$

where $\lambda_n := \frac{\lambda}{n-1}$ and λ is the ridge-regression regularization parameter. Consider two principal directions u_1 and u_2 in which the feature variance is very high and very low,

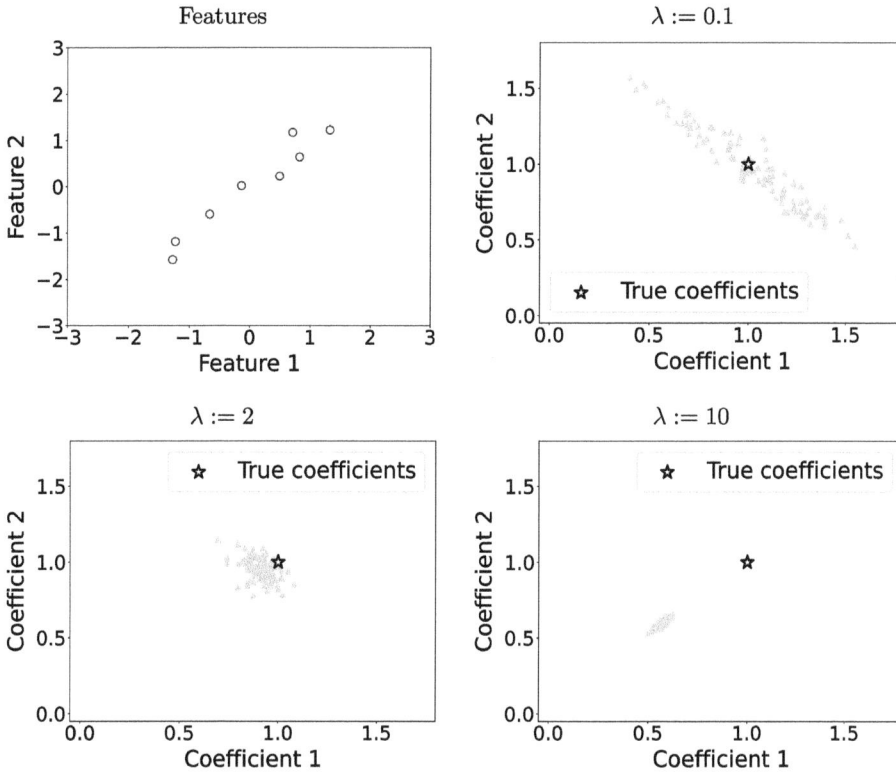

Figure 12.16 Ridge-regression coefficients for linear response with additive noise. The top left plot shows two-dimensional feature vectors corresponding to $n := 8$ training examples, which are the same as in the second row of Figure 12.9. The remaining plots show 100 ridge-regression coefficient estimates corresponding to the feature vectors for 100 different responses, generated according to Definition 12.19 using fixed-variance i.i.d. Gaussian noise and the same true linear coefficients (represented by a star). Each plot corresponds to a different value of the regularization parameter λ. As λ increases, the mean of the coefficient distribution moves toward the origin and its variance decreases, in accordance with Theorem 12.33. The variance decrease occurs more rapidly in the principal direction of minimum variance of the features, which is the direction in which the OLS coefficients have the highest variance. Consequently, the resulting coefficient estimate tends to be closer to the true coefficients than the OLS estimate for intermediate values of λ (bottom left, compare to the second row of Figure 12.9).

respectively ($\xi_1 \gg \xi_2$). Then, by (12.299), the OLS coefficient has low variance in the first direction and high variance in the second direction. If we select λ_n, so that

$$\lambda_n/\xi_2 \gg 1 \gg \lambda_n/\xi_1, \tag{12.302}$$

we suppress the variance in the second direction, while remaining unbiased in the first direction because

$$\mathbb{E}\left[\tilde{c}_{\mathrm{RR}}[1]\right] \approx c_{\mathrm{true}}[1], \tag{12.303}$$

$$\mathrm{Var}\left[\tilde{c}_{\mathrm{RR}}[2]\right] \approx 0. \tag{12.304}$$

The price to pay is bias in the second direction because

$$\mathbb{E}\left[\tilde{c}_{\mathrm{RR}}[2]\right] \approx 0. \tag{12.305}$$

The ridge-regression estimator is thus able to selectively neutralize the variance in the directions suffering from noise amplification, by biasing the corresponding components toward zero.

Theorem 12.33 (Mean and variance of the ridge-regression coefficients). *Let $x_1, \ldots, x_n$ be d-dimensional feature vectors with sample covariance matrix Σ_X, and let $\beta_{\text{true}} \in \mathbb{R}^d$ be a fixed vector of linear coefficients. The response $\tilde{y}_{\text{train}} \in \mathbb{R}^n$ is defined as in (12.148) for a noise vector $\tilde{z}_{\text{train}}$ with independent entries that have zero mean and variance σ^2. Consider the components of the true coefficients β_{true} and the ridge-regression estimator $\tilde{\beta}_{\text{RR}}$ in the basis of principal directions $u_1, \ldots, u_d$ of the features (the eigenvectors of Σ_X):*

$$c_{\text{true}}[j] := u_j^T \beta_{\text{true}}, \tag{12.306}$$

$$\tilde{c}_{\text{RR}}[j] := u_j^T \tilde{\beta}_{\text{RR}}, \qquad 1 \leq j \leq d. \tag{12.307}$$

The mean and variance of the jth ridge-regression component for $1 \leq j \leq d$ equal

$$\mathbb{E}\left[\tilde{c}_{\text{RR}}[j]\right] = \frac{c_{\text{true}}[j]}{1 + \lambda_n/\xi_j}, \tag{12.308}$$

$$\text{Var}\left[\tilde{c}_{\text{RR}}[j]\right] = \frac{1}{\left(1 + \lambda_n/\xi_j\right)^2} \frac{\sigma^2}{(n-1)\xi_j}, \tag{12.309}$$

where $\lambda_n := \frac{\lambda}{n-1}$, λ is the ridge-regression regularization parameter, and $\xi_1 \geq \xi_2 \geq \cdots \geq \xi_d$ are the sample variances of the features in each of their principal directions (the eigenvalues of Σ_X).

Proof The component of the OLS coefficients in the jth principal direction equals

$$\tilde{c}_{\text{OLS}}[j] = u_j^T \tilde{\beta}_{\text{OLS}}, \tag{12.310}$$

so by Theorem 12.31,

$$\tilde{c}_{\text{RR}}[j] = \frac{u_j^T \tilde{\beta}_{\text{OLS}}}{1 + \lambda_n/\xi_j}. \tag{12.311}$$

Consequently, by linearity of expectation (Theorem 11.8) and Theorem 12.23,

$$\mathbb{E}\left[\tilde{c}_{\text{RR}}[j]\right] = \frac{u_j^T \mathbb{E}\left[\tilde{\beta}_{\text{OLS}}\right]}{1 + \lambda_n/\xi_j} \tag{12.312}$$

$$= \frac{u_j^T \beta_{\text{true}}}{1 + \lambda_n/\xi_j} \tag{12.313}$$

$$= \frac{c_{\text{true}}[j]}{1 + \lambda_n/\xi_j}, \tag{12.314}$$

and by (12.299) and Lemma 7.36,

$$\text{Var}\left[\tilde{c}_{\text{RR}}[j]\right] = \text{Var}\left[\frac{\tilde{c}_{\text{OLS}}[j]}{1 + \lambda_n/\xi_j}\right] \tag{12.315}$$

$$= \frac{\text{Var}\left[\tilde{c}_{\text{OLS}}[j]\right]}{\left(1 + \lambda_n/\xi_j\right)^2} \tag{12.316}$$

$$= \frac{1}{\left(1 + \lambda_n/\xi_j\right)^2} \frac{\sigma^2}{(n-1)\xi_j}. \tag{12.317}$$

∎

12.4 Sparse Regression

As described in Section 12.2, linear-regression models tend to overfit when the number of data is small with respect to the number of features. In such scenarios, it is often useful to perform *variable selection*, in order to model the response using only a subset of the features. For linear models, this is equivalent to constraining the linear coefficients to be *sparse*, meaning that most of them are zero. In a linear regression model with d features, if the coefficient vector $\beta \in \mathbb{R}^d$ has k nonzero entries, then the resulting model just depends on the corresponding k features because

$$\sum_{j=1}^{d} \beta[j]x[j] = \sum_{j \in \mathcal{K}} \beta[j]x[j], \tag{12.318}$$

where $x \in \mathbb{R}^d$ is the feature vector, and $\mathcal{K}$ is the set of nonzero coefficients.

The lasso is a technique to obtain linear models with sparse coefficients, based on the insight that sparse vectors tend to have a small ℓ_1 norm. The ℓ_1 norm is equal to the sum of the absolute value of the entries of a vector. Consider the following d-dimensional vectors,

$$\beta_{\text{sparse}} := \begin{bmatrix} 1 \\ 0 \\ \cdots \\ 0 \end{bmatrix}, \qquad \beta_{\text{dense}} := \begin{bmatrix} \frac{1}{\sqrt{d}} \\ \frac{1}{\sqrt{d}} \\ \cdots \\ \frac{1}{\sqrt{d}} \end{bmatrix}. \tag{12.319}$$

The first is very sparse, only its first entry is zero. The second is completely *dense*, no entries are equal to zero (they all equal $1/\sqrt{d}$). The ℓ_2 norms of both vectors are equal to one

$$||\beta_{\text{sparse}}||_2 = \sqrt{\sum_{j=1}^{d} \beta_{\text{sparse}}[j]^2} = 1, \tag{12.320}$$

$$||\beta_{\text{dense}}||_2 = \sqrt{\sum_{j=1}^{d} \beta_{\text{dense}}[j]^2} = 1, \tag{12.321}$$

but their ℓ_1 norms are very different

$$||\beta_{\text{sparse}}||_1 = \sum_{j=1}^{d} |\beta_{\text{sparse}}[j]| = 1, \tag{12.322}$$

$$||\beta_{\text{dense}}||_1 = \sum_{j=1}^{d} |\beta_{\text{dense}}[j]| = \sqrt{d}. \tag{12.323}$$

The ℓ_1 norm of the sparse vector is much smaller, especially for large d. More generally, a unit-ℓ_2-norm vector with k nonzero entries with the same magnitude (equal to $1/\sqrt{k}$) has an ℓ_1 norm equal to $k/\sqrt{k} = \sqrt{k}$, so the sparser the vector, the smaller the ℓ_1 norm. Inspired by this, we can penalize the ℓ_1 norm to *regularize* the coefficients of a linear model, favoring sparse estimates. This approach is known as the *lasso*, originally proposed in Tibshirani (1996).

Definition 12.34 (The lasso). *Let (x_1, y_1), (x_2, y_2), ..., (x_n, y_n) be a dataset formed by n pairs of a response y_i and a corresponding vector x_i containing d features $(1 \le i \le n)$. We assume that the features and the response are centered to have zero sample mean. The lasso estimator of the response given the features is*

$$\ell_{\text{lasso}}(x_i) := \beta_{\text{lasso}}^T x_i, \quad 1 \le i \le n, \tag{12.324}$$

where the linear coefficients minimize the regularized least-squares cost function,

$$\beta_{\text{lasso}} := \arg \min_{\beta \in \mathbb{R}^d} \sum_{i=1}^{n} \left(y_i - \beta^T x_i \right)^2 + \lambda \left\| \beta \right\|_1, \tag{12.325}$$

λ is a nonnegative regularization parameter and $\left\| \beta \right\|_1 := \sum_{j=1}^{d} |\beta[j]|$.

The lasso cost function is similar to that of ridge regression (see Definition 12.29), with the crucial difference that the regularization term is designed to promote sparsity of the coefficients. As in ridge regression, the regularization parameter λ governs the trade-off between the ordinary-least-squares (OLS) data-fidelity term and the regularization term. When λ is small, the data-fidelity term dominates, so the lasso estimator is close to the OLS estimator. When λ is large, the regularization term dominates, yielding sparse coefficients. This is illustrated in Figure 12.17, which shows the results of applying the lasso to the temperature-estimation problem from Example 12.20 in a regime where the OLS estimator overfits the training data. For small λ, the coefficients have large negative and positive amplitudes, which

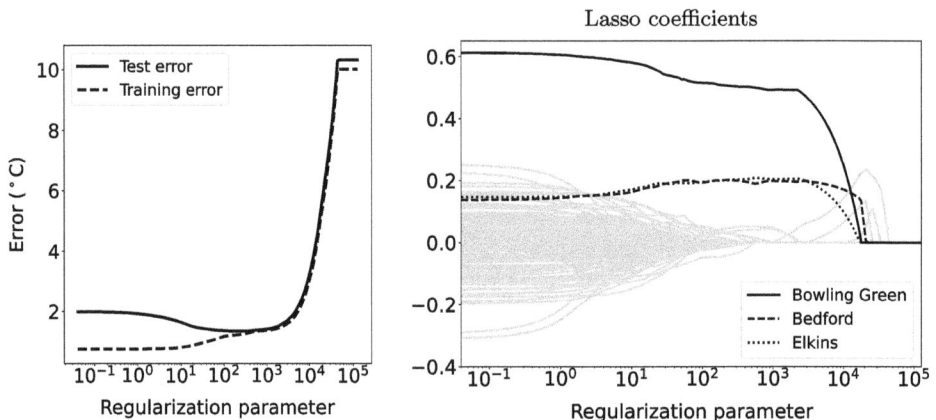

Figure 12.17 Effect of lasso regularization. The left graph shows the training and test errors of the lasso estimator applied to the temperature-estimation problem in Example 12.20 for different values of the regularization parameter λ. The number of training data is fixed to $n := 200$. The right graph shows the lasso coefficients as a function of λ. Three coefficients corresponding to locations close to Versailles (see Figure 12.8), are highlighted in black, the rest are represented by gray lines.

drown out the geographically meaningful coefficients associated with Bedford and Elkins (see Figure 12.8). This results in overfitting: The training error is small and the test error is large. As λ increases, most of the coefficients are set to zero, except for the meaningful coefficients (Bowling Green, Bedford and Elkins) and a few others. This prevents overfitting and achieves better generalization to the test data. When λ becomes too large, the regularization term dominates and all the coefficients vanish.

The plots on the right of Figures 12.17 and 12.12 show the difference between the effect of regularization in ridge regression and in the lasso. Increasing the regularization parameter λ shrinks the ridge-regression coefficients gradually toward zero.[1] By contrast, the lasso coefficients either vanish abruptly or preserve relatively large amplitudes. This is apparent in Figure 12.13, which compares OLS, ridge regression and the lasso applied to the temperature-estimation problem from Example 12.20 for different values of the number of training data n. The regularization parameter is set via cross-validation for each n. When n is small, most of the lasso coefficients are exactly zero, as opposed to ridge regression, where most coefficients have small nonzero values. In addition, the coefficients corresponding to geographically meaningful locations are larger for the lasso than for ridge regression. This results in a lower test error, which suggests that the response indeed depends mostly on the few features selected by the lasso.

Figure 12.18 compares ridge regression and the lasso under the linear-response assumption in Definition 12.19 for different realizations of the additive noise. There are two features, but the response only depends on the first one; the ground-truth coefficient corresponding to the second feature is zero. If the regularization parameter λ is large enough, the lasso estimator correctly identifies the true nonzero coefficient and sets the other to zero. By contrast, the ridge-regression coefficients are not sparse, no matter how much we increase the regularization parameter. This is consistent with Theorem 12.31: The ridge-regression estimator gradually shrinks the coefficients in the basis of principal directions of the features, which does not induce sparsity (see Exercise 12.12).

The left column of Figure 12.19 shows the contour lines of the OLS cost function (top), the lasso regularization term (center) and the lasso cost function (bottom) for a regression problem with two features. The response is a linear function of the first feature corrupted by noise, as in Figure 12.18, so the second ground-truth coefficient is zero. The regularization term reshapes the cost function, shifting its minimum to lie on the horizontal line, where the second coefficient is zero. Consequently, the lasso estimate is sparse and correctly identifies the relevant feature. The effect of the regularization term is apparent in the right column of Figure 12.19, which shows a 1D slice of the different functions restricted to the line where the first coefficient is fixed to its optimal value. Close to the origin, the linear ℓ_1-norm regularization term dominates the quadratic OLS data-fidelity term, warping the cost function so that the minimum of the cost function with respect to the second coefficient is exactly at zero.

Figure 12.20 plots the lasso coefficients for the data in Figure 12.19 as a function of the regularization parameter λ. For small λ, both coefficients are nonzero, but when λ is above a certain threshold $\lambda_{\min}$, the second coefficient vanishes. As λ increases further, the first coefficient decreases linearly and eventually vanishes too. Exercise 12.11 derives a precise

---

[1] To be precise, the regularization shrinks the components of the coefficients along the principal directions of the features, as established in Theorem 12.31.

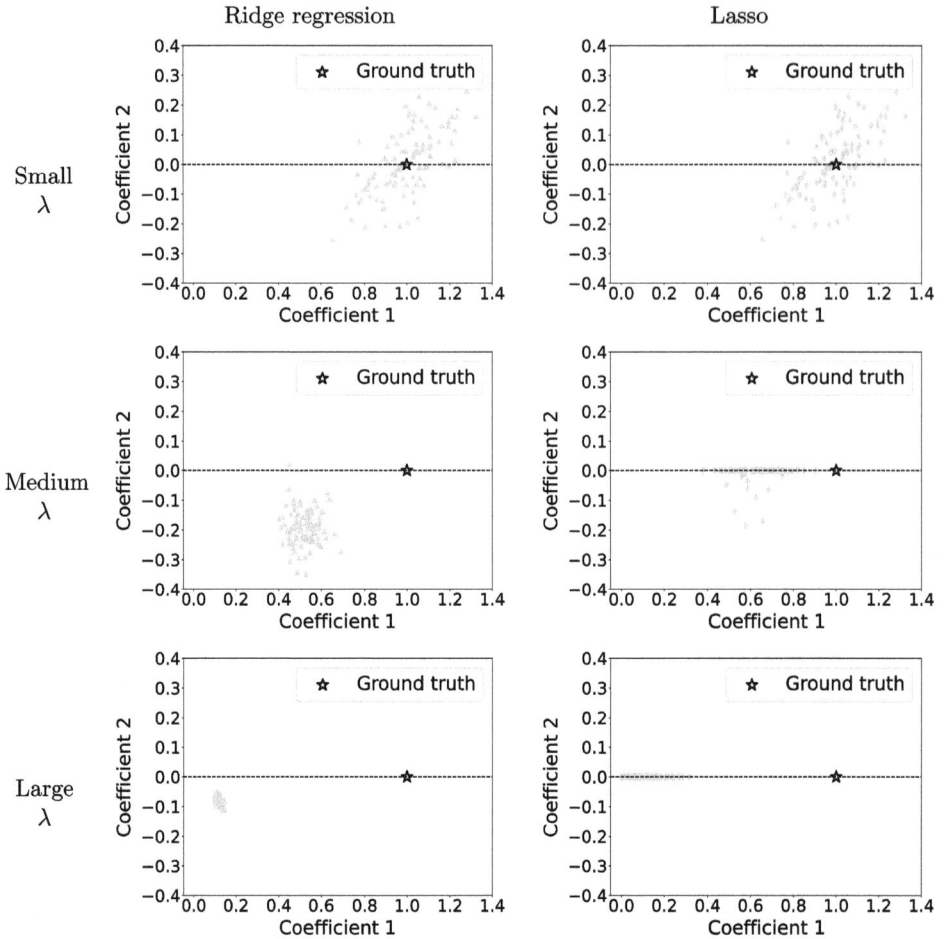

Figure 12.18 Ridge regression vs the lasso. The left column shows ridge-regression coefficients (in gray) for different values of the regularization parameter λ_{ridge}, computed from responses following Definition 12.19 with fixed-variance i.i.d. Gaussian noise. The second entry of the two-dimensional ground-truth coefficient vector, indicated by a star, is equal to zero. The right column shows the lasso coefficients (also in gray) for different values of the regularization parameter λ_{lasso}, computed from the same data. For the lasso, increasing the regularization parameter results in sparse coefficient estimates, which correctly suppress the second coefficient while preserving the first one. By contrast, ridge regression produces two nonzero coefficients, even when λ_{ridge} is large.

expression for the interval $[\lambda_{\min}, \lambda_{\max}]$ where the lasso identifies the correct feature, showing that it only depends on the correlation between the features, and the correlation between the features and the noise.

12.5 Logistic Regression

In this section, we describe logistic regression, which is a popular linear model for binary classification problems with two classes. In Section 12.6, we extend the approach to problems with more classes. Let us represent the features in a binary classification problem as a d-dimensional random vector $\tilde{x}$ and the corresponding class label as a Bernoulli random

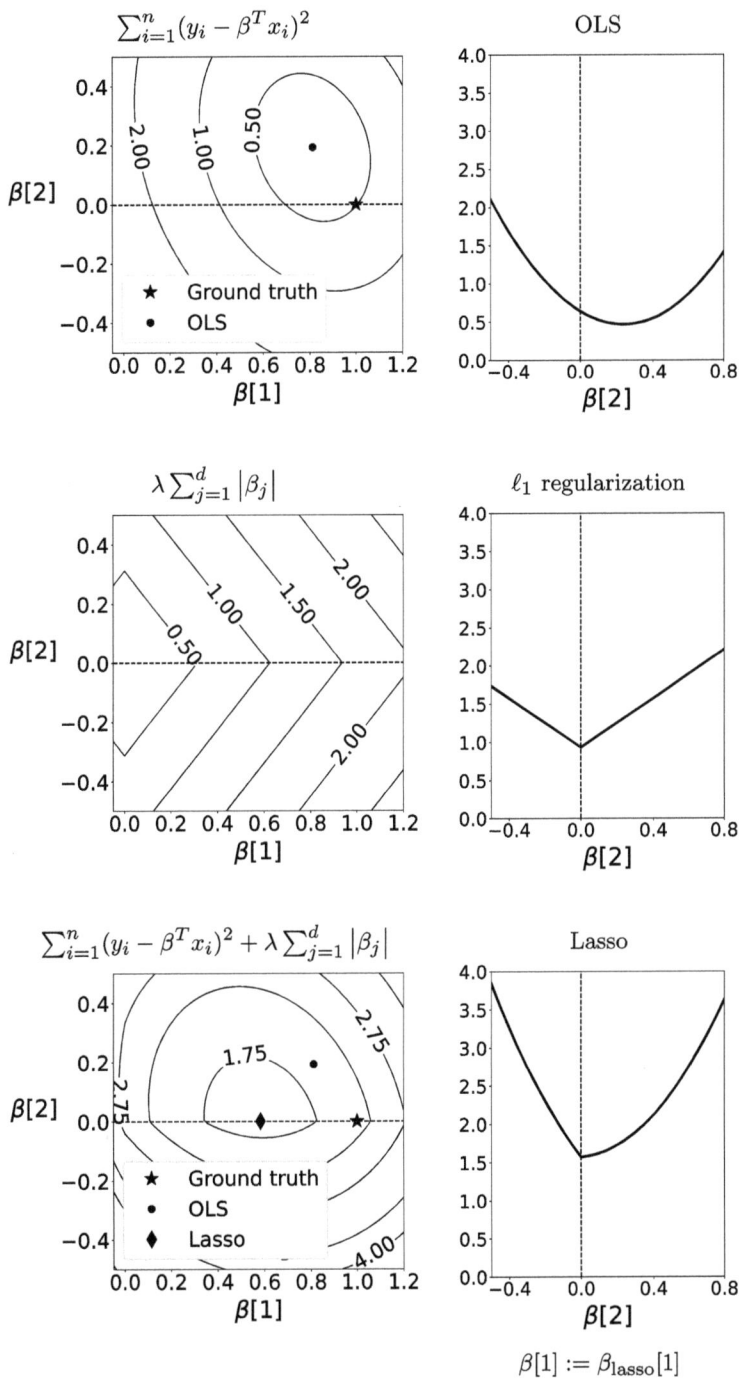

Figure 12.19 Optimization landscape of the lasso. The left column shows the contour lines of the OLS cost function (top), the lasso regularization term (center) and the lasso cost function (bottom) under the linear-response assumption in Definition 12.19, when the true second coefficient is zero (see Exercise 12.12 for the values of the feature matrix and the response). The regularization parameter is set to $\lambda := 1.6$. Even though the second true coefficient (star marker) is zero, the second OLS coefficient (circle marker) is nonzero due to the noise in the response. The ℓ_1-norm regularization shifts the minimum of the cost function to the horizontal line, so that the second lasso coefficient is zero (diamond marker). This is evident in the right column, which shows the different functions restricted to the line where the first coefficient $\beta[1]$ is equal to the first lasso coefficient $\beta_{\text{lasso}}[1]$.

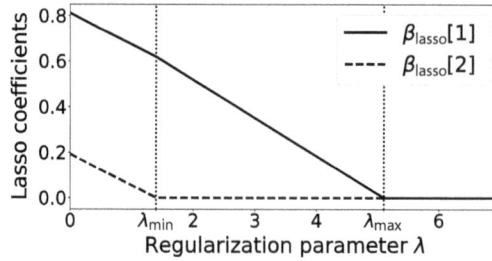

Figure 12.20 Effect of regularization in the lasso. The plot depicts the lasso coefficients for the data in Figure 12.19 as a function of the regularization parameter λ. The second coefficient decreases linearly as λ increases, until it vanishes at $\lambda := \lambda_{\min}$. From then on, it remains zero, while the first coefficient decreases linearly and eventually vanishes at $\lambda := \lambda_{\max}$. Precise expressions for $\lambda_{\min}$ and $\lambda_{\max}$ are derived in Exercise 12.11.

variable $\tilde{y}$, which equals 0 or 1. Our goal is to use a linear model to estimate the conditional probability that $\tilde{y} = 1$ given $\tilde{x} = x$, for any observed feature vector $x \in \mathbb{R}^d$. However, we cannot simply apply a linear regression model to estimate probabilities. Consider any affine function of the features,

$$\beta^T x + \alpha, \tag{12.326}$$

where $\beta \in \mathbb{R}^d$ and $\alpha \in \mathbb{R}$. For many values of x, the affine function is negative or greater than one, and consequently is not a valid probability (see Definition 1.9). To address this, we use an auxiliary function, called the *logistic function*,

$$\mathrm{lgf}(\ell) := \frac{\exp(\ell)}{1 + \exp(\ell)} = \frac{1}{1 + \exp(-\ell)}, \tag{12.327}$$

to transform the output of the affine function, so that it remains in the interval $[0, 1]$ and can be interpreted as a probability. The composition of the affine and logistic functions yields the logistic-regression model

$$\mathrm{P}\left(\tilde{y} = 1 \mid \tilde{x} = x\right) = \mathrm{lgf}(\beta^T x + \alpha). \tag{12.328}$$

Models obtained by mapping linear or affine functions of the features using one-dimensional *link* functions, such as the logistic function, are known as *generalized linear models*.

The logistic function, depicted in Figure 12.21, has several properties that make it a good link function. First, it is monotone, which means that if the affine function corresponding to a feature vector x_1 is larger than the affine function corresponding to another feature vector x_2, then so is the estimated probability,

$$\beta^T x_1 + \alpha > \beta^T x_2 + \alpha \qquad \Longrightarrow \qquad \mathrm{lgf}(\beta^T x_1 + \alpha) > \mathrm{lgf}(\beta^T x_2 + \alpha). \tag{12.329}$$

Second, the logistic function is continuous, so if two feature vectors are similar, then their respective affine functions are similar, and so are the corresponding estimated probabilities. Third, the logistic function *saturates* when its input has a large magnitude. The function tends to zero for large negative inputs, and to one for large positive inputs, so large negative and positive values of the affine function are mapped to estimated probabilities equal to almost zero or almost one, respectively. Theorem 6.7 provides additional motivation for the use of the logistic function to model probabilities.

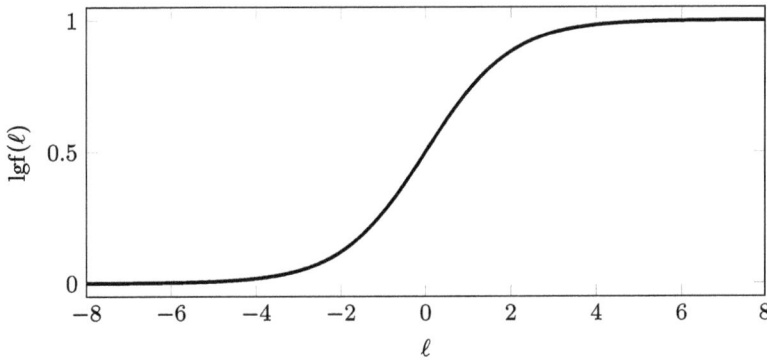

Figure 12.21 Logistic function. The logistic function is a smooth monotone function that maps the real line to the unit interval.

The inverse of the logistic function is the logit function, so its input is often called a logit. If we interpret the output of a logistic function as a probability $p := \mathrm{lgf}(\ell)$, then the associated odds (defined as the ratio between the probability of an event, and the probability of its complement) equal

$$\frac{p}{1-p} = \frac{\mathrm{lgf}(\ell)}{1-\mathrm{lgf}(\ell)} = \exp\left(\ell\right). \tag{12.330}$$

Consequently, the logit ℓ is the logarithm of the odds, or log odds. In logistic regression, the logit, and hence the log odds, are an affine function of the features:

$$\log\left(\frac{\mathrm{P}\left(\tilde{y}=1\mid\tilde{x}=x\right)}{1-\mathrm{P}\left(\tilde{y}=1\mid\tilde{x}=x\right)}\right) = \beta^{T}x + \alpha. \tag{12.331}$$

Logistic regression is a *discriminative* classification method because it seeks to directly approximate the conditional distribution of the class labels given the features. By contrast, naive Bayes (described in Section 4.8) and Gaussian discriminant analysis (described in Section 6.5) are *generative* approches because they model the full joint distribution of the labels and the features generating the data.

In order to fit a logistic-regression model to data, we maximize the corresponding likelihood function. In Sections 2.4 and 3.7, we define the likelihood of a parametric model as the joint pmf or the joint pdf of the data, interpreted as a function of the model parameters. Here we leverage a slightly different definition of likelihood tailored to discriminative models, where we condition on the available features. The likelihood is the *conditional* pmf of the observed labels, given the observed features, as a function of the model parameters. The following theorem derives the likelihood under two natural conditional-independence assumptions: (1) the labels are conditionally independent given the features, and (2) each label is conditionally independent from the features associated with other labels, given its own feature vector. As in Sections 2.4 and 3.7, we often use the log-likelihood, instead of the likelihood, because the latter tends to be very small, which can produce numerical instabilities.

Theorem 12.35 (Logistic-regression likelihood). *Let* (x_1, y_1), (x_2, y_2), ..., (x_n, y_n) *be a dataset formed by* n *pairs of a binary label* $y_i \in \{0, 1\}$ *and a corresponding feature vector* $x_i \in \mathbb{R}^d$. *We model the data as a realization from the joint distribution of random vectors* *representing the feature vectors $\tilde{x}_1, \ldots, \tilde{x}_n$ and random variables representing the labels $\tilde{y}_1, \ldots, \tilde{y}_n$. We define the likelihood of the logistic-regression model*

$$p_{\alpha,\beta}(x) := \mathrm{lgf}(\beta^T x + \alpha), \tag{12.332}$$

with parameters $\alpha \in \mathbb{R}$ and $\beta \in \mathbb{R}^d$, as the conditional pmf of the labels given the features, interpreted as a function of the parameters β and α. If the labels are conditionally independent given the features, and each label $\tilde{y}_i$ is conditionally independent from the remaining feature vectors $\{\tilde{x}_m\}_{m \neq i}$ given $\tilde{x}_i$, then the likelihood equals

$$\mathcal{L}_{XY}(\alpha, \beta) = \prod_{\{i:\, y_i=0\}} (1 - p_{\alpha,\beta}(x_i)) \prod_{\{l:\, y_l=1\}} p_{\alpha,\beta}(x_l), \tag{12.333}$$

and the corresponding log-likelihood is

$$\log \mathcal{L}_{XY}(\alpha, \beta) = \sum_{\{i:\, y_i=0\}} \log (1 - p_{\alpha,\beta}(x_i)) + \sum_{\{l:\, y_l=1\}} \log p_{\alpha,\beta}(x_l). \tag{12.334}$$

Proof   By the conditional independence assumptions (see Section 1.6),

$$\mathcal{L}_{XY}(\alpha, \beta) := \mathrm{P}\left(\tilde{y}_1 = y_1, \ldots, \tilde{y}_n = y_n \,\middle|\, \tilde{x}_1 = x_1, \ldots, \tilde{x}_n = x_n\right) \tag{12.335}$$

$$= \prod_{i=1}^{n} \mathrm{P}\left(\tilde{y}_i = y_i \,\middle|\, \tilde{x}_1 = x_1, \ldots, \tilde{x}_n = x_n\right) \tag{12.336}$$

$$= \prod_{i=1}^{n} \mathrm{P}\left(\tilde{y}_i = y_i \,\middle|\, \tilde{x}_i = x_i\right) \tag{12.337}$$

$$= \prod_{\{i:\, y_i=0\}} (1 - p_{\alpha,\beta}(x_i)) \prod_{\{l:\, y_l=1\}} p_{\alpha,\beta}(x_l) \tag{12.338}$$

because $\mathrm{P}\left(\tilde{y}_i = 0 \,\middle|\, \tilde{x}_i = x_i\right) = 1 - p_{\alpha,\beta}(x_i)$ and $\mathrm{P}\left(\tilde{y}_l = 1 \,\middle|\, \tilde{x}_l = x_l\right) = p_{\alpha,\beta}(x_l)$, according to the model.   ∎

The log-likelihood function derived in Theorem 12.35 is concave (see Section 7.1 in Boyd and Vandenberghe (2004)), but it does not have a closed-form maximum like the likelihood functions in Chapters 2 and 3. Consequently, logistic-regression models must be fit via iterative methods, such as gradient ascent (see Exercise 12.13) or faster alternatives such as Newton's method and iterative reweighted least squares.

Example 12.36 (Sex classification based on height).   In this example, we analyze the heights of 4,082 men and 1,986 women from the United States army, extracted from Dataset 5. Our goal is to fit a logistic-regression model that predicts the sex of an individual from their height. The top left graph in Figure 12.22 shows the log-likelihood function of the model. The top right plot compares the affine logits corresponding to three different choices of model parameters, one of which is the maximum-likelihood estimate. These affine functions of the feature are mapped to the estimated probabilities by the logistic function. As depicted in the plots at the bottom, the parameters maximizing the likelihood provide a much better fit to the data than the other two choices. These parameters yield the following estimator for the conditional probability of an individual being male given that their height is h:

$$p_{\alpha_{\mathrm{ML}},\beta_{\mathrm{ML}}}(h) := \frac{1}{1 + \exp(-0.29h + 48.6)}. \tag{12.339}$$

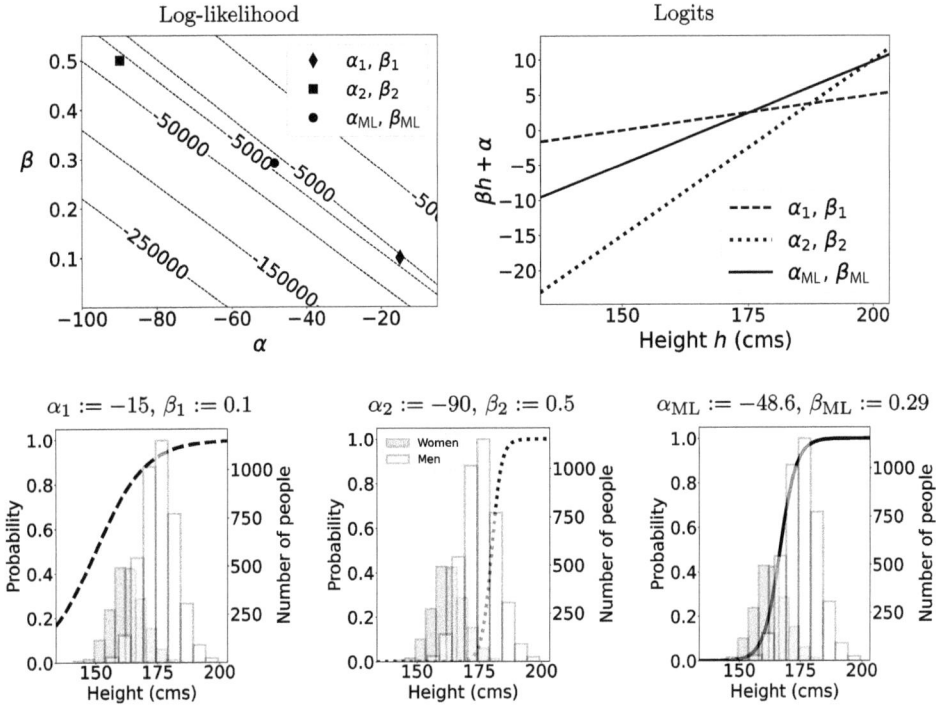

Figure 12.22 Logistic-regression model for sex. The top left graph shows the log-likelihood function of the logistic-regression model in Example 12.36. The top right plot shows the affine logits corresponding to three different choices of the parameters, including the maximum-likelihood values that maximize the log-likelihood. The plots at the bottom show the estimated conditional probability of male given height, for the three choices of parameters. The maximum-likelihood parameters provide the best discrimination between the heights of men (white histogram) and women (gray histogram) observed in the data.

The estimator is similar to the one obtained from the generative model in Example 6.3, which is depicted in Figure 6.5, but not the same (compare (12.339) to (6.35)).

. . . .

Figure 12.23 shows the result of performing automatic diagnosis of Alzheimer's disease using a logistic-regression model. As in Section 6.5, our goal is to distinguish between healthy subjects and patients with Alzheimer's using the normalized volumes of their hippocampus and entorhinal cortex. The figure shows the two-dimensional logit plane corresponding to the maximum-likelihood estimate of the model parameters. The logits are mapped to the unit interval by the logistic function. The contour lines of the resulting probability estimate are linear and orthogonal to the linear coefficients β_{ML}. A thorough evaluation of this logistic-regression model is reported in Section 12.9. The model is almost identical to the linear discriminant analysis model in Figure 6.7, which illustrates that discriminative and generative approaches can produce very similar estimators.

Just like linear regression models (see Section 12.2), logistic-regression models tend to overfit when there is multicollinearity among the features. This can be alleviated using an additive regularization term to penalize the ℓ_2 norm of the model coefficients, as in ridge regression (see Section 12.3). The resulting cost function or loss is

$$- \log \mathcal{L}_{XY}(\alpha, \beta) + \lambda \, \|\beta\|_2^2 , \qquad (12.340)$$

Logit: $-11.9\,x_{\text{hippocampus}} - 10.5\,x_{\text{entorhinal}} + 5.9$

Probability: $\text{lgf}\left(-11.9\,x_{\text{hippocampus}} - 10.5\,x_{\text{entorhinal}} + 5.9\right)$

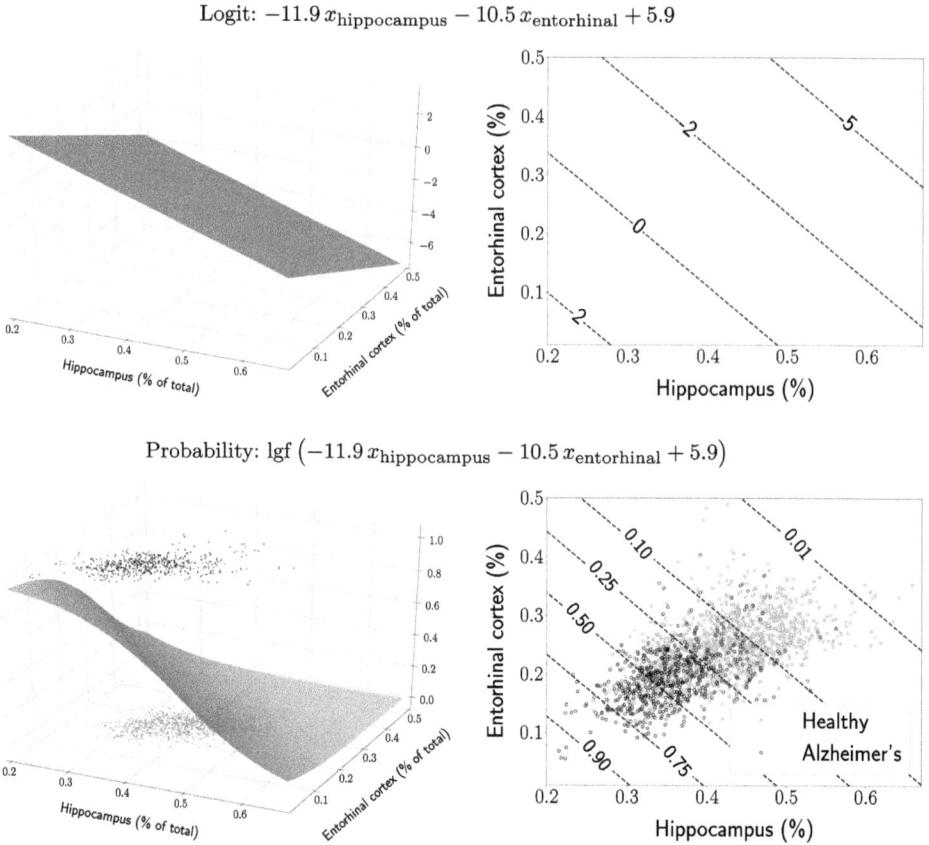

Figure 12.23 Automatic diagnosis of Alzheimer's via logistic regression. The top row shows the 3D plot (left) and contour lines (right) of the logits of a logistic-regression model to perform diagnosis of Alzheimer's from normalized volumes of the hippocampus and the entorhinal cortex. The model parameters are obtained by applying maximum-likelihood estimation to a training set of 1,926 labeled volumes, extracted from magnetic-resonance scans in the Alzheimer's Disease Neuroimaging Initiative dataset (ADNI, 2020). The logistic function maps the logits to the corresponding probability estimates. The bottom row shows the 3D plot (left) and contour lines (right) of these probabilities, superposed onto a scatterplot of the features corresponding to the healthy subjects (light gray) and Alzheimer's patients (black) in the training set.

where $\log \mathcal{L}_{XY}$ is the log likelihood derived in Theorem 12.35 and $\lambda > 0$ is a regularization parameter, which can be set using validation data. Example 12.39 illustrates the benefits of this approach for softmax regression, a generalization of logistic regression presented in Section 12.6. Alternatively, the ℓ_2 norm can be replaced by the ℓ_1 norm to promote sparse coefficients, as in the lasso (see Section 12.4).

12.6 Softmax Regression

In this section, we present a generalization of logistic regression (see Section 12.5) to classification problems with more than two classes.

Let us represent the features in a classification problem as a d-dimensional random vector $\tilde{x}$, and the corresponding class label as a random variable $\tilde{y}$ with c possible values, which we arbitrarily select to be the integers from one to c. Consider a model consisting of c different affine functions of the input feature vector $x \in \mathbb{R}^d$, one for each class:

$$\ell_k := \beta_k^T x + \alpha_k, \qquad 1 \leq k \leq c. \tag{12.341}$$

As mentioned in Section 12.5, we cannot directly use these affine functions to estimate the conditional probability that $\tilde{y} = k$ given $\tilde{x} = x$ because they are not necessarily between zero and one. Consequently, we need to transform them so that they are valid probability estimates. This can be achieved by applying an exponential function to each of the affine functions, and then normalizing the result, so that all the conditional probabilities sum to one. This operation is known as the *softmax* because it tends to push the entry with the largest magnitude toward one and the rest of entries toward zero (see Example 12.38). The resulting softmax-regression model for the conditional probability is

$$\mathrm{P}\left(\tilde{y} = k \,|\, \tilde{x} = x\right) = \frac{\exp\left(\ell_k\right)}{\sum_{l=1}^{c} \exp\left(\ell_l\right)} \tag{12.342}$$

$$= \frac{\exp\left(\beta_k^T x + \alpha_k\right)}{\sum_{l=1}^{c} \exp\left(\beta_l^T x + \alpha_l\right)}, \qquad 1 \leq k \leq c. \tag{12.343}$$

The softmax-regression model is overparametrized, meaning that the same level of expressiveness can be achieved with only $c-1$ coefficient vectors and intercepts. For instance, we can just set β_c and α_c to zero. For $c := 2$, this yields a model that is equivalent to the two-class logistic-regression model defined in Section 12.5. By analogy with logistic regression, the inputs $\ell_1, \ldots, \ell_c$ to the softmax function are often called logits.

In order to fit a softmax-regression model from data, we maximize the corresponding log-likelihood, provided by the following theorem. We omit the proof, as it is identical to that of Theorem 12.35.

Theorem 12.37 (Softmax-regression likelihood). *Let (x_1, y_1), (x_2, y_2), $\ldots$, (x_n, y_n) be a dataset formed by n pairs of a label $y_i \in \{1, \ldots, c\}$ (where $c \geq 2$ is the number of classes) and a corresponding feature vector $x_i \in \mathbb{R}^d$. We model the data as a realization sampled from a joint distribution of feature vectors $\tilde{x}_1, \ldots, \tilde{x}_n$ and labels $\tilde{y}_1, \ldots, \tilde{y}_n$. We define the likelihood of the softmax-regression model*

$$p_{\Theta}\left(x\right)_k := \frac{\exp\left(\beta_k^T x + \alpha_k\right)}{\sum_{l=1}^{c} \exp\left(\beta_l^T x + \alpha_l\right)}, \qquad 1 \leq k \leq c, \tag{12.344}$$

$$\Theta := \{\beta_1, \ldots, \beta_c, \alpha_1, \ldots, \alpha_c\}, \tag{12.345}$$

as the conditional pmf of the observed labels given the observed features. If the labels are conditionally independent given the features, and each label $\tilde{y}_i$ is conditionally independent from the remaining feature vectors $\{\tilde{x}_l\}_{l \neq i}$ given $\tilde{x}_i$, then the likelihood equals

$$\mathcal{L}_{XY}(\Theta) = \prod_{k=1}^{c} \prod_{\{i:\, y_i = k\}} p_{\Theta}\left(x_i\right)_k, \tag{12.346}$$

and the corresponding log-likelihood is

$$\log \mathcal{L}_{XY}(\Theta) = \sum_{k=1}^{c} \sum_{\{i:\, y_i = k\}} \log p_{\Theta}\left(x_i\right)_k. \tag{12.347}$$

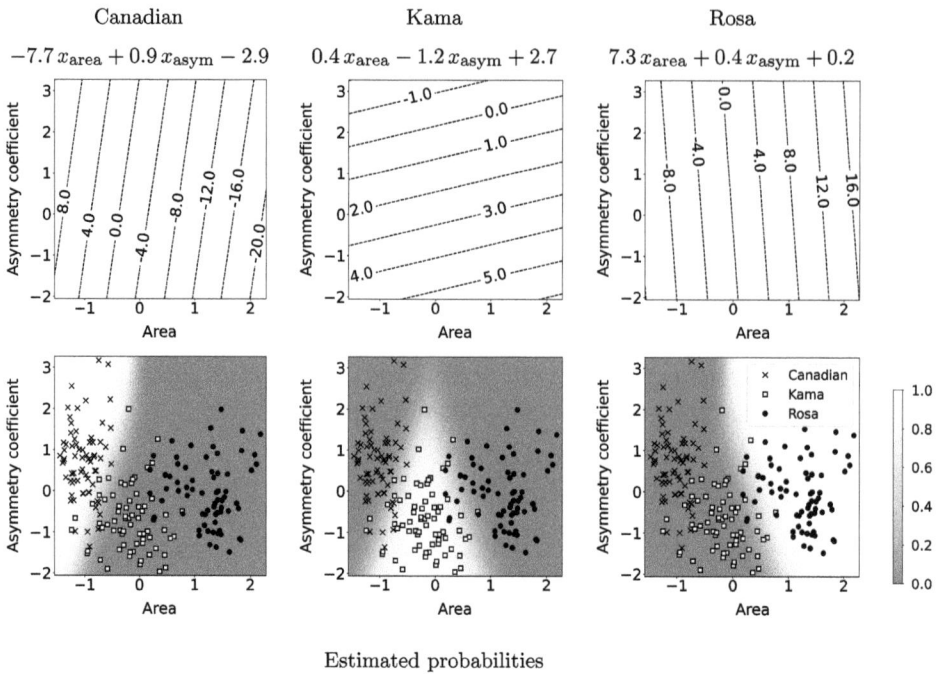

Figure 12.24 Classification of wheat seeds via softmax regression. Visualization of a softmax-regression model to classify wheat seeds into three varieties, according to their area and asymmetric coefficient. Each column corresponds to one variety (Canadian, Kama, and Rosa). The top row depicts the contour lines of the logits or softmax inputs in the model for each class. The logits are mapped via the softmax operation to the estimated probabilities, shown in the bottom row superposed onto the training data. The probabilities discriminate effectively between the three wheat varieties.

Example 12.38 (Seed classification). We consider the problem of classifying wheat seeds based on their surface area and asymmetric coefficient, using data from Dataset 21. The three classes are the wheat varieties Canadian, Kama, and Rosa. We center and normalize the features so that each has zero sample mean and unit sample variance, and then fit a softmax-regression model. The top row in Figure 12.24 shows the three logits corresponding to maximum-likelihood estimates of the coefficient vectors and intercepts. A seed with area $x_{\text{area}} := -1$ and asymmetric coefficient $x_{\text{asym}} := 2$ is assigned the following three logit values by the affine transformation in the model,

$$\begin{bmatrix} \text{Canadian} \\ \text{Kama} \\ \text{Rosa} \end{bmatrix} = \begin{bmatrix} -7.7\,x_{\text{area}} + 0.9\,x_{\text{asym}} - 2.9 \\ 0.4\,x_{\text{area}} - 1.2\,x_{\text{asym}} + 2.7 \\ 7.3\,x_{\text{area}} + 0.4\,x_{\text{asym}} + 0.2 \end{bmatrix} = \begin{bmatrix} 6.6 \\ -0.1 \\ -6.3 \end{bmatrix}. \tag{12.348}$$

The softmax operation applies an exponential function to each logit and then normalizes the three values. When the exponential is applied, it amplifies the positive logit entries and maps the negative entries to small positive numbers,

$$\exp{(6.6)} = 735.095, \quad \exp{(-0.1)} = 0.905, \quad \exp{(-6.3)} = 0.002. \tag{12.349}$$

The final probability estimates are the result of dividing these values by their sum:

$$
\begin{bmatrix} P\left(\text{Canadian} \mid x_{\text{area}}, x_{\text{asym}}\right) \\ P\left(\text{Kama} \mid x_{\text{area}}, x_{\text{asym}}\right) \\ P\left(\text{Rosa} \mid x_{\text{area}}, x_{\text{asym}}\right) \end{bmatrix} = \begin{bmatrix} \frac{735.095}{735.095+0.905+0.002} \\ \frac{0.905}{735.095+0.905+0.002} \\ \frac{0.002}{735.095+0.905+0.002} \end{bmatrix} = \begin{bmatrix} 0.999 \\ 0.001 \\ 0.000 \end{bmatrix}. \tag{12.350}
$$

The softmax pushes the largest logit entry toward one and the rest toward zero. It can thus be interpreted as a *soft maximum* of the logits, which motivates its name. The bottom row of Figure 12.24 provides a visualization of the estimated probabilities for each class, as a function of the two features. The model is able to effectively discriminate between the three varieties of wheat.

. . . .

Example 12.39 (Digit classification). In this example, we apply softmax regression to classify 28×28 images of handwritten digits, extracted from Dataset 23. The classes are the digits from 0 to 9. We separate the dataset into a training set and a test set, each containing 35,000 examples. The features are the grayscale values of the 784 pixels in each image. The parameters of the softmax-model are the linear coefficients and the intercept corresponding to each of the ten classes. The number of parameters is therefore $784 \cdot 10 + 10 = 7850$.

We fit the softmax-regression model maximizing the log-likelihood function from Theorem 12.37. The training error is just 4.3%, but the test error is 10.4%, which indicates that the model overfits the training data. In order to gain insight into why, we analyze the model coefficients. According to the model, the probability that an image $x \in \mathbb{R}^{784}$ corresponds to a digit k for $k \in \{0, 1, \ldots, 9\}$ equals

$$
P\left(\tilde{y} = k \mid \tilde{x} = x\right) = \frac{\exp\left(\beta_k^T x + \alpha_k\right)}{\sum_{l=0}^{9} \exp\left(\beta_l^T x + \alpha_l\right)}, \tag{12.351}
$$

where $\beta_k \in \mathbb{R}^{784}$ and $\alpha_k \in \mathbb{R}$ are the coefficient vector and intercept associated with digit k. The model takes inner products between the image and the ten different coefficient vectors, sums the intercepts, and then maps the resulting logits to the estimated probabilities via the softmax operation.

To determine whether the coefficient vectors capture meaningful discriminative structure, we plot β_0 and β_1 in the center column of Figure 12.25, reshaping them into 28×28 images to visualize what coefficient corresponds to each pixel. Ideally, β_0 should be such that $\beta_0^T x$ is large, if x depicts a centered image of a zero, and small otherwise.[2] This can be achieved by choosing negative coefficients for pixels that are likely to have negative values in zero images after removing the mean (e.g. at the center of the image and positive coefficients for pixels that are likely to be positive (e.g. those surrounding the center, where images of zeros have particularly large values). However, the estimated coefficients look nothing like this! They take very large negative and positive values around the edges. This is a symptom of overfitting: The coefficients achieve a higher likelihood on the training set by fitting spurious patterns in the training data, which are not present in the test data.

---

[2] To be clear, here *centered* means that we have removed the sample mean of the features, not that the digit is centered within the image.

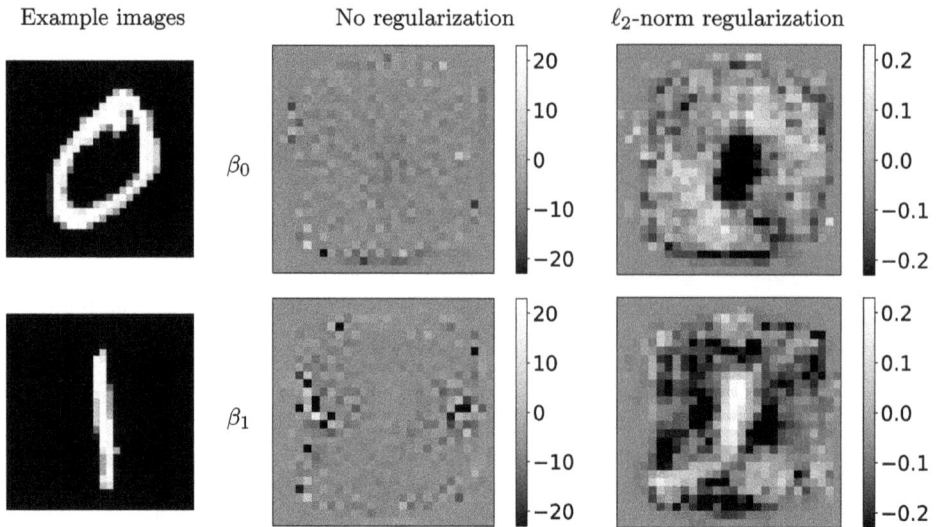

Figure 12.25 Regularization mitigates overfitting in softmax regression. The left column shows two images of handwritten digits from Dataset 23. The central column shows the coefficient vectors corresponding to zero (top) and one (bottom), obtained by fitting a softmax-regression model via maximum likelihood. The model coefficients overfit spurious patterns in the training data, which are not associated with the shape of the digits. The right column shows coefficients obtained by minimizing the regularized cost function (12.352). The regularized model provides a worse fit to the training data but generalizes better to the test data. This is reflected in the coefficients, which capture meaningful geometric features of both digits.

Section 12.2.2 reports a similar overfitting phenomenon in linear regression, which also results in noisy coefficients with large magnitudes. As explained in Section 12.3, regularization mitigates overfitting by penalizing the magnitude of the model coefficients. Motivated by this, we fit a new model, minimizing a cost function that combines the log likelihood defined in Theorem 12.37 and an additive ℓ_2-norm regularization term

$$-\log \mathcal{L}_{XY}(\alpha, \beta) + \lambda \sum_{k=0}^{9} ||\beta_k||_2^2. \qquad (12.352)$$

For simplicity, we fix the regularization parameter λ equal to 50 (in practice, we would select it using a validation set). Incorporating the regularization term increases the training error rate from 4.3% to 6.2% and decreases the test error from 10.4% to 7.8%. The regularized model does not fit the training data as well as the unregularized model, but it generalizes better to the test data. This is reflected in the model coefficients, shown in the rightmost column of Figure 12.25. The coefficients of the regularized model look much more reasonable than those of the unregularized model, clearly capturing meaningful geometric structure associated with the corresponding digits.

. . . .

12.7 Tree-Based Models

Tree-based models are popular methods for regression and classification. Sections 12.7.1 and 12.7.2 present regression and classification trees, respectively. Section 12.7.3 describes several strategies to build complex nonlinear models by combining multiple trees.

12.7.1 Regression Trees

The linear regression models described in Section 12.1 are very useful, but they have a fundamental limitation: They cannot encode any *nonlinear* dependence between the response and the features. To illustrate this, we model the temperature in Manhattan (Kansas) as a function of the day of the year and the hour of the day. The training data consist of temperatures from 2015, extracted from Dataset 9.

We begin by fitting a linear model to the data. The response is the temperature. The two features are the day $1 \leq x_{\text{day}} \leq 365$ ($x_{\text{day}} = 1$ is the first of January) and the hour $0 \leq x_{\text{hour}} \leq 23$ ($x_{\text{hour}} = 0$ is midnight). The linear-regression estimator of the temperature given the two features, obtained via ordinary least squares (see Section 12.1.2), is

$$\ell_{\text{OLS}}(x_{\text{hour}}, x_{\text{day}}) = 0.25x_{\text{hour}} + 0.03x_{\text{day}} + 5.85. \tag{12.353}$$

According to the linear model, depicted in the bottom left of Figure 12.26, if we fix the day x_{day}, then the temperature rises proportionally to the hour. As shown in the top left graph of Figure 12.26, in the data the temperature does indeed tend to rise until around noon

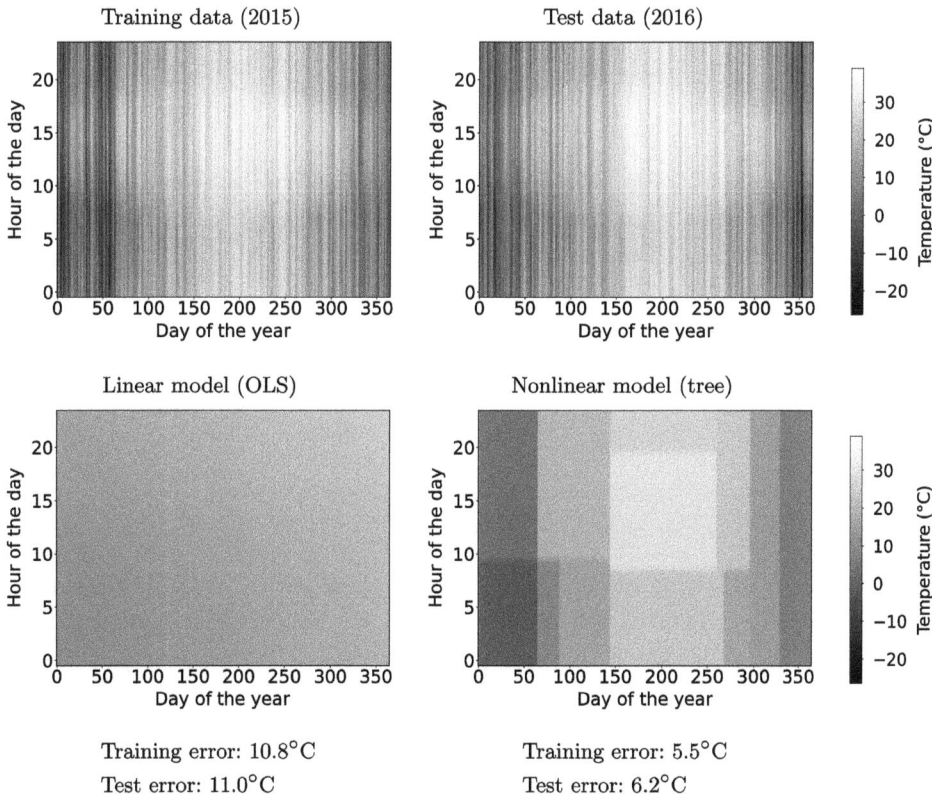

Figure 12.26 Linear vs. nonlinear temperature estimation. The top row shows hourly temperatures for 2015 (left) and 2016 (right) in Manhattan (Kansas) as a function of the day of the year $1 \leq x_{\text{day}} \leq 365$ and the hour of the day $0 \leq x_{\text{hour}} \leq 23$. The bottom row shows a linear regression model (left) and a nonlinear model (right) corresponding to the regression tree in Figure 12.27, both fit using the 2015 data. The training and test error, reported at the bottom of each graph, indicate that the nonlinear model captures the structure of the data much more effectively. The error metric is the square root of the average residual sum of squares.

($x_{\text{hour}} \approx 12$). However, from noon on ($x_{\text{hour}} \geq 12$) it *decreases* instead! For each fixed hour, the model indicates that the temperature rises continuously over the course of a year. However, in the data this is also not true. The temperature increases until around the middle of the summer ($x_{\text{day}} \approx 200$). Afterwards, it decreases during the fall and winter. The linear model is completely unable to capture this nonlinear structure, so its training and test errors are both very large.

A simple way to build a nonlinear regression model is to partition the space of possible feature vectors into multiple regions and assign a constant estimate to each region. This yields an estimator that is *piecewise constant*, as depicted in the bottom right of Figure 12.26. The following theorem establishes that the optimal piecewise-constant estimator in terms of residual sum of squares can be obtained by averaging the responses of the training data in each region.

Theorem 12.40 (Piecewise-constant regression)**.** *Let* (x_1, y_1), (x_2, y_2), $\ldots$, (x_n, y_n) *be a dataset formed by* n *pairs of a response* y_i *and a corresponding vector* x_i *containing* d *features* ($1 \leq i \leq n$)*. For any collection of* m *subsets of* $\mathbb{R}^d$, R_1, $\ldots$, R_m, *forming a partition of the feature space (so that each feature vector* x_i *belongs to only one of the sets), we define the piecewise-constant estimator*

$$n\ell_{\text{pc}}(x) := \frac{1}{n_r} \sum_{\{i:\, x_i \in R_r\}} y_i, \quad if\, x \in R_r, \tag{12.354}$$

where $n_r := |\{i \colon x_i \in R_r\}|$ *is the number of data in* R_r *for* $1 \leq r \leq m$*. In words, the estimate assigned to* R_r *is equal to the sample mean of the responses associated with feature vectors belonging to* R_r*. This estimator minimizes the residual sum of squares (RSS) among all piecewise-constant estimators associated with the partition.*

Proof Since the regions are disjoint and cover all of $\mathbb{R}^d$, the total RSS decouples into the sum of the RSS in each region. For any choice of the piecewise-constant estimates $\alpha_1, \ldots, \alpha_m$, the RSS is

$$\text{RSS} = \sum_{r=1}^{m} \sum_{\{i:\, x_i \in R_r\}} (y_i - \alpha_r)^2 \tag{12.355}$$

because each data point only belongs to one region. Consequently, we can choose the estimates to separately minimize each individual RSS. By Theorem 7.31, the minimum is achieved by averaging the data in the corresponding region,

$$\frac{1}{n_r} \sum_{\{i:\, x_i \in R_r\}} y_i = \arg\min_{\alpha_r} \sum_{\{i:\, x_i \in R_r\}} (y_i - \alpha_r)^2, \qquad 1 \leq r \leq m. \tag{12.356}$$

$\blacksquare$

Theorem 12.40 enables us to fit a piecewise-constant nonlinear estimator to data, but it requires a predefined partition of the feature space. Regression trees provide a framework to build such partitions efficiently from data. The regions in a regression tree are hyperrectangles formed by Cartesian products of intervals, which we denote by

$$R_r := \left(a_r^{[1]}, b_r^{[1]} \right] \times \left(a_r^{[2]}, b_r^{[2]} \right] \times \cdots \times \left(a_r^{[d]}, b_r^{[d]} \right] \qquad (12.357)$$

$$= \underset{j=1}{\overset{d}{\times}} \left(a_r^{[j]}, b_r^{[j]} \right], \qquad (12.358)$$

where $a_r^{[j]} < b_r^{[j]}$ for $1 \le j \le d$ and $1 \le r \le m$. To be clear, a feature vector x belongs to R_r, if its jth entry $x[j]$ is in the interval $\left(a_r^{[j]}, b_r^{[j]} \right]$ for $1 \le j \le d$. The regions are associated with the leaves (terminal nodes with no descendants) of a binary tree. Each leaf is assigned a constant estimate, obtained by averaging the response of the training data points within the corresponding region, as prescribed by Theorem 12.40. Figure 12.27 shows the regression tree corresponding to the nonlinear model depicted in the bottom right of Figure 12.26. The regression-tree model achieves a much better fit to the training data than the linear regression model because it is able to capture nonlinear patterns such as yearly seasonality (summers are warmer than winters) and daily variation (days are warmer than nights). It also generalizes better to a test set consisting of temperatures from 2016.

The regions associated with a regression tree are defined recursively. We illustrate this using the regression tree in Figure 12.27, which represents the nonlinear temperature model in Figure 12.26. We begin by assigning the whole feature space to the root node at the top of the tree. For our example this is $[0, 23] \times [1, 365]$ because the hour is between 0 and 23, and the day between 1 and 365. In a binary regression tree, a node is either a parent node with two children or a leaf with no descendants. The regions assigned to the two children are obtained by splitting the parent's region *according to a single feature*. In our example, the root-node region is split according to the day feature. The left child is assigned $[0, 23] \times [0, 65]$ (winter), and the right child is assigned $[0, 23] \times [66, 365]$ (spring, summer and fall). The left child has two children. Their regions are obtained by dividing $[0, 23] \times [0, 65]$ according to the hour feature, yielding $[0, 9] \times [0, 65]$ (winter nights) and $[10, 23] \times [0, 65]$ (winter days). These are leafs, so their regions are part of the piecewise-constant model associated with the tree. The temperature estimates assigned to each leaf are $-4.3°$C and $0.8°$C, respectively, obtained by averaging the temperature in 2015 within the corresponding regions, following Theorem 12.40.

No matter how many bifurcations there are in a regression tree, the regions associated with the leafs are always guaranteed to form a partition of the feature space (see Exercise 12.14). Consequently, the tree provides a single estimate associated with any specific feature vector, which can be retrieved by traversing the tree from the root to find the corresponding leaf. For instance, at the time of writing, it is 3 am in Kansas and the date is August 19, so $x_{\text{hour}} := 3$ and $x_{\text{day}} := 251$. To obtain the corresponding temperature estimate $n\ell_{\text{tree}}(x_{\text{hour}}, x_{\text{day}}) = 20.5°$C, we start at the root and follow the appropriate branches (represented by the thick black line in Figure 12.27):

$$x_{\text{day}} > 65 \rightarrow x_{\text{day}} \le 296 \rightarrow x_{\text{day}} > 144 \rightarrow x_{\text{hour}} \le 8 \rightarrow x_{\text{day}} \le 268 \rightarrow 20.5° \text{ C}.$$

According to the internet, the temperature in Manhattan (Kansas) is actually $22°$C, so the estimate is not too bad! The model provides an interpretable explanation of the estimate in terms of the different features and the corresponding thresholds it depends on: The temperature estimate is $20.5°$ C because $65 < x_{\text{day}} \le 268$ and $x_{\text{hour}} \le 8$ (i.e. it's a summer night).

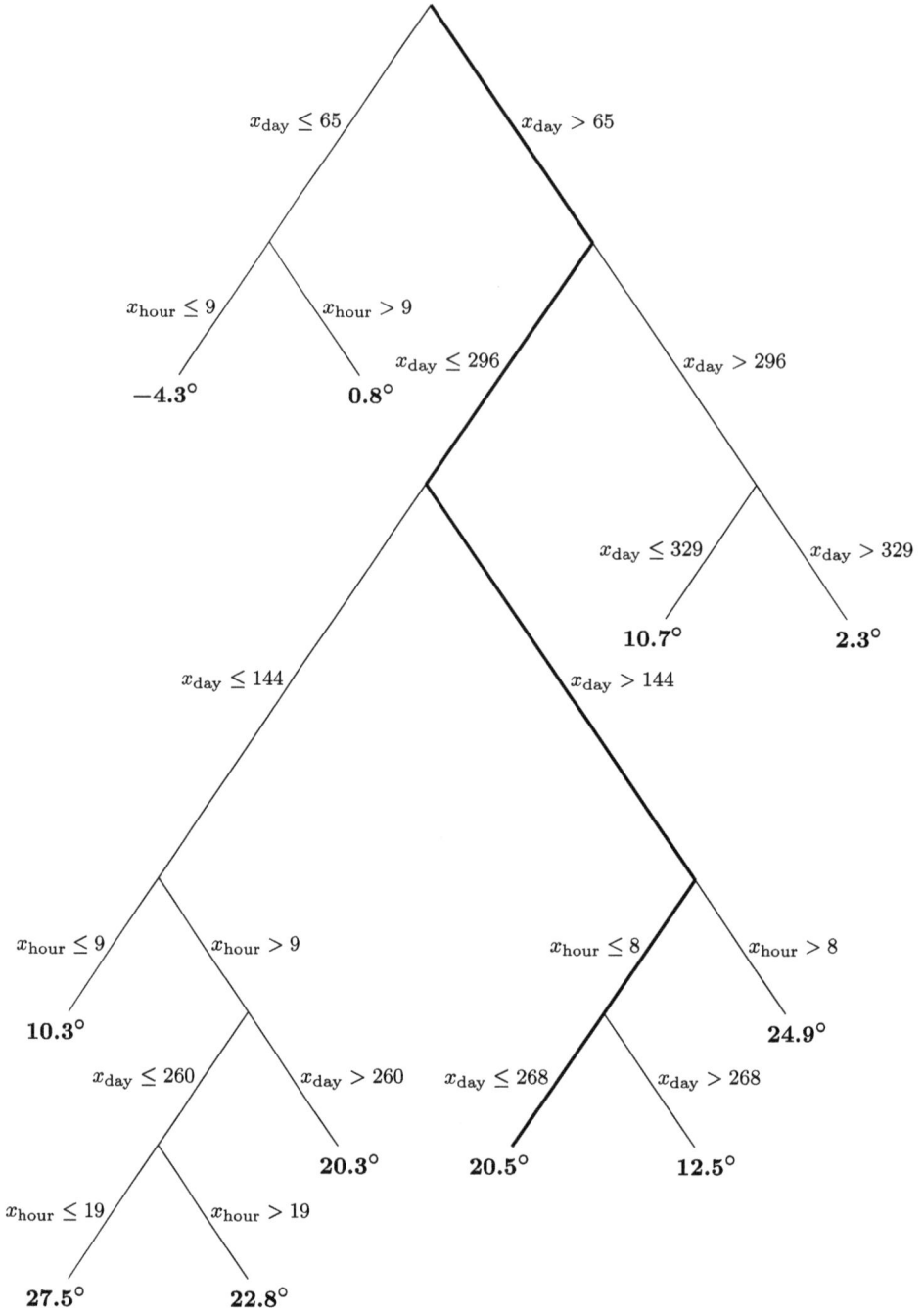

Figure 12.27 Regression tree for temperature estimation. The diagram shows the regression tree associated with the nonlinear temperature model depicted in the bottom right graph of Figure 12.26. Every node represents a region of the feature space. Each bifurcation in the tree represents a split in the parent-node region indicated on the corresponding branch, and each leaf is associated with the temperature estimate for the corresponding region. The thick black lines show the path from the root to the leaf corresponding to August 19 at 3 am ($x_{\text{day}} := 251$ and $x_{\text{hour}} := 3$).

Interpretability is an important advantage of regression trees with respect to more complicated nonlinear models, such as the tree ensembles in Section 12.7.3 or the neural networks in Section 12.8.

Ideally, the splits at each bifurcation in a regression tree should be chosen to optimize the fit to the training data. For each split, we need to determine what feature to threshold and the value of the threshold. Unfortunately, the number of possibilities to choose from quickly explodes as a function of the tree depth. This is yet another manifestation of the curse of dimensionality, discussed in Section 4.7. For simplicity, let us assume that all nodes at depth h are leaves. Then there are $b := 2^h - 1$ bifurcations in the tree (each associated with a non-leaf node). If the number of features equals d and there are t possible thresholds, then we must choose between $(dt)^b$ different trees. Consider a small depth-4 tree ($h := 4$) with 15 bifurcations. If there are $d := 10$ features and $t := 100$ possible threshold values at each bifurcation, then the total number of trees to choose from is 10^{45}, which is astronomical! It is therefore usually impossible to consider all possible regression trees in practice. Instead, regression trees are grown iteratively, adding nodes one by one to optimize the residual sum of squares in a greedy fashion. This technique is known as recursive binary splitting.

Definition 12.41 (Recursive binary splitting). *Let* (x_1, y_1), (x_2, y_2), ..., (x_n, y_n) *be a dataset formed by n pairs of a response y_i and a corresponding vector x_i containing d features ($1 \le i \le n$). Consider a partition of the feature space associated with the leafs of a binary regression tree, $R_1, \ldots, R_m$, where*

$$R_r := \underset{j=1}{\overset{d}{\times}} \left(a_r^{[j]}, b_r^{[j]} \right], \qquad a_r^{[j]} < b_r^{[j]} \quad for \; 1 \le j \le d, 1 \le r \le m. \tag{12.359}$$

For each possible split of region R_r, performed by comparing the sth feature to a threshold t, we denote the resulting two subregions of R_r by

$$\mathcal{A}_r(s, t) := \underset{j=1}{\overset{d}{\times}} \left(a_r^{[j]}, c_r^{[j]} \right], \qquad c_r^{[s]} := t, \quad c_r^{[j]} := b_r^{[j]} \quad for \; j \ne s, \tag{12.360}$$

$$\mathcal{B}_r(s, t) := \underset{j=1}{\overset{d}{\times}} \left(d_r^{[j]}, b_r^{[j]} \right], \qquad d_r^{[s]} := t, \quad d_r^{[j]} := a_r^{[j]} \quad for \; j \ne s. \tag{12.361}$$

$\mathcal{A}_r(s, t)$ *contains every data point x in R_r such that $x[s] \le t$, and $\mathcal{B}_r(s, t)$ contains every data point x in R_r such that $x[s] > t$.*

The residual sum of squares of the tree model equals

$$\text{RSS} = \sum_{r=1}^{m} \sum_{\{i: \; x_i \in R_r\}} \left(y_i - \alpha_r^{\text{tree}} \right)^2, \tag{12.362}$$

where

$$\alpha_r^{\text{tree}} := \frac{1}{n_r} \sum_{\{i: \; x_i \in R_r\}} y_i, \quad 1 \le r \le m, \tag{12.363}$$

for $n_r := |\{i \colon x_i \in R_r\}|$, following Theorem 12.40. For each possible split, we define

$$\alpha_{\mathcal{A}_r(s,t)} := \frac{1}{n_{\mathcal{A}_r(s,t)}} \sum_{\{i \colon x_i \in \mathcal{A}_r(s,t)\}} y_i, \tag{12.364}$$

$$\alpha_{\mathcal{B}_r(s,t)} := \frac{1}{n_{\mathcal{B}_r(s,t)}} \sum_{\{i \colon x_i \in \mathcal{B}_r(s,t)\}} y_i, \tag{12.365}$$

where $n_{\mathcal{A}_r(s,t)} := |\{i \colon x_i \in \mathcal{A}_r(s,t)\}|$ and $n_{\mathcal{B}_r(s,t)} := |\{i \colon x_i \in \mathcal{B}_r(s,t)\}|$ are the number of elements in $\mathcal{A}_r(s,t)$ and $\mathcal{B}_r(s,t)$. The RSS decrease after the split equals

$$\triangle \mathrm{RSS}(r,s,t) = \sum_{\{i \colon x_i \in R_r\}}^{n} \left(y_i - \alpha_r^{\mathrm{tree}}\right)^2 - \sum_{\{i \colon x_i \in \mathcal{A}_r(s,t)\}}^{n} \left(y_i - \alpha_{\mathcal{A}_r(s,t)}\right)^2$$

$$- \sum_{\{i \colon x_i \in \mathcal{B}_r(s,t)\}}^{n} \left(y_i - \alpha_{\mathcal{B}_r(s,t)}\right)^2, \tag{12.366}$$

which is always nonnegative (see Exercise 12.14). The optimal split is found by iterating over the regions, features and possible thresholds. For each region and feature, the set of possible thresholds is finite: It equals the observed values of the feature in the training examples that belong to the region. Let

$$(r^*, s^*, t^*) := \arg\max_{r,s,t} \triangle \mathrm{RSS}(r,s,t) \tag{12.367}$$

be the region, feature and threshold that achieve the greatest RSS decrease. We replace R_{r^*} by $\mathcal{A}_{r^*}(s^*, t^*)$ and $\mathcal{B}_{r^*}(s^*, t^*)$ in the original partition to complete the split and obtain a new partition with $m + 1$ regions.

To fit a regression tree using a training dataset, we begin with a single region equal to the whole feature space. Then we apply recursive binary splitting, as described in Definition 12.41 to grow the tree until it has a desired depth or number of leaves, or until another stopping criterion is satisfied. Figure 12.28 illustrates the splitting procedure for our temperature example. The top left diagram depicts a tree with four leaves. The top right image shows the four regions associated with the leaves, which roughly correspond to the four seasons (winter, spring, summer and fall). The middle row shows how much the residual sum of squares decreases, when we split each region thresholding the hour (left) or day (right) feature. The highest decrease is achieved by splitting Region 3 (summer) according to the hour feature with a threshold equal to 8. This creates two new regions, roughly corresponding to summer nights ($x_{\mathrm{hour}} \leq 8$) and summer days ($x_{\mathrm{hour}} > 8$). The resulting tree with five leaves, and the corresponding regions, are depicted in the bottom row of Figure 12.28.

The top graph in Figure 12.29 shows the training and test error of regression trees with different numbers of leaves for our temperature-estimation example. The trees are built via recursive binary splitting, as described in Definition 12.41, using the training data in Figure 12.26. The test data are also the same as in Figure 12.26. As the number of leaves in the tree increases, the fit to the training data improves and the training error decreases. At first, this is accompanied by a decrease in the test error, indicating that the model learns patterns that generalize to the test set. However, eventually (when there are more than 12 leaves) the test error starts to increase, indicating that beyond this point the model overfits spurious structure in the training set, which is not present in the test set.

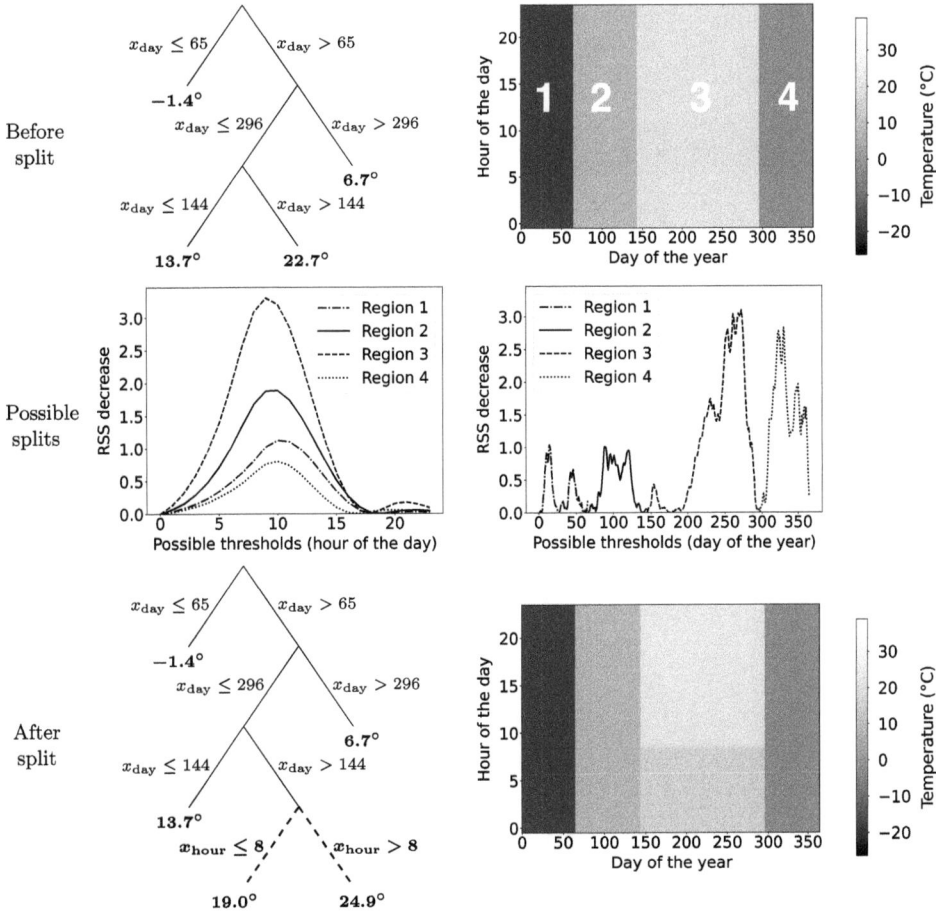

Figure 12.28 Recursive binary splitting. This figure explains how to grow a regression tree via recursive binary splitting, as described in Definition 12.41. The top left diagram depicts a regression tree with four leaves fit to the training data in Figure 12.26. The top right image shows the four regions associated with the leaves in the tree. The graphs in the middle row show the RSS decrease resulting from different splits obtained by thresholding the hour (left) and day (right) features. The highest decrease is achieved by splitting Region 3 according to the hour feature with a threshold equal to 8. The split results in two leaves that descend from the corresponding node, represented by bold dashed lines in the bottom left diagram. The corresponding piecewise-constant model is depicted on the bottom right.

The second row of Figure 12.29 shows the learned regions for the 11-leaf and 15-leaf models. Notice that the 15-leaf model includes the region $[14, 41] \times [10, 23]$ (after 10 am from mid-January to mid-February), which has an average temperature that is higher than the region $[42, 65] \times [10, 23]$ (after 10 am from mid-February to mid-March). This is surprising: One would not expect temperatures to decrease from February to March. Indeed, the pattern is not present in the test data (compare the third and fourth rows of Figure 12.29), so incorporating this region increases the test error. For tree models with regions containing a small number of training examples, such spurious patterns inevitably arise due to random fluctuations in the data. Consequently, increasing the number of regions eventually results in overfitting. To avoid this, the number of regions should be selected based on held-out validation data.

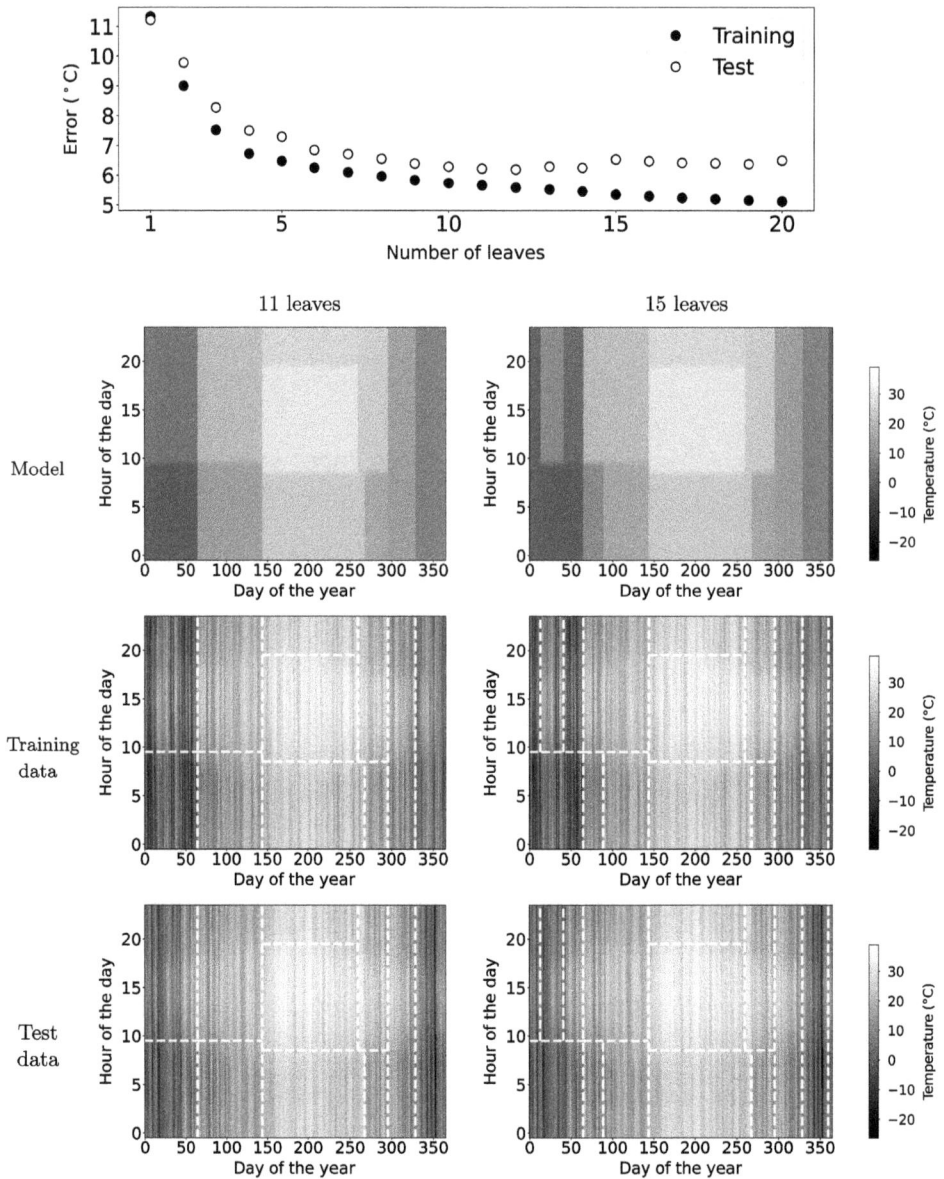

Figure 12.29 Overfitting and generalization in regression trees. The graph at the top shows the training and test error of regression trees with different numbers of leaves. The trees are built via recursive binary splitting, as described in Definition 12.41. The training and test sets consist of the 2015 and 2016 data in Figure 12.26, respectively. We observe overfitting when the number of leaves is greater than 12: The training error continues to decrease, but the test error increases. The second row shows the learned regions for the 11-leaf and 15-leaf models. The third and fourth rows show the regions superposed onto the training and test data, respectively, which reveals that the 11-leaf model generalizes better than the 15-leaf model because the latter overfits spurious structure in the training data.

12.7.2 Classification Trees

Classification trees are classification models with the same structure as regression trees. Let us represent the features in a classification problem as a d-dimensional random vector $\tilde{x}$ and the corresponding class label as a random variable $\tilde{y}$ with c possible values, which we arbitrarily select to be the integers from one to c. Classification trees allow us to learn a nonlinear estimator of the conditional probability of $\tilde{y}$ given $\tilde{x}$, based on a partition of the feature space encoded by a tree.

As in the case of regression trees, the estimator is piecewise constant. Given a partition with m regions $\mathcal{R} := \{R_1, \ldots, R_m\}$, the estimate assigned to all feature vectors in a region is the same. We denote the estimate corresponding to the region R_r as a c-dimensional vector θ_{R_r}, for $1 \leq r \leq m$. According to the model, the conditional probability of the kth class given a feature vector x is the kth entry of $\theta_{R(x)}$,

$$\mathrm{P}\left(\tilde{y} = k \mid \tilde{x} = x\right) = \theta_{R(x)}[k], \qquad 1 \leq k \leq c, \tag{12.368}$$

where $R(x)$ is the region in $\mathcal{R}$ that contains x.

In order to fit a classification tree from data, we apply maximum-likelihood estimation. As in logistic regression (see Section 12.5), the likelihood is the conditional pmf of the observed labels, given the observed features, interpreted as a function of the model parameters. In the case of classification trees the parameters are the partition of the feature space associated with the tree, and the probability estimate at each region of the partition.

Consider a classification tree associated with a partition $\mathcal{R} := \{R_1, \ldots, R_m\}$ and conditional-probability estimates $\Theta := \{\theta_{R_1}, \ldots, \theta_{R_m}\}$. Let (x_1, y_1), (x_2, y_2), $\ldots$, (x_n, y_n) be a dataset formed by n pairs of feature vectors $x_i \in \mathbb{R}^d$ and corresponding labels $y_i \in \{1, \ldots, c\}$ for $1 \leq i \leq n$. We model the data as a realization from a joint distribution of feature vectors $\tilde{x}_1, \ldots, \tilde{x}_n$ and labels $\tilde{y}_1, \ldots, \tilde{y}_n$. As in Theorems 12.35 and 12.37, we assume that the labels are conditionally independent given the features, and each label $\tilde{y}_i$ is conditionally independent from the remaining feature vectors $\{\tilde{x}_l\}_{l \neq i}$ given $\tilde{x}_i$. Then, by (12.368), the log-likelihood of the classification tree equals

$$\log \mathcal{L}_{XY}(\mathcal{R}, \Theta) = \sum_{k=1}^{c} \sum_{\{i:\, y_i = k\}} \log \theta_{R(x_i)}[k], \tag{12.369}$$

where $R(x_i)$ is the region in $\mathcal{R}$ that contains the ith feature vector x_i. We omit the derivation, as it is identical to the one in the proof of Theorem 12.35.

The following theorem shows that, if we fix the partition $\mathcal{R}$ associated with a classification tree, maximizing the log-likelihood with respect to the probabilities Θ is very simple. The maximum-likelihood estimator $\theta_{R_r}^{\mathrm{ML}}[k]$ of the conditional probability of the kth class in region R_r is the fraction of data in R_r with label equal to k.

Theorem 12.42 (Piecewise-constant classification). *For a fixed partition $\mathcal{R}$, the log-likelihood function* (12.369) *is maximized at*

$$\theta_{R_r}^{\mathrm{ML}}[k] := \frac{n_r^{[k]}}{n_r}, \quad 1 \leq r \leq m, 1 \leq k \leq c, \tag{12.370}$$

where $n_r := |\{i \colon x_i \in R_r\}|$ is the number of data points in R_r and $n_r^{[k]} := |\{i \colon x_i \in R_r, y_i = k\}|$ is the number of data points in R_r with label k.

Proof The regions form a partition, so they are disjoint and cover all of $\mathbb{R}^d$. Consequently, the log-likelihood decouples into the sum of log-likelihoods corresponding to the different regions,

$$\log \mathcal{L}_{XY}(\mathcal{R}, \Theta) = \sum_{k=1}^{c} \sum_{\{i: \, y_i = k\}} \log \theta_{R(x_i)}[k] \tag{12.371}$$

$$= \sum_{k=1}^{c} \sum_{r=1}^{m} \sum_{\{i: \, x_i \in R_r, y_i = k\}} \log \theta_{R_r}[k] \tag{12.372}$$

$$= \sum_{r=1}^{m} \sum_{k=1}^{c} n_r^{[k]} \log \theta_{R_r}[k]. \tag{12.373}$$

Since each probability estimate θ_{R_r} is only present in one of the terms, we can optimize them separately to maximize the overall log-likelihood. For each region R_r,

$$\arg\max_{\theta_{R_r}} \log \mathcal{L}_{XY}(\mathcal{R}, \Theta) = \arg\max_{\theta_{R_r}} \sum_{k=1}^{c} n_r^{[k]} \log \theta_{R_r}[k], \tag{12.374}$$

which is equivalent to maximizing the log-likelihood of a multinoulli parametric model with parameter θ_{R_r}, using only the data points with feature vectors in R_r. By Exercise 2.14 (see also Example 2.26), for $1 \leq k \leq c$, the kth entry of the maximum-likelihood estimator is equal to the fraction of data in R_r with labels equal to k, which yields (12.370). ∎

Figure 12.30 shows a tree for the classification problem in Example 12.38, where the goal is to identify wheat seeds based on two features: Surface area and asymmetric coefficient. Each node in the classification tree corresponds to a region of the feature space, exactly as in regression trees (see Section 12.7.1). The only difference is that the estimate associated with the leaf-node regions is a probability vector (instead of a single number, as in regression). Since the leafs form a partition of the feature space (see Exercise 12.14), the tree maps every possible feature vector to a single probability estimate, which can be determined by traversing the tree from the root node. To illustrate this, we use the classification tree in Figure 12.30 to classify a wheat seed with surface area $x_{\text{area}} := 0.5$ and asymmetric coefficient $x_{\text{asym}} := 0$. The path from the root to the appropriate leaf, marked by thick black lines on the diagram, is

$$x_{\text{area}} > 0.18 \rightarrow x_{\text{area}} \leq 0.78 \rightarrow x_{\text{asym}} > -0.83 \rightarrow \theta_{R_5}^{\text{ML}} = \begin{bmatrix} 0 \\ 0.25 \\ 0.75 \end{bmatrix}.$$

The feature values correspond to the fifth leaf in Figure 12.30, so we denote the corresponding region by R_5. The probability estimate is obtained from the 20 training data points in R_5, applying Theorem 12.42 to maximize the log-likelihood. 5 are from the Kama variety and 15 from the Rosa variety, so

$$\begin{bmatrix} P\left(\text{Canadian} \mid x_{\text{area}}, x_{\text{asym}}\right) \\ P\left(\text{Kama} \mid x_{\text{area}}, x_{\text{asym}}\right) \\ P\left(\text{Rosa} \mid x_{\text{area}}, x_{\text{asym}}\right) \end{bmatrix} = \theta_{R_5}^{\text{ML}} = \begin{bmatrix} \frac{0}{20} \\ \frac{5}{20} \\ \frac{15}{20} \end{bmatrix} = \begin{bmatrix} 0 \\ 0.25 \\ 0.75 \end{bmatrix}. \tag{12.375}$$

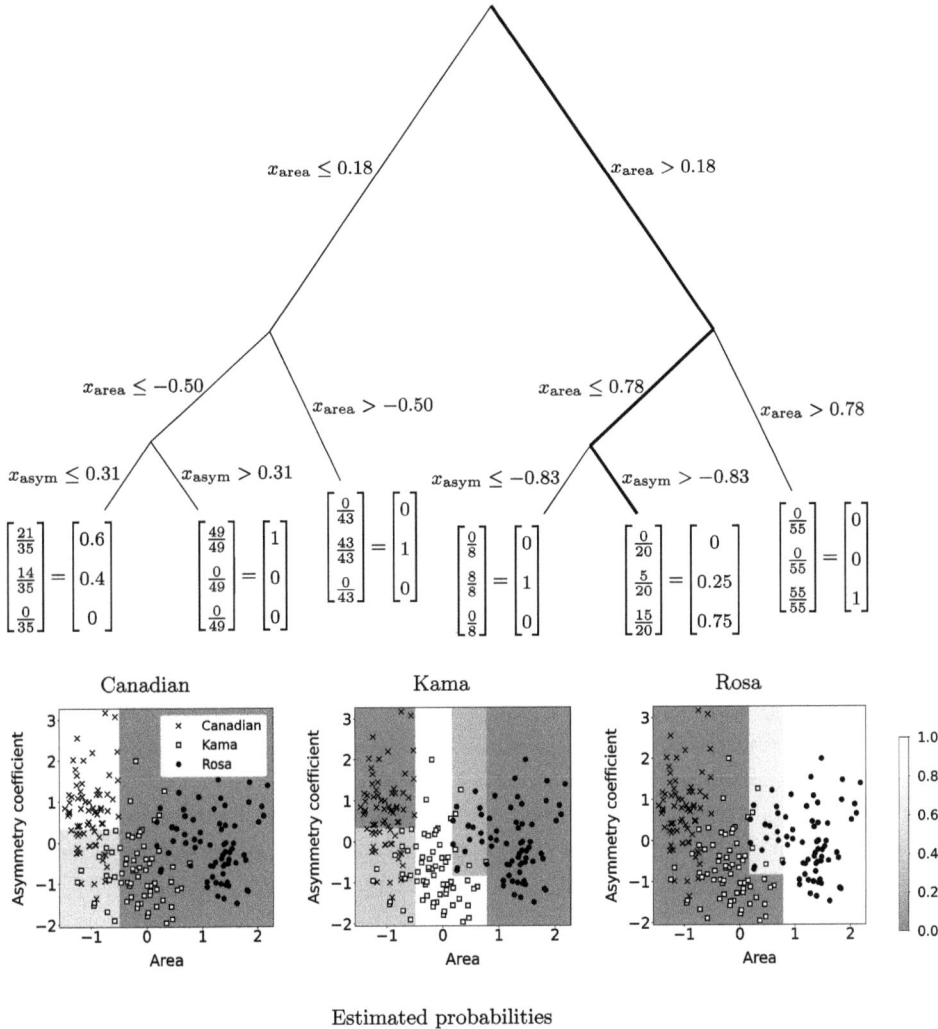

Figure 12.30 Classification tree for wheat varieties. The top diagram shows a classification tree built via recursive binary splitting, using the same data as in Figure 12.24. The tree classifies wheat seeds into three varieties according to their area and asymmetric coefficient. The thick black lines show the path from the root to the leaf corresponding to $x_{\text{area}} := 0.5$ and $x_{\text{asym}} := 0$. The plots depict the regions in feature space associated with every leaf, as well as the training data belonging to each region and the corresponding probability estimates.

As in the case of regression trees, the estimate produced by a classification tree is accompanied by an interpretable explanation, in terms of the relevant features and thresholds. In our example, the estimate is chosen because the surface area is between 0.18 and 0.78, and the asymmetric coefficient is larger than -0.83.

When fitting a classification tree to a training dataset, we need to learn the partition of regions $\mathcal{R}$. Here we again encounter the challenge discussed in Section 12.7.1: As the depth of the tree increases, the number of possible partitions explodes exponentially. It is therefore intractable to find a global maximum of the log-likelihood (12.369) with respect to $\mathcal{R}$. However, recursive binary splitting can be leveraged to optimize the log-likelihood in a greedy fashion, adding nodes one by one. The procedure is exactly as described in Definition 12.41, with the only difference that the splits are chosen to increase the log-likelihood, instead of decreasing the residual sum of squares.

In order to derive the log-likelihood increase in recursive binary splitting more precisely, let us consider a region R_r, associated with the leaf of a tree. We denote by $\mathcal{A}_r(s,t)$ and $\mathcal{B}_r(s,t)$ the two subregions of R_r resulting from thresholding the sth feature using threshold t. Following Theorem 12.42, we set the probability estimates $\theta_{R_r}^{\text{ML}}$, θ_a and θ_b associated with R_r, $\mathcal{A}_r(s,t)$ and $\mathcal{B}_r(s,t)$ equal to:

$$\theta_{R_r}^{\text{ML}}[k] := \frac{n_r^{[k]}}{n_r}, \quad n_r := |\{i\colon x_i \in R_r\}|, n_r^{[k]} := |\{i\colon x_i \in R_r, y_i = k\}|, \quad (12.376)$$

$$\theta_a[k] := \frac{n_a^{[k]}}{n_a}, \quad n_a := |\{i\colon x_i \in \mathcal{A}_r(s,t)\}|, n_a^{[k]} := |\{i\colon x_i \in \mathcal{A}_r(s,t), y_i = k\}|,$$

$$\theta_b[k] := \frac{n_b^{[k]}}{n_b}, \quad n_b := |\{i\colon x_i \in \mathcal{B}_r(s,t)\}|, n_b^{[k]} := |\{i\colon x_i \in \mathcal{B}_r(s,t), y_i = k\}|.$$

The increase in log-likelihood resulting from splitting R_r into $\mathcal{A}_r(s,t)$ and $\mathcal{B}_r(s,t)$ therefore equals

$$\triangle \mathcal{L}(r,s,t) = \sum_{k=1}^{c} \left(n_a^{[k]} \log \theta_a[k] + n_b^{[k]} \log \theta_b[k] - n_r^{[k]} \log \theta_{R_r}^{\text{ML}}[k] \right), \quad (12.377)$$

which replaces the RSS decrease in equation (12.366). During recursive binary splitting, we grow the classification tree by sequentially choosing the feature s and threshold t that maximize the log-likelihood increase across all leaves.

The log-likelihood increase in recursive binary splitting has an intuitive interpretation in terms of a fundamental quantity from information theory known as *entropy*. The entropy of a discrete random variable $\tilde{a}$ with pmf $p_{\tilde{a}}(k) = \theta[k]$ for $1 \le k \le c$, where $\theta \in \mathbb{R}^c$, is

$$H(\theta) = -\sum_{k=1}^{c} \theta[k] \log \theta[k]. \quad (12.378)$$

We refer to Cover (1999) for an in-depth introduction to the properties of entropy. For our purposes, it is sufficient to realize that the entropy of a random variable is high when the variable can take many values with similar probability (the pmf is spread out) and low when the random variable equals a few values with high probability (the pmf is concentrated). Figure 12.31 shows the entropy of a binary random variable as a function of the pmf entries:

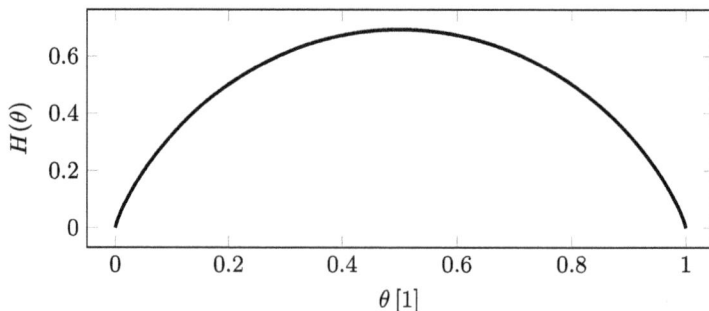

Figure 12.31 Entropy of a binary random variable. The graph shows the entropy of a binary discrete random variable equal to 1 with probability $\theta[1]$ and to 2 with probability $\theta[2] = 1 - \theta[1]$. The entropy is lowest when the probabilities are close to one or zero, and highest when they both equal 0.5.

The entropy is highest when the two entries are the same, and lowest when one entry equals zero and the other equals one.

The increase in log-likelihood (12.377) maximized during recursive binary splitting can be expressed in terms of the entropies corresponding to the different conditional-probability estimates associated with each region. By (12.378),

$$\triangle \mathcal{L}(r, s, t) = -n_r \sum_{k=1}^{c} \frac{n_r^{[k]}}{n_r} \log \theta_{R_r}^{\mathrm{ML}}[k] + n_a \sum_{k=1}^{c} \frac{n_a^{[k]}}{n_a} \log \theta_a[k] + n_b \sum_{k=1}^{c} \frac{n_b^{[k]}}{n_b} \log \theta_b[k]$$

$$= n_r H\left(\theta_{R_r}^{\mathrm{ML}}\right) - n_a H\left(\theta_a\right) - n_b H\left(\theta_b\right). \tag{12.379}$$

The increase in log-likelihood is equivalent to a decrease in entropy (weighted by the number of data in the corresponding regions). Therefore, recursive binary splitting decreases entropy as much as possible, selecting splits for which θ_a and θ_b are highly concentrated in a small number of entries. This yields discriminative partitions, where many of the labels within each region are the same, which is exactly what we want in order to achieve high classification accuracy! A similar effect can be achieved by instead maximizing the decrease in the Gini index,

$$G(\theta) := \sum_{k=1}^{c} \theta[k]\left(1 - \theta[k]\right), \qquad \theta \in \mathbb{R}^c, \tag{12.380}$$

which is an alternative metric that can be used to quantify the purity of the labels within each region (see Exercise 12.22).

Example 12.43 (Minimizing entropy yields discriminative splits). In this example, we analyze the log-likelihood splitting criterion for classification trees using a simple classification problem with one feature. The data consist of a single feature x and a binary label y, which can equal $+$ or $-$:

x	0.5	1.5	2.5	3.5	4.5	5.5
y	+	-	-	+	+	+

In Figure 12.32, each label is depicted at the position of its respective feature on the horizontal axis. We consider five possible thresholds that split the feature space into two regions A and B. The conditional-probability estimates for label $+$ in regions A and B, denoted by $\theta_a[+]$ and $\theta_b[+]$, are shown at the corresponding threshold. By Theorem 12.42 $\theta_a[+]$ and $\theta_b[+]$, are equal to the fraction of $+$ labels in each region. If the threshold is 2, then the labels in A are $\{+, -\}$ and the labels in B are $\{-, +, +, +\}$, so $\theta_a[+] = 1/2$ and $\theta_b[+] = 3/4$.

By (12.379) the increase in log-likelihood due to the split equals

$$\triangle \mathcal{L}(r, s, t) = n_r H\left(\theta_{R_r}^{\mathrm{ML}}\right) - n_a H\left(\theta_a\right) - n_b H\left(\theta_b\right), \tag{12.381}$$

where $n_r = 6$ is the total number of data and $\theta_{R_r}^{\mathrm{ML}}$ contains the overall fraction of $+$ and $-$ labels ($\theta_{R_r}^{\mathrm{ML}}[+] = 2/3$). Maximizing the log-likelihood with respect to s and t is equivalent to maximizing the weighted sum of negative entropies $-n_a H\left(\theta_a\right) - n_b H\left(\theta_b\right)$. As shown in Figure 12.31, the entropies $H\left(\theta_a\right)$ and $H\left(\theta_b\right)$ are smallest when the entries of θ_a and θ_b are close to zero or one. This occurs if the majority of the labels in region A are the same, and the majority of the labels in region B are the same. Consequently, maximizing the log-likelihood promotes homogeneous regions with high label purity, which are ideal for classification. The star markers at the bottom of Figure 12.32 represent the values of

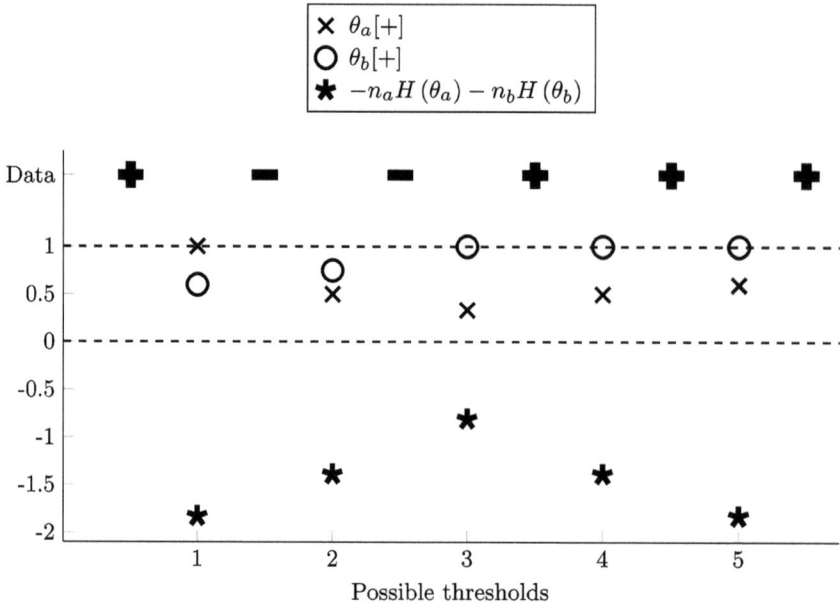

Figure 12.32 Maximizing the decrease in entropy yields discriminative splits. The diagram illustrates the process of splitting the feature space in Example 12.43, where there are two labels (+ and −) and a single feature, indexed by the horizontal axis. The top row shows the labels of the data. The cross and circle markers indicate the estimated probability of the + label in the two regions resulting from the five different possible splits. The star markers at the bottom represent the weighted sum of negative entropies corresponding to each split, which is maximized when the threshold is chosen via maximum likelihood. The maximum corresponds to the most discriminative split, which creates one region with mostly − labels on the left, and a region with only + labels on the right.

$-n_a H\left(\theta_a\right) - n_b H\left(\theta_b\right)$ for each of the thresholds. The maximum is achieved by the most discriminative split, which creates a region with mostly − labels on the left, and a region with only + labels on the right.

· ·

12.7.3 Bagging, Random Forests, and Boosting

The tree models described in Sections 12.7.1 and 12.7.2 are simple and interpretable, but their estimation performance on held-out data is often modest, due to their tendency to overfit as the tree depth increases. In this section, we explain how to build models with improved performance combining multiple trees. This brings us to the realm of *machine learning*, a data-driven approach to statistical modeling based on complex nonlinear models, which are often difficult to interpret, but achieve impressive performance in practice.

Bagging

Bagging stands for *bootstrap aggregating*. It is an *ensembling* technique, which combines the output of multiple models (called the *ensemble*) via averaging. Intuitively, if the models are different enough for their errors to be approximately independent, then averaging their outputs tends to cancel out the errors, yielding a more accurate estimate. The challenge is to ensure that the models are indeed different to each other; there is no point in combining models that are the same! Bagging promotes model diversity by fitting models on resampled versions of the dataset, obtained from the original training dataset via the bootstrap technique introduced in Section 9.9.

Definition 12.44 (Bagging). *Let* (x_1, y_1), (x_2, y_2), ..., (x_n, y_n) *be a dataset formed by* n *pairs of a response* y_i *and a corresponding vector* x_i *containing* d *features* $(1 \leq i \leq n)$. *To build a bagging ensemble:*

1 *We generate* B *bootstrap datasets of size* n. *To create the bth dataset, we sample bootstrap indices* $l_1^{[b]}, l_2^{[b]}, \ldots, l_n^{[b]}$ *independently and uniformly at random with replacement from the set of possible indices* $\{1, \ldots, n\}$, *and then select the bootstrap training data according to those indices,*

$$(x_i^{[b]}, y_i^{[b]}) := (x_{l_i^{[b]}}, y_{l_i^{[b]}}), \qquad 1 \leq b \leq B, 1 \leq i \leq n. \tag{12.382}$$

2 *We use each of the* B *bootstrap datasets to train a model* t_b, *such as a regression or classification tree* t_b $(1 \leq b \leq B)$:

$$(x_1^{[b]}, y_1^{[b]}), (x_2^{[b]}, y_2^{[b]}), \ldots, (x_n^{[b]}, y_n^{[b]}) \to t_b, \quad 1 \leq b \leq B. \tag{12.383}$$

The individual trees are typically constrained to have limited depth, in order to avoid overfitting.

3 *We average the output of the* B *models to obtain the bagging estimator,*

$$n\ell_{\text{bagging}}(x) := \frac{1}{B} \sum_{b=1}^{B} t_b(x), \tag{12.384}$$

where x *denotes an input feature vector.*

The left column in Figure 12.33 shows bagging in action for the temperature-estimation example from Section 12.7.1. The three first rows depict regression-tree models with five leaves trained on different bootstrap training sets. The bagging estimator obtained by averaging the output of 32 such trees is shown at the bottom. Figure 12.34 reports the training and test errors of bagging estimators with different numbers of trees. Bagging decreases the test error from 7.25°C for a single tree, to around 7°C for an ensemble of 30 trees or more. The price to pay is that the bagging estimate is no longer easily interpretable, as opposed to the estimate of each individual tree.

Random Forests

A key limitation of bagging is that the aggregated models may not be very diverse. Random forests are ensembles of trees that are randomized to increase diversity. This is achieved by modifying the recursive-binary-splitting procedure in Definition 12.41 to choose from a random subset of features at each split. Otherwise, the approach is identical to bagging: B randomized trees are built using bootstrap training datasets, and then combined via averaging.

Definition 12.45 (Random forest). *Let* (x_1, y_1), (x_2, y_2), ..., (x_n, y_n) *be a dataset formed by* n *pairs of a response* y_i *and a corresponding vector* x_i *containing* d *features* $(1 \leq i \leq n)$. *To build a random forest:*

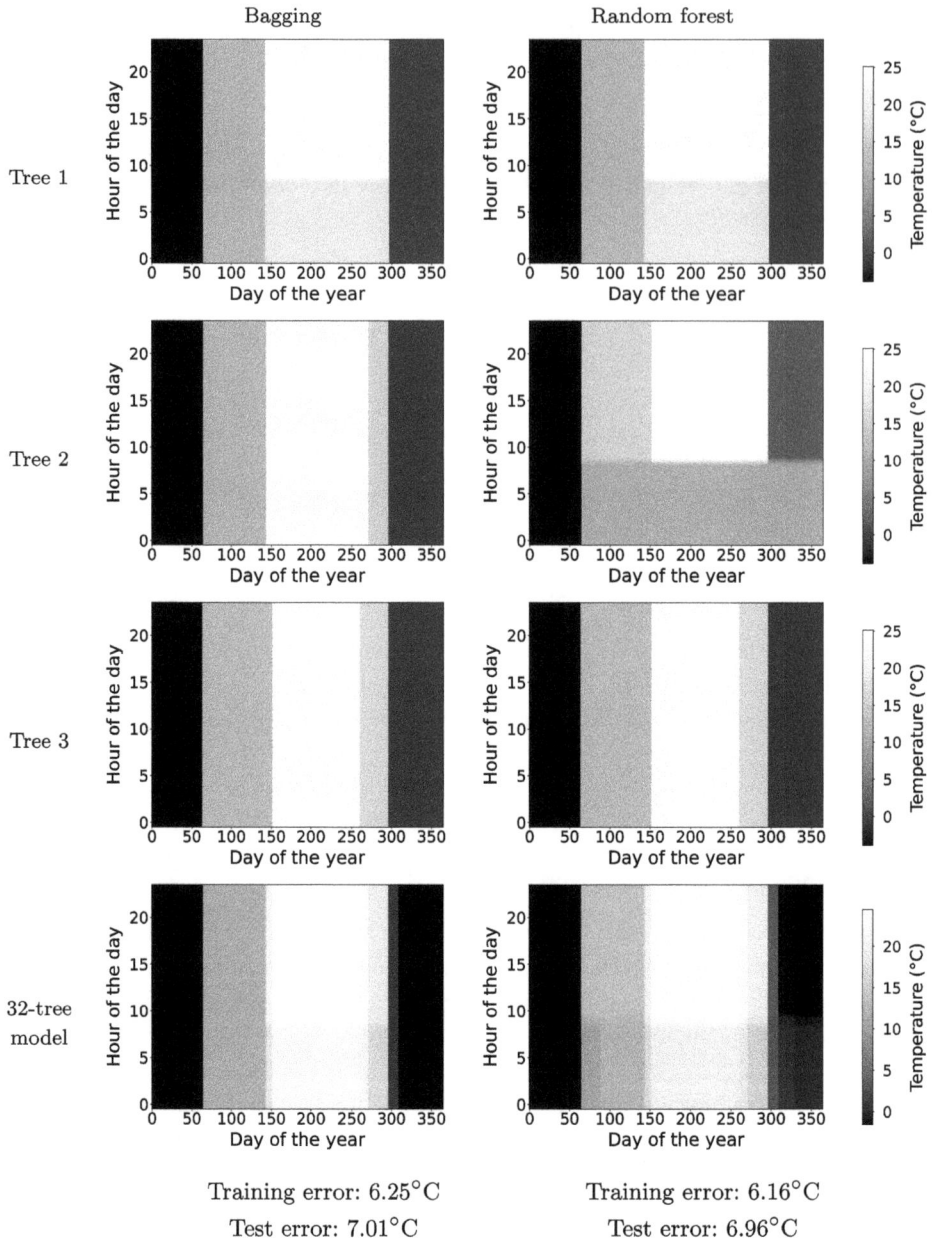

Training error: 6.25°C   Training error: 6.16°C

Test error: 7.01°C   Test error: 6.96°C

Figure 12.33 Bagging and random forests. The first three rows depict regression-tree estimators based on different bootstrap datasets sampled from the 2015 data in Figure 12.26, as
described in Definition 12.44. Each row shows a tree estimator built via standard recursive binary
splitting (left) and randomized recursive binary splitting (right) using the same bootstrap dataset.
The bottom row shows the bagging (left) and random-forest (right) estimators obtained by averaging 32 such trees. The random-forest model is more complex due to the higher diversity of the
individual trees resulting from randomized splitting. This results in a lower training and test error
(compare the values reported at the bottom of each graph). The test set consists of the 2016 data
in Figure 12.26.

1 We generate B bootstrap training sets of size n

$$(x_i^{[b]}, y_i^{[b]}), \qquad 1 \leq b \leq B, 1 \leq i \leq n, \tag{12.385}$$

as described in Definition 12.44.

2 We use the B datasets to train B trees via randomized recursive binary splitting, where each split is performed using a random subset of $d' < d$ features:

$$(x_1^{[b]}, y_1^{[b]}), (x_2^{[b]}, y_2^{[b]}), \ldots, (x_n^{[b]}, y_n^{[b]}) \to t_b^{\mathrm{rand}}, \quad 1 \leq b \leq B. \tag{12.386}$$

As in bagging, the individual trees are usually constrained to have limited depth, in order to avoid overfitting.

3 We average the B tree models to obtain the random-forest estimator,

$$n\ell_{\mathrm{rforest}}(x) := \frac{1}{B} \sum_{b=1}^{B} t_b^{\mathrm{rand}}(x), \tag{12.387}$$

where x denotes an input feature vector.

The right column of Figure 12.33 shows an application of random forests to our temperature-estimation example. Since there are only two features (hour and day), each split of the randomized recursive-binary-splitting procedure utilizes either just the day, just the hour, or both. Empirically, we observe that considering both features 80% of the time, and randomly choosing one of them 20% of the time yields good results. In problems with a larger number of features d, a popular choice is to consider $\sqrt{d}$ randomly-selected features in each split, where d is the total number of features.

Each row of Figure 12.33 shows trees constructed by applying the standard and randomized versions of recursive binary splitting to the same bootstrap training set. The pair of trees in the first row are the same, and so are the ones in the third row. However, the trees in the second row are very different because the second randomized split only considers the hour feature. By contrast, standard splitting selects the day feature, resulting in a tree that is almost identical to the third one. The ensemble models, combining 32 trees, are shown in the bottom row. Due to the higher diversity of its components, the random-forest model is more complex. Figure 12.34 reports the training and test errors of both ensembles for different numbers of trees. The random forest consistently achieves a better fit to the training data, and also better generalization to the test set.

Boosting

In bagging and random-forest ensembles, we fit each individual model independently from the rest. Boosting ensembles utilize a different strategy: The models are fit sequentially in order to ensure that they complement each other, *boosting* the overall performance.

Consider a regression problem, where the data $(x_1, y_1), (x_2, y_2), \ldots, (x_n, y_n)$ are n pairs of a response y_i and a corresponding d-dimensional feature vector x_i ($1 \leq i \leq n$). Our goal is to build a sequence of complementary regression trees. The first tree t_1 approximates the response by minimizing the residual sum of squares (RSS)

$$\sum_{i=1}^{n} (y_i - t_1(x_i))^2, \tag{12.388}$$

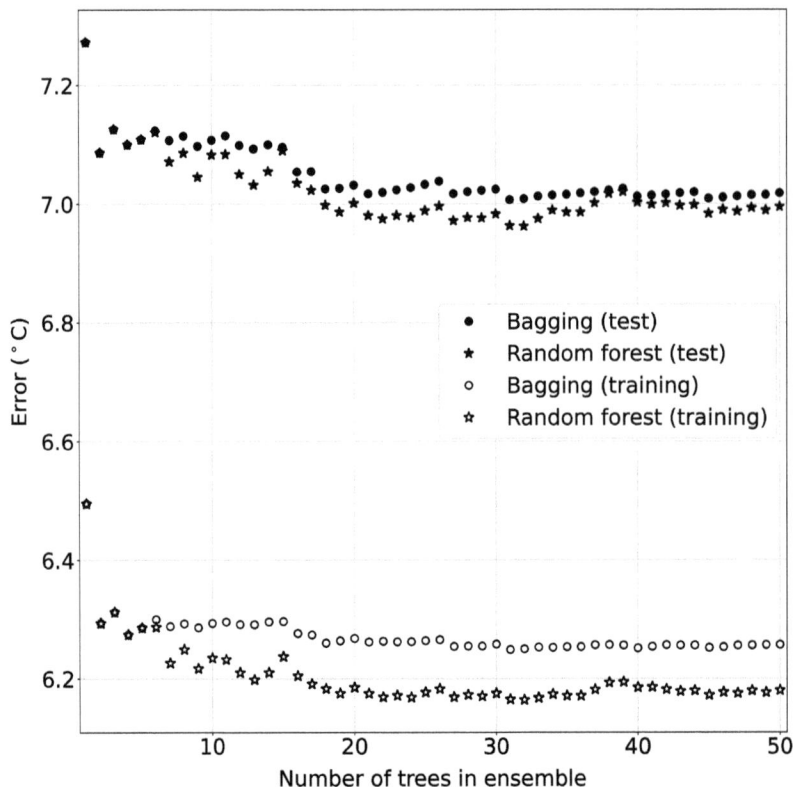

Figure 12.34 Training and test error of bagging and random-forest models. The graph shows the training and test root average squared error of bagging and random-forest ensembles with different numbers of 5-leaf trees for the temperature-estimation example in Section 12.7.1. The training and test set correspond to the 2015 and 2016 data in the top row of Figure 12.26, respectively. As the number of trees increases, the training and test error decrease, eventually plateauing. The random-forest model consistently achieves a lower training and test error than the bagging model, and both models clearly outperform the single-tree model.

as described in Definition 12.41. To obtain a bagging ensemble, we would fit the remaining trees in the same way, using different bootstrap datasets. By contrast, the second tree t_2 in our boosting ensemble *complements* the first tree. Consider the RSS of the sum of the two trees,

$$\sum_{i=1}^{n} \Big(\underbrace{y_i - t_1(x_i)}_{\text{residual}} - t_2(x_i) \Big)^2 . \tag{12.389}$$

If t_1 is fixed, fitting t_2 to the residual of the first tree minimizes the RSS of the sum. Applying the same logic, we can fit a third tree to the residuals of the sum of the two first trees, a fourth tree to the residuals of the sum of the first three trees, and so on. After fitting L trees, this yields the *naive* boosting ensemble

$$n\ell_{\text{naiveboost}}(x) := \sum_{l=1}^{L} t_l(x), \tag{12.390}$$

where x denotes an input feature vector.

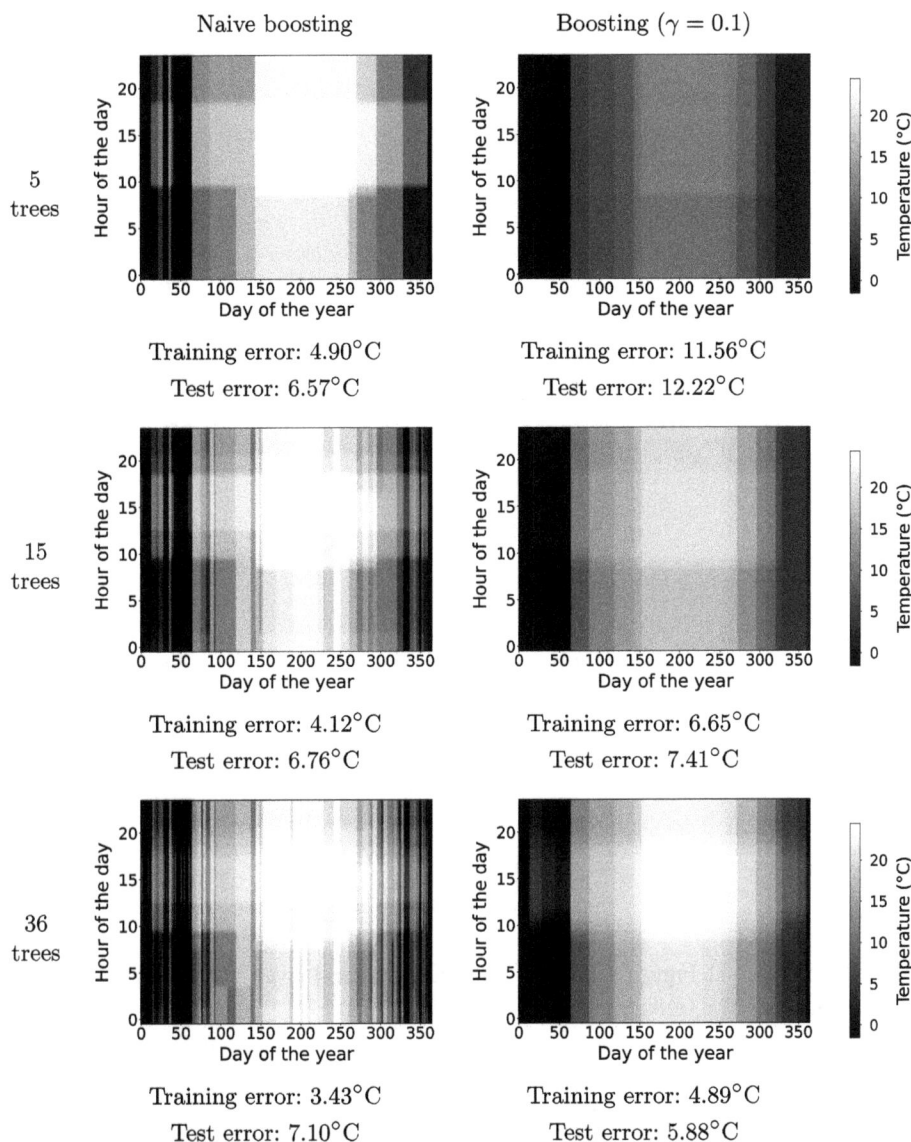

Figure 12.35 Boosting tree ensembles. The images show naive boosting (left) and boosting (right) estimators built using different numbers of 5-leaf trees for the temperature-estimation example from Section 12.7.1. The training and test error of each estimator are reported at the bottom of the corresponding image. The training and test set correspond to the 2015 and 2016 data in the top row of Figure 12.26, respectively. Naive boosting overfits the training data. By contrast, boosting achieves robust generalization to the test data.

The left column in Figure 12.35 depicts naive boosting ensembles with different numbers of 5-leaf trees for our temperature-estimation example. The training and test errors are reported at the bottom of each image. As we increase the number of trees in the ensemble, the training error decreases, but the test error increases. Naive boosting overfits the training data.

Fortunately, the overfitting tendency of our naive boosting estimator can be mitigated using a simple trick. We shrink the estimate from each tree, multiplying it by a small non-negative constant γ, in order to prevent it from fitting the training data too well. The resulting (non-naive) boosting estimator, described in the following definition, typically generalizes much better to held-out data.

Definition 12.46 (Boosting). *Let (x_1, y_1), (x_2, y_2), ..., (x_n, y_n) be a dataset formed by n pairs of a response y_i and a corresponding vector x_i containing d features $(1 \leq i \leq n)$. We initialize the residual $r_i^{[1]}$ to equal the response y_i for $1 \leq i \leq n$. To build a boosting ensemble, we set γ to equal a fixed small nonnegative constant and apply the following steps for $1 \leq k \leq L$, for some constant positive integer L, or until a stopping criterion is achieved:*

1 We train a tree via recursive binary splitting (see Definition 12.41) on a modified dataset where the response is set equal to the current residual:

$$(x_1, r_1^{[k]}), (x_2, r_2^{[k]}), \dots, (x_n, r_n^{[k]}) \rightarrow t_k. \tag{12.391}$$

As in bagging and boosting, the tree is typically constrained to have limited depth.

2 We incorporate the tree to the ensemble, and update the estimator and the residual,

$$n\ell_{\text{boost}}(x) := \gamma \sum_{l=1}^{k} t_l(x), \tag{12.392}$$

$$r_i^{[k+1]} := y_i - n\ell_{\text{boost}}(x_i). \tag{12.393}$$

The right column in Figure 12.35 shows boosting estimators built with different numbers of 5-leaf trees for our temperature-estimation example. The training and test errors are reported at the bottom of each image. The γ constant is set equal to 0.1 (in practice it should be chosen using validation data). At the beginning of the boosting sequence, the boosting estimate has small amplitude, due to the multiplication by γ, which prevents it from approximating the training data. As a result, the training error is very high, and in fact much higher than that of a single regression tree. However, as more trees join the ensemble, the boosting model is able to achieve a better approximation to the training data, which results in reduced test error. Figure 12.36 compares the performance of the random-forest, naive-boosting, and boosting ensembles for the temperature-estimation example. As the number of trees increases, naive boosting rapidly overfits. By contrast, the boosting estimator initially underfits the data, but eventually achieves much better generalization, clearly outperforming the random-forest estimator.

In this section, we have described a simplified version of boosting for regression problems. We refer to Chapter 10 in Hastie et al. (2009) for a more detailed explanation of boosting and its application to classification. In practice, boosting algorithms often achieve strong performance for regression and classification problems with a moderate number of features (on the order of hundreds). XGBoost is a particularly popular boosting method (Chen and Guestrin, 2016).

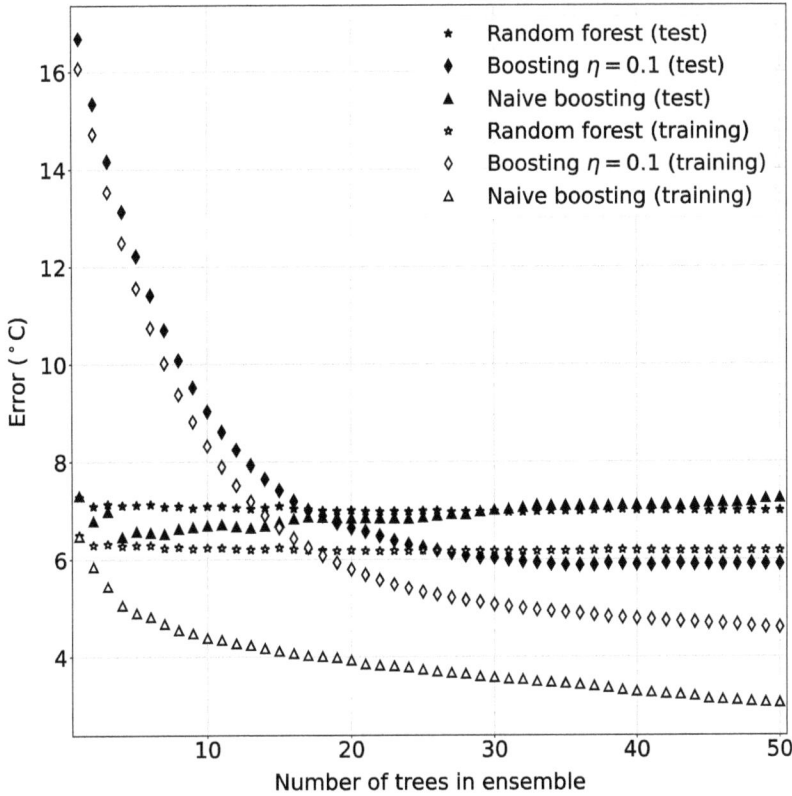

Figure 12.36 Training and test error of boosting models. The graph shows the training and test error of random-forest, naive-boosting, and boosting ensembles with different numbers of 5-leaf trees for the temperature-estimation example from Section 12.7.1. The training and test set correspond to the 2015 and 2016 data in the top row of Figure 12.26, respectively. As the number of trees increases, the naive boosting estimator overfits: The training error decreases rapidly and the test error starts increasing after a few boosting iterations. By contrast, the boosting estimator initially underfits the data, but eventually achieves much better generalization. Boosting clearly outperforms random forests in this scenario.

12.8 Neural Networks and Deep Learning

Neural networks, also known as neural nets, are machine-learning models that learn complex nonlinear functions directly from data. Fitting neural networks with many layers using large-scale datasets, an approach known as *deep learning*, has achieved spectacular success in computer vision, natural language processing, reinforcement learning, and many other applications. In Sections 12.8.1 and 12.8.2, we explain how to utilize neural networks to perform regression and classification.

12.8.1 Regression

In this section, we introduce neural networks and explain how to use them for regression. We use the temperature-estimation problem in Section 12.7.1 as a running example. Recall that a linear model is unable to capture the nonlinear relationship between the features and the response in this example, which is crucial to produce an accurate estimate (see Figure 12.26).

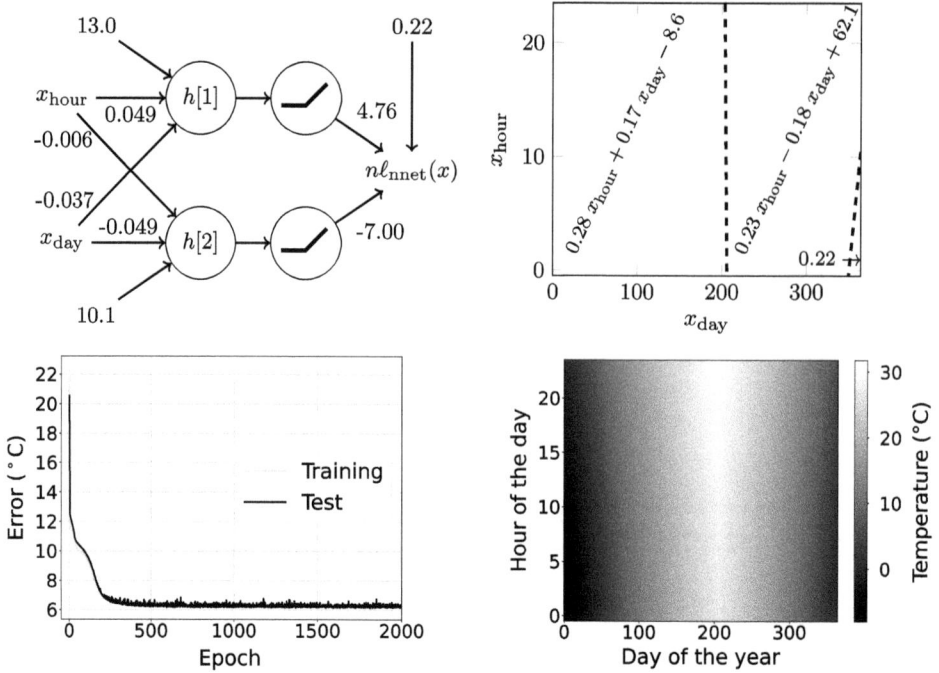

Figure 12.37 Two-layer neural network for temperature estimation. The top left diagram shows a two-layer neural network trained to estimate the temperature in Manhattan (Kansas) as a function of the hour $0 \leq x_{\text{hour}} \leq 23$ and the day $1 \leq x_{\text{day}} \leq 365$. The bottom left graph shows the evolution of the training and test error during training. The training and test set correspond to the 2015 and 2016 data in the top row of Figure 12.26, respectively. The right column shows two visualizations of the function implemented by the neural network, which reveal that the temperature estimate rises linearly during the first half of the year, and falls linearly during the second half.

Neural networks consist of *layers* that implement affine transformations, interleaved by pointwise nonlinearities. The outputs of each layer are hidden variables, typically called *feature maps* or *activation maps*. The top left diagram in Figure 12.37 shows a simple two-layer network for our temperature-estimation example, fit to the training set in Figure 12.26. The first layer transforms the input into a vector h consisting of two hidden variables,

$$h := \underbrace{\begin{bmatrix} 0.049 & -0.037 \\ -0.006 & -0.049 \end{bmatrix}}_{W_1} \begin{bmatrix} x_{\text{hour}} \\ x_{\text{day}} \end{bmatrix} + \underbrace{\begin{bmatrix} 13.0 \\ 10.1 \end{bmatrix}}_{\alpha_1} \tag{12.394}$$

$$= \begin{bmatrix} 0.049 x_{\text{hour}} - 0.037 x_{\text{day}} + 13.0 \\ -0.006 x_{\text{hour}} - 0.049 x_{\text{day}} + 10.1 \end{bmatrix}, \tag{12.395}$$

where x_{hour} and x_{day} denote the hour and day feature, respectively. The affine transformation in the first layer is parameterized by the matrix of linear weights W_1 and the additive constant or *bias*[3] α_1. The feature map is fed into a pointwise nonlinearity or *activation function* called a rectified linear unit (ReLU), which sets negative entries to zero and preserves

---

[3] Here the term bias denotes a model parameter, and should not be confused with the bias of an estimator, defined in Section 9.2.

nonnegative entries:

$$h_+ := \begin{bmatrix} h[1]_+ \\ h[2]_+ \end{bmatrix}, \tag{12.396}$$

$$a_+ := \begin{cases} a & \text{if} \quad a \geq 0, \\ 0 & \text{otherwise.} \end{cases} \tag{12.397}$$

The rectified hidden variables are combined in the second layer to generate the final regression estimate via another affine transformation, parameterized by a weight matrix W_2 and an additive bias α_2,

$$n\ell_{\text{nnet}}\left(\begin{bmatrix} x_{\text{hour}} \\ x_{\text{day}} \end{bmatrix}\right) := \underbrace{\begin{bmatrix} 4.76 & -7.00 \end{bmatrix}}_{W_2} h_+ + \underbrace{0.22}_{\alpha_2} \tag{12.398}$$

$$= 4.76 h_+[1] - 7.00 h_+[2] + 0.22. \tag{12.399}$$

The presence of nonlinearities between the layers of a neural network is crucial. It enables the network to implement nonlinear functions, which apply *different affine transformations to different inputs*. To drive this important point home, we derive the function implemented by our two-layer network.

By (12.395), the first hidden variable $h[1]$ in our network is nonnegative, as long as $x_{\text{day}} < 351 + 1.32 x_{\text{hour}}$. The second hidden variable $h[2]$ is nonnegative, if $x_{\text{day}} < 206 - 0.12 x_{\text{hour}}$. Notice that since x_{hour} is nonnegative,

$$206 - 0.12 x_{\text{hour}} < 351 + 1.32 x_{\text{hour}}. \tag{12.400}$$

Consequently, if $x_{\text{day}} < 206 - 0.12 x_{\text{hour}}$, both entries of h are nonnegative, and $h_+ = h$. The resulting network output equals

$$n\ell_{\text{nnet}}\left(\begin{bmatrix} x_{\text{hour}} \\ x_{\text{day}} \end{bmatrix}\right) = W_2 h + \alpha_2 \tag{12.401}$$

$$= W_2 \left(W_1 x + \alpha_1 \right) + \alpha_2 \tag{12.402}$$

$$= 0.28 x_{\text{hour}} + 0.17 x_{\text{day}} - 8.6. \tag{12.403}$$

However, if $351 + 1.32 x_{\text{hour}} > x_{\text{day}} \geq 206 - 0.12 x_{\text{hour}}$, the network implements a completely different affine function because of the ReLU. In that case, $h[1]$ is nonnegative, but $h[2]$ is negative. As a result, only $h[1]$ survives the ReLU, and the output equals

$$n\ell_{\text{nnet}}\left(\begin{bmatrix} x_{\text{hour}} \\ x_{\text{day}} \end{bmatrix}\right) = W_2 \begin{bmatrix} h[1] \\ 0 \end{bmatrix} + \alpha_2 \tag{12.404}$$

$$= W_2 \begin{bmatrix} 0.049 x_{\text{hour}} - 0.037 x_{\text{day}} + 13.0 \\ 0 \end{bmatrix} + \alpha_2 \tag{12.405}$$

$$= 0.23 x_{\text{hour}} - 0.18 x_{\text{day}} + 62.1. \tag{12.406}$$

Finally, if $x_{\text{day}} > 351 + 1.32 x_{\text{hour}}$, then both entries of h are negative, so the output of the network is constant,

$$n\ell_{\text{nnet}}\left(\begin{bmatrix} x_{\text{hour}} \\ x_{\text{day}} \end{bmatrix}\right) = W_2 \begin{bmatrix} 0 \\ 0 \end{bmatrix} + \alpha_2 \tag{12.407}$$

$$= 0.22. \tag{12.408}$$

In summary, the output produced by the neural network is the following piecewise-affine function of the features:

$$
\begin{aligned}
&0.28x_{\text{hour}} + \mathbf{0.17}x_{\mathbf{day}} - 8.6, && \text{if } x_{\text{day}} < 206 - 0.12x_{\text{hour}}, \\
&0.23x_{\text{hour}} - \mathbf{0.18}x_{\mathbf{day}} + 62.1, && \text{if } 206 - 0.12x_{\text{hour}} \leq x_{\text{day}} < 351 + 1.32x_{\text{hour}}, \\
&0.22, && \text{if } x_{\text{day}} > 351 + 1.32x_{\text{hour}}.
\end{aligned} \tag{12.409}
$$

The resulting estimator is depicted as a diagram and a heatmap in the right column of Figure 12.37. At the beginning of the year, the temperature estimate is directly proportional to x_{day}, and therefore rises steadily between the winter and summer. Then, after day 206 (July 25), the temperature estimate is instead *inversely* proportional to x_{day}, falling at approximately the same rate between summer and winter. The neural network is able to learn the annual seasonality of the temperature data, which is a nonlinear pattern that cannot be represented by a linear model (see Figure 12.26).

In order to train a neural network, we use a training dataset to learn its parameters, which consist of the weight matrices and additive bias parameters in all the layers of the network. This is achieved by minimizing a loss or cost function tailored to the task of interest. For regression, the most popular loss is the residual sum of squares (RSS), which approximates the mean squared error of the estimator. This is the same cost function used to train ordinary-least-squares regression models (see Theorem 12.7) and the regression trees in Section 12.7.1. In the case of neural networks, the residual sum of squares is a highly nonconvex function of the model parameters, so we cannot hope to attain a global minimum. However, this is not problematic because the goal is not to minimize the training error as much as possible, but rather to obtain a useful model.

In practice, neural networks are trained using large datasets, which are divided into disjoint sets called *batches*. The neural-network parameters are optimized by iterating through these batches. Consider a batch with n_B pairs of feature vectors $x_1, \ldots, x_{n_B}$ and corresponding responses $y_1, \ldots, y_{n_B}$. The batch RSS is the sum of the squared differences between the network outputs and the responses, interpreted as a function of the neural-network parameters Θ:

$$
\text{RSS}(\Theta) := \sum_{i=1}^{n_B} \left(y_i - n\ell_{\text{nnet}}^{[\Theta]}(x_i) \right)^2, \tag{12.410}
$$

where $n\ell_{\text{nnet}}^{[\Theta]}(x_i)$ is the network output when its input is the ith feature vector x_i and the network parameters equal Θ. Training is performed by taking a step in the direction opposite to the gradient $\nabla_\Theta \text{RSS}(\Theta)$ with respect to the model parameters,

$$
\Theta_b := \Theta_{b-1} - \eta \nabla_\Theta \text{RSS}(\Theta_{b-1}), \tag{12.411}
$$

where Θ_{b-1} and Θ_b denote the model parameters before and after processing the bth batch. The gradient is scaled by a real-valued scalar η, known as the *learning rate*. The gradient in (12.411) is obtained via backpropagation, an efficient implementation of the chain rule that propagates derivatives backwards through the network (see Section 6.5 in Goodfellow et al. (2016)).

This training procedure is known as *stochastic gradient descent* because the batch RSS can be interpreted as a stochastic approximation to the RSS of the whole training set, as long as the batches are sampled at random from the training data. A key practical consideration is how to tune the learning rate η. A small value results in very slow progress, whereas a large value can disrupt learning completely. A popular strategy is to utilize methods such as Adam (Kingma and Ba, 2014), which adapt the learning rate as training progresses, based on estimates of lower-order moments of the gradient.

Performing stochastic gradient descent over all of the batches in the training set completes an *epoch*. During training, it is advisable to monitor the model error on a separate validation set at each epoch, to evaluate how well the model generalizes to held-out data. This complements the training error, which evaluates the fit to the training set. The bottom left plot of Figure 12.37 shows the evolution of the training and test error, when training our two-layer neural network for temperature estimation via stochastic gradient descent with a batch size of 100 data points using Adam. Both the training and test error tend to decrease as the epochs proceed. The decrease is not monotonic, which is typical of stochastic gradient descent. The training error converges to $6.32°C$ and the test error to $6.25°C$.

Due to its simplicity, our two-layer network underfits the data. For instance, it clearly does not capture the dependence between the hour of the day and the temperature. In order to improve the prediction, we leverage a deeper neural network consisting of four affine layers interleaved with ReLUs:

$$h^{[1]} := W_1 x + \alpha_1, \qquad W_1 \in \mathbb{R}^{100 \times 2}, \alpha_1 \in \mathbb{R}^{100},$$

$$h^{[\ell]} := W_\ell h_+^{[\ell-1]} + \alpha_\ell, \qquad W_\ell \in \mathbb{R}^{100 \times 100}, \alpha_\ell \in \mathbb{R}^{100}, \quad \ell \in \{2,3\},$$

$$n\ell_{\text{nnet}}^{[\Theta]}(x) := W_4 h_+^{[3]} + \alpha_4, \qquad W_4 \in \mathbb{R}^{1 \times 100}, \alpha_4 \in \mathbb{R}, \tag{12.412}$$

$$\Theta := \{W_1, W_2, W_3, W_4, \alpha_1, \alpha_2, \alpha_3, \alpha_4\}, \tag{12.413}$$

where $x \in \mathbb{R}^2$ is the input to the network (containing x_{hour} and x_{day}), $h^{[\ell]}$ denotes the vector of hidden variables in layer ℓ, and $h_+^{[\ell]}$ is the result of feeding $h^{[\ell]}$ through a ReLU. Each of the intermediate layers has 100 hidden variables, so the total number of parameters in the network is 20,601, which is more than double the number of training data (8,760)! This *overparametrization* is a key property of deep neural networks, which enable them to learn highly complex functions. Neural networks for computer vision tasks have millions of parameters and large language models have billions.

The top graph in Figure 12.38 shows the training and test error of our larger network at different training epochs. Just like the smaller network, the model is trained via stochastic gradient descent with a batch size of 100 data points using Adam. The training error has a decreasing trend throughout the training epochs. By contrast, the test error initially tends to decrease, but then experiences an increasing trend instead. At 300 epochs, the four-layer model clearly outperforms the two-layer model on the test set ($6.06°C$ vs $6.25°C$), but at 5,000 epochs its test error is the same as that of the smaller model, even though the training error is much smaller ($4.78°C$ vs $6.32°C$). The temperature estimates corresponding to the four-layer model at epochs 300 and 5,000 are depicted as heatmaps in the bottom row of Figure 12.38.

The training dynamics of our four-layer model are typical of deep neural networks trained with limited data: Generalization performance tends to improve at first, but the model eventually overfits the training set. To avoid overfitting, model parameters are usually selected by monitoring performance on a held-out validation set, an approach known as *early stopping*. Exercise 12.16 shows that, in the case of linear models, early stopping has a regularizing effect on the model coefficients similar to ridge regression.

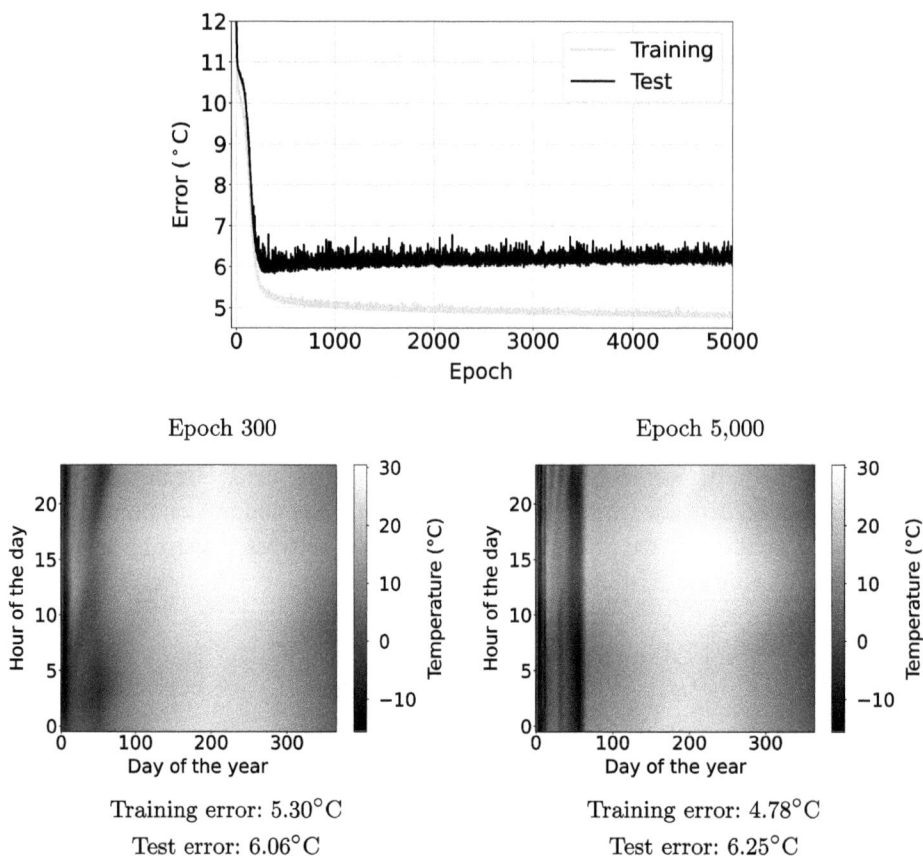

Figure 12.38 Four-layer neural network for temperature estimation. The top diagram shows the evolution of the training and test error of a four-layer neural network trained to estimate the temperature in Manhattan (Kansas) as a function of the hour $0 \leq x_{\text{hour}} \leq 23$ and the day $1 \leq x_{\text{day}} \leq 365$. The training and test set correspond to the 2015 and 2016 data in the top row of Figure 12.26, respectively. The two heatmaps in the bottom row show the temperature estimates produced by the neural network model after 300 (left) and 5,000 (right) training epochs, along with their respective training and test errors. At 5,000 epochs, the model overfits the training data, generalizing worse to the test data than at 300 epochs.

In contrast to the simpler two-layer model, it is not tractable to derive an analytical expression for the function implemented by our four-layer network because it contains 300 ReLUs (one for each hidden variable in each intermediate layer) with more than 10^{90} (2^{300}) possible states! This is the price to pay for using deep-learning models. They often achieve excellent performance due to their ability to learn complex functions, but they are notoriously difficult to interpret, precisely because of the complexity of the learned functions.

12.8.2 Classification

In this section, we explain how to perform classification using neural networks. The key idea is to use the softmax function defined in Section 12.6 to transform the network output into probability estimates. In order to illustrate the approach, we apply a two-layer neural network to the seed classification task in Example 12.38 and Figure 12.30. Recall that the goal is to identify three wheat varieties based on two features (the surface area and asymmetric coefficient of the seed).

Our two-layer neural network has 100 hidden variables in its intermediate layer. It maps an input $x \in \mathbb{R}^2$ to a three-dimensional vector as follows:

$$h := W_1 x + \alpha_1, \qquad W_1 \in \mathbb{R}^{100 \times 2}, \alpha_1 \in \mathbb{R}^{100}, \tag{12.414}$$

$$\ell := W_2 h_+ + \alpha_2, \qquad W_2 \in \mathbb{R}^{3 \times 100}, \alpha_2 \in \mathbb{R}^3, \tag{12.415}$$

where h_+ is a 100-dimensional vector obtained by applying a ReLU to the 100 hidden variables in the vector h. The dimensionality of the output ℓ is equal to the number of classes (three, in this case). We interpret the entries in ℓ as logits, borrowing the term from logistic and softmax regression (see Sections 12.5 and 12.6). The logit vector ℓ is plugged into a softmax function to generate probability estimates for each of the three classes. The probability estimate for the kth class equals

$$p_\Theta(x_i)_k := \frac{\exp(\ell[k])}{\sum_{l=1}^{3} \exp(\ell[l])}, \tag{12.416}$$

where $\Theta := \{W_1, W_2, \alpha_1, \alpha_2\}$ denotes the network parameters.

In order to train a neural-network classifier, we optimize the network parameters by maximizing the log-likelihood of the probability estimates on a training set with n examples $(x_1, y_1), (x_2, y_2), \ldots, (x_n, y_n)$, where x_i is the ith feature vector and y_i the corresponding label for $1 \leq i \leq n$. We model the training data as a realization from a joint distribution of feature vectors $\tilde{x}_1, \ldots, \tilde{x}_n$ and labels $\tilde{y}_1, \ldots, \tilde{y}_n$. As in Theorems 12.35 and 12.37, we assume that the labels are conditionally independent given the features, and each label $\tilde{y}_i$ is conditionally independent from the remaining feature vectors $\{\tilde{x}_l\}_{l \neq i}$ given $\tilde{x}_i$. Under these assumptions, the likelihood equals

$$\mathcal{L}_{XY}(\Theta) = \prod_{k=1}^{3} \prod_{\{i:\, y_i = k\}} p_\Theta(x_i)_k, \tag{12.417}$$

and the corresponding log-likelihood is

$$\log \mathcal{L}_{XY}(\Theta) = \sum_{k=1}^{3} \sum_{\{i:\, y_i = k\}} \log p_\Theta(x_i)_k. \tag{12.418}$$

The negative log-likelihood $-\log \mathcal{L}_{XY}(\Theta)$ is equal to a quantity called cross-entropy, which measures the difference between the estimated probabilities and the empirical probabilities of the labels. Because of this, it is often referred to as the *cross-entropy loss* in the deep-learning literature. To maximize the log-likelihood, we follow the same procedure used to minimize the residual sum of squares in Section 12.8.1. We perform stochastic gradient ascent (or descent if we use the negative log-likelihood) based on the gradient of the log-likelihood with respect to the network parameters Θ, computed via backpropagation on batches of training examples.

The top row in Figure 12.39 shows the logits produced by the two-layer neural network in our wheat-classification example, as well as the corresponding conditional probabilities superposed onto the training data. The neural network learns a nonlinear function of the features, which is much more complex than the function implemented by the softmax-regression model in Figure 12.24, where the logits are affine. In fact, the neural network

Logits (softmax inputs)

Estimated probabilities

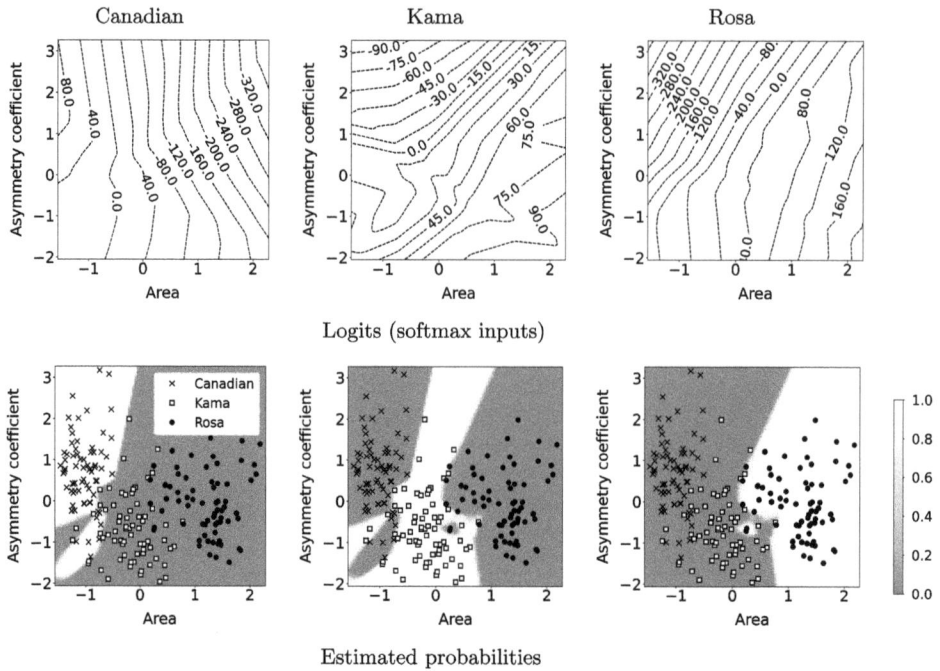

Figure 12.39 Neural-net classification of wheat seeds. Visualization of a two-layer neural-network model trained to classify wheat seeds into three varieties according to their area and asymmetric coefficient. The top row depicts the contour lines of the nonlinear logits or softmax inputs produced by the neural network for each class. These logits are mapped by the softmax operation to the estimated probabilities, shown in the bottom row. The network implements a highly intricate function of the features that is able to correctly classify every training example (compare with Figure 12.24).

is able to correctly classify every single training example. This flexibility can easily lead to overfitting the training set, so it is crucial to evaluate the generalization ability of such models on held-out data.

12.9 Evaluation of Classification Models

In this section, we discuss how to evaluate the performance of classification models, focusing on classification problems with two classes. Section 12.9.1 describes metrics to measure the discriminative ability of a classifier. Section 12.9.2 explains how to determine whether the probability estimates generated by a classifier are well calibrated.

12.9.1 Measuring Discriminative Performance

In Section 6.5 we use accuracy to evaluate the performance of our model to diagnose Alzheimer's disease. The accuracy of a classifier is the fraction of data points that it classifies correctly, which is a very reasonable measure of its discriminative ability. However, in applications such as medical diagnostics, it is often crucial to provide a more nuanced description of classification performance. For instance, we may want to determine what fraction of patients with Alzheimer's are correctly diagnosed. This metric, known as the true positive rate (TPR) or recall, can be very different from the accuracy. In our Alzheimer's example, a naive baseline that simply classifies everyone as healthy has a high test accuracy (78.4%), but an abysmal TPR (0%). Apart from the accuracy and the TPR, there exists a wide array of evaluation metrics for binary classifiers. The following definition describes some of the most popular ones.

Definition 12.47 (Evaluation metrics for binary classification). *We consider a dataset, where there are P examples with positive labels and N examples with negative labels. A classifier is applied to the data, producing an estimated positive or negative label for each example.*

Positive examples correctly classified as positive are true positives (TP). *Positive examples incorrectly classified as negative are false negatives* (FN). *Negative examples correctly classified as negative are true negatives* (TN). *Negative examples incorrectly classified as positive are false positives* (FP). *Notice that*

$$TP + FN = P, \tag{12.419}$$
$$TN + FP = N. \tag{12.420}$$

This decomposition of the data can be represented by a confusion matrix with the following format:

True Class.	Negative	Positive
Negative	**True Negatives** (TN)	**False Negatives** (FN)
Positive	**False Positives** (FP)	**True Positives** (TP)

The accuracy is the fraction of examples that are correctly classified:

$$\mathrm{Accuracy} := \frac{TN + TP}{N + P}. \tag{12.421}$$

The true positive rate (TPR), also known as the recall, sensitivity or hit rate, is the fraction of positive examples that are correctly classified:

$$\mathrm{TPR} := \frac{TP}{P}. \tag{12.422}$$

The specificity, true negative rate (TNR) or selectivity is the fraction of negative examples that are correctly classified:

$$\mathrm{Specificity} := \frac{TN}{N}. \tag{12.423}$$

The false positive rate (FPR) is the fraction of negative examples that are incorrectly classified:

$$\mathrm{FPR} := \frac{FP}{N} = 1 - \mathrm{Specificity}. \tag{12.424}$$

The precision or positive predictive value is the fraction of examples predicted as positive that are true positives:

$$\text{Precision} := \frac{\text{TP}}{\text{TP} + \text{FP}}. \qquad (12.425)$$

The F1 score is the harmonic mean[4] of the TPR and the precision:

$$F_1 := \frac{2 \cdot \text{TPR} \cdot \text{Precision}}{\text{TPR} + \text{Precision}}. \qquad (12.426)$$

All of the classification methods described in this book generate estimated conditional probabilities of the labels given the input features. This includes naive Bayes (Section 4.8), Gaussian discriminant analysis (Section 6.5), logistic regression (Section 12.5), classification trees (Section 12.7.2), and neural networks (Section 12.8). In order to perform binary classification based on these models, the estimated conditional probability of the positive class is compared to a predefined threshold. If the probability is greater than the threshold, the estimate is positive, otherwise, it is negative. The metrics in Definition 12.47 depend on the chosen threshold. Figure 12.40 illustrates this in the case of the logistic-regression model for Alzheimer's diagnosis from Section 12.5.

The top left graph in Figure 12.40 shows histograms of the probability of Alzheimer's estimated by the logistic-regression model for the negative (healthy) and positive (Alzheimer's) examples in the test set. The dashed lines represent three different thresholds, which result in different numbers of true and false positives, and true and false negatives, reported in the confusion matrices. If the threshold is set to 0.05, then the TPR is very high: We correctly identify almost all (93%) Alzheimer's cases! The bad news is that the FPR is also high; more than half (52%) of the healthy subjects are misclassified as having Alzheimer's. The precision is also bad; out of the subjects diagnosed as having Alzheimer's, only one third (33%) actually have the disease. Increasing the threshold to 0.2 reduces the FPR dramatically (13%), and increases the precision (57%). The price to pay is a reduction of the TPR to 61%. Further increasing the threshold to 0.5 almost completely eliminates false positives (the FPR is just 1%) and achieves very high precision (78%), but results in a very low TPR (19%). Ultimately, the choice of threshold should be dictated by the application of interest. Among these three options, the middle threshold achieves a reasonable trade-off between TPR and precision, as captured by the F1 score (59%, compared to 49% for the lower threshold and 31% for the higher threshold).

As illustrated by Figure 12.40, there is an inherent trade-off between the TPR and the FPR: As we increase the threshold on the estimated probabilities, the FPR decreases (because we reduce the false positives), but so does the TPR (because we also reduce the true positives). If we plot the FPR-TPR combinations achieved by every possible threshold, we obtain the graph in the top right of Figure 12.40, which is known as the receiver operating characteristic (ROC) curve of the classifier. This terminology originated in electrical engineering. The ROC curve was originally developed to perform detection based on radar signals.

The ROC curve enables us to visualize the performance of a classifier for all possible thresholds. An ROC curve that stretches towards the top left corner is ideal because it implies that there exist thresholds for which the FPR is low and the TPR is high. Equivalently, we would like the *area under the ROC curve* to be as close to one as possible. This is one of the most popular metrics for evaluation of classifiers. It can be computed via numerical

---

[4] The harmonic mean is used to compute the F1 score, as opposed to the arithmetic mean, because the TPR and the precision are fractions that share the same numerator. See Exercise 12.21.

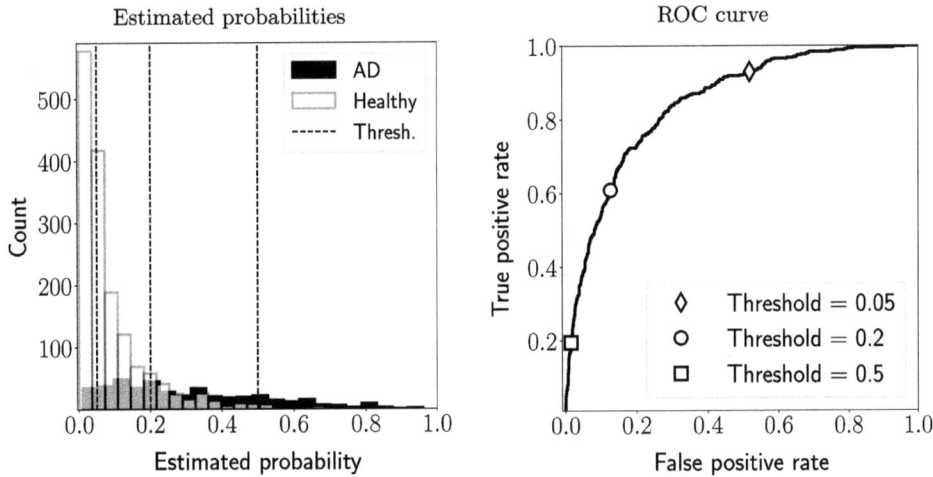

Estimated probabilities / ROC curve

Threshold = 0.05

Est.	True	Healthy	AD
Healthy		774	31
AD		829	411

Accuracy = 0.58
TPR = 0.93
FPR = 0.52
Precision = 0.33
F1 score = 0.49

Threshold = 0.2

Est.	True	Healthy	AD
Healthy		1399	173
AD		204	269

Accuracy = 0.82
TPR = 0.61
FPR = 0.13
Precision = 0.57
F1 score = 0.59

Threshold = 0.5

Est.	True	Healthy	AD
Healthy		1579	356
AD		24	86

Accuracy = 0.81
TPR = 0.19
FPR = 0.01
Precision = 0.78
F1 score = 0.31

Figure 12.40 Evaluation of a binary classifier. The top left graph shows the probability of Alzheimer's estimated by the logistic-regression model from Section 12.5 for negative (healthy, white) and positive (Alzheimer's, black) test examples, extracted from 2,046 scans in the National Alzheimer's Coordinating Center dataset (NACC, 2020). The dashed lines represent three thresholds, which result in different numbers of true and false positives, and true and false negatives, reported in the confusion matrices. The evaluation metrics in Definition 12.47 corresponding to each threshold are listed under the corresponding confusion matrix. In general, increasing the threshold decreases the FPR and increases the precision, but also decreases the TPR. The trade-off between TPR and FPR for this classifier is captured by the receiver operating characteristic curve depicted in the top right graph. The (FPR,TPR) coordinates on the curve corresponding to each of the three thresholds are indicated by different markers.

integration of the ROC curve, and is known as the AUROC, or often just AUC, for area under the ROC curve.

Beyond its obvious connection to the TPR and FPR, the AUC is equal to the concordance or c-statistic of the classifier, which directly quantifies its discriminative ability (see Exercise 12.20 for a proof). Imagine that we randomly sample a positive and a negative example from the data. The concordance is the fraction of such pairs for which the estimated probability of the positive example is higher. When the concordance approaches one, this implies that the classifier discriminates perfectly between negative and positive examples. In our Alzheimer's example, the logistic-regression classifier has good discriminative performance; its test AUC equals 0.847.

12.9.2 Calibration of Probability Estimates

A classifier is said to be well calibrated, if it generates probability estimates that adequately reflect the uncertainty in the data. Highly discriminative classifiers are not necessarily well calibrated. For example, consider a classifier for our Alzheimer's example, which assigns a probability of 0.9 to all Alzheimer's patients and a probability of 0.8 to all healthy subjects. The classifier discriminates perfectly between the classes: The estimated probability of all positive examples is higher than that of any negative example, so the concordance, and hence the AUC, equals 1. However, the probability estimates are completely misleading. If we aggregate all examples that are assigned a probability of 0.8 by the classifier, we would expect 80% of them to have Alzheimer's. Instead, none of them do!

Figure 12.41 explains how to evaluate the calibration of a classifier on a test dataset. First, we apply the classifier to all the data, in order to obtain a probability estimate for each example. Then, we separate the examples in bins according to the estimated probability assigned to the positive class, as illustrated by the top graph in Figure 12.41. For example, the first bin in the graph consists of the 1,125 subjects with estimated probabilities between 0 and 0.1. For each bin, we compute the empirical probability of the positive label. In our example, out of the 1,125 subjects in the first bin, 78 have Alzheimer's, so the empirical probability of Alzheimer's in that bin is 6.9%. We then compare the empirical probability to the midpoint of the interval, which equals 5% for the first bin. The bottom graph in Figure 12.41 shows a scatterplot of the empirical probabilities and the corresponding estimated probabilities in all bins. This type of plot is known as a reliability diagram. Perfect calibration corresponds to a diagonal reliability diagram, where the empirical and estimated probabilities are the same. In our example, the classifier is quite well calibrated, but the estimated probabilities underestimate the empirical probabilities.

Just like a discriminative classifier can be badly calibrated, a well-calibrated classifier can be nondiscriminative. Consider a simple classifier that assigns the same estimated probability to every subject in our Alzheimer's example. When we evaluate calibration, all subjects fall into the same bin. If the estimated probability is equal to the fraction of the population that has Alzheimer's (21.6% in our test set), then the empirical and estimated probabilities in the bin are the same, so the classifier is perfectly calibrated. However, it has no discriminative ability (in fact, it completely ignores the features!).

The Brier score is a metric that simultaneously evaluates the calibration and discrimination ability of a classifier, as explained in Exercise 12.22. It equals the average squared difference between the estimated probabilities and the labels.

Definition 12.48 (Brier score). *Let* (x_1, y_1), (x_2, y_2), ..., (x_n, y_n) *be a dataset formed by* n *pairs of a feature vector* x_i *and a corresponding binary label* y_i *equal to 0 or 1, for* $1 \leq i \leq n$. *The Brier score of a classifier evaluated on the dataset is*

$$\text{Brier score} := \frac{1}{n} \sum_{i=1}^{n} \left(y_i - p_{\text{est}}(x_i) \right)^2, \qquad (12.427)$$

where $p_{\text{est}}(x_i)$ *denotes the probability estimate assigned to the ith example for* $1 \leq i \leq n$.

For our Alzheimer's diagnosis test set, the Brier score of the perfectly discriminative, but uncalibrated, model that assigns 0.9 to all Alzheimer's patients and 0.8 to the healthy subjects equals

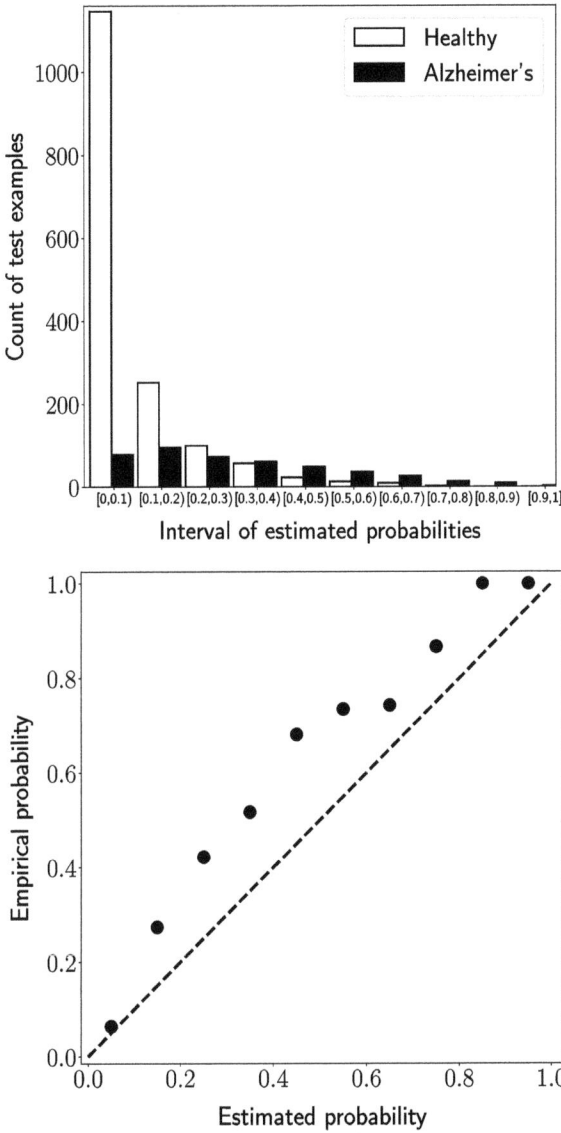

Figure 12.41 Evaluating the calibration of a classification model. In order to evaluate calibration, we separate the test data into bins according to the estimated probability assigned to the positive class. The top graph illustrates this process for the logistic-regression model from Section 12.5 applied to Alzheimer's diagnosis. It shows the number of negative (white bars) and positive (black bars) test examples in each bin. In the reliability diagram at the bottom, we plot the empirical probability of Alzheimer's in each bin against the midpoint of the corresponding estimated probabilities. Perfect calibration would correspond to the dashed diagonal line, where the empirical and estimated probabilities are equal.

$$B_{\mathrm{disc}} := \frac{1}{n} \left(\sum_{\{i:\, y_i=1\}} (1 - 0.9)^2 + \sum_{\{i:\, y_i=0\}} (0 - 0.8)^2 \right) \tag{12.428}$$

$$= 0.504. \tag{12.429}$$

The Brier score of the perfectly calibrated, but nondiscriminative, model that assigns the same probability estimate to each example is

$$B_{\mathrm{cal}} := \frac{1}{n} \left(\sum_{\{i:\, y_i=1\}} (1 - 0.216)^2 + \sum_{\{i:\, y_i=0\}} (0 - 0.216)^2 \right) \tag{12.430}$$

$$= 0.169. \tag{12.431}$$

Exercise 12.22 provides an analysis of these two Brier scores. The Brier score of our logistic-regression model equals 0.131, lower than that of the uncalibrated and the nondiscriminative models. This suggests that the probability estimates it produces are both discriminative and calibrated, which is corroborated by its high AUC (0.847) and the reliability diagram in Figure 12.41, respectively.

Exercises

12.1   (Heart beat) We are interested in estimating the heart beat of a fetus, in the presence of strong interference due to the heart beat of the baby's mother. We acquire data using two microphones: one situated near the mother's belly, and one placed away from her belly. We model the data as

$$\tilde{x}[1] = \tilde{b} + \tilde{m} + \tilde{z}_1, \tag{12.432}$$

$$\tilde{x}[2] = \tilde{m} + \tilde{z}_2, \tag{12.433}$$

where $\tilde{b}$ is a random variable representing the heart beat of the baby, $\tilde{m}$ is a random variable representing the heart beat of the mother, and $\tilde{z}_1$ and $\tilde{z}_2$ represent additive noise. All random variables are centered to have zero mean. We assume that $\tilde{b}$, $\tilde{m}$, $\tilde{z}_1$, and $\tilde{z}_2$ are all independent. The variances of $\tilde{b}$, $\tilde{z}_1$, and $\tilde{z}_2$ are equal to 1, whereas the variance of $\tilde{m}$ is much larger, it is equal to 10.

    a Compute the linear MMSE of $\tilde{b}$ given $\tilde{x}[1]$, and the corresponding coefficient of determination.

    b Compute the linear MMSE of $\tilde{b}$ given $\tilde{x}$, and the corresponding coefficient of determination.

12.2   (PCA and OLS) Consider a dataset of n 2-dimensional data points $x_1, \dots, x_n \in \mathbb{R}^2$. Assume that the data are centered to have zero sample mean. Our goal is to find a line in the 2D space that lies *closest* to the data. First, we apply PCA, and consider the line in the direction of the first principal direction. Second, we compute the OLS estimator of the second entry $x_i[2]$ given the first entry $x_i[1]$ for $1 \leq i \leq n$, and consider the line mapping $x_i[1]$ to the estimate of $x_i[2]$. Are these lines the same? Describe each line in terms of the quantity that it minimizes (for example, the sum of some squared distance from the points to the line), and draw a diagram to illustrate your description.

12.3   (Global warming) In this problem, we model temperature trends via a linear regression model, using a dataset that can be downloaded at https://github.com/cfgranda/ps4ds/blob/main/data/oxford_temperatures.txt. We focus on the second column of the dataset, which consists of the maximum temperature measured each month in Oxford from 1853 until 2014, extracted from Dataset 7. We use the first 150 years of data (the first $150 \cdot 12$ data points) as a training set, and the remaining 12 years as a test set.

In order to model the evolution of the temperature over the years, we fit the following model:

$$y[t] = a + bt + c\cos(2\pi t/T) + d\sin(2\pi t/T), \tag{12.434}$$

where $a, b, c, d \in \mathbb{R}$, and $y[t]$ denotes the maximum temperature during month t (t takes values between 1 and $162 \cdot 12$).

a What is the number of parameters in the model and how many data points are available to fit the model? Are you worried about overfitting?

b Fit the model using OLS on the training set for values of T equal to $1, 2, \ldots, 20$. Which of these models provides a better fit? Explain why this is the case. For the remainder of the problem, fix T to the value T^* that provides a better fit.

c Produce two plots comparing the observed maximum temperatures with the model predictions for $T := T^*$; one for the training set and one for the test set.

d Report the OLS coefficients and provide an intuitive interpretation of the different terms in the model.

e According to your model, are temperatures rising in Oxford? By how much?

12.4 (OLS estimator) Prove Theorem 12.7.

12.5 (Ice cream sales) Estimate the linear causal effect of advertising on sales in Exercise 8.15. What is your conclusion regarding the claim of the marketing department?

12.6 (Linear regression and maximum-likelihood estimation) Let $(x_1, y_1), (x_2, y_2), \ldots, (x_n, y_n)$ be a dataset formed by n pairs of a response $y_i \in \mathbb{R}$ and a corresponding vector $x_i \in \mathbb{R}^d$ containing d features (where $1 \leq i \leq n$), all centered to have zero sample mean. Following Definition 12.19, we model y_i for $1 \leq i \leq n$ as realizations of the random variables

$$\tilde{y}_i = x_i^T \beta_{\text{true}} + \tilde{z}_i, \qquad \beta_{\text{true}} \in \mathbb{R}^d, \tag{12.435}$$

where $\tilde{z}_1, \ldots, \tilde{z}_n$ are independent Gaussian random variables with zero mean and constant variance σ^2, which represent additive noise. What is the maximum-likelihood estimator of the linear model coefficients β_{true} given the data?

12.7 (Best linear unbiased estimator) Consider the finite-data linear response with additive noise in Definition 12.19,

$$\tilde{y}_{\text{train}} = X_{\text{train}} \beta_{\text{true}} + \tilde{z}_{\text{train}}, \tag{12.436}$$

where the noise is modeled as an n-dimensional random vector $\tilde{z}_{\text{train}}$. We assume that the noise is i.i.d. with variance σ^2. Our goal is to find an estimator of the linear coefficients β_{true} that is linear (i.e. $C\tilde{y}_{\text{train}}$ for some $d \times n$ deterministic matrix C), unbiased and has the least possible variance.

a What is the mean of the estimator $C\tilde{y}_{\text{train}}$?

b What is the covariance matrix of $C\tilde{y}_{\text{train}}$?

c Let us define $D := C - (X_{\text{train}}^T X_{\text{train}})^{-1} X_{\text{train}}^T$. What must be true of DX_{train} so that $C\tilde{y}$ is an unbiased estimator of β_{true} for any value of β_{true}?

d Let $\tilde{\beta}_{\text{OLS}} := (X_{\text{train}}^T X_{\text{train}})^{-1} X_{\text{train}}^T \tilde{y}_{\text{train}}$ denote the OLS estimator of the linear coefficients (assuming all quantities are centered to have zero mean). Show that $\tilde{\beta}_{\text{OLS}}$ is the linear unbiased estimator with the smallest variance in any direction. More precisely, for any $v \in \mathbb{R}^d$ and any $C \in \mathbb{R}^{d \times n}$,

$$\text{Var}\left[v^T C\tilde{y}_{\text{train}}\right] \geq \text{Var}\left[v^T \tilde{\beta}_{\text{OLS}}\right]. \tag{12.437}$$

12.8 (Augmented dataset) Show that ridge regression is equivalent to applying OLS on an expanded dataset with additional examples. Intuitively, what effect do these examples have on the OLS estimator?

12.9 (Correlated features) Consider a regression problem where the response vector $y \in \mathbb{R}^n$ only depends on one feature:

$$y_{\text{train}} := \beta_{\text{true}} a + z, \tag{12.438}$$

where $\beta_{\text{true}} \in \mathbb{R}$ is the true coefficient, $a \in \mathbb{R}^n$ is a feature vector, and $z \in \mathbb{R}^n$ is additive noise. We also observe a second feature $b \in \mathbb{R}^n$, which is correlated with the first. This second feature vector can be decomposed as follows:

$$b = \alpha a + \sqrt{1 - \alpha^2} a_\perp, \tag{12.439}$$

where $a_\perp$ is orthogonal to a and α is a constant between 0 and 1. The vectors a, b, $a_\perp$ and z all have unit ℓ_2 norm. In addition, we assume

$$a^T z = 0.1, \tag{12.440}$$
$$a_\perp^T z = 0.1. \tag{12.441}$$

In this problem, we study what happens if we fit a linear regression model to the response vector y_{train} using a design matrix containing both features,

$$X_{\text{train}} = \begin{bmatrix} a & b \end{bmatrix}. \tag{12.442}$$

a Derive the limit of the OLS coefficient estimate β_{OLS} when $\alpha \to 1$? Explain your result. *Hint:* Use the fact that for any c, d, g, and h such that $ch \neq dg$,

$$\begin{bmatrix} c & d \\ g & h \end{bmatrix}^{-1} = \frac{1}{ch - dg} \begin{bmatrix} h & -d \\ -g & c \end{bmatrix}. \tag{12.443}$$

b What is the limit of the OLS response estimate $y_{\text{OLS}} := X_{\text{train}}\beta_{\text{OLS}}$ when $\alpha \to 1$? Is it collinear with the relevant feature x_1?

c Derive the limit of the ridge-regression coefficient estimate β_{RR} when $\alpha \to 1$, assuming that the regularization parameter λ is positive, and compare it to the OLS coefficient estimate.

d What is the limit of the ridge-regression response estimate $y_{\text{RR}} := X_{\text{train}}\beta_{\text{RR}}$ when $\alpha \to 1$? Is it collinear with the relevant feature x_1?

12.10 (Prior knowledge) Consider a linear regression problem where we have prior information indicating that the coefficients should be close to a certain value β_{prior}.

a Suggest a modified estimator that takes into account this prior knowledge within the ridge-regression framework.

b Assume that the data are generated according to a linear model $\tilde{y}_{\text{train}} := X_{\text{train}}\beta_{\text{true}} + \tilde{z}$, where $\beta_{\text{true}} \in \mathbb{R}^d$ and $X_{\text{train}} \in \mathbb{R}^{n \times d}$ are fixed and the entries of $\tilde{z}$ are i.i.d. with zero mean and variance σ^2. Compare the mean and the covariance matrix of your proposed coefficient estimate with those of the ridge-regression coefficient estimate.

12.11 (Lasso estimation with two features) We consider a regression problem with two features, where only one of them is relevant, as in Figure 12.20. Let $x_{\text{true}} \in \mathbb{R}^n$ be a feature vector, $z \in \mathbb{R}^n$ a noise vector, $a_{\text{true}} \in \mathbb{R}$ a real-valued coefficient and $y_{\text{train}} \in \mathbb{R}^n$ the corresponding response vector

$$y_{\text{train}} := x_{\text{true}} a_{\text{true}} + z. \tag{12.444}$$

The $n \times 2$ design matrix consists of the relevant feature x_{true} and an additional spurious feature vector $x_{\text{other}} \in \mathbb{R}^n$,

$$X_{\text{train}} := \begin{bmatrix} x_{\text{true}} & x_{\text{other}} \end{bmatrix}, \tag{12.445}$$

so that

$$y_{\text{train}} = X_{\text{train}} \begin{bmatrix} a_{\text{true}} \\ 0 \end{bmatrix} + z. \tag{12.446}$$

Without loss of generality, we assume that the true coefficient is nonnegative $a_{\text{true}} \geq 0$, and that the feature vectors are centered to have zero sample mean and normalized to have unit sample variance. We denote the lasso coefficients associated with the design matrix X_{train} by

$$\beta_{\text{lasso}} := \arg \min_{\beta \in \mathbb{R}^d} ||y_{\text{train}} - X_{\text{train}}\beta||_2 + \lambda \, ||\beta||_1 \, , \tag{12.447}$$

where λ is the nonnegative regularization parameter. To alleviate notation we define $\lambda_n := \frac{\lambda}{n-1}$ and set

$$\beta := \begin{bmatrix} a \\ b \end{bmatrix}, \tag{12.448}$$

so that the lasso cost function equals

$$\mathcal{L}_{\text{lasso}}(a, b) := ||y_{\text{train}} - x_{\text{true}}a - x_{\text{other}}b||_2^2 + \lambda \, |a| + \lambda \, |b| \, . \tag{12.449}$$

This is a strictly convex function because the quadratic term is strictly convex (as long as x_{true} and x_{other} are not collinear), the ℓ_1 norm is convex (all norms are convex), and the sum of a strictly convex function and a convex function is convex. As a result, any local minimum is the unique global minimum. We refer to Boyd and Vandenberghe (2004) for a justification of these statements and background on convex functions. Our proof strategy is to show that there is a local minimum at a point where $b = 0$, which consequently is the global minimum.

a In our analysis, we assume that the sample correlation between the relevant feature vector and the noise

$$c_{\text{true,noise}} := \frac{x_{\text{true}}^T z}{n - 1} \tag{12.450}$$

is lower bounded by $-a_{\text{true}}$. Show that this is a natural assumption because otherwise y_{train} and x_{true} are negatively correlated.

b Show that the minimum of $\mathcal{L}_{\text{lasso}}(a, 0)$, interpreted as a function of a, is

$$a_{\text{lasso}} := a_{\text{true}} + c_{\text{true,noise}} - \frac{\lambda_n}{2}. \tag{12.451}$$

c Derive an expression for $\lambda_{\max}$, so that a_{lasso} is nonnegative, as long as $\lambda_n \leq \lambda_{\max}$.

d All that remains is to prove that $\mathcal{L}_{\text{lasso}}(a_{\text{lasso}}, b)$ has a local minimum at $b = 0$. Show that the derivative of $\mathcal{L}_{\text{lasso}}(a_{\text{lasso}}, b)$ with respect to b is positive when $b > 0$ for small enough b, as long as

$$\frac{\lambda_n}{2} > \frac{\lambda_{\min}}{2} := \frac{|c_{\text{other,noise}} - \rho \, c_{\text{true,noise}}|}{1 - \rho}, \tag{12.452}$$

where ρ is the sample correlation coefficient between the features and $c_{\text{other,noise}}$ is the sample correlation between the spurious feature and the noise.

e Show that the derivative of $\mathcal{L}_{\text{lasso}}(a_{\text{lasso}}, b)$ with respect to b is negative when $b < 0$ for small enough b, as long as $\lambda_n > \lambda_{\min}$. Since the lasso cost function is continuous and convex, this implies that there is a local minimum of $\mathcal{L}_{\text{lasso}}$ at

$$\beta_{\text{lasso}} = \begin{bmatrix} a_{\text{lasso}} \\ 0 \end{bmatrix}, \qquad a_{\text{lasso}} := a_{\text{true}} + c_{\text{true,noise}} - \frac{\lambda_n}{2} > 0, \tag{12.453}$$

which is the global minimum by convexity. We conclude that the lasso estimator correctly identifies the relevant feature, assigning it a positive coefficient, as long as λ_n is between $\lambda_{\min}$ and $\lambda_{\max}$.

12.12   (Ridge regression and sparsity) We consider the data model in Exercise 12.11, setting $n := 4$, $a_{\text{true}} := 1$, $\rho := 0.2$, $c_{\text{true,noise}} := -0.15$, and $c_{\text{other,noise}} := 0.155$. Derive the ridge-regression coefficients under these assumptions as a function of the regularization parameter λ. Does the ridge regression estimator set the second coefficient to zero, correctly identifying the relevant feature, for any value of λ?

12.13   (Derivative analysis of logistic regression) In this exercise we use derivatives to gain some intuition about logistic regression.

   a   Prove that the derivative of the logistic function defined in (12.327) equals

$$\frac{\mathrm{d}\,\mathrm{lgf}(\ell)}{\mathrm{d}\ell} = \mathrm{lgf}(\ell)(1 - \mathrm{lgf}(\ell)). \tag{12.454}$$

   b   Derive the partial derivative of the logistic-regression model $p_{\alpha,\beta}(x) := \mathrm{lgf}(\beta^T x + \alpha)$ with respect to the jth feature $x[j]$. Explain how the value of $p_{\alpha,\beta}(x)$ influences the sensitivity of the model to changes in the jth feature.

   c   Gradient ascent can be applied to find a maximum of the log-likelihood function (12.334). Compute the contribution of the mth data point to the gradient-ascent update of the jth linear coefficient,

$$\beta[j]^{(t)} := \beta[j]^{(t-1)} + \eta \frac{\partial \log \mathcal{L}_{XY}}{\partial \beta[j]} \left(\alpha^{(t-1)}, \beta^{(t-1)} \right), \tag{12.455}$$

   where $\eta > 0$ is the learning rate, and provide an intuitive explanation of why it improves the probability estimate for small enough η.

12.14   (Building regression trees)

   a   Prove that the leaves of a regression tree, built via recursive binary splitting as described in Definition 12.41, correspond to regions of the feature space that form a partition.

   b   Prove that the residual-sum-of-squares decrease (12.366) is nonnegative for any potential split in the recursive binary splitting procedure.

12.15   (Tree with categorical features) Tree models can be easily adapted to use categorical features. Keith is a technician at a center for biomedical imaging. He wants to predict the duration of imaging sessions to improve scheduling at the center. He has gathered the data in the following table from past sessions. Each session has either an animal (A) or human (H) subject, and between 1 and 3 scans. Fit a regression tree with three leaves to the data in order to estimate duration from the type of subject and the number of scans.

Subject	A	H	A	H	H	A	A	H
Number of scans	1	2	3	1	3	2	1	2
Duration (minutes)	30	20	50	15	40	40	40	30

12.16   (Early stopping in linear regression) In this problem, we show that early stopping applied to linear regression has an effect similar to ridge-regression regularization. Given a dataset formed by n pairs of a response y_i and a corresponding vector x_i containing d features ($1 \leq i \leq n$), we consider the following scaled version of the residual sum of squares

$$\text{RSS}(\beta) := \frac{1}{2(n-1)} \sum_{i=1}^{n} \left(y_i - \beta^T x_i \right)^2. \tag{12.456}$$

There is no intercept because the features and the response are assumed to be centered.

a Express the tth iteration of gradient descent applied to our cost function,

$$\beta^{(t)} := \beta^{(t-1)} - \eta \nabla \text{RSS} \left(\beta^{(t-1)} \right), \tag{12.457}$$

in terms of the learning rate $\eta > 0$, the sample covariance matrix of the features and the sample cross-covariance between the features and the response.

b Assume that the coefficients are all initialized to zero ($\beta^{(0)} = 0$, where 0 is the zero vector), and $\eta < 1/\xi_{\min}$, where $\xi_{\min}$ is the smallest eigenvalue of the sample covariance matrix. Show that the iterations converge to the OLS coefficient estimate as $t \to \infty$.

c Assume that the response $\tilde{y}_{\text{train}} \in \mathbb{R}^n$ is defined as in (12.148) for a noise vector $\tilde{z}_{\text{train}}$ with independent entries that have zero mean and variance σ^2. We express the true coefficients and the gradient-descent estimator in the basis of principal directions $u_1, \ldots, u_d$ of the features (the eigenvectors of the sample covariance matrix Σ_X):

$$\beta_{\text{true}} = \sum_{j=1}^{d} c_{\text{true}}[j] u_j, \tag{12.458}$$

$$\tilde{\beta}^{(t)} = \sum_{j=1}^{d} \tilde{c}_{\text{GD}}[j] u_j. \tag{12.459}$$

Derive the mean of $\tilde{c}_{\text{GD}}[j]$ as a function of $c_{\text{true}}[j]$, η, and the jth eigenvalue ξ_j of Σ_X, for $1 \le j \le d$.

d Derive the variance of $\tilde{c}_{\text{GD}}[j]$ as a function of σ^2, $c_{\text{true}}[j]$, η, and the jth eigenvalue ξ_j of Σ_X, for $1 \le j \le d$.

e Imagine that the sample covariance matrix Σ_X has a very small eigenvalue $\xi_{\min}$, which produces noise amplification in the OLS coefficients, while the rest of the eigenvalues are much larger. Explain why early stopping is effective in reducing noise amplification.

12.17 (Product) Consider a regression problem where the response $\tilde{y}$ is equal to the product of the two entries of the feature vector $\tilde{x}$. The entries are independent, and each equals 1 or -1 with probability 1/2.

a What is the linear minimum MSE estimator of $\tilde{y}$ given $\tilde{x}$? What is the corresponding MSE?
b What is the minimum MSE estimator of $\tilde{y}$ given $\tilde{x}$? What is the corresponding MSE?
c Design a regression tree that implements the minimum MSE estimator.
d Design a neural network that implements the minimum MSE estimator.

12.18 (Predicting cancer recurrence) Laura wants to build a model to predict cancer recurrence from histopathology images. She has gathered a dataset of images from a cohort of 1,000 patients. In the dataset, there are 100 images from each patient, which she labels according to whether the patient experienced recurrence or not. The model predicts recurrence from a single image.

a Laura creates a training set by selecting 80,000 images uniformly at random from the dataset. She trains a very deep neural network on these training data, and then evaluates it on the remaining images. The model has perfect accuracy on the held-out test data! However, when she tries it on data from new patients, the model performs very badly. What is going on?
b Laura trains another model on a group of patients and verifies that it performs well on held-out data from a different group of patients. Apart from the image, the model also receives

an additional input that indicates whether the patient received a certain treatment (after acquisition of the histopathology images). Since the model is working well, can Laura use it to determine whether a new patient would benefit from the treatment? If so, how?

12.19   (ROC curve of random classifier)

a   Consider a dataset with P positive examples and N negative examples, each associated with a feature vector. Let $\tilde{x}_+$ and $\tilde{x}_-$ be the features of an example chosen independently and uniformly at random from the P positive examples and from the N negative examples, respectively. We consider a classifier p_{pred} that produces a predicted probability as a function of the features. We denote the predicted probability corresponding to the randomly chosen examples by

$$\tilde{s}_+ := p_{\text{pred}}(\tilde{x}_+), \tag{12.460}$$

$$\tilde{s}_- := p_{\text{pred}}(\tilde{x}_-). \tag{12.461}$$

Let τ be a value used to threshold the predicted probability. Examples with a predicted probability above τ are classified as positive, and the rest are classified as negative. Prove that

$$\text{TPR}(\tau) = \text{P}\left(\tilde{s}_+ > \tau\right), \tag{12.462}$$

$$\text{FPR}(\tau) = \text{P}\left(\tilde{s}_- > \tau\right). \tag{12.463}$$

b   Consider a random classifier that assigns each example in a dataset a random score, which has nothing to do with the corresponding features. The scores are all independent and sampled from a uniform distribution between 0 and 1. What does the ROC curve of the classifier look like when the number of examples is large?

12.20   (The AUC is equal to the concordance) We consider a classification problem with a large number of examples, and a classifier p_{pred} that produces a predicted probability associated with each example.

a   As in Exercise 12.19, let $\tilde{x}_+$ and $\tilde{x}_-$ be the features of an example chosen independently and uniformly at random from the positive examples and from the negative examples, respectively. We denote the estimated probability of our randomly chosen examples by

$$\tilde{s}_+ := p_{\text{pred}}(\tilde{x}_+), \tag{12.464}$$

$$\tilde{s}_- := p_{\text{pred}}(\tilde{x}_-), \tag{12.465}$$

which we model as continuous random variables, because we are considering a limit where the number of examples is very large. Show that we can write the AUC as the following function of the cdf of $\tilde{s}_+$ and the pdf of $\tilde{s}_-$,

$$\text{AUC} = \int_{-\infty}^{\infty} (1 - F_{\tilde{s}_+}(\tau)) f_{\tilde{s}_-}(\tau) \, d\tau. \tag{12.466}$$

b   Conclude that the AUC is equal to the concordance or c-statistic, defined as the probability that a randomly-selected positive example has a higher predicted probability than a randomly-selected negative example,

$$\text{AUC} = \text{P}\left(\tilde{s}_+ > \tilde{s}_-\right). \tag{12.467}$$

c   Verify that (12.467) holds for the random classifier in Exercise 12.19.
d   Suggest a way to approximate the AUC via the Monte Carlo method, based on (12.467).

12.21 (F1 score) In this exercise, we justify the definition of the F1 score, as a combination of the TPR and the precision. The TPR is the fraction of true positives out of all positive examples,

$$\text{TPR} := \frac{\text{TP}}{\text{TP} + \text{FN}}. \tag{12.468}$$

The precision is the fraction of true positives out of all examples predicted as positive

$$\text{Precision} := \frac{\text{TP}}{\text{TP} + \text{FP}}. \tag{12.469}$$

Consider the *average false estimates* AFE, defined as the arithmetic mean of the false negatives and false positives,

$$\text{AFE} := \frac{\text{FN}}{2} + \frac{\text{FP}}{2}. \tag{12.470}$$

Show that we can express the F1 score as the fraction of true positives out of a fictitious set consisting of the true positives and the average false estimates,

$$F_1 = \frac{\text{TP}}{\text{TP} + \text{AFE}}. \tag{12.471}$$

12.22 (Decomposition of the Brier score) In this exercise, we show that the Brier score simultaneously evaluates the calibration and discrimination ability of a classifier.

Let $(x_1, y_1), (x_2, y_2), \ldots, (x_n, y_n)$ be a dataset formed by n pairs of a feature vector x_i and a corresponding binary label y_i equal to 0 or 1, for $1 \leq i \leq n$. We denote by $p_{\text{est}}(x_i)$ the probability estimate assigned by a classifier to the ith example. We divide the unit interval into a partition of m disjoint intervals or *bins* $\mathcal{B}_1, \ldots, \mathcal{B}_m$. Each probability estimate belongs to exactly one bin.

We assume that the bins are narrow enough that all the probabilities in a bin are approximately the same. More precisely, if the probability $p_{\text{est}}(x_i)$ assigned to the ith example is in the bth bin, then

$$p_{\text{est}}(x_i) \approx p_b, \tag{12.472}$$

where p_b is a constant (e.g. equal to the midpoint of the bin). Under this assumption, we can decompose the Brier score of the classifier on the dataset (see Definition 12.48) into a calibration term and a discrimination component,

$$\text{Brier score} \approx \underbrace{\sum_{b=1}^{m} \frac{n_b}{n} (q_b - p_b)^2}_{\text{Calibration}} + \underbrace{\sum_{b=1}^{m} \frac{n_b}{n} q_b(1 - q_b)}_{\text{Discrimination}}, \tag{12.473}$$

where n_b is the number of examples in bin b and

$$q_b := \frac{1}{n_b} \sum_{\{p_{\text{est}}(x_i) \in \mathcal{B}_b\}} y_i \tag{12.474}$$

is the empirical probability of label 1 in bin b.

a Explain why the calibration component quantifies the calibration of the classifier, and discuss its connection to the reliability diagram.

b Plot the Gini index

$$G(q) := q(1 - q) \tag{12.475}$$

for $q \in [0, 1]$, and use the plot to explain why the discrimination component quantifies the discrimination ability of the classifier.

c Prove the decomposition.

d For the Alzheimer's example in Section 12.9.2, the Brier score of the perfectly discriminative, but uncalibrated, model that assigns 0.9 to all Alzheimer's patients and 0.8 to the healthy subject equals 0.504. What are the calibration and discrimination components equal to?

e For the Alzheimer's example, the Brier score of the perfectly calibrated, but nondiscriminative, model that assigns the same probability estimate to each example is 0.169. What are the calibration and discrimination components equal to?

Appendix

Datasets

All data used in this book are available at https://github.com/cfgranda/ps4ds/tree/main/data. Here we list the original source datasets.

Dataset 1 (Congressional voting records). Voting record and political affiliation for each of the US House of Representatives Congressmen in 1984 on 16 key issues.
Source: UCI Machine Learning Repository (UCI, 1987).

Dataset 2 (NBA Free Throws). Free throws taken in NBA games between 2006 and 2016. Each free throw is annotated to indicate what player took the shot and whether it was made or not.
Source: www.kaggle.com/sebastianmantey/nba-free-throws

Dataset 3 (Call center). Database of over 440,000 individual telephone calls to an anonymous bank in Israel in 1999.
Source: Service Enterprise Engineering Center of the Technion (Mandelbaum et al., 2000).

Dataset 4 (Air traffic). Database of flight arrivals at eight large European airports in 2010 and 2011.
Source: (Lancia and Lulli, 2017).

Dataset 5 (Anthropometric Survey). Anthropometric Survey of US Army Personnel (ANSUR 2 or ANSUR II), published internally in 2012 and publicly in 2017, containing 93 body measures for over 6,000 adult US military personnel (4,082 men and 1,986 women).
Source: (Gordon et al., 2014).

Dataset 6 (Gross domestic product). Database of gross domestic products from 1970 onwards of more than 200 countries and areas of the world released by the United Nations.
Source: unstats.un.org/unsd/snaama/Downloads

Dataset 7 (Oxford weather). Weather data measured at Oxford over 150 years released by the Met Office, which is the national meteorological service for the United Kingdom.
Source: www.metoffice.gov.uk/research/climate/maps-and-data/

Dataset 8 (Movie ratings). The Movielens datasets consist of ratings collected by the GroupLens Research group from the MovieLens website. Here we use the *Small* dataset, containing 100,000 ratings from 600 users for 9,000 movies.
Source: (Harper and Konstan, 2015).

Dataset 9 (United States weather). Weather data acquired by the US Climate Reference Network (USCRN), which is a system of climate observing stations developed by the National

Oceanic and Atmospheric Administration (NOAA), with sites across the continental US, Alaska, and Hawaii.
Source: www1.ncdc.noaa.gov/pub/data/uscrn/products

Dataset 10 (NBA shooting). Data on shots taken during the 2014–2015 NBA season, including who took the shot, where on the floor the shot was taken, who was the nearest defender, time on the shot clock, etc.
Source: www.kaggle.com/datasets/dansbecker/nba-shot-logs

Dataset 11 (IMDB movie information). Database of information about movies extracted from the IMDB website.
Source: www.kaggle.com/datasets/carolzhangdc/imdb-5000-movie-dataset/

Dataset 12 (Alzheimer's disease). Dataset of brain-region volumes and surface areas corresponding to patients with Alzheimer's disease and healthy subjects in the Alzheimer's Disease Neuroimaging Initiative dataset (ADNI, 2020) and the National Alzheimer's Coordinating Center dataset (NACC, 2020).
Source: (Liu et al., 2022).

Dataset 13 (NBA players). Data on every player who was part of an NBA roster between 1996 and 2019 including demographic and biometric variables and basic box-score statistics.
Source: www.kaggle.com/datasets/justinas/nba-players-data

Dataset 14 (NBA salaries). Salaries of NBA players in the 2017/2018 season.
Source: www.kaggle.com/koki25ando/salary

Dataset 15 (Student performance). Demographic, social, and academic information, and grades in mathematics and Portuguese language, of students in two Portuguese secondary schools, collected using school reports and questionnaires.
Source: (Cortez and Silva, 2008).

Dataset 16 (Global economy). The G-Econ research project is devoted to developing a geophysically based data-set on economic activity for the world. The basic metric is the regional equivalent of gross domestic product. Gross cell product (GCP) is measured at a 1° longitude by 1° latitude resolution on a global scale.
Source: (Nordhaus et al., 2006).

Dataset 17 (NBA games). Box scores of NBA regular season games between 2012 and 2018.
Source: www.kaggle.com/datasets/pablote/nba-enhanced-stats

Dataset 18 (P-values of academic publications). P-values obtained via an exhaustive search in Open Access papers available in the PubMed database.
Source: (Head et al., 2015).

Dataset 19 (Canadian cities). Locations of cities in Canada with more than 1,000 inhabitants.
Source: simplemaps.com/data/canada-cities

Dataset 20 (Olivetti Faces). Face images taken between April 1992 and April 1994 at AT&T Laboratories Cambridge. There are 10 different images of each of 40 distinct subjects taken

against a dark homogeneous background in an upright, frontal position.
Source: (Samaria and Harter, 1994).

Dataset 21 (Wheat varieties). Measurements of seven geometrical properties of kernels belonging to three different varieties of wheat.
Source: UCI Machine Learning Repository (Charytanowicz et al., 2010).

Dataset 22 (County health). Data from the County Health Rankings & Roadmaps program of the University of Wisconsin Population Health Institute, which tracks health factors and outcomes in United States counties.
Source: (CHRR, 2022).

Dataset 23 (MNIST digits). The legendary Modified National Institute of Standards and Technology database contains 70,000 images of handwritten digits, annotated with labels that indicate the digit shown in each image.
Source: (Lecun et al., 1998).

References

ADNI. 2020. Alzheimer's Disease Neuroimaging Initiative. http://adni.loni.usc.edu.

Benjamini, Yoav, and Hochberg, Yosef. 1995. Controlling the false discovery rate: A practical and powerful approach to multiple testing. *Journal of the Royal Statistical Society: Series B (Methodological)*, **57**(1), 289–300.

Bishop, Christopher M. 2006. *Pattern recognition and machine learning*. Springer.

Boyd, Stephen P., and Vandenberghe, Lieven. 2004. *Convex optimization*. Cambridge University Press.

Candès, Emmanuel J., and Recht, Benjamin. 2009. Exact matrix completion via convex optimization. *Foundations of Computational Mathematics*, **9**(6), 717–772.

Charytanowicz, Małgorzata, Niewczas, Jerzy, Kulczycki, Piotr, Kowalski, Piotr A., Łukasik, Szymon, and Żak, Sławomir. 2010. Complete gradient clustering algorithm for features analysis of x-ray images. Pages 15–24 of: *Information Technologies in Biomedicine: Volume 2*. Springer.

Chen, Tianqi, and Guestrin, Carlos. 2016. Xgboost: A scalable tree boosting system. Pages 785–794 of: *Proceedings of the 22nd ACM SIGKDD international conference on knowledge discovery and data mining*.

CHRR. 2022. County Health Rankings National Findings Report, University of Wisconsin Population Health Institute.

Cortez, Paulo, and Silva, Alice Maria Gonçalves. 2008. Using data mining to predict secondary school student performance.

Cover, Thomas M. 1999. *Elements of information theory*. John Wiley & Sons.

Durrett, Rick. 2019. *Probability: theory and examples*. Vol. 49. Cambridge University Press.

Eckhardt, Roger. 1987. Stan Ulam, John von Neumann and the Monte Carlo method. *Los Alamos Science*, **100**(15), 131.

Fisher, Ronald A. 1915. Frequency distribution of the values of the correlation coefficient in samples from an indefinitely large population. *Biometrika*, **10**(4), 507–521.

Goodfellow, Ian, Bengio, Yoshua, and Courville, Aaron. 2016. *Deep learning*. MIT Press.

Gordon, Claire C., Blackwell, Cynthia L., Bradtmiller, Bruce, et al. 2014. 2012 anthropometric survey of us army personnel: Methods and summary statistics. Army Natick Soldier Research Development and Engineering Center MA, Technical Report.

Haider, Humza, Hoehn, Bret, Davis, Sarah, and Greiner, Russell. 2020. Effective ways to build and evaluate individual survival distributions. *Journal of Machine Learning Research*, **21**(85), 1–63.

Harper, F. Maxwell, and Konstan, Joseph A. 2015. The Movielens Datasets: History and Context. *ACM Transactions on Interactive Intelligent Systems*, **5**(4), 1–19.

Hastie, Trevor, Tibshirani, Robert, and Friedman, Jerome H. 2009. *The elements of statistical learning: Data mining, inference, and prediction*. Vol. 2. Springer.

Head, Megan L., Holman, Luke, Lanfear, Rob, Kahn, Andrew T., and Jennions, Michael D. 2015. The extent and consequences of p-hacking in science. *PLoS Biology*, **13**(3), e1002106.

Ipsos. 2020. Ipsos poll. www.ipsos.com/sites/default/files/ct/news/documents/2020-09/topline_reuters_pennsylvania_state_poll_w1_09_21_2020.pdf.

Jain, Prateek, Netrapalli, Praneeth, and Sanghavi, Sujay. 2013. Low-rank matrix completion using alternating minimization. Pages 665–674 of: *Proceedings of the forty-fifth annual ACM symposium on theory of computing*.

Jordan, M. 2009. *The multivariate Gaussian*. https://people.eecs.berkeley.edu/~jordan/courses/260-spring10/other-readings/chapter13.pdf.

Keshavan, Raghunandan, Montanari, Andrea, and Oh, Sewoong. 2009. Matrix completion from noisy entries. *Advances in Neural Information Processing Systems*, **22**.

Kingma, Diederik P., and Ba, Jimmy. 2014. Adam: A method for stochastic optimization. *arXiv preprint arXiv:1412.6980*.

Lancia, Carlo, and Lulli, Guglielmo. 2017. Data-driven modelling and validation of aircraft inbound-stream at some major European airports. *arXiv preprint arXiv:1708.02486*.

Lecun, Yann, Bottou, Léon, Bengio, Yoshua, and Haffner, Patrick. 1998. Gradient-based learning applied to document recognition. *Proceedings of the IEEE*, **86**(11), 2278–2324.

Liu, Sheng, Masurkar, Arjun V., Rusinek, Henry, et al. 2022. Generalizable deep learning model for early Alzheimer's disease detection from structural MRIs. *Scientific Reports*, **12**(1), 17106.

Mandelbaum, A., Sakov, A., and Zeltyn, S. 2000. *Empirical analysis of a call center*. Technical Report, Technion, Haifa, Israel.

Mazumder, Rahul, Hastie, Trevor, and Tibshirani, Robert. 2010. Spectral regularization algorithms for learning large incomplete matrices. *The Journal of Machine Learning Research*, **11**, 2287–2322.

NACC. 2020. National Alzheimer's Coordinating Center. https://naccdata.org.

Nordhaus, William, Azam, Qazi, Corderi, David, et al. 2006. The G-Econ database on gridded output: Methods and data. Yale University, New Haven, **6**, 11.

Paszke, Adam, Gross, Sam, Massa, Francisco, et al. 2019. PyTorch: An Imperative Style, High-Performance Deep Learning Library. Pages 8024–8035 of: *Advances in Neural Information Processing Systems 32*. Curran Associates, Inc.

Pedregosa, Fabian, Varoquaux, Gaël, Gramfort, Alexandre, et al. 2011. Scikit-learn: Machine learning in Python. *Journal of Machine Learning Research*, **12**, 2825–2830.

Polack, Fernando P., Thomas, Stephen J., Kitchin, Nicholas, et al. 2020. Safety and efficacy of the BNT162b2 mRNA Covid-19 vaccine. *New England Journal of Medicine*.

Samaria, Ferdinando S., and Harter, Andy C. 1994. Parameterisation of a stochastic model for human face identification. Pages 138–142 of: *Proceedings of 1994 IEEE workshop on applications of computer vision*. IEEE.

Shewchuk, Jonathan Richard, et al. 1994. An introduction to the conjugate gradient method without the agonizing pain.

Siroker, Dan. 2010. How Obama raised 60 million by running a simple experiment. www.optimizely.com/insights/blog/how-obama-raised-60-million-by-running-a-simple-experiment/.

Tibshirani, Robert. 1996. Regression shrinkage and selection via the lasso. *Journal of the Royal Statistical Society: Series B (Methodological)*, 267–288.

Tversky, Amos, and Kahneman, Daniel. 1983. Extensional versus intuitive reasoning: The conjunction fallacy in probability judgment. *Psychological Review*, **90**(4), 293.

UCI. 1987. Congressional voting records. UCI machine learning repository. https://doi.org/10.24432/C5C01P.

Index

A/B testing, 424
accuracy, 215, 585
activation function, 578
activation map, 578
adjust for confounders, 134, 278, 511
alternative hypothesis, 391
area under the ROC curve, 587, 596
average treatment effect, 274
averaging, 241

backpropagation, 580
bagging, 570
Bayes' rule, 18
Bayesian statistics, 225
Bernoulli distribution, 47
beta distribution, 228
bias, 329
bias (neural networks), 578
bilinear model, 465
binomial distribution, 47
Bonferroni's correction, 419
Boole's inequality, 338
boosting, 573
bootstrap, 370
bootstrap aggregating, 570
bootstrap Gaussian confidence interval, 375
bootstrap percentile confidence interval, 377
bootstrap standard error, 371
Borel set, 68
box plot, 75
Brier score, 588

calibration of a classifier, 588
categorical distribution, 64
Cauchy distribution, 347
causal effect, 127
causal inference, 127, 274, 315, 423, 510
centering, 434
central limit theorem, 349
chain rule, 15
chain rule for continuous random variables, 175
chain rule for discrete and continuous random
    variables, 208
chain rule for discrete random variables, 119
Chebyshev's inequality, 337
classification, 138, 212, 495

classification tree, 565
clustering, 217
coefficient of determination, 304, 509
collaborative filtering, 483
completing the square, 356
composite hypothesis, 392
concordance, 587
conditional continuous distribution, 174, 204
conditional cumulative distribution function, 204
conditional discrete distribution, 118, 207
conditional independence of continuous random
    variables, 180
conditional independence of discrete and
    continuous random variables, 211
conditional independence of discrete random
    variables, 125
conditional independence of events, 26
conditional mean, 262
conditional mean function, 263
conditional probability, 13
conditional probability density function, 175, 204
conditional probability mass function, 118, 207
conditional variance, 282
confidence intervals, 363
confounder, 130, 276, 318
confounding factor, 130, 276, 318
confusion matrix, 585
conjugate prior, 228
consistency, 340, 523
continuous random variable, 66
continuous variables, 66
control for confounders, 134, 278, 511
convergence in distribution, 52, 357
convergence in mean square, 335
convergence in probability, 340
convergence of Markov chains, 148
convolution, 350
correlation, 284
correlation and causation, 315
correlation coefficient, 284, 288
correlation matrix, 442
counterfactual, 128, 316
covariance, 288
covariance matrix, 436
cross-covariance, 498
cross-entropy, 583

cross-validation, 533
c-statistic, 587
cumulative distribution function, 69
curse of dimensionality, 137, 274

decile, 73
decomposition of variance, 508
deep learning, 577
design matrix, 505
deterministic variables, 37
deviation bounds, 337
dimensionality reduction, 457
directions of maximum and minimum variance, 446, 451
discrete and continuous variables, 202
discrete random variable, 37
discrete variables, 37
discriminative model, 549
domain shift, 215
double centering, 493

early stopping, 581, 594
election, 231
empirical cdf, 74
empirical joint probability mass function, 113
empirical probability, 19, 48
empirical probability mass function, 43
ensemble, 570
entropy, 568
estimator, 19
event, 7
exchangeability, 411
expectation, 241
expectation maximization, 218, 223
explained variance, 300, 507
exponential distribution, 90

F1 score, 586, 597
false negative, 399
false positive, 399
false positive rate, 585
feature, 138, 212
feature map, 578
financial crisis, 359
first moment, 241
Fisher's transformation, 383
fitting a model, 52
frequentist statistics, 329, 369
Frobenius inner product, 469
Frobenius norm, 469
function of a continuous random variable, 81
function of a discrete random variable, 42

Gaussian approximation, 357
Gaussian discriminant analysis, 212
Gaussian distribution, 95
Gaussian mixture model, 205, 217
Gaussian random vector, 186
generalization, 58, 215, 514, 562

generalized linear model, 548
generative model, 549
geometric distribution, 50
Gini index, 569
grand mean, 493

hard-impute, 486
held-out data, 58
hidden variable, 578
histogram, 44, 84
hit rate, 585
hypothesis testing, 390

independence of continuous random variables, 177
independence of discrete and continuous random variables, 211
independence of discrete random variables, 122
independence of events, 22
independent identically distributed continuous random variables, 180
independent identically distributed discrete random variables, 53, 125
inner product between random variables, 310
intercept, 496
interquartile range, 73, 75
inverse-transform sampling, 101
iterated expectation, 262, 266

joint cumulative distribution function, 163
joint probability density function, 166
joint probability mass function, 110

kernel density estimation, 86

label, 138, 212
lasso, 543
latent variable, 217, 578
law of large numbers, 339
law of total probability, 16
learning rate, 580
least squares, 256, 503
likelihood, 53
likelihood (Bayesian), 225
likelihood (continuous), 97
likelihood (discrete), 54
linear causal effect, 316, 510
linear coefficients, 496
linear dimensionality reduction, 457
linear discriminant analysis, 217
linear minimum mean-squared-error estimation, 293, 496
linear regression, 495
linearity of expectation, 248, 436
link function, 548
logistic function, 548
logistic regression, 546
logit, 549, 553, 583
log-likelihood (continuous), 97
log-likelihood (discrete), 54

long regression, 511
low-rank matrix completion, 483
low-rank model, 464

machine learning, 570, 577
marginal continuous distribution, 171
marginal discrete distribution, 115
marginal probability density function, 171
marginal probability mass function, 115
Markov chain, 141
Markov property, 159
Markov's inequality, 337
matrix completion, 483
maximum a posteriori, 138, 213
maximum likelihood, 549, 553, 565, 583
maximum likelihood (continuous), 96
maximum likelihood (discrete), 52
maximum likelihood (Gaussian random vector), 194
mean, 241
mean of a random vector, 433
mean square, 255
mean squared error, 255, 499
measurable event, 39
median, 73, 254
memoryless property, 63, 92
method of moments, 253
minimum mean-squared-error estimation, 271
mixture model, 205
model evaluation, 57
Monte Carlo method, 29
multicollinearity, 519
multidimensional density estimation, 170
multidimensional kernel density estimation, 170
multinoulli distribution, 64
multiple continuous variables, 161
multiple discrete variables, 109
multiple testing, 417
multivariate continuous distribution, 161
multivariate discrete and continuous distributions, 202
multivariate discrete distribution, 109

naive Bayes, 138
nearest neighbors, 462
neural nets, 577
neural networks, 577
noise, 59
noise amplification, 520
nonlinear classification, 565, 582
nonlinear regression, 557, 570, 577
nonnegative matrix factorization, 494
nonparametric model, 57, 59, 84, 170
nonparametric testing, 408
normal distribution, 95
null hypothesis, 391

observational study, 128, 276
one-tailed test, 396

ordinary least squares, 503
orthogonal projection, 313, 497, 525
outlier, 76
overfitting, 59, 489, 514, 555, 562, 581
overparametrization, 581

parametric model, 46, 57, 90, 186, 225
parametric testing, 392
percentile, 73
periodic Markov chain, 151
permutation test, 408
p-hacking, 427
Poisson distribution, 51
population parameter, 325
positive predictive value, 585
posterior distribution, 226
potential outcome, 127
power, 402
power function, 402
practical significance, 426
precision, 585
principal component, 447
principal component analysis, 445
principal direction, 446
prior distribution, 225
probability, 6, 7, 9
probability density function, 76
probability integral transform, 83
probability mass function, 38
probability measure, 10
probability space, 13
publication bias, 427
p-value, 393, 409
p-value function, 393

quadratic discriminant analysis, 215
quantile, 72, 74
quartile, 73

R^2 coefficient, 304, 509
random forest, 571
random sampling, 325
random variable, 37, 66
random vector (continuous), 162
random vector (discrete), 109
randomized controlled trial, 133
randomized experiment, 132, 275, 318
realization, 38
recall, 585
receiver operating characteristic curve, 596
rectified linear unit, 578
recursive binary splitting, 561
regression, 271, 495
regression tree, 557
regularization, 531, 544, 556
rejection of null hypothesis, 393, 399
rejection region, 402
reliability diagram, 588
ridge regression, 531

sample, 38
sample conditional mean, 264
sample correlation coefficient, 290
sample covariance, 290
sample covariance matrix, 441
sample cross-covariance, 503
sample mean, 247
sample size, 334
sample space, 6
sample standard deviation, 259
sample variance, 259
second moment, 255
selectivity, 585
sensitivity, 585
short regression, 511
σ-algebra, 10
signal, 59
significance level, 399
simple hypothesis, 392
simple linear regression, 293, 313
Simpson's paradox, 130
simulating continuous random variables, 101
simulating multiple continuous random variables, 184
singular-value decomposition, 467
softmax, 553, 582
softmax regression, 552
sparse, 543
sparse regression, 543
sparsity, 543
specificity, 585
spectral theorem, 445, 454
sphering, 493
spurious correlation, 315
St Petersburg paradox, 344
standard deviation, 257
standard error, 332
standardized data, 291
standardized variable, 287
state diagram, 146
state vector, 145
stationary distribution, 148

statistical significance, 399
stochastic gradient descent, 580
sum of Gaussian random variables, 354
sum of independent random variables, 350

test error, 528
test set, 58
test statistic, 392
time series, 142
time-homogenous Markov chain, 142
topic modeling, 494
training error, 523
training set, 58
transition matrix, 145
treatment, 127, 315
tree-based model, 556
true negative rate, 585
true positive rate, 585
two-sample test, 395
two-tailed test, 396
type 1 error, 399
type 2 error, 399

unbiased estimator, 329
uncertainty quantification, 225, 363
uncorrelation, 285
uncorrelation and independence, 306
underfitting, 60, 489
underpowered test, 404
uniform distribution, 71, 81
union bound, 338

validation set, 489, 533
variable selection, 543
variance, 257
vector space of matrices, 469
vector space of random variables, 310
Venn diagram, 11

whitening, 493

z test, 395